PRACTICAL CATALYTIC
HYDROGENATION

Practical Catalytic Hydrogenation

TECHNIQUES AND APPLICATIONS

Morris Freifelder
Consultant and Lecturer

Wiley-Interscience, a Division of John Wiley & Sons, Inc.
New York · London · Sydney · Toronto

Chemistry
~~Pharmacy~~

QD
281
.H8
F92

Copyright © 1971, by John Wiley & Sons, Inc.

All rights reserved. Published simultaneously in Canada.

No part of this book may be reproduced by any means, nor transmitted, nor translated into a machine language without the written permission of the publisher.

Library of Congress Catalogue Card Number: 76-123740

ISBN 0 471 27800 9

Printed in the United States of America

10 9 8 7 6 5 4 3 2 1

PREFACE

Like any writer, the author of a book on catalytic hydrogenation for the working chemist is faced with many problems.

What should be included in the book and how extensive should the coverage be? Thus the theory behind catalytic hydrogenation and the stereochemistry involved in these reactions could be subjects of individual volumes. Lack of space makes it impossible to discuss them at great length in one book. Furthermore, a writer on catalytic hydrogenation must realize that not every laboratory is so well equipped as the one in which his research was carried out. Only a few catalysts may be available for immediate use; or the chemist's only interest may be in obtaining a particular compound by a working method with the equipment and catalysts on hand in the laboratory. It is for these reasons that, in addition to specific reduction procedures, I include other workable methods (or alternate ones, while pointing out the disadvantages) even when, from the literature or from my experience, it is clear that a particular method employing a specific catalyst is the best procedure.

The problem of the "best" catalyst plagues many chemists. Often there is indeed a preferred catalyst for the hydrogenation of certain functions and reducible groups, but not infrequently this designation applies only to certain subdivisions of a particular group or function. I hope to provide a better view of the specific applicability of hydrogenation catalysts by breaking down the reduction of each function, group, or ring system into subclasses.

The use of references is always a problem in a scientific book. In the present book, references could have been cited *ad infinitum*, each chapter thus becoming a lengthy review. I have tried to supply just enough pertinent and varied examples of hydrogenations in each class, so that a chemist who has not been able to find the reduction of a specific compound may come across one of a related compound. Another reason for the large number of examples is that at times there is no "best method." Furthermore, the "best method" and "best catalyst" may not work because the structure of the substrate is somewhat different from that of the substrate cited in the example. Still more important, the substrate in question may not be so pure as the one in the example, and a less active catalyst may be more resistant to poisoning than the "best catalyst."

A primary difficulty in writing on catalytic hydrogenation is the interpretation of the work of other chemists and the decision whether a reference will be of value. Sometimes the conditions for reduction are described meagerly—the catalyst employed is noted, but the amount is not. At other times catalytic hydrogenation is only a step in the preparation of a desired intermediate with no interest in yield. The catalyst used may be one on hand; the conditions may be adapted to available equipment. When failure is cited in a reduction, the reader may not be supplied with information about the quality of the substrate or the age of the catalyst.

We also encounter other difficulties. When an investigator can carry out reactions leading to compounds to be reduced or can directly supervise these reactions, he can be confident that the substrate is in a reasonable state of purity and that one of the sources of trouble in catalytic reductions will be eliminated. Unfortunately, in some laboratories the material to be reduced is obtained from other sources. It may be purchased material that contains impurities which impede hydrogenation; it may come from another chemist who does not have or cannot supply information about its preparation; improperly cleaned equipment may also be at fault.

In the course of carrying out at least 7500 catalytic reductions, I have encountered problems from all these sources. Certain "tricks of the trade" that have been acquired through experience are imparted to the reader from time to time in the hope of aiding the chemist in overcoming difficulties. In other instances, based on this wide experience, I suggest reasons why a reaction is or is not successful and what can be done if results are unsatisfactory.

I place considerable emphasis on the use of molecular models. Some models are shown to illustrate why a particular reduction does or does not work. Although the construction of models is a fetish of mine I have tried to avoid filling the book with their pictures and, instead, often merely make the comment that a molecular model (constructed by me) does support a reported result. I hope that the examples shown in this book will encourage the reader to make greater use of molecular models. Although they are not the complete answer to the problem of catalytic hydrogenation, the construction of models of starting material and reduction products often are helpful. In some instances examination of a model of an intermediate reduction product may lead to an understanding of what is taking place.

One of the pitfalls of writing is to consider one's own work to be the best in the field. Although I have cited many of my own publications, I have tried to be objective and to point out my findings only when they seem significant. The reader will find many generalities in my unpublished work.

Abbott Laboratories, for whom I worked for many years, has generously allowed disclosure of unpublished results, but in certain instances, for security reasons, only generalities can be offered.

In the preface to his book, *Catalysis by Metals*, Dr. Geoffrey Bond points out that "One comes to see the measure of order which has been achieved in one's subject and becomes painfully aware of the many gaps and inconsistencies which exist in it." I do not search for inconsistencies, but hope that those who read this book will become inspired to fill in the gaps that exist. If my writing results in further research by other investigators into the chemistry involved in the catalytic hydrogenation of organic compounds, then the time and effort that have gone into this study will have been well spent.

Acknowledgments are due many people. First, I thank my wife, my unpaid secretary who struggled through my writing and the unfamiliar chemical names and equations. I thank also my colleagues who suggested that I tackle this task. My former co-workers at Abbott Laboratories helped in many ways, in particular Dr. John Tadanier, Dr. William Roderick, Dr. James Short, and Dr. Warren J. Close, who examined and offered constructive criticism of various parts of this book. Dr. Richard Baltzly of the Burroughs-Wellcome Company, Dr. Dale Blackburn of Smith, Kline and French, and Dr. Robert Tedeschi of the Airco Chemicals and Plastics Company were especially helpful.

I acknowledge the help of the libraries and staff at Northwestern University, Evanston, Illinois, the Franklin Institute and Philadelphia College of Pharmacy, Philadelphia, and Brandeis University, Waltham, Massachusetts. Special thanks go to Rose Lonberger of the Town Scientific Library at the University of Pennsylvania. Finally, thanks are offered to Abbott Laboratories for access to records of reductions that I carried out in their laboratories.

Without all this help the work could never have been completed.

MORRIS FREIFELDER

Framingham, Massachusetts
February 1970

CONTENTS

I.	INTRODUCTION	1
II.	FACTORS IN HYDROGENATION	3
III.	CATALYSTS AND THEIR APPLICATIONS	5
	Rhenium, 5	
	Ruthenium, 5	
	Copper Chromium Oxide, 6	
	Cobalt, 7	
	Nickel, 7	
	Platinum, 9	
	Palladium, 12	
	Rhodium, 13	
	Metal Sulfides and Sulfactive Catalysts, 15	
	Other Catalysts, 16	
	Mixed Noble Metal Catalysts, 17	
IV.	CATALYST INHIBITORS AND POISONS	23
	Metals and Metal Salts, 24	
	Halogen-Containing Compounds, 25	
	The Effect of Nitrites, 32	
	The Effect of Compounds Containing Arsenic, Antimony, Oxygen, Phosphorus, Selenium, and Tellurium, 33	
	The Effect of Sulfur, 35	
	The Effect of the Nitrogen Atom, 39	
	The Effect of Alkali Metal Bases, 47	
	Other Inhibitors, 49	
	Practical Use of Catalyst Poisons, 49	
V.	CATALYST PROMOTERS	57
	Metallic and Nonmetallic Substances, 58	

The Effect of Oxygen, 60
Effect of Noble Metals on Nickel Catalysts, 61
The Effect of Basic Additives, 63
The Effect of Acids, 66
The Effect of Water, 68

VI. OTHER EFFECTS IN HYDROGENATION — 70
Solvents, 70
The Hydrogen Acceptor, 71
Temperature and Pressure Conditions, 74
Agitation, 75
Catalyst Supports, 75

VII. PROCEDURE — 77
Equipment, 77
Reaction Conditions, 78
Choice and Amount of Catalyst, 79
Addition of Catalyst, 81
Safety in Operation, 82
Economics, 83

VIII. ACETYLENES — 84
Reductions with Platinum Catalyst, 86
Reductions with Nickel Catalyst, 91
Reductions with Palladium Catalyst, 96
  Deactivated Palladium Catalysts, 99
Use of Iron Catalyst, 109
Selective Hydrogenation in the Presence of Other Reducible Groups, 110
  Reduction in the Presence of Another Acetylenic Bond, 110
  Reductions in the Presence of Olefinic Bonds, 113
  Reductions in the Presence of Aldehyde and Ketone Groupings, 116
  Reduction in the Presence of Halogen, 118
  Reduction in the Presence of Nitro Groups, 119
Stereochemistry Resulting from Reduction, 119
Steric Effects, 122

IX. OLEFINS   127

Variables Affecting the Hydrogenation of Olefins, 127

Monoenes, 132

Di- and Polyenes, 142

Selective Hydrogenation in the Presence of Other Reducible Groups, 150

  In the Presence of Aldehydes or Ketones, 150

    α,β-Unsaturated Ketones, 151

    Isolated Double Bonds in the Presence of Ketones, 152

    α,β-Unsaturated Aldehydes, 153

    Isolated Double Bonds in the Presence of an Aldehyde Group, 154

    The Double Bond of α,β-Unsaturated Ketones in the Presence of an Isolated Double Bond, 155

Conjugated Dienones, 155

Selective Reductions in the Presence of Unsaturated Nitrogen-Containing Groups, 156

  In the Presence of Nonaromatic Nitro Groups, 156

  In the Presence of a Cyano Group, 157

  In the Presence of Other Unsaturated Nitrogen-Containing Groups, 158

Selective Reductions in the Presence of Hydrogenolyzable Functions, 158

  In the Presence of N-Benzyl and Carbobenzoxy Protecting Groups, 158

  In the Presence of Halogen, 159

X. REDUCTION OF THE NITRO GROUP   168

In Benzenoid Compounds, 168

  Conditions and Influences on Reduction, 168

  Nitrophenols, 172

  Nitrobenzoic and Nitrophenylalkanoic Acids, 173

  Aromatic Nitro Compounds with Basic Side Chains, 173

  Aromatic Polynitro Compounds, 175

  Polyaromatic Nitro Compounds, 176

  Nitrophenylarsonic, Boronic, and Phosphonic Acids and Related Compounds, 176

Partial Reduction, 177
In Heterocyclic Compounds, 182
  Nitrophenyl Substituted Heterocyclic Compounds, 182
  In the Carbocyclic Portion of Fused Heterocyclic Rings, 183
  In Unsaturated Heterocycles, 187
Selective Reduction in Aromatic Compounds Containing Other Reducible Functions, 192
  Olefinic Bonds, 192
  Aldehydes and Ketones, 194
  Unsaturated Nitrogen-Containing Groups, 195
    Nonaromatic Nitro Groups, 195
    Azomethines, 195
    Oximes, 196
    Cyano Groups, 196
  Hydrogenolyzable Groups, 198
    Benzyl Ethers, Esters, and Alcohols, 198
    Benzylamines, 199
    Nitrogen to Nitrogen Bonds, 200
    Aliphatic Halides, 200
    Aryl Halides, 201
    Halogens in Unsaturated Nitrogen-Heterocycles, 205
In Nonaromatic Compounds, 207
  Alkanes, 208
  Cyclanes, 210
  Alkanols, 211
  Partial Reduction in Saturated Aliphatic Systems, 212
  Hydrogenation in the Presence of Other Reducible Functions, 214
    Olefinic Bonds, 214
    Unsaturated Nitrogen-Containing Groups, 217
    Aldehydes and Ketones, 219
    Hydrogenolyzable Groups, 219
In *N*-Nitro Compounds, 221

XI. REDUCTION OF NITROSO, AZO, AND AZOXY GROUPS AND OTHER GROUPS RELATED TO THE AZO SYSTEM — 232

Nitroso Group, 232
  In Aromatic Systems, 232
  Nitrosamines, 232
  Halonitroso Compounds, 234
Azo and Azoxy Groups, 234
Other N=N Systems, 235

XII. NITRILES — 238

Control of Secondary Amine, 238
Aliphatic and Aromatic Mononitriles, 240
Cyano Acids and Esters, 241
Aminonitriles, 242
Cyanohydrins, 246
Dinitriles, 247
Aldehydes from Nitriles, 248
Secondary and Tertiary Amines from Reduction of Nitriles, 249
Selective Reductions, 251
  In the Presence of Aromatic Rings, 252
  In the Presence of Nonaromatic Rings, 253
  Presence of Ketones, 253
  Presence of Reducible Nitrogen-Containing Groups, 253
  Presence of Hydrogenolyzable Groups, 255
    N- and O-Benzyl Groups, 255
    Halogen, 256

XIII. REDUCTION OF OXIMES — 261

Reaction Conditions and Catalysts, 262
  Cobalt, 262
  Nickel, 262
  Platinum, 264
  Palladium, 265
  Ruthenium and Rhodium, 266

Oximino Acids, Esters, and Amides, 267
Indanone Oximes, 268
Dioximes, 269
Amidoximes to Amidines, 270
Partial Reduction, 271
Selectivity Hydrogenation in the Presence of Other Functions, 272
  Cyano Groups, 272
  Ketones, 273
  Hydrogenolyzable Groups, 274
    O- and N-Benzyl Groups, 274
    Halogens, 276
Unsaturated Heterocycles, 277

XIV. **CARBONYL GROUPS**   282

Aldehydes, 282
Monoketones, 285
  Dialkyl Ketones, 286
  Cycloalkanones, 287
  Alkyl Cycloalkyl Ketones, 289
  Dicyclic Ketones, 289
  Ketones Containing Aromatic Rings, 290
  Aminoketones, 292
  Miscellaneous Ketones, 294
Diketones, 295
Carbonyl to Methylene, 299
Selective Hydrogenation of the Carbonyl Group, 301
  In Aromatic Ring Systems, 301
    Benzenoid Rings, 301
    Unsaturated N-Heterocycles, 301
    Unsaturated O-Heterocycles, 304
  Presence of Functional Groups, 304
    Olefinic Bonds, 304
    Cyano Groups, 305

Presence of Hydrogenolyzable Groups, 305
   O-*Benzyl Groups*, 305
   N-*Benzyl Groups*, 306
   Halogen, 306

XV. **REDUCTION OF THE C═N BOND**   313
   Primary Imines, 314
   ***N*-Substituted Imines,** 314
   The Unconjugated C═N Bond in Heterocycles, 318
   The C═N Bond in Azines, Hydrazones, and Semicarbazones, 319
   In Amidines, 322
   Selective Hydrogenations, 323
     Presence of an Olefinic Bond, 323
     Presence of a Carbonyl Group, 325
     Presence of a Cyano Group, 326
     Presence of Hydrogenolyzable Groups, 326
        N- and O-Benzyl and Related Groups, 326
        Halogens, 326

XVI. **REDUCTIVE AMINATION**   333
   Aldehydes, 334
     Catalysts, 335
   Ketones, 337
     Nickel-Catalyzed Reactions, 338
     Reductions over Noble Metals, 341
     Effect of Acidic Agents, 341
   Selective Hydrogenations, 342
     Presence of Olefinic Bonds, 342
     Presence of Halogen, 343

XVII. **REDUCTIVE ALKYLATION**   346
   Procedures, 347
   Catalysts, 350
   Secondary Amine from Primary Amine, 351
     Alkylation with Aldehydes, 351
     Alkylation with Ketones, 359

Formation of Tertiary Amine, 374
   Dialkylation of Primary Amine, 374
   Alkylation of Secondary Amine, 376
Selective Hydrogenation, 380
   In the Presence of Olefinic Bonds, 380
   In the Presence of Hydrogenolyzable Groups, 381
      Hydrazines, 382
      Hydroxylamines, 382
      O- and N-Benzyl and Related Groups, 382
      Halogens, 383

XVIII. HYDROGENOLYSIS OF ALLYLIC OXYGEN- AND NITROGEN-CONTAINING COMPOUNDS   390

Allylic Oxygen Compounds, 390
Allylic Nitrogen Compounds, 394
Selective Reductions, 395
   between O- and N-Allyl Groups, 395
   In the Presence of Functional Groups, 396

XIX. DEBENZYLATION AND RELATED REACTIONS   398

Catalysts and Conditions, 398
O-Debenzylation and Related Reactions, 400
   Arylmethanols, 400
   α-Substituted Arylmethanols, 402
   Benzyl Esters of Organic Acids, 406
   Benzyl Phosphates, 408
   Benzyl Ethers, 408
   Benzyl Acetals, 412
   Arylmethyl Carbanilates, 412
N-Debenzylations, 413
   Carbobenzyloxy Groups, 414
   N-Benzylamines, 418
   N-Benzylamides, 427
Selective Reductions, 428
   Selectivity among N-Benzyl-Type Groupings, 428
   Selectivity among O-Benzyl-Type Groups, 428

N-Benzyl in the Presence of O-Benzyl, 430
Selective Reductions in the Presence of Other Reducible Groups, 433
  O-Benzyl in the Presence of Carbonyl, 433
  N-Benzyl in the Presence of Carbonyl, 435
  O- and N-Benzyl in the Presence of Nitriles, 436
  Benzyl Groups in the Presence of an Oximino Group, 436
  Benzyl Groups in the Presence of Hydroxylamine, 437
  Debenzylation in the Presence of a Hydrazine, 438
  Debenzylation in the Presence of Halogen, 438
  Miscellaneous, 440

XX. DEHALOGENATION   446
Introduction, 446
  Catalysts and Reaction Conditions, 447
  Condensation, 449
Aliphatic Halides, 449
Aryl Halides, ArX, 451
  Halogens in Benzenoid Systems, 451
  Halogens in Unsaturated Heterocycles, 452
Aralkyl Halides, 468
Halogen Attached to a Saturated Heterocycle, 471
Acyl Halides, 472
Selective Reductions, 473
  Among Dissimilar Halogens, 473
  Among Similar Halogens, 473
    In Similar-Type Halides, 474
      On the Same Carbon Atom, 474
      On Different Carbon Atoms, 475
        In Halobenzenes, 475
        In Unsaturated Heterocycles, 476
    In Dissimilar-Type Halides, 477
  In the Presence of Other Functions, 480
    Olefinic Bonds, 480
      Allylic Halides, 481
      Vinyl Halides, 481
    Carbonyl Groups, 482

Aldehydes, 482
Ketones, 483
  Allylic or Benzylic Type Haloketones, 483
  α-Haloketones, 483
  Vinyl Halides, 484
  Saturated Aliphatic Halides, 486
Cyano Groups, 487
Imino Groups, 487
N-Oxides, 488
Oximes, 488
Hydrazine Linkage, 488
Benzyl Groups, 489
  O-Benzyl, 489
  N-Benzyl, 490
Miscellaneous Ring Systems, 491

XXI. **HYDROGENOLYSIS OF CARBOXYL-CONTAINING GROUPS**   498

Acids and Esters, 498
  Selective Reductions, 500
Carboxamides, 501
Lactams, 502
Cyclic Imides, 503
Anhydrides, 504
Lactones, 506

XXII. **HYDROGENOLYSES OF ALCOHOLS, ETHERS, ACETALS, AND KETALS**   509

Alcohols, 509
Ethers, 511
  Linear Ethers, 511
  Cyclic Oxides, 511
    Ethylene Oxides, 511
      Selective Reductions, 514
        Presence of Olefinic Bonds, 514
        Presence of Carbonyl Groups, 514
        Presence of Halogen, 516
    Polymethylene Oxides, 516
Acetals and Ketals, 517

XXIII.	**MISCELLANEOUS HYDROGENOLYSES**	520
	Azides, 520	
	Hydrazines and Hydrazones, 521	
	Selective Reductions, 523	
	Peroxides and Hydroperoxides, 523	
	Nonaromatic Carbocyclic Rings, 524	
	Nonaromatic *N*-Heterocyclic Rings, 530	
	Nonaromatic *O,N*-Heterocycles, 535	
XXIV.	**AROMATIC RING SYSTEMS**	543
	Carbocycles, 544	
	Benzenes, 544	

      Alkylbenzenes, 544
      Alkoxybenzenes, 544
      Other Ethers, 546
      Benzyl Alcohols, Esters, and Ethers, 547
      Benzoic Acids, Esters, and Related Compounds, 548
        Partial Reduction, 548
      Phenylalkanoic Acids and Derivatives, 549
      Other Acids, 550
      Phenol and Substituted Phenols, 550
      Phenolic Acids and Esters, 552
      Di- and Polyhydric Phenols, 554
      Phenols to Cyclohexanones, 554
      Anilines, 556
      Phenylalkylamines, 561
      Stereochemistry, 562
      Selective Reductions, 563
        Presence of Functional Groups, 563
        Polyphenol Compounds, 564
    Indenes, 567
    Naphthalenes, 568
    Phenanthrenes, Anthracenes, and Other Fused Ring
        Systems, 572
  Heterocycles (one hetero atom), 573
    Furans, 573

Selective Reductions, 574
  In the Presence of Unsaturated Carbocycles, 574
  In the Presence of Functional Groups, 575
  In the Presence of Hydrogenolyzable Groups, 575
Pyrroles, 577
  Selective Reductions, 578
    Presence of Other Ring Systems, 578
    Presence of Functional Groups, 579
Pyrans and Pyrones, 580
  Selective Reductions, 582
Pyridines, 582
  Catalysts, 582
  Effect of the Position of Substituents, 586
  Pyridine Acids, Esters, and Amides, 586
  Hydroxypyridines and Derivatives, 589
  Aminopyridines, 590
  Pyridylalkylamines, 593
  Miscellaneous Pyridines, 593
  Quaternary Salts, 593
  Partial Reduction, 594
  Selective Reductions, 596
    Presence of Functional Groups, 596
    Presence of Hydrogenolyzable Groups, 598
    Presence of Other Rings, 600
Quinolines and Isoquinolines, 601
  Catalysts, 603
  Substituted Quinolines and Isoquinolines, 605
  Quaternary Salts, 606
  Perhydrogenation, 606
  Partial Reduction, 607
  Selective Reductions, 607
    Presence of Functional Groups, 607
    Presence of Halogen, 608
Azepines, 608
Heterocycles (two or more hetero atoms), 609
Oxazoles, 609

Pyrazoles and Related Compounds, 609
Imidazoles and Related Compounds, 610
Pyridazines and Related Compounds, 611
Pyrimidines, Quinazolines, and Related Compounds, 614
Pyrazines and Related Compounds, 618
Oxazepines, Diazepines, and Their Benzo Derivatives, 620
Triazoles, Benzotriazoles, and Triazines, 621
Miscellaneous Heterocycles (two or more hetero atoms in separate rings or in one ring with a bridgehead nitrogen), 623
Miscellaneous Polynuclear Heterocycles, 626
Three Rings with a Common Junction Point, 626
Unsymmetrically Fused Rings, 626
Symmetrically Fused Rings, 627

XXV. **HYDROGENATION OF SULFUR-CONTAINING COMPOUNDS**   641

Compounds Containing Hexavalent Sulfur, 641
Compounds Containing Tetravalent Sulfur, 642
Compounds Containing Divalent Sulfur, 643
Disulfides, 643
   Hydrogenolysis to Mercaptans, 643
   Hydrogenation of Reducible Functions, 644
Mercaptans, 645
Sulfides, RSR′, 645
   Hydrogenation of Reducible Functions, 645
     Where R and R′ are Aliphatic, 645
     Where R is Aliphatic and R′ is Aromatic, 645
     Where R is Aliphatic and R′ is Aralkyl, 647
     Where R and R′ are Aromatic, 647
     Where R is Aromatic and R′ is Aralkyl, 649
   Where RSR′ is a Ring System, 649
     Thiophenes, 649
     Benzo- and Dibenzothiophenes, 651
     Thiazoles and Related Compounds, 652

Benzothiazoles, 653
Thiadiazoles, 654
Phenothiazines, 654
Other Sulfur-Containing Ring Systems, 655
Saturation of Sulfur-Containing Rings, 655

INDEX 661

I

INTRODUCTION

The catalytic hydrogenation of organic compounds, an important tool in organic synthesis, developed from Sabatier's discovery of the reducing power of finely divided nickel in a hydrogen atmosphere [1]. This led to the investigation of other metals for similar activity.

The preparation of platinum black was reported at an earlier date [2], but it was not used for a number of years because of difficulties in obtaining a catalyst that gave reproducible results. It is worthy of mention because it was the forerunner of Adams's platinum oxide.

The development of colloidal palladium and platinum catalysts [3], not in use today, was an important finding because it led to the wide investigation of hydrogenations at low pressure.

The work of Adams with platinum oxide [4] was a very significant contribution. A catalyst, which was consistently active and whose preparation was not too involved, was now available for hydrogenations under mild conditions. The study of Hartung on the activity of palladium on carbon [5] was another important highlight. More recently supported rhodium catalysts [6] have been added to the chemist's armamentarium for reductions under a few atmospheres pressure.

High-pressure hydrogenations were first explored by Ipatieff with nickel in 1904 [7]. The development of Raney nickel [8] and the study of its activity by Adkins and his co-workers [9] made it possible to carry out reactions under less vigorous conditions. Further study on nickel on kieselguhr and on copper chromite by the same group provided more information for the chemist. The appearance of ruthenium on a support or as dioxide [10] made it possible to reduce compounds under milder high-pressure conditions (70–100° and 70 atm) than were employed in the past.

Other catalysts and combinations of them have been employed, as well as promoters to enhance the activity of known catalysts. They will be discussed in ensuing chapters.

Only heterogeneous type reductions will be covered, that is, reactions where the catalyst is not in solution. The emphasis will be on the commonly used catalysts—copper chromite, nickel, cobalt, the noble metals—and such others which may have unique properties or may have produced interesting reactions. The mechanics of the hydrogenation process will not be discussed too extensively, nor will there be a long discourse on the theory behind catalyst reductions. The physics and chemistry of metals, the energetics and kinetics of adsorption and desorption, and other phenomena related to surface chemistry are well covered by others [11, 12]. On the other hand, it is our purpose to study catalytic hydrogenation and the chemistry related to the conversion of functional groups and ring systems. These reductions will be divided into individual sections; for example, nitro groups, reduction of carbonyl function, benzenoid compounds, heterocycles, and so forth. Selective hydrogenation of one function in the presence of another reducible group, a subject of great interest to chemists, will be emphasized.

REFERENCES

[1] P. Sabatier and J. B. Senderens, *Compt. Rend.*, **124**, 1358 (1897).
[2] O. Loew, *Ber.*, **23**, 289 (1890).
[3] C. Paal and C. Amberger, *Ber.*, **37**, 124 (1904).
[4] V. Vorhees and R. Adams, *J. Am. Chem. Soc.*, **44**, 1397 (1922) and R. Adams and R. L. Shriner, *J. Am. Chem. Soc.*, **45**, 2171 (1923).
[5] W. C. Hartung, *J. Am. Chem. Soc.*, **50**, 3370 (1928).
[6] E. F. Rosenblatt, U.S. Patent 2,675,390 (1954).
[7] V. Ipatieff, *Catalytic Reactions at High Pressures and Temperatures*, Macmillan Co., New York, 1932.
[8] M. Raney, U.S. Patents 1,563,587 (1927), 1,628,190 (1927), and 1,915,473 (1933).
[9] H. Adkins, *Reactions of Hydrogen with Organic Compounds Over Copper Chromium Oxide and Nickel Catalysts*, University of Wisconsin Press, Madison, 1937.
[10] G. Gilman and G. Cohn, *Advances in Catalysis*, Vol. IX, Academic Press, New York, 1957, p. 707.
[11] G. C. Bond, *Catalysis by Metals*, Academic Press, New York, 1962, and references covered in this book.
[12] See series on *Catalysis*, P. H. Emmett, Ed., Reinhold, New York, 1954 and ensuing volumes.

II

FACTORS IN HYDROGENATION

In order to pursue the subject of catalytic reduction of organic compounds, it is necessary to become acquainted with some of the factors involved. Catalytic hydrogenation is a process by which certain metals in finely divided form or dispersed on carriers induce reaction to take place on their surface. In an oversimplification it can be stated that a reducible function or ring system of an organic compound may be converted to the desired state by subjecting it, in solution or suspension in a suitable medium, to attack by hydrogen under X atmospheres pressure at room or elevated temperature in the presence of a prescribed amount of a finely divided metal (known to have the property of catalyzing the reduction process) or such a metal on a support.

It must be borne in mind that reaction takes place at the surface of the catalyst. In order for reaction to succeed, the catalyst must have the ability to adsorb and activate the compound and hydrogen. The molecule must be held in the proper space relationship and finally the resultant product must be desorbed or removed from the catalyst. To further complicate matters it must be understood that not all of the surface of the catalyst is capable of allowing reaction to proceed. When hydrogenation does take place, it does so at certain reaction sites. This fact helps in understanding the inhibitory effect that certain elements, various impurities, and often reaction intermediates and final products have on hydrogen uptake. Indeed, very small amounts of particular inhibitors can completely prevent hydrogen uptake. These points will be discussed more fully in the section on catalyst poisons.

The possibility that certain impurities can impede the rate of hydrogen absorption makes it necessary to use reasonably pure products for reduction. It is also important to know about the preparation of the compound to be reduced so that measures can be taken to remove a possible contaminant or to overcome the effect of the presence of a reducible impurity. An illustration of the latter point occurred in this author's experience when hydrogenations of some substituted cinnamic acids to the corresponding phenylpropionic acids were to be undertaken for a colleague. In a sample run, a mixture of reduced and starting acid was obtained when the reaction was interrupted after adsorption of one molar equivalent of hydrogen. When a second reduction was carried out it was found that, if hydrogenation was not interrupted, uptake was almost 200% of theory. When it was learned that pyridine was used to prepare the cinnamic acid, it appeared obvious that some was present as the cinnamate salt and was being converted to piperidine phenylpropionate. Treatment of the reduction product with dilute hydrochloric acid yielded the desired acid in a high state of purity. In subsequent hydrogenations of other cinnamic acids, the contaminant was first removed by an acid wash. When it was no longer present, uptake of hydrogen did not proceed beyond theory for one molar equivalent.

One of the most important conditions for successful hydrogenation is good contact between that portion of the molecule to be reduced and the surface of the catalyst. The construction of molecular models of compounds to be reduced often can give an insight into what is to be expected. At times an examination of models will show whether contact will be clear and unobstructed or whether certain groups, or at times even a hydrogen atom, will prevent satisfactory contact between the catalyst and the reducible portion of the molecule. Not all of the answers will be provided. Nevertheless, models can supply useful information, even to the point of suggesting more vigorous reaction conditions. Numerous examples of their application will be found in other parts of this book.

Other factors in hydrogenation solvents, the use of promoters to speed up reaction, and the effect of acids and bases on the reduction process will be discussed in forthcoming sections.

III

CATALYSTS AND THEIR APPLICATIONS

The literature, in particular, the patent literature, is literally filled with hundreds of methods for preparing metals and their oxides or metals on supports for use as hydrogenation catalysts. Some of these materials are active only at elevated temperatures and pressures. These will be discussed first.

RHENIUM

Rhenium in the form of its oxides has been explored as a catalyst for hydrogenation. The preparation of these and the activity of each has been described [1a–f]. The oxides are not reduced to elemental rhenium but remain in the form of lower oxides after hydrogenation. Conditions for hydrogenations with the various rhenium oxides are rather vigorous; temperatures and pressures are rather high. There does not seem to be any particular advantage to their use in preference to other catalysts except in the hydrogenation of carboxylic acids to alcohols. These reactions are carried out at 150–250° and 200 atm pressure [1a]. Rhenium trioxide was reported to be unusually effective in the hydrogenation of amides.

RUTHENIUM

Ruthenium, as dioxide or as the metal on a support, has found numerous applications as a catalyst for hydrogenation. There is little need to describe its preparation since in its catalytic forms it is commercially available. It has considerable value in the reduction of functional groups and a number of ring systems. It is active at 70–100° and 60–70 atm pressure and may be the best catalyst available for these moderate high-pressure hydrogenations. At the same time it is more resistant to poisoning than other catalysts. For example, in the conversion of a number of nuclear-substituted anilines to the corresponding cyclohexylamines [2], crude starting materials were often used without any appreciable effect on the rate or extent of hydrogen uptake. A more concrete example of its resistance to poisoning can be seen in the reduction of 3,3'-dithio *bis*-(4-aminobenzenesulfonamide) to 3-mercapto-4-aminobenzenesulfonamide at 75° and less than 3 atm pressure [3]. It can be safely used in hydrogenation of pyridines in alcoholic solution without danger of *N*-alkylation. For moderate high-pressure hydrogenations it can well be the catalyst of choice for the conversion of benzene and pyridine rings to the saturated structures. Its one disadvantage in the latter reductions is the lack of selectivity when a phenyl group is a part of the pyridine compound. In such instances it is difficult to prevent some reduction of the benzene ring even though the pyridine ring is attacked first. Ruthenium catalysts have also been used for the hydrogenation of carboxylic acids, dicarboxylic acids, and hydroxy acids to the corresponding alcohols or diols. While the temperature (150°) is not excessive, pressures are generally very high (500–950 atm). The application of ruthenium catalysts for low-pressure reductions is quite limited. It has been found useful in the hydrogenation of ketones and aldehydes and is employed on commercial scale for the reduction of sugars. It will also catalyze the reduction of olefins and acetylenes. In the latter reductions there is no selectivity.

COPPER CHROMIUM OXIDE

Copper chromium oxide (copper chromite) is another catalyst useful only at elevated temperature and pressure. Its preparation has been described [4] but it is also commercially available. It is often promoted with a barium compound for stability to prevent reduction of the copper in the catalysts from its divalent state to a monovalent state, in which form it is much less active. Copper chromite is particularly active for the hydrogenation of aldehydes and ketones to alcohols without accompanying hydrogenolysis. It has little effect on the benzene ring. It is active for the hydrogenation of esters to alcohols and amides to amines. These particular reductions must be carried out at 250–300° and 250–300 atm pressure. In the laboratory reductions of esters and amides by means of lithium aluminum hydride has relegated copper chromite to a minor role.

COBALT

Cobalt in its catalytic forms is another metal that is active at higher temperatures and pressure. However, most of its use has been geared to the reduction of nitriles and dinitriles. Often these reductions can be carried out in the absence of ammonia to give high yield of primary amine. Raney cobalt was recommended for the hydrogenation of 3,4-methylenedioxybenzyl cyanide [5]. Its preparation and that of other cobalt catalysts and their uses are described. Raney cobalt was employed for the reduction of double bond and carbonyl linkages at atmospheric pressure [6] but was found to be much less active than Raney nickel.

NICKEL

Nickel catalysts have been prepared in a variety of ways by many individuals since Sabatier's discovery of its activity. Most of the work was directed to finding a better catalyst than was already available. No real success was achieved until the appearance of one developed by Raney [7]. His method produced a catalyst that was active and was convenient to use and prepare. It is commercially available and enjoys wide use today. It is a versatile catalyst capable of catalyzing hydrogenation at low pressure as well as under more vigorous conditions. It is being supplanted in certain ring reductions by rhodium and ruthenium catalysts because they can be employed under milder reaction conditions.

Essentially Raney nickel is prepared by leaching a 1:1 alloy of nickel and aluminum (Raney alloy) with sodium hydroxide solution in which most of the aluminum dissolves leaving the active catalyst. A preparation described by Mozingo [8] consists of treating the alloy with aqueous sodium hydroxide, maintaining the temperature below 25°. The mixture is then heated for 12 hr and the catalyst washed by decantation. Pavlic and Adkins [9] in a revised procedure treat the alloy with caustic at 50° and use a continuous washing process. They claim that the catalyst obtained is more active than the one prepared by Mozingo. Adkins and Billica [10] describe another procedure using a shorter digestion period at lower temperature followed by thorough washing with water in a hydrogen atmosphere. The catalysts obtained (W-6 and W-7 Raney nickel) are very active, differing only in that W-7 is alkaline. They also prepared a W-5 nickel which is less active than W-6 because it is washed at atmospheric pressure in the absence of hydrogen. An explanation of the symbols W-1 to W-7 may be seen in another article [11].

The very active catalysts, W-6 and W-7, are useful for hydrogenations at 2–3 atm pressure. Unfortunately, they retain high activity for only a short period of time. Adkins has reported that storage in the cold will maintain their activity for about two weeks. When freshly prepared, W-6 nickel not only can catalyze hydrogenation of functional groups at low pressure but also can bring about the conversion of quinoline to 1,2,3,4-tetrahydroquinoline and the conversion of naphthol to tetrahydronaphthol. Caution is necessary in using it at temperatures above 100° at high pressure because the reaction may get out of hand. At high pressure it has been used to hydrogenolyze esters to alcohols. W-7 Raney nickel gives very good results with ketones, phenols, and nitriles where the presence of alkali is beneficial.

The activity of Raney nickel is assumed by many workers in the field of hydrogenation to be due to hydrogen adsorbed on its surface. Others attribute the reducing activity to nickel hydride. Adkins, for example, suggests that the activity of W-6 Raney nickel is due to the large amount of adsorbed hydrogen. The implication for its lowered activity on standing is that hydrogen is removed. On the other hand, it could be accounted for by assuming that nickel hydride loses hydrogen: $NiH_2 \rightleftharpoons Ni + H_{2-}$ until a point where equilibrium is reached, then no further loss occurs. Bougault and co-workers [12] say nickel hydride or nickel containing nickel hydride is responsible for the activity of Raney nickel catalysts as does its discoverer [13] and Vandael [14]. On the other hand, Freidlin [15] and Sokolskii [16] attribute the activity to adsorbed and dissolved hydrogen. Smith [17] stated that there was no indication of nickel hydride, but that the catalyst surface is promoted with adsorbed hydrogen. Kokes and Emmett [18] say there is no adsorbed or dissolved hydrogen but that hydrogen is taken up in the nickel lattice by replacement of nickel atoms. However, all agree that they are able to show the presence of 40 to 100 cc of hydrogen or more per gram of Raney nickel. The amount is dependent, according to some, on the method of preparation of the catalyst. Raney and Vandael reported 380 cc per gram. It should be mentioned that the amount of hydrogen present is also dependent on the age of the catalyst.

There are a number of other nickel catalysts used for hydrogenation. Nickel on keiselguhr has been studied by Adkins. In general most other nickel catalysts require reaction at elevated temperature and pressure and usually are no more active than Raney nickel.

Japanese workers have studied the properties of a nickel catalyst discovered by Urushibara [19], which they state is easy to prepare but which requires temperatures from 70–200° and pressures of 60–100 atm to effect hydrogenation. Its preparation is relatively simple. Aqueous nickel chloride is treated with zinc dust and the precipitated nickel is digested with alkali. Work with the Urushibara catalyst, its modifications, and references to its use are summarized in a paper by Motoyama [20] who also reported that one of the catalysts was comparable in activity with Raney nickel for hydrogenation at ordinary pressures [21].

The search for improved nickel catalysts appears to be a never-ending one. Dominguez [22] describes a quick and very simple process of obtaining a catalyst designated as Raney nickel T-1. It is claimed that it is as active as the W-6 catalyst of Adkins for hydrogenation at 40–60° and 3–4 atm pressure, that it is easier to prepare, and that it retains its activity for a longer period when stored in a refrigerator. There are a number of other nickel catalysts available. Some are useful in the laboratory; others find application in the hydrogenation of fats and oils. In the ensuing work, references to different nickel forms will be given when they show some special application.

An item of interest in connection with Raney nickel catalyst is its use in acid solution. Adkins stated that acids should not be used with base metal catalysts [23]. Nevertheless, successful reductions have been carried out with nickel in acidic medium. Wetzel [24] found that a good yield of 2-chloro-5-acetaminoquinoline was obtained when 2-chloro-5-nitroquinoline was reduced in acetic acid–acetic anhydride medium in the presence of a moderate amount of Raney nickel. Kirkwood [25] also used nickel successfully in glacial acetic acid for conversion of some aromatic nitro compounds. In this reference the amount of nickel used is high. Nevertheless, in the presence of other reducible groups, the hydrogenations were selective. Even more striking were the hydrogenations by Wenner [26] of some tertiary amino aldehydes in aqueous acid solution. The pH of the solutions ranged from 3.0 to 6.0; the ratio of nickel to substrate was normal (about 10%). At 60–70° and 10–30 atm good yields were obtained. The author stated that reductions under these conditions gave better results than with noble metal catalysts.

PLATINUM

Before the advent of Adams's catalyst, liquid-phase hydrogenations had been carried out with the aid of colloidal palladium and platinum or with the so-called platinum black [27], a finely divided form of the metal containing varying amounts of oxygen. Colloidal catalysts were more active, but the difficulty in isolating the reaction products limited their use. Improvements in the method of preparation of platinum black by Willstatter [28] and further improvements by Willstatter and Waldschmidt-Leitz [29] gave a more active catalyst. Despite the fact that it was relatively easy to prepare, its activity often varied from batch to batch. As a result platinum blacks have fallen into discard. Only the catalyst developed by Adams is still on the scene. Details concerning its preparation are described [30], but in general commercially produced platinum oxide is used. The activity of Adams catalyst, the simplicity of its preparation, and the wide use it enjoyed created a demand that induced industrial scale preparation. The original process has been modified by substituting ammonium chlorplatinate for chlorplatinic acid, and certain changes in technique were also made so that large batches could be produced. The commercial process provides a catalyst that has high activity and is quite efficient for most hydrogenations.

There is a report of unsatisfactory results with platinum oxide prepared by the Adams procedure. The erratic hydrogenations of gossypol with it led Frampton and his associates [31] to modify the preparation to obtain an active catalyst which gave reproducible results in the same hydrogenation. Their modification consisted of instant heating of platinic chloride to 520°C in the presence of sodium nitrate.

Platinum oxide ($PtO_2 \cdot H_2O$) is not actually the catalyst but in reality is a precursor. Adams suggests that the actual catalyst is a lower oxide formed by reduction of platinum oxide with hydrogen. However, this is disputed by Caswell and Scheutz [32] who think the active material is the finely divided metallic platinum, as does this writer.

In general platinum oxide catalysts are quite useful in the reduction of aromatic nitro compounds and in the conversion of aldehydes to the corresponding alcohols. They will catalyze the reduction of phenols to the cyclic alcohols and also find application in the conversion of benzenoid and pyridinoid compounds to the corresponding saturated derivatives when these reductions are run in acidic media. Platinum catalysts are poisoned by nitrogen bases. Therefore, in those reductions in which strong bases are involved or where they are formed during reaction, it may be necessary to carry out hydrogenation in the presence of acid. It is for this reason that acetic acid is generally used with platinum catalyst for the reduction of pyridines to piperidines, or pyrroles to pyrrolidines and for the conversion of phenylalkylamines to cyclohexylakylamines. The hydrogenation of benzenoid compounds, excluding phenols, also requires the presence of acid [33, 34]. The attempt of Baker and Schuetz to reduce benzene under high pressure in the absence of acidic solvent is of interest [35]. When the hydrogenation was carried out at about 125 atm at room temperature in glacial acetic acid, reaction was complete in 15 min. In the absence of solvent or in dioxane or ethyl alcohol, no reduction took place at 180° and 135 atm.

It is well known that there is sodium present as an impurity in platinum oxide. Keenan [36] has found that prehydrogenation of platinum oxide in acetic acid or in methyl alcohol and thorough rinsing thereafter will give a catalyst that will hydrogenate benzene without the need to have acid present. The authors state that when the sodium component is removed, platinum oxide will catalyze the hydrogenation of benzene in the absence of additives or solvents.

Adams-type catalysts are active in the hydrogenation of carbon to carbon double bonds and for this reason are generally not used for acetylenic compounds when selective reduction to the olefin is desired. They have been used for some hydrogenolyses but are generally less effective than palladium catalysts for this purpose.

Adams-type catalysts are safely stored in the dry state and do not lose activity on standing. There is a report of a particularly active, extremely pyrophoric platinum catalyst which is described in the patent literature [37]. It is prepared by alloying platinum with a base metal and removing the metal by acid washing. The finely divided material contains adsorbed or dissolved hydrogen and must be stored in water or organic solvents. However, there is no mention of its use in the literature.

Baltzly [38] described the preparation of a platinized charcoal which has some interesting properties. It is claimed to be inactive for dehydrogenation and debenzylation and less efficient than platinum oxide for reductions of aromatic systems. It did not catalyze the hydrogenation of nitriles in the presence of hydrochloric acid. Aliphatic carbonyl compounds were reduced slowly in its presence or not at all. Simple aromatic carbonyl compounds, however, will be hydrogenated, although slowly, and uptake of hydrogen stops at one equivalent, a useful property which eliminates the danger of further conversion to the corresponding hydrocarbon. Diaryl ketones are not attacked in its presence. It is active for the reduction of aromatic nitro compounds, for azomethines, and for the reduction of carbon to carbon double bonds. It should be valuable for the selective reduction of carbon to carbon double bonds in the presence of carbonyl function because of the difference in rate which the author notes. He gives several examples of this selectivity.

All the reductions were carried out at low pressure and moderate temperature. Some of the characteristics of Baltzly's catalyst may change with more vigorous conditions. Platinum on carbon is commercially available, but slightly different from Baltzly's catalyst, which contains a trace of palladium. The commercially produced catalyst may not have quite the same properties but it does have many of those described by Baltzly. Before the publication by Baltzly there were only a few descriptions of the use of platinum on carbon. Most of them were in connection with vapor-phase reactions including aromatization of substitute hexanes.

PALLADIUM

Like platinum catalysts, low-pressure hydrogenations with palladium were carried out with this metal in colloidal form and as the black. Deficiencies noted for the same forms of platinum also apply to palladium. Much useful information about the reduction of organic compounds was gained from their use in the past, but because of their deficiencies are not used today.

Busch and Stove [39] first described the preparation of palladium on a carrier, calcium carbonate. Others followed suit using barium sulfate [40], charcoal [41], and other inert materials. A number of these and their preparations are described by Mozingo [42]. The palladium on charcoal catalyst which Ott and Schroter [41] prepared and employed for selective reduction of acetylenic compounds was further studied and more uses developed by Hartung [43]. In a later report he described a more active catalyst which was prepared in sodium acetate solution [44].

Palladium on various carriers enjoy a wide variety of uses. Their activity may be dependent on the carrier. This has been mentioned by Hartung in connection with palladium on carbon. Palladium catalyst will reduce aromatic nitro groups very readily. Schiff bases are easily converted to secondary amines. It is by far the best catalyst available for dehalogenation and debenzylation. It is used for the reduction of nitriles and oximes. When inactivated by certain inhibitors it will allow the hydrogenation of acetylenes to stop before the ethylenic linkage is reduced. It had been first reported that it could not catalyze the reduction of benzene or pyridine rings. More recent work has shown that this was dependent on reaction conditions. Aromatic compounds can be converted to cyclic structures at 80° and 3–4 atm pressure in the presence of acid. In a number of instances, reduction in neutral solution is readily effected at 70 atm pressure. Pyridines have been hydrogenated to piperidines in the presence of an acid in low-pressure reductions and in the absence of an acid at higher pressure. These will be described among the proper classifications in other sections. Mozingo and associates [45] found that when large amounts of palladium on carbon and palladium on barium sulfate were used, hydrogenation of divalent sulfur-containing compounds was successful under low-pressure conditions.

The wide use palladium catalysts have enjoyed led to preparation on commercial scale. Palladium on a number of supports can be purchased from various vendors throughout the world.

From time to time there are reports of improvements in the preparation of palladium catalysts. One describes the preparation of a catalyst at low temperature which yields a more active form of palladium on carbon [46]. This may be a point worth investigating. A private communication from a commercial producer revealed that palladium on carbon, which is heated after washing to dry the catalyst, is less active for reductions than material from the same batch which is allowed to dry with 50% of its weight as water. The loss of activity is ascribed to the long heating period. Another interesting effect of temperature is noted in an Italian patent [47] where palladium deposit is made at 10, 6, and 2°, the last figure giving the most active catalyst.

There are certain other palladium catalysts that have special uses. Lindlar catalyst, which is a lead-poisoned palladium catalyst, and others of the same type find application in selective conversions of acetylenic linkages. They will be mentioned in greater detail in the reduction of acetylenes.

Palladium on silk fibroin has been developed by Akabori [48] for asymmetric hydrogenations. It is used at 50–75° and 80–90 atm pressure. Other silk and nylon supported palladium catalysts were reported by Izumi [49], who was a member of Akabori's research team. Some of the latter catalysts were found to lose their capacity for asymmetric hydrogenation.

RHODIUM

One of the very important additions to the field of hydrogenation was the appearance of rhodium in its catalytic forms. By far its greatest value is its ability to foster hydrogenation of aromatic systems to the corresponding cyclic and saturated heterocyclic compounds at low pressure, often in the absence of acids. Earlier literature on catalytic reductions with it is concerned with colloidal rhodium [50–54] with only limited reference to ring conversions [50, 51]. The first publication of the use of rhodium on a support, carbon [55], showed its applicability in the reduction of carbon to carbon double bonds, aromatic nitro groups and intermediates, and the carbonyl function. There is also a mention of the reduction of benzonitrile to dibenzylamine in the absence of ammonia. A patent on the use of rhodium on different supports [56] gives examples for the conversion of benzenes to cyclohexanes, pyrrole to pyrrolidine, furoic acid to tetrahydrofuroic acid in water or in acetic acid, at room temperature and atmospheric pressure. Unfortunately, as is often the failing of patents, the products of reduction are not described. The only criterion appears to be completion of hydrogen uptake. Nevertheless, this work did point the way to the value of rhodium for ring reductions under very mild reaction conditions. 2,5-Dimethylpyrrole and 2,6-dimethylpyridine were converted to the corresponding saturated heterocycles with rhodium on alumina in acetic acid solution [57]. Even more interesting in the same article was the reduction of 1-amino-2,5-dimethylpyrrole to 1-amino-2,5-dimethylpyrrolidine since N-aminoheterocycles are prone to deaminate. Gilman and Cohn [58] discuss the activity of rhodium in a study of the hydrogenation of substituted benzenes, pyridine, pyrrole, and some furans in water, alcohol, or acetic acid with 1 g of 5% rhodium on alumina for 1 g of material to be reduced. Freedman and co-workers [59] report the successful conversion of phenylphosphonic and phosphanilic acids to the cyclohexyl compounds in good yield. Other investigators have studied nuclear hydrogenations of substituted benzyl alcohols and esters [60], of aniline [61], and of alkoxyanilines [62]. Rhodium on a carrier has also been explored with good results for the ring reduction of some substituted pyridines [63–66]. It has also been found worthwhile in the conversion of other heterocycles as pyran [67], quinoxaline [68], some pyrimidines [69–71] and carbazoles [72]. One of the reasons it is preferred in ring reductions, in addition to its ability to catalyze them in the absence of an acid, is the fact that it causes much less hydrogenolysis than other catalysts. This has been shown by Smith and Thompson [73] in a study of the reduction of mono-, di-, and trimethoxybenzenes with rhodium and with platinum. Previously cited references [58, 62] also point this out.

Another of its preferred uses is in the low-pressure hydrogenation of nitriles in the presence of excess ammonia [74]. In most instances good yields of primary amines are obtained, while the amount of secondary amine is kept low.

The rate of hydrogenation with rhodium is considerably influenced by changes in pH [51, 52] and by the presence of adjacent functional groups in the substrate molecule [52, 58]. Although it has been found useful in the hydrogenation of nitrogen heterocycles, the extent of reduction can be influenced by its environment. This point is dramatically illustrated in the reduction of 3-carboxy-1,2,3,4-tetrahydroisoquinoline and its ethyl ester [75] where the acid is converted to decahydro compound while in the reduction of the ester the ring remains unaffected.

$$(3.1)$$

The conversion of the 3-carboxy compound to the decahydro compound is an excellent illustration of the neutralization of the inactivating effect of the basic nitrogen on the catalyst, thereby allowing ring reduction to take place. In the hydrogenation of the ester the basic nitrogen exerts its influence as an inhibitor to prevent ring reduction. The phenomenon of inhibition of catalytic activity by nitrogen bases will be discussed more fully in the section on catalyst poisons.

Among benzenoid compounds Gilman and Cohn [58] report that the introduction of a carboxyl group lowers the rate of ring reduction more than alkyl or hydroxy groups, pointing out that the two carboxyl groups as in phthalic acid lower the rate even more. There is a surprisingly difficult reduction noted in a patent [76] in which terephthalic acid was converted to cyclohexane-1,4-dicarboxylic acid in 38% yield by reduction in water in the presence of 5% rhodium on carbon at 300° and 330 atm. An added statement indicated that the yield was lower under less drastic conditions. This is in direct contrast to low-pressure hydrogenation of some benzene polycarboxylic in aqueous solution where very high yields of the cyclic compounds were realized [77].

METAL SULFIDES AND SULFACTIVE CATALYSTS

The metal sulfides have been used mainly for the hydrogenation of carbonyl compounds in the presence of sulfur to yield thiols [78a, b]. Reaction conditions are in the range of 150° and above 70 atm pressure; the sulfides for the most part are those of cobalt, molybdenum, and nickel. Other uses have been suggested because they are resistant to poisoning. Broadbent [79] has studied the activity of rhenium heptasulfide compared with that of molybdenum sulfide and cobalt polysulfide for the reduction of a number of organic compounds but submitted that it is far less active than palladium on a carrier for the same reductions. Campaigne [80] developed what may be a worthwhile use for cobalt polysulfide when he found that the thiophene ring was reduced in its presence at 200–225° and 100–150 atm pressure. However, in the hydrogenation of the acylthiophenes which he investigated, the C=O function was reduced to CH_2 every time. In addition, when 2-acetyl-5-bromothiophene was reduced, debromination took place to give 2-ethylthiophene and 2-ethyltetrahydrothiophene. There is little doubt that loss of bromine occurred concurrently with conversion of the 2-acetyl group to 2-ethyl, followed by subsequent ring reduction. In a more recent study of base metal sulfides in the reductive alkylations of aromatic amines, some were found to be of value at 180–190° and 30–50 atm [81]. In a later publication the same authors [82] report that the platinum metal sulfides are far more useful. Hydrogenations with them can be carried out at about 100° and 35–50 atm. They have been reported to be active in the reduction of halonitrobenzenes. No dechlorination was observed with palladium, platinum, rhodium, and ruthenium sulfides. No debromination was noted with platinum sulfide but the amount was appreciable during hydrogenation with palladium sulfide. A patent [83] describes some reductions with noble metal sulfides and the sulfides on carriers in detail.

In a private communication one of the users of the noble metal sulfides reported that there was little if any evidence of hydrogen sulfide and no apparent poisoning. On the other hand, some vendors think contamination may result and suggest that vessels should be thoroughly cleaned before use for other catalytic reductions. There is also private information that vessels in which base metal sulfides are employed require extensive cleaning.

OTHER CATALYSTS

As pointed out before there is a never-ending search for hydrogenation catalysts that are more active than those in use or that have some especially desirable characteristic that a particular investigator needs. Some may be of value commercially, others may be of considerable worth for preparative purposes. Brown and Brown [84] studied the use of cobalt, iron, and nickel salts and salts of the noble metals for the hydrolysis of sodium borohydride. They found that of this group, the compounds of rhodium, ruthenium, and platinum were very effective in releasing hydrogen from an aqueous solution of sodium borohydride. As a result of their study they continued an investigation into the catalytic activity of the finely divided metals they obtained by this process [85]. They pointed out that the platinum and rhodium catalysts so obtained were more active than platinum oxide in the hydrogenation of 1-octene. This increased activity may well be due to the presence of adsorbed hydrogen on the surface of the catalyst. In another publication [86] they noted that the combination of platinum catalyst prepared *in situ* and the use of their procedure gave better results in the reduction of olefins than reaction with platinum oxide and electrolytic hydrogen. Their procedure, with the present cost of sodium borohydride, may not be commercially feasible since the borohydrides are an expensive source of hydrogen. Nevertheless, the method should be very useful for preparative scale hydrogenations. It is even possible that, where a particularly active catalyst is desired, it may be prepared according to the described procedure, washed by decantation to remove water and salts, and subsequently used in a specific reduction.

Bott and his colleagues [87] report an improved catalyst for the hydrogenation of olefins. It is obtained by reacting chloroplatinic acid with silicon hydrides. The authors claim it is more active than platinum oxide or the platinum catalyst of Brown and Brown [85].

Before the work of Brown and Brown, Paul and co-workers [88] had studied the activity of nickel, cobalt, and copper catalysts prepared by treatment of their salts with sodium borohydride solution. None was more active than Raney nickel. Some complexes from mixtures of salts were said to be much more active than Raney nickel. The authors stated that the black precipitates obtained were borides of the various metals. The chief virtue of these catalysts may be the ease of preparation and that they are neither pyrophoric nor magnetic [89]. The authors also found that they could be reused many times without significant loss of activity.

Watt had reduced a number of salts of noble metals with potassium and liquid ammonia which gave pyrophoric platinum metal that was able to hydrogenate olefins [90]. Similar treatment of ruthenium, rhodium, and palladium salts [91] yielded the metals which were reported to be exceptionally active for the hydrogenation of olefins and aromatic nitro compounds at 30° and 2 atm pressure. No other reducible groups were studied.

MIXED NOBLE METAL CATALYSTS

Young and Hartung [92] studied the effects of the addition of other noble metals to palladium on carbon. A 2% ratio of platinum, rhodium, ruthenium, osmium, or iridium to palladium metal was used in a number of reductions. The results did not show a particular pattern. A single metal activated the reduction for one substrate and inhibited the rate for another. In previous work Hartung [93] found in attempting to reduce some α-oximinoketones with palladium on carbon that a mixture of aminoketone and aminoalcohol was obtained. The addition of platinum chloride gave good yields of aminoalcohols. In the hydrogenation of oximinoamides [94] of the type

$$C_6H_5CH_2\underset{\underset{NOCH_2C_6H_5}{\|}}{C}CONHR \qquad (R = aryl)$$

with palladium on carbon reaction stopped after uptake of one molar equivalent of hydrogen to give the debenzylated oxime. At higher temperature the reaction proceeded further but never to completion. With palladium on carbon catalyst containing rhodium in a ratio of 1 part to 50 parts of palladium, hydrogenation usually led to the desired aminoamide.

Nishimura [95] in the first of a series of papers describes the preparation and properties of rhodium-platinum oxide. The catalyst consists of a 3:1 rhodium to platinum ratio by weight of metal. It is obtained by fusing a mixture of rhodium chloride and chloroplatinic acid or ammonium chloroplatinate as in the preparation of Adams catalyst [30]. The author states it is many times more active than platinum oxide. It is said to retain many of the characteristics of rhodium catalysts in its ability to hydrogenate the aromatic nucleus without accompanying hydrogenolysis. The tendency of the mixed rhodium-platinum catalyst to inhibit hydrogenolysis is illustrated, although not so markedly, in the attempted hydrogenations of cinnamyl alcohols [96].

It is further pointed out that the same catalyst will convert furfuryl alcohol to tetrahydrofurfuryl alcohol quantitatively without ring opening [97]. In the same study the advantage of it over the use of platinum oxide in the hydrogenation of pyridine and quinoline is also pointed out.

The exact structure of the mixed catalyst is not known but x-ray diffraction showed that rhodium-platinum oxide was not identical with a mechanical mixture of rhodium and platinum oxides in a 3:1 ratio by weight of the metals.

It is well known that rhodium oxide is not converted to the metal quickly at moderate temperature and pressure. The rhodium-platinum oxide on the other hand is easily reduced by hydrogen to its active state. In an effort to learn whether its activity was due entirely to the presence of rhodium, Nishimura [98] prepared other rhodium-platinum oxides of varying composition, including rhodium oxide alone, and compared their properties and activity. The longer time necessary to hydrogenate toluene with rhodium oxide than with the combinations of 90–10, 80–20, 75–25 and 70–30 rhodium-platinum oxides leads the author to suggest that the presence of platinum is advantageous. It does promote the conversion of the oxides to the catalytic form of the metals. The author further states that the presence of platinum makes the catalyst more selective as shown by the total hydrogen uptake in the conversion of acetophenone to 1-hydroxyethylcyclohexane. The figures, 4.08 molar equivalents for the reduction of 0.005 mole with rhodium oxide against 4.02–4.03 equivalents for the best of the rhodium platinum combinations, are not significant enough to support his statement.

Most of Nishimura's investigations were carried out under low-pressure conditions. As might be expected, higher pressures (100–150 atm) increased the speed of reaction in another series of reductions with rhodium-platinum oxides [99]. What is significant in this series is the generally high yields of cyclic compounds, indicative of the lower tendency toward hydrogenolysis compared with platinum or palladium. Benzyl ether, for example, was converted to cyclohexylmethyl ether in 83% yield and acetophenone was hydrogenated to the cyclic carbinol in 86% yield.

There is little question that this and other noble metal combinations display a synergistic effect in a number of catalytic reductions. How they act is an undetermined question. Rylander and Cohn [100] investigated hydrogenations of certain organic compounds with platinum-ruthenium and palladium-ruthenium compounds where the metals were co-precipitated on a carbon of high surface area. Mechanical mixtures also showed improved activity in comparison with a single metal on carbon but the co-precipitated catalysts gave somewhat better results. The authors studied the reaction rates of a number of reductions with each metal on the carrier. They concluded from their results that the improvement in rate with the binary metal systems may have been due to the hydrogenation of inhibiting intermediates or by-products with ruthenium. This led them to suggest that coprecipitation may have accomplished nothing more than supplying a second metal to reduce or remove an undesirable reactant. In another article one of the authors [101] offers the suggestion that reductions with ruthenium-platinum and ruthenium-palladium catalysts proceed either stepwise or by multiple paths and that each metal hydrogenates the substrate or intermediate which, if allowed to accumulate, would act as inhibitor for the other metal. It appears to be logical since reductions in the presence of either metal do not proceed as well as similar ones catalyzed by the binary metal system. The explanation may be an oversimplification. It is also possible that the reaction to produce the mixed metal catalysts causes a change in surface characteristics or results in more active sites for reduction to take place. Several patents [102-104] describe in detail the preparation of ruthenium containing palladium, platinum, and rhodium on a number of carriers.

Bond [105, 106] studied ruthenium-platinum oxides prepared in the same manner as the rhodium-platinum oxides of Nishimura [95, 96]. The ruthenium content varied from 3 to 30%. The catalysts had high activity in a number of reductions carried out at 30° and several atmospheres pressure. The maximum reaction rate varied with ruthenium content. In every instance, however, the increase in rate was 1.5 to 3.5 times that of the same reduction in the presence of platinum oxide. An acetylenic alcohol, maleic acid, and cyclohexene were converted to the saturated compounds. Cyclohexanone and acetophenone were reduced to the corresponding alcohols. Pyridine was also hydrogenated successfully. The nature of the reduced catalyst is still unknown; Bond thinks the metals are alloyed and not an admixture. He suggests that there is indirect evidence to support this thought. Mechanical mixtures of ruthenium and platinum oxides were not found to be highly active. Ruthenium dioxide is not readily reduced under the modest conditions used to reduce the mixed oxides, although platinum oxide is.

Some unpublished work [107] in a hydrogenation of a diphenylamino-alcohol in aqueous acid in which one aromatic ring is selectively reduced indicated that a 5% ruthenium-containing platinum oxide was somewhat more active than the corresponding amount of platinum oxide.

The greater application of these binary metal systems offers an interesting avenue of research because of their activity under moderate reaction conditions. Under more vigorous conditions their activity should be substantially increased. In the laboratory the use of these mixed catalysts may be of considerable value. Whether they could find commercial utility depends on a number of factors, one of which is the cost of separating and recovering the metals. On commercial scale a spent noble metal catalyst (often including paper filter papers or cloths) is returned to the vendor. There the mass is further processed and the metal is recovered. The value of the recovered noble metal is credited to the user's account based on the market price of the metal, less the cost of recovery. The cost of recovery is based on the ease or difficulty of obtaining pure metal. Separation of mixed metals naturally will add to the cost. According to some producers of mixed metal catalysts, the cost of separation is not high and the advantage of their use often counterbalances the cost of separation.

REFERENCES

[1] a. H. S. Broadbent, G. C. Campbell, W. J. Bartley and J. H. Johnson, *J. Org. Chem.*, **24**, 1847 (1959); b. H. S. Broadbent and J. H. Johnson, *J. Org. Chem.*, **27**, 4400 (1962); c. H. S. Broadbent and J. H. Johnson, *J. Org. Chem.*, **27**, 4402 (1962); d. H. S. Broadbent and T. G. Selin, *J. Org. Chem.*, **28**, 2343 (1963); e. H. S. Broadbent and W. J. Bartley, *J. Org. Chem.*, **28**, 2345 (1963); f. H. S. Broadbent and D. W. Seegmiller, *J. Org. Chem.*, **28**, 2347 (1963).
[2] M. Freifelder and G. R. Stone, *J. Org. Chem.*, **27**, 3568 (1962).
[3] W. H. Vinton, U.S. Patent 2,483,447 (1949).
[4] H. Adkins, R. Connor, and K. Folkers, *J. Am. Chem. Soc.*, **54**, 1138 (1932).
[5] W. Reeve and W. M. Eareckson, III, *J. Am. Chem. Soc.*, **72**, 3299 (1950).
[6] B. V. Aller, *J. Appl. Chem.*, **8**, 492 (1958).
[7] M. Raney, U.S. Patents 1,563,587 (1927), 1,628,190 (1927), and 1,915,473 (1933).
[8] R. Mozingo, *Org. Synthesis*, **21**, 15 (1941).
[9] A. A. Pavlic and H. Adkins, *J. Am. Chem. Soc.*, **68**, 1471 (1946).
[10] H. Adkins and H. R. Billica, *J. Am. Chem. Soc.*, **70**, 695 (1948).
[11] H. Adkins and A. A. Pavlic, *J. Am. Chem. Soc.*, **69**, 3039 (1947).
[12] I. J. Bougault, E. Cattelain, and P. C. Chabrier, *Bull. Soc. Chim. France*, **5**, 1699 (1938).
[13] M. Raney, *Ind. Eng. Chem.*, **32**, 1199 (1940).
[14] C. Vandael, *Ind. Chem. Belge*, **17**, 581 (1952); through *Chem. Abstr.*, **46**, 9396 (1952).
[15] L. Kh. Freidlin, *Doklady Akad. Nauk.*, **74**, 955 (1950); through *Chem. Abstr.*, **45**, 1836 (1951).
[16] D. V. Sokolskii and S. T. Bezverkhova, *Doklady Akad. Nauk.*, **94**, 493 (1954); through *Chem. Abstr.*, **49**, 10029 (1955).
[17] H. A. Smith, A. J. Chadwell, and S. S. Kinlis, *J. Phys. Chem.*, **59**, 820 (1955).
[18] R. J. Kokes and P. H. Emmett, *J. Am. Chem. Soc.*, **81**, 5032 (1959).
[19] Y. Urushibara, *Bull. Chem. Soc. Japan*, **25**, 280 (1952).
[20] I. Motoyama, *Bull. Chem. Soc. Japan*, **33**, 232 (1960).
[21] S. Nishimura, *J. Chem. Soc. Japan, Pure Chem. Soc.*, **78**, 1741 (1957).
[22] X. A. Dominguez, I. C. Lopez, and R. Franco, *J. Org. Chem.*, **26**, 1625 (1961).
[23] H. Adkins, *Reactions of Hydrogen with Organic Compounds over Copper Chromium Oxide and Nickel Catalysts*, University of Wisconsin Press, Madison, Wisconsin, 1937, p. 25.
[24] J. W. Wetzel, D. E. Weldton, J. E. Christian, G. L. Jenkins, and G. B. Bachman, *J. Am. Pharm. Assoc. Sci. Ed.*, **35**, 331 (1946).
[25] S. Kirkwood and P. H. Phillips, *J. Am. Chem. Soc.*, **69**, 934 (1947).

[26] W. Wenner, *J. Org. Chem.*, **15**, 301 (1950).
[27] O. Loew, *Ber.*, **23**, 289 (1890).
[28] R. Willstatter and D. Hatt, *Ber.*, **45**, 1471 (1912).
[29] R. Willstatter and E. Waldschmidt-Leitz, *Ber.*, **54**, 121 (1921).
[30] R. Adams, V. Vorhees, and R. L. Shriner, *Org. Syn.*, Coll. Vol. 1, Wiley, New York, 1932, p. 452.
[31] V. L. Frampton, J. D. Edwards, Jr., and H. R. Henze, *J. Am. Chem. Soc.*, **73**, 4432 (1951).
[32] L. R. Caswell and R. D. Schuetz, *Iowa Acad. Sci.*, **68**, 223 (1961).
[33] R. Adams and J. R. Marshall, *J. Am. Chem. Soc.*, **50**, 197 (1928).
[34] H. A. Smith, D. M. Alderman, and F. W. Nadig, *J. Am. Chem. Soc.*, **67**, 272 (1945).
[35] R. H. Baker and R. D. Schuetz, *J. Am. Chem. Soc.*, **69**, 1250 (1947).
[36] C. W. Keenan, B. W. Giesemann, and H. A. Smith, *J. Am. Chem. Soc.*, **76**, 229 (1954).
[37] J. S. Streicher, U.S. Patent 2,384,501 (1945).
[38] R. Baltzly, *J. Am. Chem. Soc.*, **74**, 4586 (1952).
[39] M. Busch and H. Stove, *Ber.*, **49**, 1063 (1916).
[40] E. Schmidt, *Ber.*, **52**, 400 (1919).
[41] E. Ott and R. Schroter, *Ber.*, **60**, 624 (1927).
[42] R. Mozingo, *Org. Syn.* Vol. XXVI, Wiley, New York, 1946, p. 77.
[43] W. C. Hartung, *J. Am. Chem. Soc.*, **50**, 3370 (1928).
[44] H. K. Iwamoto and W. H. Hartung, *J. Org. Chem.*, **9**, 513 (1944).
[45] R. Mozingo, S. A. Harris, D. E. Wolf, C. E. Hoffhine, Jr., N. R. Easton, and K. Folkers, *J. Am. Chem. Soc.*, **67**, 2092 (1945).
[46] M. E. Hultquist and R. P. Germann, U.S. Patent 2,520,038 (1950).
[47] Italian Patent 599,058 (1959); through *Chem. Abstr.*, **56**, 14979 (1962).
[48] S. Akabori, Y. Izumi, Y. Fujii, and S. Sakurai, *Nature*, **178**, 323 (1956).
[49] Y. Izumi, *Bull. Chem. Soc. Japan*, **32**, 932, 936, 942 (1959).
[50] G. Kahl and E. Bielaski, *Z. anorg. allgem. Chem.*, **230**, 88 (1936); through *Chem. Abstr.*, **31**, 3371 (1937).
[51] C. Zenghelis and C. Stathis, *Compt. Rend.*, **206**, 682 (1938).
[52] L. Light, *Chem. Products*, **3**, 29 (1940).
[53] L. Hernandez and F. F. Nord, *Experientia*, **3**, 489 (1947).
[54] L. Hernandez and F. F. Nord, *J. Colloid Sci.*, **3**, 363 (1948).
[55] W. P. Dunworth and F. F. Nord, *J. Am. Chem. Soc.*, **74**, 1459 (1952).
[56] E. F. Rosenblatt, U.S. Patent 2,675,390 (1954).
[57] C. G. Overberger, L. C. Palmer, B. S. Marks, and N. R. Byrd, *J. Am. Chem. Soc.*, **77**, 4100 (1955).
[58] G. Gilman and G. Cohn, *Advances in Catalysis*, Vol. IX, Academic Press, New York, 1957, p. 707.
[59] L. D. Freedman, G. O. Doak, and E. L. Petit, *J. Am. Chem. Soc.*, **77**, 4262 (1955).
[60] J. H. Stocker, *J. Org. Chem.*, **27**, 2288 (1962).
[61] R. M. Robinson, U.S. Patent 3,196,179 (1965).
[62] M. Freifelder, Y. H. Ng, and P. F. Helgren, *J. Org. Chem.*, **30**, 2485 (1965).
[63] M. Freifelder, R. M. Robinson, and G. R. Stone, *J. Org. Chem.*, **27**, 284 (1962).
[64] M. Freifelder, *J. Org. Chem.*, **28**, 602 (1963).
[65] M. Freifelder, *J. Org. Chem.*, **28**, 1135 (1963).
[66] M. Freifelder and H. B. Wright, *J. Med. Chem.*, **7**, 664 (1964).
[67] H. C. Silberman, *J. Org. Chem.*, **25**, 151 (1960).
[68] H. S. Broadbent, E. L. Allred, L. Pendleton, and C. W. Whittle, *J. Am. Chem. Soc.*, **82**, 189 (1960).

[69] W. E. Cohn and D. G. Doherty, *J. Am. Chem. Soc.*, **78**, 2863 (1956).
[70] M. Green and S. S. Cohen, *J. Biol. Chem.*, **225**, 397 (1957).
[71] M. Green and S. S. Cohen, *J. Biol. Chem.*, **225**, 601 (1957).
[72] H. Dressler and M. E. Baum, *J. Org. Chem.*, **26**, 102 (1961).
[73] H. A. Smith and R. G. Thompson, *Advances in Catalysis*, Vol. IX, Academic Press, New York, 1957, p. 727.
[74] M. Freifelder, *J. Am. Chem. Soc.*, **82**, 2386 (1960).
[75] R. T. Rapala, E. R. Lavagnino, E. R. Shepherd, and E. Farkas, *J. Am. Chem. Soc.*, **79**, 3770 (1957).
[76] H. C. Dehm and L. G. Maury, U.S. Patent 2,888,484 (1959).
[77] M. Freifelder, unpublished work.
[78] a. M. W. Farlow and F. K. Signaigo, U.S. Patent 2,402,615 (1946); b. B. W. Howk, U.S. Patent 2,402,646 (1946).
[79] H. S. Broadbent, L. H. Slaugh, and N. L. Jarvis, *J. Am. Chem. Soc.*, **76**, 1519 (1954).
[80] E. Campaigne and J. L. Diedrich, *J. Am. Chem. Soc.*, **73**, 5240 (1951).
[81] F. S. Dovell and H. Greenfield, *J. Org. Chem.*, **29**, 1265 (1964).
[82] F. S. Dovell and H. Greenfield, *J. Am. Chem. Soc.*, **87**, 2767 (1965).
[83] Netherlands Appl. 6,409,250 (1965); through *Chem. Abstr.*, **63**, 11428 (1965).
[84] H. C. Brown and C. A. Brown, *J. Am. Chem. Soc.*, **84**, 1493 (1962).
[85] H. C. Brown and C. A. Brown, *J. Am. Chem. Soc.*, **84**, 1494 (1962).
[86] H. C. Brown and C. A. Brown, *J. Am. Chem. Soc.*, **84**, 1495 (1962).
[87] R. W. Bott, C. Eaborn, E. R. A. Peeling, and D. E. Webster, *Proc. Chem. Soc.*, 337 (1962).
[88] R. Paul, P. Buisson, and N. Joseph, *Compt. Rend.*, **232**, 627 (1951).
[89] R. Paul, P. Buisson, and N. Joseph, *Ind. Eng. Chem.*, **44**, 1006 (1952).
[90] G. W. Watt, M. T. Walling, Jr., and P. I. Mayfield, *J. Am. Chem. Soc.*, **75**, 6175 (1953).
[91] G. Watt, A. Broodo, W. A. Jenkins, and S. G. Parker, *J. Am. Chem. Soc.*, **76**, 5989 (1954).
[92] J. G. Young and W. H. Hartung, *J. Org. Chem.*, **18**, 1659 (1953).
[93] W. H. Hartung and Y. Chang, *J. Am. Chem. Soc.*, **74**, 5927 (1952).
[94] J. W. Martin, Jr. and W. H. Hartung, *J. Org. Chem.*, **19**, 338 (1954).
[95] S. Nishimura, *Bull. Chem. Soc. Japan*, **33**, 566 (1960).
[96] S. Nishimura, T. Onoda, and A. Nakamura, *Bull. Chem. Soc. Japan*, **33**, 1356 (1960).
[97] S. Nishimura, *Bull. Chem. Soc. Japan*, **34**, 32 (1961).
[98] S. Nishimura, *Bull. Chem. Soc. Japan*, **34**, 1544 (1961).
[99] S. Nishimura, *Bull. Chem. Soc. Japan*, **36**, 353 (1963).
[100] P. N. Rylander and G. Cohn, *Proc. Intern. Congr. on Catalysis, 2nd, Paris, 1960*, 1, pp. 974–984 (published 1961).
[101] P. N. Rylander, *Engelhard Industries Tech. Bull.*, **1**, 133 (1961), Newark, New Jersey.
[102] J. H. Koch, Jr., U.S. Patent 3,055,840 (1962).
[103] P. N. Rylander and J. H. Koch, Jr., U.S. Patent 3,177,258 (1965).
[104] J. H. Koch, Jr., U.S. Patent 3,183,278 (1965).
[105] G. C. Bond and D. E. Webster, *Proc. Chem. Soc.*, **1964**, 398.
[106] G. C. Bond and D. E. Webster, *Platinum Metals Review*, **9**, 12 (1965), Johnson Matthey and Co., Ltd., London.
[107] Private communication from Dr. C. Forman, Abbott Laboratories, North Chicago, Illinois.

IV

CATALYST INHIBITORS AND POISONS

One of the problems in catalytic hydrogenation is the inhibiting effect which certain substances have on the reaction. Baltzly [1] attempts to distinguish between these substances. According to his definition, inhibitors diminish the reaction rate markedly when present in appreciable concentration but can be removed by washing. Poisons, on the other hand, interfere with the action of a catalyst and are not removed with a convenient amount of washing. Maxted [2] regards substances as poisons if they exert an appreciable inhibitory effect when present in small amounts. This author prefers not to differentiate but to consider, as poisons or inhibitors, materials that retard reaction or prevent it from going to completion. At best, poisoning is only a relative term. It should be sufficient that, when the term inhibitor or poison is used, the reference is to a substance that has a deterring effect on the rate and extent of hydrogenation.

As mentioned before, not all of the surface of the catalyst is capable of maintaining reduction. A few words about active sites may help toward an understanding of the effect of inhibitors or poisons on the catalyst in the reduction process.

The generally accepted theory is that all reactants are adsorbed at active sites on the surface of the catalyst and that reaction is between adsorbed molecules. The adsorption which takes place is physical in character as opposed to the chemical nature of the effect of poisons or inhibitors. In the classic view inhibition by a reactant consists of strong adsorption and correspondingly strong monopolization of the active centers of the catalyst by that reactant. The effect is essentially one of preferential chemisorptive bonding of a particular agent which allows sufficient accumulation of it so that the active sites are no longer available for the normal adsorptive-desorptive process necessary for continuing hydrogenation.

METALS AND METAL SALTS

There are a number of different types of poisons and inhibitors. The first group to be considered are certain metals, metallic compounds, and metal ions. Paal and Karl [3] studied the effects of a number of metals by digesting cleansed metals in palladium chloride solution and using the catalysts so obtained for the hydrogenation of an unsaturated ester. Of the metals tried magnesium, nickel, and cobalt did not inhibit reaction. The others, aluminum, iron, copper, zinc, silver, tin, and lead acted as anticatalysts. In a later study the authors [4] found that not only did the same metals inhibit the palladium-catalyzed reduction of cod-liver and cottonseed oils but that oxides and carbonates of these metals behaved similarly. The use of basic lead carbonate completely prevented uptake of hydrogen. Paal and Hartman [5] found that mercury and mercuric oxide also inactivated palladium catalyst. Paal and Windisch [6] investigated the effect of some metals on reduction with platinum. Aluminum, cobalt, and bismuth caused a decrease of activity, while iron, copper, zinc, silver, tin, and lead caused an increase. The increased activity that these metals caused for the platinum-catalyzed hydrogenation of cottonseed and cod-liver oil is at variance with the results of these metals combined with palladium for the same reduction [4].

Maxted [7] investigated the effect of lead, as acetate, on the reduction of oleic acid in the presence of platinum in acetic acid. He found a linear relationship. One milligram of lead inhibited the activity of 8.8 mg of platinum while the remainder of the catalyst functioned normally. Twice as much lead inhibited about 17.6 mg of platinum. When enough lead compound was added, uptake of hydrogen stopped completely. This work was extended to the influence of mercury and zinc [8]. They also had the same effect on the platinum-catalyzed reduction of oleic acid in acetic acid. The inhibitory action of a number of metals noted by Paal and co-workers [3–6] was reinvestigated by Maxted and Marsden [9] following an earlier study [7] where the metals were used in the form of the nontoxic acetate ion. The results showed that mercury, bismuth, lead, tin, zinc, cadmium, aluminum, and iron inactivated both palladium and platinum catalysts for the hydrogenation of crotonic acid. Thallium was also included as a metal having a deterring effect on reduction.

Greenfield [10] studied the effect of a number of metallic salts on the palladium-catalyzed reduction of 4-nitrotoluene. This is of interest because aromatic nitro compounds are converted so readily to the corresponding amines. Accordingly, although not necessarily so, it might be expected that compounds which act as inhibitors in these reductions could have a far greater inhibitory effect in other hydrogenations. Nickel nitrate increased reaction time 1.5-fold, while ferric chloride, ferric nitrate, cobaltous nitrate, and chromic nitrate doubled the reaction time. These and other salts were present in a 5 mole % ratio based on nitrotoluene. More severe inhibition resulted from the addition of zinc nitrate which caused a tenfold increase in reaction time. Cuprous chloride, cupric chloride, cupric nitrate, silver, aluminum, and lead nitrates completely poisoned the hydrogenation.

Very little work on the inhibition of nickel catalysts by other metals has been carried out. There are two reports by Ueno [11, 12] on the effect of metal and nonmetal soaps in the hydrogenation of soybean oil and a fish oil. Soaps of the heavy metals, including zinc, cadmium, lead, mercury, bismuth, and tin were found to be inhibitors. However, the effect of these soaps may have been due in part to mechanical coating of the catalyst since magnesium, barium, and soaps of the alkali metals were also stated to have reduced the activity of the catalyst. Some recent work [13] showed that Raney nickel was strongly affected by small amounts of mercuric chloride. Raney nickel was suspended in water and treated with an aqueous solution of mercuric chloride. The catalyst resulting from this treatment was completely poisoned for the hydrogenation of cyclohexene at a concentration of 0.05 gram atom of mercury per gram atom of nickel. Silver sulfate—Ag_2SO_4—in a ratio of 1 g atom of silver to 1 g atom of nickel caused only 50% inhibition. Cupric sulfate in the same ratio completely deactivated the nickel catalyst.

All of the inhibitory metallic substances have a common characteristic. The orbitals of the d shell of the metallic atoms have electron pairs or at least a single electron available for bonding to the catalyst [14].

HALOGEN-CONTAINING COMPOUNDS

Halogen-containing compounds have been studied for their effect on Raney nickel catalyst [15]. The investigators first added specific amounts of anhydrous hydrogen chloride to a Raney nickel catalyzed reduction of styrene. They found that a sharp break in the curve representing hydrogen uptake occurred upon the addition of 0.006 g of HCl per 0.34 g of Raney nickel. They pointed out that Taylor [16] had shown that some catalysts have more than one type of active surface. They assume from their work that Raney nickel has at least two types of active surface and that hydrogen chloride is preferentially attached to the most active one, thus causing inhibition.

In other experiments the amount of hydrogen chloride was increased until a point was reached (0.03 g per 0.34 g of nickel) when the reaction was almost completely poisoned. However, it must be pointed out that the ratio of catalyst to compound was unusually low (1 + %) in these experiments. In the presence of such a low catalyst ratio it should not take too much of an additive to depress the reaction. This not to say that hydrogen chloride is not an inhibitor. The information supplied by Pattison and Degering [15] is of value because it makes us aware that it can have an adverse effect on the hydrogenation process.

The same investigators extended their study to the effect of organic halides including alkyl and aryl halides, some polychlorides, and a few miscellaneous chloro compounds on the reduction of styrene. Among the alkyl chlorides, there was a relationship between chain length and poisoning, the shortest chain having the greatest inhibiting power. In addition the more hindered compounds exerted the weakest action. The order of activity in the aryl halides series was iodide > bromide > chloride with little real difference between bromide and chloride, the weakest inhibitor. 3-Chloroaniline and 2-amino-6-chlorotoluene were also included but these amines do not rightfully belong in the group because there is an added inhibitory effect of the amino group. Among the polychlorides and miscellaneous chlorides, carbon tetrachloride, chloroform, chloral hydrate, trichloroethanol, di- and trichloroacetic acid, alkyl chloride, benzyl chloride, and acetyl chloride were potent inhibitors. The authors suggest that the poisoning power of these compounds is due to the reactive chlorine each contains. This statement may apply to acetyl chloride, benzyl chloride, certain alkyl chlorides, and a few of the others but the halogens in chloroform and carbon tetrachloride are not especially active. It might be more to the point to attribute poisoning to the release of hydrogen chloride. This could explain why short-chain alkyl chlorides were stronger inhibitors than the long-chain alkyl chlorides which are dehalogenated less readily. It would also explain why the branched chain compound, tertiary amyl chloride, had little effect. The thought that the inhibitory effect of halogen-containing compounds is due to hydrogen chloride is suggested by some work of Baltzly and Phillips [17]. They studied the behavior of di- and trichloroacetic acids and esters in reductions with palladium on carbon. In absolute alcohol the hydrogenation of trichloroacetic acid or its ester stopped after the loss of little more than one equivalent of halogen. The reductions of the dichloro acid and ester to the monochloroacetic stage were also incomplete. Further example of the effect of hydrogen chloride was shown when the hydrogenation of ethyl dichloroacetate was suppressed by the addition of an excess of it. The results of the following experiments give added support to the suggestion that hydrogen halide, rather than organic halide, is the inhibitor.

Experimental. A solution of 16.4 g (0.2 mole) of distilled cyclohexene in 125 cc of anhydrous ethyl alcohol was hydrogenated in the presence of 0.08 g of platinum oxide at room temperature and 35 psig. Uptake of hydrogen was complete in 15 min. This was taken as a standard experiment.

A comparison was made with the addition of 0.05 mole and of 0.2 mole of *n*-propyl chloride, which from Pattison and Degering's study [15] had an inhibitory effect on the nickel-catalyzed reduction of styrene. There was no change in reduction time, nor was there any evidence of chloride ion in the solutions after hydrogen uptake was complete for the conversion of the double bond. The addition of 0.025 mole of anhydrous hydrogen chloride to the standard experiment did not cause any decrease in reaction time. However, when the amount of hydrogen chloride in the next experiments was raised to 0.125 mole and to 0.250 mole, there was a considerable increase in the times necessary for reduction (see Fig. 4.1).

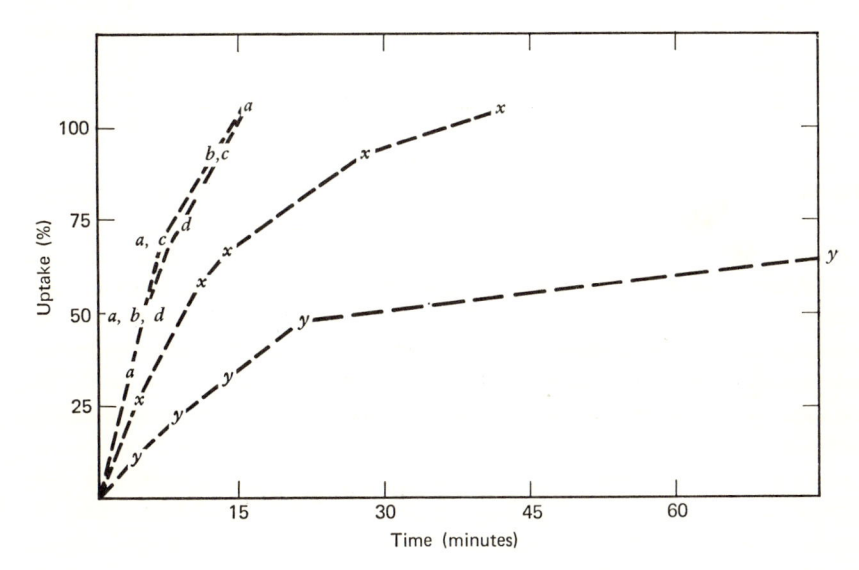

Fig. 4.1 (*a*) Hydrogenation of cyclohexene in abs EtOH with PtO$_2$; (*b*) with 0.025 mole propyl chloride; (*c*) with 0.05 mole propyl chloride; (*d*) with 0.2 mole propyl chloride; (*x*) with 0.125 mole dry HCl; (*y*) with 0.25 mole dry HCl.

5% Palladium on carbon (0.8 g) was used to hydrogenate 0.2 mole of cyclohexene in alcohol. No change in reduction time was observed in a comparison of the standard with an experiment to which *n*-propyl chloride was added. On the other hand the reaction rate was sharply decreased by the addition of 0.25 mole of dry hydrogen chloride to the palladium-catalyzed hydrogenation of cyclohexene (see Fig. 4.2).

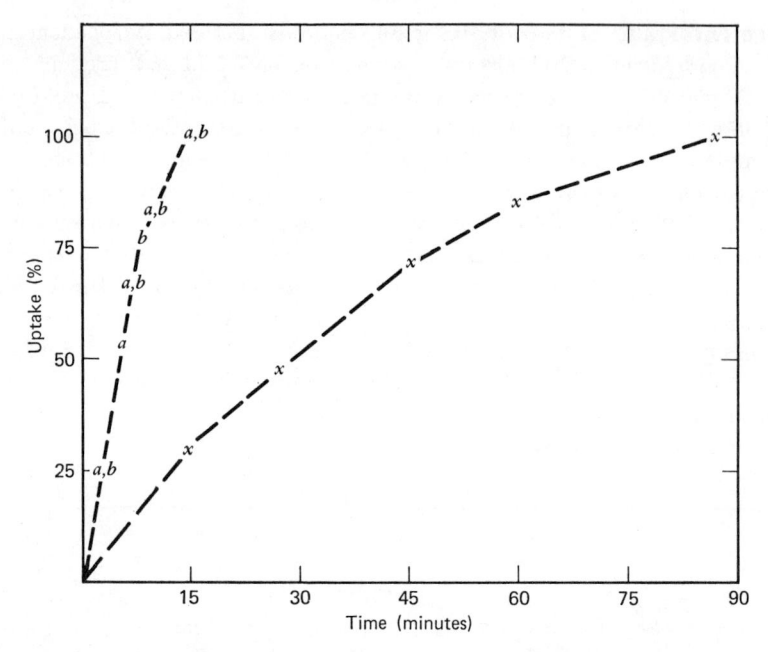

Fig. 4.2 (a) Hydrogenation of cyclohexene in abs EtOH with 5% Pd on C; (b) with 0.15 mole propyl chloride; (x) with 0.25 mole dry HCl.

Another example of the effect of hydrogen chloride may be seen in the hydrogenation and hydrogenolysis of N-methyl 4-methoxy-4'-nitrodibenzylamine hydrochloride in absolute alcohol [18]. When the reaction was rerun and 10 equivalents of hydrogen chloride were added, only reduction of the nitro group took place (Eq. 4.1). It may be argued that in another example

$$CH_3O{-}C_6H_4{-}CH_2N(CH_3)CH_2{-}C_6H_4{-}NO_2 \cdot HCl \xrightarrow{4H_2} CH_3O{-}C_6H_4{-}CH_2NHCH_3 \cdot HCl + H_2N{-}C_6H_4{-}CH_3 \quad (4.1)$$

$$\xrightarrow[\text{excess hydrogen chloride}]{3H_2} CH_3O{-}C_6H_4{-}CH_2N(CH_3)CH_2{-}C_6H_4{-}NH_2 \cdot 2HCl$$

in the same paper excess hydrogen chloride was added but it did not prevent cleavage (Eq. 4.2). It must be pointed out that the inhibitory effect of

$$CH_3O\langle\underset{}{}\rangle CH_2\overset{+}{N}(CH_3)_2 CH_2\langle\underset{}{}\rangle NH_2\cdot HCl$$
$$Cl^- \quad \downarrow H_2$$
$$CH_3O\langle\underset{}{}\rangle CH_2N(CH_3)_2\cdot HCl + H_2N\langle\underset{}{}\rangle CH_3$$
$$90\% \qquad\qquad 80\% \text{ (as } N\text{-acetyl)}$$

(4.2)

various agents is dependent, among other factors, on the ease of reducibility of the substrate. Among benzylamines, ease of hydrogenation is in descending order from quaternary amine to tertiary to secondary to primary amine. This could explain the difference in inhibition by hydrogen chloride in the examples [18]. It also helps to emphasize that hydrogen chloride is not a particularly potent inhibitor.

When Baltzly and Phillips [17] carried out reductions of di- and trichloroacetic acids and esters in aqueous solution or in aqueous alcohol, the reactions proceeded uneventfully to the monochloro stage. The authors were puzzled by the difference in the extent of reduction in aqueous media in contrast to incomplete hydrogenation of the same materials in anhydrous medium. It is likely that, in the presence of sufficient water, hydrogen chloride undergoes dissociation resulting in the nonpoisoning chloride ion. An example of the effect of water is supplied by the same authors. In absolute alcohol no uptake of hydrogen occurred during the attempted dehalogenation of 4-chloroaniline hydrochloride in the presence of palladium on carbon catalyst. When the same reaction was carried out in water, uptake of hydrogen proceeded to completion. The following experiments are a further illustration of the effect of water changing the retarding action of hydrogen chloride on the rate and extent of hydrogen uptake.

Experimental. A solution of 5.8 g (0.05 mole) of pure maleic acid in 150 cc of absolute ethyl alcohol was hydrogenated at room temperature and 35 psig in the presence of 0.05 g of platinum oxide. Uptake of hydrogen was complete in 63 min. Points were plotted on a graph. In another experiment run in the same Parr hydrogenator 0.05 mole of hydrogen chloride was present. At the end of 2 hr hydrogen uptake was only 50% and had slowed down appreciably (see Fig. 4.3).

In a second set of experiments 0.05 mole of maleic acid was hydrogenated under similar conditions in 150 cc of aqueous alcohol. Uptake of hydrogen was complete in 65–70 min. The presence of hydrogen chloride had no effect on reaction time. The points on the graph were almost identical (Fig. 4.3).

The work of Wenner [19], who achieved successful reductions with Raney

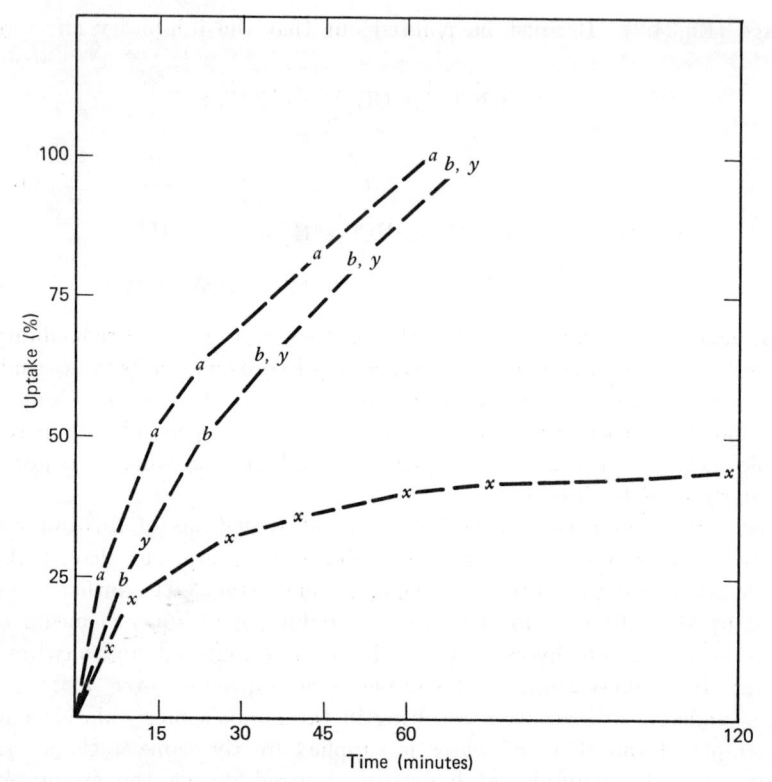

Fig. 4.3 (a) Hydrogenation of maleic acid in abs EtOH with PtO$_2$; (x) with 0.05 mole dry HCl; (b) hydrogenation in 50% aqueous EtOH; (y) with 0.05 mole dry HCl.

nickel in aqueous hydrochloric acid solution, may be another example of the effect of the presence of water.

Hydrogen bromide may act as an inhibitor as well as hydrogen chloride. Hydrogenation of 4-bromoanisole in methyl alcohol solution containing hydrogen bromide required 9 hr for complete absorption of hydrogen in the presence of palladized charcoal [17]. Unfortunately there was no comparison of reduction in aqueous medium. In an attempt to prevent debromination in the hydrogenation of 4-bromobenzylidenemethylamine, hydrogen bromide was added. Uptake was slow and never went to completion due perhaps to the effect of the additive [20]. When a similar reduction in absolute alcohol was carried out with the same catalyst in the absence of the additive, uptake of hydrogen was complete in about 1 hr.

There are other references to the inhibiting effect of organic halides. Truffault [21] studied the addition of a few of them to benzene hydrogenated in the presence of Raney nickel. Using thiophene (0.6 mM per 10 g of benzene and 3 g of Raney nickel) as a standard inhibitor he found that 1 mM of ethyl bromide, 3 mM of chlorobenzene and 15 mM of bromobenzene each decreased the reaction rate by 50%. No comment was offered as to a mechanism.

Campbell and Kemball [22] investigated the gas-phase dehalogenation of ethyl chloride and ethyl bromide on evaporated films of palladium, platinum, and nickel and to a lesser extent on those of iron, tungsten, and rhodium. They found that the reaction rate (determined under a standard set of conditions) was decreased when the temperature was raised. They reasoned that the reduction in rate was due to hydrogen halide except when nickel was used. In that case the reduction in rate, from their results, could be attributed to either hydrogen halide or alkyl halide. Experiments with added hydrogen chloride showed that there was a reduction in rate.

Greenfield [10] added octyl chloride, bromobenzene, and chlorobenzene in a 5 mole % ratio to 4-nitrotoluene. He did not observe any inhibition during the reduction of the nitro compound with 5% palladium on carbon in a 2% ratio of catalyst to substrate. It is of interest that the reactions were run in 80% aqueous isopropyl alcohol.

Pattison and Degering [15] stated that organic iodides were potent inhibitors of reduction with Raney nickel. Baltzly [1] reported that iodide ion inhibited the activity of both palladium and platinum catalysts. Hoshiai [23] found that sodium or potassium iodides were more potent inhibitors of the reduction of dipropyl ketone or 1-octene with Raney nickel than potassium bromide or chloride. In a nickel-catalyzed hydrogenation of 4-methyl-3-pentene-2-one [24], he found the following order of activity for retarding carbonyl function—$CdI_2 > BaI_2 > KI > KBr$. This particular reaction is of interest because of the retarding effect of the iodides at 25–150° and 120 atm initial hydrogen pressure. Another report on the use of the three iodides [23] indicated that the presence of KI and BaI_2 allowed selective conversion of the double bonds in mesityl oxide but that none of the three caused any selectivity in the reduction of 3,5,5-trimethyl-2-cyclohexene-1-one (isophorone).

Greenfield [10] also pointed out the potent retardation of iodide ion on reduction. He found that the addition of 2.56 mole % of sodium iodide based on 4-nitrotoluene completely poisoned the reaction. At the same time he noted that sodium fluoride, sodium bromide, and even a large amount of sodium chloride (15.4 mole %) were completely innocuous.

There are some other illustrations of the effect of iodide or iodide ion. Some, however, are not too clear-cut. The reduction of 4 g of the methiodide of methyl isoniconate with platinum oxide, 0.6 g, gave methyl 1-methyl-piperidine-4-carboxylate in good yield [26]. When the reaction was carried out for a shorter period, methyl 1-methyl-1,2,5,6-tetrahydroisonicotinate was obtained. The authors offered no suggestion about the effect of iodide, claiming only that the length of reaction governed the product. In contrast Lyle and Lyle [27] used less catalyst (0.3 g for 21 g of the methiodide) and found that hydrogen uptake stopped of its own accord before the required three equivalents were absorbed. They obtained only tetrahydro compound corresponding to that of Supniewski and Serafinovna [26]. In a later publication Lyle [28] emphasized that the formation of methyl 1-methyltetrahydroisonicotinate hydriodide is independent of the length of reaction time and that it results from poisoning of the catalyst. He showed the effect of iodide as opposed to bromide by hydrogenating the methiodide and methobromide of methyl isonicotinate with Adams's catalyst recovered from one of the reductions. In the first reduction tetrahydro compound was obtained; the second reduction yielded the hydrobromide salt of the corresponding piperidine. Grob and Ostermayer [29] reduced the methiodide of ethyl nicotinate at 120° and 120 atm in the presence of almost 50% by weight of W-5 Raney nickel and excess triethylamine. Uptake of hydrogen was complete in 8 hr. The resultant product was the piperidine. At 80° and 120 atm with the same ratio of nickel and excess triethylamine the reduction came to a standstill after 4 hr. In addition to 54% of the saturated product, 38% of a tetrahydropyridine was obtained. It may be argued that poisoning was a result of base effect or perhaps a combination of the effect of iodide and excess base (the inhibitory effect of nitrogen bases will be discussed shortly). A better answer could have been provided if the same reduction had been carried out with only a molar equivalent of triethylamine to neutralize hydrogen iodide.

It must not be assumed that all iodides are inhibitors of catalytic reduction. There are examples in the literature where no inhibiting effect is noted, although it must be pointed out that in the examples cited there was no attempt to obtain partially hydrogenated pyridines. The diethylcarboxamide of methyl pyridinium iodide was converted to 1-methylpiperidine-3-(N,N-diethyl)carboxamide [30]. Reaction was slow, 22 hr, but no other product was obtained. Similarly the methiodide of nicotinamide and isonicotinamide [31] and some alkyl iodides of nicotinamide [32] gave the corresponding 1-substituted piperidines.

THE EFFECT OF NITRITES

Nitrites have been reported to have an effect on reduction. Greenfield [10] found that the addition of sodium or potassium nitrite caused a threefold increase in reaction time in the hydrogenation of 4-nitrotoluene. There was no effect, however, with nitrate. Stevinson and Hamilton [33], investigating the effect of a number of salts on the reduction of 4-nitrophenol with Raney nickel, found that the addition of 10% by weight of sodium nitrite (based on nitrophenol) increased the time of reduction from 11 min to 80 min. In contrast, the addition of sodium nitrate did not cause any significant change in reaction time. Dart and Henbest [34] found that the addition of sodium nitrite to the platinum-catalyzed hydrogenation of some cyclic allylic alcohols led to good yields of the saturated alcohols. They also noted that, in the reduction of cholest-4-en-3β-ol, hydrogenolysis to hydrocarbon was suppressed not only by sodium nitrite but also by sodium nitrate. The addition of sodium hydroxide was as effective as nitrite in keeping the amount of hydrocarbon to a minimum. Potassium and lithium nitrate had some effect; the ratio of coprostanol and cholestanol, the saturated alcohols, to hydrocarbon was about 2 to 1. The addition of sodium chloride caused little if any change in the ratio of saturated alcohol to hydrocarbon compared to reduction without any additive.

It is unfortunate that the investigators confined their study to the use of sodium nitrite for the reduction of other cyclic allylic alcohols to saturated alcohols. Their results might have been more meaningful if they had also used the same additives that were successful in the inhibition of hydrogenolysis in the reduction of cholestenol.

THE EFFECT OF COMPOUNDS CONTAINING ARSENIC, ANTIMONY, OXYGEN, PHOSPHORUS, SELENIUM, AND TELLURIUM

By far the most important group of inhibitors of catalytic activity and the type most frequently encountered are compounds containing elements of the periodic group Vb—nitrogen, phosphorus, arsenic, and antimony—and VIb—oxygen, sulfur, selenium, and tellurium. Of these elements oxygen, except as carbon monoxide, rarely acts as a retardant of reduction. The most important members, sulfur and nitrogen, will be dealt with separately.

It should be borne in mind that poisoning by these elements depends on their electronic structures. The geometry of the molecules of which they are a part is another factor. Maxted and Moorish [35] have suggested that the poisoning elements of the two groups are not toxic to catalysts when they possess a completely shared electron octet. In such cases Maxted [2] chooses to refer to these elements as existing in a shielded or nontoxic state. Opposed to this are the elements in the unshielded or toxic state. Here the element has an unshared pair of electrons with which it can be strongly bound (chemisorbed) to the active centers of the catalyst. The sulfate ion—nonpoisoning—and the sulfite ion—unshielded—are good examples of the two states shown in 4.3. The authors [35] point out that the phosphate ion with a filled

$$\begin{bmatrix} \text{O} \\ \text{O:}\ddot{\text{S}}\text{:O} \\ \ddot{\text{O}} \end{bmatrix}^{2-} \qquad \begin{bmatrix} \text{O} \\ \text{O:}\ddot{\text{S}}\text{:O} \end{bmatrix}^{2-} \qquad (4.3)$$

Sulfate ion (shielded)   Sulfite ion (unshielded)

electron octet about the phosphorus atom is not a poison. Phosphine (PH_3) is an example of a phosphorous-containing compound which is a catalyst inhibitor (4.4).

$$\begin{bmatrix} \text{O} \\ \text{O:}\ddot{\text{P}}\text{:O} \\ \ddot{\text{O}} \end{bmatrix}^{3-} \qquad \begin{matrix} \text{H} \\ \text{H:}\ddot{\text{P}}\text{:H} \end{matrix} \qquad (4.4)$$

Phosphate ion   Phosphine

Phosphite and hypophosphite ions are inhibitors in spite of their shielded structure (4.5). The authors attribute the poisoning effect to the weakness

$$\begin{bmatrix} \text{O} \\ \text{H:}\ddot{\text{P}}\text{:O} \\ \ddot{\text{O}} \end{bmatrix}^{2-} \qquad \begin{bmatrix} \text{O} \\ \text{H:}\ddot{\text{P}}\text{:O} \\ \ddot{\text{H}} \end{bmatrix}^{-} \qquad (4.5)$$

Phosphite ion   Hypophosphite ion

of the phosphorus-hydrogen bond which is not strong enough to prevent hydrogen displacement and formation of the toxic phosphorus-platinum linkage. Another example of the difference in reduction in phosphorous-containing compounds of different oxidation states is seen in the work of Freedman and co-workers [36] who successfully converted phenylphosphonic acid to cyclohexylphosphonic acid with rhodium but were unable to reduce phenylphosphinic acid under the same conditions.

Horner and his associates [13] studied the effect of a number of phosphorous-containing compounds in different valence states on the hydrogenation of cyclohexene in the presence of Raney nickel. They found that substituted phenylphosphines were usually strongly inhibitory, while phenylphosphine oxide, $(C_6H_5)_3PO$, had little effect. Another example of the difference in the toxic effect of shielded and unshielded states of the phosphorus atom can be seen in the reduction of cyclohexene in the presence of ethyl phosphate, $(H_5C_2O)_3\,PO$, and ethyl phosphite, $(H_5C_2O)_3\,P$. There was no poisoning with the former up to the addition of 1 mM of it to the reduction, while the addition of 0.25 mM of ethyl phosphite caused almost complete inhibition of reaction. Based on Maxted's concept of nonpoisoning shielded structures, Horner's finding that phenylphosphonic acid was an inhibitor is unexpected. However, a large amount of nickel was used in the reduction, 3.0 to 0.087 g (0.5 mM) or 0.174 g (1.0 mM) of phenylphosphonic acid per

2.05 g (25 mM) of cyclohexene. It might be assumed that phenylphosphonic acid was reduced to a compound with phosphorus in a lower valence state and in that form the resulting compound was the inhibitor.

The effect of structure on the activity of a potentially poisoning phosphorus atom may be seen in Horner's experiments. Diphenylphosphinic acid, because of the added bulk of a second benzene ring, was only a mild deterrent. Freedman [36] also showed the effect of bulk on a poisoning phosphorus atom. He and his co-workers were able to convert diphenylphosphinic acid to dicyclohexylphosphinic in contrast to the previously noted failure to hydrogenate phenylphosphinic acid.

Arsine, AsH_3, and stibine, SbH_3, are toxic to hydrogenation catalysts. Each is structurally similar to phosphine with an unshared pair of electrons. Arsenates and antimonates when stable do not affect reductions adversely. Arsenites, in a lower oxidation state, do have anticatalytic activity. Stevinson and Hamilton [33] compared the effect of adding sodium arsenite and sodium arsenate to the reduction of nitrophenol with Raney nickel. The reaction treated with arsenite required 70 min for completion of uptake of hydrogen against 11 min for the arsenate-treated reduction. A standard hydrogenation was also complete in 11 min. Freedman [36] reported on an unsuccessful attempt to convert phenylarsonic acid to cyclohexylarsonic, reasoning that the arsono group, AsO_3H_2, was reduced to the trivalent poisoning form. The finding of Maxted and Moorish [35] was cited as a reference that reduction from arsenate to arsine does indeed take place. Horner and his group [13] investigated the effect of a few arsenicals on the nickel-catalyzed hydrogenation of cyclohexene. They found that triphenylarsine, $(C_6H_5)_3As$, and tribenzylarsine, $(C_6H_5CH_2)_3As$, poisoned the catalyst. The triphenyl ester of arsonic acid was somewhat inhibitory upon the addition of 1 mM but its poisoning effect could have been due to reduction of the arsenic atom to the inhibiting trivalent state.

Selenium and tellurium in their lower valence states as selenites and tellurites are inhibitors but are not too often encountered. Horner [13] showed the inhibitory effect of selenium as benzyl selenide, $(C_6H_5CH_2)_2Se$, and as diphenyl selenide, $(C_6H_5)_2Se$.

THE EFFECT OF SULFUR

Consistent with the concept of the inhibitory effect of the unshielded forms of the poisoning atoms of groups Vb and VIb it is found that sulfides, mercaptans, disulfides, and thiosulfates are potent catalyst inhibitors. For example, thiophene, a cyclic sulfide and a contaminant of benzene, is known

36   *Catalyst Inhibitors and Poisons*

to have considerable anticatalytic activity (4.6). On the same basis it should

$$RC\!:\!\ddot{S}\!:\!CR \qquad RC\!:\!\ddot{S}\!:\!H \qquad RC\!:\!\ddot{S}\!:\!\ddot{S}\!:\!CR \qquad \left[\begin{array}{c} O \\ O\!:\!\ddot{S}\!:\!S\!: \\ O \end{array}\right]^{2-} \qquad (4.6)$$

Sulfide   Thiol   Disulfide   Thiosulfate ion

be expected that compounds containing a sulfur atom in the shielded form such as sulfonamides, sulfonic acids, and sulfones will not inhibit the course of catalytic hydrogenation (4.7). In contrast to most arsenic, antimony,

$$\begin{array}{ccc} O & O & O \\ RC\!:\!\ddot{S}\!:\!NH_2 & RC\!:\!\ddot{S}\!:\!OH & RC\!:\!\ddot{S}\!:\!CR \\ O & O & O \end{array} \qquad (4.7)$$

Sulfonamide   Sulfonic acid   Sulfone

and phosphorous compounds in the higher oxidation states, sulfur-containing organic compounds such as sulfonic acids and sulfonates, sulfonamides, and sulfones are reasonably stable substances. They do not undergo breakdown during hydrogenation except under severe conditions. As a rule other functional groups present in these compounds may be reduced catalytically with the same ease as similar functional groups of other organic compounds.

Horner and his group [13] studied the effect of a number of sulfur-containing compounds in the reduction of cyclohexene with nickel. Almost all of the compounds, in which the sulfur atom had an available pair of electrons to bond with the catalyst, were definitely catalyst poisons. Among the inhibiting compounds were a number of sulfides, thiocyanates, thioureas, thioacids, and thiophenols. Thiophene, thiophane, and another cyclic sulfide were also shown to be highly anticatalytic. Oddly enough sodium sulfide was not found particularly potent although sodium polysulfide was an effective catalyst poison. The investigators did show that the shielded type of sulfur-containing compounds had little inhibitory effect. They also studied the effect of the addition of some sulfoxides. Dodecyl methyl sulfoxide, $C_{12}H_{25}SOCH_3$, did not appear to be a retardant. On the other hand dibenzylsulfoxide, $(C_6H_5CH_2)_2SO$, seemed to be a mild inhibitor. The sulfur atom in sulfoxides has available an unshared pair of electrons in contrast to the atom in sulfones which has a shared octet. In agreement with the current concept,

$$\begin{array}{c} O \\ RC\!:\!\ddot{S}\!:\!CR \end{array} \qquad (4.8)$$

Sulfoxide

it may be assumed that sulfoxides are potentially poisoning substances. As yet there is insufficient knowledge of the effect of the sulfur atom in these compounds. There is also too little information on the reduction of functional groups present in sulfoxides. It is an area in which further study is warranted.

Horner and co-workers also showed that sodium phenylsulfinate, $C_6H_5SO_2Na$, was an inhibitor. This is not unexpected since organic sulfinates are closely related to inorganic sulfites which are known to be catalyst poisons. The effect of inorganic sulfites has been demonstrated by Greenfield [10]. He reported that sodium sulfite caused complete poisoning of the reduction of 4-nitrophenol with palladium on carbon after slow uptake of one-third of the necessary amount of hydrogen. He also showed that sodium bisulfite as well as sodium sulfide caused complete inhibition. The effect of sodium sulfide is of interest in view of Horner's finding [13] noted on a previous page. Greenfield also reported that phenyl disulfide was strongly poisoning at very low concentrations.

Tonnies and Kolb [37] found that the addition of N-acetylmethionineallyl-sulfonium bromide

$$[\text{HOOCCH}(\text{NHCOCH}_3)\text{CH}_2\overset{+}{\text{S}}(\text{CH}_3)(\text{CH}_2\text{CH}=\text{CH}_2)]\text{Br}^-$$

or other related sulfonium salts poisoned the palladium on barium sulfate and platinum oxide reductions of maleic acid. They concluded from their results that the sulfonium ion was a strong poison for noble metal catalysts.

Deem and Kaveckis [38] followed the inhibitory effect of sulfur-containing compounds, in varying oxidation states, on the hydrogenation of phenol, quinoline, and naphthalene at 120–130° and 50–60 atm initial pressure in the presence of Raney nickel. They found that thiophenol, benzyl sulfide, and benzyl sulfoxide inhibited all the reductions. Methyl p-toluenesulfonate had little effect as did sodium benzenesulfonate. Phenyl sulfone appeared to have a slight deterring effect but this could have been due to the reaction conditions which may have led to another sulfur-containing compound which was the poisoning agent. When the temperature was raised to 190° to convert quinoline to decahydroquinoline and over 200° to convert naphthalene to decahydronaphthalene, the sulfone and the sulfonates did exert an inhibiting effect. The authors agree that the higher temperatures and pressures could have caused some breakdown of these materials with formation of other sulfur-containing inhibitors.

Horner and his group [13] point out that poisoning also depends on the substrate as well as the catalyst inhibitor. They showed that the reduction of acetophenone and cyclohexanone to the respective alcohols was as profoundly effected by poisoning sulfur-containing compounds as was cyclohexene to cyclohexane during hydrogenation with nickel. In another comparison they showed that *bis*-4-chlorophenyl sulfide caused only about 30% inactivation of the nickel catalyst in the reduction of nitromethane, while the corresponding amount of the same sulfide, 1 mM, strongly inhibited the conversion of cyclohexene to cyclohexane. Thiophene, although normally a potent catalyst inhibitor, had even less effect on the nitromethane reduction. Thiourea, phenylthiourea, and naphthylthiourea, which were all highly inhibitory for the cyclohexene reduction at very low concentration, caused only mild inactivation in the nitromethane reaction except when 1 mM of inhibitor was employed. Thiourea, strongly inhibitory for the reduction of cyclohexene, was far less effective in retarding the hydrogenation of nitrobenzene. The difference is due no doubt to the greater ease of reducing an aromatic nitro group.

The manner in which a catalyst is prepared can also influence its susceptibility to poisoning agents. Kelber [39], for example, noted differences with nickel catalysts which had been reduced under hydrogen at temperatures varying in range from 300–450°. A catalyst reduced at 450° was poisoned more substantially for the hydrogenation of sodium cinnamate by the addition of hydrogen sulfide or carbon disulfide than were those prepared at lower temperatures.

Maxted and Evans [40] studied the relative poisoning power of a number of sulfur compounds in the nickel-catalyzed hydrogenation of olive oil and in the platinum reduction of crotonic acid. They reported that the retarding effect of the sulfur compounds increased with their molecular weight. Hydrogen sulfide, which was taken as a standard with a relative poisoning value of one per gram atom of sulfur, was less of a poison than sulfur or carbon disulfide. Kelber [39] had also shown that carbon disulfide was a stronger inhibitor than hydrogen sulfide. The relative toxicity of thiophene was more than four times that of the standard; cysteine was five times more potent than hydrogen sulfide. The authors reasoned that the inhibiting effect per unit of sulfur increased with the complexity of the chain or the ring attached to the sulfur atom.

They carried out their study still further [41] and reported that the inhibiting effect of a series of sulfides increased with chain length. Hydrogen sulfide was the weakest poison and cetyl sulfide, with the longest chain of the compounds investigated, the most potent inhibitor. The thiols followed the same pattern. Ethyl thiol was the least effective poison of the group under study and cetyl thiol was the most effective of the group. Their findings also appear to indicate that sulfides are more potent inhibitors than the corresponding thiols. This is of interest because an examination of molecular models of phenyl sulfide and thiophenol suggests that the sulfur atom in the smaller molecule may be less obstructed and closer to the surface of the catalyst than the corresponding atom flanked on each side by a benzene ring.

In a further extension of the effect of the sulfur atom in its unshielded state, Maxted and Evans [42] compared the effect of a thiol with a dithiol of equal hydrocarbon chain length on the platinum-catalyzed reduction of crotonic acid. They found that in spite of the greater amount of sulfur the toxicity of propylenedithiol was considerably less than that of n-propylthiol or of n-butylthiol, which would be the equal of propylenedithiol in chain length with its additional SH group. They attribute the difference in effect to anchoring of the two sulfur atoms to the catalyst surface thereby restricting mobility. In the case of a single sulfur atom they assume that the chain as a whole is free to move, except for permanent attachment of the terminal sulfur atom, and that its area of influence is a circle whose radius is equal to the length of the chain. They found, however, that when two sulfur atoms were adjacent, as in ethyl disulfide, these compounds were little different in effect from that of the related thiol. This is probably due to conversion of the disulfide to thiol during reduction. The effect of unsaturation was also found negligible. There was little difference in inhibition, caused by allyl and n-propyl sulfide, due probably to conversion of the unsaturated compound to n-propyl sulfide. A straight chain sulfide was not significantly more inhibitory toward the hydrogenation of crotonic acid than a branched chain sulfide with the same number of carbons. That part of the study, however, was limited to a comparison of propyl and isopropyl sulfides. There was no comparison of the effect of aromatic sulfide against an aliphatic sulfide with the same number of carbons. They did, however, study a few compounds containing an aromatic ring. They found that thiophenol was more toxic than 2-phenylethylthiol which, in turn, was found to be more inhibitory than thiophene toward platinum catalyst in the hydrogenation of crotonic acid.

THE EFFECT OF THE NITROGEN ATOM

The effect of the nitrogen atom in organic compounds on catalysts used for hydrogenation is governed by electron effects similar to those of other atoms of groups Vb and VIb. Compounds containing unshielded nitrogen atoms act as inhibitors, but the retarding action on catalysts can be neutralized by conversion of these compounds to a form in which the nitrogen atom is shielded. The inhibiting effect of the nitrogen atom is limited to those instances where, as part of an organic compound, it exists as a base capable of donating electrons. Nonbasic nitrogen-containing compounds have little, if any, deterring effect on the catalytic process. Horner [13] reported that cyanoacetamide had an anticatalytic effect on the hydrogenation of cyclohexene. The inhibition, however, is not caused by the amide nitrogen but by the nitrile function, the effect of which will be discussed subsequently. Ureas do not appear to be detrimental to catalytic reduction. However, semicarbazide, $H_2NCONHNH_2$, was shown to be a deterrent [13]. It must be

40   Catalyst Inhibitors and Poisons

pointed out that semicarbazide is a basic compound capable of forming a salt. On the same basis it might be assumed that guanidines, which are strongly basic, would be inhibitors. Horner does not report on their effect, nor is there too much mention of inhibition by them in the literature. Nevertheless, their effect can be inferred from the large amount of catalyst used for reduction when the guanidine moiety was part of an organic compound. Merrifield [43] reported that to cleave the protecting nitro groups from nitro-L-arginyl-L-prolyl-L-prolylglycyl-L-phenylalanyl-L-seryl-L-prolyl-L-phenylalanyl-nitro-L-arginine to yield the peptide, bradykinin, three times as much catalyst as compound was used. Nicolaides and DeWald [44] also used more catalyst than compound for the same purpose. This author had a similar experience. When the dinitro derivative of bradykinin was hydrogenated with an equal weight of palladium on carbon, uptake of hydrogen stopped completely at about 50%. A second addition of catalyst was required for completion of the reaction. Schröder and Lübke [45] note that peptides containing nitroarginine residues require long hydrogenation time to convert the N-nitro function to N-amine or to reduce it completely to the arginine grouping.

$$\text{RHNC}(=\text{NH})\text{NHNO}_2 \begin{array}{c} \xrightarrow{3\text{H}_2} \text{RHNC}(=\text{NH})\text{NHNH}_2 \\ \xrightarrow{4\text{H}_2} \text{RHNC}(=\text{NH})\text{NH}_2 \end{array} \quad (4.9)$$

R = amino acid residue

The following experiments were performed which clearly indicate that guanidine is an inhibitor. One millimole of guanidine carbonate was added to a solution of 8.2 g (0.1 mole) of cyclohexene in 150 cc of 90% aqueous ethyl alcohol. The mixture was hydrogenated in the presence of 0.042 g of platinum oxide under 2 atm pressure. Uptake of hydrogen was not very rapid. After about 40 min uptake of hydrogen stopped completely at 25% of the amount necessary to reduce cyclohexene to cyclohexane. In contrast, in the absence of the inhibitor, reduction was complete in 15–20 min.

Oximes *per se* are not poisoning agents but amines that result from their reduction are capable of inactivating catalysts. The effect of oximes is probably dependent on the ease with which they are reduced. Horner found that, in contrast to the pronounced inhibition caused by benzaldoxime, cyclopentanone oxime had little effect on the nickel-catalyzed hydrogenation of cyclohexene. Schiff bases, imines, azines, hydrazones, and related compounds containing an azomethine linkage are essentially nonpoisoning. Although Horner did report that a number of compounds of this type were inhibitors, the effect is due largely or perhaps totally to the resulting amines. In one example in the study it was shown that isopropylidenesemicarbazone,

$H_2NCONHN=C(CH_3)_2$, did not retard the hydrogenation of cyclohexene. Apparently, because of steric effects, either it was reduced too slowly to the base, isopropylsemicarbazide, to interfere with the hydrogenation of cyclohexene, or not enough of the base was produced to retard the reduction of cyclohexene before it was converted to cyclohexane. The following comparison shows the effect of a nitrogen base resulting from hydrogenation of the C=N bond [46]. When N-benzylbutylamine was added to the rhodium-catalyzed reduction of N-benzylidenebutylamine, there was a significant increase in the time necessary for complete conversion of the azomethine to secondary amine compared with the time required for reduction without the added nitrogen base.

Horner [13] reported on additions of a number of organic nitriles to nickel-catalyzed reductions of cyclohexene. Most of them had an anticatalytic effect. Organic nitriles themselves may have some inhibitory properties because of the combination of the effect of the aromatic ring and the C≡N bond. More likely the inhibition is caused by the resulting bases as in the case of imines and related compounds. A series of experiments were undertaken to determine whether the nitrile or the base resulting from reduction of it was the more potent inhibitor.

Experimental. A solution of 16.4 g (0.2 mole) of cyclohexene in 150 cc of absolute ethyl alcohol was hydrogenated in the presence of 0.050 g of platinum oxide at 36 psig. Uptake of hydrogen was complete in 20 min.

A similar experiment was carried out to which 0.001 mole of benzonitrile was added and in another run 0.001 mole of benzylamine was a part of the reduction mixture.

Hydrogen uptake for the mixture containing benzonitrile was 90% complete in 63 min, while the mixture containing benzylamine was only 50% complete in 88 min after an initial rapid uptake. The initial uptake followed the pattern of benzonitrile for 10 min (12–15% uptake) after which it decreased considerably. At the end of 138 min hydrogen uptake was 62% of the amount necessary to reduce cyclohexene.

The reduction solution, to which benzonitrile was added, was definitely basic after the reaction was interrupted indicating that some amine was formed. Two other hydrogenations containing the same amount of benzonitrile were carried out. One was interrupted after 0.05 mole of hydrogen was absorbed, the second after 0.1 mole of uptake. A test of basicity of the first reduction solution showed perhaps a trace of base while a sample of the reduction solution interrupted at 0.1 mole uptake contained titratable nitrogen base.

In another set of experiments phenylacetonitrile and the corresponding reduction product, phenethylamine, were the additives. Phenylacetonitrile was much less inhibitory. Uptake of hydrogen was complete in 35 min against about 100 min for the reduction retarded by the nitrogen base.

The results seem to indicate that the bases are more responsible than the parent nitriles for the inhibiting effect noted in the preceding experiments. The results also make it appear that benzylamine is a more potent inhibitor of platinum catalyst than phenethylamine. If there is a correlation in the difference in reducibility among benzonitrile, phenylacetonitrile, and other aralkyl nitriles with longer methylene bridges between the benzene ring and the nitrile function it might be explained by a difference in inhibitory effect among the corresponding amines.

Another example of inhibition resulting from conversion of a nonbasic form of nitrogen to one which has some basicity may be seen in Baltzly's investigations of the reducing capabilities of platinum on carbon [1]. In one of his experiments it was shown that conversion of the —N=N— bond in azobenzene took place rapidly in contrast to the much slower rate for the second stage, the hydrogenolysis of hydrazobenzene to aniline. It can be inferred that the difference in reduction rate between the first and second stage is caused by the resulting hydrazobenzene or by aniline as Baltzly suggests. In another experiment the addition of two molar equivalents of acetic acid had little effect on the first stage but greatly aided the second state. In this instance it appears likely that the presence of acid neutralized the retarding effect of either hydrazobenzene or aniline.

The inorganic counterpart of organic nitriles, the cyanides, are known to retard the activity of hydrogenation catalysts. Maxted [2] states that the cyanide ion is a poison of moderate intensity. Kelber [39], studying the effect of potassium cyanide and hydrogen cyanide on the reduction of sodium cinnamate with nickel, reported that hydrogen cyanide was less inhibiting than potassium cyanide in the reduction at room temperature and atmospheric pressure. He reasoned that the difference is due to the rapid reaction which takes place with hydrogen cyanide and nickel catalyst and hydrogen. Hydrogen cyanide may attack the catalyst or be converted to a nitrogen base. While it might be expected to yield methylamine on hydrogenation, Kelber reported that no amine was formed, only ammonia. Baltzly [1] found that the cyanide ion inhibited the platinized charcoal reduction of cyclohexene at a low concentration but the rate rose progressively from uptake of 3.3 mM of hydrogen per minute to 12 mM in 5 min. He attributes this to the rapid reduction of adsorbed cyanide ion to methylamine, which he states is a weaker inhibitor. Horner [13] found that potassium cyanide inhibited the nickel-catalyzed reduction of cyclohexene but to no greater degree than a number of organic nitriles. Stevinson and Hamilton [33] added potassium cyanide, 10% by weight of 4-nitrophenol, to the reduction in the presence of

Raney nickel. Reaction time was three times as long as in the hydrogenation of the nitro compound without the additive. When 4-nitrotoluene was reduced with a low ratio of palladium on carbon, the addition of potassium cyanide resulted in complete poisoning of the catalyst [10].

Pyridine is reported to be an inhibitor. Hamilton and Adams [47] were the first to show its failure to undergo hydrogenation under low-pressure conditions in neutral solvent. Maxted and Walker [48] have suggested that it is self-poisoning or, more likely, that the poisoning effect is due to the long adsorbed life of the reduction product, piperidine, on the catalyst. According to Maxted's theory of the inhibitory effect of unshielded structures, both pyridine and piperidine should adversely affect the course of hydrogenation. The nitrogen atom in each compound has an unshared pair of electrons which can bond to the catalyst.

$$\text{Pyridine} \qquad \text{Piperidine} \tag{4.10}$$

However, there is reason to believe, as suggested by Maxted and by this author that the more basic piperidine is responsible for the strong inhibitory action against the catalyst.

Comparative reductions of cyclohexene with platinum oxide in the presence of pyridine and in the presence of piperidine show that piperidine is by far a more powerful inhibitor than pyridine [49]. When piperidine was added to the attempted reduction, complete poisoning resulted. The addition of an equal amount of pyridine to the cyclohexene reduction caused uptake of hydrogen to proceed at a slow rate. The reaction required 7 hr and was incomplete compared with the reduction time of 20 min for the standard. Another example of the difference in inhibitory effect of pyridines and piperidines may be seen in the same reference. The addition of 2-methylpyridine caused a 50% increase in reaction time compared with the standard. The addition of an equal amount of 2-methylpiperidine to the cyclohexene reduction resulted in a 350% increase in reaction time.

It is likely that pyrrole, a much weaker base than pyridine, is difficult to reduce because it is converted to the saturated strong base pyrrolidine which causes inactivation of the catalysts used in the reaction. The poisoning effect of another saturated nitrogen heterocycle, morpholine, was reported

44   Catalyst Inhibitors and Poisons

by Kindler [50], who found that the addition of it prevented the hydrogenolysis of α-methylbenzyl alcohol to ethylbenzene in a reduction with palladium oxide. Another example of the poisoning power of piperidine is seen in work of Ungnade and Nightingale [51] who showed that a very small amount of it retarded the reduction of phenol to cyclohexanol in the presence of nickel catalyst at 100° and 130–150 atm.

The assumption that saturated heterocycles, which are strong bases, cause catalyst poisoning rather than the weakly basic unsaturated compounds is given credence by the work of Devereux, Payne, and Peeling on the kinetics of the hydrogenation of a number of unsaturated amines with platinum oxide [52]. Their results seem to indicate that weakly basic aromatic amines such as pyridine, N,N-dimethylaniline, m-toluidine, p-toluidine, and pyrrole are not inhibitory toward platinum oxide, but that strong nitrogen bases caused catalyst poisoning. Baltzly [1] also showed a difference in poisoning effect between a weakly basic aromatic amine, p-toluidine and a much stronger base, N-benzyldimethylamine. To inhibit the reduction of cyclohexene in the presence of platinum on carbon, 7 mM of aromatic amine were needed to obtain the same effect produced by 1 mM of the aralkylamine. Five to 6 mM of p-toluidine were required to cause similar retardation when palladium was the catalyst.

Based on Maxted's concept, conversion of the nitrogen atom in pyridine or piperidine to a shielded form should neutralize its poisoning effect. The ease of reducibility of pyridines where the free electron pair is shielded was reported by Hamilton and Adams [47] some time before Maxted had advanced his theory. They found that acid salts and quaternary salts of

$$\begin{array}{cc} \text{Pyridinum ion} & \text{Piperidinum ion} \end{array} \qquad (4.11)$$

pyridines and quinolines could be reduced readily under low-pressure conditions if a suitable solvent was used. In contrast, as noted before, no uptake of hydrogen occurred during the attempted reduction of the base in neutral solvent. In work prior to that of Hamilton and Adams, Skita and Brunner [53] carried out hydrogenations of pyridine and quinoline bases in acid solution in the presence of colloidal platinum but there is no record of their attempts to effect similar reductions in the absence of acid.

Devereux, Payne, and Peeling [52] investigated the effect of acid in their hydrogenation experiments. They found that when less than an equivalent of hydrochloric acid was added to the reduction of N,N-dimethylaniline, for example, the reaction, which was rapid initially, reverted to a much slower rate when the amount of hydrogen absorbed corresponded to the formation of an amount of N,N-dimethylcyclohexylamine equal to the amount of added acid. The slower rate was similar to the rate of hydrogen uptake observed in reduction without acid. The authors reasoned that the cyclic base displaced the weaker aromatic base from its hydrochloride until the displacement was complete, after which time the rate of hydrogenation decreased indicating that the catalyst had been poisoned. They pointed out at this time that free N,N-dimethylcyclohexane had formed and this substance, a strong base, was the poison. They suggested that similar results should occur whenever a weak base is converted to a strong base by hydrogenation.

The same authors indicated from some unpublished work in their laboratory that piperidine, pKa 11.12, was a more potent inhibitor than N,N-dimethylcyclohexylamine, pKa 10.48. This is compatible with their suggestion that poisoning is dependent on base strength. In this instance, however, another factor is involved. Molecular models of the two compounds appear to show that the nitrogen atom in piperidine is less sterically hindered than that in N,N-dimethylcyclohexylamine. Another example of the same phenomenon can be seen in the comparative effects of piperidine, 2-methylpiperidine, and 2,6-dimethylpiperidine on the platinum-catalyzed reduction of cyclohexene [49]. Piperidine caused complete poisoning, 2-methylpiperidine caused a 50% increase in reduction time, while the addition of 2,6-dimethylpiperidine did not produce any significant difference in reduction time compared with the standard. The difference in anticatalytic activity among the three compounds, which are all strong bases, must be due to the presence of the methyl groups adjacent to the nitrogen atom. In the case of piperidine there is nothing to interfere with bonding between the nitrogen atom and the active centers of the catalyst. In 2-methylpiperidine contact between the offending atom and the catalyst is weakened while the two methyl groups in 2,6-dimethylpiperidine make contact between the catalyst and the nitrogen atom difficult and poisoning does not take place.

Comparable steric effects were noted in the pyridine series [49]. The addition of pyridine affected the reduction of cyclohexene so that hydrogen uptake was only 65% after 7 hr. The addition of 2-methylpyridine increased the reaction time from 20 min for a standard reduction to 28–30 min. The addition of 2,6-dimethylpyridine appeared to have little or no poisoning effect on the reaction. Another example of steric effects on catalyst inhibition is seen among the many compounds studied by Horner [13]. 2-Methylpyridine N-oxide caused about 60% inhibition at a 1 mM level when added to the reduction of cyclohexene with nickel, whereas the addition of a similar amount of 2,6-dimethylpyridine N-oxide resulted in only 2% inhibition.

Devereux and his colleagues [52] suggested that N,N-dimethylphenethylamine, pKa 8.6, and the reduction product, N,N-dimethyl-2-cyclohexylethylamine, pKa 8.92, because of their basicity should be catalyst poisons. By analogy, related compound should not reduce with ease in neutral medium. Conversion of aralkylamines to cyclohexylalkylamines, in general, requires conditions of elevated temperature and pressure when the reductions are carried out in a neutral solvent. With certain catalysts, reaction conditions often induce hydrogenolysis. Metayer [54] pointed out that in the reduction of phenethylamines with nickel incidental hydrolysis and cracking occurred. King and co-workers [55] experienced difficulty in hydrogenating N,N-dimethylphenethylamine with nickel. At approximately 80 atm pressure of hydrogen, the high temperature (200°) required for hydrogen uptake caused extensive hydrogenolysis. However, the investigators were able to obtain the corresponding cyclohexyl compound by reducing the amine in acid solution in the presence of palladium on carbon catalyst. Acid conditions were used by another group who reduced a number of aralkylamines with platinum catalyst under mild conditions [56].

In a continuation of his study on the poisoning effect of nitrogen-containing compounds Maxted, with one of his associates, reported that completely anhydrous ammonia was a more effective inhibitor than butylamine or cyclohexylamine in poisoning the hydrogenation of dry cyclohexene [57]. Its inhibiting effect was lost in the presence of water because of the change of the nitrogen atom from an unshielded form to the nonpoisoning shielded state. Baltzly [1] also reported on the effect of ammonia on the platinum-catalyzed reduction of cyclohexene. From his work the effect may be dependent on the amount of ammonia present. An 0.1 molar solution of ammonia in methanol completely inhibited the reaction. At a concentration of 0.005 mole per 0.05 mole of substrate, 25% of the normal activity of the catalyst was restored. When the amount of ammonia was decreased to 0.002 mole, the catalyst had about half of its usual activity.

Of interest in the Maxted and Biggs study [57] was the very low relative poisoning effect of dicyclohexylamine, which from its pKa value, 10.3, should be almost as inhibitory as n-butylamine, pKa 10.4. Nevertheless, on a molar basis butylamine was 80 times more inhibitory toward the reduction of cyclohexene. This is another example which illustrates that the geometry of a potential inhibitor can affect its anticatalytic activity. Despite the basicity of dicyclohexylamine, the inhibiting effect of the nitrogen atom was greatly diminished, because the bulky cyclohexane rings did not allow good contact between that atom and the active sites of the catalyst.

Horner [13] noted the effect of some nitrogen-containing aromatic compounds on the reduction of cyclohexene with Raney nickel. All were fairly weak bases. In view of their weakly basic character they might not be expected to be inhibitors [52]. However, some including pyrazine, 5-aminotetrazole, benzimidazole, some aromatically substituted oxazolones, and 4-methyl- and 4-methyl-5-nitrouracils had a potent adverse effect.

Poisoning by the above compounds is due to the strong bonding effect of aromatic rings which enables the ring systems of such compounds to compete with cyclohexene for active sites on the catalysts and for the most part to be preferentially adsorbed.

It was surprising to find that quinoline, often used to inhibit the activity of noble metal catalysts, was a weak inhibitor toward the nickel reduction of cyclohexene, particularly since pyridine was a potent one for the same reaction [13]. The difference in effect among 2-, 3-, and 4-methylpyridine-N-oxides is also of interest. The 3-methyl compound was far more inhibitory than the 2- or 4-methylpyridine-N-oxides. This difference may have been due to a difference in purity among them. More likely it is due to the more facile removal of oxygen from 3-methylpyridine-N-oxide than the 2- and 4-methylpyridine-N-oxides to the corresponding pyridines.

THE EFFECT OF ALKALI METAL BASES

Strong bases, such as sodium hydroxide and potassium hydroxide, at times inhibit the activity of hydrogenation catalysts. Adams and co-workers [58] reported that sodium hydroxide, 0.025 mole per 0.1 mole of substrate, had a distinct anticatalyst effect on the hydrogenation of nitrobenzene with platinum oxide. Sodium carbonate, in contrast to sodium hydroxide, did not prevent the uptake of hydrogen. It merely doubled the reaction time necessary for complete reduction. The investigators tried to make use of the inhibiting effect of sodium hydroxide to selectively reduce the nitro group and the double bond in (3-nitrobenzylidene)acetophenone and (3-nitrobenzylidene)acetone. The addition of 0.1 mole of sodium hydroxide to the hydrogenation of 0.1 mole of each nitro compound resulted in complete poisoning of both reactions. Less alkali, 0.003 mole per 0.1 mole of nitro compound, completely prevented the reduction of nitrobenzene in the presence of 3 g of Raney nickel [59]. The poisoning effect was limited to nitrobenzenes with neutral substituents. In another study [60] it was shown that excess sodium hydroxide poisoned the nickel-catalyzed hydrogenation of sodium 3- and 4-nitrobenzoates. The addition of acetic acid allowed the inhibited reduction to continue; the presence of ammonium chloride as a buffer had the same effect. This suggested to the authors that alkalinity poisoned the hydrogenations. Another group of investigators [61] reported that nitrobenzenes containing basic or acidic substituents were adversely affected by the presence of sodium hydroxide as well as nitrobenzenes with neutral substituents. An analysis of their data, however, is not in complete agreement with their statement. The reductions of 3,5-dinitrosalicylic acid, 2-amino-6-nitrophenol, 2,4-dinitrophenol, and 2-amino-4,6-dinitrophenol did not go to completion under any of the described reaction conditions, among which were reductions in the presence or absence of alkali.

The effect of alkali on the hydrogenation of nitrobenzenes may depend on the catalyst used. All inhibitors do not necessarily poison every catalyst. Baltzly [1] found that nitrobenzene was rapidly reduced to aniline in methanolic sodium hydroxide solution in the presence of palladium on carbon or platinum on carbon. The lack of poisoning by sodium hydroxide in the reduction with platinum on carbon catalyst is of interest in view of Adams's report of poisoning when platinum oxide was used in the hydrogenation of nitrobenzene [58]. Baltzly also noted that alkali, which had no effect on the reduction of azobenzene to hydrazobenzene, prevented any further uptake of hydrogen. In the absence of alkali, the reduction with platinum on carbon proceeded to yield aniline and hydrazobenzene.

Tedeschi [62] found that small amounts of sodium or potassium hydroxide, 0.05–0.1 g per mole of substrate, inhibited hydrogenolysis in the hydrogenation of acetylenic glycols with palladium on carbon. The addition of base did not affect the conversion of acetylene to olefin, but it greatly inhibited hydrogenolysis of the tertiary hydroxyl function and retarded the reduction of olefin to the saturated compound to some degree. Sodium hydroxide was a stronger inhibitor than potassium hydroxide. Potassium carbonate was about one-tenth as effective as potassium hydroxide; potassium bicarbonate had no effect. Sodium methoxide was also an inhibitor but the effect was also limited to prevention of hydrogenolysis. Only potassium metal, 0.05 g per mole of acetylenic glycol, inhibited hydrogenation to such an extent that uptake of hydrogen did not go beyond formation of the olefinic diol. Another investigation [63] showed that addition of alkali metal bases acted as inhibitors on the hydrogenation of tertiary ethynyl carbinols in the presence of palladium, platinum, or rhodium on carriers. The addition of the bases in a 1:1 or 1:2 ratio of base to catalyst caused the reaction to come to a halt after the olefinic carbinol was formed. In the absence of base there was no selectivity.

Another example of the deterring effect of strong base may be seen in the reduction of 2,5-dimethyl-3-hexyne-2,5 diol to 2,5-dimethyl-3-hexene-2,5-diol in the presence of a Raney-type nickel catalyst obtained by refluxing a 1:1 nickel silicon alloy with sodium hydroxide [64]. The secret of success appeared to lie in the retention of alkali after washing the catalyst with only enough water to remove silicate ion but not residual sodium hydroxide.

Uptake of hydrogen stopped of its own accord after 1 hr when the reaction was carried out for 18 hr under about 200 psig.

Foresti [65], studying the effect of pH of the hydrogenation of olefins with platinum catalyst, found that a sharp drop in rate occurred in alkaline media. The same decrease in rate was observed in the reduction of benzene and a few other aromatic hydrocarbons under similar conditions. In a later investigation [66], the velocity of the hydrogenation of benzene decreased with increasing pH to about 10% of the rate at neutrality. The rate of reduction of cyclohexene also decreased radically in alkaline media. Foresti suggested that the increase in alkalinity of the medium has the same effect as a selective and reversible poison.

OTHER INHIBITORS

There is very little mention of the effect of oxygen as an inhibiting element towards hydrogenation catalysts, in contrast to that of the other poisoning elements of groups Vb and VIb. It appears to have little, if any, effect except in carbon monoxide. The poisoning power of carbon monoxide is probably due to preferential adsorption on the active sites of the catalyst. More carbon monoxide than hydrogen is adsorbed on most of the commonly used hydrogenation catalysts [66]. However, carbon monoxide should be of little concern to the chemist involved in hydrogenation because it is very seldom encountered except in the Fischer-Tropsch reaction which will not be covered in this book.

Greenfield [10] reported that the addition of 4-nitroso-N-methylaniline and 4-nitrosodiphenylamine caused a twofold and sixfold increase respectively in reaction time in the hydrogenation of 4-nitroluene with palladium on carbon. The effect of the addition of N-nitrosodiphenylamine was much more striking. A 1 hr induction period ensued which was followed by hydrogen uptake at one-third of the rate of the uninhibited reaction. There was no explanation nor any attempt by the author to investigate the effect of possible reduction products of the additives on the hydrogenation of 4-nitroluene.

PRACTICAL USE OF CATALYST POISONS

The number of catalyst poisons mentioned in this text need not alarm the reader. It should be understood that potential poisoning agents do not have the same effect on every catalyst. What is of value is to be aware of the poisoning potential. Many of these agents can be and have been used to good advantage to inactivate certain catalysts. By this means the reaction rate becomes sufficiently depressed so that it becomes possible to interrupt a

50   Catalyst Inhibitors and Poisons

reduction to obtain a particular product or, by using the proper amount of inhibitor, the reaction stops of its own accord at the desired point. In a number of instances an inhibitory material is used as a solvent. There are many examples of what Maxted calls "beneficial poisoning" [2]. The Rosenmund reduction [67] makes use of retardants or, to coin a phrase, selectivity agents. The reaction is essentially a hydrogenolysis or dehalogenation of an acid chloride to an aldehyde. However, the reaction does not always stop at this point but continues stepwise to hydrocarbon as seen in Eq. 4.12. Rosenmund pointed out that success was often achieved but in

$$C_6H_5COCl \xrightarrow{H_2} C_6H_5CHO + HCl$$
$$C_6H_5CHO \xrightarrow{H_2} C_6H_5CH_2OH \xrightarrow{H_2} C_6H_5CH_3 + H_2O \quad (4.12)$$

later investigations with pure starting materials yields of aldehyde were often insignificant. This led to the conclusion that extraneous impurities in the previously used starting materials were responsible for affecting the catalyst to such an extent that further reduction of the aldehyde did not take place. It also led to a study of poisons of which quinoline or quinoline containing dissolved sulfur appeared to be best suited for use in the presence of palladium on barium sulfate. Weygand and Meusel [68] found that benzaldehyde could be obtained in high yield—96%—from benzoyl chloride when platinum oxide was used as the catalyst and thiourea was the poisoning agent. Ammonium thiocyanate could be substituted for thiourea, but it resulted in a somewhat lower yield of benzaldehyde—80%. Quinoline has been used successfully to poison palladium on calcium carbonate so that partial hydrogenation of β-carotene would take place [69]. Neither quinoline, quinoline-sulfur, nor pyridine poisoned palladium on carbon catalyst sufficiently to allow reduction of the double bond of 2-phenyl-3-(2'furyl) acrylonitrile without attack on the furan ring (see Eq. 4.13). However, the addition of thiophene, 8–8.5 mg per gram of catalyst, gave the desired furan compound.

Teichmann [71] employed a variety of poisons to accomplish a purpose. He used triphenylphosphine, 0.001 mole per mole of compound, to prevent

(4.13)

reduction of carbonyl function while dehalogenating trichloroacetaldehyde. Thiophene, thiophenol and thiourea, 0.001 mole per mole of substrate, served the same purpose, as did N,N-dimethylaniline, acetonitrile, and pyridine, 0.001 mole per mole of trichloroacetaldehyde. He also added copper sulfate, mercuric chloride, or silver sulfate at a concentration of 0.005 mole per mole to prevent reduction of the aldehyde group. Only the reduction inhibited with triphenylphosphine gave a small amount of trichloroethanol.

Pyridine has been found a useful agent for selective poisoning. Sokol'skii and Biesekov [72] reported that the addition of it, 0.005–0.02 g, to a catalyst prepared from 1 g of Raney nickel aluminum alloy decreased the reaction rate to insure reduction of cinnamaldehyde to 3-phenylpropionaldehyde. Masaki and his co-workers [73] used pyridine to attain debenzylation without concomitant hydrogenolysis of the resulting hydroxamic acid to amide (Eq. 4.14).

$$(H_3C)_2CHCH_2CH(R)CONHOCH_2C_6H_5 \quad \xrightarrow{2H_2} (H_3C)_2CHCH_2CH(R)CONH_2$$
$$R = H \text{ or } OH \qquad \xrightarrow[\text{pyridine present}]{H_2} (H_3C)_2CHCH_2CH(R)CONHOH$$

Pd on C, 40 atm (4.14)

When the hydrogenation of 2,3-dimethoxy ω-nitrostyrene was carried out in pyridine as solvent in the presence of palladium on carbon, the corresponding phenylacetaldoxime was obtained [74]. References 75 and 76 are further examples of this technique which leads to high yields of substituted phenylacetaldoximes.

In the hydrogenation of acetylenic compounds to ethylenic structures the catalyst must be inactivated for selectivity to be realized. The catalyst developed by Lindlar [77], palladium on calcium carbonate poisoned by lead acetate, is often used for the reaction. It is not only useful for preventing further reduction of the ethylenic linkage but it can also inhibit the hydrogenation of other functional groups. A fine example is seen in the selective reduction of an acetylenic bond to an olefinic one, without attacking the aliphatic nitro group [78]. Furthermore, the patent points out that the nitro group of the ethylenic compound can be reduced to the amine in the presence of Lindlar catalyst without affecting the double bond. The examples are seen in Eq. 4.15.

$$C_{13}H_{27}C{\equiv}CCH(OH)CH(NO_2)CH_2OH \xrightarrow{H_2}$$
$$C_{13}H_{27}CH{=}CHCH(OH)CH(NO_2)CH_2OH$$
Lindlar catalyst (4.15)

$$C_{13}H_{27}CH{=}CHCH(OH)CH(NO_2)CH_2OH \xrightarrow{3H_2}$$
$$C_{13}H_{27}CH{=}CHCH(OH)CH(NH_2)CH_2OH$$
Lindlar catalyst

Fukuda and Kusama [79] studied the effect of Lindlar catalyst on the reduction of 2-butyne-1,4-diol. They found that it had little effect on the conversion of the acetylenic diol to the corresponding butenediol, but that it retarded the next stage of the reduction to 1,4-butanediol. They further observed that the rate of hydrogenation of the butenediol decreased with the increasing amount of lead acetate present in the catalyst. In a later publication Fukuda [80] showed that larger amounts of lead acetate also affected the first step.

Other catalysts and other inhibitors have been employed in order to attain selectivity in the reduction of acetylenic compounds. Oroshnik, Karmas, and Mebane [81] were successful in converting the triple bond to an ethylenic bond without affecting either an isolated double bond or a conjugated double bonded system which were a part of the compound in question. Two inhibitors, piperidine and zinc acetate, were added to the Raney nickel reduction. During the major part of the reaction hydrogen was absorbed at a rate of 25 cc per minute. Near the end of the reaction it had dropped to 2 cc per minute. Grenet [82] found that palladium on calcium carbonate could be inactivated by the addition of stannous chloride or stannous acetate. The catalyst, thus obtained, was suitable for selective hydrogenation of butynols to butenols. In the reduction of 1,6-dihydroxy-3,7-dimethyl-9-(2,6,6-trimethyl-1-cyclohexenyl)-2,7-nonadien-4-yne with palladium on carbon, quinoline was added. The reaction was interrupted after one equivalent of hydrogen was absorbed [83]. The high yield of the corresponding cyclohexenyl 2,4,7-nonatriene, 97%, showed that the reduction was indeed selective. The selectivity is probably attributable to the addition of quinoline. Another example of its effect is illustrated in Cram's work on the reduction of the following cyclophane in which quinoline was the retarding agent for palladium on barium sulfate catalyst [84] (Eq. 4.16).

$$\text{(cyclophane with } (CH_2)_4 \text{ bridges and } C{\equiv}C) \xrightarrow{H_2} \text{(cyclophane with } (CH_2)_4 \text{ bridges and } CH{=}CH)$$

(4.16)

In a later publication Cram and Allinger [85] reported that the addition of chemically pure quinoline to the reduction of acetylenic compounds retarded the activity of palladium on barium sulfate catalyst so that ethylenic compounds were obtained. They claimed that coal tar quinoline was unsuitable.

Fukuda [79, 80], in addition to his work on the action of lead acetate on palladium on calcium carbonate, studied the effect of addition of quinoline, pyridine, and piperidine during reductions of 2-butyne-1,4-diol with the same catalyst. He reported that the reduction of the triple bond was only slightly retarded by the addition of quinoline, but it was not affected by the presence of either pyridine or piperidine. On the other hand the hydrogenation of 2-butene-1,4-diol was inhibited to some degree by the addition of pyridine, more markedly so by piperidine, and poisoned almost completely by quinoline.

Papa and co-workers [86] achieved success in reducing some acetylenic carbinols to the corresponding vinyl carbinols by carrying out the reaction in pyridine in the presence of palladium on calcium carbonate. Ruzicka and Muller [87] inactivated palladium on calcium carbonate catalyst by pretreating it in pyridine in a hydrogen atmosphere. 17-Ethinyltestosterone dissolved in piperidine was added and was selectively reduced to 17-vinyltesterone.

Elsner and Paul [88] reported that a poisoned palladium catalyst was not selective for the hydrogenation of some octadecynes, but pyrophoric Raney nickel which was treated with copper acetate gave a catalyst which was useful for their purpose. When octadec-4-yne was reduced in its presence, uptake for one molar equivalent was complete in 3 min. No further uptake occurred within 10 more min. When octadec-1-yne was similarly hydrogenated, uptake for the first equivalent was complete in 4 min. In this instance the catalyst did not poison the reaction completely but retarded it so that uptake of the second equivalent of hydrogen required 40 min.

A flow system was used to hydrogenate nitrobenzene to aniline over a nickel catalyst. Nickel sulfate, described as an inhibitor in a previous section in this chapter, was added to prevent further reduction of aniline to cyclohexylamine [89]. Thiophene also poisoned the ability of the nickel catalyst to reduce the aromatic ring in a flow system at 180° but it did not prevent reduction of the nitro group [90].

Attempts to prepare cyclohexylhydroxylamine from nitrocyclohexane were dependent on the presence of a nitrogen base when the reduction was carried out with palladium on carbon in neutral solvent [91]. Hydrogenation could be carried out in alcohol containing various bases which included pyridine, piperidine, amylamine, or morpholine; or the nitrogen base could act as solvent and inhibitor at the same time.

The examples that have been cited illustrate how poisoning agents may serve a useful purpose. Many more examples will be seen in the ensuing chapters.

REFERENCES

[1] R. Baltzly, *J. Am. Chem. Soc.*, **74**, 4586 (1952).
[2] E. B. Maxted, *Advances in Catalysis*, Vol. III, Academic Press, New York, 1951, p. 129.

[3] C. Paal and A. Karl, *Ber.*, **44**, 1013 (1911).
[4] C. Paal and A. Karl, *Ber.*, **46**, 3069 (1913).
[5] C. Paal and W. Hartman, *Ber.*, **51**, 711 (1918).
[6] C. Paal and E. Windisch, *Ber.*, **46**, 4010 (1913).
[7] E. B. Maxted, *J. Chem. Soc.*, **117**, 1501 (1920).
[8] E. B. Maxted, *J. Chem. Soc.*, **119**, 225 (1921).
[9] E. B. Maxted and A. Marsden, *J. Chem. Soc.*, **1940**, 469.
[10] H. Greenfield, *J. Org. Chem.*, **28**, 2431 (1963).
[11] S. Ueno, *J. Chem. Ind. Japan*, **21**, 898 (1918); through *Chem. Abstr.*, **13**, 383 (1919).
[12] S. Ueno, *J. Chem. Ind. Japan*, **23**, 845 (1920); through *Chem. Abstr.*, **15**, 1226 (1921).
[13] L. Horner, H. Reuter, and E. Hermann, *Ann.*, **660**, 1 (1962).
[14] E. B. Maxted, *J. Chem. Soc.*, **1949**, 1987.
[15] J. N. Pattison and E. Degering, *J. Am. Chem. Soc.*, **73**, 611 (1951).
[16] H. S. Taylor and S. C. Liang, *J. Am. Chem. Soc.*, **69**, 1306 (1947).
[17] R. Baltzly and A. P. Phillips, *J. Am. Chem. Soc.*, **68**, 261 (1946).
[18] R. Baltzly and P. B. Russell, *J. Am. Chem. Soc.*, **72**, 3410 (1950).
[19] W. Wenner, *J. Org. Chem.*, **15**, 301 (1950).
[20] M. Freifelder, unpublished work.
[21] R. Truffault, *Bull. Soc. Chim. France*, **1935**, 44.
[22] J. S. Campbell and C. Kemball, *Trans. Faraday Soc.*, **57**, 809 (1961).
[23] K. Hoshiai and J. Miyata, *Kôgyô Kagaku Zasshi*, **60**, 253 (1957); through *Chem. Abstr.*, **53**, 7974 (1959).
[24] K. Hoshiai, *Kôgyô Kagaku Zasshi*, **60**, 40 (1957); through *Chem. Abstr.*, **53**, 6067 (1959).
[25] K. Hoshino and J. Miyata, *Kôgyô Kagaku Zasshi*, **59**, 236 (1956); through *Chem. Abstr.*, **51**, 10368 (1957).
[26] J. V. Supniewski and M. Serafinovna, *Arch. Chem. Farm.*, **3**, 109 (1936); through *Chem. Abstr.*, **33**, 7301 (1939).
[27] R. E. Lyle and G. G. Lyle, *J. Am. Chem. Soc.*, **76**, 3536 (1954).
[28] R. E. Lyle, E. F. Perlowski, H. J. Troscianiec, and G. G. Lyle, *J. Org. Chem.*, **20**, 1761 (1955).
[29] C. A. Grob and F. Ostermayer, *Helv. Chim. Acta*, **45**, 1119 (1962).
[30] A. Lasslo, W. M. Marine, and P. D. Waller, *J. Org. Chem.*, **21**, 958 (1956).
[31] S. Sugasawa and Y. Degueli, *J. Pharm. Soc. Japan*, **76**, 968 (1956); through *Chem. Abstr.*, **51**, 2771 (1957).
[32] K. Tomita, *J. Pharm. Soc. Japan*, **71**, 220 (1951) in English.
[33] M. R. Stevinson and C. S. Hamilton, *J. Am. Chem. Soc.*, **57**, 1298 (1935).
[34] M. C. Dart and H. B. Henbest, *J. Chem. Soc.*, **1960**, 3563.
[35] E. B. Maxted and R. W. D. Moorish, *J. Chem. Soc.*, **1940**, 252.
[36] L. D. Freedman, G. O. Doak, and E. L. Petit, *J. Am. Chem. Soc.*, **77**, 4262 (1955).
[37] G. Toennies and J. J. Kolb, *J. Am. Chem. Soc.*, **67**, 1141 (1945).
[38] A. G. Deem and J. E. Kaveckis, *Ind. Eng. Chem.*, **33**, 1373 (1944).
[39] C. Kelber, *Ber.*, **49**, 1868 (1916).
[40] E. Maxted and H. E. Evans, *J. Chem. Soc.*, **1937**, 603.
[41] E. Maxted and H. E. Evans, *J. Chem. Soc.*, **1937**, 1004.
[42] E. B. Maxted and H. C. Evans, *J. Chem. Soc.*, **1938**, 455.
[43] R. B. Merrifield, *Biochemistry*, **3**, 1385 (1964).
[44] E. D. Nicolaides and H. A. DeWald, *J. Org. Chem.*, **26**, 3872 (1961).
[45] E. Schröder and K. Lübke, *The Peptides*, Vol. I, Academic Press, New York, 1965, p. 169.

References

[46] M. Freifelder, *J. Org. Chem.*, **26**, 1835 (1961).
[47] T. S. Hamilton and R. Adams, *J. Am. Chem. Soc.*, **50**, 2260 (1928).
[48] E. B. Maxted and A. G. Walker, *J. Chem. Soc.*, **1948**, 1093.
[49] M. Freifelder, *Advances in Catalysis*, Vol. XIV, Academic Press, New York, 1963, p. 203.
[50] K. Kindler, H. G. Helling, and E. Sussner, *Ann.*, **605**, 200 (1957).
[51] H. Ungnade and D. V. Nightingale, *J. Am. Chem. Soc.*, **66**, 1218 (1944).
[52] J. M. Devereux, K. R. Payne, and E. R. A. Peeling, *J. Chem. Soc.*, **1957**, 2845.
[53] A. Skita and W. Brunner, *Ber.*, **49**, 1597 (1916).
[54] M. Metayer, *Bull. Soc. Chim. France*, **1952**, 276.
[55] F. E. King, J. A. Barltrop, and R. J. Walley, *J. Chem. Soc.*, **1945**, 277.
[56] B. L. Zenitz, E. B. Macks, and M. L. Moore, *J. Am. Chem. Soc.*, **69**, 117 (1947).
[57] E. B. Maxted and M. S. Biggs, *J. Chem. Soc.*, **1957**, 3844.
[58] R. Adams, F. L. Cohen, and O. W. Rees, *J. Am. Chem. Soc.*, **49**, 1093 (1927).
[59] J. R. Reasenberg, E. Lieber, and G. B. L. Smith, *J. Am. Chem. Soc.*, **61**, 384 (1939).
[60] S. S. Scholnik, J. R. Reasenberg, E. Lieber, and G. B. L. Smith, *J. Am. Chem. Soc.*, **63**, 1192 (1941).
[61] G. S. Samuelson, V. L. Garik, and G. B. L. Smith, *J. Am. Chem. Soc.*, **72**, 3872 (1950).
[62] R. J. Tedeschi, *J. Org. Chem.*, **27**, 2398 (1962).
[63] R. J. Tedeschi and G. Clark, Jr., *J. Org. Chem.*, **27**, 4323 (1962).
[64] B. Foresti, *Boll. soc. Eustachiana*, **38**, 19 and 29 (1940); through *Chem. Abstr.*, **36**, 6400 (1942).
[65] B. Foresti, *Ann. Chim.* (Rome), **41**, 425 (1951); through *Chem. Abstr.*, **46**, 330 (1952).
[66] G. C. Bond, *Catalysis by Metals*, Academic Press, New York, 1962, Section 5.6, p. 90.
[67] K. W. Rosenmund and F. Zetsche, *Ber.*, **54**, 425 (1921).
[68] C. Weygand and W. Meusel, *Ber.*, **76**, 503 (1943).
[69] H. H. Inhoffen, F. Bohlmann, K. Bartran, G. Rummert, and H. Pommer, *Ann.*, **570**, 54 (1950).
[70] M. Pesson, S. Dupin, M. Antoine, D. Humbert, and M. Joannic, *Bull. Soc. Chim. France*, **1965**, 2262.
[71] B. Teichmann, *J. Prakt. Chem.*, **301**, 51 (1965).
[72] D. V. Sokol'skii and K. Z. Beisekov, *Vestn. Akad. Nauk Kaz SSR*, **21**, 71 (1965); through *Chem. Abstr.*, **63**, 6906 (1965).
[73] M. Masaki, J. Ohtake, M. Sugiyama, and M. Ohta, *Bull. Chem. Soc. Japan*, **38**, 1802 (1965).
[74] A. Lindemann, *Helv. Chim. Acta*, **32**, 69 (1949).
[75] B. Reichert and W. Koch, *Arch. Pharm.*, **273**, 265 (1935).
[76] B. Riechert and H. Marquardt, *Pharmazie*, **5**, 10 (1939).
[77] H. Lindlar, *Helv. Chim. Acta*, **35**, 446 (1952).
[78] C. Grob and E. Jenny, U.S. Patent 3,118,946 (1964).
[79] T. Fukuda and T. Kusama, *Bull. Chem. Soc. Japan*, **31**, 339 (1958).
[80] T. Fukuda, *Bull. Chem. Soc. Japan*, **32**, 420 (1959).
[81] W. Oroshnik, G. Karmas, and A. D. Mebane, *J. Am. Chem. Soc.*, **74**, 295 (1952).
[82] J. B. Grenet, U.S. Patent 3,192,168 (1965).
[83] O. Isler, W. Huber, A. Ronco, and M. Koffler, *Helv. Chim. Acta*, **30**, 1911 (1947).
[84] D. J. Cram and M. Cordin, *J. Am. Chem. Soc.*, **77**, 4090 (1955).
[85] D. J. Cram and N. L. Allinger, *J. Am. Chem. Soc.*, **78**, 2518 (1956).

[86] D. Papa, F. J. Villani, and H. F. Ginsberg, *J. Am. Chem. Soc.*, **76**, 4246 (1954).
[87] L. Ruzicka and P. Muller, *Helv. Chim. Acta*, **22**, 755 (1939).
[88] B. B. Elsner and P. F. M. Paul, *J. Chem. Soc.*, **1953**, 3156.
[89] K. Yoshikawa, *Sci. Papers Inst. Phys. Chem. Research Tokyo*, **25**, 235 (1934); through *Chem. Abstr.*, **29**, 5005 (1935).
[90] K. Yoshikawa, T. Yamanaka, and B. Kubota, *Bull. Inst. Chem. Research Tokyo*, **14**, 313 (1935); through *Chem. Abstr.*, **29**, 7957 (1935).
[91] Unpublished work.

V

CATALYST PROMOTERS

Pease and Taylor [1] define catalyst promoters as substances which are capable of producing a greater catalytic effect than can be accounted for if each component, catalyst or additive, acted independently and in proportion to the amount present. The authors did not confine catalytic effect to hydrogenation alone but included other catalytic reactions. "Producing a greater catalytic effect" is rather a broad term as applied to hydrogenation. A definition that might be more applicable to their effect in the hydrogenation process would consider those substances as promoters which, when present in small amounts, would increase the speed of reaction or would aid reduction to proceed to completion. The limiting factor of low concentration of promoter would differentiate between their action and that seen, for example, in the effect of acidic media on the hydrogenation of pyridines [2] or that of the addition of equivalent amounts of acids of varying pKa values in conversion of N-benzylidenebutylamine to N-benzylbutylamine [3].

Additives used for promoting catalytic activity include a variety of materials. Some of these substances are active for hydrogenation, as some of the noble metals, although it is doubtful in most instances that they would have any effect in the amount used for promotion. For the most part additives which enhance the activity of catalysts in hydrogenations are themselves inactive for catalytic reduction.

It must be borne in mind that promoters are often specific in their action. A substance which may increase the activity of a catalyst in one type of hydrogenation may act as a retardant or even as a complete poison in another. Indeed, there is a fine line of distinction. It may be noted that some of the additives to be discussed have already been described in Chapter IV as poisons. Furthermore, the amount of additive may determine whether it is promoter or poison.

METALLIC AND NON-METALLIC SUBSTANCES

Covert, Connor, and Adkins [4] added a number of salts, in a ratio of one part to ten parts of nickel nitrate, and incorporated these materials into nickel on kieselguhr for the hydrogenation of acetone, furaldehyde diethylacetal, and toluene at 175° and 125 atm pressure. Their results indicated some promotion, although not striking, from the presence of boron, chromium, cobalt, molybdenum, and tin in the catalyst in the reduction of acetone. The presence of barium, cerium, iron, manganese, molybdenum, and silicon appeared to have improved the activity of the catalyst for the hydrogenation of the acetal. Cerium distinctly promoted the reduction of toluene. The authors appeared to be unenthusiastic about the promoter effect of the various additives, pointing out, as had been previously mentioned, that promoters are too specific in their action.

A patent [5] describes an increase in activity of palladium and platinum-palladium catalysts for the reduction of nitro compounds by the addition of various metallic and nonmetallic substances. The foreword of the patent states that oxides or hydroxides of iron, cobalt, nickel, magnesium, aluminum, chromium, vanadium, or tungsten or combinations of them are activators. Boron and silicon fluorides are also mentioned. These may be added before, during, or after precipitation of the noble metal on a specified carbon black. However, there is no description of their use in the examples except for iron as ferric chloride, ferrous chloride, or ferric hydroxide in a 1:1 ratio of iron to noble metal. The use of the iron-promoted catalysts resulted in a 400% increase of hydrogen absorption per minute.

Carothers and Adams [6] found that the addition of ferric chloride, 0.1 mM per 0.001 g atom of platinum, induced the reduction of benzaldehyde and heptaldehyde to proceed at a rapid rate in the presence of platinum oxide, in contrast to the extremely slow uptake of hydrogen in the absence of the promoter. Larger amounts of ferric chloride caused poisoning. In other experiments ferrous chloride was substituted. Optimum activity resulted from 0.1 mM of the additive. Increasing the amount did not increase the reaction rate, nor did it cause inhibition. Hydrogenation without the additive required a number of aerations before uptake was complete. The investigators studied the effect of other salts on the platinum-catalyzed reduction of aldehydes and found that manganese, nickel, and cobalt acetates and chlorides, 0.1 mM, accelerated the reaction. Nickel and cobalt salts were almost as effective as iron salts and chromic chloride was in the range of the cobalt salt. Uranium as a salt also had some promoting activity. Increasing the amount of these additives had no effect on the reaction rate.

On the other hand, the activity of zinc salt was completely dependent on concentration. Above a certain concentration zinc lost its promoting power and became a poison. It is of interest that metallic iron, not intimately mixed with the catalyst but simply added to the reduction mixture, showed considerable accelerating effect.

Tuley and Adams [7] used ferrous chloride to promote the hydrogenation of cinnamaldehyde with platinum. They pointed out that most of the substances used as promoters by Carothers and Adams [6] acted as poisons for the reduction of olefins. Nevertheless, they consider the additives as promoters because of the increase in the reaction rate and elimination of aeration of the catalyst that has been found necessary for the reduction of aldehydes. Another example of the use of an iron salt can be seen in the hydrogenation of geranial, $(H_3C)_2C{=}CHCH_2CH_2C(CH_3){=}CHCHO$, to geraniol, $(H_3C)_2C{=}CHCH_2CH_2C(CH_3){=}CHCH_2OH$ [8].

Maxted and Akhtar [9] reported that stannous chloride was the most effective promoter among a series of chlorides in the platinum-catalyzed reduction of valeraldehyde, more effective than ferric chloride. Stannic chloride was found to be a much less potent accelerator than stannous chloride. This is in contrast to the similarity in promoting action seen in the addition of ferrous and ferric chlorides to the catalyst in the reduction of heptaldehyde [6]. Cobaltous chloride was more active than stannic chloride but less active than ferric chloride. Other salts, the chlorides of aluminum, copper, cerium, chromium, and zinc were about half as active as stannic chloride.

Appreciable promotion with stannous chloride and, to a lesser degree, with ferric chloride was observed in the hydrogenation of ethyl cinnamate with platinum oxide. The authors noted that the reduction of ethyl cinnamate did not require aeration. The addition of stannous chloride increased the rate in the reduction of acetone with platinum but was of little value when added to the cyclohexene reduction.

Rylander and Kaplan [10] noted that stannous chloride promoted the reduction of heptaldehyde with 5% platinum on carbon as well as with platinum oxide. The addition of it to the reduction in the presence of ruthenium on carbon eliminated the long induction period often seen in reductions with ruthenium.

There is a report on the use of ferric chloride in the hydrogenation of some nitroalkanes in ethyl or methyl alcohol with Raney nickel at 40–50° and 6–110 atm pressure [11]. The authors make no claim for the salt as a promoter, nor do they indicate any increase in the rate of absorption of hydrogen due to its use. They do, however, consider it responsible for the increased yields of alkylamines.

THE EFFECT OF OXYGEN

The necessity for aeration of certain platinum metal catalysts in the hydrogenation of aldehydes in the absence of promoters raises the question whether oxygen is a promoter. It has been pointed out that unless the catalyst is subjected to aeration uptake of hydrogen is extremely slow after an initial absorption [6–10]. In addition, if aeration is not resorted to, it is found necessary to use a rather large amount of platinum catalyst; Vavon [12] used 12 g of platinum catalyst per mole of aldehyde.

Vorhees and Adams [13] state that revival with oxygen only applies to the reduction of readily oxidizable compounds (most of the references in the literature pertain to the reduction of aldehydes).

Willstätter and Jacquet [14] supplied another example of the necessity of aeration. They reported that it was necessary to remove hydrogen and shake the mixture of platinum catalyst in the solution of phthalic anhydride with air for several minutes. Despite the high catalyst ratio, 5 g of platinum black to 20.3 g of compound, the initial hydrogen uptake was only 2.5%. Twenty-one aerations were required. The need for aeration in this reduction has been corroborated [9].

Rylander and Kaplan [10] have indicated an improvement in the activity of platinum and ruthenium on carbon supports in the hydrogenation of heptaldehyde by intermittent shaking with air or by the addition of 1% oxygen as air in the hydrogen. They also reported an improvement in the performance of platinum oxide in the same reduction, but it did not appear to be as striking as the increase noted in the addition of air to the reductions with the supported catalysts.

A number of reasons have been offered for the need for oxygen. Adams and co-workers have suggested that the active catalyst resulting from the treatment of platinum oxide is a mixture of platinum metal and a lower oxide. When the oxygen content is depleted its activity is lowered. Maxted [9] attributes the decrease in activity to too rapid conversion of platinum oxide to platinum and theorizes that promoters like stannous chloride prolong the period of conversion.

Rylander [10] suggests that the decrease in activity of platinum oxide, which has been reported to take place as it is reduced, may not necessarily be deactivation but may reflect only a gradual flocculation of the catalyst and effective removal from the system. Platinum generally settles in thick flocks as the oxide is reduced. This has often been observed by this author and has been reported before [15]. Rylander suggests further that effective promoters for aldehyde reductions, as stannous chloride, may function by stabilizing the platinum oxide dispersion.

Whatever the reasons, it appears likely that the presence of oxygen does aid the reduction of aldehydes with platinum catalyst or with ruthenium. It is not a case of periodic flushing or venting as is often practiced to remove an accumulation of poisoning vapors. Nitrogen has been substituted for air in the aeration procedure but it did not change the catalytic activity [10].

Examination of the study of the effect of oxygen on the rate of hydrogenation of heptaldehyde raises an interesting point. Platinum on carbon and ruthenium on carbon were far superior to a corresponding amount of platinum oxide. In the presence of ruthenium on carbon uptake of hydrogen was about 80% of theory in over 5 hr, against approximately 45% for the reduction with platinum on carbon and 10% with platinum oxide. Exclusive of the superiority of the supported catalysts due to the much greater surface, their effect in the reduction of heptaldehyde may be due to the presence of oxygen in activated carbon. It is well known that activated carbons adsorb air. It is not an uncommon occurrence for the organic chemist to note that when a solution of oxidizable material is treated with decolorizing carbon darkening results because of air adsorbed by the carbon.

The beneficial effect of oxygen is not general and is also dependent on the amount. Rylander reported that the addition of 3% oxygen as air decreased the activity of the catalysts. He also found that the addition of 0.5 and 1% oxygen to the hydrogen was without effect in the reduction of cyclohexanone with either platinum on carbon or ruthenium on carbon. There was a slight decrease in rate with 1-octene and nitropropane and an improvement in rate for the reduction of benzaldehyde, but the increase was not as great as in the oxygen-promoted hydrogenation of heptaldehyde with the same catalysts.

EFFECT OF NOBLE METALS ON NICKEL CATALYSTS

Delepine and Horeau [16] reported that the hydrogenation of ketones in the presence of Raney nickel was improved by treatment of the catalyst with a platinum salt. In another study on the reduction of ketones and aldehydes they found that a ketone derived from cholesterol, which resisted hydrogenation with platinum catalyst, was readily reduced with nickel containing a small amount of platinum [17]. They noted that among the noble metals, platinum, osmium, and iridium had the greatest promoting effect for Raney nickel reductions. Rhodium and ruthenium had some effect but did not increase the activity of the catalyst to the same extent as did the other noble metals. Lieber and Smith [18] found that a small amount of platinic chloride, added to Raney nickel catalyst just before the start of the reduction, produced considerable enhancement of the reaction rate. The rate of absorption in the hydrogenation of nitroguanidine increased by a factor of almost 2, castor oil, 3.5, benzaldehyde, 6.5, and nitrobenzene by a factor of 9. The difference between the catalyst of Delepine and Horeau and that of Lieber and Smith is that the former is prepared from a 30–70% nickel-aluminum alloy and the resulting catalyst is plated with platinum. The Raney nickel catalyst used by Lieber and Smith was obtained from a 1:1 nickel-aluminum alloy and may or may not have been plated with platinum since the promoter was added as hydrogenation was to begin. In an extension of the work of Lieber and Smith, Reasenberg [19] found that the addition of platinic chloride to Raney nickel reductions of alkyl-substituted nitrobenzenes and of nitrophenols and nitroanilines resulted in a marked increase in hydrogen uptake during the reactions. There was only a 30% increase in uptake in the promoted hydrogenation of maleic acid but a substantial one, over 500%, in the promoted reduction of benzaldehyde. In a continuation of the work, Scholnik [20] found that Raney nickel catalyst promoted by the addition of platinic chloride was particularly active for the hydrogenation of sodium, ethyl, and methyl nitrobenzoates as well as azo, hydrazo, and azoxybenzene and phenylhydroxylamine.

After the development of the highly active W-6 Raney nickel catalyst of Adkins and Billica [21], Levering and Lieber [22] found that the addition of 0.22 mM of platinic chloride to the hydrogenation of 0.05 mole of benzaldehyde in the presence of 3 g of the wet active catalyst resulted in conversion to benzyl alcohol within 17 min against 170 min for the unpromoted reduction. They also reported very marked promoter effect for the hydrogenation of nitro and nitrile groups. In a more extensive study [23] the group reported that the reduction of diisobutylene with W-6 nickel and platinic chloride was only slightly improved. The promoter also had little effect on the reduction of cinnamic acid. The reaction time for the promoted reduction of phenylacetonitrile was cut in half; nitroethane which was not completely reduced with the W-6 catalyst was converted to ethylamine within 27 min by the addition of platinic chloride.

Samuelsen [24] states that chloroplatinic acid is a true promoter since through its action compounds which are hydrogenated with Raney nickel are reduced more readily. Samuelsen and his colleagues found that the absorption of hydrogen resulting from the addition of the promoter was complete in half the time of that of unpromoted reductions. In the reduction of alkyl-substituted mononitrobenzenes there was a pronounced improvement due to the additive. The reaction rate was increased up to ten times that of the rate in its absence. The addition of chloroplatinic acid also resulted in promotion of the hydrogenation of some dinitrobenzenes with nickel but the increase in activity was not as substantial as with the mononitro compounds. The increase in rate in the reduction of the dinitro compounds was also dependent on the presence of other substituents on the ring.

Reasenberg and Goldberg [25] used Raney nickel promoted with platinic chloride in the reductive alkylation of aminoalcohols with cyclohexanone. They found that the catalyst could be reused many times with very little diminution of activity.

THE PROMOTING EFFECT OF BASIC ADDITIVES

Delepine and Horeau [16] reported that the addition of sodium hydroxide promoted the hydrogenation of ketones in the presence of Raney nickel. It appeared to be as effective for the purpose as the addition of platinic chloride to Raney nickel. Forrest and Tucker [26] added potassium hydroxide to the reduction of fluorenone to 9-hydroxyfluorene with Raney nickel. Uptake of hydrogen was complete in 5 min. In a similar reduction at room temperature without alkali, uptake of hydrogen was still incomplete after 55 min. The addition of platinic chloride to Raney nickel had no effect in the reduction without base. The addition of sodium hydroxide, 2 mM per 3 g of Raney nickel, was found especially useful by Reasenberg [19] in promoting the hydrogenation of ethyl methyl ketone. It also had some activating effect in the reduction of benzaldehyde and certain nitroanilines and nitrophenols. The promoter was added to the suspension of catalyst, in the solution to be reduced, just before the start of the hydrogenation.

The addition of a small amount of triethylamine to the hydrogenation of aldehydes and ketones with the very active W-6 Raney nickel cut the reaction time in half [21]. Scholnik [20] reported that the Raney nickel-catalyzed hydrogenations of azobenzene and azoxybenzene were aided by the addition of a small amount of sodium hydroxide. These reductions were not activated to the same extent as was seen in the promotion of the catalyst by platinic chloride. The reaction rate was considerably increased during the hydrogenation of azoxybenzene by the combined addition of platinum and alkali to the nickel catalyst.

Ungnade and McLaren [27] found that the addition of a small amount of 40% aqueous sodium hydroxide to the nickel-catalyzed reduction of acylphenols, phenylphenol, and some disubstituted alkylphenols made it possible to carry out the reaction at lower temperature and in less time than hydrogenations without the additive. Reduction of the simpler phenols gave anomalous results. Ungnade and Nightingale [28] later determined that the addition of water was responsible for the anomalies obtained in the promoted reduction of those phenols. They showed that promotion was gained by adding a small amount of sodium or solid sodium hydroxide to the hot melted phenol before reduction with Raney nickel.

The rate and extent of reduction of nitriles with Raney nickel is often affected by the presence of alkali. Fluchaire and Chambret [28] had reported only 65% yield of phenethylamine from a Raney nickel-catalyzed reduction of phenylacetonitrile (they made no mention of the amount of secondary amine obtained) in contrast to high yield of primary amine in similar reductions of nitriles by Paty [30]. Paty suggested that the low yield may have been due to mechanical loss or thermal decomposition. Fluchaire and Chambret later discovered the reason for the low yield. In a second publication [31] they carried out a hydrogenation with Raney nickel which had been thoroughly washed with water until it was neutral, as it was in their previously reported reduction. They obtained 51% of primary amine and 37.5% of secondary amine. The addition of sodium hydroxide to a second reduction with neutral Raney nickel yielded 92.5% of phenethylamine. The increase in the yield of primary amine in this instance at the expense of secondary amine is actually an example of promotion by base. According to the reaction scheme by which nitriles are converted to primary amines, the yield of primary amine is dependent to a great degree on the reaction rate. The mechanism will be discussed in detail in the reduction of nitriles but for the present it will suffice to say that rapid conversion of the intermediate imino compound to primary amine prevents or minimizes secondary and tertiary amine formation, particularly in the absence of an agent which cuts down side reaction. Although Fluchaire and Chambret did not report the reaction time in either experiment, it is possible that the presence of sodium hydroxide caused such rapid conversion of the intermediate imine to primary amine that very little imine was present to react with primary amine to give the undesirable side products. The difference in effect of Raney nickel catalysts due to alkali content is illustrated in a report on the application of the highly active W-6 catalyst of Adkins and Billica [21]. They stated that W-7 Raney nickel [32], which is an alkaline catalyst, is more active than W-6 Raney nickel for ketones, phenols, and nitriles. In the preparation of W-6 Raney nickel the catalyst, obtained after treatment of Raney alloy with alkali, is thoroughly washed with water in a hydrogen atmosphere. Apparently all or most of the occluded alkali is removed. Adkins and Billica also reported that the alkaline W-7 catalyst was more active for the reduction of acetophenone than W-6 promoted with triethylamine.

Other strong bases have been used as promoters for the reduction of nitriles with nickel catalyst. A patent [33] gives examples of the addition of barium hydroxide, potassium hydroxide, and methyltriethylammonium hydroxide as well as sodium hydroxide which led to increased yields of primary amines.

Decombe [34] stated that difficult hydrogenations take place quantitatively by the use of platinized nickel in alkaline medium. He gave the reduction of triphenylacetonitrile to triphenylethylamine as an example.

Paul [35] reported that the addition of sodium hydroxide to Raney nickel containing small amounts of chromium, cobalt, or molybdenum increased the rate of reduction of furfural to furfuryl alcohol.

Chabrier and Sekera [36] showed that the rate of hydrogenation of semicarbazones of α- and γ-keto acids with Raney nickel was four times faster in 0.1 N alkali than in neutral solution.

Lieber [18], who studied the promoter effect of platinic chloride, after noting and confirming the effect of triethylamine on W-6 Raney nickel, reported that the combination of triethylamine and platinic chloride with it produced "promotions far exceeding any activity previously known" at very low concentrations of platinum chloride [22]. They found that the reaction time for the hydrogenation of 0.05 mole of benzaldehyde to benzyl alcohol was decreased from 170 to 6 min by the addition of 2 cc of triethylamine and 0.22 mM of platinum chloride to 3 g of W-6 Raney nickel. They found that the combination of promoters enabled the hydrogenation of acetone to proceed rapidly. In an extension of the investigation [23] the authors found that in some reductions the combination of promoters was not as effective as the addition of triethylamine to W-6 nickel. The combination of base and platinum salts appeared most promising in the reduction of aldehydes and ketones, nitroethane, and aromatic nitro groups. The authors conclude that the active promoter is triethylamine chloroplatinate, $[(H_5C_2)_3N]_2 \cdot H_2PtCl_6$.

Blanche and Gibson [37] studied the effect of the same group of promoters, investigated by Lieber and his associates for the hydrogenation of a number of ketones, with a nickel catalyst prepared by the method of Adkins and Pavlic [38]. Their results indicated that the combination of triethylamine, platinum salt, and sodium hydroxide with nickel catalyst was the best means of obtaining rapid reduction. They also reported lithium hydroxide could replace sodium hydroxide.

Most of the reports on the use of basic promoters have been concerned with reductions with nickel. Dunworth and Nord [39] found that, in the reduction of nitrobenzene to aniline in the presence of rhodium on carbon, the addition of a small amount of potassium hydroxide solution more than tripled the amount of hydrogen absorbed in 30 min compared with the unpromoted standard. In the reduction of 2-nitroaniline with palladium on carbon or platinum on carbon, the reaction was carried out in water containing 1% of sodium hydroxide based on the amount of nitroaniline. Yields of 90–96% of o-phenylenediamine were obtained. In a comparative reduction with palladium on carbon without sodium hydroxide the resulting product was discolored and difficult to purify.

PROMOTER EFFECT OF ACIDS

Kindler, Oelschlager, and Heinrich [41] found that strong acid aided the reduction of 2-acetyl-4-fluorophenol to 2-ethyl-4-fluorophenol. When the reaction was carried out in glacial acetic acid in the presence of palladium black, uptake of hydrogen was complete in 15 hr. Addition of sulfuric acid as an activator decreased the reaction time to 4.5 hr. The reaction rate in the reduction of 2-acyl-4-chlorophenols with palladium black to the corresponding 2-alkyl-4-chlorophenols was increased considerably and dehalogenation kept to a low level when sulfuric acid carried was added to the reaction mixture. In these reductions the acid was added in a molecular equivalent to the amount of substrate.

In later work Kindler and his associates [42] found that lesser amounts of strong acids, 1 and 0.1 mM per 0.05 mole of substrate, decreased reaction times very substantially in the reductions of phenyl methyl carbinol to ethylbenzene with either palladium or platinum catalyst. In the hydrogenation of acetophenone to ethylbenzene in methanol with palladium black, only 5% hydrogen uptake took place in 5 hr. The addition of hydrogen chloride, 1 mM, resulted in 90% uptake in 1.5 hr. Other strong acids, in small amounts, caused substantial promotion, except phosphoric acid, 10 mM per 0.05 mole of substrate, which required over 13 hr to achieve 90% uptake. The addition of 0.1 mole of acetic acid was without effect.

There are other examples in work by Kindler and his associates in which the hydrogenation rate was increased because of the presence of strong acids. However, the type of compounds subjected to hydrogenation were those which contained a basic nitrogen atom or which resulted in the formation of a strong nitrogen base, and the amount of acid present was equivalent to the molar amount of substrate. These compounds included O-substituted mandelonitriles, benzoylcyanides, ephedrine, ω-aminoaceto, and propiophenones where the presence of acid should cause an increase in rate because of neutralization of the inhibiting action of the basic nitrogen atom. Other effects of acids noted in these reductions will be discussed in the sections on hydrogenolysis of alcohols and on the reduction of aralkylamines.

Hershberg [43] reported difficulty in obtaining cholestanol from cholesterol following a described procedure [44]. Purification of starting material, and pretreatment of the solution with Raney nickel and the use of a large amount of catalyst had little effect on the very slow uptake of hydrogen. However, reduction in ethyl acetate with platinum oxide was highly successful upon the addition of a small amount of perchloric acid, 2 cc of 72% acid for 1250 g of cholesterol. Uptake of hydrogen was complete in 30–45 min. Other acids with pKa values lower than 3 were also promoters. They included sulfuric, maleic, oxalic, hydrochloric, phosphoric, citric, and p-toluenesulfonic acids.

$$\text{Cholesterol} \xrightarrow{H_2} \text{Cholestanol} \quad (5.1)$$

Reductions to which acetic acid or benzoic acid was added were slow and incomplete.

The effect of strong acid increasing the rate of hydrogen uptake in the platinum-catalyzed reductions of a number of ketones may be seen in the work of Peterson and Casey [45]. They showed that in reductions in trifluoroacetic acid solution, the amount of hydrogen absorbed per minute was from two to six times that of similar reactions in glacial acetic acid. Its effect was dependent on the structure of the ketone and on the concentration in trifluoroacetic acid.

Dunworth and Nord [39], who reported considerable activation from the addition of dilute potassium hydroxide to the rhodium-catalyzed hydrogenation of nitrobenzene, noted far less promoting effect from the addition of a small amount of acetic acid to a similar reaction.

Smith, Fuzek, and Meriwether [46] showed that the promoting effect of a small amount of palmitic acid, 0.4 mM per gram of nickel catalyst, was very pronounced in the hydrogenation of some terpenes in the absence of solvent at varying pressures from 20–100 atm. In a second investigation Smith and Fuzek [47] studied the effect of a number of fatty acids on the nickel-catalyzed reduction of some of the same terpenes. Acids from C_{10} to C_{25} had some promoting activity but the greatest effect came from the addition of palmitic acid. It is of interest that with a very active nickel catalyst, W-6 Raney nickel, the best additive caused retardation. With a normally active nickel catalyst the addition of palmitic acid caused a considerable increase in reaction rate in the reduction of d-limonene, for example. Other additives, which like palmitic acid may be classed as surfactants, also influenced catalytic activity. These compounds included sodium oleate, certain long-chain amines, sulfated alcohols as sodium lauryl sulfate, and quaternary ammonium salts. The authors insisted that the effect is not simply a wetting action. They reasoned that if acceleration was due to wetting action then promotion should have taken place with either normally active nickel or with the high active W-6 Raney nickel.

It is of interest that in the first study [46] the addition of methyl alcohol to the nickel-catalyzed hydrogenation of d-limonene caused far greater activation under the reaction conditions than the addition of promoter to the reduction without solvent. It would have been worthwhile in the two investigations to have studied the effect of the addition of palmitic acid, the best promoter of those tried, to a nickel-catalyzed reduction in methyl alcohol.

Smith and Fuzek [47] observed that α-pinene was the only one of the terpenes studied where the hydrogenation was not promoted by the addition of palmitic acid. It was the only terpene in the series which did not possess an exocyclic double bond.

PROMOTER EFFECT OF WATER

The promoting effect of water appeared to be limited to ruthenium-catalyzed hydrogenations. Reduction at low pressure and room temperature would not take place unless the solvent contained water [48]. This was found to apply even when the substrate was essentially water-insoluble as in the hydrogenation of olefins [49].

The effect of water was shown in reduction of benzoic acid in hexahydrobenzoic acid with 5% ruthenium on carbon at 130° and 150 atm [50]. In 30 min hydrogen uptake was only 10%. In a similar reaction in hexahydrobenzoic acid solution containing water equal to the weight of benzoic acid, uptake was 100% in the same period. Little promotion was obtained by the addition of water to similar palladium on carbon or rhodium on carbon reductions.

There was also an increase in rate in the ruthenium hydrogenations of pyridine and aniline in water as compared to reductions in other solvents.

The addition of small amounts of water decreased the time for hydrogenation in the reduction of toluene with ruthenium hydroxide [51]. It also decreased the reaction time in the hydrogenation of hydroquinone dimethyl ether but did not promote the reduction in the presence of rhodium hydroxide. A very small amount of glacial acetic acid was added to these reactions to counteract the effect of the possible presence of alkali in the catalysts.

On the other hand there was little or no promotion by water in the ruthenium-catalyzed reductions of a number of aromatic nitro compounds except with 4-nitrophenol [52].

REFERENCES

[1] R. N. Pease and H. S. Taylor, *J. Phys. Chem.*, **24**, 241 (1920).
[2] T. S. Hamilton and R. Adams, *J. Am. Chem. Soc.*, **50**, 2260 (1928).
[3] M. Freifelder, *J. Org. Chem.*, **26**, 1835 (1961).
[4] L. W. Covert, R. Connor, and H. Adkins, *J. Am. Chem. Soc.*, **54**, 1651 (1932).

References

[5] D. P. Graham and L. Spiegler, U.S. Patent 2,823,235 (1958).
[6] W. H. Carothers and R. Adams, *J. Am. Chem. Soc.*, **45**, 1071 (1923).
[7] W. F. Tuley and R. Adams, *J. Am. Chem. Soc.*, **47**, 3061 (1925).
[8] R. Adams and B. S. Garvey, *J. Am. Chem. Soc.*, **48**, 477 (1926).
[9] E. B. Maxted and S. Akhtar, *J. Chem. Soc.*, **1959**, 3130.
[10] P. N. Rylander and J. Kaplan, *Englehard Industries Tech. Bull.*, **2**, 48 (1961).
[11] K. Johnson and E. F. Degering, *J. Am. Chem. Soc.*, **61**, 3195 (1939).
[12] G. Vavon, *Compt. Rend.*, **154**, 359 (1912).
[13] V. Vorhees and R. Adams, *J. Am. Chem. Soc.*, **44**, 1397 (1922).
[14] R. Willstätter and D. Jacquet, *Ber.*, **51**, 767 (1918).
[15] R. Willstätter and E. Waldschmidt-Leitz, *Ber.*, **54**, 113 (1921).
[16] M. Delepine and A. Horeau, *Compt. Rend.*, **201**, 1301 (1935).
[17] M. Delepine and A. Horeau, *Compt. Rend.*, **202**, 995 (1936).
[18] E. Lieber and G. B. L. Smith, *J. Am. Chem. Soc.*, **58**, 1417 (1936).
[19] J. R. Reasenberg, E. Lieber, and G. B. L. Smith, *J. Am. Chem. Soc.*, **61**, 384 (1939).
[20] S. S. Scholnik, J. R. Reasenberg, E. Lieber, and G. B. L. Smith, *J. Am. Chem. Soc.*, **63**, 1192 (1941).
[21] H. Adkins and H. R. Billica, *J. Am. Chem. Soc.*, **70**, 695 (1948).
[22] D. R. Levering and E. Lieber, *J. Am. Chem. Soc.*, **71**, 1515 (1949).
[23] D. R. Levering, F. L. Morritz, and E. Lieber, *J. Am. Chem. Soc.*, **72**, 1190 (1950).
[24] G. S. Samuelsen, V. L. Garik, and G. B. L. Smith, *J. Am. Chem. Soc.*, **72**, 3872 (1950).
[25] J. R. Reasenberg and S. D. Goldberg, *J. Am. Chem. Soc.*, **67**, 933 (1945).
[26] J. Forrest and S. H. Tucker, *J. Chem. Soc.*, **1948**, 1137.
[27] H. E. Ungnade and A. D. McLaren, *J. Am. Chem. Soc.*, **66**, 118 (1944).
[28] H. E. Ungnade and D. V. Nightingale, *J. Am. Chem. Soc.*, **66**, 1218 (1944).
[29] Fluchaire and F. Chambret, *Bull. Soc. Chim. France*, **1942**, 189.
[30] M. Paty, *Bull. Soc. Chim. France*, **1938**, 1276 and **1940**, 55.
[31] Fluchaire and F. Chambret, *Bull. Soc. Chim. France*, **1944**, 22.
[32] H. Adkins and A. A. Pavlic, *J. Am. Chem. Soc.*, **69**, 3039 (1947).
[33] M. Grunfeld, U.S. Patent 2,449,036 (1948).
[34] J. Décombe, *Compt. Rend.*, **222**, 90 (1946).
[35] R. Paul, *Bull. Soc. Chim. France*, **1946**, 208.
[36] P. Chabrier and A. Sekera, *Compt. Rend.*, **226**, 818 (1948).
[37] R. B. Blanche and D. T. Gibson, *J. Chem. Soc.*, **1954**, 2487.
[38] H. Adkins and A. A. Pavlic, *J. Am. Chem. Soc.*, **68**, 1471 (1946).
[39] W. P. Dunworth and F. F. Nord, *J. Am. Chem. Soc.*, **74**, 1459 (1952).
[40] J. Levy, U.S. Patent 3,230,259 (1966).
[41] K. Kindler, H. Oelschläger, and P. Heinrich, *Ber.*, **86**, 501 (1953).
[42] K. Kindler, H. G. Helling, and E. Sussner, *Ann.*, **605**, 200 (1957).
[43] E. B. Hershberg, E. Olivetto, M. Rubin, H. Staeudle, and L. Kuhlen, *J. Am. Chem. Soc.*, **73**, 1144 (1951).
[44] W. F. Bruce, *Org. Syn.*, Vol. 17, Wiley, New York, 1937, p. 45.
[45] P. E. Peterson and C. Casey, *J. Org. Chem.*, **29**, 2325 (1964).
[46] H. A. Smith, J. F. Fuzek, and H. T. Meriwether, *J. Am. Chem. Soc.*, **71**, 3765 (1949).
[47] H. A. Smith and J. F. Fuzek, *J. Am. Chem. Soc.*, **72**, 3454 (1950).
[48] P. N. Rylander, *Engelhard Ind. Tech. Bull.*, **1**, 133 (1961).
[49] L. M. Berkowitz and P. N. Rylander, *J. Org. Chem.*, **24**, 708 (1959).
[50] P. N. Rylander, N. Rakoncza, D. Steele, and M. Bolliger, *Engelhard Ind. Tech. Bull.*, **4**, 95 (1963).
[51] Y. Takagi, T. Naito, and S. Nishimura, *Bull. Chem. Soc. Japan*, **38**, 2119 (1965).
[52] K. Taya, *Sci. Papers Inst. Phys. Chem. Res. Japan*, **56**, 285 (1962).

VI

OTHER EFFECTS IN HYDROGENATION

Excluding catalysts, promoters, and poisons discussed in previous chapters, there are other variables that can have an influence on the hydrogenation process. They include the solvent system, the nature of the hydrogen acceptor and the concentration in solution, temperature, and pressure conditions, agitation, and perhaps the carrier upon which the metal catalyst is deposited. The choice and amount of catalyst will be covered in the following chapter on procedures.

SOLVENTS

Reductions can be carried out in the absence of solvent when the material to be reduced is a liquid or a low melting solid. However, reaction generally proceeds more rapidly in most solvents. Their presence helps to modify and allows better control of exothermic reactions. They are usually chosen for the ability to dissolve the compound to be reduced. Often, however, the starting material is poorly soluble but the resulting reduction product dissolves so readily that the reaction is not impeded by physical covering of the catalyst.

There are reports in the literature which emphasize the superiority of the one solvent over another upon its influence on the rate and extent of hydrogenation. Within limits this may be true. The results of the work of Carothers and Adams [1], who studied the platinum-catalyzed reductions of some aromatic and aliphatic aldehydes in a number of solvents, are probably more to the point. They concluded that the effect of solvent on the reaction rate was specific for each system.

In discussing the effect one must qualify whether the solvent is acidic, basic, or neutral. At the same time it must be borne in mind that acidic and basic solvents often serve a dual purpose. In the platinum-catalyzed reductions of pyridines and quinolines, acetic acid acts not only as a solvent but its prime purpose is to overcome catalyst poisoning caused by the starting materials or resulting hydrogenated bases. Conversely, nitrogen bases can act as inhibitor as well as solvent to induce a reaction to stop at a desired point. This may be seen in the work of Forman and Freifelder [2] who employed cyclohexylamine, triethylamine, piperidine, and other amines in the selective reduction of nitrocyclohexane to cyclohexylhydroxylamine.

Acidic and basic media have a distinct influence on the rate and extent of reaction and on stereochemistry resulting from reduction. They can also change the course of reaction. Some examples have been cited in the chapters on poisons and promoters. More will be seen in subsequent chapters on the reduction of functional groups and ring systems.

The effect of neutral solvents is unpredictable. There may be some dependency in certain systems based on the relative acidity or basicity of the solvent. There may be some correlation between polarity and effect. Kindler and his co-workers [3] attempted selective hydrogenation of 2-allyl-4-chlorophenol with palladium catalyst. They reported that in benzene or in cyclohexane the double bond was reduced without the release of halide ion. In the presence of methyl, ethyl, and isopropyl alcohols the amount of chloride ion was 6.2, 8.6, and 3.5% respectively. Ham and Coker [4] obtained much higher yields of 1,3-dichloropropane from the rhodium-catalyzed reduction of 1,3-dichloropropene in cyclohexane, 47.9% and ether, 30.8%, than in ethyl alcohol or acetic acid, 2.5%. There may be differences in effect between hydroxylated and aprotic solvents. All of these points are worthy of further investigation. Too often conclusions are drawn from a limited study or from a comparison of work of different investigators. At times, too, the effect may be due to impurities in either the solvent, catalyst, or substrate. A more reliable assessment of the effect of neutral solvents on the hydrogenation reaction would necessitate rigid purification of the solvents involved, a thorough cleaning of the apparatus before and after each use, purification of the various substrates, and finally reductions under a controlled set of conditions. Only then would the findings be valid.

THE HYDROGEN ACCEPTOR

The nature of the hydrogen acceptor is a point to be considered as well as the concentration of it in the proper medium. The structure of the compound to be reduced will play a part in the ease or difficulty of reduction. It should be obvious that a highly hindered functional group or ring system will

72 Other Effects in Hydrogenation

be more difficult to hydrogenate than one which has fewer and/or smaller groups about the portion of the molecule undergoing reaction. Smith [5] has noted that in the ring reduction of benzenoid compounds containing two substituents there is a relationship of proximity of the groups to ease of reduction. The order is *para*, *meta*, and *ortho* in increasing reaction time. He has also reported that symmetrical trisubstitution in the aromatic system favors more rapid rate than reduction of similar 1,2,3- and 1,2,4-trisubstituted compounds.

Freifelder [6], by means of molecular models, has shown examples of the effect of structure on the course of reduction. The hydrogenation of ethyl 2-cyano-3-methyl-2-hexenoate over platinum oxide yielded the expected saturated cyano ester. When ethyl 2-cyano-2-ethyl 3-methyl-3-hexenoate was

$$CH_3CH_2CH_2C(CH_3)=C(CN)COOC_2H_5 \xrightarrow[PtO_2]{H_2}$$
$$CH_3CH_2CH_2CH(CH_3)CH(CN)COOC_2H_5 \quad (6.1)$$

similarly hydrogenated, the saturated ester was not obtained. Instead attack took place at the nitrile function, not at the double bond, yielding a mixture of primary and secondary amino compounds with the double bond intact. The model of this unsaturated ester shows that the double bond

$$CH_3CH_2CH=C(CH_3)C(C_2H_5)(CN)COOC_2H_5 \quad (6.2)$$

→ Saturated cyano ester

→ Unsaturated primary and secondary amines

is well removed from the surface of the catalyst while the nitrile function rests upon it (see Fig. 6.2), whereas in Fig. 6.1 the model of the readily reduced unsaturated ester clearly indicates the reducibility of the double bond.

Acidic and basic substituents in a molecule also exert some influence on the reduction rate. The hydrogenation of compounds containing basic groups or functions which lead to basic groups may be retarded when certain catalysts are used unless an equimolar amount of acid is present to convert the nitrogen base to an ionic or noninhibiting form. On the other hand while the poisoning effect of the pyridine or piperidine nitrogen on platinum catalyst is well known, it has been shown that 2- and 4-pyridinecarboxylic acids are readily reduced with this catalyst in aqueous solution [7].

The concentration of the hydrogen acceptor can be important depending on reaction conditions and the substrate. Its effect is usually an indirect one. In concentrated solutions the heat of reaction may rise to a point where

Fig. 6.1 Model of ethyl 2-cyano-3-methyl-2-hexenoate.

Fig. 6.2 Note position of double bond in ethyl 2-cyano-2-ethyl-3-methyl-3-hexenoate.

undesirable side reactions can take place. In the hydrogenation of α-aminonitriles the development of too high a temperature can lead to cleavage. When potentially exothermic reactions are to be carried out, such as the reduction of aromatic nitro compounds, the use of a concentrated solution may be hazardous because the heat developed cannot be dissipated rapidly enough.

TEMPERATURE AND PRESSURE CONDITIONS

The reaction temperature will affect the rate and extent of hydrogenation as it does any chemical reaction. At times an increase in temperature can be detrimental to a reduction because it may lead to loss of selectivity if a second functional group is present. In the semihydrogenation of acetylenes control of the temperature is extremely important. It has been reported that such reductions started at 22–23° showed a rise in temperature to about 40° which enabled the investigators to achieve selective conversion to the corresponding olefins [8]; on the other hand those started at 30° reached a temperature of 80° and were not selective. The effect of temperature on the reduction of α-aminonitriles has already been noted. The temperatures at which a particular reduction is to be carried out is often dependent on the catalyst. In the case of certain catalysts reductions must be carried out under elevated temperature and pressure conditions.

Hydrogen pressure, another form of energy, can be beneficial as well as detrimental. At a fixed temperature there is not usually a very significant difference between reactions run at atmospheric pressure and those at 2–3 atm. There is a report of the detrimental effect of increase in pressure from atmospheric to 45 psig in the reduction of acetylenes. At atmospheric pressure selectivity was obtained, at 45 psig it was destroyed [9].

An increase in pressure from several atmospheres to higher pressure will decrease reaction time considerably. The hydrogenation of N,N-diethylnicotinamide

$$\text{[pyridine ring]}\text{CON}(C_2H_5)_2$$

required 20 hr for conversion to the corresponding piperidine in the presence of rhodium catalyst [10]. In a comparison at a later date, reduction at room temperature and 70 atm with less catalyst was complete in 30 min [11a].

In general with the wide variety of catalysts available today most reductions can be carried out under mild conditions. There are some hydrogenations that will require elevated temperature and pressure conditions. There are others on commercial scale, which for economic reasons, are carried out at higher temperature and increased pressure. However, except for the hydrogenation of esters to alcohols and amides to amines with copper chromite, most reductions should not require pressures above 2000 psig or temperatures beyond 200°.

AGITATION

Some form of movement of the catalyst in the reaction medium is necessary in order for reduction to take place. Except for micro-scale reductions where the pressure of hydrogen is sufficient to cause movement, there will be no uptake of hydrogen without some form of agitation. It can be demonstrated in all reductions (except on micro-scale) that hydrogen absorption comes to a halt when agitation by any means is stopped. Cessation of mixing will keep an exothermic reaction under control. The effect of agitation on the speed of reaction can be readily seen in reductions carried out in the low-pressure Parr hydrogenator. A difference of 50 rev/min between two similar reductions can cause a dramatic difference in reaction time [116]. In batch hydrogenations on commercial scale, baffle plates are usually installed in the reaction vessel and hydrogen is introduced under the surface to produce adequate agitation during reduction.

The volume of material in a particular piece of apparatus also has a relationship to good mixing. While there is also some relationship of the amount of hydrogen present to reduction too little free space in the reaction vessel results in poor agitation and prolonged reaction times and at times incomplete reduction especially in low-pressure hydrogenations [11b].

CATALYST SUPPORTS

The effect of the catalyst support on hydrogenation is as difficult to evaluate as the effect of the solvent. A support results in more catalytic surface area and allows more economical use of the metal. On commercial scale mechanical loss of metal can be minimized when supported catalysts are employed. The financial saving can be substantial when noble metals are used, since the recovered catalysts are returned to the manufacturer for credit. There are reports which show, on an equal weight of metal basis, that reduction proceeds more rapidly with supported catalyst than with the metal itself and others which report failure with the supported catalyst, although at times the reported failure is due to an improper comparison of the amount of catalyst rather than the amount of metal.

There are many examples in the literature covering a variety of reactions in which one catalyst support is preferred to another. The ensuing chapter on the reduction of acetylenes contains examples of palladium on a variety of supports. Observations on the effect of supports for the most part are incidental and are isolated instances. The support at times may have considerable influence on certain reductions but no serious study has been made on their effect. Several things must be kept in mind. Certain supports such as carbon and alumina are excellent adsorbents and are used in chromatography. As a result of their adsorptive power they may hold impurities more strongly than other supports and be less efficient for this reason. On the other hand, because of the way in which catalysts are prepared, they can retain traces of acid or base which can affect a particular hydrogenation adversely or advantageously. Other supports, as the alkaline earth carbonates, may have some influence because of their basicity.

The support may have some other function. Fukuda [12] studied the role of calcium carbonate as a carrier for palladium catalyst in the hydrogenation of 1,4-butynediol poisoned by lead acetate and quinoline. He suggested that the carrier neither increased the activity of palladium nor had an effect on selectivity. Instead it appeared to suppress polymerization as a side reaction.

It is difficult to assess the effect of supports without an intensive study involving pure substrates, pure solvents, and a standard method of depositing the particular metal on the various supports. Even then there is the added factor of the relative acidity or basicity of the support.

REFERENCES

[1] W. H. Carothers and R. Adams, *J. Am. Chem. Soc.*, **46**, 1675 (1924).
[2] J. C. Forman and M. Freifelder, unpublished work.
[3] K. Kindler, H. Oelschlager, and P. Henrich, *Ber.*, **86**, 167 (1953).
[4] G. E. Ham and W. P. Coker, *J. Org. Chem.*, **29**, 194 (1964).
[5] H. A. Smith, *Ann. N.Y. Acad. Sci.*, **145**, 72 (1967).
[6] M. Freifelder, *Platinum Metals Review*, **9**, 38 (1965).
[7] M. Freifelder, *J. Org. Chem.*, **27**, 4046 (1962).
[8] G. F. Hennion, W. A. Schroeder, R. P. Lu, and W. B. Scanlon, *J. Org. Chem.*, **21**, 1142 (1956).
[9] N. A. Khan, *J. Am. Chem. Soc.*, **74**, 3018 (1952).
[10] M. Freifelder, R. M. Robinson, and G. R. Stone, *J. Org. Chem.*, **27**, 284 (1962).
[11] a. M. Freifelder, *Ann. N.Y. Acad. Sci.*, **145**, 5 (1967); b. M. Freifelder, unpublished work.
[12] T. Fukuda, *Bull. Chem. Soc. Japan*, **31**, 343 (1958).

VII

PROCEDURE

Since the object of this book is to concentrate on the hydrogenation process, the factors related to it, and the chemistry involved, the discussion of equipment will be brief. Although the results of reductions from vapor-phase reactions and continuous process hydrogenation may be included in subsequent chapters, that type equipment will not be described. Coverage will be limited to batch-type apparatus for use at atmospheric to high pressures.

EQUIPMENT

Most workers in the field are familiar with the many pieces of apparatus in use today and their operation. For others there is considerable information available from the manufacturers of the various low-pressure and high-pressure units as well as information in the chemical and chemical engineering literature.

Low-pressure Apparatus

Any piece of glass equipment should be suitable for laboratory-scale reductions at or slightly above atmospheric pressure and at moderate temperature if it meets the following criteria. It should be capable of withstanding the pressure, it should have both a mixing device and a means of measuring hydrogen consumption, and finally it should be leakproof. An all-glass unit that can be of value on laboratory scale is the Brown hydrogenator [1]. The catalyst is prepared *in situ;* reductions are carried out at atmospheric pressure and are readily self-controlled. Hydrogen is generated from the decomposition of sodium borohydride with acid. The hydrogenator is reported to be useful as an analytical tool and for reductions up to 1 kg of material. As valuable as this technique may be, it must be pointed out that the cost of hydrogen from sodium borohydride will be prohibitive for most commercial applications. There may be times, however, when the hydrogen cost would only be an incidental part of the total in the production of a particular compound.

By far the most universally used piece of equipment for low-pressure reductions (up to 4 atm) is the Burgess-Parr hydrogenator [2] or modifications of it. It is simple to operate, the reactor can be heated, and it is relatively safe to use within the prescribed conditions. Recently in addition to the glass reactor bottle larger stainless steel flasks have been made available for low-pressure reductions.

For very small-scale work the microhydrogenators of Clauson-Kaas and Limborg [3], Ogg and Cooper [4], and Vanderheuvel [5] are among the examples of units described in the literature.

High-pressure Reactors

Reductions at elevated temperature and pressure are carried out in sealed, thick-walled metal reactors which are commercially available [2, 6].

REACTION CONDITIONS

Solvent

Solvent is usually chosen for its ability to dissolve the starting material and/or reduction product. Ethyl and methyl alcohols are most commonly used. Water, the cheapest and safest solvent for hydrogenation, is less often employed as are ethyl acetate, cyclohexane, methylcyclohexane, and benzene or other aromatic hydrocarbons. Benzene has some disadvantages because it attacks the rubber stoppers in the Parr hydrogenator and it is itself strongly bound to the catalyst surface which results in decreased rate of hydrogen absorption during reaction. However, the latter effect can be used to advantage at times to insure selectivity in the hydrogenation of a compound containing two reducible groups. Certain petroleum fractions as Skelly B are reported to be of value in the semihydrogenation of acetylenes. Methylcellosolve (ethyleneglycol monomethyl ether) is an excellent solvent. Its low volatility as well as the fact that it does not form a peroxide makes it particularly safe to use. It is not too difficult to remove after reduction, boiling point 124° (760 mm.), and it is completely miscible with water. Dimethylformamide, boiling point 177°, due to its excellent solubilizing power can also be of value in the hydrogenation process. It, too, is completely miscible with water. Any solvent that is pure may be considered for use if it is known that it will not react with the substrate or products of reduction or will not, itself, undergo attack under the conditions of reaction. For example, acetone or chloroform, either of which can be reduced, may be used at room temperature and several atmospheres pressure, but they are unsatisfactory under more vigorous conditions. Most of the others are employable under high pressure if the reaction temperature is kept below 200–220°. Above this temperature range the vapor pressure of the solvent will cause an increase in pressure during reaction (the International Critical Tables should be consulted for specific data). When methyl alcohol, ethyl alcohol, or other lower aliphatic primary alcohols are solvents in the nickel-catalyzed reduction of pyridines, N-alkylation takes place at 150° [7]. If copper chromite is used as catalyst in the same reductions, N-alkylation takes place at 200° [8]. Mozingo [9] cautions against the use of dioxane with Raney nickel at 210° because of the formation of explosive mixtures. The writer does not favor it even for low-pressure reductions unless it is purified and distilled before use.

Acidic and Basic Conditions

When hydrogenation is to be carried out in acid or basic solution, the investigator must be cognizant of the effect of these media on the activity of the catalyst. Their action has been pointed out in Chapters IV and V on catalyst poisons and promoters. The investigator must also have knowledge of the effect of each medium on the starting material and product or products of reduction. Information of their effect in certain reductions may be gleaned from the investigations of McQuillin and Ord [10], Augustine [11], and Nishimura and co-workers [12].

CHOICE AND AMOUNT OF CATALYST

There is no particular catalyst that is best for all hydrogenations. A number can be used for a variety of reductions; one may be the catalyst of choice for certain type reactions as palladium on carbon, under the proper conditions, is for the removal of chlorine, bromine, and iodine and for O- and N-debenzylation. Each has advantages and disadvantages.

Some, as noted in Chapter III, are active only at elevated temperature and pressure. Therefore the choice of catalyst can vary with the type of compound and the reaction conditions. A study of examples of the hydrogenation of functional groups and ring systems in the ensuing chapters should aid in selecting the proper catalyst.

The amount of catalyst is important and can vary with the compound to be reduced, with the type of reaction, and with the size of the experiment. Reduction at high pressure will require less catalyst than a corresponding reaction at low pressure. A reduction on very small scale will usually require a much higher than normal ratio of catalyst to compound. On the other hand a scale-up of laboratory size experiments will need less catalyst than usual. There is another point worth mentioning. An increase in the amount of catalyst in a reduction has a much greater than linear effect on the rate. Doubling the amount of catalyst can cause a five- to tenfold increase in the reaction rate.

Table 7.1 shows what may be an average weight ratio of the more commonly used catalysts to substrate for the reduction of functional groups. The table is offered only as a guide line. In general more catalyst is required for the reduction of ring systems unless more energy, in the form of heat and/or higher pressures, is applied. There will be instances when it is desirable to increase the amount of catalyst in order to attain more rapid reaction and suppress the formation of an undesirable side product. Such an example is seen in the reduction of a number of nitriles with rhodium on a support [13]

Table 7.1

Catalyst	Ratio[a] (%)
Copper chromium oxide (copper chromite)	10–20
Palladium, platinum, or rhodium on support	10
Platinum oxide	1–2
Raney nickel	10–20[b,c]
Cobalt[d]	10–20
Ruthenium dioxide	1–2
Ruthenium on supports	10–25

[a] The ratio of catalyst to substrate is based on the weight of catalyst not on the percentage of metal to substrate. [b] It is often convenient to measure the volume of wet Raney nickel. When the catalyst has settled, 1 cc is equal to about 1.5 g. [c] Normally a 10% ratio may be sufficient. A larger ratio should be used if the catalyst has been standing for 6 months or more. Dry stabilized nickel catalysts are also available, with which 15–20% weight ratios of catalyst to substrate should be satisfactory. [d] Use in the same manner as nickel with wet or dry catalyst.

where increasing the catalyst ratio to almost 20% induced complete hydrogen uptake within two hours. This rapid reaction contributed to the high yield of primary amine. The method of preparation and age of the catalyst can be a factor in the amount used for reaction. In reductions with the freshly prepared, highly active Raney nickel of Adkins and Billica [14], the catalyst ratio may be decreased for reactions under low-pressure conditions. For reductions at higher pressure it should be drastically decreased. After several weeks storage, when it loses its high activity, a normal ratio is employed. When potentially exothermic reactions, such as the hydrogenation of aromatic nitro compounds, are carried out less catalyst is usually required even under low-pressure conditions. In batch-type reactions on a large scale, such as are carried out in the plant or pilot plant, the level for palladium on carbon for example may drop to 1% or less for the same hydrogenation.

The amount of catalyst can also affect selectivity. When the hydrogenation of 1-(2- and 4-bromobenzoyl)-2-isopropylidenehydrazine was carried out in the presence of an 8% ratio of 5% platinum on carbon, loss of bromine resulted. Although the catalyst had been reported to be inactive for dehalogenation [15], loss of bromine was not controlled until a 2.5% ratio was adopted for the reduction of the 2-bromo compound and a 4% ratio for the 4-bromo compound. Under these conditions the desired products were obtained [16]. There are other examples of selectivity resulting from the use of low catalyst ratios. A number will be seen in the section on reduction of acetylenes.

ADDITION OF CATALYST

There is always some hazard attached to the addition of most of the noble metal catalysts or Raney nickel to certain solvents used for hydrogenation. The hazard is due to the fact that these materials also act as catalysts for oxidation and can cause solvent vapors to ignite. Therefore, whenever possible the more volatile solvents, as methyl alcohol, should be avoided particularly when the metal is on a carbon support. These are usually less dense than catalysts on other supports and do not sink as rapidly into the solvent. Raney nickel can be added with comparative safety. Most of the catalysts that are active only at higher pressure are nonpyrophoric and present no problem.

Any catalyst can be added to methyl cellosolve or dimethylformamide with almost complete safety from ignition. It is probably true to a lesser degree with glacial acetic acid, dependent perhaps on whether a large amount of catalyst is added at once. The safest solvent obviously is water. After addition of catalyst to any solvent, the reactor should be mixed in some manner so that the catalyst becomes covered. The catalyst can usually be added to ethyl alcohol with reasonable safety by adding it in portions and swirling it about until it is covered with the solvent. An added precaution is to draw off any vapors by attachment to a vacuum line and/or add nitrogen to the reactor before the next portion of catalyst. The same procedure can be used for the addition of catalyst to methyl alcohol. In many instances the presence of water or of methylcellosolve will have no adverse effect on reduction in methyl alcohol. In such cases the catalyst should be added to either solvent and, when wetted, covered with methyl alcohol. It might further be mentioned that the reactor should be covered after the addition of catalyst. If, in rare cases, ignition does take place, covering the reactor will usually smother the burning solvent.

SAFETY IN OPERATION

Catalytic hydrogenation must be regarded as a potentially hazardous chemical reaction. However, an awareness of its potential and an adherence to some simple precautions can minimize the hazards attached to it so that the process will be as safe as any other chemical reaction.

Peroxide-forming solvents should be avoided. If it is necessary to employ one, it should be rigidly purified and distilled immediately before use. However it is a rare occasion when another solvent cannot be substituted. This author can never see any reason for the use of diethyl ether as a solvent for hydrogenation. Its low boiling point and high volatility makes the addition of catalysts, except the most inactive ones such as copper chromite, a considerable hazard.

After hydrogenation is complete many catalysts are more pyrophoric than before reduction because of adsorbed hydrogen. Care must be taken during filtration. The filter must be covered with liquid at all times and not allowed to go dry or ignition may occur. This can be avoided by filtering under nitrogen or keeping the filter covered with a watch glass. There is usually no hazard with the less active catalysts but it must be remembered that most of the hydrogen catalysts are also capable of promoting oxidation. If the spent catalyst is to be saved and returned for credit, as is often the case with the noble metals, it should be stored under water. If used nickel catalysts are to be destroyed, they can be digested with concentrated hydrochloric acid. The same procedure can be used for cobalt catalysts. Most of the other base metal catalysts can be discarded without danger.

No hydrogenation, however small, should be in operation without a safety shield. When a higher pressure reaction is carried out it should be behind a substantial barricade or isolated behind heavy walls. No pressurized unit should be in operation without a rupture disk or a venting system. When it is necessary to carry out reactions which are known or suggested to be exothermic in heavy equipment, it is very wise to use only about 50% of the necessary amount of hydrogen for the first vigorous stage of the reduction and then add the remaining amount of hydrogen when the reaction begins to moderate. Most important of all is that no attempt should ever be made to exceed the rated capacity of any piece of equipment, small or large. By adhering to simple precautions catalytic hydrogenations can indeed be a safe as well as a very valuable procedure.

ECONOMICS

In the laboratory there is usually little concern about the economics of the process unless a particular reduction is to be scaled up to pilot plant or manufacturing scale. In that case it is not only essential to choose the proper catalyst, but also to find conditions which will enable the reaction to proceed at a rapid rate without hazard. At the same time reuse of the catalyst is important. There are times when a less active catalyst may be chosen over a more active one because it is less sensitive to poisoning and can be reused more often. The solvent in the process may be chosen because of its low cost and ease of removal. It may also be more economical to use high-pressure equipment because of more rapid reaction.

REFERENCES

[1] Available from Delmar Scientific Laboratories, Inc., Maywood, Illinois.
[2] Available from Parr Instrument Co., Moline, Illinois.
[3] N. Clauson-Kaas and F. Limborg, *Acta Chem. Scand.*, **1**, 884 (1947).
[4] C. L. Ogg and F. J. Cooper, *Anal. Chem.*, **21**, 1400 (1949).
[5] F. A. Vanderheuvel, *Anal. Chem.*, **24**, 847 (1952).
[6] Autoclave Engineers, Erie, Pennsylvania and American Instrument Co., Silver Springs, Maryland are among the manufacturers of this type of equipment.
[7] H. Adkins and H. Cramer, *J. Am. Chem. Soc.*, **52**, 4349 (1930).
[8] Y. Sawa, K. Inoue, and S. Kitamura, *J. Pharm. Soc. Japan*, **63**, 619 (1943); through *Chem. Abstr.*, **45**, 2940 (1951).
[9] R. Mozingo, *Org. Syn. Coll. Vol.* **3**, 181 (1955).
[10] F. J. McQuillin and W. O. Ord, *J. Chem. Soc.*, **1959**, 2902.
[11] R. L. Augustine, *Ann. N.Y. Acad. Sci.*, in press.
[12] S. Nishimura, M. Shimahara, and M. Shiota, *J. Org. Chem.*, **31**, 2394 (1966).
[13] M. Freifelder, *J. Am. Chem. Soc.*, **82**, 2386 (1960).
[14] H. Adkins and H. R. Billica, *J. Am. Chem. Soc.*, **70**, 695 (1948).
[15] R. Baltzly, *J. Am. Chem. Soc.*, **74**, 4586 (1952).
[16] M. Freifelder, W. B. Martin, G. R. Stone, and E. L. Coffin, *J. Org. Chem.*, **26**, 383 (1961).

VIII

ACETYLENES

The acetylenic bond is readily attacked by hydrogen under mild conditions in the presence of a variety of catalysts. An important application for the reduction of acetylenes is the selective conversion to the corresponding ethylenes and in the case of disubstituted acetylenes, the stereochemistry involved. Industrial hydrogenation of acetylene itself is chiefly concerned with reduction to ethylene, and with the synthesis of hydrocarbons formed by polymerization, which occurs during the hydrogenation reaction.

Only a few side reactions accompany the catalytic reduction of acetylenes, especially when the reaction is carried out at moderate temperatures. Polymerization does occur occasionally. Acetylene especially is most prone to this. With tertiary acetylenic carbinols and diols some cleavage may take place [1], particularly under basic conditions (Eq. 8.1). N-(1-Propynyl)phenothiazine, when hydrogenated in the presence of palladium on carbon

$$RR'C(OH)C\equiv CH \xrightarrow{\Delta} RR'C=O + HC\equiv CH$$
$$\uparrow \Delta \qquad (8.1)$$
$$RR'C(OH)C\equiv CC(OH)RR' \xrightarrow{\Delta} RR'C(OH)C\equiv CH + RR'C=O$$

at low pressure, yielded only phenothiazine [2]. Hydrogenation in the presence of platinum oxide gave N-propylphenothiazine. The yield (about 40%), however, suggested that hydrogenolysis may have accompanied this reduction (Eq. 8.2). Hennion and DiGiovanna [3] reported extensive hydrogenolysis in the platinum-catalyzed hydrogenation of some crowded acetylenic amines in an attempt to prepare ditertiary amylamine. However, the reductions were successful in the presence of Raney nickel (Eq. 8.3). In an extension of the effect of hydrogenation on sterically crowded acetylenic amines [4], reduction of the quaternary salts of the following compounds in the presence of Raney nickel, platinum oxide, or palladium on carbon

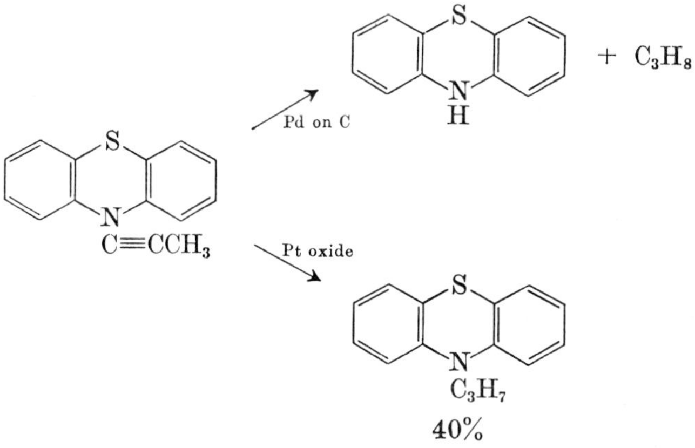

$$\text{(phenothiazine-N-C}\equiv\text{CCH}_3\text{)} \xrightarrow[\text{Pd on C}]{} \text{(phenothiazine-NH)} + \text{C}_3\text{H}_8$$

$$\xrightarrow{\text{Pt oxide}} \text{(phenothiazine-N-C}_3\text{H}_7\text{)} \quad 40\%$$

(8.2)

resulted in hydrogenolysis (Eq. 8.4). In an earlier study, Hennion and Hanzel [5] reported hydrogenolysis in a few similar reductions with a mildly active nickel catalyst. They suggest that the amount of catalyst and

$$[\text{HC}\equiv\text{CC}(\text{CH}_3)_2]_2\text{NH}\cdot\text{HCl} \xrightarrow[\text{Pt}]{} \text{hydrogenolysis}$$

$$\Bigg\downarrow 4\text{H}_2$$

$$\xrightarrow{\text{Raney nickel}} [\text{H}_5\text{C}_2\text{C}(\text{CH}_3)_2]_2\text{NH}\cdot\text{HCl} \quad (8.3)$$

$$\Bigg\uparrow 2\text{H}_2$$

$$\text{H}_5\text{C}_2\text{C}(\text{CH}_3)_2\text{NHC}(\text{CH}_3)_2\text{C}\equiv\text{CH}\cdot\text{HCl} \xrightarrow[\text{Pt}]{} \text{hydrogenolysis}$$

its activity are critical. In the reduction in which hydrogenolysis took place they quoted a very rapid rise in temperature. Huggill and Rose [6] found that when 4-morpholino-1,1-diphenyl-2-pentyn-1-ol was hydrogenated with

$$\text{RR'C}\big[\overset{+}{\text{N}}(\text{CH}_3)_3\big]\text{C}\equiv\text{CH}(\text{Cl}^-) \rightarrow (\text{H}_3\text{C})_3\text{N}\cdot\text{HCl}$$

(8.4)

R	R'
CH$_3$	CH$_3$
CH$_3$	C$_2$H$_5$
CH$_3$	$i-$C$_4$H$_9$
CH$_2$(CH$_2$)$_3$CH$_2$	

palladium on calcium carbonate at room temperature and atmospheric pressure, the desired olefinic compound was obtained but when the same reaction was carried out under pressure (not indicated), considerable hydrogenolysis resulted. Reduction under pressure in the presence of Raney nickel also gave cleavage products. Attempts to completely reduce the acetylenic amino alcohol with platinum oxide gave morpholine and a low yield of the ethylenic aminoalcohol.

REDUCTIONS WITH PLATINUM CATALYST

The addition of hydrogen to the triple bond takes place in stepwise fashion and leads first to the olefin and then to the saturated hydrocarbon. Campbell and Campbell [7], Burwell [8], and Bond [9] each state that selectivity is not generally obtained with platinum catalysts. When a platinum-catalyzed reduction of an acetylene is interrupted at the halfway point, the products of reaction usually consist of fairly large amounts of saturated hydrocarbon and unreacted acetylenic material. Crombie [10], for example, found that when the hydrogenation of undec-10-ynoic acid with platinum oxide was interrupted at about one molar equivalent of hydrogen, 21.5% of starting material, 35.5% of undecenoic acid, and 43% of saturated acid were obtained. Markman [11], conducting a polarographic investigation of the hydrogenation of acetylenedicarboxylic acid, showed that in the presence of platinum the acid was directly converted to succinic acid and only very little maleic acid was formed simultaneously. Most of the examples in the literature seem to indicate that platinum catalysts are of value only as an analytical tool in the reduction of acetylenic compounds.

Nevertheless, there are some examples in the more recent literature where selectivity with these catalysts is obtained. The work of Braude and Coles [12a] is of interest in this respect. They reported that the inadvertent use of platinum oxide in the reduction of 1-phenyl-2-butyn-1-ol,

$$C_6H_5CH(OH)C{\equiv}CCH_3,$$

gave the corresponding phenylbutenol when the reaction was interrupted after uptake of one molar equivalent of hydrogen. Selectivity may be attributed to the very low ratio of catalyst to substrate (less than 0.2%). In contrast the use of an even smaller amount of platinum oxide (less than 0.1% per weight of substrate) was not effective in the semihydrogenations of some secondary and tertiary acetylenic alcohols [12b].

In the platinum-catalyzed semihydrogenation of hex-1-en-4-yn-3-ol, interrupted after one equivalent was absorbed, the triple bond was reduced but the dienol was not obtained as the terminal double bond was also affected [12a], Eq. 8.5.

$$CH_2{=}CHCHOHC{\equiv}CCH_3 \begin{array}{l} \xrightarrow{/\!/\!/} CH_2{=}CHCH(OH)CH{=}CHCH_3 \\ \\ \longrightarrow CH_3CH_2CH(OH)CH{=}CHCH_3 \end{array} \quad (8.5)$$

The hydrogenation of *bis*-(1-hydroxycyclohexyl)acetylene may be another example of selectivity in the presence of platinum [13]. The reduction was interrupted at one equivalent uptake after about 20 to 30 min (reaction had slowed down perceptibly with precipitation of the product), Eq. 8.6. The resulting substituted ethylene was also obtained by reduction of the acetylene

$$\underset{\text{Cy}}{\text{Cy}}{-}\underset{\text{OH}}{\overset{\text{OH}}{\text{C}}}{\equiv}\underset{\text{HO}}{\overset{\text{HO}}{\text{C}}}{-}\underset{\text{Cy}}{\text{Cy}} \xrightarrow{\underset{\text{Pt}}{\text{H}_2}} \underset{\text{Cy}}{\text{Cy}}{-}\underset{\text{OH}}{\overset{\text{OH}}{\text{CH}}}{=}\underset{\text{HO}}{\overset{\text{HO}}{\text{CH}}}{-}\underset{\text{Cy}}{\text{Cy}} \quad (8.6)$$

in the presence of palladium on barium sulfate. This might be expected since palladium usually gives better results in semihydrogenations of acetylenes. It is possible that the selectivity obtained in the platinum-catalyzed hydrogenation of this acetylene resulted from purely mechanical covering of catalyst sites with precipitated material.

Nikitin and Timofeeva [14] reported that neither platinum oxide nor palladium on carbon were capable of causing reduction of a number of acetylenic tertiary polyols to the saturated compounds. In acetic acid, hydrogenation continued beyond the olefinic stage with loss of OH but without reduction of the double bond. The acetylenic compounds included 2,5,6-trimethyl-3-heptyne-2,5,6-triol, 2-(1-hydroxycyclohexyl)-5-methyl-3-hexyne-2,5-diol, and others of similar structure where the carbon atom on each side of the acetylenic bond was tertiary. In a later publication [15] other acetylenic triols of similar structure were also selectively reduced with platinum oxide or palladium on carbon. The uptake of hydrogen was reported to be more rapid with platinum catalyst but selective conversion to the corresponding olefins did take place.

It seems likely that any catalyst that is capable of reducing the triple bond would be unable to induce further reduction of the olefins reported in References 14 and 15 because of steric effects. It appears that the unusual selectivity commented upon by the author is actually a result of complete substitution of the carbon atom on each side of the double bond which prevents contact between the catalyst and the olefinic linkage. From construction of a molecular model of 2,5,6-trimethyl-3-heptene-2,5,6-triol,

88 Acetylenes

which illustrates a typical structure of the olefins resulting from the reduction of the corresponding acetylenes [14, 15], it appears that the double bond is too far from the surface of the catalyst to undergo further attack by hydrogen (see Fig. 8.1).

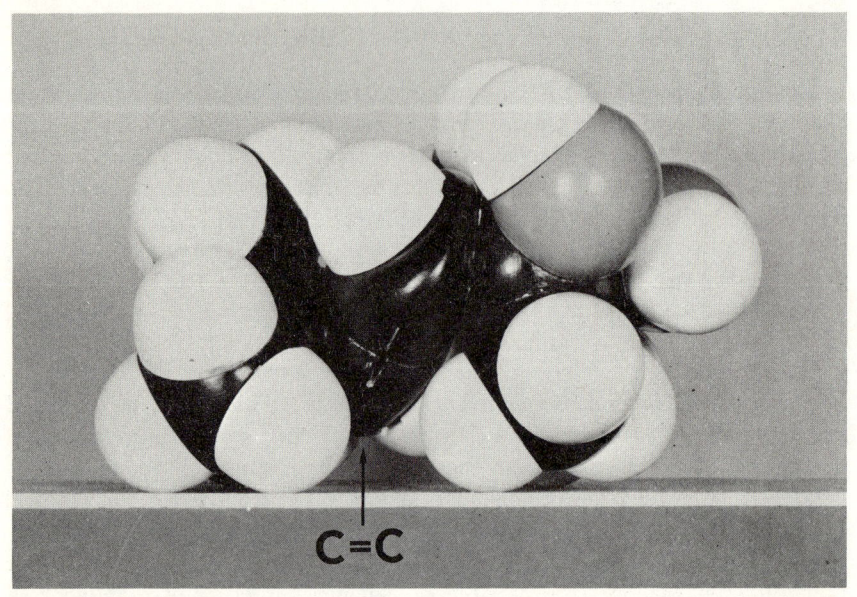

Fig. 8.1 2,5,6-Trimethyl-3-heptene-2,5,6-triol; note distance of double bond from the surface of the catalyst.

In more recent work [15a] 2,3,6,7-tetramethyl-4-nonyne-2,3,6,7-tetralol was converted to the corresponding olefin in the presence of either platinum or palladium. On the other hand, while 2,3-dimethyl-4-nonyne-2,3,6-triol could be selectively reduced to the olefinic compound, further reduction to the saturated triol would also take place in the presence of either catalyst, although at a slower rate than for the first molar equivalent.

A model of the dimethylnonenetriol (Fig. 8.2) shows that further reduction is possible since the double bond appears to be in contact with the surface of the catalyst. A comparison of the two models clearly demonstrates how substitution of a hydrogen atom for one of the substituents on the tertiary carbon adjacent to the double bond causes a complete change in the ability of the double bond to add hydrogen.

Karrer reported on semihydrogenation of acetylenic carbinols with platinum catalyst [16]. He did not describe the type of catalyst or the ratio

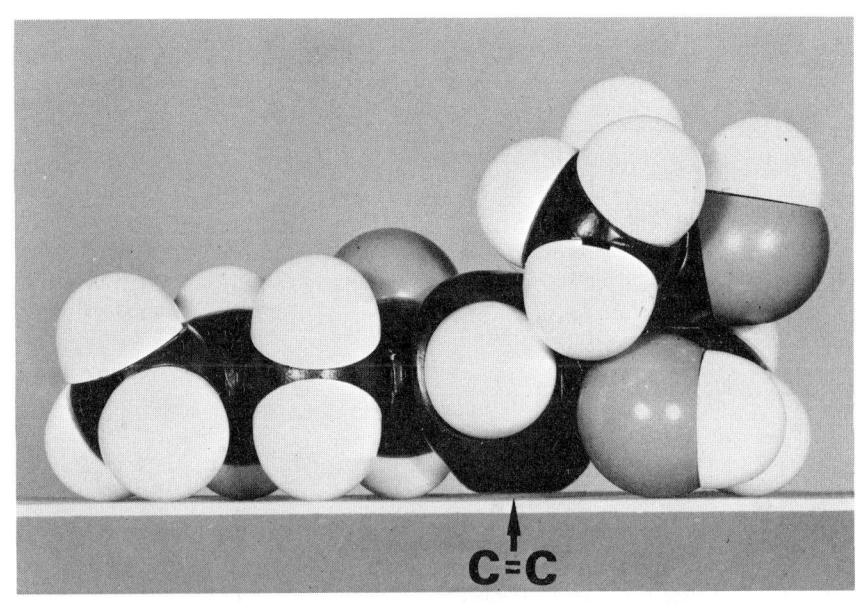

Fig. 8.2 2,3-Dimethyl-4-nonene-2,3,6-triol; note proximity of double bond to the catalyst surface.

of catalyst to compound. He simply stated that 3,7,11,15-tetramethyl-1-hexadecen-3-ol was obtained when the platinum-catalyzed reduction of the acetylenic carbinol was interrupted after absorption of one molar equivalent

$$(H_3C)_2CH(CH_2)_3CH(CH_3)(CH_2)_3CH(CH_3)(CH_2)_3C(OH)(CH_3)C\equiv CH$$
$$\downarrow H_2 \qquad (8.7)$$
$$(H_3C)_2CH(CH_2)_3CH(CH_3)(CH_2)_3CH(CH_3)(CH_2)_3C(OH)(CH_3)CH=CH_2$$

(Eq. 8.7). Another example is given in the semihydrogenation of 3,7-dimethyl-1-octyn-3-ol, $(H_3C)_2CH(CH_2)_3C(OH)(CH_3)C\equiv CH$ [17]. Other partial reductions in connection with his work on homologs of vitamin E and the synthesis of phytol were all carried out in the same manner [18–20]. The reductions were run in absolute alcohol in the presence of platinum and the reaction was interrupted after absorption of one equivalent of hydrogen. It is of particular interest that all of the selective reductions reported by Karrer were those of terminal acetylenes. Other investigators using Raney nickel generally reported lack of selectivity and no noticeable difference in reduction rate between the first and second equivalent of hydrogen when the semihydrogenation of terminal acetylenes was attempted. A number of examples will be seen in the ensuing section on reductions with nickel.

90   Acetylenes

Selectivity does not occur because acetylenes are reduced more rapidly than olefins. On the contrary Bourguel [21] has shown that the olefins from the corresponding terminal acetylenes add hydrogen as rapidly or even more rapidly than the parent acetylene. However, it is well recognized that acetylenes are preferentially chemisorbed at the active centers of the catalyst and that little or no olefin can be adsorbed until the acetylene is completely converted [22].

An examination of a model of 3,7-dimethyl-1-octen-3-ol (Fig. 8.3), the product of one of the partial reductions reported [18], does not explain the selectivity of the hydrogenation. It would appear from the model that there

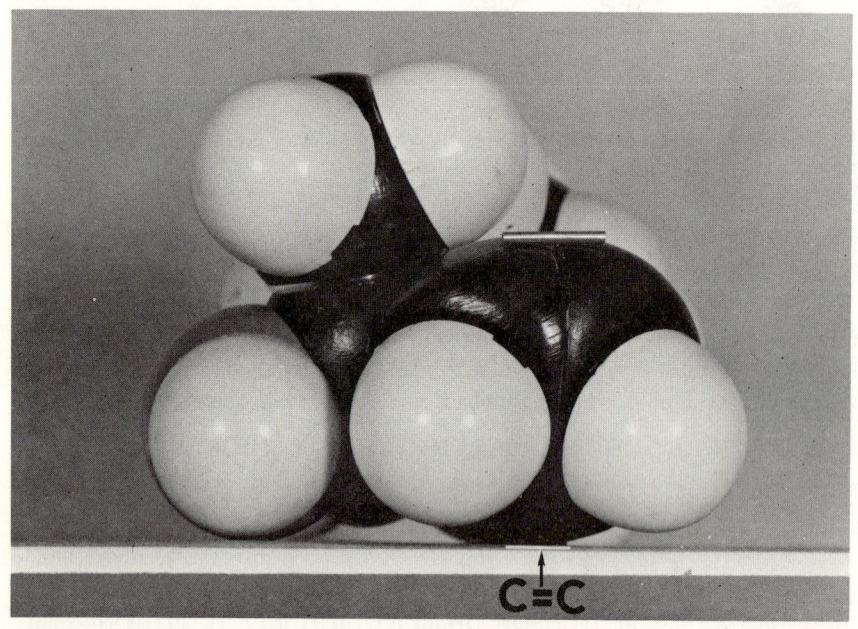

Fig. 8.3 3,7-Dimethyl-1-octen-3-ol; note contact of the double bond with the surface of the catalyst.

are no serious obstructions preventing the double bond from making good contact with the catalyst and that it should be reduced rather readily. The model is typical of those of the resulting terminal olefinic compounds described in the Karrer work.

These facts and the selectivity observed by Karrer suggest that probably hydrogen uptake was very rigidly controlled and/or that a very low catalyst ratio was used or that the catalyst was of low activity.

It is conceivable that platinum could be useful in semihydrogenations under the proper conditions, such as rigid control of hydrogen uptake, use of a small amount of catalyst, and control of the reaction temperature. It has been noted in a number of reductions of acetylenic compounds that there is a rise in temperature during the uptake of the first equivalent of hydrogen. If the exothermic reaction could be controlled it might increase selectivity. Another means of controlling reductions with platinum could involve the employment of inhibiting agents.

REDUCTIONS WITH NICKEL CATALYST

Nickel catalysts have been reported to give selectivity in the reduction of acetylenes. Yet there is much discussion on the ability of certain nickel catalysts to cause reaction to stop at the end of the first stage. Selectivity appears to be dependent on the catalyst and on the compound. Freshly prepared Raney nickel catalyst was not selective in the semihydrogenation of monoalkylacetylenes [23]. It was more selective with dialkylacetylenes. With octadec-3-yne and octadec-9-yne a break in the curve occurred at 1.5 molar equivalents of hydrogenation; with 1-octyne there was no break. The authors stated that aged Raney nickel was far more selective. The catalyst, when no longer pyrophoric on exposure to air, was sufficiently active to promote rapid reduction of dialkylacetylenes. The rate of uptake dropped very sharply after rapid absorption of the first molar equivalent. With 1-octadecyne there was little difference but in the other octadecynes, from 2- to 9-octadecdne, the difference in rate was from 6 to 24 times more rapid for the first mole than for the second. However, they also pointed out that selectivity with aged nickel had obvious disadvantages since the method of preparation of the catalyst and the period of storage were factors.

The apparent difference in selectivity in the reduction of substituted acetylenes with Raney nickel had been reported by Dupont [24], who found there was little change in velocity in the hydrogenation of 1-heptyne or phenylacetylene after one molar equivalent was absorbed but there was a difference in the case of 2-octyne and 1-methoxy-2-nonyne.

The hydrogenations of a series of mono- and dialkylacetylenes with Raney nickel could be interrupted after absorption of one equivalent to give generally good selectivity and good yield of ethylenes except in the reduction of 1-hexyne which yielded only 27% of 1-hexene [25].

Freidlin [26] reported on the difference in selectivity between the terminal acetylene, 1-hexyne, and the two disubstituted acetylenes, 2- and 3-hexyne in nickel-catalyzed reductions. The first equivalent of hydrogen was absorbed at almost the same rate in each case. In the reduction of 1-hexyne the addition of the second equivalent was accelerated, but it was retarded with the 2- and 3-isomers. Hexane was found in the early stages of the reduction of 1-hexyne. The reduction was nonselective and a 3:1 ratio of hexene to hexane was maintained until the reaction was stopped. In contrast, when 2- and 3-hexynes were selectively reduced, no more than 1–2% of hexane was found.

The variables associated with Raney nickel catalysts, the method of preparation, its activity, its age, and possibly the amount of occluded alkali present, may make it difficult to obtain consistent results in the semihydrogenation of acetylenes with it.

75–90% yields of 5-decene, 4-octene, 3-octene, and 3-hexene were obtained from reductions of the corresponding acetylenes [27] with the same Raney nickel [28] that had been employed in another semihydrogenation [25]. A reduction, employing the procedure described in Reference 27, gave only 49% of 2,2,5,5-tetramethylene-3-hexene from the corresponding hexyne [29]. It may be significant that, from an almost quantitative yield of residue, several fractional distillations were required to effect high purity. No comment was offered to indicate whether the yield was due to mechanical loss or whether the product, after reduction of the hexyne, contained other components which required fractionation for their removal.

Other examples of differences in similar catalysts are cited in the literature. Knight and Diamond [30] reported that, in the hydrogenation of some octenoic acids in thiophene-free benzene with W-5 Raney nickel [31] which had been aged 5 to 6 months, there was not the sharp decrease in rate that had been reported by others to take place after absorption of one molar equivalent. They noted that, after the approximate theoretical volume of hydrogen was absorbed, there was a gradual decrease in the rate. In contrast, Howton and Davis [32] reported that in the complete hydrogenation of 5-octynoic acid with W-5 Raney nickel the first equivalent of hydrogen was taken up in 10 min while the second mole required an additional 4 hr. In a controlled hydrogenation with the same catalyst, they discontinued the reaction when the sharp decrease in rate was noted or when one molar equivalent was absorbed. They reported that the yield of 5-octenoic acid was essentially quantitative and essentially free of starting material and caprylic acid. There was no notation of the age of the catalyst. At the same time the authors found that there was no break in the reduction of the methyl ester of 5-octynoic acid.

Good results have been obtained in some semihydrogenations with Raney nickel. Huggill and Rose [6] noted that there was a difference in time of uptake in the Raney nickel-catalyzed reduction of 1,1-diphenyl-4-dimethylamino-2-pentyn-1-ol to the saturated amino alcohol. They reported that the reduction proceeded in two stages. Absorption of the first equivalent of hydrogen required 1 hr while absorption of the second mole of hydrogen was complete in 12 hr. A similar difference in rate was noted in the reduction of the corresponding 4-diethylamino compound. It is of interest that in the hydrogenation of the 4-morpholino compound successful reduction took place at atmospheric pressure but increased pressure resulted in low yield of the ethylenic amine with much hydrogenolysis.

Valette [33] hydrogenated 2-butyn-1,4-diol in absolute ethyl alcohol with a rather large amount of Raney nickel. The reaction was interrupted at 92% uptake for one mole of hydrogen. Valette reported that it was simple to separate the starting material and saturated diol from the olefinic compound.

Gouge [34] carried out hydrogenations of a number of 2-alkyn-1,4-diols to the ethylenic diols in the presence of 10 g of Raney nickel to 0.5 mole of substrate. The reductions were stopped at 95% uptake for one equivalent of hydrogen; good yields of products were reported.

A graph of two compounds employed in Gouge's study (Fig. 8.4) shows that there is a difference in rate between absorption of the first and second mole of hydrogen. The rate curve suggests the advisability of interrupting reduction at less than one equivalent of hydrogen. The break in the curve, while it is definite, may not be sharp enough to insure complete selectivity.

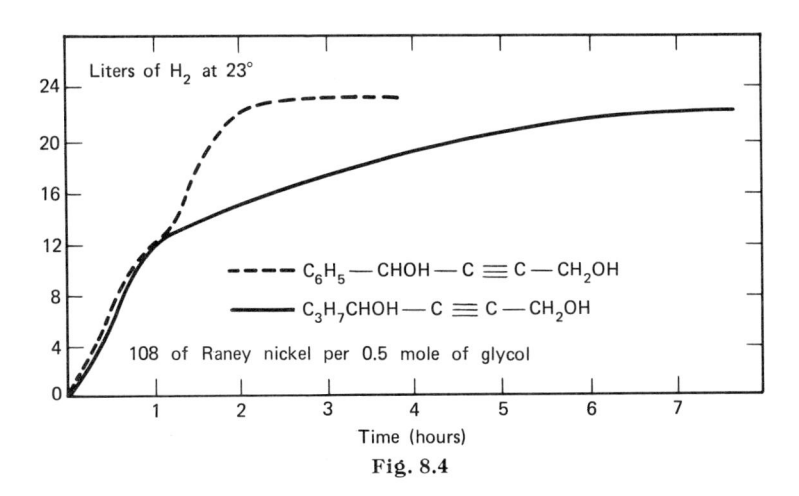

Fig. 8.4

While there are examples of successful partial reductions of acetylenic compounds with Raney nickel, the chemist should be aware of its limitations because of the variables affecting its activity.

Bernard and Cologne [35] reported that the semihydrogenation of some 3,4-dialkyl-1-alkyne-3,4-diols could be effected by judicious use of Raney nickel. In a selective hydrogenation the reaction was found to be complete when the temperature rose to a maximum. The speed of hydrogenation for the first stage was found to be greater than for the second stage of reduction to the saturated diols. In a second study [36] they noted that 3-ethyl-4-methoxy-1-butyn-3-ol could be converted in 70% yield to the corresponding 1-buten-3-ol. In each study they were able to achieve selectivity with terminal acetylenes in contrast to the lack of selectivity in the reduction of monoalkylacetylenes. This difference has also been noted by others [23–25] with Raney nickel.

Another variable which may complicate the action of Raney nickel catalysts in the hydrogenation of acetylenes may be the presence of alkali and aluminum. Khan [37] used a specially prepared Raney nickel catalyst which had been thoroughly digested and washed to remove almost all of the alkali and aluminum. The resulting catalyst was covered with dioxane and some of the solvent was distilled off. Stearolic acid (9-octadecynoic acid) and its methyl ester were hydrogenated with this specially prepared catalyst and also with W-2 and W-3 Raney nickel. With the alkali-free catalyst uptake of hydrogen fell to an almost negligible rate after reduction of the triple bond to the double bond, while other catalysts tended to saturate the double bond. The composition of the crude products after reduction of methyl stearolate with the new catalyst showed 94.2% of methyl octadecanoate, 0.7% methyl stearate, and 2.1% of starting material. The author reported that the other nickel catalysts tended not only to reduce the double bond but also to partially reduce the ester group.

Other investigators, apparently bothered by the variables attendant in the preparation of Raney nickel catalysts, adopted what they hoped were better means of attaining greater selectivity. Henne and Greenlee [38] reported that nickel on kieselguhr, notably less active than Raney nickel under low-pressure conditions, appeared to have little tendency toward hydrogenating certain acetylenes beyond the olefinic stage. They found that at 30–80° and 1–3 atm pressure uptake of hydrogen almost halted as the amount absorbed approached one molar equivalent. Nevertheless, reduction with this catalyst may not be as selective as reported since hydrogenations were interrupted at 80% uptake for the first stage to obtain pure alkenes and minimize alkane formation.

Vaughan [39] reported that a catalyst prepared from a nickel-silicon alloy was selective for the conversion of 2,5-dimethyl-3-hexyn-2,5 diol to the ethylenic diol. It was pointed out that the reaction is dependent on the preparation of the catalyst with sufficient sodium hydroxide present. It is of interest that, in a reduction in an autoclave with this catalyst at 200°, uptake of hydrogen stopped at one equivalent after 1 hr although reaction was carried out for 18 hr and 94% of 2,5-dimethyl-3-hexen-2,5-diol was obtained. In contrast to the results obtained above, another investigator reported nonselective reductions [12b].

Successful semihydrogenations were carried out with Raney nickel containing copper [23]. Reduction of a number of octadecynes to octadecenes took place rapidly with no further uptake of hydrogen, except in the case of 1-octadecyne where one molar equivalent was absorbed rapidly (4 min) but uptake did not stop. The second equivalent was absorbed in 40 min.

2-Butyn-1,4-diol was hydrogenated with Raney nickel or nickel containing 10% of copper, to 2-buten-1,4-diol (86% yield) in aqueous or alcoholic solution at 25–100° and 1–21 atm in the presence of bases such as ammonia, pyridine, and piperidine [40], the bases apparently acting as inhibitors. From unpublished experiments carried out in his laboratory, Tedeschi stated that although Raney nickel at low to moderate pressures did not cause complete hydrogenation of acetylenic glycols to the saturated glycols, the addition of bases did not make it selective for semihydrogenation.

Shcheglov, Sokol'skii, and Ishchenki [41], who reported good selectivity in the reduction of phenylacetylene with skeletal nickel or nickel promoted with titanium, found that the addition of pyridine had no effect on the rate of reduction of the triple bond but did decrease the rate for the reduction of the double bond. From unpublished experiments Oroshnik, Karmas, and Merbane [43] used two inhibiting materials, piperidine and zinc acetate, with Raney nickel and obtained an excellent yield (90%) of cis ethylenic compound from the reduction of 1-(2-cyclohexenylidene)-3,7-dimethyl-9-methoxy-2,7-nonadien-4-yne (Eq. 8.8). The inhibitors may have had little

$$\text{C}_6\text{H}_9=\text{CHCH}=\text{C}(\text{CH}_3)\text{C}\equiv\text{CCH}_2\text{C}(\text{CH}_3)=\text{CHCH}_2\text{OCH}_3$$
$$\downarrow \text{H}_2 \qquad (8.8)$$
$$\text{C}_6\text{H}_9=\text{CHCH}=\text{C}(\text{CH}_3)\text{CH}=\text{CHCH}_2\text{C}(\text{CH}_3)=\text{CHCH}_2\text{OCH}_3$$

effect on the reduction of the triple bond but their purpose probably was to prevent double-bond reductions.

Sulfur-poisoned Raney nickel was employed for the selective reduction of 3-methyl-1-butyn-3-ol to the corresponding butenol at 100° and 17 atm [44].

The work of Hoff, Greenlee, and Boord [45] is of interest because of the means they employed to keep alkanes at a minimum level during the reduction of some alkynes with Raney nickel. Hydrogenation was carried out at 4–5 atm pressure in a bomb equipped with a cooling coil through which ice water was passed during the reaction period. High yields of very pure alkenes were obtained.

Although the preferred catalyst for semihydrogenation of 3-alkylamino-3-alkyl-1-pentynes in nonpolar solvents was palladium, it was claimed that Raney nickel could be used [46]. It was employed for the reduction of 3-t-butylamino-3-ethyl-1-pentyne to the corresponding ethylenic amine. In that reaction hydrogen uptake stopped after absorption of one molar equivalent.

The use of W-6 Raney nickel [31], a rather active catalyst, also has been reported for successful semihydrogenations. Ahmad and Strong [47] converted 6-hendecynoic acid to 6-hendecenoic acid with this catalyst in a reduction at 3 atm pressure. The reaction was stopped immediately after absorption of one equivalent of hydrogen. In a later report [48] 86% yield of 11-octadecenoic acid was obtained from reduction of the acetylenic acid. In this work the ratio of catalyst to compound was about 7%. However, the catalyst was weighed wet, so the ratio was much lower. Uptake for one equivalent took from 10 to 60 min. Taylor and Strong [49] following these two methods obtained pure 6-heptenoic, 7-octenoic, 7-dodecenoic, 6- and 7-tridecenoic, and 7-tetradecenoic acids from the acetylenic acids. Huber [50] also used the same method with good results.

REDUCTIONS WITH PALLADIUM CATALYST

Campbell [7] and Bond [9] both state that, while palladium black does not bring about selective conversion of acetylenes to olefins, other palladium catalysts have been found useful. Campbell reports a number of papers which describe interrupted hydrogenations of acetylenes in the presence of colloidal palladium. The products noted in these references are largely or entirely olefinic. Colloidal catalysts are seldom if ever used today; more recent work deals with palladium on various carriers. In these forms palladium is more or less accepted as the catalyst of choice for the semihydrogenation of acetylenes. Nevertheless, from reports in the literature palladium on a specific carrier falls short of being the ideal catalyst for a wide variety of acetylenes.

Hennion and co-workers [51] describe the use of 5% palladium on barium carbonate in the low-pressure semihydrogenation of some tertiary ethinyl carbinols and their acetates. They suggest that petroleum ether is the best solvent for these reductions. It is possible that the solvent and carrier do play a part in selectivity. It would appear, however, that selectivity is more likely due to control of the exothermic reaction which is known to occur during reductions of acetylenic compounds. Hennion and his group found that by starting the hydrogenation at 22–23°, in the presence of a low weight ratio of catalyst to substrate (0.15–0.2%), the temperature did not rise above

40°; the reaction was clean and self-terminating. On the other hand, starting the reduction at 30° caused a rise to 80° and a loss of selectivity. In the same article the authors reported that in the reduction of some tertiary acetylenic amines it was necessary to use twice as much catalyst as in the reaction with the carbinols. They suggested that preliminary experiments should be run to ascertain the necessary amount of catalyst required to allow the consumption of one equivalent of hydrogen to take place in 30 to 60 min. Under these conditions they stated that the heat of reaction was dissipated rapidly enough to avoid extensive overhydrogenation.

The use of palladium on other carriers has been reported to give good results. Ashley and MacDonald [52] hydrogenated 1,4-*bis*-(4-amidinophenoxy)-2-butyne dihydrochloride in the presence of 10% palladium on calcium carbonate at room temperature. They interrupted the reaction after uptake of one molar equivalent. From analytical data it appears that the desired symmetrical olefinic compound was obtained.

Crombie [53] used about 4% by weight of 5% palladium on calcium carbonate to convert 1,1-diethoxy-2-octyne to the corresponding octene in 70% yield. The reaction was interrupted at about 1.05 molar equivalents. He also obtained ethyl 3-hydroxy-4-decenoate by interrupting the reduction of the acetylenic compound. Campbell, Fatora, and Campbell [54] reported success in the semihydrogenation of 1-dimethylamino-2-heptyne with the same catalyst but found that there was no break in the reduction of 1-dimethylamino-2-propyne. Milas and Priesing [55] hydrogenated commercially available 1-ethynylcyclohexanol selectively with palladium hydroxide deposited on calcium carbonate. These are only a few of the examples of selectivity resulting from the use of this catalyst. In every instance reduction was stopped at one molar equivalent. A rather remarkable example of semihydrogenation with it is seen in the reduction of 2,5-dimethyl-3-hexyne-2,5-diol at 80 atm pressure to yield 94% of the ethylenic diol [56]. Uptake of hydrogen was complete in 3 min. Based on experience with other triple-bond reductions it is surprising that there was such selectivity in the absence of control of the reaction exotherm. However, the low catalyst ratio, 0.35% by weight of starting acetylene, and the use of an exact amount of hydrogen apparently were responsible for the high selectivity.

It is claimed that palladium on calcium carbonate is the catalyst of choice for reducing certain acetylenic glycols to the corresponding ethylenic glycols [57]. In contrast, Harper and Smith [58] stated that palladium on barium sulfate was more selective. They reduced 3-hexyn-1-ol with it at 15° until one equivalent of hydrogen was absorbed and obtained a 79% yield of 3-hexen-1-ol. The selectivity was very likely due to the control of the temperature, a phenomenon noted before. There is a report in the patent literature on the use of palladium on the same carrier or on calcium sulfate

98   Acetylenes

[59] in the semihydrogenation of 2-methyl-1-butyn-3-ol. The resultant product was obtained in about 97% yield and contained less than 3% of saturated alcohol. The very small amount of catalyst probably was responsible for the excellent results. Another patent [60] reports selective reduction of 1-cyclohexyl-1-phenyl-4-dimethylamino-2-butyn-1-ol with 0.4% palladium on barium sulfate but it points out that two equivalents will be absorbed if the reaction is allowed to run long enough.

Palladium on carbon was employed successfully for semihydrogenations of acetylenic amines to ethylenic amines [61]. The reduction of 3-ethylamino-3-phenyl-1-butyne in petroleum ether in the presence of palladium on carbon was a typical example. Meinwald, Dicker, and Danieli [62] reported that the reduction of 6-(2-tetrahydropyranyloxy)-3-hexyn-1-ol with it proceeded more smoothly in a mole-size experiment in ethyl acetate solution than did a previous semihydrogenation with Lindlar catalyst (Eq. 8.9).

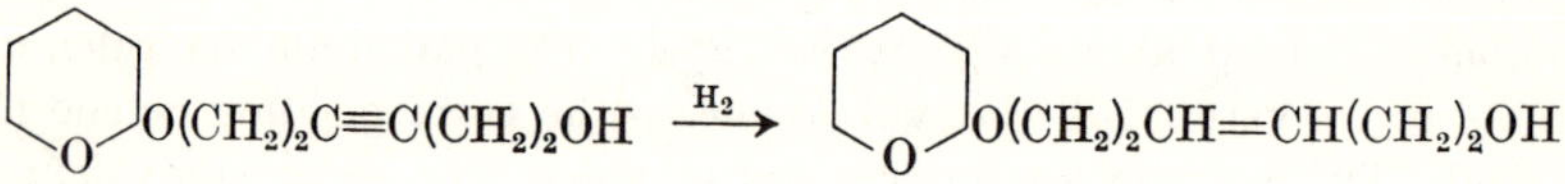

$$\text{(THP)}O(CH_2)_2C{\equiv}C(CH_2)_2OH \xrightarrow{H_2} \text{(THP)}O(CH_2)_2CH{=}CH(CH_2)_2OH \quad (8.9)$$

Heuberger and Owen [63] found that the conversion of 3-butyn-1,6-diol to the ethylenic glycol took 4 hr in the presence of palladium on carbon. Gensler and Schlein [64] found that with pure 9-octadecyne-1,18-dioic acid there was no break in the curve with palladium on carbon but with less pure material the reduction stopped automatically at one molar equivalent and gave pure octadecenedioc acid in 90% yield.

In the semihydrogenation of 2-methyl-3-butyn-2-ol, it is suggested [44] that palladium on carbon is the preferred catalyst for partial reduction of the acetylenic linkage of alkynols in water and methyl alcohol. In each instance the ratio of catalyst to acetylenic alcohol was 1% or less (the ratio is based on weight of catalyst, not metal content). In several experiments reduction was carried out at 18 and 50° psig; in one run the temperature was held at 50°. The end of the reaction was determined by treatment of aliquots withdrawn from the reaction mixture with ammoniacal silver nitrate. When a precipitate no longer formed, the reduction was stopped immediately. The close check on the reduction and the low catalyst ratio were probably responsible for the reported selectivity.

Marszak and Olomucki [65] achieved selectivity in the semihydrogenation of some ω-dimethylaminoalkynoic acids of the type,

$$(H_3C)_2N(CH_2)_nC{\equiv}CCOOH$$

where $n = 1, 2, 5,$ and 9, with palladium on alumina using 1 g of catalyst for 2 g of substrate. It is of interest that, in spite of the high ratio of catalyst to starting material, selective conversion of acetylene to olefin was obtained. Two factors may have been responsible. Because of the dilute solution, 2 g of substrate in 70 cc of 93% ethyl alcohol, it is likely that the heat of the exothermic reaction, typical of acetylene reductions, was dissipated quickly. This control of exothermicity probably decreased the tendency for hydrogen uptake to continue to complete saturation. It may also be assumed that strict monitoring of hydrogen uptake was exercised, since the authors showed that reduction to the saturated compounds would also take place in the presence of the same catalyst.

Rigid control of the amount of hydrogen may have been more responsible for selectivity than the small amounts of catalyst in the reductions of butynediols to butenediols with supported palladium catalysts or with similar quantities of zinc acetate deactivated ones [66]. Hydrogen was passed into an apparatus containing substrate, solvent, and catalyst (appropriately stirred) and the end point of the reaction was determined by a sharp drop in electrochemical potential.

Control of the amount of hydrogen employed in semihydrogenations of acetylenes is a point worth investigating. In many instances interruption of attempted partial reductions in the presence of excess hydrogen does not always result in good selectivity. It may be more advisable to add hydrogen gradually to a system until one equivalent has been absorbed.

Deactivated Palladium Catalysts

Lindlar catalyst, lead-poisoned palladium on calcium carbonate, is one of a number of palladium catalysts modified by transition metal salts which are used in the semihydrogenation of acetylenes. Lead salts are generally employed [67, 68] although bismuth salts or a combination of bismuth and lead salts are claimed to be effective [67].

Lindlar compared the selectivity of the lead-treated catalyst with palladium on carbon [68]. Reduction with Lindlar catalyst was reported to stop after the required amount of hydrogen was absorbed but the palladium on carbon-catalyzed reduction continued beyond that point without any apparent break in the curve. Lindlar also suggested that petroleum ether was the preferred solvent for acetylenic semihydrogenations and in addition pointed out that, when other reducible functions were present in the compound to be reduced, the presence of a small amount of quinoline in the reaction mixture might give better selectivity.

Because it has been more widely studied than most deactivated catalyst systems a general impression has been created that Lindlar catalyst is a general, specific one for all semihydrogenations of acetylenics. However, personal experience [2] and that of another investigator [12b] and reports from knowledgeable chemists in the perfume and flavor industry indicate that this point is somewhat controversial.

It is known to be useful in the semihydrogenation of vitamin A intermediates and other internal triple-bonded systems. Excellent results have been obtained with it, 5% by weight of substrate, in reductions of tertiary acetylenic alcohols to allylic alcohols carried out at atmospheric pressure [69]. The rate of hydrogen uptake decreased markedly after absorption of one equivalent, and yields of 90–95% were obtained (Eq. 8.10).

$$RC{\equiv}CC(OH)(CH_3)R' \xrightarrow{H_2} RCH{=}CHC(OH)(CH_3)R'$$

R	R'
CH_3	H
C_2H_5	H
$(CH_3)_2CHCH_2$	H
$-(CH_2)_4-$	H
C_6H_5	H
CH_3	C_2H_5
CH_3	$n\text{-}C_6H_{13}$

(8.10)

Success in the above series may have been due to reaction at room temperature and atmospheric pressure; related reductions at 25–40° and 1–4 atm were not too selective [12b].

Lindlar catalyst gave good results in the reduction of a Mannich-type acetylenic alcohol to the corresponding ethylenic amino alcohol [70] (Eq. 8.11).

$$(H_5C_2)_2NCH_2C{\equiv}CCHOHCH_3 \xrightarrow{H_2} (H_5C_2)_2NCH_2CH{=}CHCHOHCH_3 \quad (8.11)$$

On the other hand, Libman and Kuznetsov [71] reported that reduction of some related acetylenic amino alcohols was not as selective as expected with it, even when quinoline was added to the reduction mixture. They stated that uptake of hydrogen did not decrease sharply but only gradually after absorption of one equivalent. They found it necessary to stop the reaction after one equivalent uptake and recrystallize the product of reduction a number of times to obtain pure material; yields were not recorded.

Iwai and associates [72] claimed selectivity in the hydrogenation of 1,4-bis(tetrahydro-2-pyranyloxy)-2-butyne to the butene in 76% yield in ethyl acetate with Lindlar catalyst (Eq. 8.12). However, purity of the material was

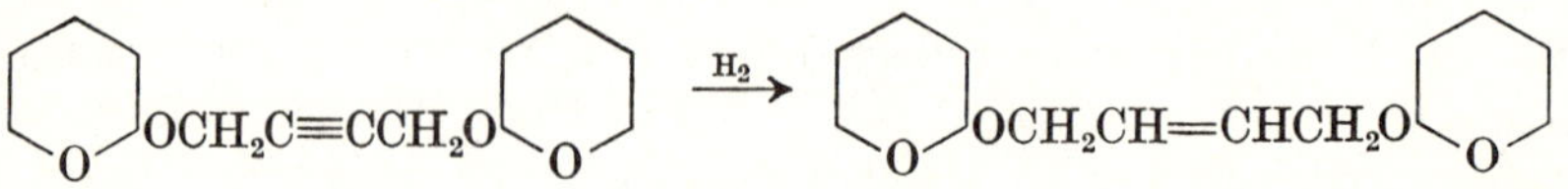

(8.12)

based on elemental analysis which would not show great differences between the butene, butane, and butyne.

Quantitative yield of 6-(2-tetrahydropyranyloxy)-3-hexen-1-ol was reported in the reduction of the acetylenic compound with quinoline-poisoned Lindlar catalyst [73]. Another group stated that in a mole-size experiment reaction proceeded more smoothly with palladium on carbon [62].

Lindlar catalyst appeared to give good results in the partial reduction of some 2- and 4-alkynyl-3,5-dioxopyrazolidines [74]; only 0.95–1.02 equivalents of hydrogen were absorbed. It was also employed in the semihydrogenation of 1-(2-pyridyl)-2-(2-, 3- and 4-pyridyl)acetylenes [75].

Despite the controversy about the merits of Lindlar catalyst, it has given good results in acetylenic semihydrogenations but there has never been an adequate study of a comparison of it with supported or other deactivated supported palladium catalysts for partial reduction of a wide variety of acetylenics.

Selectivity in reductions with it is often improved by the use of quinoline as an inhibitor. Blomquist and Goldstein [76] carried out the hydrogenation of cyclodecyne with Lindlar catalyst and quinoline but apparently resorted to chromatography to remove quinoline to obtain pure cyclodecene. It is possible that contamination by quinoline may have led to the use of the catalyst without the base in later work in the preparation of pure cyclononene [77]. Baker, Linstead, and Weedon [78] converted stearolic acid (octadec-9-ynoic acid) to oleic acid with quinoline-treated catalyst. Other reductions in the same solvent (ethyl acetate) or in other solvents with untreated catalyst always gave products containing stearic acid. Chase and Galender [79] described the use of palladium on calcium carbonate modified by lead acetate for the conversion of 2-methyl-3-butyn-2-ol to the ethylenic alcohol in the presence of 150 g of quinoline for 1000 g of substrate. The authors reported that the catalyst could be reused. The conditions under which the hydrogenation was carried out were applicable to the semihydrogenation of terminal acetylenes which are usually more difficult to control than other acetylene reductions. They reported the preparation of 3-methyl-1-penten-3-ol and 3,5-dimethyl-1-hexen-3-ol from partial reduction of the corresponding acetylenes, but did not give yields. They also pointed out that the reductions they carried out with the deactivated catalysts were exothermic and specified that the preferred reaction temperature is between 15 and 20°.

Control of temperature is a variable that has been cited before in acetylene reductions with other catalysts. Other investigators regard control of the reaction temperature important even in the presence of a partially inactivated catalyst. Payne [80] maintained control of the reaction temperature at 25–27° by directing an air jet against the moving bottle during the reduction

102   Acetylenes

of 2-butyn-1-ol with lead-poisoned palladium on calcium carbonate. A 95% yield of 2-buten-1-ol was obtained. In another study Kimel and associates [81] cooled a 50% solution of 3-methyl-1-butyn-3-ol in petroleum ether to 10° before reduction with Lindlar catalyst inactivated by the presence of quinoline. Absorption of the necessary amount of hydrogen required 3 to 5 hr; hydrogen uptake declined markedly thereafter. 3-Methyl-1-buten-3-ol was obtained in 94% yield after distillation through a packed column.

The addition of certain sulfur-containing compounds is said to increase selectivity in the reduction of acetylenes with Lindlar catalyst [82]. 3-Methyl-1-butyn-3-ol was converted to 3-methyl-1-buten-3-ol of greater than 99% purity, in 98–100% yield in the presence of 14 g of thiodiethyleneglycol and 50 g of catalyst per 1000 g of acetylenic alcohol. The rate of hydrogen uptake decreased almost to zero after absorption of one equivalent. The inhibitors were of the type, RS_nCH_2R' where $n = 1$ or 2 and R and R' were aliphatic or aromatic groups. It is of interest that reductions were carried out at 20–22°.

Other supported lead-treated palladium catalysts have been found useful. Some 6-phenethinylnicotinic acids and amides were hydrogenated with lead acetate-deactivated palladium on carbon further inactivated by the addition of quinoline [83] (Eq. 8.13). Yields of the 6-styryl compounds were not particularly high (40–69% after recrystallization) but this was probably due to mechanical losses because of the scale of the reductions, 2 mM.

$$ROC\text{-pyridyl-}C{\equiv}CC_6H_5 \xrightarrow{H_2} ROC\text{-pyridyl-}CH{=}CHC_6H_5 \quad (8.13)$$

$$R = OH, NH_2, NHCH_3 \text{ and } N\text{-morpholinyl}$$

Another Lindlar-type catalyst, lead-containing palladium on barium sulfate inactivated with quinoline, was employed with excellent results for the semihydrogenation of some α, ω-substituted acetylenic diamines [84] (Eq. 8.14).

$$RCH_2C{\equiv}C(CH_2)_nN(CH_3)_2 \xrightarrow{H_2} RCH_2CH{=}CH(CH_2)_nN(CH_3)_2 \quad (8.14)$$

Yield, 80%, $n = 1$; 92%, $n = 2$

R = 1-(1,2,3,4-tetrahydroquinolyl); $n = 1$ or 2

Other metal-poisoned palladium catalysts similar to the Lindlar catalyst have been reported to be effective in semihydrogenation of acetylenics.

Palladium on calcium carbonate containing tin, 2.25 parts to 1 of palladium based on metal content, was effective in the reduction of 1-butyn-3-ol to 1-buten-3-ol [85]. Hydrogen uptake stopped automatically at one equivalent. In one instance quinoline was added to the reduction mixture.

Zinc, cadmium, mercury, or thallium-containing supported palladium catalysts were employed for selective reduction of acetylenic compounds to olefins [86]. At times ammonia, quinoline, pyridine, piperidine, or methylamine was also added to the reaction mixture. Most of the reductions came to a complete stop after uptake of one equivalent. In some reactions the temperature was held at 10–20°, in others it rose to 30–40°; very high yields were reported.

The weight ratio of deactivating metal to palladium can be important. Normally the ratio of lead to palladium in Lindlar catalyst is about 1.25:1 [67, 68]. Nevertheless, at times it may be necessary to decrease the amount of lead salt. It was found, in an attempted reduction of 1,1-diphenyl-4-dimethylamino-2-butyn-1-ol with the catalyst described by Lindlar, that uptake of hydrogen was slow and incomplete. When the preparation of the catalyst was modified by using half the amount of lead acetate, the same weight ratio of catalyst to substrate employed in the unsuccessful reaction led to the formation of the desired olefinic compound [87] (Eq. 8.15).

$$(C_6H_5)_2C(OH)C\equiv CCH_2N(CH_3)_2 \xrightarrow{H_2} (C_6H_5)_2C(OH)CH=CHCH_2N(CH_3)_2$$
$$\text{Modified Lindlar catalyst}$$

(8.15)

Not only is the metal content of the catalyst important but there may be instances when the weight ratio of the catalyst to substrate is a controlling factor in semihydrogenation of acetylenes. In general, 5–10 g of Lindlar catalyst per 100 g of acetylene will prove satisfactory; more often than not lower ratios give better results. However, when this author attempted to convert 3-amino-3-methyl-1-butyn in pentane solution to the corresponding olefin, no uptake was observed with a 10% ratio of catalyst to compound and only 37.5% uptake with a 20% ratio [2]. There was no question of the purity of the starting material since it was readily converted to the saturated amine in the presence of a small amount of platinum oxide. When the amount of Lindlar catalyst was increased to 30–40%, absorption of one equivalent was complete in 3–5 hr. Vapor-phase chromatography showed 96% of olefinic amine and a mixture of starting material, 1.5%, and saturated amine, 2.5%.

Before the advent of Lindlar calalyst supported palladium catalysts were deactivated with nitrogen bases, as pyridine and quinoline, to achieve selectivity in acetylene reductions. Hershberg and his group found that

17-ethinylandrost-5-en-3β,17β-diol was selectively hydrogenated in pyridine solution in the presence of palladium on calcium carbonate to the 17-vinylandrostenediol in 95% yield [88] (Eq. 8.16). Schechter, Green, and

(8.16)

LaForge [89] deactivated the same catalyst with quinoline to reduce 2-(2-butynyl)-3-methyl-4-hydroxy-2-cyclopenten-1-one to the corresponding 2-butenyl compound without affecting the other reducible groups. They stated that the rate of hydrogenation could be varied at will by adjusting the proportion of quinoline to catalyst. In their described experiment they used 0.8 cc of quinoline to 2.4 g of catalyst per 9.45 g of compound. They further stated that if quinoline was not used uptake, which was extremely rapid, tended to continue beyond absorption of one equivalent of hydrogen. They pointed out that in the presence of quinoline the hydrogenation slowed down to such an extent that it appeared difficult to carry reduction further without taking an excessively long time. Isler and his group [90] in their work on the synthesis of vitamin A added quinoline to a palladium on carbon-catalyzed semihydrogenation of an acetylenic compound which contained other reducible groups. The desired ethylenic compound was obtained in 95% yield.

Elsner and Paul [23], who studied the partial hydrogenation of octadecynes with a variety of catalysts, reported that pyridine-poisoned palladium on calcium carbonate showed selectivity in the reduction of 6-octadecyne. No hydrogen uptake took place in the quinoline-poisoned palladium on carbon reduction of 2-octadecyne when the amount of quinoline, 0.2 g, was equal to the weight of catalyst. However, addition of 0.3 g of catalyst to the same reaction enabled reduction to take place with absorption of only one equivalent of hydrogen. The authors stated that certain other quinoline- and pyridine-deactivated palladium catalysts reported by other investigators were not active enough for the reduction of octadecynes. They noted that at the time of their investigation information about Lindlar catalyst was not available.

Papa, Villani, and Ginsburg [91] reduced the isomeric 2,3- and 4-methyl-1-ethinycyclohexanols in pyridine solution with 5% palladium on calcium carbonate to the corresponding vinyl compounds in 68, 72, and 63% yields respectively. They made no comment except to state that the theoretical amount of hydrogen was absorbed in 10 min. There are other semihydrogenations reported which are catalyzed by supported catalysts deactivated with pyridine or quinoline. In one instance [92] the successful conversion of 1-ethinyl-2,6,6-trimethylcyclohexanol to the 1-vinyl compound may be attributed to the small amount of palladium on carbon used for the reduction and control of the reaction temperature (22°) rather than the fact that the catalyst was deactivated with quinoline.

After the appearance of Lindlar catalyst some investigators continued to use and prefer pyridine- or quinoline-deactivated palladium catalysts. In some instances comparisons were made with Lindlar catalyst and reasons were stated for the author's preference. Cram and Allinger [93] reported that palladium on barium sulfate with a trace of added quinoline had received extensive use in their laboratory and was far superior to Lindlar catalyst in ease of preparation and reproducibility. They added that hydrogenation with their catalyst was highly specific for acetylenes giving cleanly *cis*-olefins. They converted cyclododeca-1,7-diyne and the dimethyl ester of deca-5-yn-1,10-dioic acid to the corresponding *cis*-olefins in very high yield. Other acetylenic compounds were similarly reduced. Biel and Di Pierro [94], citing Cram and Allinger, used 5% palladium on barium carbonate to obtain a number of *cis* 1,5-*bis*-substituted amino alkenes from the corresponding alkynes. They noted that hydrogen absorption stopped abruptly after uptake of one molar equivalent. It should be noted that a small amount of catalyst was used, 0.1 g to 21 g of substrate.

Newman [96] hydrogenated 1,1-diethoxy-2-hexyn-5-ol to the corresponding hexenol in 80% yield in the presence of palladium on carbon. He also used Lindlar catalyst and stated that both catalysts behaved erratically leading in some instances to 2-ethoxy-6-methyl-5,6-dihydro-2H-pyran, which he suggested was due to traces of acid somewhere in the system (Eq. 8.17).

He noted that the desired hexenol was consistently obtained when a trace of quinoline was added to the Lindlar-catalyzed reduction of the acetylenic

$$CH_3CHOHCH_2C{\equiv}CCH(OC_2H_5)_2 \xrightarrow{H_2} CH_3CHOHCH_2CH{=}CHCH(OC_2H_5)_2$$

$$\xrightarrow[H_2]{H^+}$$

[structure: 2-ethoxy-6-methyl-5,6-dihydro-2H-pyran with H_3C and OC_2H_5 substituents on the O-containing ring]

(8.17)

106   *Acetylenes*

compound. Newman was of the opinion that palladium on carbon, similarly deactivated, would perform as well but did not attempt to prove his statement by experimentation. It is more than likely that the real function of quinoline in this instance was to counteract the effect of traces of acid.

Fukuda and Kusama [96] studied the effect of the addition of lead acetate, quinoline, pyridine, and piperidine to palladium on calcium carbonate in the partial hydrogenation of 2-butyn-1,4-diol. They obtained selectivity with both the lead- and quinoline-treated catalyst, but not with the other deactivated catalysts.

Skowronski and colleagues [97], influenced by Fukuda and Kusama, attempted reductions of some mono- and diacetylenic quinols in tetrahydrofuran with palladium on calcium carbonate deactivated with quinoline. Their results made it appear to them that treatment of palladium catalyst with lead was unnecessary. The amount of catalyst, 50–100 g per mole of compound of average molecular weight of 300, may appear excessive but it contained only 0.5% palladium against the usual palladium content, 5%. In this instance the ratio was actually low and could have been responsible in part for selectivity. However, the authors emphasized the importance of the presence of quinoline, 35 cc per mole of substrate. They noted that in its absence there was no break in the curve after absorption of one equivalent of hydrogen and that the reduction proceeded to saturation. On the other hand, uptake came to a complete stop at one molar equivalent using the quinoline-treated catalyst (Eq. 8.18).

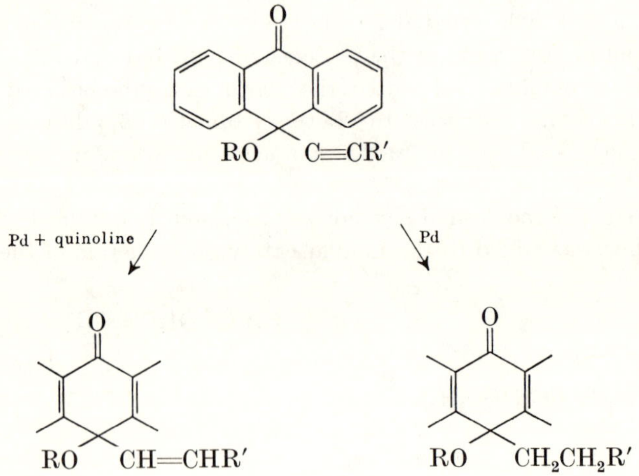

$$R = H \text{ or } CH_3$$
$$R' = H \text{ or } C_6H_5$$

(8.18)

The following example describes a successful semihydrogenation employing a pyridine-deactivated catalyst [98]. The investigators carried out reduction of the acetylenic compound in 10% pyridine in dimethylformamide to obtain a 90% yield of ethylenic alcohol (Eq. 8.19).

$$\text{acetylenic compound} \xrightarrow[\text{Pd on CaCO}_3]{H_2} \text{ethylenic alcohol}$$

(8.19)

It may be desirable at this point to focus attention on the role of deactivators in the palladium-catalyzed reduction of acetylenes. Many references note, in semihydrogenations with Lindlar-type catalysts, that the reaction must be interrupted after the required amount of hydrogen is absorbed. On this basis it may be assumed that the treated catalyst does not completely inhibit the second stage of reduction, saturation of the olefinic bond. The function of the metal additive apparently is to poison the catalyst sufficiently so that hydrogenation will not go beyond the first stage. Too large an amount of metal can poison reduction completely [86] or in part [87]. Fukuda and Kusama [96] found that pyridine and piperidine had no effect and quinoline was only slightly retarding toward the first stage of the palladium-catalyzed reduction of 2-butyn-1,4-diol. However, attack on the resultant butenediol was slowed down somewhat by pyridine, more markedly by piperidine, and almost completely by the quinoline-treated catalyst.

It is interesting to speculate on the effect of quinoline in the hydrogenations with palladium catalyst. The citation of Chase and Galender [79], that quinoline-treated palladium catalyst could be reused in subsequent reductions, bolsters the suggestion of Fukuda and Kusama that its presence has little effect on the acetylenic linkage. Its retarding effect is due, in all probability, to the fact that it is more strongly bonded to the active sites of the catalyst than the olefin. The stronger bonding power stems from a combination of forces, namely, the effect of the aromatic pi field and that of the unshared electron pair of the nitrogen atom. It is possible that the actual retardant is 1,2,3,4-tetrahydroquinoline from reduction of quinoline during the reaction. Tetrahydroquinoline, because of its greater basicity, should exert an even stronger adsorptive effect than quinoline and should completely inhibit adsorption by the C═C bond.

The stronger inhibitory effect of a more basic nitrogen compound was shown by Fukuda and Kusama [96] in the more pronounced retarding effect of piperidine over pyridine on the second stage of the reduction of acetylenes and by Freifelder [99] who demonstrated the effect resulting from differences in basicity of the two retardants on a carbon to carbon double bond in the platinum-catalyzed reductions of cyclohexene, which was completely poisoned by piperidine and far less retarded, although eventually poisoned, by pyridine. It would be of interest to study the effect of 1,2,3,4-tetrahydroquinoline on a number of catalysts in the semihydrogenation of acetylenes.

Tedeschi and Clark [100] claimed that use of the method of Hennion [51] or reduction with Lindlar catalyst was unsatisfactory in the semihydrogenation of some tertiary acetylenic carbinols of the type RR′C(OH)C≡CH, where R = CH_3 and R′ = CH_3, C_2H_5, i-C_4H_9, C_6H_{13}, and C_6H_5 or where RR′C = 1-cyclohexyl. They pointed out that the products of reduction were always contaminated with acetylenic and saturated compounds. However, in similar reductions with 5% palladium on a variety of carriers, the addition of a small amount of potassium hydroxide, in a base to catalyst ratio of 1:1 or 2:1 per 0.15–0.3 g of catalyst per mole of substrate, caused the reactions in most instances to stop selectively after adsorption of one equivalent of hydrogen. Products of high purity were obtained in good yield.

Addition of the base to hydrogenations with palladium on carbon did not halt selectively but reactions could be interrupted to give vinyl carbinols of 96–97% purity in 77–92% yields. On the other hand the use of the base with platinum on carbon or rhodium on carbon did cause cessation of hydrogen uptake after absorption of one equivalent.

The completely selective reduction of 3-methyl-1-butyn-3-ol could not be effected by the addition of potassium hydroxide to either palladium or platinum catalysts. The addition of sodium hydroxide, which has a far greater effect on the reaction rate in reductions of this type with palladium or platinum catalyst [42], also failed to achieve selectivity, but the reduction could be interrupted manually with good results. For this particular reduction the use of 5% rhodium on carbon and an equal amount of sodium methoxide resulted in selective semihydrogenation. However, the rhodium-sodium methoxide system was not studied further.

All hydrogenations in the absence of alkali hydroxide gave vinyl carbinols of inferior purity. Other basic materials, sodium carbonate, triethylamine and quinoline, were ineffective.

The authors suggested that the alkali metal hydroxide exerted its effect through formation of a 1:1 complex with the acetylenic carbinol [100]. They noted that a 1:1 complex was formed quantitatively in reaction between finely divided potassium hydroxide and 3-methyl-1-butyn-3-ol in dry inert solvent. The solid complex, when used as a substitute for potassium hydroxide in the same concentration employed for semihydrogenation, showed the same selectivity in the reduction of 3-methyl-1-pentyn-3-ol and 1-ethynylcyclohexan-1-ol giving products of high purity in good yield.

Potassium metal was responsible for selectivity in the partial reduction of 3-methyl-1-pentyn-3-ol with palladium on carbon [100]. The rate, however, was about one-third of that noted in reductions in the presence of potassium hydroxide. The inhibitor is probably the potassium acetylide or the potassium tertiary alkoxide derivative or a combination of both. These species are believed to account for the much slower hydrogenation rate compared to the use of potassium hydroxide. The catalytic semihydrogenation of 4,7-dimethyl-5-decyn-4,7-diol aided by the presence of potassium (0.044 g per mole of diol) halted selectively at the end of the first stage [42].

The original intent in using alkali hydroxides was to prevent serious hydrogenolysis side reactions in the reduction of acetylenic 1,4-diols with palladium on carbon catalyst [42]. Complete reduction of these alkyndiols gives saturated diols in low yields (9–33%); the major products are saturated mono alcohol and saturated hydrocarbon. However, when bases such as potassium or sodium hydroxides, potassium metal, or triethylamine (0.025–0.1 g/mole of diol) were used, complete elimination of hydrogenolysis was realized. Hydrogenolysis was shown to arise mainly from the intermediate olefinic diol and was not due to traces of acidity in the catalyst support [42].

Alkali hydroxide appears to have a further use in the semihydrogenation of acetylenic glycols. In general, partial reduction of disubstituted acetylenes give excellent yields of *cis* olefins [101]. However, the reactive acetylenic-1,4-diols and particularly 2,5-dimethyl-3-hexyn-2,5-diol appear to be exceptions [102]. Tedeschi [103] has shown that it is possible to obtain *cis*-olefinic diols of 96–99% purities in high yields by reducing acetylenic-1,4-diols with palladium on carbon at 15–20° in the presence of small amounts of potassium or sodium hydroxides. The intermediate base-diol complex is believed to be responsible for stereochemical control of the reduction as well as elimination of hydrogenolysis.

USE OF IRON CATALYST

Paul and Hilly [104] described the preparation of Raney iron and in a later publication [105] reported that terminal acetylenes, RC≡CH where R = C_5H_{11}, C_6H_{13}, or C_6H_5, were converted to the corresponding olefins at 100–110° and 50 atm in the presence of that catalyst. Other acetylenes with internal triple bonds were also reduced selectively. They stated that the

110 Acetylenes

catalyst was unable to effect the reduction of the olefins. Thompson and Wyatt [106] tried the same catalyst in the reduction of 1,1-diphenylacetylene at 100° and 1000 psig but obtained only the saturated compound. However, they were successful in hydrogenating 2,5-dimethyl-3-hexyn-2,5-diol to the olefinic glycol in near quantitative yield but required more vigorous conditions, 150° and 1400 psig pressure. Elsner and Paul [23] found that Raney iron did not catalyze the hydrogenation of octadec-9-yne at atmospheric pressure; at 110° and 50 atm reduction was not selective, giving octadecane. A patent disclosed that compounds of the type shown in Eq. 8.20 can be selectively hydrogenated with Raney iron at 100° and 200 atm pressure [107].

$$C_6H_{11}C(C_6H_5)(OH)C{\equiv}CCHRNR'R'' \xrightarrow{H_2}$$

$$C_6H_{11}C(C_6H_5)(OH)CH{=}CHCHRNR'R''$$

R = H or alkyl
R' = H, alkyl or cycloalkyl
R" = alkyl or cycloalkyl
NR'R" = $N(CH_2)_n$, where n = 4, 5, or 6.   (8.20)

In more recent years there has been little mention of the use of iron catalysts for acetylene reductions, except for some work by Tairi [108] with Urushibara iron catalysts which have been reported to give good selectivity in the conversion of 2-butyn-1,4-diol to the olefin. The preparation of the catalysts is comparatively simple. All contain varying amounts of zinc; in a few cases there is a very small amount of nickel or cobalt present. The ratio of catalyst to compound is high, about 50% based on iron content; hydrogenations are carried out at 100° and 50 atm pressure. It is difficult to evaluate the worth of these catalysts without more experimentation on different acetylenic compounds.

SELECTIVE HYDROGENATIONS IN THE PRESENCE OF OTHER REDUCIBLE GROUPS

Reduction in the Presence of Another Acetylenic Bond

When diacetylenes or polyacetylenes are to be hydrogenated selectively, Lindlar catalyst or Lindlar catalyst deactivated with quinoline is generally the catalyst of choice although other deactivated palladium catalysts have been employed. In general, if the diacetylenes are symmetrical as in diethyl nona-2,7-diyn-1,9-dioate, the corresponding diene is obtained [109]. Another example is found in the reduction of cyclododeca-1,7-diyne to cyclodeca-1,7-diene with quinoline-inactivated palladium on barium sulfate [94]. In di- and polyacetylenes, where primary carbon atoms are adjacent to the acetylenic linkages, di- and polyenes are obtained. This is illustrated in the reductions of tetradeca-5,8-diyn-1-ol[110], nonadeca-6,9,12,15-tetraynoic acid [111], and octadeca-9,12,15-triynoic acid [112].

Conjugated diacetylenes usually follow the same pattern unless a carbon atom adjacent to one of the acetylene linkages is highly substituted. Crombie [53] converted methyl deca-2,4-diynoate to the conjugated dienoate in 65% yield. The work of Audier, Dupont, and Dulou [113] is of special interest because it shows a difference in results obtained from deactivated and untreated palladium catalysts. 2,7-Dimethylocta-3,5-diyn-2,7-diol was hydrogenated with Lindlar catalyst inactivated with quinoline to obtain 70% yield of the corresponding conjugated dienediol. However, with palladium on barium sulfate, reduction was not selective. After interruption of the reaction at absorption of two equivalents of hydrogen, recovered starting material was obtained along with a mixture of two other components. One was the expected diolefinic diol. The other was identified as the *trans* form of 2,7-dimethyloct-4-en-2,7-diol (Eq. 8.21). The formation

$$(CH_3)_2C(OH)C{\equiv}CC{\equiv}CC(OH)(CH_3)_2 \begin{cases} \xrightarrow[\text{Lindlar catalyst}]{2H_2} (CH_3)_2C(OH)CH{=}CHCH{=}CHC(OH)(CH_3)_2 \\ \\ \xrightarrow{\text{Pd on BaSO}_4} \textit{trans-}(CH_3)_2C(OH)CH_2CH{=}CHCH_2C(OH)(CH_3)_2 \end{cases} \quad (8.21)$$

of the latter product apparently resulted from reduction of the *trans* form of the conjugated diene by 1,4-addition of hydrogen. The formation of *cis* and *trans* olefins from reduction of acetylenes and the difference in reducibility will be discussed in another section.

Dobson and his group [114] were able to show the stepwise reduction of a diacetylene containing a terminal and an internal triple bond. The results of separate experiments in the hydrogenation of undeca-1,7-diyne with palladium on carbon catalyst, where the reductions were stopped at uptake of one, two, and three molar equivalents of hydrogen, were clear-cut. The products were respectively undec-1-en-7-yne, undeca-1,7-diene, and undec-4-ene. Each experiment gave essentially one product (Eq. 8.22). Each of the

$$C_3H_7C{\equiv}C(CH_2)_4C{\equiv}CH \begin{cases} \xrightarrow{H_2} C_3H_7C{\equiv}C(CH_2)_4CH{=}CH_2 \\ \\ \xrightarrow{2H_2} C_3H_7CH{=}CH(CH_2)_4CH{=}CH_2 \\ \\ \xrightarrow{3H_2} C_3H_7CH{=}CH(CH_2)_5CH_3 \end{cases} \quad (8.22)$$

112   Acetylenes

reduction products was identified by various physical and chemical methods but not by gas-liquid chromatography. The precedence of reduction was established as terminal C≡C, internal C≡C, and terminal double bond. Dobson and his group were able to substantiate the preferential hydrogenation of the terminal triple bond over the internal triple bond by reducing an equimolar mixture of 1- and 4-octynes. After uptake of 50% of one equivalent of hydrogen, the products were 1-octene and unchanged 4-octyne. When one molar equivalent was absorbed equal amounts of 1- and 4-octene were obtained. The results of the two reductions were determined by gas-liquid chromatography. The latter result may also be considered an illustration of the selective reduction of an acetylene bond in the presence of an olefinic bond. This point will be discussed at greater length in the next section.

The above reductions were carried with a 5% weight ratio of catalyst to substrate. When palladium on calcium carbonate treated with quinoline was used, the rate of hydrogen uptake slowed down appreciably but did not stop at $2H_2$. Quinoline-treated Lindlar catalyst caused the reduction to come to a halt at this point.

Petrov and Forost [115] also noted the difference in reduction between a terminal and what they chose to call a disubstituted acetylenic linkage. They found that the terminal bonds of hepta-1,5-diyne and octa-1,5-diyne were selectively converted to hept-1-en-5-yne and oct-1-en-5-yne in the presence of palladium on calcium carbonate. The influence of substituents on selectivity between the two internal bonds has been shown by Zakharova and Ilína [116]. They found that absorption of hydrogen stopped at two molar equivalents in the reduction of 4,4,7,7-tetramethylocta-2,5-dyne with palladium on calcium carbonate. They identified the product by chemical and physical means, including nuclear magnetic resonance spectroscopy, as 2,2,5,5-tetramethyloct-3-yne (Eq. 8.23).

$$(CH_3)_3CC{\equiv}CC(CH_3)_2C{\equiv}CCH_3 \quad \begin{array}{l} \xrightarrow{2H_2} (CH_3)_3CCH{=}CHC(CH_3)_2CH{=}CHCH_3 \\ \\ \xrightarrow{2H_2} (CH_3)_3CC{\equiv}CC(CH_3)_2C_3H_7 \end{array} \qquad (8.23)$$

A framework molecular orbital model [117] of the diyne clearly shows that the difference in reducibility between the two acetylene bonds is due to the effect of the terminal tertiary butyl group. It prevents contact between the bond to which it is attached and the catalyst surface. The terminal methyl group attached to the other triple bond, on the other hand, has no blocking effect. A model of the intermediate, 2,2,5,5-tetramethyl-3-yn-6-ene, where the less hindered acetylenic bond is converted to the olefin shows that the acetylenic bond is well removed from the catalyst surface while the double bond can make contact with it and should be readily reduced as was reported. It is a clear case of steric hindrance preventing reduction of one of the acetylenic bonds.

Reductions in the Presence of Olefinic Bonds

As in the semihydrogenation of di- and polyacetylenes, Lindlar catalyst deactivated with quinoline finds the greatest use in selective low-pressure reductions of acetylenes which also contain an olefinic bond. A sulfur-treated Lindlar catalyst has been found useful [82]. Base-deactivated as well as untreated palladium catalysts also find application. Raney iron at elevated temperature and pressure has been reported to give selectivity because it has little activity for reduction of double bonds [106].

Selective reduction of an acetylenic bond over an ethylenic bond is favored because of its preferential adsorption to the active portions of the catalyst [22]. Dobson and his colleagues [114] showed that the order of reduction in an unsubstituted nonconjugated diacetylenic system was terminal acetylene, internal acetylene, and finally terminal double bond. Based on their findings it may be assumed that semihydrogenation of a nonconjugated enyne will lead to the corresponding diene, except if the triple bond is so surrounded by substituents that it cannot make contact with the surface of the catalyst. The previously mentioned work of Zakharova and Il'ina [116] is an illustration of this point.

Shackleford [118] suggests that while there is good selectivity in the reduction of an acetylenic compound which also contains a nonconjugated double bond, it may not be as complete as one would hope. In order to get over 99% purity in the reduction of hex-1-yn-4-en-3-ol it was necessary to use 5% palladium on calcium carbonate that was previously deactivated by the addition of 40% by weight of lead acetate. Without the additive, reduction gave material which was 96–98% pure. Palladium on carbon was only 92% selective. The heavily lead-treated catalyst gave very pure hexa-1,4-dien-3-ol in 94% yield.

It is possible in systems where the acetylenic bond is not conjugated with an olefinic system that greater selectivity could be achieved if strict control of the reaction temperature as well as hydrogen consumption is maintained. This is illustrated in the hydrogenation of 3-methylocta-4,6-dien-1-yn-3-ol with palladium on calcium carbonate which was held at 22° by periodic cooling [119]. Interruption at the proper time gave 95% yield of 3-methyl-octa-1,4,6-trien-1-ol distilling over a narrow range.

Semihydrogenation of the acetylenic bond in a conjugated enyne system may not be completely selective because the resultant conjugated diene may be almost as firmly bound to the active sites of the catalyst as the triple bond. In addition, selectivity can be affected by substituents at any part of the unsaturated linkages.

Theoretically, hydrogenation of a conjugated enyne can proceed (a) by addition of hydrogen across the acetylenic bond to yield a conjugated diene, (b) by 1,4-addition and rearrangement to form an allene, or (c) by preferential attack on the double bond to yield an acetylene. Zonis [120] and Zal'kind and Khudekova [121] reported that semihydrogenation of substituted vinylacetylenes with palladium catalyst gave the corresponding conjugated dienes.

Crombie, Harper, and Newman [122] stated that, although there was a tendency toward overhydrogenation, their reductions of oct-7-en-5-yn-2-one and hept-6-en-4-ynoic in the presence of deactivated Lindlar catalyst showed that the acetylenic bond in each instance was converted before the terminal olefinic bond was reduced. The corresponding respective conjugated dienes obtained were contaminated with oct-5-en-2-one and hept-4-ynoic acid from overhydrogenation of the terminal double bond of the resultant dienes.

Balyan and associates [123] noted in palladium reductions of conjugated enynes of the type, $RC{\equiv}CCH{=}CH_2$ (R = alkyl), that addition of hydrogen took place first at the triple bond. They stated however that there was no break in the rate until two molar equivalents of hydrogen were absorbed. When compounds of the type, $HC{\equiv}CCH{=}CHR$ or $HC{\equiv}CC(R){=}CH_2$ (R = alkyl), were hydrogenated there was a break in the curve at one equivalent and reduction was reasonably selective giving 90% conjugated diene and 10% olefin.

In later work Balyan [124] reported that greater selectivity was obtained in the palladium-catalyzed hydrogenations of the conjugated enynes, hept-1-en-3-yne and oct-1-en-3-yne, by addition of 4-chlorophenylthiocynate to the reaction mixture. However, the poor material balance (53% of diene, 20–25% of unreacted enyne, and no mention of the remainder) leaves much to be desired. There do not appear to be any references to formation of allenes from catalytic reduction of conjugated enynes.

In the attempted hydrogenation of an internal triple bond in conjugation with a terminal double bond, that of 5,6-dimethyloct-1-en-3-yn-5,6-diol, the investigators reported they obtained 5,6-dimethyloct-3-yn-5,6-diol resulting from attack at the terminal double bond [125] (Eq. 8.24). Infrared examination showed the presence of conjugated diene. However, the authors pointed out that the weak intensity of the bands in the spectrum and other tests convinced them that the reduction product was the acetylenic glycol.

Selective Hydrogenations 115

$$\text{C}_2\text{H}_5\text{C(OH)(CH}_3\text{)-}\text{C(OH)(CH}_3\text{)C}\equiv\text{CCH}=\text{CH}_2 \begin{cases} \xrightarrow{\cancel{H}} \text{C}_2\text{H}_5\text{C(OH)(CH}_3\text{)C(OH)(CH}_3\text{)CH}=\text{CHCH}=\text{CH}_2 \\ \\ \xrightarrow{\text{H}_2} \text{C}_2\text{H}_5\text{C(OH)(CH}_3\text{)C(OH)(CH}_3\text{)C}\equiv\text{CC}_2\text{H}_5 \end{cases} \quad (8.24)$$

A molecular model of the starting enyndiol (Fig. 8.5) shows that the 5-methyl and 5-hydroxy groups on the carbon atom adjacent to the triple bond do not prevent the bond from making contact with the catalyst and

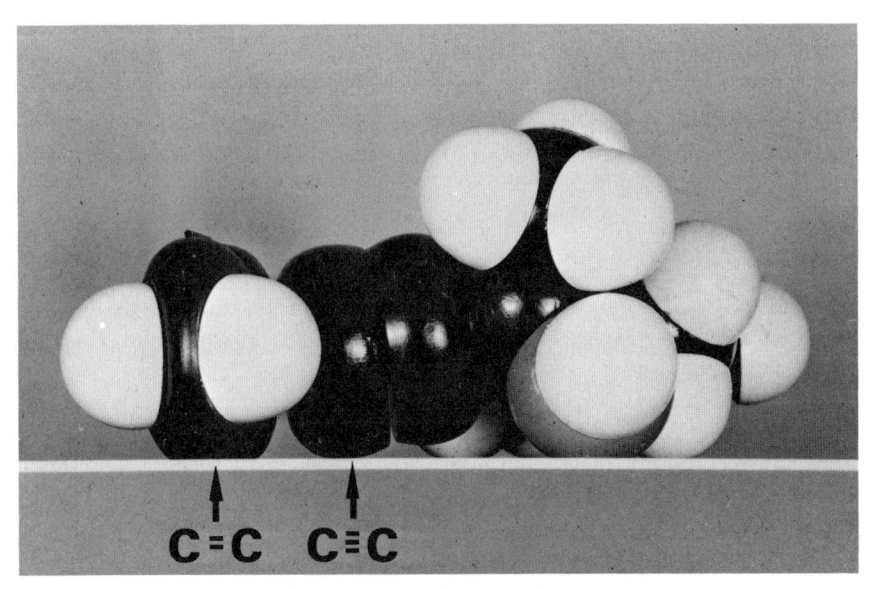

Fig. 8.5 5,6-Dimethyloct-1-en-3-yn-5,6-diol.

indicates that it as well as the terminal double bond should be subject to reduction. Based on the preferential adsorption of the acetylenic linkage [22] it would be assumed that the triple bond would be attacked first. The finding of the authors [125] instead of the normally expected diene is of considerable interest. It is unfortunate that they did not offer stricter proof of purity and structure of the reduction product by subjecting it to examination by gas-liquid chromatography and nuclear magnetic resonance spectroscopy.

One of the problems in selective hydrogenation, particularly in semi-hydrogenation of conjugated enynes, is proper identification of the products of reduction. Marvell and Tashiro [126] brought out a very pertinent and important point when they stated that in too many instances quantitative data on the products of reduction were not supplied. Gas-liquid chromatography and other modern techniques now allow investigators in the field to determine accurately the amounts of the various components obtained from attempted selective reductions and to identify those components.

Marvell and Tashiro reported that hydrogenation of some acetylenes of the type, $C_6H_9C{\equiv}CR$ where C_6H_9 was 1-cyclohexenyl and R is hydrogen, methyl, or ethyl, were not especially selective in the presence of Lindlar catalyst deactivated with quinoline. It was necessary to interrupt these and other reductions after uptake of one molar equivalent of hydrogen because there was no break in the curve. Analysis of the reduction products by gas chromatography showed mixtures of unreacted starting material and alkylcyclohexenes (10–22%). In the reduction where R was a vinyl group 54% of the conjugated triene was obtained in addition to 8% of unreacted material, 25% of overhydrogenated side chain products, 5% of unidentified material, and 8% of 1-(1-butinyl)cyclohexene. The last named compound may be an example that selective hydrogenation of a terminal olefin bond in a conjugated enyne system, as claimed by Nogaideli and Rtveliashvili [125], is possible.

When R was 1-cyclohexenyl the reduction was also not particularly selective. Starting material and overhydrogenation products were obtained in addition to the desired conjugated triene. Schlatmann and Havinga [127] reported similar difficulties in the attempted reduction of 1-(2-methylcyclohexanyl)-2-cyclohexenylacetylene.

The study of Marvel and Tashiro showed that the cyclic system was not affected except in one instance where R was cyclohexenyl. There they found that in one of the overhydrogenated products one of the cyclic double bonds was reduced.

It is noteworthy, in spite of the difficulties sometimes encountered in selective hydrogenation of conjugated enynes, that semihydrogenation of an internal acetylenic bond in conjugation with a long multiple double-bonded system is reported to take place in very high yield. This is well illustrated in the carotinoid work of Isler and associates [128], of Akhtar and Weedon [129], of Schwieter, Rüegg, and Isler [130], and in the reductions of long-chain aliphatic conjugated enynes by Bohlmann and Mannhardt [131].

Reductions in the Presence of Aldehyde and Ketone Groupings

In general, with the proper catalyst, conversion of triple bond to a double bond should take place selectively in hydrogenations of acetylenes containing aldehyde or ketone functions. An example of selectivity in the presence of an isolated carbonyl group is seen in the palladium-catalyzed reduction of 1-ethinyl-3,7a-dimethyl-5-ketoperhydroindan-1-ol to the corresponding 1-vinyl-5-keto compound [132] (Eq. 8.25).

$$\text{[structure: 1-ethinyl-3,7a-dimethyl-5-ketoperhydroindan-1-ol]} \xrightarrow{H_2} \text{[structure: 1-vinyl-5-keto compound]} \quad (8.25)$$

An excellent example of selectivity in the presence of a conjugated carbonyl system, not conjugated with the triple bond, is seen in the reduction of 1-ethinyl-3,7a-dimethyl-1,2,3,6,7,7a-hexahydroindan-3a,4-en-5-one-1-ol to the 1-vinyl compound without attack at the conjugated carbonyl system [133]. Similar examples can be seen in the palladium on carbon reduction of ethinyltestosterone to vinyl testosterone in 95% yield [88] and in the reduction of 19-nor 17α-ethinyltestosterone with palladium on calcium carbonate to the 17α-vinyl compound [134]; in each instance the 3-keto-Δ^4-ene portion of the molecule remained intact.

The hydrogenation of 1-(1-cyclohexyl)but-1-yn-3-one with palladium on calcium carbonate to 1-(1-cyclohexyl)but-1-en-3-one in high yield is an example of selectivity where the carbonyl group is part of a conjugated enynone system [135]. However, 1-(1-hydroxycyclohexyl)but-1-yn-3-one did not undergo reduction with either palladium or platinum catalyst. The compound was shown not to be a poison since the dehydration product from it, 1-(cyclohex-1-enyl)but-1-yn-3-one, was readily hydrogenated in its presence. Nonreducibility is apparently due to the effect of the hydroxyl group. A comparison of molecular models shows that the acetylenic linkage of the hydroxy compound cannot make contact with the catalyst while the triple bond of the dehydration product is unhindered.

In a conjugated yn-one system where the triple bond was terminal,

$$\text{R}-\overset{\overset{\text{O}}{\|}}{\text{C}}\text{C}\equiv\text{CH}$$

(R = 3-cyclohexenyl or 2- or 4-methyl-3-cyclohexenyl), reasonably good yields (57–65%) of the corresponding pure vinyl ketones were obtained after reduction with palladium on calcium carbonate [136].

Pure *trans*-3-methylpenta-2,4-dienal was obtained from the reduction of *trans*-3-methylpent-2-en-4-ynal in the presence of a partially poisoned palladium catalyst further deactivated with quinoline [137].

When overhydrogenation occurs in the reduction of acetylenes containing various conjugated or nonconjugated aldehyde or ketone groups, it usually involves further attack on the resultant double bonds. Edgar, Harper, and Kazi [138] noted that the reduction of 3-methyl-2-(penta-*cis*-2-en-4-ynyl)-cyclohex-2-enone with Lindlar catalyst was not fully selective although in the main the desired *cis*-2,4-diene was obtained. In the hydrogenation of the *trans* compound with palladium on calcium carbonate, the ultraviolet spectrum of the reduction product showed low conjugated diene content.

Reduction in the Presence of Halogen

The reduction of the triple bond in the presence of a chlorine atom may be expected to be selective and lead to the corresponding ethylenic compound without loss of halogen if the halogen is not attached to one of the acetylenic carbon atoms. 1-Chlorononadeca-4,7-10,13-tetrayne was converted to 1-chlorononadeca-4,7,10,13-tetraene by two groups of investigators. One used Lindlar catalyst to obtain 69% yield [139], the other also used Lindlar catalyst but deactivated with quinoline to obtain an 80% yield [140]. Olomucki [141] reduced 6-chlorohex-2-ynoic, 7-chlorohept-2-ynoic, and 8-chlorooct-2-ynoic acids to the corresponding chloroenoic acids in high yield with palladium on alumina. They were able to show that the chlorine atom was not affected when they reduced the chlorine-containing acetylenes to the saturated chloro acids with the same catalyst.

1-(3-Chloro-4-methoxyphenyl)prop-1-yne was reduced with palladium on calcium carbonate and pyridine as a solvent [142]. Gas-liquid chromatography showed 12.6% of unreacted material, 58.6% of the expected propene, and 24.9% of another component which was believed to be the corresponding chlorophenylpropane. Another possibility is a dehalogenation product since the reduction was carried out in pyridine. The effect of base on loss of halogen will be examined more thoroughly in the section on dehalogenation.

McLamore, P'an, and Bavley [143] reduced 1-chloro-3-ethylpent-1-en-4-yn-3-ol to the chloropentadienol without apparent loss of halogen. Mechanical loss after distillation through a rather long-packed distilling column may be largely responsible for the modest yield, 51.2%.

The same authors prepared bromo- and iodoacetylenic carbinols but did not attempt reduction. Eğe [140] also prepared a bromine-containing acetylene, 1-bromoundeca-2,4-diyne, but did not subject it to hydrogenation. It is unfortunate that these investigators did not study the effect of the larger halogens.

Dobson [114] found that bromine at the end of an acetylene chain could act as a means of almost completely blocking reduction of the linkage to which it was attached with little loss of bromine. When 1-bromoocta-1,7-diyne and 1-bromoundeca-1,7-diyne were hydrogenated with palladium on carbon, two molar equivalents were rapidly absorbed to give 1-bromooct-1-yne and 1-bromoundec-1-yne respectively. When the reduction of 1-bromoundeca-1,7-diyne was stopped at one molar equivalent, 89% of impure

1-bromoundec-7-en-1-yne was obtained. The authors suggested, from analytical data and the infrared spectrum of the reduction product, that it was contaminated with about 10% of undec-1-en-7-yne. They did not indicate an order of reduction where hydrogenolysis occurred but it could be assumed that the following competing side reaction took place yielding the contaminant:

$$\text{BrC} \equiv \text{C(CH}_2)_4\text{C} \equiv \text{CC}_3\text{H}_7 \xrightarrow[\text{H}_2]{-\text{Br}} \text{HC} \equiv \text{C(CH}_2)_4\text{C} \equiv \text{CC}_3\text{H}_7 \xrightarrow{\text{H}_2} \text{CH}_2 = \text{CH(CH}_2)_4\text{C} \equiv \text{CC}_3\text{H}_7$$

When they hydrogenated 1-chloroundeca-1,7-diyne, hydrogenolysis occurred but the extent of it and the products of reduction were not investigated.

1-Bromohept-1-yne was used as a model to study the difficulty in reducing this blocked 1-yne system. Uptake of hydrogen was found to be slow and a mixture of products resulted. A molecular model of the bromoheptyne offers no steric explanation for the difficulty in hydrogenation. One position shows the triple bond and the bromine atom in the same plane, indicating that either can be subject to attack by hydrogen. The model can also be rotated so that the triple bond is raised from the surface of the catalyst. These points might explain the lack of selectivity. It is also possible that release of hydrogen bromide caused poisoning. However, since there was no attempt to identify the product or products of reduction, the effect of the bromine atom in this system must remain a matter for speculation and further study.

Reduction in the Presence of Nitro Groups

An example of the selective reduction of the triple bond in the presence of an aliphatic nitro group is described by Grob and Jenny [144] who reduced 2-nitrooctadec-4-yn-1,3-diol with Lindlar catalyst to obtain 2-nitrooctadec-4-en-1,3-diol. However, little is known about hydrogenation of acetylenes containing an aromatic nitro group. Each function is readily reduced under mild conditions; the triple bond and, as will be shown later, the aromatic nitro group are both reduced preferentially in the presence of an olefinic bond. A study of the hydrogenation of compounds containing a triple bond and an aromatic nitro group should be one worthy of investigation.

STEREOCHEMISTRY RESULTING FROM REDUCTION

It has been generally assumed that catalytic hydrogenation of acetylenes at room temperature and low pressure leads exclusively to *cis* olefins [7, 101].

Burwell [8] has pointed out that earlier results were too often dependent on isolation techniques and that analyses employing more modern methods have shown the presence of 5–20% of *trans* olefin.

Too often the presence or absence of *trans* isomer was based on the absence or presence of particular bands in the infrared spectrum of the reduction product. More accurate methods would involve nuclear magnetic resonance and gas-liquid chromatography. For example, gas-liquid chromatography, a more quantitative procedure than infrared examination, showed that the hydrogenation of deca-5,9-diyn-1-ol with Lindlar catalyst deactivated with quinoline yielded a 4:1 *cis, trans* mixture of deca-5,9-dien-1-ol [145].

A number of factors may influence the formation of *cis* and *trans* isomers. Stereoselectivity may be dependent on the catalyst, the support, on the preparation of the catalyst, on acid or base effect, temperature, possibly on the speed of reaction, and even on the structure of the acetylenic compound.

Alkali-free Raney nickel [37], Raney nickel deactivated with piperidine and zinc acetate [43], and supported palladium catalysts deactivated with quinoline [68, 90, 146] have been reported useful for stereoselective reductions [126]. Others have indicated that only Lindlar catalyst is substantially stereoselective for the preparation of *cis* ethylenes from acetylenes [111, 114, 131, 147–149].

Cram and Allinger [93] consider palladium on barium sulfate treated with a small amount of quinoline a superior and more selective catalyst than Lindlar catalyst. They stated that reductions with it gave cleanly *cis* products in the reductions they attempted. On the other hand Dobson [114] found that, while reduction of undec-4-yne with Lindlar catalyst gave *cis* undec-4-ene, the use of palladium on barium sulfate as well as palladium on other supports gave large amounts of *trans* isomer. Loev and Dawson [146] also noted a striking difference in the reduction of 1-chlorotetradec-7-yne with Lindlar catalyst which gave essentially *cis* 1-chlorotetradec-7-ene; with palladium on carbon or on calcium carbonate the olefin showed intense absorption in its infrared spectrum indicative of considerable amounts of the *trans* form.

Dobson suggested that the presence of large amounts of *trans* compound resulted from stereomutation of the initially formed *cis* undec-4-ene. It has been shown that such stereomutation takes place only in an atmosphere of hydrogen [150]. Dobson did show that the *cis* olefin could be converted to about 70% *trans* within 1 hr when the amount of catalyst used was too small to induce uptake of hydrogen.

We can never be sure that trace amounts of impurities (sometimes acidic ones) from the preparation of the catalyst on different supports may not be responsible for isomerism. The effect of acid was shown by Dobson's group in a comparison of the reduction of undec-4-yne with palladium on carbon which gave 32% *trans* ene; in the reduction to which a small amount of acetic acid was added the amount of *trans* ene rose to 49%.

To accurately evaluate the effect of different supported catalysts on stereoselectivity, prepared or purchased catalysts should be thoroughly washed to insure removal of any substance which may have an undesirable effect. The chemist must remember that some of the supports are known to have strong adsorptive powers and are used in column chromatography. Therefore, considerable washing may be necessary to insure removal of even small amounts of impurity.

It is possible that stereoselectivity ascribed to quinoline-deactivated Lindlar catalyst by Dobson is due to neutralization of a small amount of acidic material occluded in the supported catalyst although it may actually be an effect of nitrogen base. When quinoline was added to the reduction of undec-4-yne with Lindlar catalyst only 4% of *trans* undec-4-ene was found. Unfortunately a comparative run without quinoline was not carried out. However, an example of the effect of quinoline is seen in the hydrogenation of stearolic acid. Doubling the amount of it added to a previously deactivated Lindlar-catalyzed reduction decreased the *trans* content in *cis* octadec-9-enoic acid from 5% to 1–2% [78].

The effect of a small amount of another amine, triethylamine, was shown in the palladium on carbon hydrogenation of undec-4-yne [114]; 17% of *trans* undecene was obtained against 32% in its absence. Base was not added to a reduction in the presence of palladium on barium sulfate which gave 40% of the *trans* isomer or to the reduction with palladium on calcium carbonate which gave 63% of the same isomer.

The effect of a small amount of potassium or sodium hydroxide on *cis* formation during palladium on carbon reductions of acetylenic-1,4-diols has already been noted. However maintenance of the reaction temperature at 15–20° must also have been responsible for minimizing formation of *trans* isomers in those experiments.

It was suggested that the speed of hydrogenation did not appear to have an effect on formation of *trans* isomer [114]. The lower amount of *trans* undecene (31%) found after hydrogenation of undec-4-yne with 17.4% by weight of palladium on carbon against 68% *trans* isomer from the use of a 10% weight ratio of the same catalyst was attributed to the amount of catalyst. Based on experience that reduction with the larger amount of catalyst should proceed more rapidly than one with less catalyst, it may be assumed that decreasing the reaction time could lower *trans* isomer formation in semihydrogenations of acetylenes.

Isomer formation during hydrogenation of diynes and enynes may give complex mixtures. Undeca-1,7-diyne gave essentially *cis* diene upon reduction with palladium on carbon [114]. Marvel and Tashiro [126] obtained predominantly *cis* dienes in the reduction of some conjugated cyclohexenylacetylenes with Lindlar catalyst, although some saturated side chain compounds were also obtained. They also hydrogenated 1-(1-cyclohexenyl)-but-3-en-1-yne to obtain the *cis*, (47%) and *trans* triene (7%). Reduction was accompanied by a number of other products. Based on separation difficulties the authors concluded that synthetic use of semihydrogenation to prepare trienes with a central *cis* double bond was unsatisfactory. Similar difficulties have been noted by others [127, 151]. Oddly enough in long conjugated systems with multiple bonds not only is reduction selective but *cis* polyenes are also obtained [128–131].

STERIC EFFECTS

Substituents on the acetylene bond can influence reducibility. A terminal triple bond, because it is less hindered than an internal one, is selectively reduced in the presence of an internal or disubstituted triple bond [114]. It is generally assumed, perhaps not unequivocally proven, that the acetylenic bond in vinylacetylenes is selectively reduced [120–122]. Selective hydrogenation of the terminal vinyl groups of 5,6-dimethyloct-1-en-3-yn-5,6-diol reported by Nogaideli and Rtveliashvili [125] in contrast to the expected attack at the acetylene bond may be due to steric effects. Although a molecular model of the enyne shows that the triple bond can be reduced (see Fig. 8.5), the reported result may be due to the difference in the size of the substituents on the double bond and on the triple bond.

The unexpected selectivity noted by Nikitin and Timofeeva [14, 15] in the reduction of a number of acetylenic tertiary polyols with platinum oxide is certainly due to steric effects as shown by a molecular model which is typical of the structure of the resultant polyols (see Fig. 8.1).

An accumulation of aromatic groups hinders or completely inhibits reduction of the acetylenic bond. Wieland and Kloss [152] found that 1,1,1,4,4,4-hexaphenylbut-2-yne was not reduced with palladium or platinum catalyst and that 1,3,3,3-tetraphenylprop-1-yne was hydrogenated very slowly. Zalkind [153] reported that 1,4-diphenyl-1,4-di-α-naphthylbut-2-yn-1,4-diol was hydrogenated more slowly than 1,1,4,4-tetraphenylbut-2-yn-1,4-diol.

Jardine and McQuillen [154] noted some rather interesting results in a study of the effect of increasing substitution in a series of acetylenic diols. They found as substitution increased on each side of the triple bond that the rate of hydrogen uptake increased. The order of hydrogen uptake was 2,5-dimethylhex-3-yn-2,5-diol > hex-3-yn-2,5-diol > pent-2-yn-1,4-diol > but-2-yn-1,4-diol. They reasoned that substitution assisted hydrogenation by reducing chemisorption of the acetylene, which because of its strong bonding to the catalyst excludes adsorption of hydrogen and retards reaction. The effect of an aromatic group was not examined by the investigators.

REFERENCES

[1] W. Sung, *Ann. Chim.*, [10] **1**, 372 (1924).
[2] M. Freifelder, unpublished results.
[3] G. F. Hennion and C. V. Di Giovanna, *J. Org. Chem.*, **30**, 2645 (1965).
[4] G. F. Hennion and C. V. Di Giovanna, *J. Org. Chem.*, **30**, 3696 (1965).
[5] G. F. Hennion and R. S. Hanzel, *J. Am. Chem. Soc.*, **82**, 4908 (1960).
[6] H. P. W. Huggill and J. D. Rose, *J. Chem. Soc.*, **1950**, 335.
[7] K. N. Campbell and B. K. Campbell, *Chem. Rev.*, **31**, 76 (1942), see Chapter IV, Section D.
[8] R. L. Burwell, Jr., *Chem. Rev.*, **57**, 895 (1957).
[9] G. C. Bond, *Catalysis*, Reinhold, New York, 1955, Chapter IV, p. 143.
[10] L. Crombie, *J. Chem. Soc.*, **1955**, 3510.
[11] A. L. Markman, *J. Gen. Chem. (USSR) (Eng. translation)*, **24**, 63 (1954).
[12] a. E. A. Braude and J. A. Coles, *J. Chem. Soc.*, **1951**, 2078; b. Private communication, Dr. R. J. Tedeschi, Airco Chem., Middlesex, New Jersey.
[13] J. D. Chanley, *J. Am. Chem. Soc.*, **71**, 829 (1949).
[14] V. I. Nikitin and I. M. Timofeeva, *Zhur. Obshchei Khim.*, **25**, 1334 (1955); through *Chem. Abstr.*, **50**, 4929 (1956).
[15] V. I. Nikitin and I. M. Timofeeva, *J. Gen. Chem. (USSR)*, **27**, 1880 (1957).
[15a] V. I. Nikitin, E. M. Glazunova, I. M. Putapova, and A. B. Zegelman, *Zhur. Organ. Khim.*, **1**, 2123 (1965); through *Chem. Abstr.*, **64**, 11074 (1966).
[16] P. Karrer and B. H. Ringier, *Helv. Chim. Acta*, **22**, 610 (1939).
[17] P. Karrer and K. S. Yap, *Helv. Chim. Acta*, **23**, 581 (1940).
[18] P. Karrer and K. S. Yap, *Helv. Chim. Acta*, **24**, 639 (1941).
[19] P. Karrer and F. Kehrer, *Helv. Chim. Acta*, **25**, 29 (1942).
[20] P. Karrer, A. Geiger, H. Rentschler, E. Zbinden, and A. Kugler, *Helv. Chim. Acta*, **26**, 1741 (1943).
[21] M. Bourguel, *Bull. Soc. Chim. France*, **1932**, 253.
[22] A. Farkas, *Trans. Faraday Soc.*, **35**, 906 (1934).
[23] B. B. Elsner and P. F. M. Paul, *J. Chem. Soc.*, **1953**, 3156.
[24] G. Dupont, *Bull. Soc. Chim. France*, **1936**, 1030.
[25] K. N. Campbell and M. F. O'Connor, *J. Am. Chem. Soc.*, **61**, 2897 (1939).
[26] L. Kh. Freidlin, Yu. Yu. Kaup, E. F. Litvin, and T. I. Ilomets, *Dokl. Akad. Nauk SSSR*, **143**, 883 (1962); through *Chem. Abstr.*, **57**, 3258 (1962).
[27] K. N. Campbell and L. T. Eby, *J. Am. Chem. Soc.*, **63**, 216 (1941).
[28] L. W. Covert and H. Adkins, *J. Am. Chem. Soc.*, **54**, 4116 (1932).
[29] G. F. Hennion and T. B. Banigan, Jr., *J. Am. Chem. Soc.*, **68**, 1202 (1946).
[30] J. A. Knight and J. H. Diamond, *J. Org. Chem.*, **24**, 400 (1959).
[31] H. Adkins and H. R. Billica, *J. Am. Chem. Soc.*, **70**, 695 (1948).
[32] D. R. Howton and R. H. Davis, *J. Org. Chem.*, **16**, 1405 (1951).
[33] A. Valette, *Ann. Chim.*, [12] **3**, 644 (1948).
[34] M. Gouge, *Ann. Chim.*, [12] **6**, 648 (1951).
[35] G. Bernard and J. Cologne, *Bull. Soc. Chim. France*, **12**, 345 (1945).
[36] G. Bernard and J. Cologne, *Bull. Soc. Chim. France*, **1945**, 356.

[37] N. A. Khan, *J. Am. Chem. Soc.*, **74**, 3018 (1952).
[38] A. L. Henne and K. W. Greenlee, *J. Am. Chem. Soc.*, **65**, 2020 (1943).
[39] T. H. Vaughan, U.S. Patent 2,157,365 (1939).
[40] E. V. Hort and D. E. Graham, Ger. Patent 1,139,832 (1962).
[41] N. I. Shcheglov, D. V. Sololskii, and A. A. Ishchenko, *Izv. Akad. Nauk Kaz. SSSR, Ser. Khim.*, **1962**, 59; through *Chem. Abstr.*, **57**, 1605 (1962).
[42] R. J. Tedeschi, *J. Org. Chem.*, **27**, 2398 (1962).
[43] W. Oroshnik, G. Karmas, and A. D. Mebane, *J. Am. Chem. Soc.*, **74**, 295 (1952).
[44] Br. Patent 595,459 (1947).
[45] M. C. Hoff, K. W. Greenlee, and C. E. Boord, *J. Am. Chem. Soc.*, **73**, 3329 (1951).
[46] N. R. Eaton and E. C. Kornfeld, U.S. Patent 3,067,101 (1962).
[47] K. Ahmad and F. M. Strong, *J. Am. Chem. Soc.*, **70**, 1699 (1948).
[48] K. Ahmad, F. M. Bumpus, and F. M. Strong, *J. Am. Chem. Soc.*, **70**, 3391 (1948).
[49] W. R. Taylor and F. M. Strong, *J. Am. Chem. Soc.*, **72**, 4263 (1950).
[50] W. F. Huber, *J. Am. Chem. Soc.*, **73**, 2730 (1951).
[51] G. F. Hennion, W. A. Schroeder, R. P. Lu, and W. B. Scanlon, *J. Org. Chem.*, **21**, 1142 (1956).
[52] J. N. Ashley and R. D. MacDonald, *J. Chem. Soc.*, **1957**, 1668.
[53] L. Crombie, *J. Chem. Soc.*, **1955**, 1007.
[54] K. N. Campbell, F. C. Fatora, Jr., and B. K. Campbell, *J. Org. Chem.*, **17**, 1141 (1952).
[55] N. A. Milas and C. P. Priesing, *J. Am. Chem. Soc.*, **79**, 6295 (1957).
[56] G. F. Woods and A. Viola, *J. Am. Chem.*, Soc., **78**, 4380 (1956).
[57] A. I. Nogaideli and Ts. N. Vardosanidze, *J. Gen. Chem. USSR Eng. translation*, **33**, 372 (1963).
[58] S. H. Harper and R. J. D. Smith, *J. Chem. Soc.*, **1955**, 1512.
[59] Belgian Patent 614,501 (1962); through *Chem. Abstr.*, **59**, 1487 (1952).
[60] German Patent 1,003,209 (1957).
[61] N. R. Easton, D. R. Cassidy, and R. D. Dillard, *J. Org. Chem.*, **27**, 2746 (1962).
[62] J. Meinwald, D. W. Dicker, and N. Danieli, *J. Am. Chem. Soc.*, **82**, 4087 (1960).
[63] O. Heuberger and L. N. Owen, *J. Chem. Soc.*, **1952**, 910.
[64] W. J. Gensler and H. N. Schein, *J. Am. Chem. Soc.*, **77**, 4846 (1955).
[65] I. Marszak and M. Olomucki, *Bull. Soc. Chim. France*, **1959**, 182.
[66] F. Beck and L. Schuster, Belgian Patent 647,708 (1964); through *Chem. Abstr.*, **63**, 9806 (1965).
[67] H. Lindlar, U.S. Patent 2,681,938 (1954).
[68] H. Lindlar, *Helv. Chim. Acta*, **35**, 446 (1952).
[69] W. Kimel, J. D. Surmatis, J. Weber, G. O. Chase, N. W. Sax, and A. Ofner, *J. Org. Chem.*, **23**, 1611 (1957).
[70] C. B. Clarke and A. R. Pinder, *J. Chem. Soc.*, **1958**, 1967.
[71] N. M. Libman and S. G. Kuznetsov, *J. Gen. Chem. U.S.S.R.*, **33**, 23 (1963) (English translation).
[72] I. Iwai, T. Iwashige, M. Asai, K. Tomita, and J. Ide, *Chem. Pharm. Bull.*, **11**, 188 (1963).
[73] J. Meinwald and H. Nozaki, *J. Am. Chem. Soc.*, **80**, 3132 (1958).
[74] K. E. Schulte and J. Witt, *Arch. Pharm.*, **291**, 404 (1958).
[75] D. Jerchel and W. Melloh, *Ann.*, **622**, 53 (1959).
[76] A. T. Blomquist and A. Goldstein, *J. Am. Chem. Soc.*, **77**, 1001 (1955).
[77] A. T. Blomquist and P. R. Taussig, *J. Am. Chem. Soc.*, **79**, 3505 (1957).
[78] B. W. Baker, R. P. Linstead, and B. C. L. Weedon, *J. Chem. Soc.*, **1955**, 2218.
[79] G. O. Chase and J. Galender, U.S. Patent 2,883,431 (1959).

[80] G. B. Payne, *J. Org. Chem.*, **27**, 3819 (1962).
[81] W. Kimel, N. W. Sax, S. Kaiser, G. G. Eichmann, G. O. Chase, and A. Ofner, *J. Org. Chem.*, **23**, 153 (1958).
[82] Neth. Appl., 6,506,928 (1965); through *Chem. Abstr.*, **64**, 14109 (1966).
[83] J. Lehrfeld, A. M. Burkman, and J. E. Gearien, *J. Med. Chem.*, **7**, 150 (1964).
[84] J. L. Neumeyer, J. G. Cannon, and J. F. Buckley, *J. Med. Chem.*, **5**, 784 (1962).
[85] J. B. Grenet, U.S. Patent 3,192,168 (1965).
[86] British Patent 871,804 (1961).
[87] M. Freifelder, *J. Pharm. Sci.*, **56**, 903 (1967).
[88] E. B. Hershberg, E. P. Oliveto, C. Gerold, and L. Johnson, *J. Am. Chem. Soc.*, **73**, 5073 (1951).
[89] M. S. Schechter, N. Green, and F. B. LaForge, *J. Am. Chem. Soc.*, **74**, 4902 (1952).
[90] O. Isler, W. Huber, A. Ronco, and M. Kofler, *Helv. Chim. Acta*, **30**, 1911 (1947).
[91] D. Papa, F. J. Villani, and H. F. Ginsberg, *J. Am. Chem. Soc.*, **76**, 4446 (1954).
[92] T. Miki and Y. Hara, *Chem. Pharm. Bull. Tokyo*, **4**, 85 (1956).
[93] D. J. Cram and N. L. Allinger, *J. Am. Chem. Soc.*, **78**, 2518 (1956).
[94] J. H. Biel and F. Di Pierro, *J. Am. Chem. Soc.*, **80**, 4614 (1958).
[95] H. Newman, *J. Org. Chem.*, **29**, 1461 (1964).
[96] T. Fukuda and T. Kusama, *Bull. Chem. Soc. Japan*, **31**, 339 (1958).
[97] R. Skowronski, W. Chodkiewicz, and P. Cadot, *Compt. Rend.*, **249**, 552 (1959).
[98] J. Martel, E. Toromanoff, and C. Huynh, *J. Org. Chem.*, **30**, 1752 (1965).
[99] M. Freifelder, *Advances in Catalysis*, Vol. XIV, Academic Press, New York, 1963, p. 203.
[100] R. J. Tedeschi and G. Clark, Jr., *J. Org. Chem.*, **27**, 4323 (1962).
[101] R. A. Raphael, *Acetylenic Compounds in Organic Synthesis*, Academic Press, New York, 1955, p. 22–28 and references therein.
[102] R. J. Tedeschi, H. C. McMahon, and M. S. Pawlak, *Annals New York Acad. Sciences*, **145**, 91 (1967).
[103] R. J. Tedeschi, to be published.
[104] R. Paul and G. Hilly, *Bull. Soc. Chim. France*, **1936**, 2330.
[105] R. Paul and G. Hilly, *Compt. Rend.*, **206**, 608 (1938).
[106] A. F. Thompson, Jr. and S. B. Wyatt, *J. Am. Chem. Soc.*, **62**, 2555 (1940).
[107] Brit. Pat. 749,156 (1956).
[108] S. Tairi, *Bull. Chem. Soc. Japan*, **35**, 840 (1962).
[109] W. Parker, R. A. Raphael, and D. L. Wilkinson, *J. Chem. Soc.*, **1959**, 2433.
[110] Ger. Pat. 1,111,615.
[111] H. J. J. Pabon, D. Van Der Steen, and D. A. Van Dorp, *Rec. Trav. Chim.*, **84**, 1319 (1965).
[112] A. A. Kraevskii and N. A. Preobrazhenskii, *J. Gen. Chem. USSR.*, **35**, 620 (1965) (English translation).
[113] L. Audier, G. Dupont, and R. Dulou, *Bull. Soc. Chim. France*, **1957**, 248.
[114] N. A. Dobson, G. Eglinton, M. Krishnamurti, R. A. Raphael, and R. G. Willis, *Tetrahedron*, **16**, 16 (1961).
[115] A. A. Petrov and M. P. Forost, *J. Gen. Chem. USSR*, **34**, 3331 (1964) (English translation).
[116] A. I. Zakharova and G. D. Il'ina, *J. Gen. Chem. USSR*, **34**, 1391 (1964) (English translation).
[117] Framework molecular orbital models are available from Prentice-Hall, Inc., Englewood Cliffs, New Jersey.
[118] J. M. Shackleford, W. A. Michalowicz, and L. H. Schwartzman, *J. Org. Chem.*, **27**, 1631 (1962).

[119] G. W. H. Cheeseman, I. Heilbron, E. R. H. Jones, S. F. Sondheimer, and C. L. Weedon, *J. Chem. Soc.*, **1949**, 2031.

[120] S. Zonis, *J. Gen. Chem. USSR*, **9**, 2191 (1939); through *Chem. Abstr.*, **34**, 4052 (1940).

[121] Yu S. Zal'kind and N. D. Khudekova, *J. Gen. Chem. USSR*, **10**, 435 (1940); through *Chem. Abstr.*, **34**, 7847 (1940).

[122] L. Crombie, S. H. Harper, and F. C. Newman, *J. Chem. Soc.*, **1956**, 3963.

[123] K. V. Balyan, A. A. Petrov, and Yu. I. Porfiryeva, *J. Gen. Chem. USSR*, **27**, 409 (1957) (English translation).

[124] K. V. Balyan, Z. A. Lerman, and L. A. Merkur'eva, *J. Gen. Chem. USSR*, **28**, 110 (1958) (English translation).

[125] A. I. Nogaideli and N. A. Rtveliashvili, *J. Gen. Chem. USSR*, **34**, 1751 (1964) (English translation).

[126] E. N. Marvell and J. Tashiro, *J. Org. Chem.*, **30**, 3991 (1965).

[127] J. L. M. A. Schlatmann and E. Havinga, *Rec. Trav. Chim.*, **80**, 1101 (1961).

[128] O. Isler, H. Lindlar, M. Montavon, R. Rüegg, and P. Zeller, *Helv. Chim. Acta*, **39**, 249 (1956).

[129] M. Akhtar and B. C. L. Weedon, *J. Chem. Soc.*, **1959**, 4058.

[130] U. Schwieter, R. Rüegg, and O. Isler, *Helv. Chim. Acta*, **49**, 992 (1966).

[131] F. Bohlmann and H. Mannhardt, *Ber.*, **89**, 1307 (1956).

[132] I. N. Nazarov, L. N. Terkhova, and L. D. Bergelson, *J. Gen. Chem. USSR*, **20**, 697 (1950) (English translation).

[133] P. S. Venkataramani, J. P. John, V. T. Ramakrishnan, and S. Swaminothan, *Tetrahedron*, **22**, 2021 (1966).

[134] A. Sandoval, G. H. Thomas, C. Djerassi, G. Rosenkranz, and F. Sondheimer, *J. Am. Chem. Soc.*, **77**, 148 (1955).

[135] M. S. Newman, I. Waltcher, and H. F. Ginsberg, *J. Org. Chem.*, **17**, 962 (1952).

[136] G. P. Kugatova-Shemyakina and V. I. Vidugirene, *J. Gen. Chem. USSR*, **34**, 1742 (1964) (English translation).

[137] E. E. Boehm and M. C. Whiting, *J. Chem. Soc.*, **1963**, 2541.

[138] A. J. B. Edgar, S. H. Harper, and M. A. Kazi, *J. Chem. Soc.*, **1957**, 1083.

[139] A. I. Rachlin, N. Wasyliw, and M. W. Goldberg, *J. Org. Chem.*, **26**, 2688 (1961).

[140] S. N. Ege, R. Wolovsky, and W. J. Gensler, *J. Am. Chem. Soc.*, **83**, 3080 (1961).

[141] M. Olomucki, *Bull. Soc. Chim. France*, **1963**, 2067.

[142] H. U. Daeniker, *Helv. Chim. Acta.*, **49**, 1543 (1966).

[143] W. M. McLamore, S. Y. P'an, and A. Bavley, *J. Org. Chem.*, **20**, 109 (1955).

[144] C. Grob and E. Jenny, U.S. Patent 3,118,946 (1964).

[145] W. S. Johnson and J. K. Crandall, *J. Org. Chem.*, **30**, 1785 (1965).

[146] D. J. Cram and M. Cordin, *J. Am. Chem. Soc.*, **77**, 4090 (1955).

[147] B. Loev and C. R. Dawson, *J. Org. Chem.*, **24**, 980 (1959).

[148] M. Svoboda and J. Sicher, *Chem. & Ind.*, **1959**, 290.

[149] A. T. Blomquist, M. Passer, C. S. Schollenberger, and J. Wolinsky, *J. Am. Chem. Soc.*, **79**, 4972 (1957).

[150] S. Siegel and G. V. Smith, *J. Am. Chem. Soc.*, **82**, 6087 (1960).

[151] W. Oroshnik and A. D. Mebane, *J. Am. Chem. Soc.*, **76**, 5719 (1954).

[152] H. Wieland and H. Kloss, *Ann.*, **470**, 201 (1929).

[153] Yu. S. Zalkind and S. V. Nedzvetskii, *J. Gen. Chem. USSR*, **3**, 573 (1933); through *Chem. Abstr.*, **28**, 2707 (1934).

[154] I. Jardine and F. J. McQuillin, *J. Chem. Soc. (C)*, **1966**, 458.

IX

OLEFINS

VARIABLES AFFECTING THE HYDROGENATION OF OLEFINS

The alkene linkage is readily reduced under mild hydrogenation conditions (low pressure and room temperature) with a variety of catalysts. In general, unless the double bond is highly substituted, elevated temperature and/or pressure will not be necessary for reduction to take place. In addition, unless steric effects play a part, a double bond will be preferentially hydrogenated in the presence of other reducible functions; triple bonds and aromatic nitro groups are the exceptions.

For reductions of olefins under mild conditions the catalysts most often used are palladium on a carrier, platinum oxide, and Raney nickel. At times rhodium and ruthenium on specific supports find application. In this author's opinion there is no catalyst of choice for the reduction of the ethylenic bond. Selection of the catalyst will depend on the type of compound to be reduced, on the presence of other reducible groups and, among other things, on whether the particular catalyst has a tendency to cause isomerization or bond migration.

Obviously the structure of the olefinic compound will influence the reducibility of the double bond. Lebedev and associates [1] found that monosubstituted olefins were reduced more rapidly with platinum catalyst than more highly substituted ethylenes. They were not able to determine the difference in ease of reduction of symmetrical and unsymmetrical disubstituted ethylenes, RCH=CHR, R_2C=CH, or trisubstituted ethylenes, R_2C=CHR, but they did find that tetrasubstituted ethylenes were reduced at the slowest rate. Dupont [2] found that Raney nickel gave results similar to those with platinum, although symmetrical disubstituted ethylenes, RCH=CHR, were reduced before either the unsymmetrical or the trisubstituted ethylenes. Dobson and his group [3] established in the hydrogenation of undeca-1,7-diene with palladium on carbon that the external double bond was reduced in preference to the internal one. In another study 1-octene was reduced preferentially in the presence of 2-octene with ruthenium on carbon catalyst and 4-methyl-1-pentene was selectively reduced in the presence of 2-methyl-2-pentene [4]. The latter example is one of a monosubstituted olefin reducing more rapidly than a trisubstituted one.

Kern, Shriner, and Adams [5] reported that isolated double bonds were more easily reduced with platinum oxide than were the double bonds of α,β-unsaturated carbonyl compounds. They, as well as Lebedev, noted the deactivating effect of aryl groups on the platinum-catalyzed reductions of ethylenes. Kazanskii and Tatevosyan [6] reported the following order of reduction confirming the effect of aryl groups on reduction with platinum catalysts:

$$CH_3CH{=}C(C_2H_5)_2 > CH_3C(C_6H_5){=}CHCH_3$$
$$> CH_2{=}C(C_6H_5)_2 > (C_6H_5)_2C{=}CHC_6H_5$$

Gauthier [7] also showed the effect of the double bond conjugated with an aromatic ring in a comparison of the reduction of eugenol and isoeugenol with Raney nickel. The nonconjugated ethylenic compound, eugenol, was hydrogenated more readily and at lower temperature than isoeugenol. In contrast to the effect on nickel and platinum reductions, aryl groups were reported to increase the ease of hydrogenation with palladium [8]. A comparison of the effect of phenyl groups on palladium and platinum-catalyzed reductions was studied [9]. Results showed that as phenyl groups were substituted at the ethylenic linkage hydrogenations with platinum became increasingly slow; tetraphenylethylene could not be reduced. Reductions with palladium were not retarded except in the hydrogenation of tetraphenylethylene where absorption of hydrogen was slow. The difference in behavior may be ascribed to the strong pi-bonding of the aromatic ring to the active sites of platinum so that the reduction product cannot be desorbed. In contrast the ring is less strongly bound to the sites of palladium so that reduction product is desorbed and more olefin is adsorbed and reduced until saturation of the double-bonded material is completed.

The size and degree of branching also has an influence on the hydrogenation of olefins. The following sequences of reducibility were shown by Lagerev in the hydrogenation of isomeric hexenes over platinum [10, 11]:

$$C_4H_9CH{=}CH_2 > (CH_3)_3CCH{=}CH_2$$
$$> C_2H_5C(CH_3){=}CHCH_3 > C_2H_5CH{=}C(CH_3)_2$$
$$C_3H_7CH{=}CCH_3 > (CH_3)_2CHCH{=}CHCH_3 > (CH_3)_2C{=}C(CH_3)_2$$

Based on Dupont's findings [2] the order in the two series is not unexpected.

Increasing the size and number of substituents on the double bond interferes with its ability to bond to the catalyst. In such instances because of the slow rate of hydrogen uptake, increased temperature and/or pressure are necessary to induce reaction to proceed at a respectable rate.

In certain instances steric effects are so pronounced that reduction does not take place. Hennion and Hazy [12] reported that some alkenylamines of the type $R^1CH=C(CH_2R^2)CH(NR^3R^4)CH_3$ could not be reduced with palladium, platinum, or Raney nickel when $R^1 = CH_3$, R^2 and $R^3 = H$, and $R^4 = CH(CH_3)_2$, when R^1 and $R^2 = CH_3$, $R^3 = H$ and $R^4 = CH(CH_3)_2$, and when R^1, $R^2 = -CH_2CH_2CH_2-$, and R^3 and $R^4 = CH_3$. Molecular models of these compounds show that the double bond cannot make contact with the catalyst because of the surrounding groups. Other less hindered compounds were reduced. Another example of steric effects may be seen in the attempt to hydrogenate the olefinic bond in 2,2,3-trimethyl-4-(1-methyl-2-naphthyl)-but-3-enoic acid with platinum in acetic acid [13]. Instead the less substituted naphthalene ring was hydrogenated (see Eq. 9.1).

$$(9.1)$$

R,R' = methyl

When R was hydrogen instead of methyl and R^1 was methyl [14] or hydrogen [15a], the ethylenic bond was readily reduced. A molecular model shows that the 1-methyl group attached to the naphthalene ring restricts movement of the highly substituted side chain so that contact between the ethylenic bond and the catalyst is made difficult while the less substituted portion of the naphthalene ring is open to attack by hydrogen [15b]. In contrast, models with the 1-methyl group replaced by hydrogen show that the double bond is in good position for contact with the catalyst when R' is methyl and is even less hindered as expected when it is hydrogen.

Another factor in the reduction of ethylenes is the effect of geometrical configuration. There appears to be a general opinion that *cis* isomers are reduced more rapidly than the corresponding *trans* isomers. This point is open to qualification. Results could depend on the type of compound used, on substitution at the double bond, and possibly on the catalyst.

According to Platonov [16] the difference in reduction between *cis* and *trans* isomers is dependent on their absorption to active portions of the catalyst surface. In a comparison of the reduction of some unsaturated acids with platinum black he noted that *cis* acids were more strongly adsorbed than the *trans* acids and were also reduced more rapidly. The relative rates of reduction of isocrotonic acid (*cis*) and crotonic acid (*trans*) and their esters were studied [17]; *cis* isomers were hydrogenated substantially more rapidly with palladium on barium sulfate or with platinum in alcohol or in acetic acid.

Maleic, citraconic, and oleic acids were more rapidly reduced than the corresponding *trans* acids (fumaric, mesaconic, and elaidic) with palladium on barium sulfate or palladium on calcium carbonate [18a]. An extension of the study to other *cis* and *trans* unsaturated acids gave similar results [18b]. Markman and Zinkova [19] reported preferential saturation of the *cis* form of some unsaturated acids with palladium or platinum on alumina. They found that use of the palladium catalyst generally gave higher percentages of saturated acids when mixtures of the *cis* and *trans* forms were hydrogenated.

In another group of experiments Weygand [20] reported that *cis* 1,4-*bis*(4-methylphenyl)- and 1,4-*bis*(4-butylphenyl)-2-butene-1,4-dione were reduced in a shorter time than the *trans* isomers but *trans* 1-(4-methylphenyl)-4-phenyl and the *trans* 1-(4-butylphenyl)-4-phenylbutenediones were hydrogenated more rapidly than the *cis* isomers. It is difficult to understand how the absence of an alkyl substituent on one benzene ring could cause this change. Possibly the difference in the two sets of experiments was due to a difference in purity of the substrates.

Dobson [3], who found that stereomutation of *cis* undec-4-ene to the *trans* form took place with insufficient 10% palladium on carbon to cause uptake of hydrogen, also found that the *trans* isomer was rapidly saturated during hydrogenation with the identical amount of catalyst. This led to a suggestion that the pure *trans* isomer was reduced more rapidly than the *cis* but that in mixtures of the two, the *cis* isomer was preferentially bound to the active catalyst sites, and as a result inhibited reduction of the *trans*.

Jardine and McQuillin [21], studying a series of symmetrical butenes, showed that the *cis* isomers were generally hydrogenated more rapidly than the *trans* forms with palladium on carbon, but that the *cis* series was more sensitive to the size of the group adjacent to the double bond. With large groups the *trans* isomer reacted more rapidly. Molecular models of highly substituted symmetrical butenes do show that the *trans* form is usually in better contact with the catalyst than the *cis* form.

Double-bond migration and *cis-trans* isomerization are complications which arise during hydrogenation of the double bond, dependent on the catalyst and the reaction medium.

Bond [22] has stated that platinum and iridium catalysts show a distinct preference for hydrogenation while iron, nickel, rhodium, and palladium have a marked tendency to effect isomerization during reduction of the ethylenic bond. In a study of group VIII metals, osmium, iridium, and platinum caused little isomerization during hydrogenation [23]. In contrast, isomerization occurred to a substantial degree in the presence of ruthenium, rhodium, and palladium.

Young and associates [24] reported that double-bond migration and isomerization took place rapidly in the hydrogenation of 1-butene in alcohol in the presence of palladium on barium sulfate or Raney nickel. When reduction was interrupted after partial uptake of hydrogen, separation and examination of the butene fraction from palladium showed over 90% of 2-butene in a 4:1 ratio of *trans* to *cis*; more 1-butene (21%) was found in the fraction from a nickel hydrogenation but the ratio of *trans* to *cis* 2-butene was almost the same as with palladium (3.5:1). Although there was far less change with platinum catalyst (72% 1-butene, 18.5% *trans*, and 9.5% *cis* 2-butenes), some double-bond migration did take place.

The investigators also established that rearrangement took place only in a hydrogen atmosphere. The same observation was made by Stavely and Bollenback [25] who noted that the double bond in 3β-acetoxy-5α-cholest-7-ene was completely isomerized in ethyl acetate solution to the 8,14-position in the presence of palladium black saturated with hydrogen (Eq. 9.2).

3β-Acetoxy-5α-cholest-7-ene   3β-Acetoxy-5α-cholest-8(14)-ene

In the absence of hydrogen or in the presence of nitrogen no change occurred.

It is well known that the 7,8- and 8,9-bond in steroidal compounds can migrate to the 8,14-position. Wieland and Benend [26] showed that migration took place with palladium catalyst in either acid or neutral medium and with platinum catalyst in acetic acid. They found that in neutral solvent platinum catalyst did not cause migration.

Bladon and co-workers [27] reported that the use of Raney nickel in neutral solvent gave more reliable and reproducible results than platinum under the same conditions. They were able to saturate the side chain double bond in 3β-hydroxy-5α-ergosta-7,22-diene at normal temperature and pressure without causing migration of the 7,8-bond (Eq. 9.3).

3β-Hydroxy-5α-ergosta-7,22-diene → 3β-Hydroxy-5α-ergost-7-ene (9.3)

The 9,11-bond in the C ring of a steroid was reduced in ethyl acetate solution with platinum oxide leaving the 7,8-bond intact [28]. In a hydrogenation of a steroidal 5,7-diene the 5,6-bond could be selectively reduced without causing migration of the 7,8-bond in the presence of palladium on carbon when piperidine or pyridine was added to the reaction mixture. These reactions will be discussed at greater length in the section on the reductions of di- and polyenes.

MONOENES

Monosubstituted ethylenes are less complicated and more easily reduced than other more substituted ones [1, 2]. Hydrogenation should not lead to problems, unless double-bond migration or *cis-trans* isomerization takes place. There are literally hundreds of references available on saturation of monosubstituted ethylenic compounds as well as references to the reduction of more substituted ethylenes, such as the stilbenes. The reduction of similarly substituted aliphatic double-bonded compounds is also well documented.

The systems to be discussed will include cinnamic acids and related compounds, certain nitrogen-containing double-bonded compounds, and endocyclic and exocyclic alkenes.

α,β-Unsaturated Acids and Related Compounds

Kern, Shriner, and Adams [5] note that the presence of a carboxyl group has a deterring effect on the reduction of olefinic bonds in conjugation with it. Its effect is of little consequence in the aliphatic series. Most unsaturated acids, esters, and amides are easily converted to the saturated compounds. Kern and others showed that conjugation with an aromatic ring also had some adverse effect on reduction of the double bond in the presence of platinum and nickel catalysts. On this basis it might be expected that cinnamic acids and compounds related to it might not be readily hydrogenated because of conjugation of the double bond with an aromatic ring and with a carboxyl group. However, such is not the case. Various cinnamic acids have been reduced as sodium salts in water at room temperature and slight hydrogen pressure [29]. Palladium on carbon gave the best results; Raney nickel was satisfactory. The acids, if soluble, can also be hydrogenated in neutral organic solvent with platinum oxide [29] or Raney nickel or palladium on carbon [30]. Compounds like 3,4-dihydroxycinnamic acid (caffeic acid) are best reduced as the free acid in a neutral solvent because the resultant dihydroxyhydrocinnamic acid is water soluble and will be difficult to isolate after acidification when reduction is carried out as sodium salt in water. Nickel, palladium, or platinum catalysts may be employed for reductions of the esters or amides in neutral solution, or platinum or palladium catalysts for reductions in acetic acid. In acidic medium the chemist must be aware of the possibility of ring hydrogenation in case a reduction must be carried out overnight. Supported rhodium catalysts may also be used in neutral or acid solution but with them the reaction should be watched more closely since rhodium is able to catalyze ring reductions under mild conditions.

The author would add a further precaution. Substituted cinnamic acids, which have been prepared by a modified Doebner synthesis, should be checked to make sure there is no contaminating pyridine (used as condensing agent) which might be present as the salt of the acid. It would not poison the catalyst as a salt but its presence could be a problem in knowing when reduction was complete (see Chapter II).

α,β-Unsaturated acids related to cinnamic acid, RCH=CHCOOH where R is an aromatic nitrogen heterocycle, are also readily hydrogenated under mild conditions. Although Katritzky [31] hydrogenated 3-(4-pyridyl)acrylic acid in alkaline solution with Raney nickel at 100° and 50 atm pressure to obtain 3-(4-pyridyl)propionic acid, he did point out that Walter and co-workers [32] carried out the same reduction at room temperature and 2–3 atm pressure. Freifelder [33] used palladium on carbon to convert 3-(2-, 3- and 4-pyridyl)acrylic acids in dilute alkaline solution to the corresponding pyridinepropionic acids in good yield. Walker [34] employed palladium on strontium carbonate to reduce 3-(6-methoxy-4-quinolyl)acrylic acid in alkaline solution to the corresponding quinolylpropionic acid under low-pressure conditions. Jones and colleagues [35] reduced a series of acids, where R was 2-quinoxalyl, 2-pyrazinyl, 3-pyridazinyl, and 2- and 4-pyrimidyl, in alkaline solution at low pressure with a very low weight ratio of Raney nickel to substrate (2%) to give 75–93% yields of the substituted propionic

134   Olefins

acids. Ried and Keller [36] also used a very small amount of Raney nickel (*spatelspitzen*) to carry out similar reductions. In their series R was 2-pyridyl, 2-quinolyl, 2-quinoxalyl, 2-quinazolyl, 2-benzoxazolyl, and 2-benzothiazolyl. Taylor and McKillop [37] also reduced 3-(2-quinoxalyl)-acrylic acid in alkaline solution but employed palladium on calcium carbonate. When they used Raney nickel for the same reduction they obtained the corresponding propionic acid admixed with another product which was identified as the lactam of tetrahydroquinoxalinepropionic acid. The authors noted an unusual dependence on the weight ratio of nickel catalyst to lactam. They found that when they used as much catalyst as compound only the lactam was obtained (see Eq. 9.4). It is interesting that ring reduction took place so readily under such mild conditions even with an equal weight of catalyst since the catalyst used is not known to be especially active.

$$\text{quinoxalyl-CH=CHCOOH} \xrightarrow{\text{Ni(R)}} \text{lactam} \quad (9.4)$$

Alkenylamines

The inhibiting effect of a basic nitrogen atom, with an unshared pair of electrons, on activity of the catalyst has been discussed in Chapter IV. The presence of such a nitrogen atom in an olefinic compound can have a profound effect on the reaction rate, unless the unshared electrons are paired. Elderfield, Pitt, and Wempen [38] reported that the hydrogenation of 3-methoxymethylpent-2-enylamine with platinum oxide was slow (18 hr) under low-pressure conditions. Unfortunately they did not carry out another reduction on the amine salt for comparison but resorted to more vigorous conditions (75–100°, 100 atm) to obtain very rapid reduction of the base with Raney nickel in cyclohexane. A molecular model of the starting alkenylamine shows that the double bond is not prevented from contacting the catalyst, but it also indicates that the basic nitrogen atom is in position to bond strongly to the catalyst by means of its unshared electron pair. It appears that more rapid hydrogen uptake under mild conditions would occur if the nitrogen base were neutralized. This point was illustrated by Adams and Billinghurst [39] who used palladium on carbon to reduce the double bond of 2-(1-phenyl-3-dimethylamino-1-propenyl)pyridine as base in acetic acid or as hydrochloride salt in alcohol. Katz and Karger [40] reduced the 1-(4,4-diphenylbut-3-enyl)piperidine hydrochloride in methyl alcohol within 2 hr in the presence of platinum oxide. In the reduction of 2-(1-cyclohexenyl)-pyridine in acetic acid in the presence of platinum oxide, Lochte and associates [41] noted a tendency toward slow hydrogenation of the pyridine ring. In such instances, it would be advisable to substitute a supported palladium catalyst which will have less tendency to reduce the pyridine ring at room temperature and low pressure. This technique appeared to be used successfully by Suzuki [42] who reduced 2- and 4-(1-hydroxy-3-butenyl)-pyridine to the corresponding hydroxybutylpyridines in 10% hydrochloric acid with palladium on carbon.

Reduction of the double bond in alkenylamines can be effected as free amine in neutral solvent. Hennion, Price, and Wolff [43] used a 15% weight ratio of Raney nickel to saturate the ethylenic bond of 2-phenyl-4-dimethyl-amino-1-butene in 2 hr under 4 atm pressure. Whether reduction as amine salt would be more rapid was not studied.

Cyclo and N-Heterocycloalkenes

Endocyclic Double Bonds. In cyclic systems double bonds without substituents generally reduce under very mild conditions. Crane, Boord, and Henne [44] carried out the hydrogenation of 3-methyl, 3-isopropyl, and 3-*tert*-butylcyclopentene at room temperature and low pressure with platinum oxide. Fieser and Novello [45] reduced 3-(α-hydroxybenzyl)-cyclohexene-4-carboxylic acid in alcohol and 1,2,3,6-tetrahydrophthalic anhydride in ethyl acetate over platinum oxide. Baumgarten and colleagues [46] converted 4-methylcyclohexene-4-carboxylic acid to the corresponding cyclohexane with platinum oxide in alcohol. Bailey and Lawson [47] hydrogenated *endo*-2,3-dicarbethoxy-5-bicyclo[2,2,2]octene with Raney nickel at room temperature and 70 atm pressure (Eq. 9.5). No reason for the use of high pressure was given although it is possible that the reduction was carried out at this pressure to provide sufficient hydrogen for a 3-mole-size reaction.

$$\text{[bicyclic diene diester]} \xrightarrow[\text{Ni(R)} \atop \text{70 atm}]{H_2} \text{[bicyclic diester]} \tag{9.5}$$

The presence and nature of a substituent on an endocyclic double bond may affect hydrogen uptake. The presence of the long-chain aliphatic group of diethyl 1-octylcyclohexene-4,5-dicarboxylate did not appear to have too profound an effect on its reduction. Bailey and Klein [48] carried out the reaction with Raney nickel at 100° and 170 atm, but stated that hydrogenation would take place with platinum in alcohol at room temperature and low pressure. Arnold and Dowdall [49] reported a slow reduction of sodium cyclohexenemethylsulfonate with palladium on carbon at 3 atm pressure. There was no explanation for the lengthy reaction time but since the substrate was prepared from methylenecyclohexane and sulfur trioxide, the possibility of catalyst poisoning by a sulfur-containing reduction product derived from sulfur trioxide cannot be ignored.

Bachmann and co-workers [50] attempted the hydrogenation of 2-(4-methoxyphenyl)cyclohexenylacetic acid with platinum oxide in methyl alcohol using a rather high catalyst ratio (7.7%). No uptake of hydrogen took place until a small amount of concentrated hydrochloric acid was added, whereupon reduction occurred within 8 hr (Eq. 9.6). It is possible that acid activated the catalyst. A molecular model of the starting material shows that the double bond is not able to make contact with the catalyst. On the other hand, it is possible that the function of the acid is to cause migration

$$\text{(9.6)}$$

R = 4-methoxyphenyl

of the tetrasubstituted double bond to a position where it becomes trisubstituted, facilitating its approach to the catalyst surface. This author has reduced cyclohexenylacetic acid in alcohol with platinum catalyst in less than 2 hr and 7-methoxy-3,4-dihydronaphthaleneacetic acid in ethyl acetate with palladium on carbon within 45 min [30]. Both are trisubstituted ethylenic compounds. On the other hand, Linstead, Whetstone, and Levine [51] found that the reduction of 2-phenylcyclohexenylacetic acid in acetic acid with platinum oxide was slow (18 hr).

It would be of interest to study whether double-bond migration was a factor in the reduction of the hindered tetrasubstituted double-bonded compounds. It could be determined by interruption of a reduction before hydrogen uptake was complete and examination of the reduction solution by nuclear magnetic resonance spectroscopy for vinyl protons.

The effect of structure on the reduction of endocyclic double bonds was shown by Klemm and Hodes [52]. They suggested that reduction of 1-(1-naphthyl)cyclohexene was slower than that of 1-1(naphthyl)cyclohexene because of the effect of the hydrogen atoms at the 2-position in the cyclohexene ring and the 8-position in the naphthalene ring. A model of 1-(1-naphthyl)cyclohexene shows that contact between the double bond and the catalyst surface is inhibited by the two aforementioned atoms whereas a model of 1-(1-naphthyl)cyclopentene shows there is no such problem with the smaller ring. 1-(2-Naphthyl)cyclohexene, a model of which does not show any unusual steric effects, was hydrogenated at a more rapid rate than 1-(1-naphthyl)cyclohexene.

In a continuation of this work Klemm and Mann [53] showed that 1-(1-naphthyl)cycloheptene was reduced more slowly than the corresponding cyclohexene. In the less hindered 2-naphthalene series the order of the ease of reduction was 1-cyclopentenyl > 1-cyclohexenyl > 1-cycloheptenyl. The nonconjugated 3-(1-naphthyl)cyclohexene was hydrogenated more readily than the conjugated 1-(1-naphthyl)cyclohexene. A molecular model of the nonconjugated compound points out the accessibility of the cyclic double bond. Klemm and Mann gave one other example of the effect of substitution in the slower hydrogenation of 6-methyl-1-(2-naphthyl)cyclohexene compared with that of 1-(2-naphthyl)cyclohexene.

A double bond which forms the junction between two rings, such as the 9,10-bond in octahydronaphthalene, is less readily reduced than other endocyclic bonds. A molecular model shows the difficulty the double bond has in making contact with the catalyst and suggests that a long reaction period or more virogous conditions will be required for reduction [15b].

Reduction of endocyclic double bonds in nitrogen-containing cyclic systems is complicated by the inhibiting effect of the ring nitrogen, particularly when it is strongly basic. The hydrogenation of 1-aza-4,5-cyclopenteno-6,7-benzocyclohept-4-ene required 24 hr for reduction [54] in spite of the use of 50% by weight of platinum oxide to compound (Eq. 9.7).

(9.7)

1-Aza-4,5-cyclopenteno-6,7-benzocyclohept-4-ene

Reduction of the free base in acid solution or hydrogenation of the hydrochloride salt might have reduced reaction time. In the cited reduction the long reaction period may result from the inhibiting effect of the basic nitrogen atom and the poor contact between the double bond and the catalyst. From a framework molecular orbital model it appears that the bond representing the unshared pair of electrons of the nitrogen atom is in contact with the catalyst and the double bond joining the 5-membered and the 7-membered ring is in an unfavorable position for reduction [15b].

The double bond in 2-trifluoromethyl-3,6-dihydro-1,2-oxazine was reduced in a neutral medium with palladium on carbon without difficulty [55]. In this reduction the electronegativity of the trifluoromethyl group decreased the basicity of the nitrogen atom. Nevertheless, neutralization of the effect of the basic nitrogen in such compounds does aid reduction of the double bond.

Some reductions of the 3,4-bond in 1,2-dihydroquinolines and isoquinolines show that the basicity of the nitrogen atom affects reaction time. Johnson and Buell [56] found that hydrogenation of 1,2-dihydroquinoline in neutral solvent with palladium on carbon was slow (17.5 hr). Bersch [57], on the other hand, noted that reduction of a more highly substituted compound, 2-methyl-3-[2-(1-hydroxyethyl)-4,5-methylenedioxyphenyl]-7,8-dimethoxy-1,2-dihydroisoquinoline, to the tetrahydroisoquinoline in aqueous acetic acid with the same catalyst took place in 30 min. This author has successfully carried out hydrogenations of 1,2-dihydroquinolines and isoquinolines as hydrochloride salts with palladium or platinum catalysts since hydrogen uptake was slow during reduction of the free bases [30].

Exocyclic Double Bonds. Molecular models of alicyclic and heterocyclic compounds with exocyclic double bonds show that those bonds are generally unhindered and should be reduced under moderate conditions. Siegel and Dmuchovsky [58] hydrogenated 1-methylene-4-*tert*-butylcyclohexane with platinum oxide in acetic acid; Baggett and his co-workers [59] reduced 2-phenyl-4-methylene-1,3-dioxolane in methyl alcohol with platinum oxide; and Anderson and his associates [60] hydrogenated the same compound under similar conditions and also used platinum on carbon in methyl or ethyl acetate or chloroform. Each obtained the corresponding methyl compound. Easton, Cassady, and Dillard [61] hydrogenated a number of 2-methylenemorpholines in alcohol with palladium on carbon to the corresponding methylmorpholines in 80–90% yield. Reaction times were not reported, nor was the reduction of any salt studied.

Molecular models of trisubstituted exocyclic double-bonded systems show that the double bond is still in a relatively favorable position for making contact with the catalyst. Bachman and Polansky [62] reduced 9-(1-heptenyl) fluorene in alcohol with platinum oxide. Plesek [63] used similar conditions to obtain 6-cyclohexylhexanoic acid from the corresponding cyclohexylidene compound. Sugimoto and associates [64, 65] obtained 5- and 8-benzyl-5,6,7,8-tetrahydroisoquinolines by reduction with palladium on carbon. 5-(Substituted benzylidene)hydantoins were hydrogenated under mild conditions in alkaline solution with Raney nickel [66] or with palladium on carbon [67, 68]. The latter compounds are related structurally to cinnamic acids and reduce with similar ease.

The ease or difficulty in reducing a tetrasubstituted exocyclic double bond will depend on the size of the substituents. A typical example was shown by Anthony [69] who found that the double bond of 3-isopropylidenoxindole was reduced in acetic acid with a 10% weight ratio of palladium on carbon within 5 min. When they attempted the reduction of 3,3-diphenylmethylene-1-methyloxindole, they used a combination of 10% by weight of platinum oxide and 25% by weight of palladium on carbon (Eq. 9.8).

$$\text{indole}=C(CH_3)_2 \xrightarrow[\text{Pd on C, 10\% by weight, 5 min}]{H_2} \text{indole}-CH(CH_3)_2$$

$$\text{1-methylindole}=C(C_6H_5)_2 \xrightarrow[\substack{\text{Pd on C, 25\% by weight} \\ \text{PtO}_2\text{, 10\% by weight}}]{H_2} \text{1-methylindole}-CH(C_6H_5)_2$$

(9.8)

Steroidal Double Bonds

The ease of reduction of double bonds in steroids is greatly affected by their position and degree of substitution. The construction of molecular models greatly facilitates the understanding of when and how catalytic hydrogenation can take place. A model of the naturally occurring steroid molecule showing the β-oriented angular methyl groups (Fig. 9.1) suggests that attack

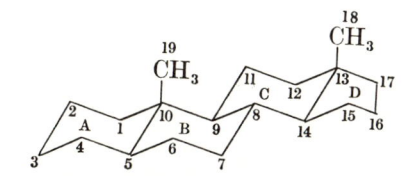

Fig. 9.1 *cis* and *trans* ring fusion is based on the relationship of hydrogen at the ring junction to the angular β-methyl groups; α bonds (below the ring) – – –; β bonds (above the ring) ———.

by hydrogen is generally favored from the less hindered α side of the molecule. From the structural formula and from the order of reducibility as shown by Dupont [2], it can be assumed that symmetrically disubstituted double bonds, 1,2-, 2,3-, 3,4-, 6,7-, 11,12-, and 15,16-, should be readily reduced. Double bonds, when present in the side chain, should be saturated with the same facility as any aliphatic double bond. The ease of reducibility of these ring and side-chain bonds may be illustrated by the construction of molecular models which will show that the double bonds can make contact with the catalyst surface.

The tetrasubstituted 8,14-bond is said to be resistant to hydrogenation [70]. It can be isomerized to the reducible 14,15-position. An example may be seen in the treatment of 8,14-ergostenol in chloroform solution with dry hydrogen chloride resulting in isomerization to 14,15-ergostenol which is then reduced in ethyl acetate with palladium to ergostanol [71].

Recently, Johns [72] found, in contrast to the reported inertness of the 8,14-bond to hydrogenation, that he could convert 3-methoxyestra-1,3,5(10), 8(14)-tetraen-17β-ol to 3-methoxy-8α,9α,14α-estra-1,3,5(10)-trien-17β-ol in 15 hr with an equal weight of palladium on carbon in alcohol (Eq. 9.9).

(9.9)

Framework molecular orbital models and Drieding models of the tetraene suggest that the presence of the aromatic ring appears to lead to a conformation in which the 8,14-bond seems to be susceptible to attack by hydrogen at the α-face of the molecule.

An isolated 8,9-double bond has not been hydrogenated *per se*, but it will isomerize during hydrogenation in the presence of acid or in the presence of palladium in neutral solvent to the 8,14-position and then be subsequently reduced after further treatment.

Of the trisubstituted double bonds, saturation of the 14,15-bond has been noted. An example of the reduction of the 16,17-bond is seen in the conversion of 3α-hydroxyandrost-16-ene in alcohol in the presence of platinum catalyst to the corresponding androstane [73].

When the 7,8-bond is the only double bond present in the molecule, it will not be reduced as such. However, under acidic hydrogenation conditions with palladium or platinum catalyst or under neutral conditions in the presence of palladium, it has been shown that it is converted without hydrogen uptake to the 8,14-position [70] where it may be further isomerized to the 14,15-position and then reduced. Mazur and Sondheimer [74] carried out a hydrogenation of 3β-acetoxy-4α-methyl-5α-cholest-7-ene with platinum in acetic acid containing hydrochloric acid. The investigators made no mention of hydrogen uptake but interrupted the reaction after 1 hr to add another small portion of strong acid, then continued hydrogenation for 24 hr to obtain the corresponding cholestane. Although the authors made no comment, this writer suggests that the reduction followed the pattern described above.

Reduction of the 5,6-bond appears to be best carried out in acidic medium. Although Shriner and Ko [75] claimed that ethyl acetate was the best solvent for the reduction of cholesterol, it must be noted that they used an enormous amount of platinum oxide (40% by weight). Hershberg and associates [76] reported that they could not duplicate the result of Bruce [77] who carried out the reduction of cholesterol at 65–75° in glacial acetic acid with platinum oxide to 3β-hydroxy-5α-cholestane (cholestanol) in several hours. They found that although the 5,6-bond in stigmasterol and sitosterol could be reduced under Bruce's reaction conditions, the hydrogenation of cholesterol required the presence of a strong acid (pKa < 3.0). A high yield of cholestanol was obtained by Hershberg's group by reduction of cholesterol in ethyl acetate with platinum oxide when the reaction was promoted by the presence of acids such as perchloric, sulfuric, and other strong acids. Acetic acid was unsatisfactory. Nace [78] suggested that the problem in the reduction is due to the insolubility of cholestanol and cholestanyl acetate (formed by esterification during reaction) which precipitate at about 75% uptake of hydrogen and impede complete reduction. He obtained good results, however, by keeping the reactants and reduction products in solution with acetic acid and cyclohexane. Nace also suggested that the procedure of Ralls [79], in which cholesteryl acetate is hydrogenated in ether and acetic acid, eliminated the difficulty noted by Hershberg.

Lewis and Shoppee state that catalytic hydrogenation of the 3β-substituted 5,6-double bond of steroids leads mainly to 5α-H, A/B-*trans* saturated compounds [80 and references therein]. They found that reduction of 3α-substituted 5,6-unsaturated steroids in methyl alcohol or ethyl acetate containing perchloric, sulfuric, or hydrobromic acid in the presence of platinum oxide gave varying amounts of A/B *cis* saturated steroids. The course of reduction was dependent of the bulk of the 3α-substituent. 3α-Hydroxycholest-5-ene gave 50% of 3α-hydroxycoprostane, the 5β-form, and 3α-hydroxy-5α-cholestane, 40%. Reduction of 3α-methoxy, acetoxy, methylamino, acetylamino, and dimethylaminocholest-5-enes gave over 90–100% of the corresponding coprostanes.

Shoppee, Agashe, and Summers [81] studied the reduction of 3α- and 3β-substituted-4,5-double-bonded steroids with platinum oxide in ethyl acetate and ethyl acetate containing acetic, sulfuric, or perchloric acids. They found that the influence of substituents was less marked than in the 5,6-series. Reduction was complicated by extensive hydrogenolysis of the allylic 3-substituents during reduction of 3α- and 3β-hydroxy, 3α- and 3β-acetoxy, and 3β-methoxycholest-4-enes in the presence of sulfuric and perchloric acid. Reduction in the presence of acetic acid notably decreased hydrogenolysis.

DI- AND POLYENES

General

The major interest in the hydrogenation of di- and polyenes is the selective reduction of one double bond in the presence of another. In general, saturation of all the multiple double bonds of nonaromatic compounds can be carried out with any catalyst suitable for low-pressure reductions. Saturation of conjugated double bonds in a cyclic system with certain catalysts does not require the presence of acid or the vigorous reaction conditions that are generally necessary for the hydrogenation of the aromatic rings. Aumüller and Muth [82] reduced N^1-(2,4,6-cycloheptatrienyl)-N^2-(3-and 4-substituted benzenesulfonyl)ureas to the cycloheptyl compounds within 10 min with palladium black. Akiyoshi and Matsuda [83] converted bicyclo-[6.1.0]-3,5,7-nonatriene-1-carboxylic acid to the saturated cyclic structure with Raney nickel, and Helferich and colleagues [84] used Raney nickel at low pressure to reduce a series of sultams to the corresponding 2-substituted-tetrahydro-1,2-thiazane-1, 1-dioxides in high yield (Eq. 9.10).

$$\underset{SO_2}{\overset{H_3C\quad CH_3}{\diagup\diagdown NR}} \xrightarrow{2H_2} \underset{SO_2}{\overset{H_3C\quad CH_3}{\diagup\diagdown NR}} \qquad (9.10)$$

R = alkyl, benzyl, phenyl, and substituted phenyl

Selective hydrogenation of a single bond in multiple-bonded cyclic systems is possible with certain catalysts. Evans, Morris, and Shokal [85] were able to preferentially hydrogenate cyclopentadiene and 1,3-cyclohexadiene and their methyl and ethyl derivatives to the cyclopentene and cyclohexene compounds with Raney nickel at 0–40° and 2–5 atm pressure using 65–85% of the required amount of hydrogen. Friedlin and Polkovnikov [86] noted a break in the curve in the reduction of cyclopentadiene with palladium black at room temperature and atmospheric pressure, but not with platinum black or platinum on barium sulfate. They did find that Raney nickel could be used but in later work [87] they reported it was not selective in the reduction of 1,3-cyclohexadiene unless the reaction was carried out in cyclohexene. Better results were obtained when pyridine was used as solvent. In that instance hydrogen uptake would stop at the cyclohexene stage. In this reference the authors noted that benzene and cyclohexene were formed from cyclohexadiene by disproportionation during the reaction.

Gardner and Narayana [88] reduced 1,5-cyclooctadiene to *cis*-cyclooctene in 87% yield with a small amount of 5% palladium on calcium carbonate.

Control of the reaction temperature may also have aided the selectivity. Moore and associates [89] described the use of molybdenum and nickel sulfides to reduce cyclic and aliphatic diolefins to monoolefins at 100 atm initial hydrogen pressure and 150°. They stated that monoenes did not reduce under similar conditions. They suggest that the reaction only takes place with conjugated systems or with systems which become conjugated through double-bond migration in the presence of the catalyst and hydrogen. Seefelder and Raskob [90] reported that the reduction of conjugated or nonconjugated cyclic polyenes stopped at the monoene stage in the presence of palladium catalyst poisoned with heavy metals. The polyenes included cyclopentadiene, 1,3-cyclohexadiene, and cyclooctatetraene.

Allenes

The literature on the hydrogenation of allenes is rather meager. Bond and Sheridan [91] showed that allene resembles acetylene and methylacetylene in its ease of hydrogenation. They suggested that it is selectively adsorbed and held more strongly by the catalyst than 1-propene. In separate experiments allene, acetylene, and methylacetylene, each were selectively hydrogenated with palladium, platinum, and nickel in the presence of 1-propene without further reduction of it. Balyan and associates [92] studied the hydrogenation of some long-chain 2,3- and 3,4-alkadienes with colloidal palladium. They noted that uptake of the first molar equivalent of hydrogen was rapid and was considerably slower for the second equivalent, an effect observed in acetylene reductions. Balyan's group reported that after half hydrogenation, allene bands were not seen in the infrared spectra of the resultant products; reduction of 2,3-dienes gave 2- and 3-alkenes and reduction of 3,4-dienes gave 3- and 4-alkenes as shown by ozonolysis of the products of reduction. Infrared spectra indicated that reduction gave *trans* alkenes. The investigators also hydrogenated some 1,3,4-trienes. Uptake of hydrogen was rapid for two equivalents, then slow for the third as in the reduction of vinylacetylenes. Interruption of the reduction of 8-methylnona-1,3,4-triene at one molar equivalent and subsequent examination of the infrared spectrum of the reduction product revealed the absence of the 1,2-double bond, the disappearance of the allene band in the spectrum, and the presence of bands indicative of a conjugated diene system. The authors suggested that the conjugated diene resulted from 1,4-addition and rearrangement, as seen in Eq. 9.11, or from the known isomerizing effect of palladium catalyst.

$$RCH=C=CHCH=CH_2 \xrightarrow{H_2} \left[\begin{array}{c} RCH=C=CHCH=CH_2 \\ H \qquad\qquad H \end{array} \right] \longrightarrow$$

$$RCH=CHCH=CHCH_3$$

$$R = (CH_3)_2CH(CH_2)_2- \tag{9.11}$$

Based on the above references it should be possible to learn if isomerization occurs by interruption of the hydrogenation at 50–75% uptake of one molar equivalent and to determine if allenic bands are still present and if migration of the terminal bond took place as noted by Young in the interrupted reductions of 1-butene [24].

Hennion and Sheehan [93] found that hexa-1,2-diene was completely reduced with platinum oxide within 30 min. With Raney nickel, however, they found that when the reaction was interrupted after uptake of one molar equivalent of hydrogen a mixture of 1- and 2-hexene was obtained with the latter predominating.

Hennion and DiGiovanna [94] selectively hydrogenated 3-ethylpenta-1,2-dienyl trimethylammonium chloride with Raney nickel in ethyl alcohol to obtain only 3-ethylpent-2-enyl trimethylammonium chloride. The authors proved the structure by preparing a quaternary salt from 3-ethylpent-2-enyl chloride and trimethylamine which was identical in all respects to the reduction product. When they carried out the half hydrogenation of 3-ethylocta-1,2-dienyl trimethylammonium chloride, they obtained a product with a wide melting range which they suggested was a mixture of the two possible isomers. Examination by nuclear magnetic resonance of that reduction product might have proved useful.

Selective Partial Reduction of Unconjugated Di- and Polyenes

Partial hydrogenation of unconjugated di- and polyenes leading to a single product is possible depending on the structure of the compound and the effect of catalyst. In many instances platinum catalysts are not suitable because they are not selective. When they can be used their one virtue is that they have the least tendency to cause isomerization or double-bond migration.

When symmetrically substituted dienes are hydrogenated, as in the work of Horner and Grohmann [95], a single alkene may result with the proper catalyst. They reduced a series of dienes, $CH_2=CH(CH_2)_nCH=CH_2$ where $n = 1$ to 8, with Raney nickel to obtain 1-alkenes. In the same reductions, supported palladium catalysts caused double-bond migration leading to a mixture of 1- and 2-alkenes.

Selective reductions of nonconjugated dienes are seen in the hydrogenations of dextropimaric acid and isodextropimaric acid (with the position of the vinyl and methyl groups reversed) with palladium on carbon [96], Eq. 9.12, and in reductions of related pimaric acids [97]. Other examples include the hydrogenation of 4-(2-methyl-5-isopropenyl-cyclopent-1-enyl)-2-butanone to the isopropylcyclopentenylbutanone with Lindlar catalyst in hexane at room temperature and atmospheric pressure [98], the reduction of 1-methyl-4-isopropenylcyclohexene (limonene) to 1-methyl-4-isopropylcyclohexene with a very small amount of 5% platinum on carbon [99] and the hydrogenation of 3β-acetoxy-5,20-pregnadiene in dioxane containing some acetic acid in the presence of platinum oxide to obtain 3β-acetoxy-20-methyl-5-pregnene [100]. All the above are examples of the selective reduction of a terminal double bond (monosubstituted ethylene) in the presence of an endocyclic double bond which is trisubstituted.

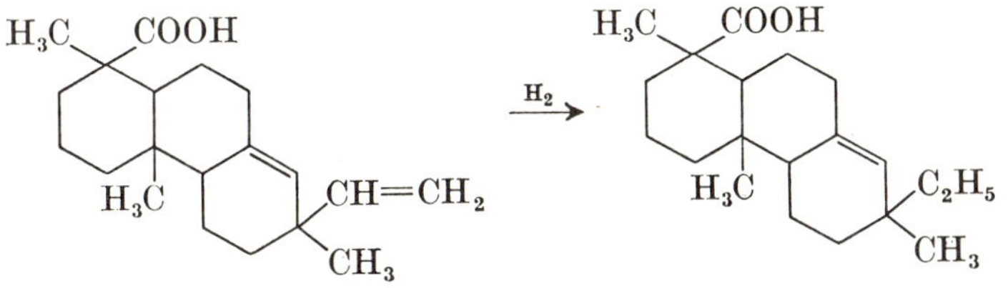

(9.12)

Dextropimaric acid

Sanderman and Bruns [101] preferentially reduced the vinyl linkage in larixol, a terpenic diol, in the presence of an exocyclic double bond with platinum oxide in alcohol. This is an example of the preferential reduction of a monosubstituted double bond in the presence of an unsymmetrical disubstituted double bond (Eq. 9.13).

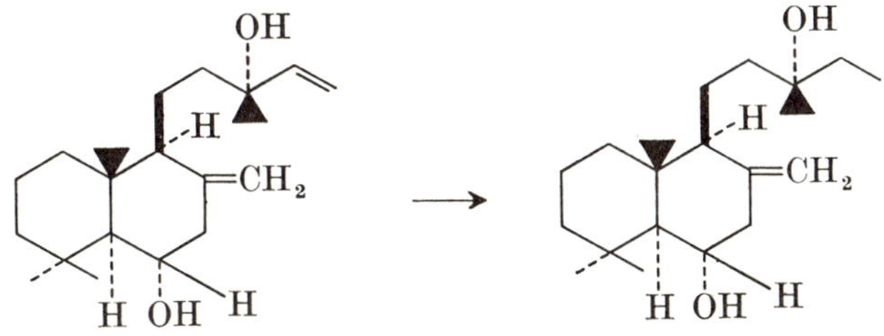

(9.13)

Larixol

Kupchan, Flouret, and Matuszak [102a] reduced the 8,9-bond in 1,3,4,6,7,10,11,11a-octahydro-2H-benzo[b]quinolizine with platinum oxide to obtain the decahydro-2H-benzo[b]quinolizine. Kupchan and DeGrazia [102b] obtained similar results in the same series. These are examples of the favored hydrogenation of a symmetrical disubstituted double bond over a tetrasubstituted one (Eq. 9.14). A similar example is seen in the work of

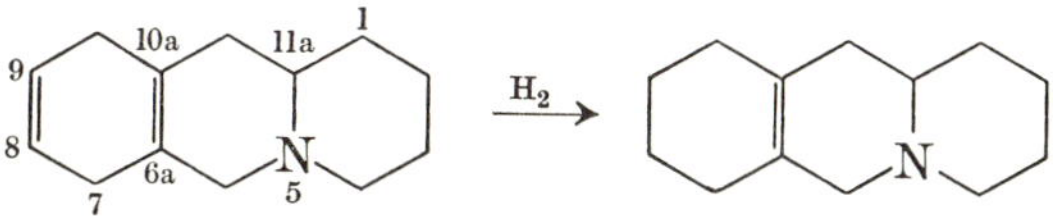

(9.14)

Octahydrobenzoquinolizine

Inhoffen's group [103] who converted 3,4-*bis*(3-cyclohexenyl)-3-hexene to 3,4-*bis*-cyclohexyl-3-hexene in the presence of palladium on carbon.

It should be pointed out that the above examples follow the pattern suggested by Dupont [2] and, unless the bulk of the substituent is an adverse factor, selectivity, if desired, can frequently be accomplished with many unconjugated olefinic systems.

Prediction of results in the reduction of unconjugated bonds of steroids is more difficult. For example, Hershberg and associates [104] reduced 3β-acetoxypregna-5,17(20)-diene to the corresponding 5-pregnene with palladium on carbon in dioxane. In this instance the difference in selectivity in reduction of the trisubstituted double bonds may be due to the fact that the 5,6-bond does not reduce readily in neutral solvent. A molecular model suggests that, while the 5,6-bond is open to attack at the α-face, the 17,20-double bond may be slightly less hindered. Selective reduction of another trisubstituted double bond in the presence of the 5,6-bond in neutral solvent may be seen in the patent of Freifelder and Kurath [105] who reduced 3β,17β-diacetoxyandrost-5-en-16-ylideneacetic acid to the androst-5-enyl-acetic acid with platinum oxide in methanol containing some water. Hydrogen uptake stopped at one molar equivalent. However, if acetic acid was substituted for water, reduction was no longer selective.

Another example of selectivity in the reduction of two similar double bonds in a steroidal diene may be seen in the conversion of 3β-acetoxy-5α-cholesta-6,8(14),22-triene to the corresponding 8(14),22-diene with platinum oxide in dioxane [106]. Both the 6,7- and 22,23-double bond are symmetrically disubstituted, but differ in that the 6,7-double bond is conjugated with the 8,14-double bond.

Kern [5] has suggested that the presence of a carboxyl group conjugated with a double bond retards the rate of hydrogenation of the double bond. The selective reduction of an isolated double bond in the presence of such a conjugated double bond may be due to that effect. This author prefers to consider the palladium-catalyzed reduction of fusidic acid and its isomer,

(9.15)

Fusidic acid   24,25-Dihydrofusidic acid

lumifusidic acid, to the corresponding 24,25-dihydro acids [107 and references] as examples of the selective hydrogenation of a trisubstituted double bond in the presence of a tetrasubstituted one (see Eq. 9.15). Jones and associates [108] preferentially reduced a symmetrically disubstituted double bond in the presence of a tetrasubstituted one in the conversion of 2-methylcyclohexa-1,4-dienecarboxylic acid with palladium on calcium carbonate to 2-methylcyclohexenecarboxylic acid.

Selective Reduction Among Conjugated Di- and Polyenes

Lebedev and Yakubchik [109a, b, c] studied the reduction of a number of conjugated diolefins with platinum catalyst. From rate curves of the absorption of hydrogen they suggested that during uptake of one equivalent 1,2-, 1,4-, and 3,4-addition all took place. Their work was disputed by some workers because of the lack of verification by chemical analysis, although others also reported partial hydrogenation of butadiene derivatives [24 and included references].

Young and associates [24] studied the half hydrogenation of butadiene with platinum from platinum oxide, palladium from palladium oxide, palladium on barium sulfate, and Raney nickel. Their results indicated that palladium gave the best results for partial hydrogenation to monoene but since palladium also caused isomerism and/or double-bond migration, the authors did not say whether 1,4-addition actually took place. With platinum there was very little bond migration and it appeared that 1,2- or 3,4-addition was the major reaction.

Gostunskaya and colleagues [110] studied the half hydrogenation of isoprene, $CH_2{=}CHC(CH_3){=}CH_2$, and of 2,5-dimethylhexa-2,4-diene,

$$(CH_3)_2C{=}CHCH{=}C(CH_3)_2,$$

with palladium, platinum, and nickel catalysts, as well as the isomerizing effect of each catalyst. As in the work of Young, palladium caused the most isomerization; nickel caused the double bond to shift only to a slight extent, while platinum had no apparent isomerizing effect. From the results of the reduction of isoprene it appears that 1,4-addition as well as 1,2- and 3,4-addition took place. With platinum 31% of 2-methyl-2-butene was obtained, with nickel 42%, and with palladium 43%. The result with palladium may be discounted because of the known isomerizing effect of the catalyst. In the partial reduction of the dimethylhexadiene, 1,4-addition was a very minor reaction. Friedlin and associates [111], who showed that the reduction of isoprene in pyridine or quinoline would stop at one molar equivalent, obtained a mixture of isopentenes in their reactions.

Inhoffen, Müller, and Brendlar [112] hydrogenated 2,3-*bis*-(1-methyl-4,4-ethylenedioxycyclohexyl)-1,3-butadiene with palladium on carbon to obtain the corresponding 2-butenyl compound (Eq. 9.16), the structure of which was

$$\left[\begin{array}{c} RC \\ \parallel \\ CH_2 \end{array} \right]_2 \xrightarrow[\text{Pd on C}]{H_2} RC(CH_3)\!=\!C(CH_3)R \qquad (9.16)$$

established by nuclear magnetic resonance spectroscopy. The butenyl compound might result from a 1,4-addition of hydrogen or from 1,2-addition and double-bond migration during reduction since palladium is known to effect such a change. The authors, who obtained a 31% yield of the 2-butenyl compound after recrystallization, do not account for the remainder of the material, thus the extent of 1,2-addition is not known.

From a study of the literature on the hydrogenation of conjugated carbon to carbon double bonds it appears that most of the work points to 1,2-addition. Ohloff, Farnow, and Schade [113] reported 1,2-addition in the semihydrogenations of 1-vinylcyclohexene and of some conjugated 1-vinylbicyclo[3.1.0]heptenes to give very good yields of 1-ethylcyclo- and bicycloheptenes respectively. Wilcox and Neely [114] obtained an 82% yield of 1,2,3,4-tetramethylcyclobuten-3-ol benzoate from the reduction of 4-methylene-1,2,3-trimethylcyclobutene-3-ol benzoate with palladium on carbon. The authors stated that results of NMR spectroscopy showed that the reduction product was a mixture of geometrical isomers. Reductions of conjugated di- and polyenes among steroids also point to 1,2-addition [28, 106].

Ruyle and associates [115] reduced the 5,6-bond in 3β-acetoxy-5,7,22-ergosterol acetate; Fieser and Herz [116] selectively hydrogenated the 9,11-bond of 7,9-dehydrocholesterol (3,4-addition). In the hydogenation of 9α-lumisterol, a steroid of unnatural configuration, Castells [117] and his group reported that the 5,6-double bond was selectively saturated in the presence of Raney nickel (Eq. 9.17).

In some platinum-catalyzed low-pressure reductions of some highly substituted conjugated dienynes of the structure, RCH=CHC≡CCH=CHR,

9α-Lumisterol

Di- and Polyenes   149

symmetrical monoenes were obtained [118]. The monoenes were identified by results of oxidation. It was suggested [119] that after reduction of the acetylene bond to a conjugated triene further reduction by 1,6-addition and rearrangement gave RCH$_2$CH=CHCH=CHCH$_2$R which then underwent 1,4-addition and rearrangement to give the symmetrical monoene. Partial reduction of the dienynes with Raney nickel gave mixtures of symmetrical dienes and symmetrical monoenes which were identified by oxidation.

Conjugated Polyenic Acids, Esters, and Lactones

Early work on the selective catalytic reduction of compounds containing two or more double bonds in conjugation with a carboxyl group gave contradictory results [120]. There is little doubt that, under the proper conditions and with the proper catalyst, partial reduction can be achieved.

Stork [121] showed that methyl 4-(6-methoxy-1,2,3,4-tetrahydronaphthylidene)crotonate was semihydrogenated successfully with Raney nickel at room temperature and atmospheric pressure to yield methyl 4-(6-methoxy-1,2,3,4-tetrahydronaphthyl)but-2-enoate (reduction with platinum oxide or palladium on barium sulfate showed no discernible change in the rate of uptake). Stork suggested that when two structures may be expected from catalytic reduction, the favored one will be the one more stabilized by resonance. He stated that the work of Farmer [122], who obtained 2-hexenoic acid from sorbic acid, CH$_3$CH=CHCH=CHCOOH, was an illustration of that principle. Another illustration may be found in the selective reduction of cyclohexa-2,4-diene-1,2-dicarboxylic acid with platinum oxide to 1,4,5,6-tetrahydrophthalic acid [123], although this may be the result of the favored reduction of a symmetrical disubstituted double bond over the more hindered trisubstituted one. Another example of the retarding effect of greater substitution in this series is shown in the selective reduction of a trisubstituted bond in the presence of a tetrasubstituted double bond. Martell [124] hydrogenated ethyl 3-[2,3,4-trimethoxy-5-(2-carbethoxyethyl)-benzocyclohept-5-en-6-yl]-2-cyano-2-propenoate to the corresponding ethyl benzocycloheptenylcyanopropionate (Eq. 9.18).

Blake and co-workers [125] studied the hydrogenation of the lactone of

(9.18)

R^1, R^2, R^3 = OCH$_3$; X = CN, R^4 = COOC$_2$H$_5$, R^5 = (CH$_2$)$_2$COOC$_2$H$_5$

6-hydroxymethyl-7-carboxymethylene-1,2,4,7-tetrahydroazepino[3.2.1-hi]-indole with platinum, Raney nickel, and palladium on carbon and on barium sulfate. In all cases complex mixtures resulted. However, in one reduction with palladium on barium sulfate the investigators were able to isolate by column chromatography 14% of apo-β-erythroidine, 20% of isoapo-β-erythroidine, and 13% of unreacted lactone (see Eq. 9.19). The postulate of Stork [121] predicts the presence of isoapo-β-erythroidine; the formation of

Lactone → Apo-β-erythroidine + Isoapo-β-erythroidine (9.19)

apo-β-erythiodine may result from 1,4-addition and rearrangement. However, the known isomerizing effect of palladium catalyst must also be considered before one suggests 1,4-addition.

Literature on the catalytic hydrogenation of muconic acid (2,4-hexadien-1,6-dioic acid) is scanty and at times contradictory [see Ref. 24]. Elvidge, Linstead, and Smith [126] studied the reduction of the *cis-cis*, *trans-trans*, and *cis-trans* acids. They showed that reduction of the acids in aqueous solution neutralized with sodium bicarbonate in the presence of palladized charcoal was more or less selective leading mainly to 2-hexenedioic acids. 3-Hexenedioic acid (11–13%) was found which the authors suggested was formed by direct 1,4-addition.

The question of whether 1,4-addition does take place in the partial reduction of conjugated polyenic acids, esters, and lactones is still unanswered. The possibility that the products of apparent 1,4-addition of hydrogen may not have been formed by 1,2-addition followed by double-bond migration has not been excluded.

SELECTIVE HYDROGENATION IN THE PRESENCE OF OTHER REDUCIBLE GROUPS

In the Presence of Aldehydes or Ketones

There are literally hundreds of references on the selective hydrogenation of the olefinic bond in the presence of aldehyde or ketone groups. In general, selective conversion of the double bond may be expected in hydrogenations with most catalysts under mild conditions. An example can be seen in the Raney nickel-catalyzed reduction of 2,5-diphenyl-4-acetoxycyclopent-4-en-1,3-dione [127]. The uptake of two equivalents of hydrogen resulted in saturation of the double bond and hydrogenolysis of the acetoxy group to form 2,4-diphenylcyclopentan-1,3-dione. Bourdiol and associates [128] used more vigorous conditions, 70° and 40 atm pressure, to reduce 4-methylpent-3-en-2-one with Raney nickel in the presence of a small amount of water and potassium hydroxide to obtain 4-methylpentan-2-one.

α,β-Unsaturated Ketones. Covert, Connor, and Adkins [129] selectively reduced 4-phenyl-3-butene-2-one (benzalacetone) with nickel on kieselguhr at 45–85° and 125 atm pressure to obtain a 93% yield of 4-phenylbutan-2-one. In an attempt to obtain 4-(2-hydroxy-3-methoxy and 3-methoxy-4-hydroxyphenyl)butan-2-ones from the corresponding unsaturated ketones, over-hydrogenation occurred with Raney nickel and with platinum oxide at 2–3 atm pressure [30]. The reduction was controlled by using a slight excess of hydrogen (105–110% of one molar equivalent) and a lower than normal amount of catalyst. Nomura [130] reported success with platinum black in a slow reduction (9 hr) of 3-methoxy-4-hydroxybenzalacetone in ether at room temperature and atmospheric pressure. Mannich and Merz [131], on the other hand, found that the isomeric 3-hydroxy-4-methoxy compound absorbed 1.5 molar equivalents in the presence of palladium on carbon in alcoholic solution giving the saturated ketone as the major product and 25.5% of the saturated alcohol.

Examples of selectivity in reductions of other α,β-unsaturated ketones may be seen in the palladium oxide hydrogenation of 2-methylene-3-carbethoxycyclopentanone to 2-methyl-3-carbethoxycyclopentanone [132], in the reduction of 4-(2-methyl-5-isopropenylcyclopentenyl)butan-2-one to the corresponding isopropylcyclopentylbutanone with palladium on calcium carbonate at 125° and 35 atm pressure [98] and in the reduction of 2-benzylidene-1-tetralone to 2-benzyl-1-tetralone with platinum oxide [133]. The Raney nickel hydrogenation of 2-methylpent-1-en-3-one stopped at one molar equivalent to yield 2-methyl-3-pentanone [134]. In another example, the conversion of 3-(4-methoxybenzoyl)acrylic acid to the corresponding benzoylpropionic acid was accomplished with 1% by weight of platinum oxide to substrate or with a 10% weight ratio of 5% palladium on carbon, perhaps the better catalyst for the reduction [30].

Sam and associates [135] selectively reduced the ethylenic bond in some 2-pyridylmethylen-1-indanones in ethyl alcohol and in acetic acid and as their hydrochloride salts in aqueous alcohol in the presence of palladium on carbon at 3.5 atm pressure. With Raney nickel, reduction of 5,6-dimethoxy-2-(2-pyridylmethylene)-1-indanone under the same conditions gave 58% of the corresponding pyridylmethylindanone plus 12% of 5,6-dimethoxy-2-(2-pyridylmethyl)indan-1-ol.

Nakata [136] suggested that the method of Brown and Brown [137] might be useful in the reduction of conjugated double bonds in steroidal compounds. He obtained very high yield of 3β-hydroxy-5α,17α-pregnan-20-one from the hydrogenation of 3β-hydroxy-5α-pregn-16-en-20-one with platinum catalyst from chlorplatinic acid and sodium borohydride. Maksimov [138] reduced the 4,5-bond of 3-keto-24-hydroxy-24,24-diphenylchol-4-ene in pyridine with 2% palladium on calcium carbonate in a reaction that was interrupted after uptake of one molar equivalent of hydrogen. Mazur [139] noted that uptake of hydrogen stopped at one equivalent in the conversion of 4-methylstigmast-4-en-3-one in ethyl alcohol with palladium on carbon to yield 4α-methylstigmastan-3-one. Robinson and associates [140] cited an example of the reduction of the 1,2-double bond in 5,9-cyclo-11β,21-diacetoxypregn-1-en-17α-ol-3,20-dione in ethyl acetate in the presence of palladium on carbon.

Isolated Double Bonds in the Presence of Ketones. Reductions of isolated double bonds in the presence of a ketone group have been carried out in the presence of platinum, palladium, and nickel catalysts. Fieser and Novello [45] used platinum oxide to saturate the double bond in 2-(1-naphthoyl)cyclohex-4-enylcarboxylic acid; Conroy and Firestone [141] used the same catalyst to convert 2,6-dimethyl-2-propenylcyclohexanone to 2,6-dimethyl-2-propylcyclohexanone in 92.6% yield. Other examples may be found in the work of Beerebom [142], in that of Foote and Woodward [143], who reduced bicyclo[3.2.0]oct-3-en-8-one to the saturated ketone in methyl alcohol, and in the high yield of saturated ketone obtained by hydrogenation of 2-carbethoxy-2-(4-carbethoxy-1-butenyl)cyclopentanone [144].

Newman [145] used platinum on kieselguhr to effect an excellent conversion of 10-methylenespiro[4.5]decan-6-one and 10-methylspiro[4.5]dec-9-en-6-one to the 10-methyldecanone. Schechter and Kirk [146] found Raney nickel useful in reducing 2-allylcyclohexanone to 2-propylcyclohexanone, and Stetter [147] used the same catalyst in aqueous sodium hydroxide to reduce 2-allyl-5,5-dimethylcyclohexan-1,3-dione to 2-propyl-5,5-dimethylcyclohexane-1,3-dione. Many investigators have found palladium catalysts selective for reduction of isolated double bonds in the presence of a ketone function even in instances where the amount of catalyst was larger than normal. Brown and co-workers [148] obtained only the saturated ketone in the reduction of N-[2-(3-methoxyphen)ethyl]-7-azaperhydroind-1(8)-ene-3,6-dione with 20% by weight of 5% palladium on carbon (see Eq. 9.20). Dinwiddie and McManus [149] successfully reduced the isomeric 2-acetylbicyclo[2.2.1]hept-5-enes with 15% by weight of 5% palladium on carbon to the corresponding 2-acetylbicycloheptanes; House and Carlson [150] converted 1,1a,4,4a-tetrahydrofluoren-9-one with the same catalyst to 1,1a,2,3,4,4a-hexahydrofluoren-9-one. 2,2-Dimethyl-5-phenylcyclohex-4-enone was reduced to the saturated ketone in the presence of palladium on carbon [151]. Hydrogenation of 4-acetylcyclohexene to acetylcyclohexane

$$\text{(9.20)}$$

took place rapidly with a small amount of the same catalyst [30]. Uptake of hydrogen never went beyond one molar equivalent. Sands [152] found palladium on barium sulfate useful in reduction of 1-carbethoxybicyclo-[4.3.1]dec-3-en-10-one to the saturated bicyclic ketone.

Supported palladium catalysts may be best for the reduction of unconjugated double bonds in steroidal ketones. Bowers and Ringold [153] used palladium on carbon to selectively reduce the 5,6-double bond in a 3-keto steroid system in methyl alcohol, and Dodson [154] converted 1α, 3β-dihydroxyandrost-5-en-17-one to the dihydroxyandrostan-17-one with the same catalyst in ethyl alcohol. Plattner [155] used platinum oxide to hydrogenate pregnenolone acetate in acetic acid solution. Despite a sharp break in the rate of hydrogen uptake at one molar equivalent, it was found necessary to allow overhydrogenation to take place and to reoxidize to convert the C-20 hydroxyl group to the ketone. In the platinum-catalyzed reduction of pregnenolone in acetic acid containing 1–2% of water, it was found that the carbonyl group was hydrogenated almost as readily as the 5,6-double bond [30].

α,β-*Unsaturated Aldehydes.* Since an aldehyde group is less hindered than a ketone group, it is therefore more open to attack by hydrogen in the presence of a hydrogenation catalyst. Nevertheless an isolated double bond in the presence of an aldehyde or a double bond in conjugation with an aldehyde can be reduced selectively.

In the reduction of double bonds conjugated with an aldehyde, carbonyl palladium may be the catalyst of choice. Moe, Warner, and Buckley [156] reduced the double bond of 4,4-dicarbethoxycyclohexene-2-carboxaldehyde with palladium on carbon. In some experiments concerned with the reduction of 2-methylpent-2-enal [30], it was found that 5% palladium on carbon could be used to saturate the double bond without affecting the aldehyde group. When a low weight ratio of catalyst to substrate was used (2.5%), it appeared virtually impossible to cause overhydrogenation at 60° and 3 atm pressure.

In the same reductions, uptake of hydrogen using a normal weight ratio (1%) of platinum oxide was very slow at 60° and 3 atm pressure. Raney nickel could be used but it was necessary to carefully follow hydrogen uptake. The tendency toward overhydrogenation in the nickel-catalyzed reduction of double bonds conjugated with an aldehyde group has been noted by others. Woods and Sanders [157] in the reduction of penta-2,4-dienol found that amyl alcohol as well as veleraldehyde was obtained, and Palfray [158] reported that an analysis of the products of reduction of cinnamaldehyde with Raney nickel at 15° showed 78% of aldehyde and 22% of alcohol. As the reaction temperature was increased, the amount of alcohol also increased.

Isolated Double Bonds in the Presence of an Aldehyde Group. Reduction of an isolated double bond in the presence of an aldehyde is generally selective. The double bond of 3-methoxy-4-hydroxy-5-allylbenzaldehyde was saturated in glacial acetic acid in the presence of palladium on barium sulfate [159]; 1,2,3,6-tetrahydrophthalaldehyde was converted to hexahydrophthalaldehyde with palladium on carbon [160]; Hennis and Trapp [161] used a very small amount of palladium on carbon to hydrogenate cyclohex-3-enaldehyde at 75–80° and 14 atm pressure. The temperature was not allowed to rise above 80°. High yield of cyclohexanecarboxaldehyde (99% pure by gas-liquid chromatography) was obtained.

Nazarov and his group [162] used palladium on calcium carbonate to reduce 4-methyl, 2,6-dimethyl, and 4,6-dimethylcyclohex-3-enecarboxaldehydes to the saturated aldehydes. In each reduction uptake was incomplete. In order to complete saturation of the double bond, the catalyst was removed and hydrogenation continued with platinum oxide catalyst. It is of interest that palladium on calcium carbonate, which was used successfully by Nazarov in reducing an isolated double bond in some cyclohexenyl ketoalcohols, was less efficient in the hydrogenation of the double bond in the presence of an aldehyde group.

When there is a tendency toward overhydrogenation in the reduction of double bonds in the presence of aldehyde or ketone functions, the use of a smaller than normal amount of catalyst and the use of a slight excess of hydrogen might be helpful in controlling the reduction. A modified Raney nickel catalyst reported by Fukawa [163] to deactivate reduction of the carbonyl group in unsaturated compounds might also prove useful. In some cases it may be necessary to protect the carbonyl function as a ketal or acetal before subjecting the compound to reduction.

The Double Bond of α,β-Unsaturated Ketones in the Presence of an Isolated Double Bond. In the case of α,β-unsaturated ketones, conjugation with the carbonyl may make the double bond easier to reduce. The reduction of such a conjugated double bond in the presence of an isolated double bond may not always be predictable but in many instances it can be rationalized. Edgar, Harper, and Kazi [164] hydrogenated 3-methyl-3-allylcyclohex-2-enone in methyl alcohol with palladium on barium sulfate to obtain an 81% yield of methylpropylcyclohexenone. Based on steric factors alone, the result could be expected since it can be viewed as the reduction of a monosubstituted double bond over a tetrasubstituted one. The results are also consistent with Stork's suggestion [121] that when two reducible double bonds are present the one stabilized by resonance is less likely to be reduced.

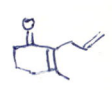

In contrast to this suggestion, the reductions of (a) 3β-acetoxypregna-5,16-dien-12,20-dione with palladium on barium sulfate in ethyl acetate [165], (b) of 3β-acetoxy-6,16-dimethylpregna-5,16-dien-20-one in ethyl alcohol with Raney nickel (washed until it was neutral) [166], and (c) of 3β-acetoxy-16-acetyl-17-acetylaminoandrosta-5,16-diene in ethyl acetate and acetic acid in the presence of platinum oxide [167], all showed that the 16,17-double bond was preferentially hydrogenated. The first two results may be rationalized on the basis of the known difficulty of the 5,6-bond to undergo reduction in neutral solvent but this anology does not apply to the result obtained in the third hydrogenation. In the above examples the 16,17-bond may be more accessible to the catalyst than the 5,6-bond. 24,24-Diphenylchola-4,23-dien-3-one was reduced to the corresponding 4-en-3-one in pyridine in the presence of palladium on calcium carbonate [138]. In this reduction the presence of the aromatic rings in the side chain may allow the double bond conjugated with them to be preferentially adsorbed to the catalyst and also preferentially reduced.

Conjugated Dienones

Krakower and Van Dine [168] found that 1α,2α-methylene-16α,17α-(dimethylmethylenedioxy)pregna-4,6-diene was selectively hydrogenated to the corresponding pregn-4-en-3,20-dione in methyl alcohol containing potassium hydroxide in the presence of palladium on carbon. Shepherd and his group [169] used a similar procedure for the reduction of ergosta 4,6,22-trien-3-one to ergosta-4,22-dien-3-one in 80% yield. They suggested that the yield was dependent on the concentration of potassium hydroxide in the solvent and on prehydrogenation of the catalyst. The authors reported that the catalyst exhibited a fair degree of selectivity in neutral solvents but that reduction in the presence of potassium hydroxide gave the best results. Kinetic studies indicated that reduction of the trienone went stepwise to the 4,22-dien-3-one and then to ergost-22-en-3-one. The first step may be

156   Olefins

expected since to 22,23-double bond in steroids is reported to reduce slowly. It is possible that reduction by way of the enol or enolate anion formed in the basic medium may have been responsible for selective reduction of the conjugated 4,5-double bond over the 22,23-double bond. In another example of selectivity [171] 1,14-dimethyl-2-keto-3,4,5,8,12,13-hexahydrophenanthrene in methyl alcohol containing sodium hydroxide was hydrogenated to 1,14-dimethyl-2-keto-3,4,5,8,9,10,12,13-octahydrophenanthrene in the presence of 2% palladium on strontium carbonate (Eq. 9.21).

$$\text{(structures)} \quad (9.21)$$

Selective Reductions in the Presence of Unsaturated Nitrogen-Containing Groups

The olefinic bond will be preferentially reduced in the presence of most unsaturated nitrogen-containing groups except an aromatic nitro group. However there is some question about selectivity in the presence of a C=N bond, see chapter XV.

In the Presence of Nonaromatic Nitro Groups. Reduction of the double bond in substituted nitrocyclohexenes with platinum oxide is reported by Drake and Ross [172]. They noted that there was a sharp break in the rate of uptake at one equivalent in the hydrogenation of 1,2,4-trimethyl-5-nitrocyclohexene and 1,2-dimethyl-4-(2-ethoxyethyl)-5-nitrocyclohexene. Other references cited in the same article also report reductions of nitrocycloalkenes with platinum catalyst in acetic acid to yield the corresponding nitrocycloalkanes. An exception, in which inconsistent results were reported, was the hydrogenation of 2-nitrobicyclo[2.2.1]hept-5-ene [173]. Minami and Uyeo [174] used platinum oxide in the reduction of the double bond of 4-nitro-5-(2-methoxy-3-ethoxyphenyl)cyclohexene in dioxane solution to the corresponding nitrocyclohexane in high yield. Palladium on carbon was reported by Drake and Ross to be less selective in these reductions. Apparently the nitro group was also affected. Huitric [175] reported its application in the hydrogenation of the olefinic bonds of 4-nitro-5-(4-acetoxyphenyl) and 4-(3,4,5-trimethoxyphenyl)-5-nitrocyclohexene in neutral solution.

Reduction of the double bond of 1,2-dimethyl-4-nitro-5-(3,4,5-trimethoxyphenyl)cyclohexene could not be carried out selectively with palladium on carbon in acetic acid [30]. A mixture of products was obtained. A study of the reduction showed that there was no break in the rate of uptake of one equivalent of hydrogen. Reduction in acetic acid in the presence of Raney nickel resulted in preferential hydrogenation of the nitro group to give the unsaturated amine in 85% yield. This unexpected result can be understood by comparing a molecular model of the starting material with a model of a nitrocyclohexene containing an unsubstituted double bond. The model of the latter compound shows that the double bond and the nitro group are in position to contact the catalyst. Selective reduction of the double bond in that type of compound has already been reported [175]. The model of the 1,2-dimethyl substituted compound shows that the double bond is in a less favorable position for reduction than the nitro group.

Other reductions of the double bond in the presence of an aliphatic nitro group have been reported [176a, b]; reduction of a nitrodihydropyran gave the nitrotetrahydropyran [177]. High yields were obtained in these palladium-catalyzed hydrogenations in neutral solvents.

In the Presence of a Cyano Group. An ethylenic bond is usually selectively reduced in the presence of a cyano group. Cope [178] found that palladium on carbon and platinum oxide catalyzed selective reductions of alkylidene cyanoacetic esters to the corresponding alkyl cyanoacetic esters in yields of 90% or better. In some instances the reduced ester was washed with dilute acid to remove a small amount of amine formed by reduction of the cyano group.

Substituents on the double bond may change the course of reduction. It was shown [30] that the ethylenic bonds in ethyl 2-cyano-2-ethyl-3-methylhex-3-enoate and ethyl 2-cyano-2,3-diethylpent-4-enoate were unaffected by reductions in ethyl alcohol with palladium on carbon or platinum oxide; reaction led to mixtures of primary and secondary amines. On the other hand, the double bond of ethyl-2-cyano-3-methylhex-2-enoate was readily saturated without affecting the cyano group. A comparison of molecular models (Figs. 6.1 and 6.2) suggests that the atypical results are probably due to steric effects.

Wawzonek and Smolin [179] reported that the double bond of 2,3-diphenylcinnamic acid failed to reduce with the usual catalysts, but platinized nickel led to 2,3,3-triphenylpropylamine.

Poor selectivity appears to have been obtained in the reduction of 1-cyanocyclohexene with platinum oxide [180] which gave both cyanocyclohexane (31% yield) and di(cyclohexylmethyl)amine, $(C_6H_{11}CH_2)_2NH$.

Pesson and his group [181] noted that in the reduction of 2-phenyl-3-(2-furyl)acrylonitrile in alcohol with palladium on carbon there was no change in the rate of hydrogenation until two equivalents were absorbed. The products of reduction consisted of a mixture of the substituted propionitrile and 2-phenyl-3-(tetrahydro-2-furyl) propionitrile. The addition of thiophene (7–8.5 mg/g of catalyst) to a similar reduction resulted in the formation of only 2-phenyl-3-(2-furyl)propionitrile.

Bergman and Ikan [182] carried out the selective hydrogenation of diethyl cyclopentylidenecyanoacetate in ethyl alcohol to the corresponding cyclopentyl compound. It is of interest that in this reduction the stereochemistry of the product of reduction is highly dependent on the catalyst. Reduction with platinum oxide gave the *cis* hydrogenated cyanodiester, while hydrogenation in the presence of palladium on carbon gave the corresponding *trans* compound.

In the Presence of Other Unsaturated Nitrogen-Containing Groups.

Although Ohno and associates [183] carried out the reduction of 2-morpholino and 2-piperidinocyclodedeca-5,9-dienonoxime under conditions in which oximes are frequently converted to amines (acetic acid with a large amount of palladium catalyst), the double bonds were selectively reduced.

Cope and LeBel [184] selectively reduced some alkenyl N-methylhydroxylamines in methyl alcoholic solution with hydrogen-pretreated palladium on calcium carbonate to obtain high yields of N-alkyl-N-methylhydroxylamines.

Nace and Goldberg [185] found that the hydrogenation of 2-(2-ethylhex-2-enylideneamino)-1-butanol with Raney nickel in dry methyl alcohol stopped after uptake of one equivalent of hydrogen to give 2-(2-ethylhexylideneamino)-1-butanol. Studying the effect of substitution on the olefinic bond they found that in the reduction of $C_2H_5CH(CH_2OH)N=CHCR=CHR'$ when $R = H$ and $R' = CH_3$ and $R = C_2H_5$ and $R' = C_3H_7$, the rate of reduction of the C=C bond was twenty times that of the C=N bond [186]. When $R = H$ and $R' = C_6H_5$ the rate was only four times more rapid than the reduction of the azomethine group.

Selective Reductions in the Presence of Hydrogenolyzable Functions

As a rule, catalytic hydrogenolysis of N- and O-benzyl and carbobenzoxy groups and related types occurs less readily than does reduction of ethylenic linkages.

In the Presence of N-Benzyl and Carbobenzoxy Protecting Groups.
4-Styryl-2-benzylaminopyrimidine was selectively hydrogenated to 4-(2-phenethyl)-2-benzylaminopyrimidine [187] with palladium on carbon which is the catalyst of choice for debenzylation of amines. The ethylenic bond and the aromatic nitro group in 1-(3-carbobenzoxyamino-5-carbomethoxy-2-pyridyl)-2-(2-nitrophenyl)ethylene were both reduced in the presence of Raney nickel without removal of the carbobenzoxy protecting group [188].

In contrast to these reductions Bailey and Lutz [189] found that cis-1,2-dibenzoyl-1-(morpholinomethyl)ethylene as base or hydrochloride underwent deamination in ethyl alcohol in the presence of platinum oxide or palladium on barium sulfate to give 1,2-dibenzoylpropane. Reduction in acetic acid gave the expected 1-morpholine-2,3-dibenzoylpropane. Since two molar equivalents of hydrogen were rapidly absorbed in the reduction in alcohol and it was found that 1-morpholino-2,3-dibenzoylpropane was relatively stable to reduction, it was concluded that deamination took place before or concurrently with reduction of the double bond. The authors suggested several mechanisms for the reaction, one of which was direct 1,2-reductive deamination analogous to debenzylation. Selective reductions of double bonds in the presence of benzyloxy groups are exemplified in the hydrogenation of 1,2-*bis* benzyloxy-3-(1-pentadecenyl)benzene with palladium on carbon [190] and in the reduction of 3-benzyloxy-4-methoxycinnamic acid with nickel or platinum catalyst in neutral or alkaline solution [30]. In contrast, with the isomeric 3-methoxy-4-benzyloxycinnamic acid and 2-benzyloxy-3-methoxycinnamic acid hydrogenolysis always occurred leading to a mixture of products [30]. However, reduction of the double bond of 2-benzyloxy-3-methoxycinnamic acid could be effected in ethyl alcohol with copper chromite at 125–150° and 80 atm pressure without debenzylation [30].

In the Presence of Halogen. Conditions most conducive to hydrogenolysis of halogen atoms in organic compounds will be discussed at greater length in a later chapter. However, under the proper conditions selective hydrogenation of double bonds in the presence of chlorine and bromine can be accomplished. The literature concerning selective reductions of olefinic iodides is very meager and further study is needed.

To prevent possible loss of halogen it may be advisable to avoid basic conditions, particularly in the presence of palladium catalyst or to choose a less active catalyst when the compound to be reduced contains a basic side chain. Nevertheless, despite the presence of a basic side chain, 2-aminomethyl-3-chloromethylbicyclo[2.2.1]hept-5-ene was reduced to the bicycloheptane with palladium on carbon without loss of halogen [191]. There are also reports of the use of palladium in selective reductions of alkenyl halides. However, since it is the most active of the catalysts for hydrogenolyses, other catalysts capable of reducing the double bond should be selected.

In general, saturation of the olefinic bond, in compounds containing a halogen atom bound to an aromatic ring, will be the preferred reaction. Platinum oxide gave good selectivity in the reduction of some halophenylethylenes [192a, b] although in one instance loss of bromine occurred during hydrogenation of 1,1-di(4-methoxyphenyl)-2-(3-bromophenyl)ethylene at 65° and 3 atm pressure when the reaction was allowed to continue beyond uptake of one equivalent of hydrogen [192b]. Raney nickel catalyst has also been found effective for reduction of the double bond without loss of chlorine or bromine [175a, 193, 194]. Huitric [175a] noted that when 6-(4-iodophenyl)cyclohex-3-enone was hydrogenated, only a fraction of the calculated amount of hydrogen was absorbed. Deiodination may have occurred, releasing hydrogen iodide which either destroyed some of the catalyst or poisoned it. Palladium catalysts have been used without hydrogenolysis of chloride or bromide [82, 175a, 195, 196]. Selective hydrogenation of 3-bromo-4,5-dimethoxycinnamic acid in methyl alcohol to 3-bromo-4,5-dimethoxyhydrocinnamic acid in the presence of palladium on barium sulfate has been reported [197]. In contrast, attempts to reduce the double bond of 3-bromo-4-hydroxy-5-methoxycinnamic acid in neutral solvent with platinum, nickel, or palladium catalysts without loss of bromine were unsuccessful [30].

Kindler and associates [198], in studies on the hydrogenation of 2-allyl-4-chlorophenol with palladium black to 2-propyl-4-chlorophenol, found that dehalogenation, when it occurred, was dependent on the solvent. In reductions with the lower alcohols as solvents the amount of ionic halide formed ranged from 3.0–8.6% and was 20% in 80% ethyl alcohol and only 2.8% in dioxane and glacial acetic acid. In contrast no ionic halogen was found on hydrogenation in thiophene-free benzene or in cyclohexane. In reductions in ethyl alcohol the addition of nicotinamide, diethylnicotinamide, or thiophene as inhibitors limited loss of chlorine to less than 1%. The corresponding 4-bromo compound underwent considerable loss of bromine on reduction in ethyl alcohol, but in thiophene-free benzene no bromine ion was detected after uptake of one equivalent of hydrogen; 2-propyl-4-bromophenol was obtained in 90% yield. This study was confined to the use of palladium catalyst. It would be of value to extend it to the use of nickel and platinum catalysts.

It may be expected that reductions of double bonds in aliphatic structures, which also contain a chlorine or bromine atom, will be selective when the double bond is separated from the halogen atom so that an allylic or vinylic type halide cannot be formed. While it is known that vinyl halides are not particularly reactive, the possibility of isomerization and double-bond migration to form an allylic halide in the presence of certain catalysts cannot be ignored.

In examples of successful selective double-bond hydrogenations, platinum oxide was employed to reduce 1-chlorononadeca-4,7,10,13-tetraene to nonadecyl chloride; 1-chlorotetradec-7-ene was hydrogenated to myristyl chloride in quantitative yield with palladium on carbon catalyst [200], and 5-(4-chlorobutylidene)hydantoin was reduced with platinum oxide in alcohol to 5-(4-chlorobutyl)hydantoin [201]. Lewis and Shoppee [80] not only reduced the 3α-chlorocholest-5-ene to 3α-chlorocoprostane in ethyl acetate containing perchloric acid with platinum oxide but also found that the corresponding 3α-bromo compound was hydrogenated without loss of bromine.

1,3,5-Triphenyl-7,7-dichloro-2-oxanorcarene was reduced in poor yield (19%) with Raney nickel in tetrahydrofuran [202]. The investigators, however, reported 90% yield of 7,7-dichloronorcarane from reduction of 7,7-dichloronorcarene with platinum by means of the method of Brown and Brown [137].

McMullen and associates [203] stated that their experiments on the hydrogenation of some vinyl bromides appeared to indicate that palladium catalyst was specific for conversion of the —CC(Br)=C— grouping to —CCH₂CH—, although they did show that in the reduction of 2,3-dibromopropene only the double bond was reduced.

Ham and Coker [204] suggested that rhodium on alumina was a superior catalyst for the hydrogenation of a vinyl or allyl chloride to a haloalkane. Reductions were carried out at 100° and 30–40 atm pressure, but they made no comparison with other catalysts. Allyl chloride and 1-chloropropene gave 57.8% and 42.4% respectively of propyl chloride. 5-Chlorohexene, where the double bond and chlorine atom are further separated than in allyl chloride, was reduced to 2-chlorohexane in almost 97% yield under similar conditions. 1,3-Dichloropropene was hydrogenated with rhodium on alumina to yield a mixture of 47.9% of 1,3-dichloropropane and 35.4% of propyl chloride. While the yield of propyl chloride was about the same with palladium and with platinum on alumina, the yield of 1,3-dichloropropane was much lower with palladium (19.2%) and platinum (6.1%). The authors found the use of a less polar solvent as cyclohexane, as noted by Kindler [198], gave better yield than reduction in alcohol or acetic acid, but the addition of thiophene did not act as a deterrent to hydrogenolysis.

Loss of halogen as well as rupture of the C—N bond was reported in the hydrogenation of 1-benzoyl and 1-phenylacetyl-2,2-di(2-chloroallyl)hydrazines with rhodium in neutral solvent and with palladium on carbon or platinum on carbon in alcoholic hydrogen chloride [205]. The authors implied that loss of halogen occurred before reduction of the double bond and rupture of the C—N bond but did not stop hydrogen uptake at two equivalents to determine whether the double bond was reduced or whether dehalogenation took place.

REFERENCES

[1] S. V. Lebedev, G. G. Kobliansky, and A. O. Yakubchik, *J. Chem. Soc.*, **172**, 417 (1925).
[2] G. Dupont, *Bull. Soc. Chim. France*, **1936**, 2021.
[3] N. A. Dobson, G. Eglinton, M. Krishnamurti, R. A. Raphael, and R. G. Willis, *Tetrahedron*, **16**, 16 (1961).
[4] L. M. Berkowitz and P. N. Rylander, *J. Org. Chem.*, **24**, 708 (1959).
[5] J. W. Kern, R. L. Shriner, and R. Adams, *J. Am. Chem. Soc.*, **47**, 1147 (1925).
[6] B. A. Kazanskii and G. T. Tatevosyan, *J. Gen. Chem. USSR*, **9**, 1458 (1939); through *Chem. Abstr.*, **34**, 2783 (1940).
[7] B. Gauthier, *Compt. Rend.*, **217**, 28 (1943).
[8] B. A. Kazanskii and G. T. Tatevosyan, *J. Gen. Chem. USSR*, **9**, 2256 (1939); through *Chem. Abstr.*, **34**, 4731 (1940).
[9] N. K. Yurashevskii, *J. Gen. Chem. USSR*, **8**, 438 (1938); through *Chem. Abstr.*, **32**, 7908 (1938).
[10] S. P. Lagerev and S. F. Babek, *J. Gen. Chem. USSR*, **7**, 1661 (1937); through *Chem. Abstr.*, **31**, 8342 (1937).
[11] S. P. Lagerev and M. M. Abramov, *J. Gen. Chem. USSR*, **8**, 1682 (1938); through *Chem. Abstr.*, **33**, 4957 (1939).
[12] G. F. Hennion and A. C. Hazy, *J. Org. Chem.*, **30**, 2650 (1965).
[13] A. Horeau and J. Jacques, *Compt. Rend.*, **230**, 1667 (1950).
[14] J. Julia, J. Jacques, and A. Horeau, *Compt. Rend.*, **230**, 660 (1950).
[15] a. J. Jacques and A. Horeau, *Bull. Soc. Chim. France*, **1948**, 711; b. Models were constructed by this author.
[16] M. S. Platonov, *J. Russ. Phys. Chem. Soc.*, **61**, 1055 (1929); through *Chem. Abstr.*, **24**, 539 (1940).
[17] A. K. Plisov and A. V. Bogatsky, *J. Gen. Chem. USSR*, **27**, 403 (1957) (English translation).
[18] a. C. Paal and H. Schiedwitz, *Ber.*, **60**, 1221 (1927); b. *Ber.*, **63**, 766 (1930).
[19] A. L. Markman and E. V. Zinkova, *J. Gen. Chem. USSR*, **32**, 346 (1962), English translation.
[20] C. Weygand, A. Werner, and W. Lanzendorf, *J. Prakt. Chem.*, **151**, 231 (1938).
[21] I. Jardine and F. J. McQuillin, *J. Chem. Soc. (C)*, **1966**, 458.
[22] G. C. Bond, *Catalysis by Metals*, Academic Press, London and New York, 1962, Section 21.33.
[23] G. C. Bond, G. Webb, P. B. Wells, and J. M. Winterbottom, *J. Catalysis*, **1**, 74 (1962).
[24] W. G. Young, R. L. Meier, J. Vinograd, H. Bollinger, L. Kaplan, and S. L. Linden, *J. Am. Chem. Soc.*, **69**, 2046 (1947).
[25] H. F. Stavely and G. N. Bollenback, *J. Am. Chem. Soc.*, **65**, 1600 (1943).
[26] H. Wieland and W. Benend, *Ann.*, **554**, 1 (1943).
[27] P. Bladon, J. M. Fabian, H. B. Henbest, H. P. Koch, and G. W. Wood, *J. Chem. Soc.*, **1951**, 2402.
[28] C. Djerassi, J. Romo, and G. Rosenkranz, *J. Org. Chem.*, **16**, 754 (1951).
[29] M. B. Moore, H. B. Wright, M. Vernsten, M. Freifelder, and R. K. Richards, *J. Am. Chem. Soc.*, **76**, 3656 (1954).
[30] M. Freifelder, unpublished results.

References

[31] A. R. Katritzky, *J. Chem. Soc.*, **1955**, 2581.
[32] L. A. Walter, W. H. Hunt, and R. J. Fosbinder, *J. Am. Chem. Soc.*, **63**, 2771 (1941).
[33] M. Freifelder, *J. Org. Chem.*, **28**, 602 (1963).
[34] J. Walker, *J. Chem. Soc.*, **1947**, 1684.
[35] R. G. Jones, E. C. Kornfeld, and K. C. McLaughlin, *J. Am. Chem. Soc.*, **72**, 3539 (1950).
[36] W. Ried and H. Keller, *Ber.*, **89**, 2578 (1956).
[37] E. C. Taylor and A. McKillop, *J. Am. Chem. Soc.*, **87**, 1984 (1965).
[38] R. C. Elderfield, B. M. Pitt, and I. Wempen, *J. Am. Chem. Soc.*, **72**, 1334 (1953).
[39] D. W. Adamson and J. W. Billinghurst, *J. Chem. Soc.*, **1950**, 1039.
[40] L. Katz and L. S. Karger, *J. Am. Chem. Soc.*, **74**, 4085 (1952).
[41] H. H. Lochte, P. F. Kruse, Jr., and E. N. Wheeler, *J. Am. Chem. Soc.*, **75**, 4480 (1953).
[42] I. Suzuki, *Chem. Pharm. Bull. Japan*, **6**, 479 (1956).
[43] G. F. Hennion, C. C. Price, and V. C. Wolff, Jr., *J. Am. Chem. Soc.*, **77**, 4633 (1955).
[44] G. Crane, C. E. Boord, and A. L. Henne, *J. Am. Chem. Soc.*, **67**, 1237 (1945).
[45] L. F. Fieser and F. C. Novello, *J. Am. Chem. Soc.*, **64**, 802 (1942).
[46] H. E. Baumgarten, F. A. Bower, and T. T. Akomoto, *J. Am. Chem. Soc.*, **79**, 3145 (1956).
[47] W. J. Bailey and W. B. Lawson, *J. Am. Chem. Soc.*, **79**, 1444 (1957).
[48] W. J. Bailey and W. A. Klein, *J. Am. Chem. Soc.*, **79**, 3124 (1957).
[49] R. T. Arnold and J. F. Dowdall, *J. Am. Chem. Soc.*, **70**, 2590 (1948).
[50] W. E. Bachmann, G. I. Fujimoto, and L. B. Wick, *J. Am. Chem. Soc.*, **72**, 1995 (1950).
[51] R. P. Linstead, R. R. Whetstone, and P. Levine, *J. Am. Chem. Soc.*, **64**, 2014 (1942).
[52] L. H. Klemm and W. Hodes, *J. Am. Chem. Soc.*, **73**, 5181 (1951).
[53] L. H. Klemm and R. Mann, *J. Org. Chem.*, **29**, 900 (1964).
[54] A. Bertho, *Ber.*, **90**, 29 (1957).
[55] R. E. Banks, M. G. Barlow, and R. N. Haszeldine, *J. Chem. Soc.*, **1965**, 4714.
[56] W. S. Johnson and B. G. Buell, *J. Am. Chem. Soc.*, **74**, 4517 (1952).
[57] H. W. Bersch, *Arch. Pharm.*, **284**, 217 (1951).
[58] S. Siegel and B. Dmuchovsky, *J. Am. Chem. Soc.*, **84**, 3132 (1962).
[59] N. Baggett, J. M. Duxbury, A. B. Foster, and J. M. Webber, *J. Chem. Soc. (C)*, **1966**, 208.
[60] J. E. Anderson, F. G. Riddell, J. P. Fleury, and J. Morgen, *Chem. Commun.*, **1966**, 128.
[61] N. R. Easton, D. R. Cassady, and R. D. Dillard, *J. Org. Chem.*, **28**, 448 (1963).
[62] G. B. Bachmann and S. Plansky, *J. Org. Chem.*, **16**, 1690 (1951).
[63] J. Plesek, *Coll. Czechoslov. Chem. Commun.*, **21**, 902 (1956).
[64] N. Sugimoto, S. Ohshiro, H. Kugita, and S. Saito, *Chem. Pharm. Bull. Japan*, **5**, 62 (1957).
[65] N. Sugimoto and H. Kugita, *Chem. Pharm. Bull. Japan*, **5**, 67 (1957).
[66] W. P. Ward, *J. Am. Chem. Soc.*, **74**, 4212 (1952).
[67] E. C. Britton and H. C. White, U.S. Patent 2,605,282 (1952).
[68] M. Freifelder, unpublished results.
[69] W. C. Anthony, *J. Org. Chem.*, **31**, 77 (1966).
[70] L. F. Fieser and M. Fieser, *Steroids*, Reinhold, New York, 1959, Sec. 4.8 and 4.9.
[71] I. M. Heilbron and D. G. Wilkinson, *J. Chem. Soc.*, **1932**, 1708.
[72] W. F. Johns, *J. Org. Chem.*, **31**, 3780 (1966).

[73] H. L. Mason and J. J. Schneider, *J. Biol. Chem.*, **184**, 593 (1950).
[74] Y. Mazur and F. Sondheimer, *J. Am. Chem. Soc.*, **80**, 6296 (1958).
[75] R. L. Shriner and L. Ko, *J. Biol. Chem.*, **80**, 1 (1928).
[76] E. Hershberg, E. Oliveto, M. Rubin, H. Staeudle, and L. Kuhlen, *J. Am. Chem. Soc.*, **73**, 1144 (1951).
[77] W. F. Bruce, *Organic Synthesis*, Coll. Vol. II, John Wiley and Sons, New York, 1943, p. 191.
[78] H. R. Nace, *J. Am. Chem. Soc.*, **73**, 2379 (1951).
[79] J. A. Ralls, ref. 76, p. 191.
[80] J. R. Lewis and C. W. Shoppee, *J. Chem. Soc.*, **1955**, 1365.
[81] C. W. Shoppee, B. D. Agashe, and G. H. R. Summers, *J. Chem. Soc.*, **1957**, 3107
[82] W. Aumüller and K. Muth, U.S. Patent 3,150,178 (1964).
[83] S. Akiyoshi and T. Matsuda, *J. Am. Chem. Soc.*, **77**, 2476 (1955).
[84] B. Helferich, R. Dhein, K. Geist, H. Junger, and D. Wiehle, *Ann.*, **646**, 32 (1961).
[85] T. W. Evans, R. C. Morris, and E. C. Shokal, U.S. Patent 2,360,555 (1944).
[86] L. Kh. Friedlin and B. D. Polkovnikov, *Doklady Akad. Nauk. USSR*, **112**, 83 (1957); through *Chem. Abstr.*, **51**, 11259 (1957).
[87] L. Kh. Friedlin and B. D. Polkovnikov, *Bull. Acad. Sci., Div. Chem. Sci.*, **1959**, 691 (English translation).
[88] P. D. Gardner and M. Narayana, *J. Org. Chem.*, **26**, 3518 (1961).
[89] R. J. Moore, R. A. Trimble, and B. S. Greensfelder, *J. Am. Chem. Soc.*, **74**, 373 (1952).
[90] M. Seefelder and W. Raskob, U.S. Patent 3,251,892 (1966).
[91] G. C. Bond and J. Sheridan, *Trans. Faraday Soc.*, **48**, 658 (1952).
[92] Kh. H. Balyan, A. A. Petrov, N. A. Borovikova, V. A. Kormer, and T. V. Yakovleva, *J. Gen. Chem. USSR*, **30**, 3217 (1960) (English translation).
[93] G. F. Hennion and J. J. Sheehan, *J. Am. Chem. Soc.*, **71**, 1964 (1949).
[94] G. F. Hennion and C. V. Di Giovanna, *J. Org. Chem.*, **30**, 3696 (1965).
[95] L. Horner and I. Grohmann, *Ann.*, **670**, 1 (1963).
[96] G. C. Harris and T. F. Sanderson, *J. Am. Chem. Soc.*, **70**, 2081 (1948).
[97] J. W. ApSimon, P. V. Demarco, and J. Lemke, *Can. J. Chem.*, **43**, 2793 (1965).
[98] W. Kimel, N. W. Sax, S. Kaiser, G. H. Eichmann, G. O. Chase, and A. Ofner, *J. Org. Chem.*, **23**, 153 (1958).
[99] W. F. Newhall, *J. Org. Chem.*, **23**, 1274 (1958).
[100] R. T. Blickenstaff, *J. Am. Chem. Soc.*, **82**, 3673 (1960).
[101] W. Sandermann and K. Bruns, *Ber.*, **99**, 2835 (1966).
[102] a. S. M. Kupchan, G. R. Flouret, and C. A. Matuszak, *J. Org. Chem.*, **31**, 1707 (1966); b. S. M. Kupchan and C. G. De Grazia, *ibid.*, p. 1716.
[103] H. H. Inhoffen, R. Jones, H. Krösche, and U. Eder, *Ann.*, **694**, 19 (1966).
[104] E. B. Hershberg, E. P. Oliveto, C. Gerold, and L. Johnson, *J. Am. Chem. Soc.*, **73**, 5073 (1951).
[105] M. Freifelder and P. Kurath, U.S. Patent 3,192,237 (1965).
[106] G. D. Laubach and K. J. Brunings, *J. Am. Chem. Soc.*, **74**, 705 (1952).
[107] W. O. Godtfredsen, W. von Daehne, L. Tybring, and S. Vangedal, *J. Med. Chem.*, **9**, 15 (1966).
[108] E. R. H. Jones, G. H. Mansfield, and M. C. Whiting, *J. Chem. Soc.*, **1956**, 4073.
[109] a. S. V. Lebedev and A. O. Yakubchik, *J. Chem. Soc.*, **1928**, 823; b. p. 2190; c. **1929**, p. 220.
[110] I. V. Gostunskaya, N. B. Dobroserdova, and B. A. Kazansky, *J. Gen. Chem. USSR*, **27**, 2458 (1957) (English translation).

[111] L. Kh. Friedlin, A. A. Balandin, I. F. Zhukova, and I. P. Yakovlev, *Bull. Acad. Sci. USSR, Div. Chem. Sci.*, **1959**, 1576 (English translation).
[112] H. H. Inhoffen, K. D. Müller, and O. Brendler, *Ann.*, **694**, 31 (1966).
[113] G. Ohloff, H. Farnow, and G. Schade, *Ber.*, **89**, 1549 (1956).
[114] C. F. Wilcox, Jr. and D. L. Nealy, *J. Org. Chem.*, **28**, 3446 (1963).
[115] W. V. Ruyle, E. M. Chamberlin, J. M. Chemerda, G. E. Sita, L. M. Aliminoso, and R. L. Erickson, *J. Am. Chem. Soc.*, **74**, 5929 (1952).
[116] L. F. Fieser and J. E. Herz, *J. Am. Chem. Soc.*, **75**, 121 (1953).
[117] J. Castells, E. R. H. Jones, G. D. Meakins, S. Palmer, and R. Swindells, *J. Chem. Soc.*, **1962**, 2907.
[118] M. Tuot and M. Guyard, *Bull. Soc. Chem. France*, **1947**, 1086.
[119] M. Tuot and M. Guyard, *Compt. Rend.*, **225**, 809 (1947).
[120] K. N. Campbell and B. K. Campbell, *Chem. Rev.*, **31**, 77 (1942).
[121] G. Stork, *J. Am. Chem. Soc.*, **69**, 2936 (1947).
[122] E. H. Farmer and L. A. Hughes, *J. Chem. Soc.*, **1934**, 1929.
[123] K. Alder and M. Schumaker, *Ann.*, **564**, 96 (1949).
[124] J. Martel, E. Toromanoff, and C. Huynh, *J. Org. Chem.*, **30**, 1753 (1965).
[125] J. Blake, J. R. Tretter, G. J. Juhasz, W. Bonthrone, and H. Rapoport, *J. Am. Chem. Soc.*, **88**, 4061 (1966).
[126] J. A. Elvidge, R. P. Linstead, and J. F. Smith, *J. Chem. Soc.*, **1953**, 708.
[127] P. Ruggli and J. Schmidlin, *Helv. Chim. Acta*, **29**, 383 (1946).
[128] M. Bourdiol, G. Calcagni, and J. Ducasse, *Bull. Soc. Chim. France*, **1941**, 375.
[129] L. W. Covert, R. Connor, and H. Adkins, *J. Am. Chem. Soc.*, **54**, 1651 (1932).
[130] H. Nomura, *J. Chem. Soc.*, **1917**, 769.
[131] C. Mannich and W. Merz, *Arch. Pharm.*, **265**, 15 (1927).
[132] K. Toki, *Bull. Chem. Soc. Japan*, **30**, 450 (1957).
[133] A. Hassner, N. H. Cromwell, and S. J. Davis, *J. Am. Chem. Soc.*, **79**, 230 (1957).
[134] O. E. Curtis, Jr., Ph.D. Thesis, Michigan State University, 1955.
[135] J. Sam, D. W. Alwani, and K. Aparajithan, *J. Het. Chem.*, **2**, 366 (1965).
[136] H. Nakata, *Bull. Chem. Soc. Japan*, **38**, 500 (1965).
[137] H. C. Brown and C. A. Brown, *J. Am. Chem. Soc.*, **84**, 1495 (1962).
[138] V. I. Maksimov, F. A. Lur'i, and G. P. Krupina, *J. Gen. Chem. USSR*, **28**, 2910 (1958).
[139] Y. Mazur, A. Weizmann, and F. Sondheimer, *J. Am. Chem. Soc.*, **80**, 6293 (1958).
[140] C. H. Robinson, D. Gnoj, E. P. Oliveto, and D. H. R. Barton, *J. Org. Chem.*, **31**, 2749 (1966).
[141] H. Conroy and R. A. Firestone, *J. Am. Chem. Soc.*, **78**, 2290 (1956).
[142] J. J. Beerebom, *J. Org. Chem.*, **30**, 4230 (1965).
[143] C. S. Foote and R. B. Woodward, *Tetrahedron*, **20**, 687 (1964).
[144] H. Pommer, *Arch. Pharm.*, **291**, 23 (1958).
[145] M. S. Newman, V. DeVries, and R. Darlak, *J. Org. Chem.*, **31**, 2171 (1966).
[146] H. Schechter and J. C. Kirk, *J. Am. Chem. Soc.*, **73**, 3089 (1951).
[147] H. Stetter, R. Engl, and H. Rauhut, *Ber.*, **91**, 2882 (1958).
[148] R. E. Brown, D. M. Lustgarten, R. J. Stanaback, and R. I. Meltzer, *J. Org. Chem.*, **31**, 1489 (1966).
[149] J. G. Dinwiddie, Jr. and S. P. McManus, *J. Org. Chem.*, **30**, 766 (1965).
[150] H. O. House and R. G. Carlson, *J. Org. Chem.*, **29**, 74 (1964).
[151] M. Budzikiewicz and H. Janda, *Monotoh.*, **91**, 1043 (1960).
[152] R. D. Sands, *J. Org. Chem.*, **29**, 2488 (1964).
[153] A. Bowers and H. J. Ringold, *J. Am. Chem. Soc.*, **81**, 424 (1959).

[154] R. M. Dodson, A. H. Goldkamp, and R. D. Muir, *J. Am. Chem. Soc.*, **82**, 4027 (1960).

[155] Pl. A. Plattner, H. Heusser, and E. Angliker, *Helv. Chim. Acta*, **29**, 468 (1946); see also A. Ruff and T. Reichstein, *Helv. Chim. Acta*, **34**, 70 (1951).

[156] O. A. Moe, D. T. Warner, and M. I. Buckley, *J. Am. Chem. Soc.*, **73**, 1062 (1951).

[157] G. F. Woods and H. Sanders, *J. Am. Chem. Soc.*, **69**, 2926 (1947).

[158] L. Palfray, S. Sabetay, and B. Gauthier, *Compt. Rend.*, **218**, 553 (1944).

[159] K. Freudenberg and H. Richtzenhain, *Ber.*, **76**, 997 (1943).

[160] D. L. Hufford, D. S. Tarbell, and T. R. Koszalka, *J. Am. Chem. Soc.*, **74**, 3014 (1952).

[161] H. E. Hennis and W. B. Trapp, *J. Org. Chem.*, **26**, 4678 (1961).

[162] I. N. Nazarov, G. P. Kugatova, and G. A. Laumenskas, *J. Gen. Chem. USSR*, **27**, 2511 (1958) (English translation).

[163] H. Fukawa, Y. Izumi, S. Komatsu, and S. Akabori, *Bull Chem. Soc. Japan*, **35**, 1703 (1962).

[164] A. J. B. Edgar, S. H. Harper, and M. A. Kazi, *J. Chem. Soc.*, **1957**, 1083.

[165] R. E. Marker, *J. Am. Chem. Soc.*, **71**, 2656 (1949).

[166] B. Ellis, S. P. Hall, V. Petrow, and D. M. Williamson, *J. Chem. Soc.*, **1962**, 22.

[167] J. Romo and A. R. De Vivar, *J. Am. Chem. Soc.*, **81**, 3446 (1959).

[168] G. W. Krakower and H. A. Van Dine, *J. Org. Chem.*, **31**, 3467 (1966).

[169] D. A. Shepherd, R. A. Donia, J. A. Campbell, B. A. Johnson, R. P. Holysz, G. Slomp, Jr., J. E. Stafford, R. L. Pederson, and A. C. Ott, *J. Am. Chem. Soc.*, **77**, 1212 (1955).

[170] D. H. R. Barton, J. D. Cox, and N. J. Holness, *J. Chem. Soc.*, **1949**, 1771.

[171] L. B. Barkley, M. W. Farrar, W. S. Knowles, H. Raffelson, and Q. E. Thompson, *J. Am. Chem. Soc.*, **76**, 5014 (1954).

[172] N. L. Drake and A. B. Ross, *J. Org. Chem.*, **23**, 719 (1958).

[173] W. C. Wildman and C. H. Hemminger, *J. Org. Chem.*, **17**, 1641 (1952).

[174] S. Minami and S. Uyeo, *Chem. Pharm. Bull. Japan*, **12**, 1012 (1964).

[175] a. A. C. Huitric and W. D. Kumler, *J. Am. Chem. Soc.*, **78**, 614 (1956); b. W. F. Trager and A. C. Huitric, *J. Pharm. Sci.*, **54**, 1552 (1965).

[176] a. J. Yoshimura, H. Komoto, H. Ando, and T. Nakagawa, *Bull. Chem. Soc. Japan*, **39**, 1775 (1960); b. J. C. Sowden and H. O. L. Fischer, *J. Am. Chem. Soc.*, **69**, 1047 (1947).

[177] H. H. Baer and F. Kienzle, *Can. J. Chem.*, **43**, 4074 (1965).

[178] A. C. Cope, C. M. Hoffman, C. Wyckoff, and E. Hardenbaugh, *J. Am. Chem. Soc.*, **63**, 3452 (1941).

[179] S. Wawzonek and E. M. Smolin, *J. Org. Chem.*, **16**, 746 (1951).

[180] L. Bauer and J. Cymerman, *J. Chem. Soc.*, **1950**, 1826.

[181] M. Pesson, S. Dupin, M. Antoine, D. Humbert, and M. Joannic, *Bull. Soc. Chim. France*, **1965**, 2262.

[182] E. D. Bergmann and R. Ikan, *J. Am. Chem. Soc.*, **78**, 1482 (1956).

[183] M. Ohno, S. Torimitsu, N. Naruse, M. Okamotu, and I. Sakai, *Bull. Chem. Soc. Japan*, **39**, 1129 (1966).

[184] A. C. Cope and N. A. LeBel, *J. Am. Chem. Soc.*, **82**, 4656 (1960).

[185] Pl. R. Nace and E. H. Goldberg, *J. Am. Chem. Soc.*, **75**, 3646 (1953).

[186] E. H. Goldberg and H. R. Nace, *J. Am. Chem. Soc.*, **77**, 359 (1955).

[187] T. Matsukawa and K. Shirakawa, *J. Pharm. Soc. Japan*, **72**, 909 (1952); through *Chem. Abstr.*, **47**, 6425 (1953).

[188] H. Plieninger and M. S. von Wittenau, *Ber.*, **91**, 1905 (1958).

[189] P. S. Bailey and R. E. Lutz, *J. Am. Chem. Soc.*, **67**, 2232 (1945).
[190] B. Loev and C. R. Dawson, *J. Am. Chem. Soc.*, **78**, 6095 (1956).
[191] H. Christol, A. Donche, and F. Ple'nat, *Bull. Soc. Chim. France*, **1966**, 2536.
[192] a. O. C. Grummitt, A. C. Buck, and E. I. Becker, *J. Am. Chem. Soc.*, **67**, 2265 (1945); b. R. E. Allen, E. L. Schumann, W. C. Day, and M. G. Van Campen, Jr., *J. Am. Chem. Soc.*, **80**, 591 (1958).
[193] B. I. Mikhant'ev and E. I. Federov, *J. Gen. Chem. USSR*, **33**, 851 (1963) (English translation).
[194] J. Klein and E. D. Bergmann, *J. Org. Chem.*, **22**, 1019 (1957).
[195] T. W. Campbell and R. N. McDonald, *J. Org. Chem.*, **24**, 1246 (1959).
[196] G. A. Coppens, M. Coppens, D. N. Kevill, and N. H. Cromwell, *J. Org. Chem.*, **28**, 3267 (1963).
[197] H. Kataoka, *Chem. Abstr.*, **51**, 16502 (1957).
[198] K. Kindler, H. Oelschlager, and P. Heinrich, *Ber.*, **86**, 167 (1953).
[199] S. N. Eğe, R. Wolovsky, and W. J. Gensler, *J. Am. Chem. Soc.*, **83**, 3080 (1961).
[200] B. Loev and C. R. Dawson, *J. Org. Chem.*, **24**, 980 (1959).
[201] H. Conroy, U.S. Patent 2,786,850 (1957).
[202] K. Dimroth, W. Kinzebach, and M. Soyka, *Ber.*, **99**, 2351 (1966).
[203] E. J. McMullen, H. R. Henze, and B. W. Wyatt, *J. Am. Chem. Soc.*, **76**, 5636 (1954).
[204] G. E. Ham and W. P. Coker, *J. Org. Chem.*, **29**, 194 (1964).
[205] B. T. Gillis and R. E. Kadunce, *J. Org. Chem.*, **32**, 91 (1967).
[206] R. P. Linstead, W. E. Doering, S. B. Davis, P. Levene, and R. R. Whetstone *J. Am. Chem. Soc.*. **64**, 1985 (1942).
[207] H. O. House, *Modern Synthetic Reactions*, Benjamin, New York, 1966, p. 21.

X

REDUCTION OF NITRO GROUPS

IN BENZENOID COMPOUNDS

Conditions and Influences on Reduction

The nitro group attached to a benzene ring, with the exception of the acetylenic linkage, is the most amenable of all reducible systems to catalytic hydrogenation. Not enough work has been carried out on the reducibility of the triple bond and the aromatic nitro group to state authoritatively that one will be selectively hydrogenated in the presence of the other. In most other instances an aromatically bound nitro group will be preferentially reduced in the presence of another reducible function.

In general aromatic nitro compounds can be reduced at room temperature and under slight hydrogen pressure with rhodium, palladium, or platinum on carriers, with palladium or platinum oxide, or with Raney nickel to give high yields of amines. Ruthenium catalyst has been shown to be much less efficient at low pressure [1]. Methods that employ a hydrogenation catalyst and sodium borohydride in water or alcohol as the source of hydrogen [2] or a catalyst and hydrazine as the source of hydrogen [3] are worthwhile laboratory procedures for reduction of nitro groups. However, both sodium borohydride and hydrazine are expensive sources of hydrogen.

Hydrogenation of the nitro group in aromatic compounds takes place in the following manner:

$$ArNO_2 \xrightarrow{H_2} ArNO \xrightarrow{H_2} ArNHOH \xrightarrow{H_2} ArNH_2$$

None of the intermediates or side reaction products (azo, azoxy, or hydrazobenzenes resulting from condensation of the intermediates or intermediates and reduction products) are generally encountered. The aromatic nitro group, which is reduced more rapidly than any of the intermediates or side products, goes directly to the amine. This is probably due to the exothermicity of the reaction which supplies enough energy to aid rapid reduction of the intermediates to the amine.

Because of the heat of reaction, care must be taken not to reduce very concentrated solutions unless the reactions can be cooled. The amount of catalyst normally used for reduction of other groups can be drastically decreased for the hydrogenation of nitrobenzenes. On commercial scale, where active catalysts are used under moderate conditions or where less active catalysts may be employed under more drastic conditions, dissipation of the heat of reaction can be a problem.

As a rule methyl, ethyl, and propyl alcohols are excellent solvents for reductions. Methyl cellosolve (ethylene glycol monomethyl ether) and dimethylformamide are especially useful when the starting materials are poorly soluble in other solvents. Each is completely miscible with water; either solvent can be removed by concentration under reduced pressure. After removal of catalyst the amine can be precipitated as a solid or thrown out of solution as an oil by the addition of water. It is not necessary for the nitro compound to be completely dissolved. If it is only slightly soluble and the amine is soluble in the solvent, reduction will usually proceed as rapidly as the amine goes into solution. Acetic acid is also useful. It has been reported to decrease the rate of reaction in reductions with platinum oxide in comparison with similar reductions of nitro compounds in alcohol [4]. The decrease in rate does not always apply nor does it appear to apply to reductions with rhodium or palladium. Normally it is expected that the presence of acid (unless it is a large excess of strong acid) will neutralize the inhibitory effect of the amino nitrogen and will help to accelerate reduction rather than retard it.

Kornblum and Iffland [5] used acetic acid as solvent in the platinum-catalyzed reduction of N-(3-nitrobenzyl)phthalimide to N-(3-aminobenzyl)-phthalimide as did Kloetzel and Abidir [6] with platinum oxide to reduce 2,5-dihydroxy-3-nitrobenzoic acid to the corresponding amine. Eastman and Detert [7] carried out the hydrogenation of 2-nitro-4-methylphenyl-acetic acid in acetic acid with platinum oxide to give high yield of 6-methyl-oxindole. Loudon and Ogg [8] found acetic acid useful in the reduction of 2-nitrophenoxyacetic acid with palladium in obtaining a quantitative yield of 3-oxo-2H-1,4-benzoxazine.

One disadvantage of acetic acid as a solvent in the reduction of nitrobenzenes with platinum oxide is the possibility of ring reduction if the reaction is allowed to run unattended for too long a period. The same danger applies to its use in the presence of rhodium catalyst and to a lesser extent in the presence of palladium catalyst.

Adams [4] suggested that the presence of more than 10% of water in ethyl alcohol decreased the rate in platinum-catalyzed reductions of aromatic nitro compounds. This may be due to a decrease in solubility and possible suspension of the nitro compound which could physically coat and inactivate the catalyst. Adams' report should not deter the chemist from employing water as a solvent for reduction of nitro compounds. It was used by Kloetzel and Abidir [6] for the reduction of 2,5-dihydroxy-3-nitrobenzoic acid. Benner and Stevenson [9] hydrogenated 2,4-dinitrotoluene by adding the dinitro compound in hot water portionwise to palladium catalyst until each portion was reduced to obtain high yield of the corresponding diamine. 1,3-Phenylenediamine was obtained in a similar manner from 1,3-dinitrobenzene.

Adams [4] pointed out that excess hydrochloric acid caused inhibition in the hydrogenation of nitrobenzenes. The effect of excess hydrogen chloride on reduction has already been noted (Chapter IV). The addition of an equivalent of acid had little effect on the reductions carried out by Adams. In contrast Moffett [10] stated that time of reduction of 4-nitropyrogallol in ethyl alcohol with platinum oxide was greatly decreased by the addition of an equivalent of hydrogen chloride.

Adams also noted that the presence of sodium hydroxide acted as an inhibitor in the reduction of nitrobenzenes with platinum catalyst in alcohol. Shmonina and colleagues [11] found that the reduction of nitrophenols was accelerated in alkaline medium with Raney nickel but the presence of alkali hindered reduction with platinum. Samuelson, Garik, and Smith [12] pointed out that the addition of alkali often retarded hydrogenation of aromatic nitro compounds. The retarding effect of excess base was noted in the hydrogenation of (2-carboxyphenyl)4-nitrobenzoate [13]. In a reduction in aqueous solution containing a slight excess of sodium hydroxide (pH-10.0), no uptake of hydrogen took place in the presence of Raney nickel. However, if reduction was carried out in water containing an equivalent of sodium bicarbonate (pH of the solution was below 8.0), uptake of hydrogen in the presence of the same catalyst was complete in about 2 hr. Baltzly [14] reported that nitrobenzene was rapidly reduced with platinum on carbon or palladium on carbon in 0.8% methyl alcoholic sodium hydroxide or 1% sodium methoxide. Levy [15] reported that the presence of sodium hydroxide was necessary in the reduction of 1,2-dinitrobenzene to 1,2-phenylenediamine in water in the presence of palladium on carbon.

Adams [4] also studied the effects of substitution in reduction with platinum. He stated that *ortho* substituents had no effect on reduction. A series of reductions in which the bulk of the 2-alkoxy groups was increased appeared to substantiate this observation, at least with palladium on carbon [13]. There appeared to be little difference, if any, in the hydrogenation of 2-substituted nitrobenzenes in ethyl alcohol whether the substituent was methoxy, propoxy, butoxy, phenoxy, or cyclohexyloxy. With pure compounds hydrogen uptake was complete in less than 0.5 hr in each instance. It is possible that with a low weight ratio of catalyst a difference in time might be observed between the reduction of 2-methoxynitrobenzene and 2-cyclohexyloxynitrobenzene. The reductions of 2-, 3-, and 4-tertiary butylnitrobenzenes were reported [16] but there was no comparison of the reaction times to determine if there was an effect of the branched chain in 2-position. In contrast, Reasenberg and his group [17] reported that 2-substituted nitrobenzenes were reduced more readily than the corresponding 3- or 4-substituted compounds with Raney nickel.

Ortho substituents on both sides of the nitro group can influence the rate of hydrogenation. For example, 2,3-, 3,4-, and 2,6-dimethoxynitrobenzenes were hydrogenated in the presence of palladium on carbon in ethyl alcohol [13]. Reduction of the first two compounds was extremely rapid (15–30 min) and exothermic at room temperature and 2 atm pressure. Reduction of 2,6-dimethoxynitrobenzene, while satisfactory, took at least four times as long for complete uptake of hydrogen. Unless substituents on each side of the nitro group are large, 2,6-disubstituted nitrobenzenes should be hydrogenated without much difficulty although at a slower rate than a mono *ortho* substituted nitrobenzene with one of the same substituents.

In another comparison of reaction times of some trimethoxynitrobenzenes with palladium on carbon [13], it was found that 2,3,6- and 2,4,6-trimethoxynitrobenzenes reduced very slowly (12 hr); with half the amount of catalyst 2,3,5- and 2,4,5-trimethoxynitrobenzenes were converted to the corresponding amines in less than 1 hr.

Adams [4] reported that there was little difference in the rate of hydrogenation between nitrobenzene and substituted nitrobenzenes in reduction with platinum oxide. Samuelson and associates [12], reporting on the influence of structure in the reduction of nitrobenzenes in alcohol with Raney nickel, stated that aliphatic side chains contributed to complete uptake of hydrogen and that functional groups as amino or hydroxy groups usually, although not always, interfered with reduction. Henderson and Nord [18] noted that when palladium on polyvinyl alcohol was the catalyst there was no difference in rate between the reduction of nitrobenzene and substituted nitrobenzenes. Substituents in 3- and 4-position appeared to have a distinct effect in the presence of rhodium on polyvinyl alcohol. However, any relationship between an increase or decrease in rate due to the substituent cannot seriously be considered in view of the conflicting results obtained in reductions of the same compounds in the presence of rhodium on carbon [18a].

The particular interest of Samuelson and his associates [12] in the reduction of aromatic nitro compounds with Raney nickel was in the promoter effect of chloroplatinic acid. It would appear that if promoters are necessary with

Raney nickel for low-pressure hydrogenation of these compounds it might be more advantageous to use another catalyst in its place because of the vagaries of Raney nickel whose activity is dependent on the method of preparation, the age of the catalyst, and perhaps the presence or absence of occluded alkali.

From the literature it may be seen that palladium catalyst is less affected by the medium used for hydrogenation and by substituents on the ring in the reduction of nitrobenzenes. A combination of these facts and unpublished work [13] makes palladium on carbon this investigator's personal choice for most low-pressure reductions of the nitro group in aromatic compounds. If another reducible function is also a part of the compound to be hydrogenated, the choice of catalyst may be governed by the reducibility of the particular function.

Nitrophenols

Nitrophenols have been hydrogenated under a variety of conditions. Some typical examples of low-pressure hydrogenations can be seen in Table 10.1.

Table 10.1

OH	R	Catalyst	Solvent	Time (hr)	Yield of Amine (%)	Reference
2-	5-CH$_2$CH$_2$N(CH$_3$)COCH$_3$	Ni(R)[a]	EtOH	5	>90	19
2-	3-OH	Ni(R)	MeOH	1	[b]	20
2-	3-C$_3$H$_7$	Ni(R)	EtOH	1	83	13
2-	6-COOCH$_3$	Ni(R)	EtOH	4	[b]	21
2-	4-OCH$_3$	Ni(R)	acetone	[b]	90	22
2-	4-OCH$_3$	PtO$_2$	EtOH	[b]	[b]	23
2-	4-CH$_3$	PtO$_2$	EtOH	[b]	[b]	24
2-	5-OCH$_3$	Pd on C	EtOH	2	71	13
4-	2-CH$_3$	Pd on C	EtOH	1	85	13
2-	4-CH$_3$	Pd on C	EtOH	1	85	13
2-	3-CH$_3$	Pd on C	EtOH	[b]	85–90	13
3-	6-C$_2$H$_5$	Pd	EtOH	3	>95	25

[a] Ni(R) = Raney nickel. [b] Not listed.

Palladium and sodium borohydride as the source of hydrogen gave good results in the reduction of nitrophenol [2]. 4-Aminophenol has been prepared from 4-nitrophenol by reduction with palladium on carbon in water containing an equivalent of hydrochloric acid or a slight excess of acetic acid [13]. A patent [26] describes the effect of catalyst, temperature, and other factors in the reductions of 4-nitrophenol in aqueous solution with palladium and platinum catalysts.

Nitrobenzoic and Nitrophenylalkanoic Acids

Nitrobenzoic acids and nitrobenzenes with a carboxyl group in an attached side chain have also been easily hydrogenated to the corresponding amino acids under low-pressure conditions (see Table 10.2).

The reduction of 4-nitrosalicylic acid has given poor results in certain cases. Wenis and Gardner [34] reported only 20% yield of 4-aminosalicylic acid from reduction of the nitro compound with palladium in alcoholic hydrogen chloride. Huckel and Janecka [35] reduced a suspension of the nitro acid in methyl alcohol with Raney nickel to obtain 72% yield of 4-aminosalicylic acid. Hydrogenation of the nitro acid was slow with platinum oxide in alcohol but the use of Raney nickel, 25% by weight of nitro compound, in a reduction in water containing an equivalent of sodium bicarbonate gave 80% yield of 4-aminosalicyclic acid in a 2-hr reaction.

The hydrogenation of 2-nitrophenyl and 2-nitrophenoxy-alkanoic acids can lead to cyclic structures. 2-Nitrophenoxyacetic acid was reduced to the lactam with platinum oxide in alcohol [36]; 2-nitrophenylacetic acid was hydrogenated in glacial acetic acid with platinum oxide in a rapid reaction (20 min) to give 2-oxindole in 88% yield [37] and 3-nitro-4,5-dimethoxyphenylacetic acid was reduced in glacial acetic acid with palladium on carbon and simultaneously cyclized to 5,6-dimethoxy-2-oxindole in 75% yield [38].

Aromatic Nitro Compounds with Basic Side Chains

The hydrogenation of an aromatic nitro compound containing a basic side chain is best carried out on a salt. A useful method is seen in reductions in aqueous solution with palladium on carbon of compounds of the type, $NO_2C_6H_4COO(CH_2)_2NHR \cdot HCl$, where R = alkyl, aralkyl, cyclohexyl, or substituted cyclohexyl [39]; reduction may also be carried out in alcohol or aqueous alcohol. Platinum oxide has been employed as catalyst for the reduction of related compounds [40]. Hydrogenation of these compounds as bases can be carried out in neutral solvent but hydrogen uptake may be slow because the presence of the basic nitrogen atom with its unshared pair of electrons

Table 10.2

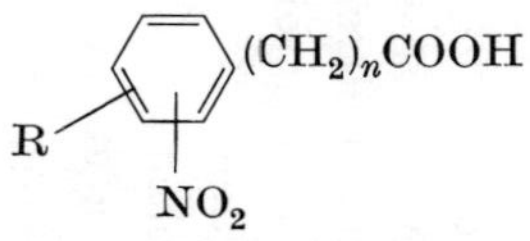

$$R\text{—}\underset{NO_2}{\underset{|}{\bigcirc}}\text{—}(CH_2)_n COOH$$

NO_2	R	n	Catalyst	Solvent	Time (hr)	Yield of Amine (%)	Reference
2-	3-OCH$_3$, 4-CH$_3$	0	Ni(R)	MeOH	a	81	27
2-	4-(CH$_3$)$_3$C-	0	PtO$_2$	abs EtOH	0.75	71b	28
2-	5-CH$_3$	0	Ni(R)	EtOH	a	92	29
4-		0	Pd on C	H$_2$O + NaHCO$_3$	<0.5	90	13
3-	4-OC$_4$H$_9$	0	Pd on C	EtOH	<0.1	96	13
3-	4-OCH$_3$	1	{Ni(R) / Pd on C}	EtOH / MeOH	a / 1	90 / a	30 / 31
4		1	{PtO$_2$ / Ni(R)}	EtOH^c	2–3	80	13
4		1		H$_2$O + NaOH	No reduction		13
4		2	Pd on C	EtOH	a	a	32
4		4^d	Ni(R) or PtO$_2$	EtOH	a	quantitative	33

^{a} Not listed. ^{b} After two recrystallizations. ^{c} Methyl cellosolve also used with similar results. ^{d} The side chain is (CH$_3$)$_2$CHCH$_2$–.

174

can have an inactivating effect on the catalyst. The inhibiting effect of the basic side chain may be the reason that a pressure of 80 atm was used for the palladium on carbon-catalyzed hydrogenation of compounds of the type, $NO_2C_6H_4O(CH_2)_nR$, where $n = 2$ or 3 and R = diethylamino, morpholino, or piperidine, [41], or for the rather high catalyst ratio employed (2 g of 10% palladium on carbon for 5 g of substrate) in the reduction of a substituted aminoethyl 4-nitrobenzoate as base [42]. Nevertheless, reduction, as bases, of N-(2-diethylaminoethyl)-2-nitroaniline [43] and N,N-diethyl-1-(4-nitrophenyl)nipecotamide [44] has been reported with Raney nickel at low pressure.

Aromatic Polynitro Compounds

The reduction of di- and polynitro compounds can be extremely exothermic. In most instances it is advisable to reduce the amount of catalyst used for hydrogenation and to carry out the reaction at room temperature and slight pressure. The following references show examples of reductions of such compounds to di- and polyamines under a variety of conditions with a number of catalysts [9, 45–51].

Although hydrogen uptake in the reduction of dinitro compounds often is very rapid, it is possible to reduce only one nitro function. Brunner and Halasz [52a, b] used only three equivalents of hydrogen with platinum or palladium on carbon in acidic media to effect selective reduction of one nitro group in a series of 2,4-dinitro-N-substituted anilines. The unhindered group in 4-position was selectively reduced. A difference in reduction rate between a hindered and unhindered nitro group led to selective reduction of 3-(3-nitrobenzyl)nitromesitylene in benzene solution with platinum oxide to 2,4,6-trimethyl-3-(3-aminobenzyl)nitrobenzene [53]. In some symmetrical dinitro compounds, 2,6-dinitrophenetole and 2,6-dinitropropoxybenzene, one nitro group was selectively reduced in benzene solution in the presence of platinum oxide [54]. The authors reported that when they carried out reductions in alcohol they did not obtain the desired product. Benzene, which is strongly adsorbed to most catalysts, probably retarded the activity of the catalyst sufficiently so that the reaction could be interrupted after hydrogenation of only one nitro group. Frisch and Bogart [55] used palladium black for the reduction of 4,5-dinitroveratrole in ethyl alcohol to 4-amino-5-nitroveratrole. When the same conditions were employed in the reduction of 1,2- and 1,3-dinitrobenzene and 2,4-dinitroaniline, both nitro groups were hydrogenated in each instance. The selective reduction of 4,5-dinitroveratrole to 4-amino-5-nitroveratrole in good yield can be carried out with a lower than normal amount of palladium on carbon in alcohol in the presence of some excess hydrogen if the reaction is interrupted after consumption of three equivalents [13]. Raney nickel and hydrazine as a source of hydrogen gave a 71% yield of 3-amino-5-nitrobenzoic acid from the reduction of 3,5-dinitrobenzoic acid [56].

Polyaromatic Nitro Compounds

Reduction of the nitro group in polyaromatic systems is carried out in the same manner as in the reduction of nitrobenzenes. There is such a difference between the reducibility of the nitro group and the benzene ring that there should be no concern about ring reduction if reaction conditions are not too vigorous. For example, nitronaphthalenes have been converted to naphthylamines by reduction in alcohol at 2–3 atm pressure with platinum oxide [57, 58], with a small amount of palladium on carbon [59], and with palladium on pumice (pumice was used as the support because the reduction product was strongly adsorbed by carbon) [60]. Palladium on carbon was used in a reduction of a nitronaphthalene at room temperature and 60 atm pressure [61]. Raney nickel was used under similar conditions. Reaction was exothermic to 70°C. In order to obtain the tetrahydro compound it was necessary to raise the temperature to 100° and pressure to 100 atm.

1-Methoxy-2-nitro and 3-methoxy-2-nitrofluorene were converted to the corresponding aminofluorenes with palladium in acetic acid [62]; 1-methyl-2-nitrofluorene was reduced to the corresponding amine with palladium in alcohol at low pressure [63].

Nitrophenylarsonic, Boronic, and Phosphonic Acids and Related Compounds

As pointed out in Chapter IV, arsenic is not a catalyst poison in the pentavalent state. If the compounds are stable, nitrophenylarsonic acids are reduced to aminophenylarsonic acids in aqueous alkaline solution in the presence of Raney nickel under 2–3 atm pressure [64, 65]. In the cited references the nitro group was always *meta* to the arsonic acid. It has been reported that 2-nitrophenylarsonic acid could not be reduced catalytically to 2-aminophenylarsonic acid [66]. Hydrogen uptake was observed but the desired compound was not obtained [13]. Freedman and Doak [67] reported that 4-(2-nitro-4-arsonophenyl)butyric acid was not hydrogenated successfully with Raney nickel but was reduced chemically. Tri-(3-nitro-4-ethoxyphenyl)arsenoxide in alcohol was reduced with Raney nickel under 3 atm pressure to tri-(3-amino-4-ethoxyphenyl)arsenoxide [68]. At 80° and 70 atm reduction led to tri-(3-amino-4-ethoxyphenyl)arsine. It must be noted that attempted reductions of nitro groups in arsenicals are successful only with Raney nickel. Palladium and platinum catalysts cause cleavage of the carbon-arsenic bond.

Boron-containing aromatic compounds appear to be more stable to reduction than similar type arsenicals and since boron is not a poisoning element, catalytic reduction of nitrophenylboronic acids takes place more or less readily. Raney nickel and platinum oxide have been used in alcoholic solution in reductions at 2–3 atm pressure [69a, b, c, 70]. In one instance reduction was carried out with platinum in acetic acid [71]. In general, yields are high and the position of the nitro group has little effect on its reducibility. In some instances the amine has been isolated as the N-acetyl derivative and in only one reaction, the hydrogenation of 2-nitro-4-aminophenylboronic acid, was there a report of cleavage to boric acid [69a].

A series of 3-nitro-4-substituted amino phenylphosphonic acids were reduced as sodium salts in water with Raney nickel under 2.7 atm pressure [72]. 3-Nitro-4-hydroxyphenylphosphonic acid was converted to 3-amino-4-hydroxyphenylphosphonic acid under similar conditions. In another study [73] when palladium on carbon was used for the hydrogenation of 3-nitro-4-hydroxy-phenylphosphonic acid several additions of catalyst were required before hydrogen uptake was complete. On the other hand, 3-nitro-4-aminophenylphosphonic acid was readily reduced to the diamino acid with palladium on carbon in alkaline solution [74]. Doak and Freedman [75] reported that reduction of 3- and 4-nitrophenylphosphonic acids as potassium salts with Raney nickel was slow and incomplete unless the pH of the solution was adjusted to 6.7. *Bis* (3-nitrophenyl)phosphinic acid was dissolved in aqueous sodium hydroxide but the solution was acidified to pH 5.0 before reduction. In all instances the addition of chloroplatinic acid increased the rate of reduction. The same authors [76] employed Raney nickel in alcohol to reduce a series of N-phenyl or N-substituted-4-nitrophenylphosphodiamides to the corresponding amino compounds. The reduction of 2-hydroxy-4-nitrophenylphosphonic acid in solution at pH 6.0 with Raney nickel resulted in cleavage to 3-aminophenol and phosphoric acid [77]. On the other hand, 2-methoxy-4-nitrophenylphosphonic acid was reduced without cleavage in alkaline solution with Raney nickel. 2-Hydroxy-4-aminophenylphosphonic acid was obtained from reduction of the nitro compound in dilute hydrochloric acid with platinum oxide. The same catalyst was used to reduce methyl *bis*-(4-nitrophenyl)phosphate to the amine in alcoholic solution containing a small amount of hydrochloric acid [78].

Reduction of sulfur-containing aromatic nitro compounds will be discussed separately in the section on sulfur-containing compounds.

Partial Reduction

There is no record of partial catalytic reduction of nitrobenzene to nitrosobenzene. Although phenylhydroxylamine is formed during hydrogenation of

178   Reduction of Nitro Groups

nitrobenzene, its isolation is not always possible after uptake of two equivalents. Nevertheless, use has been made of the fact that it is an intermediate in the reduction. 4-Aminophenols have been obtained during partial reduction of nitrobenzenes in acidic medium with a variety of catalysts. There are other patents [80 a, b] which describe similar procedures with platinum oxide in which quaternary salts are added to improve yields of aminophenols.

Formation of aminophenol is rationalized as involving acid-catalyzed rearrangements of the intermediate hydroxylamine through a series of steps in which the conjugate acid of phenylhydroxylamine is the active electrophilic intermediate [79a] (see Eq. 10.1). The hydrogenations of nitrobenzene

$$C_6H_5\overset{+}{N}HOH_2 \xrightarrow[-H_2O]{Y^{(-)}} \underset{Y}{\overset{H}{\diagup}}\!\!\!=\!\!NH \longrightarrow Y\!-\!C_6H_4\!-\!NH_2 \qquad (10.1)$$

Y = OH

and 2-nitrotoluene in concentrated hydrochloric acid with platinum oxide to yield 4-chloroaniline and 4-chloro-2-methylamine respectively [79] are also examples of the same type of reaction. The preparation of 4-fluoroaniline from the reduction of nitrobenzene in anhydrous hydrogen fluoride with platinum oxide [81] is another example. In this reduction some aniline was always obtained. Palladium oxide and palladium or platinum on carbon were less effective than platinum oxide; other catalysts led to aniline or did not reduce nitrobenzene.

Reductions of 4-substituted nitrobenzenes, which were carried out in anhydrous hydrogen fluoride with palladium on carbon in the presence of phenol, aniline, and anisole to give 4,4'-substituted diphenylamines [81a], could proceed through the mechanism shown in Eq. 10.1. The author, who described the action of hydrogen fluoride as a condensing agent, stated that the mechanism of action in the reductions he carried out remained uncertain.

The reduction of β-2-nitrophenyl-α-alanine hydrochloride in aqueous methyl alcohol with platinum catalyst to 1-hydroxy-3-amino-3,4-dihydroisoquinoline [82] presumably proceeds through a hydroxylamine intermediate (see Eq. 10.2). The reductions of 2-nitrophenylacetic acid to

$$\underset{NO_2}{C_6H_4}\!\!-\!\!CH_2CH(NH_2)COOH \xrightarrow{2H_2} \left[\underset{NHOH}{C_6H_4}\!\!-\!\!CH_2CH(NH_2)COOH\right] \xrightarrow{-H_2O} \text{1-hydroxy-3-amino-3,4-dihydroisoquinoline} \qquad (10.2)$$

1-hydroxy-2-oxindole with platinum [37] and of butyl 4,6-dichloro-2-nitrophenoxyacetate to 6,8-dichloro-4-hydroxy-3-oxo-2H-1,4-benzoxazine with palladium on carbon [82a] also may form through cyclization of a hydroxylamino intermediate.

The hydrogenation of 4-(2-nitrobenzylidene)-2-pyrazole-5-ones to 3a,4,9,9a-tetrahydro-9-hydroxy-1H-pyrazolo-[3,4-b]quinolines with sodium borohydride and palladium on carbon or preferably with the same catalyst and cyclohexene as the hydrogen donor probably proceeds through the hydroxylamine and the lactim form of the cyclic amide [83] as in Eq. 10.3.

$$\text{(structures)} \quad (10.3)$$

R = C_6H_5; R^1 = CH_3 or C_6H_5; R^2, R^3 = H or OCH_3 (in one instance R^2 = Cl).

The isolation of phenylhydroxylamines by hydrogenation of nitrobenzenes has been observed in a number of cases. A number of factors may be involved in allowing these selective reductions to take place, one of which may be the activity of the catalyst. The presence of inorganic impurities, either acidic or basic, may have an effect. Brand and Steiner [84] suggested that the neutrality of the solution, a decrease in hydrogen pressure, and the choice of a suitable catalyst should moderate the reduction of nitrobenzenes so that phenylhydroxylamines could be isolated. Interruption of a hydrogenation in alcohol at atmospheric pressure with 2% palladium on carbon [84a] after uptake of two equivalents resulted in an 80% yield of phenylhydroxylamine. If the reaction was not interrupted, aniline was obtained. Freshly prepared and thoroughly washed Raney nickel also gave good yields of phenylhydroxylamine in interrupted reductions [85]. Hydrogenation with iridium, usually a catalyst of limited activity, also gave phenylhydroxylamine in an interrupted reaction [86]; a combined iridium-platinum catalyst caused a two- to threefold increase in reaction rate but gave a lower yield of phenylhydroxylamine. A patent describes a Raney nickel-catalyzed reduction leading to phenylhydroxylamine [87]. Hydrogenation of nitrobenzene in isopropyl alcohol at −2° and 2 atm pressure when interrupted at one equivalent gave 47% of phenylhydroxylamine, 2% of aniline, and 51% of nitrobenzene which could be recycled. Higher temperature or greater hydrogen uptake increased the yield of aniline at the expense of phenylhydroxylamine.

It was never possible to obtain phenylhydroxylamine from interrupted reductions of nitrobenzene with low weight ratios of commercially available palladium or platinum on carbon catalysts [13]. The use of benzene as solvent and the addition of pyridine, quinoline, or piperidine to overcome possible traces of acid in the catalysts and to act as inhibitors failed to inactivate the catalysts sufficiently to obtain phenylhydroxylamine; only aniline and unreacted nitrobenzene were obtained from the reaction.

The reduction of nitrobenzenes to phenylhydroxylamines also appears to be affected by the presence of certain substituents. Sugimori [85] found that ethyl 4-nitrobenzoate was reduced to the hydroxylamine in 60% yield with neutral Raney nickel but noted that reduction of 4-nitroanisole under the same conditions gave only 22% of the hydroxylamine and 40% of *para*-anisidine. He pointed out that this significant difference might be related to the difference in the effect of an electron-withdrawing and an electron-donating group.

There are other examples of reductions of substituted nitrobenzenes in which the corresponding phenylhydroxylamines are isolated. Interrupted reductions of 1,3-dinitrobenzene and 2,6-dinitrotoluene gave the corresponding 3-nitrophenylhydroxylamines [84]. 2-Nitro- and 4-nitrobenzoic acids were hydrogenated to the hydroxylamines with palladium black in ether [88]. 3-Nitrobenzaldehyde and 1,2-dinitrobenzene gave 3-hydroxylaminobenzaldehyde and 2-nitrophenylhydroxylamine respectively in similar reductions. The high yields of 4-hydroxylaminobenzamide, 94%, and 4-hydroxylaminobenzoic acid, 98%, from the palladium on carbon-catalyzed reductions of the corresponding nitro compounds may have been due to the combined effect of the substituent and the use of pyridine and quinoline, respectively, as inhibitors [89]. 2-Chloro-4-nitrobenzoic acid was converted to 2-chloro-4-hydroxylaminobenzoic acid in dilute aqueous ammonia containing quinoline with platinum on carbon as catalyst. The chloronitro acid was also hydrogenated to the hydroxylamine in very high yield in alcoholic solution containing quinoline with 5% palladium on carbon (3 parts/100 parts of substrate) without loss of halogen. Raney nickel reduction under similar conditions was also successful.

2-Chlorocyclohexyl 4-chloro-2-nitrophenyl sulfone was reduced to the hydroxylamine in 77% yield [90]. It was possible to isolate the 4-hydroxylamines from the palladium-catalyzed reductions of some substituted 4-nitrobenzenesulfonamides [13].

The cited examples suggest that partial hydrogenation of aromatic nitro

(10.4)

compounds to the corresponding hydroxylamines is facilitated by the presence of electron-withdrawing substituents on the aromatic ring. In the case of *ortho* substituents steric effects may also play a part.

Another example of partial reduction, due possibly to a combination of the effect of the electron-withdrawing trifluoromethyl group and steric effects, may be noted in the attempted hydrogenation of di-(2-nitro-4-trifluoromethylphenyl) sulfone with Raney nickel [90a]. Instead of the diamine three cyclic products were obtained resulting from reduction of the nitro compound to the intermediate hydroxylamino followed by condensation and reduction (Eq. 10.4). The major product was the cyclic hydrazo compound.

IN HETEROCYCLIC COMPOUNDS

Reductions of the nitro group of heterocyclic compounds may be classified as (a) the reduction of the nitro function on a benzene ring which is a substituent on the heterocyclic ring as in (nitrophenyl)pyridine or *N*-(nitrophenyl)piperidine, (b) the reduction of the nitro group attached to the carbocyclic ring of a fused polycyclic heterocycle as in 5-nitroquinoline, and (c) the hydrogenation of the nitro group attached to an aromatic heterocyclic ring as in 3-nitropyridine.

Nitrophenyl Substituted Heterocyclic Compounds

In general reductions of the nitro group among these compounds are readily carried out under mild reaction conditions. More vigorous conditions may be employed but they are usually unnecessary. Examples of reductions to amines are given in Table 10.3.

When the nitrophenyl group is a substituent on a saturated nitrogen heterocycle such as a piperidine or piperazine, reductions are similar in nature to those discussed in the section on aromatic nitro compounds. In such cases the method of Cope [39] (hydrogenation in aqueous dilute acid with palladium on carbon or modifications using alcohol as solvent or platinum oxide as catalyst) may be the one of choice. It has been applied successfully in rapid reductions of 1-methyl-4-(4-nitrophenyl)piperazine and 1-(2- and 4-nitrophenyl)piperazines to the aminophenylpiperazines with palladium [13] and in the reduction of 3-methyl-2-(4-nitrophenyl)morpholine hydrochloride in alcohol with platinum oxide [101]. There is one report of rapid uptake of hydrogen in the reduction of 1-(3,4-dimethoxyphenyl)-4-(2-nitrophenyl)piperazine with platinum oxide in alcohol [102].

Table 10.3 Heterocyclic Compounds Containing Nitrophenyl Groups[a]

Np[b]	Catalyst	Solvent	Yield of Amine (%)	Reference
3-(4-Np)piperidine-2,6-dione-3-COOH	PtO$_2$	EtOH	75	91
3-Et-3-phenyl-(2-Np)azetidin-2-one	PtO$_2$	EtOH	49–75	92
3-Et-3-(4-Np)azetidin-2-one	Pd on C	abs EtOH	70	93
5-(3- and 4-Np)hydantoins	Pd on C	DMF	84	94
3-Me-3-(4-Np)succinimide	Pd on C	—	—	95
N-(4-Nb)-3-morpholone	Pd on C	abs EtOH	89	96
4,5-Di Me-2-(2- and 4-Np)-1,2,3-triazole	NiR[c]	abs EtOH (hot)	91–97	97
1-(2,4-di Np)benzotriazole	PtO$_2$	AcOH	89	98
1-(Np)-5-alkyl-1,2,3,4-tetrazoles	PtO$_2$	AcOH	90[d]	99
5-(Np)-1,2,3,4-tetrazoles	Pd on C	Methyl cellosolve	83[d]	13
(Npo)pyridines	NiR[e]	MeOH	78–91	100
3-(4-Np)pyridine	Pd on C	EtOH	90	13

[a] All reductions were carried out at room temperature and 1–3 atm pressure. [b] Np = nitrophenyl, Nb = nitrobenzyl, Npo = nitrophenoxy. [c] Reduction was slow, 24 hr. [d] No evidence of cleavage of the tetrazole ring. [e] Reduction with platinum in acidified alcohol required a longer reduction period.

In the Carbocyclic Portion of Fused Heterocyclic Rings

Reduction of the nitro group on an aromatic ring fused to an oxygen-containing heterocyclic ring generally poses no problem unless conditions are vigorous enough to reduce the heterocyclic ring. Any catalyst capable of reducing the aromatic nitro group under moderate conditions is normally satisfactory. In general the same conditions should apply when the benzene ring is fused to an N-heterocyclic ring. Examples of such reductions are listed in Table 10.4.

Catalytic reductions of 5,6- and 7-nitroindazoles were first carried out with Raney nickel in alcohol at 50% and 20 atm pressure [115]. Other nitroindazoles were reduced to the amines in high yield under similar conditions [116]. More recently low-pressure reductions with palladium on carbon were reported [117, 118]. 1-Acetyl-5-nitroindazole in glacial acetic acid and 1-acetyl-6-nitroindole and 1-acetyl-6-nitroindazolidin-3-one in methyl cellosolve solution containing acetic acid were hydrogenated very rapidly in the presence of platinum oxide at 1–2 atm pressure to give the corresponding amines [119]. It is likely that palladium or platinum on a carrier could be substituted for platinum oxide in these reductions.

Reduction of Nitro Groups

Table 10.4 Nitrobenzo O- and N-heterocycles[a]

Substrate	Catalyst	Solvent	Yield of Amine (%)	Reference
Nitrodibenzofurans	NiR	95% EtOH	92	103a,b
	Pd on C	EtOH	—	104
5-Nitrobenzofuran	NiR	EtOH	92	105
4-Nitrobenzofurans	PtO$_2$[b]	EtOH	43	106
6-Nitrobenzo-1,4-dioxan-7-COOH	NiR	97.8% EtOH	65	107
4,7-Di MeO-5-nitrobenzimidazole	Pd on C	EtOH	95	108
2-Benzyl-5-nitrobenzimidazole	NiR	EtOH	100	109
2-Me-4-nitrobenzoxazole	NiR	EtOH	92	110
Nitrobenzoxazolones	Pd on C	70% and 95% EtOH	—	111
Nitrobenzoxazoles	PtO$_2$	AcOH[c]	67	112
6-Nitrobenzotriazoles	Pd on C	95% EtOH	63–70	113
1-Me-4- and 7-nitrobenzotriazoles	Pd on C	EtOH	72–78	114

[a] All reductions were carried out at room temperature and 1–3 atm pressure. [b] Reduction was slow. Most of the compounds were highly substituted. [c] Reduction was very rapid; it was not tried in neutral solvent.

In the reduction of nitroindoles hydrogen absorption is often slow. In many instances larger than normal amounts of Raney nickel have been used to reduce 4-, 5-, and 6-nitroindoles [120–125]. Some reactions have been carried out using nickel catalyst with hydrazine as the hydrogen donor [120, 122]. In certain instances high yields of aminoindoles resulted from the use of this procedure [124, 126]. Hydrogenations of some 3-substituted-5-nitroindoles, where the substituent was —COOC$_2$H$_5$, —CH$_2$COOC$_2$H$_5$, and —CH$_2$C(NHCOCH$_3$)(COOC$_2$H$_5$)$_2$, were carried out in neutral solvent with platinum oxide [127]. Yields of amines were high but large amounts of catalyst (6–14% by weight of substrate) were required. The reduction of nitroindoles to aminoindoles might be improved by carrying out the reaction in acidified solvent in the presence of palladium or platinum catalyst.

When the benzene ring is fused to a six-membered hetero ring and the nitro group is in 6- or 7-position, reduction is relatively uncomplicated. In that position, as can be seen from a molecular model, the nitro group is not prevented from making contact with the catalyst. The ease of reduction among such compounds is illustrated in the hydrogenation of 6-nitroquinoline in alcohol with a very small amount of 2% palladium on calcium carbonate [128]. Reduction was carried out at room temperature and atmospheric pressure. The reaction was exothermic and the yield of 6-aminoquinoline was almost quantitative. Other examples are seen in the rapid reductions of the nitro groups in 3-aryl-2-methyl-6- and 7-nitro-3H-4-quinazolones [129] and of 6-nitro-1,2,3,4-tetrahydroquinoxaline with Raney nickel at atmospheric pressure [130]. 5-Methoxy-7-nitro- and 5-methoxy-2,3-dimethyl-7-nitroquinoxalines in alcohol were reduced with palladium on carbon at atmospheric pressure [113]; 6-nitro-3-(4-phenyl-1-piperazinyl)propyl-2,4-(1H,3H)-quinazolinedione in acetic acid was reduced to the amine in almost quantitative yield with palladium under 3 atm pressure [131].

When a benzene ring is fused to a six-membered heterocyclic ring, molecular models show that the nitro group in 5-position, as in 5-nitroquinoline or isoquinoline or in 5-nitrocinnoline or in 5-nitroquinazoline, is hindered by the hydrogen atom in 4-position. The hydrogen atom prevents the nitro group of these compounds from making as firm contact with the catalyst as does the nitro group in nitrobenzene where it is co-planar with the benzene ring. In 8-nitroisoquinoline the hydrogen atom in 1-position also interferes with contact between the nitro group and the active portions of the catalyst. It is assumed that reaction only takes place at active sites on the surface of the catalyst. In reductions of these partially hindered nitro compounds it may also be assumed that fewer sites will be available because of somewhat unfavorable contact between the substrate and catalyst. It is therefore particularly important in these reactions that the active sites still available are not poisoned by impurities in the substrates.

In addition, difficulty in reduction of some of these compounds is compounded by copious precipitation of the poorly soluble intermediate hydroxylamine. Investigators have argued the merits of various solvent systems. The solvent, while important, is probably less of a factor than the purity of the substrate and the activity of the catalyst. Molecular models of the hydroxylamines show that the NHOH group is co-planar with the ring and is in far better position for contact with the catalyst than the corresponding nitro compound. It would appear that, if reaction could be induced to proceed rapidly enough, precipitation of the hydroxylamine would not take place because the intermediate would be reduced to amine almost as quickly as it was formed.

More rapid uptake may be obtained by increasing the amount of catalyst used for reduction. 5-Nitroisoquinoline was hydrogenated to the amine with about 35% by weight of Raney nickel [132]. However, it was reported later than unless the catalyst was freshly prepared and very active, uptake of hydrogen stopped at two equivalents [133]. The same investigators found that reduction in acetic acid with palladium on carbon was more rapid and complete, giving high yields of 5-aminoisoquinoline. Hydrogenation of the same nitro compound in alcohol with palladium was reported [134] with particular emphasis on the correlation between the purity of the starting material and its reducibility.

In general, reductions of the partially hindered 5- and 8-nitro compounds do not proceed as readily with Raney nickel as with palladium under similar conditions. In a direct comparison 5-nitroisoquinoline was rapidly reduced in methyl cellosolve solution with palladium on carbon (10% by weight) to give 93% yield of amine. In contrast, hydrogenation with a higher weight ratio of Raney nickel to substrate (15%) required a second addition of catalyst after an 18-hr reaction period. The yield of the amine was also lower (72.5%) [13]. 1-Hydroxy-5-nitroquinoline and 6-acetylamino-4-methoxy-2-methyl-5-nitroquinoline were reduced with palladized carbon in the same solvent to give over 90% yield of each amine in very short reaction periods [13].

In contrast to the reductions of 5-nitroquinolines and isoquinolines with Raney nickel, no difficulties were reported in the hydrogenations of 3-aryl-2-methyl-5-nitro-3H-4-quinazolones [129]. When 5-nitrocinnoline was hydrogenated with palladium on carbon in alcoholic solution 27% of 5-aminocinnoline was obtained as well as 37% of 5-amino-1,2- or 1,4-dihydrocinnoline [135]. Since as much palladium metal as substrate was present in the catalyst prepared *in situ*, it is likely that the relative rates of reduction between the nitro group and the hetero ring were so changed that it was no longer possible to achieve selectivity.

In the reduction of 8-nitroquinolines and similar compounds which have a nitro group in 8-position and the ring nitrogen in 1-position the rate of hydrogen uptake may be decreased because of the inhibitory effect of the ring nitrogen. Inactivation of the catalyst is probably responsible for precipitation of the poorly soluble intermediate 8-hydroxylaminoquinoline which occurs in the reduction of 8-nitroquinoline because the intermediate cannot be reduced to amine at the same rate that it is formed.

Although Raney nickel has been used for reduction of 8-nitroquinolines, the results indicate that in general it is less satisfactory than palladium or platinum catalysts. 6-Methoxy-4-methyl-8-nitroquinoline was hydrogenated in ethyl acetate with Raney nickel at 60° and 3 atm [136a]. However, it was reported that reductions did not proceed well on large scale because of catalyst poisoning [136b]. The hydrogenation of 6-methoxy-5-methyl-8-nitroquinoline with Raney nickel was carried out at 75–80° and 130 atm presumably to overcome catalyst inactivation [137]. Reductions under such conditions may lead to saturation of the nitrogen-containing ring. In the reduction of 6-methoxy-8-nitroquinoline at 60–65° and 20 atm the authors pointed out that if the reaction was not interrupted after the uptake of three equivalents of hydrogen, 8-amino-6-methoxy-1,2,3,4-tetrahydroquinoline was formed [138]. The same nitro compound on the other hand was reduced at atmospheric pressure with a very small amount of palladium on barium sulfate [128] to yield 95% of 8-amino-6-methoxyquinoline. A similar reduction with platinum oxide gave a quantitative yield [139].

There are a number of examples of the reduction of the 8-nitro group in similar fused systems where the nitrogen atom in the heterocyclic ring is in 1-position. 3-Ethyl-8-nitroquinazolin-4-one in methyl cellosolve solution was converted to the amine with a substantial amount of palladium on carbon [140]. 4-Methoxy- and 4,6-dimethoxy-8-nitroquinazoline in methyl alcohol were readily reduced to the corresponding amines with a moderate amount of 4% palladium on calcium carbonate [141]. 3-Carbethoxy-8-nitro-2-(2-pyridyl) and 8-nitro-2,3-di(2-pyridyl)quinoxaline in methyl alcohol solution were hydrogenated to the corresponding amines in very good yields with palladium on carbon [142] but, unfortunately, the amount of catalyst was not reported. The reductions of 1,3-dinitrophenoxazine in glacial acetic acid with platinum oxide [143] and of 1,6- and 1,9-dinitrophenazines in acetone with palladium on carbon [144] are other examples of hydrogenation in fused aromatic-heterocyclic systems where the nitro group in the benzene ring and the ring nitrogen are adjacent to one another.

In reductions where the ring nitrogen and the nitro group are in 1- and 8-positions, respectively, the addition of an equivalent of acid should increase the rate of hydrogen uptake in the presence of palladium or platinum catalysts. This procedure should be preferable to reductions with these catalysts in acetic acid. Under the latter conditions saturation of the hetero ring may take place if reaction is allowed to run unattended for too long a period.

In Unsaturated Heterocycles

Furans. There is no record of the catalytic hydrogenation of an unsubstituted nitrofuran to the amine. The only unequivocal reductions among the group are those of esters of 5-nitro-2-furoic acid with platinum oxide [144a, b]. Later it was reported that ethyl 4-acetylamino-2-furoate was obtained in reasonably good yield from the nitro ester by hydrogenation with platinum in acetic anhydride at room temperature and 45–65 atm or with nickel in acetic anhydride at 60–80° and 60–90 atm pressure [144c]. The reduction of methyl 4-nitro-2-furoate was carried out at room temperature and 2–3 atm pressure with platinum oxide in acetic acid or ethyl acetate containing acetic anhydride; uptake of hydrogen was extremely rapid but a crystalline product was never obtained until the reaction was carried out in acetic anhydride at 60 atm pressure [13]. The only other reported reductions were those in which 5-nitro-2-furaldehyde derivatives were hydrogenated, but difficulty due to instability of the reduction products prevented characterization of the compounds except by infrared analysis. Recently it has been shown that 5-nitro-2-furaldehyde derivatives can be reduced to amines and that these products are well characterized. Hydrogenations were carried out in ethyl acetate solution with palladium on carbon; the aldehyde function was protected or converted to a semicarbazide or an ylidene derivative or the same type compounds were reduced in the presence of acetic anhydride to give the corresponding 5-acetylaminofurans [145a, b].

Pyrroles. Low-pressure reductions of 1-alkyl-3-nitropyrrole-5-carboxylic acid, ester, and substituted amides are reported to take place rapidly with palladium on carbon [146]. 3-Nitro-2,5-diphenylpyrrole has been reduced to the corresponding amine with Raney nickel [147]. Any of the catalysts useful at low pressure—nickel, palladium, or platinum—should reduce nitropyrroles to aminopyrroles in neutral solvent without causing saturation of the ring. Supported rhodium catalysts probably could be used but they are generally less efficient for reduction of the nitro group and may induce ring reduction if the reaction is not interrupted at the proper point.

Pyrazoles and Related Ring Systems. Reductions of nitropyrazoles to aminopyrazoles under moderate conditions have been carried out with palladium on carbon [148, 149] and with Raney nickel [150]. In the hydrogenation of some 1,3-dimethyl-4-nitro-5-substituted pyrazoles, when the 5-substituent was a bulky group as phenyl, phenoxy, or anilino, palladium on carbon and Raney nickel were employed [13]. In these reductions methyl cellosolve was found to be an especially useful solvent. In a comparison of the hydrogenation of 1,3-dimethyl-4-nitro-5-(1-piperidino)pyrazole in alcohol with palladium on carbon and platinum oxide, it appeared that the platinum catalyst was less active [13]. This was probably due to inhibition by the basic piperidine nitrogen since platinum was found to be useful in other reductions where the 5-substituent was not a basic one.

5-Amino-3-hydroxy-4-nitrosopyrazole was simultaneously hydrogenated and acylated in 98% formic acid in the presence of palladium on carbon [151]. *The addition of catalyst to formic acid can be extremely hazardous because of release of hydrogen from decomposition of formic acid.* A safe procedure is to add the catalyst to sufficient water to wet it and then add formic acid when it is seen that the catalyst is completely covered.

It should also be noted that, when formic acid is used as solvent in reduction, hydrogen uptake may appear incomplete because nascent hydrogen from decomposition of the acid has already caused partial reduction. This has been observed in numerous hydrogenations of 5-amino-1,3-dimethyl-4-nitropyrazole with palladium in formic acid when uptake was 40–50% of the calculated amount and 75–80% yields of 4,5-diformylamino-1,3-dimethylpyrazole were obtained [13].

Imidazoles. Mild conditions are used for the hydrogenation of the nitro group on the imidazole ring. 4-Nitroimidazole-5-carboxamide was reduced to the corresponding amine with platinum oxide catalyst [152]. Good yields of amines have resulted from reductions of 1-methyl, ethyl, or phenyl-4-nitroimidazole-5-carboxamides or 5-carboxylic esters with either Raney nickel or palladium on carbon [153]. The hydrogenation of 5-cyano-1-methyl-4-nitroimidazole was successfully carried out to the amine using almost as much Raney nickel as compound [154]. Another group reported that reduction of the same compound was variable with nickel depending on the activity of the catalyst. They used platinum oxide in alcohol or in alcoholic hydrogen chloride [155]. However, while reduction was very rapid, the product obtained was the 4-hydroxylamino compound which was further reduced to 4-amino-5-cyano-1-methylimidazole with an equal amount of Raney nickel.

Pyridines and Related Ring Systems. The reduction of the nitro group in the pyridine ring or in the pyridine portion of the quinoline ring is readily carried out at low pressure with nickel, platinum, or palladium catalysts. At times there may be some inhibiting effect because of the unshared pair of electrons of the nitrogen atom in the ring but, in general, uptake of hydrogen is accompanied by a rise in temperature so that reaction proceeds more or less readily.

In reductions with Raney nickel larger than normal amounts of catalyst were often used. Whether or not it was necessary was not studied. 2-Methoxy-5-nitropyridine in alcohol was hydrogenated with an equal amount of nickel [156]. A similar reduction with about 20% by weight of 5% palladium on carbon to substrate was just as rapid. In each instance yield was quantitative. Nickel can be used advantageously at higher pressure because it will not reduce the ring too readily. 2,6-Dimethyl-3-nitropyridine was hydrogenated at 70° and 100 atm pressure to yield 3-amino-2,6-dimethylpyridine [157]. Platinized nickel was used at room temperature and atmospheric pressure in the reductions of 2- and 4-N(pyrrolidino, piperidino, and morpholino)-3-nitropyridines [158]. Raney nickel gave good results in reductions of 2-anilino or 2-phenoxy-3-nitropyridines in methyl cellosolve solution, but it was necessary to use about 30% by weight of catalyst to substrate to get complete hydrogen uptake in a reasonable length of time [13].

2-Methoxy-3,5-dinitropyridine was reduced in alcoholic solution with platinum oxide [159]. A number of 1-methyl-3-nitro-2-pyridones [160] and some 5-nitro-3-pyridones [161] were also hydrogenated in the presence of the same catalyst; the amounts appear to be rather high (6–10% of the weight of the substrate).

4-Methylamino-3-nitropyridine in methyl alcohol was reduced to 3-amino-4-dimethylaminopyridine in 90% yield in the presence of 1% palladium on calcium carbonate and 3-nitro-4-(4-toluenesulfonamido)pyridine in acetic acid was hydrogenated with 5% palladium on carbon [162]. 3-Methyl-4-nitropyridine was reduced to the corresponding amine with palladium on carbon in neutral solution [163].

Adjacence of the ring nitrogen to the nitro group may have a greater effect on reduction than when the nitro group is *meta* or *para* to it. There did not appear to be any difficulty in the hydrogenation of 3-hydroxy-2-nitropyridine in methyl alcohol with palladium on carbon; a moderate amount of catalyst gave the corresponding amine in 85% yield [164]. In the reductions of 3-amino- and 3-substituted amino-2-nitropyridines relatively large amounts of the active Raney nickel catalyst were used [165]. In the hydrogenation of 2-nitropyridine, in alcohol with 10% by weight of platinum oxide to substrate, absorption decreased markedly after uptake of two equivalents to give 2-hydroxylaminopyridine [166].

The nitro group in nitropyridine- and nitroquinoline-1-oxides can be selectively reduced without removing the N-oxide if the reaction is carried out in neutral solvent in the presence of palladium. Nickel- or platinum-catalyzed reduction or reduction in acidic medium gives aminopyridine or aminoquinoline. 4-Nitropyridine-1-oxide and 4-nitroquinoline-1-oxide and related nitropyridine-1-oxides have been hydrogenated to the corresponding amine-1-oxides with palladium in neutral solvent [167, 168a, b]. Reduction with palladium on strontium carbonate at 60° practically stopped after absorption of three equivalents to give 78% of 4-aminopyridine-1-oxide [169]. 4-Nitro-2-picoline-1-oxide was reduced to the amine oxide in neutral solvent, but the same reaction in 10% hydrochloric acid gave 4-amino-2-picoline [170]. In the palladium-catalyzed reduction of 4,4'-dinitro-3,3'-picolyl-1,1'-dioxide in alcohol containing concentrated hydrochloric acid hydrogen uptake stopped at four equivalents; 4,4'-dihydroxylamino-3,3'-picolyl-1,1-dioxide was obtained [171]. 2,6-Dimethyl-3-nitropyridine-1-oxide was selectively reduced to amine-1-oxide in 98% yield with palladium on carbon [172]. However, when 2-nitropyridine-1-oxide was hydrogenated with the same catalyst only 2-aminopyridine was obtained [173].

Pyridazines. Aminopyridazines are more often prepared by amination of chloropyridazines than by catalytic reduction of nitropyridazines. There are few reported hydrogenations by catalytic means. The conversion of 3,5-diamino-4-nitropyridazine with Raney nickel to the triamine [174] and the reduction of 4-amino-2,6-dimethoxy-5-nitropyridazine to the diamino compound with palladium on carbon [175] are examples. A series of 1-substituted-5-amino-4-nitro-6-pyridazones in alcohol have been hydrogenated to the corresponding diamines with Raney nickel, platinum oxide, and palladium on carbon at 100° and 20 atm pressure [176]. Some of the reductions were carried out in tetrahydrofuran or, at times, in the same solvent containing ammonia.

As with nitropyridine-1-oxides, selective reduction of nitropyridazine-1-oxides is limited to the use of palladium in neutral solvent [177a, b, c]. The palladium-catalyzed reduction of 3-nitropyridazine-1-oxide in alcohol is of interest because 3-hydroxylaminopyridazine-1-oxide can be obtained by interrupting the reaction after absorption of two equivalents of hydrogen; 3-aminopyridazine-1-oxide is obtained after absorption of three equivalents [178].

Pyrazines. Since it is difficult to nitrate the pyrazine ring and since aminopyrazines are more easily prepared by other methods, there are almost no reported reductions of nitropyrazines. 2-Hydroxy-3-nitro-5,6-diphenylpyrazine was reduced to the amine by means of Raney nickel and hydrazine hydrate [179]. The same compound was hydrogenated in methyl cellosolve solution with palladium on carbon catalyst under 2 atm pressure [13].

Pyrimidines and Related Ring Systems. With few exceptions the nitro and nitroso groups are found only in 5-position on the pyrimidine ring. In general, hydrogenation with Raney nickel, palladium on carbon, and platinum oxide in neutral solvent is straightforward, leading to the corresponding aminopyrimidine. In one instance, with "old" Raney nickel, reduction of 5-nitro-2-phenylpyrimidine in dioxane gave 2,2'-diphenyl-5,5'-azoxypyrimidine [180].

It is difficult to evaluate the effect of the catalysts in this series because in too many instances the amount employed for reduction is not reported. In other experiments excessive amounts are used as seen in some reductions of 5-nitropyrimidines with Raney nickel [181–185]. In some of these reductions the presence of an amino group in 4-position may have made the use of more catalyst necessary because of the basicity of the amino group. On the other hand, a very modest amount of nickel (1 g/25 g of substrate) catalyzed the reduction of 4,6-diamino-5-nitropyrimidine to the triamine, in a reaction in which hydrogen was passed into a warm, vigorously stirred alcoholic solution of the nitro compound [186]. 4-Methoxy-5-nitro-6-(2-carbomethoxyphenoxy)-pyrimidine was hydrogenated in a conventional manner in a Parr hydrogenator with a 20% weight ratio of commercially available Raney nickel [13].

There are only a few reductions reported with platinum oxide which indicate the amount of catalyst used. Very rapid hydrogen absorption and good yield of triamine was noted in the platinum-catalyzed reduction of 2,4-diamino-6-trifluoromethyl-5-nitropyrimidine [187].

There are also a number of platinum-catalyzed reductions where the weight of catalyst is not recorded or an abnormally large amount of catalyst is employed without explaining the need for such an amount. About 50%

by weight of 5% palladium on carbon was used to reduce 2,4- and 4,6-dialkoxy-5-nitropyrimidines to the amines [188]. Hydrogenations of 1,3,6-trimethyl- and 1,3-diethyl-6-methyl-5-nitrouracils were carried out in the presence of a normal amount of the same catalyst but at 70–90° and 50 atm pressure [189]. 4-Amino-6-dimethylamino-5-nitropyrimidine was reduced to 4,5-diamino-6-dimethylaminopyrimidine in methyl cellosolve solution with only 10% by weight of the same catalyst at room temperature and 2 atm pressure [13].

Reduction of 2-nitro-5-phenylpyrimidine to the amine could only be accomplished in the presence of palladium on carbon [190]. Other procedures, both chemical and catalytic, failed to give the amine.

Summary. A review of this section might suggest that palladium is the preferred catalyst for hydrogenation of the nitro group in unsaturated heterocycles. Performance of the other catalysts, as well as palladium, under low-pressure conditions might be improved by carrying out reactions in the presence of a molar equivalent of acid to overcome possible inhibition by nitrogen base, except in situations where the presence of acid is contraindicated.

Hydrogenations, in which the nitro group is attached to fused two-ring unsaturated systems containing a hetero atom in each ring, are very sparsely reported. In such instances reduction may be compared to related compounds in these series in order to find comparable conditions. For example, the reduction of 3-nitro-7-azaindole may be compared with that of 3-nitroimidazole and the hydrogenation of 5-nitro-7-azaindole related to that of 3-nitropyridine. These reductions have been reported [191] but they illustrate a point.

SELECTIVE REDUCTION IN AROMATIC COMPOUNDS CONTAINING OTHER REDUCIBLE FUNCTIONS

Olefinic Bonds

There is usually a significant difference between the rate of hydrogenation of a nitro group on an aromatic ring and that of a nonaromatic double bond which is a part of the same molecule. This is due to the preferential adsorption of the nitro group to the active portions of the catalyst. As a result reduction of the double bond does not take place until the nitro group is reduced and the amine is desorbed from the catalyst. The amine contributes to selectivity by inhibiting the activity of the catalyst. It is also known that an aromatic ring inhibits the activity of hydrogenation catalysts for reduction of ethylenic bonds. These factors and the general ease of reducing an aromatic nitro compound to an amine make it possible to achieve preferential reduction of the nitro group under the proper conditions. Uptake of hydrogen may continue beyond three equivalents but the reaction can be interrupted to obtain the amine with the double bond intact. Reductions are carried out at room temperature and at 1–3 atm pressure with Raney nickel or platinum oxide or with palladium on a carrier.

The only reported catalytic hydrogenation of an aromatic nitro compound containing a monosubstituted double bond is that of 3-nitrostyrene. Here 3-aminostyrene was obtained in 17% yield with rhenium catalyst at 135° and 200 atm pressure [192].

There are many examples of selective reduction in the presence of symmetrical disubstituted double bonds. Hydrogenations of nitrostilbenes to aminostilbenes have been carried out with Raney nickel [193–195]. The reduction of 3-nitrostilbene with palladium on barium sulfate gave 3-aminobibenzyl; the authors reported no change in the hydrogenation curve with palladium or Raney nickel [196]. In contrast, 3-aminostilbene was obtained from Raney nickel reduction of 3-nitrostilbene [194]. In the reduction of 1,4-bis-(3-nitrostyryl)benzene in dimethylformamide overhydrogenation occurred with Raney nickel; a similar reduction with ammonia added gave the desired bis-(aminostyryl)benzene [197].

2-, 3-, and 4-Nitrocinnamic acids and esters were selectively reduced with Raney nickel [198]. In the reduction of the 2-nitro compounds it was necessary to stop the reaction after absorption of three equivalents to prevent overhydrogenation. In the other reductions there was a distinct difference in the rate of hydrogen uptake after absorption of three equivalents. 3,4-Dimethoxy-5-nitrocinnamic acid was selectively reduced in alcohol with Raney nickel [199]. Attempts to obtain the corresponding aminohydrocinnamic acid by hydrogenation of the nitro compound with nickel in the presence of base failed. Other examples of reductions of the nitro group in the presence of a symmetrically disubstituted double bond are seen in the hydrogenation of 4-(2,6-dinitrostyryl)pyridine to the diaminostyrylpyridine with nickel [200] and in that of 2-chloromethyl-2-(4-nitrostyryl)-1,3-dioxolane with platinum to the corresponding aminostyryl compound [201]. In the latter reduction uptake of hydrogen was complete within 15 min. Reduction of both the nitro group and the double bond required 2 days.

Selective reduction of the nitro group in the presence of a trisubstituted double bond should be no problem. Examples are seen in the hydrogenations of 2- and 3-nitro-α-(4-nitrophenyl)cinnamic acids to the corresponding diamines with a large amount of Raney nickel [193, 202]. In other examples 5-(4-nitrobenzylidene)hydantoin was reduced to 5-(4-aminobenzylidene)-hydantoin with palladium on carbon in dimethylformamide, with platinum oxide in acetic acid and with Raney nickel in aqueous alkaline solution [94], and 4-(4-nitrobenzylidene)-2-phenyl-5-oxazolone was reduced to the aminobenzylidene with palladium on alumina [203].

In the hydrogenation of the nitro group of 1,1-dichloro-2,2-*bis*-(4-nitrophenyl)ethylene in acetic acid with nickel catalyst absorption stopped of its own accord at $3H_2$ [204]. Such selectivity might be expected in this instance because the double bond is tetrasubstituted and therefore difficult to reduce. High yield of the dichloro-*bis*-(aminophenyl)ethylene was obtained.

Selective reduction of the nitro group to amine in the presence of an α,β-unsaturated carbonyl group has been carried out with platinum oxide [4] and with palladium on carbon [205] without attack on the other reducible groups. Similarly, 2(4-methoxyphenyl)-3-(3-nitrophenyl)acrylnitrile was reduced in alcohol to the aminophenylacrylonitrile with platinum oxide [206] and some 2-(nitrophenyl)-3-phenylacrylonitriles were hydrogenated in ethyl acetate with palladium on carbon [207, 208]. In the latter case [208] it was necessary to raise the temperature to 70–80° to reduce the double bond.

Aldehydes and Ketones

Since selectivity is obtained in the reduction of aromatic nitro compounds also containing a nonaromatic double bond and since reduction of such double bonds has been shown to be favored over that of a carbonyl group, it is expected that an aromatic nitro group will also be preferentially reduced in the presence of an aldehyde or ketone function. 2-Nitrobenzaldehyde and 3- and 4-nitrobenzaldehydes have been reduced to aminobenzaldehydes with supported palladium catalysts [205, 209] and with Raney nickel [210]. 2'- and 3'-Nitroacetophenones were rapidly hydrogenated in the presence of platinum oxide to the aminoacetophenones [211]; nickel was found to be less efficient in similar reductions. At higher pressure with nickel there was a tendency to reduce the carbonyl group, but selectivity was obtained if the reaction was interrupted after absorption of $3H_2$. Others, however, have reported good results with nickel at 50° and 1 atm [212] and at 35–40° and 25 atm pressure [213]. Satisfactory yield was obtained in the reduction of 3-nitroacetophenone in ethyl acetate with palladium on carbon, but the authors stated that Raney nickel was more suitable for larger scale reductions where 94.7% yield of 3'-aminoacetophenone was obtained [214]. Selective reductions of 3'-nitroacetophenone and 3'-nitropropiophenone were carried out with palladium on carbon in methyl cellosolve [214a]. Hydrogenations of 2'- and 3'-nitropropiophenone in benzene were rather slow with palladium on carbon [215], perhaps due to inhibition by the solvent. 6-Acetyl-2-nitrobenzoic acid was reduced in dilute aqueous ammonia with palladium on carbon to 6-acetylanthranilic acid in 73% yield [214a]; the yield was only 18% when alcohol was the solvent for hydrogenation. 4,4'-Dinitrobenzophenone was selectively reduced to the diamine in glacial acetic acid with Raney nickel [204]; 4-nitrobenzophenone was selectively hydrogenated with Adams catalyst [216] and a reagent-grade benzene solution of 3-nitrobenzophenone, after a carbon pretreatment, was selectively reduced with palladium [217].

It is possible that platinum oxide (or platinum on a carrier) may be the catalyst of choice in selective reductions of an aromatic nitro group in the presence of a ketone group. Reduction with palladium may be just as selective and as rapid in the presence of acid if the hydrogenation of 5-(4-nitrophenyl)cyclohexen-4-one can be used as an example. Reduction of the double bond and the nitro group was complete in a very short period in a hydrogenation with palladium on carbon in alcohol containing hydrochloric acid [218]. Yield of 2-(4-aminophenyl)cyclohexanone was 93%. A similar effect has been noted in reduction with Raney nickel. 2-(4-Nitrophenyl)-propiophenone was selectively hydrogenated in ethyl acetate to the amine with nickel at 100° and 90 atm pressure in 3–4 hr. Reduction in acetic acid was complete in 1 hr at ordinary temperature and pressure to give 70–80% yields of 2-(4-aminophenyl)propiophenone [194].

Unsaturated Nitrogen-Containing Groups

Nonaromatic Nitro Groups. There appears to be no record of attempts at selective reduction of 2,β- and 4,β-dinitrostyrenes. In the hydrogenation of 2,β-dinitrostyrenes, reduction and cyclization were carried out in one step to form indoles.

Some 4-nitro-(2-nitroalkyl)benzenes have been hydrogenated under moderate high-pressure conditions with Raney nickel (60–70°, 80 atm) but no observations were made regarding selectivity since the investigator's only interest was in obtaining 4-(2-aminoalkyl)anilines.

The reduction of 2,2-dimethyl-5-nitro-5-(4-nitrobenzyl)-1,3-dioxane in alcohol with palladium on carbon catalyst at 25° and 4 atm pressure to 5-(4-aminobenzyl)-2,2-dimethyl-5-nitro-1,3-dioxane [219] illustrates the difference in reducibility between an aromatic and nonaromatic nitro group. Reduction of both nitro groups was carried out at 180° and 115 atm with Raney nickel.

Azomethines. Rapid and selective reductions of (nitrobenzylidene)-anilines and other nitrobenzylideneamines have been carried out with Raney nickel [210], with palladium on carbon [220], with platinum oxide [221], and with platinum on carbon [13]. Other examples of selectivity have been noted in reductions of 5-nitro-2-furaldehydesemicarbazone to the corresponding amine [145a, b]. When the C=N bond is an internal one, as in 7-nitrobenzo-1,4-diazepines, where the nitro group was selectively reduced with Raney nickel [222], any catalyst suitable for low-pressure hydrogenations should be capable of selectively reducing the nitro group without affecting the C=N linkage.

Oximes. There are a few selective hydrogenation among nitrobenzaldoximes. 2-Nitrobenzaldoxime was reduced to 2-aminobenzaldoxime in 90% yield with Adams catalyst [223]. 4-Chloro-2-nitrobenzaldoxime was similarly hydrogenated to 2-amino-4-chlorobenzaldoxime [224]. Ethyl 4-nitrophenyl-α-hydroxyiminoacetate was also hydrogenated with platinum to obtain ethyl 4-aminophenyl-α-hydroxyiminoacetate [225]. Reduction with Raney nickel gave ethyl 4-aminophenyl-α-aminoacetate, but it was not made clear whether there had been an attempt at selective reduction.

Cyano Groups. Selective reduction of the aromatic nitro group may also be expected in the presence of a cyano group. In the reduction of nitrobenzonitriles to aminobenzonitriles or in reductions in other aromatic systems, reaction is relatively uncomplicated except when the two groups are adjacent to one another. 4-Amino-3-nitrobenzonitrile was reduced very rapidly under 3 atm pressure with platinum oxide to 3,4-diaminobenzonitrile [226]. 4-Nitrobenzonitrile was hydrogenated under 10 atm pressure with a small amount of palladium on strontium carbonate [227]. Hydrogen uptake was slow but reduction was selective; 88% of 4-aminobenzonitrile was obtained. 5-Nitro-1-naphthonitrile and 5-nitro-2-naphthonitrile were hydrogenated in a mixture of solvents containing water to the corresponding aminonaphthonitriles with a reduced nickel catalyst [228]. 5-Cyano-1-methyl-4-nitroimidazole was reduced with Raney nickel to 4-amino-5-cyano-1-methylimidazole [154].

2-Nitrobenzonitrile was hydrogenated with a reduced nickel catalyst in an alcohol-ethyl acetate mixture. Only 2-aminobenzamide was obtained which the investigators attributed to the effect of water (from reduction of the nitro group) in a lengthy reaction period. In reductions of 2-nitrobenzonitrile, in methyl alcohol with palladium and with platinum, the aminoamide was obtained in 90% yield [230]. As a result of a similar reduction to which O^{18} water was added, the authors concluded that since amide did not contain any O^{18} tagged material, its formation could not arise from water resulting from reduction of the nitro group. In a later publication [231a] they postulate that, in hydrogenations in alcohol with nickel or platinum, oxygen transfer proceeds intramolecularly after formation of 2-hydroxylaminobenzonitrile to 3-amino-1,2-benzisoxazole, which undergoes rupture at the nitrogen-oxygen bond to give 2-aminobenzamide (see Eq. 10.5).

$$\begin{array}{c} \text{C}_6\text{H}_4(\text{CN})(\text{NO}_2) \xrightarrow{2\text{H}_2} \\ \left[\text{C}_6\text{H}_4(\text{CN})(\text{NHOH}) \rightarrow \text{benzisoxazole intermediate} \right] \xrightarrow{\text{H}_2} \text{C}_6\text{H}_4(\text{CONH}_2)(\text{NH}_2) \end{array} \qquad (10.5)$$

In the presence of palladium catalyst reduction of the nitronitrile in alcohol or in 3:1 dioxane-alcohol yielded 2-aminobenzonitrile because, according to the authors, cyclization did not take place.

In most of the reactions carried out by this group, *ortho* aminoamides were obtained from reductions of 1,2-nitroarylnitriles in alcohol with nickel, palladium, or platinum catalysts [231a, b, c]. In an early study by another group of the hydrogenation of nitronaphthonitriles, 1-nitro-2-naphthonitrile was reduced with a nickel catalyst in a mixture of organic solvents and water [228]. The product of reduction was claimed to be 1-amino-2-naphthonitrile. In the later work [231c], it was proved that the product was not the aminonitrile but was 1-aminonaphthalene-2-carboxamide, which was obtained by reductions of the nitronaphthonitrile in alcohol with Raney nickel, palladium on barium sulfate, or platinum oxide or by chemical reduction with zinc and acetic acid.

In the hydrogenations of *ortho* nitronitriles [231a, b, c], the combination of dioxane and palladium catalyst appeared to favor formation of aminonitrile. In a reduction of 1-nitro-2-naphthonitrile in *very pure dry* dioxane, 1-amino-2-naphthonitrile was obtained when palladium on barium sulfate was used [231c]. Reduction in unpurified dioxane with platinum oxide gave the aminoamide and less than 1% of aminonitrile. It was not determined whether reduction in pure dry dioxane with nickel or platinum also led to aminonitrile.

In some hydrogenations of 4-trifluoromethyl-2-nitrobenzonitrile in absolute methyl alcohol with commercially available Raney nickel packed in water the corresponding aminoamide was obtained in 62–73% yield [13]. The addition of about 1 mole of water per 0.15 mole of substrate to the reduction mixture increased the yield to 83–95%. Reduction in alcohol with palladium on carbon gave 2-amino-4-trifluoromethylbenzonitrile and 2-amino-4-trifluoromethylbenzamide in a 2:1 ratio.

Selective hydrogenation of the nitro group in 2-, 3-, and 4-nitrophenylacetonitriles is readily accomplished, although there is a tendency in reduction of the 2-nitro compounds toward continued uptake to form indoles. In such cases the amount of catalyst can be reduced substantially since the nitro group is reduced so readily. Indeed it is a practice which should be followed in attempted selective hydrogenations whenever there is an indication of continuing hydrogen absorption.

Palladium on carbon in ethyl acetate has been found useful in hydrogenations of 2-nitrophenylacetonitriles [232]. 4-Methoxy-3-nitrophenylacetonitrile was reduced rapidly to the corresponding aminonitrile in quantitative yield in its presence [233]. Raney nickel also gave good results in reduction of the same compound [13]. Platinum oxide was employed in the hydrogenation of 4-nitrophenylacetonitrile to 4-aminophenylacetonitrile [234]. In the palladium-catalyzed reduction of 2-methoxy-4-nitrophenyl-acetonitrile, in alcohol containing hydrochloride acid, hydrogen absorption came to a stop at $3H_2$ but continued in the presence of platinum [235]. Palladium on a carrier may be the catalyst of choice in selective reductions of an aromatic nitro group in the presence of a cyano group. 5% Palladium on carbon proved to be especially adaptable to the selective reduction of N-(4-nitrobenzoyl)amino- and N-methyl-N-(4-nitrobenzoyl)aminoacetonitriles in methyl cellosolve solution [236]. Hydrogen uptake was fairly rapid and yields of the corresponding 4-aminobenzoylaminoacetonitriles were about 80%.

Hydrogenolyzable Groups

Benzyl Ethers, Esters, and Alcohols. Reduction of 4-benzyloxynitrobenzenes illustrates the difference in reducibility between the nitro group and the O-benzyl group. Hydrogenation of 4-benzyloxynitrobenzene with Raney nickel has been carried out at 60° and 10–15 atm [13] and 70° and 75 atm [237] to give 90% yields of 4-benzyloxyaniline. Reductions with nickel at low pressure are often slow but are usually selective. Palladium catalysts have been used but are less preferred because they are so active for hydrogenolyses in general. The experience of this writer suggests that platinum on carbon may be the catalyst of choice for reductions of this type. As low as 1% by weight 5% platinum on carbon gave extremely rapid and selective conversion of 2- and 4-benzyloxynitrobenzenes to the corresponding anilines in low-pressure hydrogenations. Platinum oxide was used to obtain a quantitative yield of 3,4,5-tribenzyloxyaniline from 3,4,5-tribenzyloxynitrobenzene [238]. It was also used in the hydrogenation of 5-amino-1-benzyl-3-benzyloxy-4-nitropyridazin-6-one to 4,5-diamino-1-benzyl-3-benzyloxypyridazin-6-one [176].

Platinum oxide was chosen because of its "nondebenzylating" character for the reduction of 3- and 4-nitrobenzyl alcohol to the 3- and 4-aminoalcohols [239]. 3-Aminobenzyl alcohol was obtained in quantitative yield but the product of reduction of the 4-nitro compound was a polymer. 2-, 3-, and 4-Nitrobenzhydrols were reduced with platinum oxide to the aminobenzhydrols in 80–92% yield [240]. However, in one of the reductions of 2-nitrobenzhydrol, 2-aminobenzhydrol was obtained as a minor product. The major product was 2,2'-di(α-hydroxybenzyl)azoxybenzene. The formation of the latter product may have been due to steric effects which caused incomplete reduction and condensation (see Eq. 10.6).

$$RC_6H_4\text{-}2\text{-}NO_2 \xrightarrow{H_2} RC_6H_4\text{-}2\text{-}NO \xrightarrow{H_2} RC_6H_4\text{-}2\text{-}NHOH$$

$$RC_6H_4\text{-}2\text{-}NHOH + RC_6H_4\text{-}2\text{-}NO \xrightarrow{-H_2O} RC_6H_4\text{-}2\text{-}N\!\!=\!\!\overset{+}{\underset{O^-}{N}}\text{-}2\text{-}C_6H_4R$$

(10.6)

$R = C_6H_5CHOH$

Palladium catalyst was employed without causing hydrogenolysis in the reductions of methyl 4-methoxy-3-nitromandelate [241] and 2-nitromandelic acid [242].

Selective hydrogenation of the aromatic nitro group should also be expected in the presence of a benzyl ester and also when the O-protecting group is carbobenzoxy.

Benzylamines. Selective reduction of the nitro group can be readily carried out without accompanying cleavage of the carbon to nitrogen bond. The favored catalysts in these reductions are Raney nickel, platinum oxide, and platinum on carbon. Other supported platinum catalysts should be just as useful. Raney nickel is such a poor catalyst for debenzylation that a high catalyst ratio can be employed for reduction without effect on the C—N bond. More catalyst than compound was used to hydrogenate methyl 5-carbobenzoxyamino-6-(2'-nitrostyryl)nicotinate to the amine without reducing the double bond or removing the N-benzyl group [243]. Nickel may be particularly useful in the selective reduction of a nitro group on a heterocyclic ring containing an N-benzyl group because it has little or no effect on the ring under low-pressure conditions. It was employed to convert 1-benzyl-5-cyano-4-nitroimidazole to 4-amino-1-benzyl-5-cyanoimidazole in 99% yield [244]. Nickel catalyst was also used to selectively reduce the nitro group in some 4-benzylamino-5-nitropyrimidines [245a, b] and in the reduction of 2-(N-benzyl-N-methyl)amino-3'-nitroacetophenone hydrochloride to the corresponding 3'-amino compound [214].

Palladium catalysts have been used successfully at times in selective reductions. N-Methyl-2-nitrobenzylamine was reduced to N-(2-aminobenzyl)methylamine with 50% of its weight of 5% palladium on carbon

[246]. [3-Carbobenzoxyamino-3-(4-nitrophenyl)propionyl]taurinamide was selectively reduced to the amine with palladium black in a long reaction period under 200 atm pressure [247]. On the other hand, when palladium was substituted for nickel in the selective hydrogenation of 2-(N-benzyl-N-methyl)amino-3'-nitroacetophenone hydrochloride, cleavage of the benzyl group as well as conversion of the nitro group took place [214]. It must be noted, however, in that reduction that the ratio of catalyst to substrate was high (about 50%). Palladium on carbon was not selective in the hydrogenation of N-bis-(2-chloroethyl)-4-nitrobenzylamine [248]. Cleavage of the benzyl group occurred under both acidic and neutral conditions [248]. By using an aged catalyst the 4-aminobenzyl nitrogen mustard was obtained. N-bis-(2-hydroxyethyl)-4-nitrobenzylamine was cleaved to p-toluidine during reduction with palladium or Raney nickel [249].

Platinum oxide was judged to be preferable to palladium on carbon for the reduction of N-benzyl-2-nitroaniline because it decreased debenzylation [250]. Platinum oxide catalyzed the reduction of 4-(3-nitrobenzylamino)-benzoic acid [251] and was used to obtain a 96% yield of 3-amino-N,N-dimethylbenzylamine from the 3-nitro compound [252]. In the reduction of 10-benzyl-3-nitrophenoxazine with platinum oxide in acetic acid, uptake of hydrogen halted at three molar equivalents to give 3-amino-10-benzylphenoxazine in quantitative yield [253].

Nitrogen to Nitrogen Bonds. The reduction of 4-hydrazino-6-hydroxy-5-nitropyrimidine with palladium on carbon to 5-amino-4-hydrazino-6-hydroxypyrimidine has been described [254]. It was reported that in larger runs some 4,5-diamino compound was obtained which may have been due to the use of half as much catalyst as compound. In a related reduction with Raney nickel the yield of 4,5-diamino-6-hydrazinopyrimidine was only 45% in a reaction carried out with more catalyst than compound. 3-Nitro-N-nitrosodibenzylamine was reduced in the presence of a low ratio of palladium on carbon to yield 92% of 3-amino-N-nitrosodibenzylamine [255].

Aliphatic Halides. The nitro group attached to the aromatic ring of compounds also containing an aliphatic halide has been reduced to amine in neutral solution under low-pressure conditions with nickel [25b], palladium [90], and platinum [257] and in acidic solution with platinum catalysts [257, 258] without accompanying dehalogenation.

When the nitro compound also contains an activated halogen, such as a phenacyl bromide, reduction to amine without loss of halogen may be difficult. The hydrogenation of 2-bromo-4'-chloro-3'-nitrophenylbutyrophenone (4'-Cl,3'-NO$_2$C$_6$H$_3$COCHBrCH$_2$CH$_3$) with a large amount of Raney nickel gave 3'-amino-4'-chlorophenylbutyrophenone [259]. There was no effort to obtain selectivity in this and similar reductions of related compounds.

It is unlikely in the presence of such a large amount of catalyst (35–40%) by weight of substrate) that selective reduction of the nitro group could be effected.

When 2-chloro-2'-nitro-N-phenylacetanilide was hydrogenated in alcohol with platinum oxide, three molar equivalents of hydrogen were absorbed [260]. The reduction product was shown to be 2-methyl-1-phenylbenzimidazole-3-oxide hydrochloride. It was suggested that 2-chloro-2'-hydroxylamino-N-phencylacetanilide was the intermediate which led to the 3-oxide. In contrast, the platinum-catalyzed reduction of 2'-nitro-N-phenylacetanilide in alcohol gave the corresponding 2'-amino compound, while reduction in alcohol containing 0.2 molar equivalent of hydrogen chloride yielded 2-methyl-1-phenylbenzimidazole. Platinum reductions of the same nitro compound with 1.0 and 2.0 molar equivalents of hydrogen chloride gave the 3-oxide as obtained from reduction of the 2-chloro-2'-nitro compound. As a result of these experiments it was concluded that ring closure was acid catalyzed and that, in the hydrogenation of the 2-chloro-2'-nitro compound, dehalogenation of the intermediate 2-chlor-2'-hydroxylamino-N-phenylacetanilide occurred before ring closure, as seen in Eq. 10.7.

$$\underset{NO_2}{\text{C}_6\text{H}_4}\text{N(R)COCH}_2\text{Cl} \xrightarrow{2H_2} \underset{NHOH}{\text{C}_6\text{H}_4}\text{N(R)COCH}_2\text{Cl} \xrightarrow{H_2}$$

(10.7)

$$\left[\underset{NHOH}{\text{C}_6\text{H}_4}\text{N(R)COCH}_3 + \text{HCl} \right] \longrightarrow \text{benzimidazole-N-oxide with CH}_3$$

R = C₆H₅

Aryl Halides. Based on the ability of the halogen atoms to make contact with the catalyst surface, examination of molecular models of halonitrobenzenes suggests that the chlorine, bromine, and iodine atoms may be cleaved during reduction while the fluorine atom will not undergo hydrogenolysis. Nevertheless, it has been shown that loss of fluorine has taken place during reduction. When 4'-fluoro-3'-nitroacetophenone was hydrogenated to 5-ethyl-2-fluoroaniline, in acetic acid with palladium black under 2.5 atm pressure, some 3-ethylaniline was obtained [261]. This finding suggests that factors other than steric ones have an influence on the relative ease of hydrogenolysis in the reduction of halonitrobenzenes.

It must be pointed out that chemical reactivities of the halogens in halonitrobenzenes should not be equated with their reducibility. Although the fluorine atom in 2,4-dinitrofluorobenzene is extremely active toward nucleophilic substitution, 2,4-dinitrofluorobenzene has been reduced to 4-fluoro-1,3-phenylenediamine in over 90% yield in the presence of Raney nickel or palladium black [262]. In the hydrogenation of 2,4-dinitrochlorobenzene with Raney nickel at 50° and 100 atm pressure, complete loss of halogen did occur [263]; only 1,3-phenylenediamine was obtained.

It was shown that both nitro groups in 2,4-dinitrochlorobenzene are reduced before hydrogenolysis takes place [13]. Reductions with a 5% weight ratio of platinum on carbon at 25° and 3 atm pressure were interrupted at various stages to determine the amount of ionic chloride. At absorption of 1H_2 it was about 0.5%, at 3H_2 it was 1.25%, and at 6H_2 it was 2.45%. With less catalyst the amount of ionic halide decreased during reduction. Extensive dehalogenation occurred only after both nitro groups were reduced. Partial reduction of 2,4-dinitrochlorobenzene with palladium on carbon also showed that little dehalogenation had taken place.

These experiments suggest that the dehalogenation which occurred in the reduction of 2,4-dinitrochlorobenzene which led to 1,3-phenylenediamine [263] may have been due to the presence of two amino groups, one of which is adjacent to the halogen atom, and to the reaction conditions.

Catalytic hydrogenolyses of the various type halogen-containing compounds will be discussed at greater length in another section. It may be pointed out, however, that the reactions are usually carried out in the presence of base. Therefore, in the reductions of halonitrobenzenes the effect of the resulting amines must be considered. It has been shown that chloro- and bromoanilines can be dehalogenated in the presence of palladium on carbon [264]. In the hydrogenation of 2-iodonitrobenzene with Raney nickel only 23% of 2-iodoaniline was obtained [263]. In this instance, reduction to amine was accompanied by considerable loss of iodine.

The proximity of the resulting amino group to the halogen may also be a factor in causing dehalogenation in the hydrogenation of halonitrobenzenes, but this point has not been investigated other than with too large an amount of catalyst to determine whether there is any effect [264].

Probably the most important factor in the selective reduction of the nitro group in halonitrobenzenes and related compounds is the activity of the catalyst for dehalogenation.

In general selective reduction of the nitro group in halonitrobenzenes can be achieved by carrying out hydrogenations in neutral solvent under several atmospheres pressure in the presence of nickel, platinum, and rhodium catalysts. The presence of mineral acid or acetic acid will help to neutralize the effect of the nitrogen base if dehalogenation does takes place [264]. Raney nickel, which essentially is a poor catalyst for dehalogenation, has been used for reductions under more vigorous conditions [263]. The specially prepared more active Raney nickel catalysts may not be selective under the same conditions. Palladium catalysts may be employed in these reductions, particularly when either chlorine or fluorine is present. One rather remarkable example is the very good yield of 3-chloro-4-(2-N-morpholinoethoxy)-aniline obtained from the reduction of the corresponding nitro compound with palladium on carbon under 80 atm pressure [264a]. Nevertheless, of all the useful low-pressure hydrogenation catalysts, palladium is the one most likely to cause dehalogenation.

Noble metal sulfide catalysts, inactive at low pressure, have been used for selective reduction of chloro- and bromonitrobenzenes at 85° and 35–50 atm pressure [265a, b]. It was pointed out that reduction of bromonitrobenzenes could not be effected with palladium sulfide without loss of bromine.

Hydrogenations of fluoronitrobenzenes have been carried out with palladium [262, 266] and with nickel [262, 267]. In the latter reference 2-fluoroaniline was obtained in 44% yield and 4-fluoroaniline was obtained in 84% yield. 4-Fluoroaniline was obtained in 72% yield when 4-fluoronitrobenzene was hydrogenated at 85° and 80 atm pressure with a 1% ratio of Raney nickel in a reaction that was interrupted after absorption of two molar equivalents [268]. There was no notation that interruption of the reduction was a precaution to prevent loss of fluorine. The use of platinum oxide catalyst or Raney nickel promoted with platinum chloride gave comparable results in the hydrogenation of 3-fluoro-6-methoxynitrobenzene [269]. In a similar comparison with 4-fluoronitrobenzene, promoted nickel gave better yield of 4-fluoroaniline than did platinum oxide [270], 92–95% against 65–70%.

There are many examples of selective reduction of the nitro group in halonitroaromatic compounds. When 3-5-dichloro-2-methoxynitrobenzene was hydrogenated in the presence of Raney nickel, no dehalogenation occurred [13]. The use of 3% by weight of 5% platinum on carbon gave equally good results in the same reduction. Raney nickel was used in a series of reductions of N-(2-diethylaminoethyl)-2-chloro-4-nitro and 4-chloro-2-nitro and related chloronitrobenzamides [271]. In reductions of these compounds as hydrochloride salts, dehalogenation did not take place in the presence of platinum on carbon but did occur in the presence of palladium on carbon [13]. Platinum oxide has also been used. High yields have been reported in reductions of 2- and 4-chloronitrobenzene [4], although it was noted that some cyclohexylamine hydrochloride was obtained when the reaction was allowed to continue to uptake of 4H_2. Other references report high yields without loss of chlorine. A specially prepared platinum catalyst modified by magnesium oxide was used to reduce chloro- and bromonitrobenzenes at 70–90° and 34 atm pressure with little or no dehalogenation [272]. The secret of success may have been the very small amount of catalytic metal used for reduction of the substrate, 1 part to 10,000 parts of nitro compound. Rhodium on alumina has been used in reductions of chloronitrobenzenes at 80–120° and 4–14 atm without causing much loss of chlorine [273]. If palladium catalyst is to be employed, dehalogenation may be avoided by the use of a very low catalyst ratio, as seen in the hydrogenation of 2-chloro-5-nitrotoluene with 10% palladium on carbon, 1 g/150 g of substrate [274]. In another group of reductions with 5% palladium on carbon or on alumina a 0.1% weight ratio of catalyst to substrate was used [275].

The reduction of 5-chloro-2-nitromandelic acid in alkaline solution was carried out at low pressure in the presence of palladium on carbon without dehalogenation [242] but loss of chlorine resulted when the reaction was allowed to continue. Raney nickel gave the best results and was most selective in the reduction of 2-chloro-4-nitrobenzoic acid in alcohol or in aqueous sodium bicarbonate or ammonium hydroxide solution [276]. Reduction was less selective in the presence of platinum on carbon unless the ratio of catalyst to compound was about 2%. The use of a similar amount of palladium on carbon resulted in loss of chlorine even when the reduction was interrupted at exactly three molar equivalents. Poisoning the reaction with a small amount of quinoline completely prevented dehalogenation in reduction with platinum on carbon but only caused reduction with palladium to stop at $2H_2$. When more catalyst was used the inhibitor had no effect. Uptake of hydrogen continued until 4-aminobenzoic acid was formed.

The high-pressure reduction of 4-bromonitrobenzene with Raney nickel was pointed out previously [263]. Other reductions with nickel have yielded bromoanilines without loss of halogen [277a, b]. 5-Bromo-3-nitrobenzhydrol was reduced rapidly to the corresponding aminobromo compound with platinum oxide [240]. 5% Rhodium on alumina was employed in the hydrogenation of 3-bromonitrobenzene and 3-bromo-4-nitrophenol [278]. The addition of calcium hydroxide to the reduction mixture increased the speed of reaction.

Bromo- and iodonitrobenzenes were converted to the corresponding bromo- and iodoanilines by the use of Raney nickel and hydrazine hydrate as the source of hydrogen [279a]. Dehalogenation was not observed when ruthenium on carbon and hydrazine hydrate were used in similar reductions [279b].

Methyl α-acetylamino-3-iodo-5-nitro-4-(methoxyphenoxy) cinnamate was hydrogenated in the presence of Raney nickel to the corresponding methyl iodoaminocinnamate [280] and 5′-iodo-2′-nitroacetophenone was reduced to 2′-amino-5′-iodoacetophenone with platinum oxide in reasonably good yield [211].

Any of the successful procedures for selective reduction of halonitrobenzenes should also apply to the hydrogenations of halonitronaphthalenes. There may be problems when the halogen and nitro group are in 4,5- or 1,8-positions. 5-Chloro-1-nitronaphthalene in benzene solution was reduced with a Raney-type nickel catalyst at atmospheric pressure to the aminochloronaphthalene in almost quantitative yield [280a]. In a reaction at 4 atm pressure the reduction product contained about 1% of 1-naphthylamine. Hydrogen uptake in the reduction of 5,8-dichloro-1-nitronaphthalene was slow and incomplete despite further additions of catalyst. Chromatography showed that 5,8-dichloro-1-naphthylamines, 5-chloro-1-naphthylamine, and starting nitro compound were present.

The hydrogenation of halonitroaromatic compounds is still not clear-cut. In the references in this section there are so few in which the reduction products are examined by means of chromatography that there are still unsolved questions. It has been generally assumed that the fluorine atom is not hydrogenolyzed during the reduction of the nitro group but it has not been proved. In too many instances the product of reduction is assumed to be pure because physical constants approximate described physical constants. Yet the products are rarely examined by gas-liquid chromatography.

Halogens in Unsaturated Nitrogen-Heterocycles

It is known that nitrogen bases can aid hydrogenolysis of the halogen atom during catalytic reduction. It may be assumed that there will be greater danger of dehalogenation in the hydrogenation of the nitro function in unsaturated nitrogen heterocycles which contain a halogen atom than in the reduction of similar carbocyclic compounds. Nevertheless, reduction without dehalogenation can be accomplished.

Most of the following examples show that selectivity may be expected. Raney nickel gave high yields of amines without apparent loss of chlorine in the reduction of 2-chloro-3- and 5-nitro-4-aminopyridines [281]. Good yields of aminochloropyridines have been obtained in reductions of chloronitropyridines with platinum oxide [282]. It was observed that in similar reductions in the presence of platinum oxide uptake of hydrogen stopped at $3H_2$ [161]. In the hydrogenation of 6-chloro-5-nitro-2-picoline with this catalyst uptake of hydrogen decreased markedly after $3H_2$; 65% yield of 5-amino-6-chloro-2-picoline was obtained along with some 5-amino-2-picoline [283]. The reduction of 2-amino-5-chloro-3-nitropyridine at low temperature (5°) with platinum on carbon gave only 20–30% of 2,3-diamino-5-chloropyridine [284]. It is difficult to draw any conclusions from this result since the amount of catalyst and substrate was not reported, nor was there a mention of other products of reduction.

Methyl 6-chloro-5-cyano-2-methyl-3-nitroisonicotinate was reduced with platinum oxide or with palladium on carbon in aqueous solution containing a large excess of hydrochloric, hydrobromic, or sulfuric acid [285]. During the reduction process the ester and cyano groups were hydrolyzed but dehalogenation did not take place. Very high yield of 3-amino-6-chloro-2-methylpyridine-4,5-dicarboxylic acid was obtained. It was rather striking to find that the ethyl ester of the chloronitro compound was reduced in neutral solution with twice its weight of palladium on carbon to give 73% of ethyl 3-amino-6-chloro-5-cyano-3-methylisonicotinate without any indication of dehalogenation [286].

When the halogen and nitro group of quinolines or isoquinolines are attached to the carbocyclic ring, reductions should be treated as those of halonitrobenzenes. In the reduction of 5-chloro-8-nitroisoquinolines the 5-chloro group could not be removed during hydrogenation with Raney nickel in alkaline solution; reduction with palladium on carbon in the presence of sodium acetate also did not cause loss of halogen but resulted in high yield of 8-amino-5-chloroisoquinoline [287].

If dehalogenation does occur in reductions where the halogen atom is attached to the heterocyclic ring, the use of lead-poisoned catalyst and/or acidic conditions should prevent or minimize hydrogenolysis.

Preferential reductions of 1-chloro-5-nitro-, 4-bromo-5-nitro and 4-bromo-8-nitroisoquinolines to the corresponding aminohaloisoquinolines have been obtained with Raney nickel in neutral solution [288]. 4-Chloro-3-nitroquinoline was converted to 3-amino-4-chloroquinoline in a like manner [289]. Dehalogenation did occur when potassium hydroxide was added before reduction. 4-Chloro-6-methoxy-8-nitroquinoline was reduced in neutral solvent to 8-amino-4-chloro-6-methoxyquinoline with platinum oxide [290].

The reduction of 4,6-dichloro-2-ethyl-5-nitropyrimidine with Raney nickel is typical of those among halonitropyrimidines [291 and references therein]; the yield of 5-amino-4,6-dichloro-2-ethylpyrimidine was 80.5%. 5% Platinum on carbon was used for the hydrogenation of 4,6-dichloro-5-nitropyrimidine in alcohol to 5-amino-4,6-dichloropyrimidine. It appeared to be more useful than Raney nickel because reduction with it was more rapid and the ratio of catalyst to substrate for the particular reduction was much lower—10% by weight of platinum on carbon to 40–50% by weight of Raney nickel [13].

In other heterocyclic systems 4-chloro-5-nitro- and 4-chloro-7-nitroquinazolines were selectively reduced to the corresponding aminochloroquinazolines with Raney nickel [292] and the resultant products were subsequently dehalogenated with palladium. 2,6-Dichloro-3-(2-nitrophenoxy)pyridazine was reduced in the presence of Raney nickel [293].

Since the condensation product, 2-chloro-3,4-diazaphenoxazine, was obtained in good yield, it appears obvious that dehalogenation of the intermediate did not take place (Eq. 10.8).

$$\text{(10.8)}$$

When 1-(chlorophenyl)-4-nitropyrazole was reduced with ruthenium on carbon and hydrazine as the source of hydrogen, 4-amino-1-(chlorophenyl)-pyrazole was obtained [294], but when palladium on carbon was used, dehalogenation occurred.

IN NONAROMATIC COMPOUNDS

Nonaromatic nitro groups are not reduced to amines as readily as nitro groups which are bound to an aromatic ring. An example of the difference in rate of reduction between the two types is shown in the hydrogenation of nitrobenzene and nitropropane with supported palladium and platinum catalysts [295]; hydrogen uptake in the reduction of nitrobenzene in neutral solvent was at least twenty times more rapid than that of nitropropane under similar conditions. The difference was more pronounced in reductions in acetic acid. The difference in reducibility between the two types is probably due to weaker bonding of the nonaromatic nitro function to the active portions of the catalyst and to the resulting amine, which is more basic than aniline and which exerts a greater inhibitory effect on the activity of the catalyst.

More side reaction takes place in the reduction of aliphatic nitro compounds to amines. At times there is interaction between the amino and nitro groups to form complex derivatives. Often more catalyst is required for reduction of the nonaromatic nitro function; many times reductions are carried out at elevated pressure to insure rapid absorption of hydrogen and to minimize side reaction.

Alkanes

Nitroparaffins, from C_1 to C_4, were reduced in alcohol to amines in 82–94% yield with 25–40% by weight of Raney nickel at 40–50° and 6–110 atm pressure [296]. Longer chain nitro compounds have been reduced with nickel at room temperature and atmospheric pressure [297a, b]. Hydrogenation of C_1 to C_4 nitroalkanes with platinum oxide gave 48–91% yields of amines [298], the yield increasing with chain length. In another study, supported platinum catalysts were found ineffective for the reduction of nitroparaffins at room temperature and atmospheric pressure but palladium oxide, palladium on carbon, and rhodium on carbon were useful under the same conditions [295].

Dinitroalkanes have been reduced to diamines in reasonably good yield in the presence of Raney nickel at 20–75° and 40–75 atm pressure [299a, b, c]. In the nickel-catalyzed reduction of 2,2-dimethyl-1,3-dinitropropane in alcohol, in addition to 67% yield of diamine, 5% of 2,2-dimethylmalonamide was obtained [300]. The latter product resulted from partial reduction of the dinitro compound to dioxime and rearrangement of the dioxime as seen in Eq. 10.9.

$$(R)_2C(CH_2NO_2)_2 \rightleftharpoons (R)_2C(CH{=}NOH)_2 \xrightarrow{2H_2}$$
$$\downarrow O$$

$$(R)_2C(CH{=}NOH)_2 \longrightarrow R_2C(CONH_2)_2 \quad (10.9)$$

$R = CH_3$

Very good yields of diamines have resulted from nickel reductions of dinitroalkanes at 40–50° and 75 atm pressure in the presence of an excess of organic acid [301]. Reductions of 2-methyl-2,4-dinitrohexane and 3,5-dinitroheptane with platinum oxide at 25° and 3 atm pressure gave only 22% and 48% of the corresponding diamines [302]. Uptake of hydrogen could not be brought to completion in either reduction. The use of acid to overcome possible catalyst inhibition was not attempted.

Yields of amines in the low-pressure reductions of 2-nitroalkylamines of the type —RCH(NH$_2$)CH(NO$_2$)R′— with Raney nickel appeared to depend on the basicity of the nitroamine or diamine [303]. When R and R′ = alkyl, yields were variable but when R was phenyl, yields were about 90%. Reductions of related compounds were carried out as hydrochloride salts in aqueous solution in the presence of palladium on carbon at 80–90° and 100–120 atm [304]. Hydrogenations of 1-(tertiary amino)-2-methyl-2-nitropropanes with Raney nickel at 30–50° and 33 atm [305a] and at room temperature and 70 atm [305b] were reported in good yield when the tertiary amino nitrogen was in a cyclic system. Reductions of 1-dimethylamino- and

1-dibutylamino-2-methyl-2-nitropropanes with nickel at elevated pressure gave good yields of the corresponding diamines [305c]. On the other hand, in the reduction of 1-diethylamino-2-methyl-2-nitropropane only 20% yield of the diamine was obtained [305b]. Low-pressure hydrogenation of the 1-diethylamino compound was studied in detail in the presence of Raney nickel, platinum oxide, palladium on carbon, and rhodium on carbon in neutral solvent and with the latter three catalysts in glacial acetic acid [306]. In all instances considerable cleavage took place. In addition a number of N-methylated amines were obtained. It was suggested that water resulting from reduction of the nitro group caused hydrolysis that led to the many reaction products obtained during reduction (see Eq. 10.10) and that reductive methylation of the amines with the formaldehyde obtained from hydrolysis gave N-methyl compounds. In order to obtain the desired

$$(R)_2NCH_2C(R')_2NO_2 \xrightarrow{3H_2} (R)_2NCH_2C(R')_2NH_2 + H_2O$$
$$(R)_2NCH_2C(R')_2NO_2 + H_2O \longrightarrow (R)_2NH + HCHO + CH(R')_2NO_2$$
$$CH(R')_2NO_2 \xrightarrow{3H_2} CH(R')_2NH_2 \qquad (10.10)$$

$R = C_2H_5$, $R' > CH_3$

N',N'-diethyl-2-methylpropane-1,3-diamine, it was found necessary to reduce freshly prepared nitro compound immediately after distillation and to carry out the reduction in the presence of Raney nickel at 75 atm pressure. When the 1-tertiary amino group was part of a cyclic system, reduction of the nitroamines was carried out with Raney nickel at low pressure to give N-methyl-free products.

In most examples of the reduction of nitroalkanes of the type-RYNO$_2$, where R is an aromatic ring or an unsaturated heterocyclic ring and Y is a straight or branched alkyl chain, Raney nickel finds the most application, although platinum oxide has also been employed. Conditions governing Raney nickel are often dependent on activity of the catalyst. This is exemplified in the hydrogenation of 3-(2-nitropropyl) indoles with W-7 Raney nickel, a highly active catalyst, at 20° and 1 atm and with W-2 Raney nickel at 50–100° and 50 atm [307]. Although platinum oxide also was employed for the same reductions at 1 atm nickel appeared to be preferred. Platinum oxide was used to give 78% yield of 3-(2-aminoethyl)indole from reduction of the corresponding nitro compound under 3 atm pressure [308]. Fairly high yields of 1-amino-1-phthalidylalkanes (C$_1$–C$_6$) were obtained from palladium on carbon reductions of the nitro compounds in alcohol containing an equivalent of hydrochloric acid at 50–70° and 3 atm pressure [309a]. Reduction was also carried out in acetic acid [309b]. Palladium on carbon was employed in another series in neutral solvent at 3 atm and 80° but 50–100% ratios of catalyst were used [309c].

Deamination was reported in the Raney nickel reduction of ethyl 2-carbethoxy-3-(3-indolyl)-2-nitropropionate at 180° and 150 atm [310] and at 100° and 25 atm [13]. When ethyl 2-carbethoxy-3-methyl-2-nitrobutyrate was hydrogenated at 25° and 3 atm with an active nickel catalyst, the corresponding nitrogen-free diester was obtained [311]. Deamination also occurred in low-pressure reductions of related nitro esters with palladium on carbon and with platinum catalyst [312, 313]. Deamination appears to be limited to the reduction of nitro esters where the nitro group is attached to a tertiary carbon atom.

Cyclanes

Raney nickel and platinum oxide appear to be the preferred catalysts for the reduction of nitro groups attached to saturated ring systems. Nitrocyclobutane was hydrogenated with platinum oxide at low pressure to give 85% of aminocyclobutane [314]. Nitrocyclopropane was reduced in alcohol to the amine at 25–30° and 2.7 atm with W-6 Raney nickel [315]. 1-Methyl-1-nitrocyclopentane, 1-methyl-1-nitrocyclohexane, and 1,2-, 1,3-, and 1,4-dimethyl-1-nitrocyclohexanes in alcohol containing excess acetic acid were rapidly reduced to amines in good yields in the presence of commercially available Raney nickel at 75° and 30–50 atm [316]. Reductions of the nitro compounds with the same catalyst at 2–3 atm pressure were interminably long [13]. Long reaction periods have been reported in low-pressure hydrogenations of nitrocyclanes with palladium catalyst. The reduction of nitrocyclododecane with palladium on calcium carbonate promoted with ferric chloride required almost three days for complete hydrogen uptake [317]. In the absence of the promoter only cyclododecylhydroxylamine was obtained. In order to convert the nitro compound to amine in the absence of ferric chloride, 50% by weight of catalyst was used. The hydrogenation of nitrocyclohexane to cyclohexylamine was also slow even in the presence of a 30% weight ratio of 5% palladium on calcium carbonate. The reduction of 2-(3,4,5-trimethoxyphenyl)nitrocyclohexane in acetic acid with 10% palladium on carbon was extremely slow [317b]. Ruthenium oxide could not catalyze the reduction of nitrocyclohexane in alcohol or acetic acid at room temperature and 3.3 atm pressure; uptake of hydrogen was slow in this reduction in acetic acid at 50° [13]. However, conversion to cyclohexylamine took place in alcohol at 60–70° and 80 atm in the presence of 2% by weight of the catalyst; with 1% by weight of catalyst uptake of hydrogen did not begin until the reaction temperature reached 90–100°.

Examples of reductions of the nitro group attached to saturated heterocyclic rings may be seen in the high-pressure reductions of some 5-nitro-1,3-dioxanes with Raney nickel [318a, b] and in the hydrogenation of

3-nitro-2-piperidone to the corresponding amine with Raney nickel and with palladium on carbon at 80–100° and 150–160 atm [318c].

1,2-Dinitrocyclohexanes have been reduced in a reaction chamber containing nickel or cobalt catalyst through which the substrate, ammonia (or various amines), and hydrogen are passed at 70–110° and 250 atm [319]. Good yields of 1,2-diaminocyclohexanes or 1-amino-2-secondary or 1-amino-2-tertiaryaminocyclohexanes were obtained.

Reduction with 75–90% retention of configuration has been reported in the hydrogenations of cis and trans-1,3- and 1,4-dinitrocyclohexanes in acetic acid with platinum oxide [320]. Reasonably good yields of diamines were obtained. Cis- and trans-2-(3,4,5-Trimethoxyphenyl)nitrocyclohexanes have been converted to the corresponding cis and trans amines with palladium on carbon [317b]. trans-2-Phenylnitrocyclohexane was reduced to the trans amine in 80% yield by means of Raney nickel at 3.7 atm [321]; the cis-nitro compound was reduced to amine with retention of configuration with nickel at 0 to 5°. When the reduction was carried out at 60–70°, a mixture of cis and trans amine resulted. Hydrogenation of the cis-nitro compound with platinum oxide also led to a mixture of cis and trans amine; reduction with palladium on carbon in methyl alcohol containing sulfuric acid gave only cis amine in 45% yield.

Alkanols

Aliphatic nitro alcohols are readily prepared by base-catalyzed condensations of nitroparaffins with aldehydes. Their hydrogenation is complicated by reversal of the nitroalcohol to its original components. It is generally assumed, although not proved, that reversal is caused by the resulting amine.

The earliest catalytic reductions of nitroalcohols were carried out with an equal weight of palladium on barium sulfate to substrate in solution containing oxalic acid [322]. Excellent yields of aminoalcohols were obtained. Reduction in neutral or alkaline solution or in the presence of mineral acid gave poor results.

Very high yields of aminoalcohols and aminoglycols have been obtained from reductions of the corresponding nitro compounds with Raney nickel at 30° and 33–130 atm pressure [323]. In the reduction of nitroglycols some reversal took place during reduction yielding a small amount of aminoalcohol in addition to the major product, the aminoglycol. In earlier work reductions of nitroalcohols with Raney nickel were carried out in the presence of excess carbon dioxide to give good yields of aminoalcohols [324a]. The same procedure, applied to the reduction of 2-nitro-1-phenyl-1-propanol, gave good yield of the racemic aminoalcohol [324b]. About 5% of N-ethylbenzylamine was also obtained. When the reduction of the nitroalcohol was carried out without carbon dioxide, 45% of N-ethylbenzylamine was obtained probably through the following series of reactions (Eq.10.11).

$$C_6H_5CHOHCH(NO_2)CH_3 \rightleftarrows C_6H_5CHO + C_2H_5NO_2$$

$$C_2H_5NO_2 \xrightarrow{3H_2} C_2H_5NH_2 \tag{10.11}$$

$$C_6H_5CHO + C_2H_5NH_2 \xrightarrow[H_2]{-H_2O} C_6H_5CH_2NHC_2H_5$$

Apparently the buffering action of carbon dioxide on the basic products of reduction prevented reversal of the nitroalcohol.

In most instances reductions of nitroalcohols with nickel are best carried out in acetic acid although formic and butyric acids also have been employed [325a, b]. Reduction may also be run in acidified solvent. Hydrogenations have been carried out in neutral solvent at 1–3 atm pressure [326a, b]. However, a comparison of the reductions of 1-(1-nitrocyclohexyl)methanol as a model compound with Raney nickel in alcohol and in alcohol containing 1.1 equivalents of acetic acid showed that while comparable good yields of aminoalcohol were obtained, uptake of hydrogen in the reduction in acidified solvent was three to four times more rapid [13].

Palladium on carbon was reported to be ineffective in the reduction of nitroalcohols [325b]. While palladium has catalyzed hydrogenations of nitroalcohols to aminoalcohols, in general uptake of hydrogen was slow and/or a large amount of catalyst was necessary. The reduction of 2-methyl-2-nitro-1-phenyl-1-propanol in alcohol containing acetic acid [327] and that of 1-(1-nitrocyclohexyl)methanol with palladium on carbon [328] are exceptions to the general trend.

Failures have been reported with platinum oxide and with platinum on carbon [328, 329]. Other examples note the use of these catalysts in the reduction of nitroalcohols but do not report details. Ethyl 2-acetylamino-2-carbethoxy-5-hydroxy-6-nitrohexanoate was hydrogenated to the corresponding amino compound at 3 atm with a modest amount of platinum oxide, although hydrogen uptake was slow [330]. 1-(2-Furyl)-2-aminoethanol was obtained in 84% yield from a platinum oxide reduction of the nitro compound in alcohol saturated with ammonia [331].

5% Rhodium on carbon was employed in a reduction at 3–4 atm pressure to give 77% yield of 1-(1-aminocyclohexyl)methanol from the nitroalcohol [328].

Partial Reduction in Saturated Aliphatic Systems

Hydrogenation of the nitro group in aliphatic compounds proceeds in a stepwise manner from nitro to nitroso to hydroxylamine to amine. When the nitro group is attached to a primary or secondary carbon atom, the intermediate is the oxime

$$RCHNO_2 \xrightarrow{H_2} RCHNO \rightleftharpoons RC{=}NOH$$

Acetone oxime was isolated in 30% yield from the reduction of 2-nitropropane at 110–130° and 30 atm initial pressure with a lead-treated palladium catalyst [332a, b]. Reductions of nitrocyclanes in a similar manner gave good yields of oximes although these were accompanied by cyclic amine, cyclic ketone, and recovered starting material. In the absence of lead compounds little or no oxime was obtained; the major product was cyclic amine contaminated with a small amount of cyclic ketone. Other catalysts were employed under similar reaction conditions but lead-treated palladium catalyst containing some magnesium gave the best results for the preparation of the oximes. Cyclohexanone oxime was obtained in high yield from the reduction of nitrocyclohexane in acidic medium with palladium on carbon [333]. In a tube-type reaction nitrocyclohexane and hydrogen were passed over a special silver-containing catalyst to give a 44% conversion to the oxime [334a]. Other silver or silver-containing catalysts have been used at elevated temperature and pressure to give fairly good yields of oxime from nitrocyclohexane [334b, c]. Alkaline promoters, sodium hydroxide, sodium salt of nitrocyclohexane, and benzyltrimethylammonium chloride, decreased reaction time in the reduction [334c]. In the hydrogenation of the *aci* form of nitrocyclohexane, with platinum in concentrated hydrochloric acid solution diluted with methanol, at 200 atm pressure, 70% yield of cyclohexanoneoxime was obtained along with 5% of cyclohexanone [335a]. A similar reduction with palladium on carbon gave 90% of cyclohexylhydroxylamine and 5% of cyclohexylamine. A platinum catalyst containing basic oxides was used to reduce nitrocyclohexane in aqueous ammonia to cyclohexanone oxime at 105° and 100 atm pressure [335b]. Yield was 65%, but in the absence of the oxides it dropped to 15%. Similar results were obtained with nickel; with basic oxides yield of oxime was 70%, otherwise it was 10%.

Reductions of aliphatic nitro compounds to hydroxylamines appear to be more straightforward than reductions of the nitro compounds to oximes. Nitroalkanes have been reduced to the corresponding hydroxylamines in aqueous solution containing oxalic acid in the presence of palladium on barium sulfate at low pressure. Hydroxylaminoalcohols were prepared in the same manner from nitroalcohols [336]. Cyclohexylhydroxylamine was readily obtained by reduction of nitrocyclohexane in aqueous acetic acid with palladium on alumina at 20° and 1–5 atm [337], and also from reduction of nitrocyclohexane in methyl alcohol with 1% palladium on calcium carbonate or on alumina [338]. In the latter reaction hydrogen uptake was very slow, possibly due to physical inactivation of the catalyst caused by precipitation of the hydroxylamine. It was pointed out that palladium catalysts containing more than 5% of the metal tended to produce cyclohexylamine.

In general hydrogenations of nitrocyclanes to cycloalkylhydroxylamines are most successful with palladium catalysts. Other catalysts are usually not selective and tend to carry the reductions to amines. When platinum catalyst was substituted for palladium in the hydrogenation of nitrocyclohexane, cyclohexylamine was obtained [338]. A sharp break after uptake of four equivalents of hydrogen was noted in the reductions of 1,3- and 1-4-dinitrocyclohexane and in one instance hydrogen uptake stopped [320]. The products were not isolated or identified. Attempted low-pressure reductions of nitrocyclohexane to cyclohexylhydroxylamine with platinum oxide in acidic and neutral solution and with ruthenium dioxide in glacial acetic acid all gave cyclohexylamine and nitrocyclohexane [13].

When 3-(2-nitropropyl)indole was hydrogenated in the platinum oxide in alcohol containing an equivalent of hydrochloric acid, the products of reaction consisted of 17% starting material, 12% of 3-(2-nitropropyl)-octahydroindole, and 39% of 3-(2-nitropropyl)indoline [339]. It was suggested that the formation of indoline may have occurred through the indoleninium ion (Eq. 10.12).

$$\text{indole-CH}_2\text{CH}_2\text{NO}_2 + \text{HCl} \rightleftharpoons \text{indoleninium-CH}_2\text{CH}_2\text{NO}_2$$

(10.12)

There are precedents which indicate that in heterocyclic systems the rings can be reduced before a functional group when the functional group is separated from the nucleus by a methylene bridge and the ring nitrogen is protonated. This will be discussed at greater length in the hydrogenation of nitrogen-containing ring systems.

Hydrogenation in the Presence of Other Reducible Functions

Olefinic Bonds. Selective reductions of the nitro group in unconjugated nitroolefins have been reported with Raney nickel. 1,2-Dimethyl-4-nitro-5-phenyl and 4-nitro-3,5,6-triphenylcyclohexenes were converted to the cyclohexenylamines in its presence at 50–60° and 200 atm pressure [340]. The hydrogenation of 4-nitro-5-phenylcyclohexene to the cyclohexenylamine could only be accomplished at room temperature and 3 atm pressure. 5-Ethyl-5-nitro-2-(2-pentenyl)- and 5-ethyl-5-nitro-2-(2-hexenyl)-1,3-dioxanes and related compounds with di- and trisubstituted alkenyl groups in the

2-position were reduced preferentially to the unsaturated amines with Raney nickel at 30–50° and 70–100 atm [341].

Selective reduction among these compounds may be dependent on the catalyst as well as the degree of substitution of the double bond. In the hydrogenation of 1,2-dimethyl-4-(2-ethoxyethyl)-5-nitrocyclohexene in acetic acid with platinum oxide, uptake of hydrogen stopped at one equivalent, which suggests selective reduction of the double bond [342]. When the same reduction was carried out with palladium on carbon, three equivalents of hydrogen were absorbed in less than 1 hr while absorption of the fourth equivalent took an additional 4 hr. Since the reaction was not interrupted after three equivalents uptake and the products of reduction were not investigated at that point, definite conclusions about selectivity cannot be made. In the reduction of 4-ethyl-1,2-dimethyl-5-nitrocyclohexene with palladium, the reaction was interrupted after three equivalents of hydrogen were absorbed. In this instance only saturated amine was obtained. The authors suggested that in the reduction of nitrocyclohexenes there was a tendency toward formation of nitrocyclohexane prior to attack of the nitro group, (342 and references therein; see also Chapter IX). The use of platinum oxide in the hydrogenation of 2-nitrooctadec-4-en-1,3-diol gave a mixture of saturated and unsaturated amines [343]. When the corresponding nitroacetylenic acid was reduced with Lindlar catalyst, rapid uptake of one equivalent of hydrogen occurred with conversion of the triple bond to a double bond. After uptake of additional hydrogen for reduction of the nitro group the absorption stopped; 2-aminooctadec-4-en-1,3-diol was obtained.

When 3-ethyl-5-nitro-3-hexene was hydrogenated with Raney nickel at 3 atm pressure 4-ethyl-2-hexanone was obtained [344]. It is likely that formation of the ketone resulted from partial reduction of the nitro group followed by hydrolysis of the resultant oxime from the water obtained from reduction (see Eq. 10.13).

$$(C_2H_5)_2C{=}CHCH(R)NO_2 \xrightarrow{2H_2} (C_2H_5)_2CHCH_2CH(R)NO \rightleftharpoons$$
$$(C_2H_5)_2CHCH_2C(R){=}NOH \xrightarrow[-NH_2OH]{H_2O} (C_2H_5)_2CHCH_2COR \quad (10.13)$$
$$R = CH_3$$

Reductions of conjugated nitroolefins are nonselective; both the nitro group and double bond are reduced. When the double bond is also conjugated with a benzene ring, it is possible to obtain arylnitroalkanes in hydrogenations of arylnitroolefins by control of the temperature [18–38°] and by use of small amounts of palladium on carbon or platinum oxide at 33 atm pressure [345]. Although reductions are not completely selective, yields of arylnitroalkanes are reasonably good (29.5–63%). Ketones (5–17%) and oximes (10–42%) are obtained as by-products.

It is likely in these reductions and in the hydrogenations of conjugated nitroolefins in general that 1,4-addition takes place to give the *aci* form of the nitroalkane which may tautomerize to the nitro form:

$$RCH=C(R')NO_2 \xrightarrow{H_2} [RCH_2C(R')=N(OH) \to O] \to RCH_2CH(R')NO_2$$

R = H or aryl, R' = H or alkyl.

At times considerable side reactions are encountered in the reductions of conjugated nitroolefins. Dimerization occurred in the hydrogenation of nitrostyrene in alcohol with platinum black; 1,4-dinitro 2,3-diphenylbutane was obtained [346]. Reduction of nitrostyrene in acetic acid gave the dinitro compound together with phenylacetaldoxime and a mixture of products which were not characterized. Hydrogenation of 2-nitro-1-butene in acetic acid with platinum led to formation of oxime, saturated amine, and polymer [347]. Ketones have often been obtained in the reduction of conjugated nitroolefins under certain reaction conditions. In the hydrogenation of 2-nitroolefins, in water containing a weak acid, with Raney nickel at 70° and 75 atm, 53–72% of ketones were obtained together with small amounts of amines [348]. On the other hand, in nickel reductions carried out in solutions of methanol containing acetic acid, the primary amines were the major product [349].

Hydrogenation of nitrostyrenes to phenethylamines are readily carried out under mild conditions in acetic acid containing mineral acid in the presence of palladium or platinum catalysts. In some instances the nitro compound in acetic acid is added to the catalyst in an acetic-mineral acid mixture while stirring under hydrogen. Because of the dropwise addition of substrate the catalyst ratio is very high. This procedure results in rapid hydrogen uptake and minimizes side reactions. The following references illustrate the variety of conditions used for these reductions (350a–f).

3,4-Methylenedioxy-α-nitrostilbene was hydrogenated to the corresponding diphenylethylamine in acetic acid-sulfuric acid medium with platinum oxide [351]. In other instances α-nitrostilbenes were hydrogenated with palladium on carbon in alcohol at 50–60° and 3.5 atm to give 52–87% yields of 1,2-diphenylethylamines [352a, b].

Early attempts to reduce 2-nitro-1-phenyl or substituted phenyl-1-propenes catalytically were unsuccessful; electrolytic methods were used instead. However, other related compounds have been hydrogenated with nickel in alcohol at 20–50° and 50–80 atm [353a]. It was previously mentioned that amines were obtained from reductions of similar nitro compounds but that ketones were by-products [349]. In another process the yield of amine was increased and amount of ketone minimized by raising the reaction temperature to 120° after hydrogen uptake appeared to slow down or stop [353b]. This procedure gave a 74% yield of 1-phenylisopropylamine from 2-nitro-1-phenyl-1-propene.

It is of interest that, in the reduction of 1-nitro-1-alkenes with palladium on carbon, the presence of water was essential for good yields of primary amine [354] when in other reductions it appeared to be responsible for formation of ketones. When 1-nitro-1-alkenes were hydrogenated in the absence of water the products of reduction consisted of oximes, ketones, aldehydes, and polymeric compounds; very little primary amine was formed.

Products resulting from them during catalytic reduction of nitroolefins to saturated amines has been discussed. Oximes can be prepared in very high yield from partial reduction of ω-nitrostyrenes and related compounds. They also can be reduced readily to amines in good yield (this will be discussed in a subsequent chapter). It may be possible to increase yields of saturated amine from reductions of nitroolefins and eliminate or minimize by-products by a two-step synthesis.

Phenylacetaldoximes were obtained in over 90% yield from hydrogenations of ω-nitrostyrenes in pyridine solution with palladium on carbon under low-pressure conditions [355a]. High yields of desoxybenzoin oximes resulted from similar reductions of α-nitrostilbenes [351, 355b]. The same procedure used in the hydrogenations of 1-nitro-1-octadecene and 1-nitrocyclooctene gave only 1-nitrooctadecane and nitrocyclooctane [356]. However, when the reductions were carried out in methyl alcoholic hydrogen chloride with palladium on carbon, the corresponding oximes were obtained.

Cyclohexylhydroxylamine has been prepared by hydrogenation of 1-nitrocyclohexene with palladium or platinum on carbon at 20–30° and 35–70 atm [357].

Unsaturated Nitrogen-Containing Groups. The products obtained from reductions of aliphatic nitro compound which contain cyano groups appear to be dependent on the relative positions of the two groups, the catalyst and reaction conditions.

The hydrogenation of 3-(α-cyano-α-nitromethyl)-2,2-dimethylbicyclo-[2.2.1]heptane with platinum black in cyclohexane at 50° and 79 atm gave the aminocyano compound [358].

In the hydrogenation of 2-methyl-2-nitropropionitrile, $(CH_3)_2C(NO_2)CN$, in methyl alcohol with Raney nickel at atmospheric pressure, 2-amino-2-methylpropionamide was obtained in 15% yield [359]. When 2-methyl-3-nitrobutyronitrile was reduced under similar conditions, 3-amino-2-methylbutyronitrile and 3-amino-2-methylbutyramide were obtained. When the same nitrile was hydrogenated with nickel and ammonia at 100° and 100 atm, the products of reduction were aminoamide and 2-methylpropane-1,3-diamine. 2,2-Dimethyl-3-nitropropionitrile was hydrogenated with Raney nickel at room temperature and atmospheric pressure to yield small amounts of aminonitrile, diamine, 50% of aminoamide, and 20% of a product, identified as 2-(5,5-dimethylhexahydro-2-pyrimidyl)-2,2-dimethylacetamide. When palladium on carbon was substituted for nickel in the reduction, 89% yield of 3-amino-2,2-dimethylpropionamide was obtained. Formation of aminoamides in the hydrogenations of 2-nitroalkylcyanides appear to follow the pattern shown in the hydrogenations of 2-nitrobenzonitriles [230, 231a, b, c], that is, reduction of the nitro group to hydroxylamine followed by cyclization to an isoxazole and ring cleavage to yield aminoamide as shown in Eq. 10.5.

Attempts to reduce 4-methyl-4-nitropentanenitrile to the aminonitrile with palladium or platinum in the presence of mineral acids were unsuccessful [360]. Reduction of the nitronitrile with Raney nickel at 20° and 1 atm pressure yielded 5-amino-2,2-dimethylpyrroline-1-oxide or the tautomeric 1-hydroxy-5-imino-2,2-dimethylpyrrolidine. Reduction of other 3-nitroalkylcyanides with Raney nickel or with palladium on calcium carbonate at 100 atm or at atmospheric pressure gave analogous cyclic products. Reductions with platinum catalyst gave poor yields. It was suggested that the cyclic products were obtained through reduction of the nitro compound to hydroxylaminoalkanenitrile followed by ring closure.

Some of the work on the reduction of 4-methyl-4-nitropentanenitrile to the cyclic amidine was duplicated with Raney nickel at 90° and 80 atm [361]. Under more vigorous conditions (240°, 185 atm) the nitronitrile was converted to 5,5-dimethyl-2-pyrrolidone.

In the hydrogenation of aliphatic nitro groups in the presence of other unsaturated nitrogen-containing functions, 6-hydroxyimino-6-nitrohexanoic acid was reduced in aqueous ammonia with Raney nickel or with palladium on carbon at 20° and 80–100 atm [362a]. The resulting product was the amidine of 6-hexanoic acid presumably formed by selection reduction of the nitrooxime to amidoxime and subsequent reduction to the amidine. This appears to have been borne out by the reduction of the nitrooxime to the amidoxime in glacial acetic acid with platinum on carbon at 25° and 1 atm pressure [362b].

Hydrogenations of hydrazones of α-nitrobenzaldehyde,

$$C_6H_5C(NO_2)=NNHR$$

(R = alkyl or substituted phenyl), to the corresponding amidrazones have been carried out with palladium on barium sulfate or with Raney nickel at low pressure [363].

Reduction of 3-nitro-4-iminohexane with platinum oxide at 2–3 atm gave 3,4-diaminohexane in 63% yield [364]. Some ammonia was noted indicating about 6% hydrolysis or hydrogenolysis. A similar reduction with nickel at

35° and 200 atm gave 15% of ammonia. When 3-nitro-4-(N-phenylimino)-hexane was reduced with platinum oxide at 3 atm, complete hydrolysis or hydrogenolysis to aniline took place.

Aldehydes and Ketones. 6-Nitro-4-ketohexanoic acid has been selectively reduced to 6-amino-4-ketohexanoic acid in glacial acetic acid with palladium on barium sulfate, in alcohol with the same catalyst, and in water with palladium on alumina at atmospheric pressure [356a]; selective reduction of the keto nitro acid was carried out in water or in 2% hydrochloric acid with palladium on alumina at 70–100 atm [356b]. 4-Nitrohexaldehyde was hydrogenated to 4-aminohexanol with platinum oxide at low pressure [366]. No attempt was made to determine whether the nitro group was converted to amine before the carbonyl group was reduced. 4-(2-Nitro-1-hydroxy)ethyltropolone was selectively reduced to 4-(2-amino-1-hydroxy)ethyltropolone with palladium on carbon in alcohol containing a mole of acetic acid [367a]. 1-Nitro-4-phthalimido-2-butanone in hydrochloric acid in acetic acid was also reduced with palladium to the corresponding aminobutanone [367b]. 2-Acetyl-4-nitro-3-phenylbutyric acid and its ethyl ester have been reduced to the corresponding 2-acetyl-4-amino compounds with Raney nickel in methyl alcohol at low pressure [368].

In general, reductions of γ-nitroketones when carried out with Raney nickel give pyrrolines or pyrrolidines dependent on reaction conditions and whether the catalyst is freshly prepared [369a–d]. Palladium on carbon has also been used to reductively cyclize γ-nitroketones to pyrrolines [370]. 5-Nitro-5-phenyl-2-pentanone, when hydrogenated in alcohol in the presence of palladium on carbon at 2 atm pressure, gave 5-phenyl-2-pentanone [13]. It was assumed that the nitro group was partially hydrogenated to oxime and underwent hydrolysis to 1-phenylpentan-1,4-dione. The latter product was selectively reduced to 5-hydroxy-5-phenyl-2-pentanone and subsequently hydrogenolyzed to give 5-phenyl-2-pentanone.

Hydrogenolyzable Groups. The nitro group in an aliphatic compound which also contains a benzyl ester can be selectively reduced when catalysts other than palladium are used. Typical examples are seen in the hydrogenations of 7-benzyloxy-3-(2-nitroethyl)indole and related compounds with platinum oxide which yielded the corresponding 7-benzyloxytryp-tamines [371a, b]. In the presence of palladium, concomitant reduction of both functions occurred.

4-(1-Methoxy-2-nitroethyl)anisole, another benzyl ether, was readily reduced to 2-methoxy-2-(4-methoxyphen)ethylamine with platinum oxide in alcoholic oxalic acid [372a]. Related amines were similarly prepared from the corresponding 1-methoxy-2-nitroethyl compounds. The same nitro compounds were converted to amines without cleavage of the methoxy group in reductions with palladium black in acetic acid at 60–70° and 2 atm pressure [372b]. The presence of sulfuric acid in the palladium black reductions resulted in hydrogenolysis of the methoxy group.

1,3-Dibenzyl-5-nitrohexahydropyrimidine has been reduced to the corresponding aminodibenzyl compound with Raney nickel at 75–100° and 70 atm pressure [373]. However, this is a poor example of selective reduction because it was shown that the resulting aminodibenzyl compound could not be debenzylated in the presence of palladium on carbon at 100 atm and up to 150°. Nevertheless, it is likely that with nickel or platinum catalysts preferential reduction of an aliphatic nitro group should be expected in the presence of an N-benzyl group.

The hydrogenations of aliphatic nitro compounds which also contain alkyl, aryl, or cycloalkyl halides have been studied very meagerly. Some 2-nitro-1,1-di(chlorophenyl)propanes have been reduced to the corresponding 1,1-di(chlorophenyl)isopropylamines with Raney nickel at 50° and 100 atm [374]. It is difficult to draw any conclusions about selectivity in these reductions since yields were reported only in a few instances. However, it is likely that the nitro group in these compounds could be hydrogenated without causing loss of halogen if reduction was carried out in acidic medium with platinum catalyst. It may also be possible to use Raney nickel in acetic acid for the same purpose.

Low-pressure reduction of 3,3,3-trichloro-1-nitro-2-propanol with 45% by weight of freshly prepared Raney nickel was reported to give 1-amino-3,3,3-trichloro-2-propanol [375a]. Duplication of this work indicated that the product of reduction was 1-amino-3,3-dichloro-2-propanol [375b]. The large amount of catalyst used may have been responsible for the loss of chlorine. In the hydrogenation of 2-chloronitrocyclohexane to cyclohexylamine with Raney nickel, no observations were made regarding selectivity [376a]. When 1-chloronitrocyclohexane was reduced in alcohol with palladium on carbon at 3 atm or with a very small amount of catalyst at 33 atm pressure, 80% of cyclohexylhydroxylamine was obtained [376b]. 2-Chloro-2-nitropropane, similarly hydrogenated, gave acetone oxime in only 35% yield. Reduction of 1-chloronitrocyclohexane with platinum oxide resulted in 65% yield of cyclohexylhydroxylamine. Hydrogenation of 1-chloronitrocyclohexane in aqueous solution with palladium in water gave cyclohexanone and nitrocyclohexane. Reduction of 3-chloro-3-nitropentane-2,4-diol and related compounds gave alkanolones [377]. It is likely that loss of halogen is the first stage in the hydrogenation of compounds which have a nitro group and a halogen attached to the same carbon. Thereafter further hydrogenation, if it proceeds, does so through the *aci* form of the nitro compound. Loss of chlorine is also seen in the reduction of 2-chloro-2-nitroethanol to 2-nitroethanol in 71% yield in aqueous pyridine with palladium on barium sulfate

[378]. The hydrogenation of 1-chloronitrocyclohexane with palladium on carbon in aqueous sodium hydroxide gave the sodium salt of *aci*-nitrocyclohexane in 75% yield [376b].

IN N-NITRO COMPOUNDS

Hydrogenation of the N-nitro group usually results in cleavage to give the amine; reduction proceeds through the hydrazine:

$$\text{RNHNO}_2 \xrightarrow{3\text{H}_2} \text{RNHNH}_2 \xrightarrow{\text{H}_2} \text{RNH}_2 + \text{NH}_3$$

Cleavage of the nitro group during reduction with Raney nickel has been employed to prove structures of some nitraminopyridazines [379]. In polypeptides containing arginine, the N-nitro group has been used to prevent reaction on the guanidine portion of the molecule. The protecting group is removed by hydrogenation with palladium catalyst [380a, b]. The reaction is not always clear-cut. Four equivalents of hydrogen should be absorbed but uptake does not proceed beyond three equivalents [13, 381]. Reaction often stops at the aminoarginine stage [381]:

$$\text{HOOCCH}(\text{NH}_2)(\text{CH}_2)_3\text{NHC}(=\text{NH})\text{NHNO}_2 \rightarrow -\text{NHC}(=\text{NH})\text{NHNH}_2$$

REFERENCES

[1] K. Taya, *Sci. Papers Inst. Phys. Chem. Res. Tokyo*, **56**, 285 (1952).
[2] T. Neilson, H. C. S. Wood, and A. G. Wylie, *J. Chem. Soc.*, **1962**, 371.
[3] D. Balcom and A. Furst, *J. Am. Chem. Soc.*, **75**, 4334 (1953).
[4] R. Adams, F. L. Cohen, and O. W. Rees, *J. Am. Chem. Soc.*, **49**, 1093 (1927).
[5] N. Kornblum and D. C. Iffland, *J. Am. Chem. Soc.*, **71**, 2137 (1949).
[6] M. C. Kloetzel and B. Y. Abidir, *J. Am. Chem. Soc.*, **77**, 3823 (1955).
[7] R. H. Eastman and F. L. Detert, *J. Am. Chem. Soc.*, **73**, 4511 (1951).
[8] J. D. Loudon and J. Ogg, *J. Chem. Soc.*, **1953**, 739.
[9] R. G. Benner and A. C. Stevenson, U.S. Patent 2,619,503 (1952).
[10] R. B. Moffett, *J. Med. Chem.*, **9**, 475 (1966).
[11] V. P. Shmonina, G. P. Temnikova, and D. V. Sokol'skii, *J. Gen. Chem. USSR*, **31**, 681 (1961) (English translation).
[12] G. S. Samuelson, V. I. Garik, and G. B. L. Smith, *J. Am. Chem. Soc.*, **72**, 3872 (1950).
[13] M. Freifelder, unpublished results.
[14] R. Baltzly, *J. Am. Chem. Soc.*, **74**, 4586 (1952).
[15] J. Levy, U.S. Patent 3,230,259 (1966).
[16] H. J. B. Bieckart, H. B. Dessens, P. E. Verkade, and B. M. Wepster, *Rec. Trav. Chim.*, **71**, 321 (1952).

[17] J. R. Reasenberg, E. Lieber, and G. B. L. Smith, *J. Am. Chem. Soc.*, **61**, 384 (1939).
[18] L. Hernandez and F. F. Nord, *J. Colloid Sci.*, **3**, 363 (1948).
[18a] W. P. Dunworth and F. F. Nord, *J. Am. Chem. Soc.*, **74**, 1459 (1952).
[19] U. A. Corta, *Helv. Chim. Acta*, **32**, 681 (1949).
[20] L. Horner and K. Sturm, *Ann.*, **608**, 128 (1957).
[21] S. J. Angyal, E. Bullock, W. G. Hanger, W. C. Howell, and A. W. Johnson, *J. Chem. Soc.*, **1957**, 1593.
[22] J. Hill and G. R. Ramage, *J. Chem. Soc.*, **1964**, 3711.
[23] E. E. Smissman, J. B. Lapidus, and S. D. Beck, *J. Org. Chem.*, **22**, 220 (1957).
[24] R. Adams and K. R. Brower, *J. Am. Chem. Soc.*, **79**, 1950 (1957).
[25] C. Hansch, B. Schmidhalter, F. Reiter, and W. Saltonstall, *J. Org. Chem.*, **21**, 265 (1956).
[26] L. Spiegler, U.S. Patent 2,947,781 (1960).
[27] H. Brockman and H. Muxfeldt, *Ber.*, **89**, 1397 (1956).
[28] G. S. Skinner and H. C. Zell, *J. Am. Chem. Soc.*, **77**, 5441 (1955).
[29] T. P. C. Mulholland and G. Ward, *J. Chem. Soc.*, **1954**, 4676.
[30] A. Burger and S. Avakian, *J. Org. Chem.*, **5**, 606 (1940).
[31] O. Hromatka, *Ber.*, **75**, 123 (1942).
[32] W. A. W. Cummings, *J. Chem. Soc.*, **1958**, 2058.
[33] J. Corse and E. Rohrmann, *J. Am. Chem. Soc.*, **70**, 370 (1948).
[34] E. Wenis and T. S. Gardner, *J. Am. Pharm. Assoc. Sci. Ed.*, **38**, 9 (1949).
[35] W. Hückel and K. Janecka, *Arch. Pharm.*, **284**, 341 (1951).
[36] R. W. Holley and A. D. Holley, *J. Am. Chem. Soc.*, **74**, 3069 (1952).
[37] F. J. DiCarlo, *J. Am. Chem. Soc.*, **66**, 1420 (1944).
[38] G. N. Walker, *J. Am. Chem. Soc.*, **77**, 3844 (1955).
[39] A. C. Cope and E. M. Hancock, *J. Am. Chem. Soc.*, **66**, 1448 (1944).
[40] S. L. Shapiro, H. Soloway, E. Chodos, and L. Freedman, *J. Am. Chem. Soc.*, **81**, 203 (1959).
[41] D. Chabrier, H. Najer, and R. Guidicelli, *Bull. Soc. Chim. France*, **1955**, 1353.
[42] G. Cignarella, E. Ocelli, G. Maffii, and E. Testa, *J. Med. Chem.*, **6**, 387 (1963).
[43] S. Bell, R. Foster, and W. E. B. Soutar, *J. Chem. Soc.*, **1959**, 2316.
[44] J. Sam, W. F. Minor, and Y. G. Perron, *J. Am. Chem. Soc.*, **81**, 710 (1959).
[45] A. E. Blood and C. R. Noller, *J. Org. Chem.*, **22**, 711 (1957).
[46] G. R. Clemo and A. F. Daglish, *J. Chem. Soc.*, **1950**, 1481.
[47] G. Gaertner, A. Gray, and F. G. Holliman, *Tetrahedron*, **18**, 1105 (1962).
[48] W. V. Wirth and S. E. Krahler, U.S. Patent 2,765,341 (1956).
[49] H. Gilman and H. S. Broadbent, *J. Am. Chem. Soc.*, **70**, 2619 (1948).
[50] S. K. Freeman and P. E. Spoerri, *J. Org. Chem.*, **16**, 438 (1951).
[51] British Patent 805,249 (1958).
[52] a. W. H. Brunner and A. Halasz, U.S. Patent 3,088,978 (1963); b. U.S. Patent 3,274,249 (1966).
[53] R. Adams and K. R. Brower, *J. Am. Chem. Soc.*, **78**, 663 (1956).
[54] P. E. Verkade and P. H. Witjens, *Rec. Trav. Chim.*, **65**, 361 (1946).
[55] K. C. Frisch and M. T. Bogert, *J. Org. Chem.*, **8**, 331 (1943).
[56] D. Pitre and E. Lorenzotti, *Chimia*, **19**, 462 (1965).
[57] R. Adams and R. A. Wankel. *J. Am. Chem. Soc.*, **73**, 131 (1951).
[58] D. E. Spalding, E. C. Chapin, and H. S. Mosher, *J. Org. Chem.*, **19**, 357 (1954).
[59] D. H. Rosenblatt, M. M. Lachlas, and A. M. Seligman, *J. Am. Chem. Soc.*, **80**, 2463 (1958).

[60] C. F. Kelly and A. R. Day, *J. Am. Chem. Soc.*, **67**, 1074 (1945).
[61] F. Kuiban, Z. Dumitresku, and I. Ambrush, *J. Gen. Chem. USSR*, **34**, 1592 (1964) (English translation).
[62] H. T. Nagasawa and H. R. Guttman, *J. Med. Chem.*, **9**, 719 (1966).
[63] E. O. Arene and D. A. H. Taylor, *J. Chem. Soc.*, C, **1966**, 481.
[64] M. R. Stevinson and C. S. Hamilton, *J. Am. Chem. Soc.*, **57**, 1298 (1935).
[65] W. F. Holcomb and C. S. Hamilton, *J. Am. Chem. Soc.*, **61**, 1236 (1939).
[66] G. O. Doak, H. G. Steinman, and H. Eagle, *J. Am. Chem. Soc.*, **63**, 99 (1941).
[67] L. D. Freedman and G. O. Doak, *J. Am. Chem. Soc.*, **71**, 779 (1949).
[68] J. R. Vaughan and D. S. Tarbell, *J. Am. Chem. Soc.*, **67**, 144 (1945).
[69] a. K. Torsell, *Arkiv Kemi*, **10**, 473 (1957); b. K. Torsell, H. Meyer and B. Zacharias, p. 497; c. K. Torsell, p. 513.
[70] D. N. Butler and A. H. Soloway, *J. Med. Chem.*, **9**, 362 (1966).
[71] A. H. Soloway, *J. Am. Chem. Soc.*, **82**, 2442 (1960).
[72] G. B. Arnold and C. S. Hamilton, *J. Am. Chem. Soc.*, **63**, 2637 (1941).
[73] V. R. Bell, Jr. and G. M. Kosolapoff, *J. Am. Chem. Soc.*, **75**, 4901 (1953).
[74] R. W. Bost and L. D. Quin, *J. Org. Chem.*, **18**, 358 (1953).
[75] G. O. Doak and L. D. Freedman, *J. Am. Chem. Soc.*, **74**, 753 (1952).
[76] G. O. Doak and L. D. Freedman, *J. Am. Chem. Soc.*, **76**, 1621 (1954).
[77] G. O. Doak and L. D. Freedman, *J. Am. Chem. Soc.*, **75**, 6307 (1953).
[78] J. G. Moffatt and H. G. Khorana, *J. Am. Chem. Soc.*, **79**, 3741 (1957).
[79] C. O. Henke and J. V. Vaughan, U.S. Patent 2,198,249 (1940).
[79a] C. K. Ingold, *Structure and Mechanism in Organic Chemistry*, Cornell, Ithaca, 1953, p. 621–624.
[80] a. British Patent 856,366 (1960); b. French Patent 1,338,899 (1963).
[81] D. A. Fidler, J. S. Logan, and M. M. Boudakian, *J. Org. Chem.*, **26**, 4014 (1961).
[81a] V. Weinmayr, *J. Am. Chem. Soc.*, **77**, 1762 (1955).
[82] A. L. Davis, O. H. P. Choun, D. E. Cook, and T. J. McCord, *J. Med. Chem.*, **7**, 632 (1964).
[82a] G. W. Cavill and D. L. Ford, *J. Chem. Soc.*, **1954**, 565.
[83] R. T. Coutts and J. B. Edwards, *Can. J. Chem.*, **44**, 2009 (1966).
[84] K. Brand and J. Steiner, *Ber.*, **55**, 875 (1922).
[84a] C. Mannich and E. Thiele, *Ber. Dtsch. Pharm. Ges.*, **26**, 36 (1916).
[85] A. Sugimori, *Bull. Chem. Soc. Jap.*, **33**, 1599 (1960).
[86] K. Taya, *Chem. Commun.*, **1966**, 464.
[87] French Patent 1,200,050 (1959).
[88] G. Cusmano, *Ann. Chim. Applicata*, **12**, 123 (1919); through *Chem. Abstr.*, **14**, 1314 (1920).
[89] M. Freifelder, to be submitted.
[90] O. Hromatka, E. Flieder, and J. Augl, *Monatsh.*, **91**, 1016 (1960).
[90a] H. H. Szmant and R. L. Lapinski, *J. Am. Chem. Soc.*, **75**, 6338 (1953).
[91] C. F. Koelsch, *J. Org. Chem.*, **25**, 164 (1960).
[92] B. J. R. Nicolaus, E. Bellasio, G. Pagani, L. Mariani, and E. Testa, *Helv. Chim. Acta*, **48**, 1867 (1965).
[93] L. L. Fontanella and E. Testa, *Ann.*, **622**, 117 (1959).
[94] T. A. Connors, W. C. J. Ross, and J. G. Wilson, *J. Chem. Soc.*, **1960**, 2994.
[95] G. F. Woods, T. L. Heying, L. H. Schwartzman, S. M. Grenell, W. F. Gasser, E. W. Rowe, and N. C. Bolgiano, *J. Org. Chem.*, **19**, 1290 (1954).
[96] A. R. Surrey, S. O. Winthrop, M. K. Rukwid, and B. F. Tullar, *J. Am. Chem. Soc.*, **77**, 633 (1955).

[97] R. F. Coles and C. S. Hamilton, *J. Am. Chem. Soc.*, **68**, 1799 (1946).
[98] B. Stárkova, A. Vystrcil, and L. Stárka, *Coll. Czechoslov. Chem. Commun.*, **22**, 1019 (1957).
[99] R. M. Herbst, C. W. Roberts, H. T. F. Givens, and E. K. Harvill, *J. Org. Chem.*, **17**, 262 (1952).
[100] D. Jerchel and L. Jakob, *Ber.*, **92**, 724 (1959).
[101] M. J. Kalm, U.S. Patent 2,832,777 (1958).
[102] H. Ratouis, J. R. Boissier, and C. Dumont, *J. Med. Chem.*, **8**, 104 (1965).
[103] a. H. Gilman and J. Swiss, *J. Am. Chem. Soc.*, **66**, 1884 (1944); b. H. Gilman and S. Avakian, *J. Am. Chem. Soc.*, **68**, 580 (1946).
[104] S. O. Onyiriuka and A. H. Rees, *J. Chem. Soc. (C)*, **1966**, 504.
[105] H. Erlenmeyer, W. Grubenmann, and H. Block, *Helv. Chim. Acta*, **31**, 75 (1948).
[106] A. N. Grinev and A. P. Terent'ev, *J. Gen. Chem. USSR*, **28**, 80 (1958) (English translation).
[107] P. M. Heertjes, B. J. Knape, H. C. A. van Beek, and K. van den Boogaart, *J. Chem. Soc.*, **1957**, 3445.
[108] L. A. Weinberger and A. R. Day, *J. Org. Chem.*, **24**, 1451 (1959).
[109] B. N. Feitelson and R. Rothstein, *J. Chem. Soc.*, **1958**, 2426.
[110] Ch. Sannie and H. Lapin, *Bull. Soc. Chim. France*, **1952**, 369.
[111] R. L. Clark and A. A. Pessolano, *J. Am. Chem. Soc.*, **80**, 1662 (1958).
[112] F. F. Stephens and J. D. Bower, *J. Chem. Soc.*, **1949**, 2971.
[113] H. B. Gillespie, M. Engelmann, and S. Graff, *J. Am. Chem. Soc.*, **76**, 3531 (1954).
[114] M. Kamel, M. I. Ali, and M. M. Kamel, *Tetrahedron*, **22**, 3351 (1966).
[115] C. E. Kwartler and P. Lucas, *J. Am. Chem. Soc.*, **65**, 1804 (1943).
[116] R. R. Davies, *J. Chem. Soc.*, **1955**, 2412.
[117] F. Hunziker, H. Lehner, O. Schindler, and J. Schmutz, *Pharm. Acta Helv.*, **38**, 538 (1963).
[118] M. Kamel, M. A. Allam, F. I. Abdel Hay, and M. O. Osman (in part), *J. Prakt. Chem.*, **303**, 100 (1966).
[119] Freifelder and Wright, to be published.
[120] R. K. Brown and N. A. Nelson, *J. Am. Chem. Soc.*, **76**, 5149 (1954).
[121] R. K. Brown and R. A. Garrison, *J. Am. Chem. Soc.*, **77**, 3839 (1955).
[122] R. K. Brown, R. F. Synder, and M. D. Stevenson, *J. Org. Chem.*, **21**, 261 (1956).
[123] M. S. Melzer, *J. Org. Chem.*, **27**, 496 (1962).
[124] W. E. Noland and K. Rush, *J. Org. Chem.*, **28**, 2921 (1963).
[125] S. P. Hiremath and S. Siddappa, *J. Ind. Chem. Soc.*, **41**, 357 (1964).
[126] Z. J. Vejdělek, *Coll. Czechoslov. Chem. Commun.*, **22**, 1852 (1957).
[127] J. DeGraw and L. Goodman, *J. Org. Chem.*, **27**, 1395 and 1728 (1962) and *J. Med. Chem.*, **7**, 213 (1964).
[128] L. Haskelberg, *J. Org. Chem.*, **12**, 434 (1947).
[129] French Patent 1,412,615 (1965); through *Chem. Abstr.*, **64**, 5113 (1966).
[130] P. Clarke and A. Moorhouse, *J. Chem. Soc.*, **1963**, 4763.
[131] S. Hayao, H. J. Havera, W. G. Strycker, J. J. Leipzig, R. A. Kulp, and H. E. Hartzler, *J. Med. Chem.*, **8**, 807 (1965).
[132] J. J. Craig and W. E. Cass, *J. Am. Chem. Soc.*, **64**, 783 (1942).
[133] F. Misani and M. T. Bogert, *J. Org. Chem.*, **10**, 347 (1945).
[134] P. K. Jospeh and M. M. Joullié, *J. Med. Chem.*, **7**, 801 (1964).
[135] I. Suzuki, T. Nakashima, and N. Nagasawa, *Chem. Pharm. Bull. (Tokyo)*, **14**, 816 (1966).

[136] a. K. N. Campbell, A. H. Sommers, J. F. Kerwin, and B. K. Campbell, *J. Am. Chem. Soc.*, **68**, 1556 (1946); b. *J. Am. Chem. Soc.*, **69**, 1465 (1947).
[137] M. C. Carmack, L. W. Kissinger, and I. Von, *J. Am. Chem. Soc.*, **68**, 1551 (1946).
[138] H. J. Barber and W. R. Wragg, *J. Chem. Soc.*, **1946**, 610.
[139] R. S. Tipson and M. A. Clapp, *J. Org. Chem.*, **11**, 292 (1946).
[140] Neth. Appl. 6,403,115 (1965); through *Chem. Abstr.*, **64**, 5114 (1966).
[141] R. C. Elderfield, T. A. Williamson, W. J. Gensler, and C. B. Kremer, *J. Org. Chem.*, **12**, 405 (1957).
[142] F. R. Pfeifer and F. H. Case, *J. Org. Chem.*, **31**, 3384 (1966).
[143] B. Boothroyd and E. R. Clark, *J. Chem. Soc.*, **1953**, 1504.
[144] a. H. Gilman and G. Wright, *Iowa State College J. Sci.*, **5**, 85 (1931); b. I. J. Rinkes, *Rec. Trav. Chim.*, **51**, 349 (1932); c. A. Ponomarev and M. D. Lipanova, *J. Gen. Chem. USSR*, **31**, 897 (1961) (English translation).
[145] a. F. F. Ebetino, J. J. Carroll, and G. Gever, *J. Med. Chem.*, **5**, 513 (1962); b. F. F. Ebetino and G. Gever, U.S. Patent 3,154,543 (1964).
[146] C. W. Waller, M. J. Weiss, and J. S. Webb, U.S. Patents 2,785,181, 2,785,183, and 2,797,228 (1957).
[147] A. Kreutzberger and P. A. Kalter, *J. Org. Chem.*, **26**, 3791 (1961).
[148] R. K. Robins, L. B. Holum, and F. W. Furcht, *J. Org. Chem.*, **21**, 835 (1956).
[149] R. K. Robins, F. W. Furcht, A. D. Grauer, and J. W. Jones, *J. Am. Chem. Soc.*, **78**, 2419 (1956).
[150] M. J. S. Dewar and F. E. King, *J. Chem. Soc.*, **1945**, 114.
[151] E. C. Taylor, J. W. Barton, and T. S. Osdene, *J. Am. Chem. Soc.*, **80**, 421 (1958).
[152] W. Shive, W. W. Ackermann, M. Gordon, M. E. Getzendaner, and R. E. Eakin, *J. Am. Chem. Soc.*, **69**, 725 (1947).
[153] F. G. Mann and J. W. G. Porter, *J. Chem. Soc.*, **1945**, 751.
[154] R. N. Prasad and R. K. Robins, *J. Am. Chem. Soc.*, **79**, 6401 (1957).
[155] E. C. Taylor and P. K. Loeffler, *J. Org. Chem.*, **24**, 2035 (1959).
[156] Y. Ahmad and D. H. Hey, *J. Chem. Soc.*, **1954**, 4516.
[157] J. C. Clayton and J. Kenyon, *J. Chem. Soc.*, **1950**, 2952.
[158] O. Meth-Cohn, R. K. Smalley, and H. Suschitzky, *J. Chem. Soc.*, **1963**, 1666.
[159] J. Barycki and E. Plazek, *Roczniki Chem.*, **37**, 1443 (1963); through *Chem. Abstr.*, **60**, 7987 (1964).
[160] A. Dornow and H. von Plessen, *Ber.*, **99**, 244 (1966).
[161] W. F. Bruce and L. A. Perez-Medina, *J. Am. Chem. Soc.*, **69**, 2571 (1947).
[162] J. W. Clark-Lewis and R. P. Singh, *J. Chem. Soc.*, **1962**, 2379.
[163] W. Herz and L. Tsai, *J. Am. Chem. Soc.*, **76**, 4184 (1954).
[164] J. Fraser and E. Tittensor, *J. Chem. Soc.*, **1957**, 4625.
[165] J. W. Clark-Lewis and M. J. Thompson, *J. Chem. Soc.*, **1957**, 443.
[166] G. T. Newbold and F. S. Spring, *J. Chem. Soc.*, **1949**, S 133.
[167] E. Ochiai, *J. Org. Chem.*, **18**, 534 (1953).
[168] a. J. A. Benson and T. Cohen, *J. Org. Chem.*, **20**, 1461 (1955); b. G. Tacconi and S. Pietra, *Ann. Chim. Rome*, **55**, 810 (1965).
[169] J. N. Gardner and A. R. Katritzky, *J. Chem. Soc.*, **1957**, 4375.
[170] T. Kato and F. Hamaguchi, *Chem. Pharm. Bull. Japan*, **4**, 174 (1956).
[171] E. C. Taylor, A. J. Crovetti, and N. E. Boyer, *J. Am. Chem. Soc.*, **79**, 3549 (1957).
[172] T. Kato and T. Niitsuma, *Chem. Pharm. Bull. Japan*, **13**, 963 (1965).
[173] E. V. Brown, *J. Am. Chem. Soc.*, **79**, 3565 (1957).
[174] W. D. Guither, D. G. Clark, and R. N. Castle, *J. Heterocyclic Chem.*, **2**, 67 (1965).
[175] T. Itai and S. Suzuki, *Chem. Pharm. Bull. Japan*, **8**, 999 (1960).

[176] F. Reicheneder and K. Dury, Belg. Patent 660,637 (1965); through *Chem. Abstr.*, **64**, 5108 (1966).
[177] a. M. Ogata and H. Kano, *Chem. Pharm. Bull. Japan*, **11**, 29 (1963); b. T. Itai and S. Natsume, p. 83; c. S. Sako, p. 337.
[178] T. Itai and S. Natsume, *Chem. Pharm. Bull. Japan*, **11**, 342 (1963).
[179] G. Karmas and P. E. Spoerri, *J. Am. Chem. Soc.*, **75**, 5517 (1953).
[180] P. E. Fanta and T. R. Hughes, *J. Am. Chem. Soc.*, **72**, 5343 (1950).
[181] D. J. Brown, *J. Appl. Chem.*, **7**, 109 (1957).
[182] W. Pfleiderer and H. Mosthaf, *Ber.*, **90**, 738 (1957).
[183] K. L. Dille, M. L. Sutherland, and B. E. Christensen, *J. Org. Chem.*, **20**, 171 (1955).
[184] D. E. O'Brien, C. W. Noell, R. K. Robins, and C. C. Cheng, *J. Med. Chem.*, **9**, 121 (1966).
[185] Y. F. Shealy and C. A. O'Dell, *J. Org. Chem.*, **29**, 1235 (1964).
[186] D. Söll and W. Pfleiderer, *Ber.*, **96**, 2977 (1963).
[187] C. Kaiser and A. Burger, *J. Org. Chem.*, **24**, 113 (1959).
[188] R. Urban and O. Schnider, *Helv. Chim. Acta*, **41**, 1806 (1958).
[189] V. Papesch and R. M. Dodson, *J. Org. Chem.*, **28**, 1329 (1963).
[190] P. E. Fanta and E. A. Hedman, *J. Am. Chem. Soc.*, **78**, 1434 (1956).
[191] M. M. Robison, B. L. Robison, and F. P. Butler, *J. Am. Chem. Soc.*, **81**, 743 (1959).
[192] H. S. Broadbent and T. G. Selin, *J. Org. Chem.*, **28**, 2345 (1963).
[193] P. Ruggli and A. Dinger, *Helv. Chim. Acta*, **24**, 173 (1941).
[194] G. Benoit and D. Marinopoulos, *Bull. Soc. Chim. France*, **1950**, 829.
[195] G. Drefahl and J. Ulbricht, *Ann.*, **598**, 174 (1956).
[196] F. Bergmann and D. Schapiro, *J. Org. Chem.*, **12**, 57 (1947).
[197] T. W. Campbell and R. N. McDonald, *J. Org. Chem.*, **24**, 1246 (1959).
[198] E. K. Blout and D. C. Silverman, *J. Am. Chem. Soc.*, **66**, 1442 (1944).
[199] V. Boekelheide and A. P. Michels, *J. Am. Chem. Soc.*, **74**, 256 (1952).
[200] R. Roger, *J. Chem. Soc.*, **1947**, 560.
[201] B. R. Baker and J. H. Jordaan, *J. Med. Chem.*, **8**, 33 (1965).
[202] D. S. Morris, *J. Chem. Soc.*, **1950**, 1913.
[203] A. Pedrazzoli, *Helv. Chim. Acta*, **40**, 80 (1957).
[204] S. Kirkwood and P. H. Phillips, *J. Am. Chem. Soc.*, **69**, 934 (1947).
[205] W. Borsche and W. Ried, *Ber.*, **76**, 1011 (1943).
[206] J. H. Burckhalter, J. Sam, and L. A. R. Hall, *J. Am. Chem. Soc.*, **81**, 394 (1959).
[207] J. T. Suh and B. M. Puma, *J. Org. Chem.*, **30**, 2253 (1965).
[208] G. N. Walker, *J. Med. Chem.*, **8**, 583 (1965).
[209] W. Borsche and F. Sell, *Ber.*, **83**, 78 (1950).
[210] W. A. Ried, A. Berg, and G. Schmidt, *Ber.*, **85**, 204 (1952).
[211] N. J. Leonard and S. N. Boyd, Jr., *J. Org. Chem.*, **11**, 405 (1946).
[212] D. Amico, L. Bertolini, and C. Monreale, *Chimica e Industria*, **38**, 43 (1956); through *Chem. Abstr.*, **50**, 13800 (1956).
[213] E. Profft and A. Jumar, *Arch. Pharm.*, **289**, 90 (1956).
[214] S. I. Sergievskaya and G. A. Ravdel, *J. Gen. Chem. USSR*, **22**, 559 (1952) (English translation).
[214a] R. B. Baker, R. E. Schaub, J. P. Joseph, F. J. McEvoy, and J. H. Williams, *J. Org. Chem.*, **17**, 164 (1952).
[215] B. L. Zenitz and W. H. Hartung, *J. Org. Chem.*, **11**, 444 (1946).
[216] W. Theilacker, H. M. Muller, and V. Blumencron, *Ber.*, **89**, 984 (1956).
[217] B. R. Baker and B. Ho, *J. Heterocyclic Chem.*, **2**, 72 (1965).
[218] A. C. Huitric and W. D. Kumler, *J. Am. Chem. Soc.*, **78**, 614 (1956).

References

[219] E. Schipper, E. Chinery, and J. Nichols, *J. Org. Chem.*, **26**, 4145 (1961).
[220] L. Krbechek and H. Takimoto, *J. Org. Chem.*, **29**, 1150 (1964).
[221] S. Wawzonek and E. Yeakey, *J. Am. Chem. Soc.*, **82**, 5718 (1960).
[222] E. Reed and L. H. Sternbach, U.S. Patent 3,222,359 (1965).
[223] W. L. F. Armarego, *J. Chem. Soc.*, **1961**, 2697.
[224] Neth. Appl., 6,413,744 (1965); through *Chem. Abstr.*, **64**, 3570 (1966).
[225] W. Davis, J. S. Roberts, and W. C. J. Ross, *J. Chem. Soc.*, **1955**, 890.
[226] F. F. Stephens and J. D. Bower, *J. Chem. Soc.*, **1950**, 1722.
[227] B. H. Chase, J. P. Thornton, and J. Walker, *J. Chem. Soc.*, **1951**, 3439.
[228] H. Rupe and A. Metzger, *Helv. Chim. Acta*, **8**, 838 (1925).
[229] H. Rupe and H. Vogler, *Helv. Chim. Acta*, **8**, 832 (1925).
[230] M. Moll, H. Musso, and H. Schroeder, *Angew. Chem. Internat. Edit.*, **2**, 212 (1963).
[231] a. H. Musso and H. Schroeder, *Ber.*, **98**, 1562 (1965); b. H. Musso and H. Schroeder, *Ber.*, **98**, 1577 (1965); c. H. Schroeder, U. Schwabe, and H. Musso, *Ber.*, **98**, 2556 (1965).
[232] G. N. Walker, *J. Am. Chem. Soc.*, **77**, 3844 (1955).
[233] H. W. Bersch, *Arch. Pharm.*, **277**, 271 (1939).
[234] S. Wawzonek, *J. Am. Chem. Soc.*, **68**, 1157 (1946).
[235] J. Harley-Mason and A. H. Jackson, *J. Chem. Soc.*, **1954**, 1158.
[236] M. Freifelder, work to be published.
[237] B. Heath-Brown and P. G. Philpot, *J. Chem. Soc.*, **1965**, 7185.
[238] T. A. Geissman and T. G. Halsall, *J. Am. Chem. Soc.*, **73**, 1280 (1951).
[239] A. P. Phillips and A. Maggiolo, *J. Org. Chem.*, **15**, 659 (1950).
[240] R. T. Puckowski and W. A. Ross, *J. Chem. Soc.*, **1959**, 3555.
[241] P. Pratesi, A. LaManna, A. Campiglio, and V. Ghislandi, *Farmaco*, I (Pavia), *Sci. Ed.*, **12**, 993 (1957).
[242] E. J. Alford and K. Schofield, *J. Chem. Soc.*, **1952**, 2102.
[243] H. Plieninger and M. S. von Wittenau, *Ber.*, **91**, 1905 (1958).
[244] N. J. Leonard, K. L. Carraway, and J. P. Helgeson, *J. Heterocyclic Chem.*, **2**, 291 (1965).
[245] a. W. S. Fidler and H. C. S. Woods, *J. Chem. Soc.*, **1957**, 3980; b. H. Goldner and F. Carstens, *J. Prakt. Chem.*, **12**, 242 (1961).
[246] A. R. Osborn and K. Schofield, *J. Chem. Soc.*, **1956**, 3977.
[247] M. M. Rapport, J. F. Mead, J. T. Maynard, A. E. Senear, and J. B. Koepfli, *J. Am. Chem. Soc.*, **69**, 2561 (1947).
[248] M. Ishidate, Y. Sakurai, and S. Owari, *Chem. Pharm. Bull. (Tokyo)*, **5**, 199 (1957).
[249] A. Chizhov, K. V. Levshina, and S. I. Sergievskaya, *J. Gen. Chem. USSR*, **30**, 3659 (1960) (English translation).
[250] K. H. Taffs, L. V. Prosser, F. B. Wigton, and M. M. Jouillé, *J. Org. Chem.*, **26**, 462 (1961).
[251] B. R. Baker, T. J. Schwan, J. Novotny, and B. Ho, *J. Pharm. Sci.*, **55**, 295 (1966).
[252] J. N. Ashley and S. S. Berg, *J. Chem. Soc.*, **1959**, 3725.
[253] B. Boothroyd and E. R. Clark, *J. Chem. Soc.*, **1953**, 1499.
[254] C. Temple, Jr., R. L. McKee, and J. A. Montgomery, *J. Org. Chem.*, **30**, 829 (1965).
[255] C. G. Overberger and N. P. Marullo, *J. Am. Chem. Soc.*, **83**, 1378 (1961).
[256] G. B. Marini-Bettolo and R. L. Vittory, *Gazz. Chim. Ital.*, **87**, 1038 (1957).
[257] R. S. Hanslick and W. F. Bruce, U.S. Patent 2,784,229 (1957).
[258] S. Akaboshi and S. Ikegami, *Chem. Pharm. Bull. (Tokyo)*, **14**, 622 (1966).
[259] L. N. Nikolenko, A. V. Chistyakova, E. N. Karpova, and S. A. Kabanova, *J. Gen. Chem. USSR*, **34**, 4091 (1964) (English translation).

[260] J. W. Schulenberg and S. Archer, *J. Org. Chem.*, **30**, 1279 (1965).
[261] H. Oelschlager and P. Schmersahl, *Arch. Pharm.*, **296**, 324 (1963).
[262] N. N. Vorozhtsov, Jr., G. G. Yakobson, and L. I. Denisova, *J. Gen. Chem. USSR*, **31**, 1137 (1961) (English translation).
[263] C. F. Winans, *J. Am. Chem. Soc.*, **61**, 3564 (1939).
[264] R. Baltzly and A. P. Phillips, *J. Am. Chem. Soc.*, **68**, 261 (1946).
[264a] H. Najer, D. Chabrier, and R. Guidicelli, *Bull. Soc. Chim. France*, **1956**, 106.
[265] a. F. S. Dovell and H. Greenfield, *J. Am. Chem. Soc.*, **87**, 2767 (1965); b. Neth. Appl. 6,409,250 (1965); through *Chem. Abstr.*, **63**, 11428 (1965).
[266] E. D. Bergmann and M. Bentov, *J. Org. Chem.*, **26**, 1480 (1961).
[267] N. Lofgren, B. Lundqvist, and S. Lindstrom, *Acta Chem. Scand.*, **9**, 1079 (1955).
[268] F. Bennington, E. V. Shoop, and R. H. Poirier, *J. Org. Chem.*, **18**, 1506 (1953).
[269] A. F. Helin and C. A. Vanderwerf, *J. Org. Chem.*, **17**, 229 (1952).
[270] H. L. Bradlow and C. A. Vanderwerf, *J. Am. Chem. Soc.*, **70**, 654 (1948).
[271] R. K. Richards and M. R. Vernsten, U.S. Patent 3,170,955 (1965).
[272] L. Spiegler, U.S. Patent 3,073,865 (1963).
[273] F. C. Trager, U.S. Patent 2,772,313 (1956).
[274] E. D. Bergmann and R. Barshai, *J. Am. Chem. Soc.*, **81**, 5641 (1959).
[275] D. E. Graham, J. B. Normington, and H. B. Freyermuth, U.S. Patent 3,149,161 (1964).
[276] M. Freifelder and D. A. Dunnigan, to be published.
[277] a. A. E. Senear, M. M. Rapport, J. F. Mead, J. T. Maynard, and J. B. Koepfli, *J. Org. Chem.*, **11**, 378 (1946); b. F. Dallacker, L. Doyen, and G. Schmets, *Ann.*, **694**, 117 (1966).
[278] A. J. Dietzler and T. R. Keil, U.S. Patent 3,051,753 (1962).
[279] a. B. E. Leggetter and R. K. Brown, *Can. J. Chem.*, **38**, 2363 (1960); b. S. Pietra, *Ann. Chim.* (Rome), **52**, 727 (1962).
[280] H. Nahm and W. Siedel, *Ber.*, **96**, 1 (1963).
[280a] R. W. Dennison and F. Scheinmann, *J. Chem. Soc.*, **1965**, 4206.
[281] R. J. Rousseau and R. K. Robins, *J. Heterocyclic Chem.*, **2**, 196 (1965).
[282] S. A. Harris and K. Folkers, *J. Am. Chem. Soc.*, **61**, 1245 (1939).
[283] E. D. Parker and W. Shive, *J. Am. Chem. Soc.*, **69**, 63 (1947).
[284] J. R. Vaughan, Jr., J. Krapcho, and J. P. English, *J. Am. Chem. Soc.*, **71**, 1885 (1949).
[285] C. E. Larrabee, U.S. Patent 2,860,141 (1958).
[286] L. Vellus and G. Amiard, *Bull. Soc. Chim. France*, **1947**, 136.
[287] A. R. Osborn and K. Schofield, *J. Chem. Soc.*, **1956**, 4191.
[288] B. Elpern and C. S. Hamilton, *J. Am. Chem. Soc.*, **68**, 1436 (1946).
[289] A. R. Surrey and R. A. Cutler, *J. Am. Chem. Soc.*, **73**, 2413 (1951).
[290] C. C. Price, E. W. Maynert, and V. Boekelheide, *J. Org. Chem.*, **14**, 484 (1949).
[291] C. Temple, Jr. and J. A. Montgomery, *J. Org. Chem.*, **31**, 1417 (1966).
[292] M. B. Naff and E. R. Christensen, *J. Am. Chem. Soc.*, **73**, 1372 (1951).
[293] V. G. Nyrkova, T. V. Gortinskaya, and M. N. Shchukina, *Zh. Organ. Khim.*, **1**, 1688 (1965); through *Chem. Abstr.*, **64**, 2084 (1966).
[294] C. Alberti and C. Tironi, *Farmaco, I. (Pavia) Ed. Sci.*, **21**, 883 (1966).
[295] Br. Patent 925,458 (1963).
[296] K. Johnson and E. F. Degering, *J. Am. Chem. Soc.*, **61**, 3194 (1939).
[297] a. G. D. Buckley, *J. Chem. Soc.*, **1947**, 1494; b. G. D. Buckley and E. Ellery, *J. Chem. Soc.*, **1947**, 1497.
[298] D. C. Iffland and F. A. Cassis, Jr., *J. Am. Chem. Soc.*, **74**, 6284 (1952).

[299] a. A. Lambert and A. Lowe, *J. Chem. Soc.*, **1947**, 1517; b. A. Lambert and H. A. Piggott, p. 1489; c. G. S. Skinner and P. R. Wunz, *J. Am. Chem. Soc.*, **73**, 3815 (1951).
[300] J. Rockett and F. C. Whitmore, *J. Am. Chem. Soc.*, **71**, 3249 (1949).
[301] M. Senkus, U.S. Patent 2,418,237 (1947).
[302] H. R. Snyder and W. E. Hamlin, *J. Am. Chem. Soc.*, **72**, 5082 (1950).
[303] R. L. Heath and J. D. Rose, *J. Chem. Soc.*, **1947**, 1486.
[304] T. Tsuji and T. Ueda, *Chem. Pharm. Bull. (Tokyo)*, **12**, 946 (1964).
[305] a. M. Senkus, U.S. Patent 2,426,375 (1947); b. G. H. Butler and F. N. McMillan, *J. Am. Chem. Soc.*, **72**, 2978 (1950); c. H. G. Johnson, *J. Am. Chem. Soc.*, **68**, 12 (1946).
[306] M. Freifelder and Y. H. Ng, *J. Org. Chem.*, **30**, 4370 (1965).
[307] B. Heath-Brown and P. G. Philpott, *J. Chem. Soc.*, **1965**, 7165.
[308] W. E. Noland and P. J. Hartman, *J. Am. Chem. Soc.*, **76**, 3227 (1954).
[309] a. G. E. Ullyot, J. J. Stehle, C. L. Zirkle, R. L. Shriner, and F. J. Wolf, *J. Org. Chem.*, **10**, 429 (1945); b. G. E. Ullyot, U.S. Patent 2,480,105 (1949); c. G. N. Walker, R. T. Smith, and B. N. Weaver, *J. Med. Chem.*, **8**, 626 (1965).
[310] D. I. Weisblat and D. A. Lyttle, *J. Am. Chem. Soc.*, **71**, 3079 (1949).
[311] E. E. van Tamelen and G. Van Zyl, *J. Am. Chem. Soc.*, **72**, 2979 (1950).
[312] M. Hamana and M. Yamazaki, *Chem. Pharm. Bull. (Tokyo)*, **11**, 415 (1963).
[313] P. M. G. Bavin, *J. Med. Chem.*, **9**, 52 (1966).
[314] D. C. Iffland, G. X. Criner, M. Koral, F. J. Lotspeich, Z. B. Papanastassiou, and S. M. White, Jr., *J. Am. Chem. Soc.*, **75**, 4044 (1953).
[315] H. B. Hass and H. Schechter, *J. Am. Chem. Soc.*, **75**, 1382 (1953).
[316] K. E. Hamlin and M. Freifelder, *J. Am. Chem. Soc.*, **75**, 369 (1953).
[317] a. H. Meister, *Ann.*, **679**, 83 (1964); b. W. F. Trager and A. C. Huitric, *J. Pharm. Sci.*, **54**, 1552 (1965).
[318] a. M. Senkus, U.S. Patent 2,413,250 (1946); b. G. H. Morey, U.S. Patent 2,415,021 (1947); c. J. H. Ottenhaym and P. L. Kerkhoffs, U.S. Patent 3,048,580 (1962).
[319] H. Scholz and P. Guenthert, U.S. Patent 2,828,131 (1958).
[320] A. T. Nielsen, *J. Org. Chem.*, **27**, 1998 (1962).
[321] F. Bordwell and R. L. Arnold, *J. Org. Chem.*, **27**, 4426 (1962).
[322] E. Schmidt and R. Wilkendorf, *Ber.*, **52**, 389 (1919).
[323] K. Johnson and E. F. Degering, *J. Org. Chem.*, **8**, 7 (1943).
[324] a. B. M. Vanderbilt, U.S. Patent 2,157,391 (1939); b. F. W. Hoover and H. B. Hass, *J. Org. Chem.*, **12**, 506 (1947).
[325] a. J. B. Tindall, U.S. Patent 2,347,621 (1944); b. W. C. Gakenheimer and W. H. Hartung, *J. Org. Chem.*, **9**, 85 (1944).
[326] a. W. E. Noland, J. E. Kneller, and D. E. Rice, *J. Org. Chem.*, **22**, 695 (1957); b. F. Zymalkowski, *Arch. Pharm.*, **289**, 52 (1956).
[327] B. L. Zenitz, E. B. Macks, and M. L. Moore, *J. Am. Chem. Soc.*, **70**, 955 (1948).
[328] P. E. Fanta, R. J. Smat, L. F. Piecz, and L. Clemens, *J. Org. Chem.*, **31**, 3113 (1966).
[329] P. M. Ruoff and J. R. Miller, *J. Am. Chem. Soc.*, **72**, 1417 (1950).
[330] G. Van Zyl, E. E. van Tamelen, and G. D. Zuidema, *J. Am. Chem. Soc.*, **73**, 1765 (1951).
[331] W. Herz and S. Tocker, *J. Am. Chem. Soc.*, **77**, 3554 (1955).
[332] a. R. E. Foster and A. F. Kirby, U.S. Patent 2,967,200 (1961); b. C. B. Flack and D. O. Halvorson, U.S. Patent 3,157,702 (1964).

[333] F. Möller and O. Bayer, Ger. Patent 966,201 (1957).
[334] a. G. von Schuckmann and H. Danziger, Ger. Patent 959,644 (1957); b. Belg. Patent 626,468 (1963); through *Chem. Abstr.*, **60**, 8688 (1964); c. J. Christian, U.S. Patent 2,711,427 (1955).
[335] a. P. Guyer and H. J. Merz, *Chimia*, **18**, 144 (1964); b. J. Weise, Ger. Patent 917,426 (1954).
[336] E. Schmidt, A. Ascherl, and L. Mayer, *Ber.*, **58**, 2430 (1925).
[337] G. G. Joris and J. Vitrone, Jr., U.S. Patent 2,829,163 (1958).
[338] H. Meister and W. Franke, U.S. Patent 2,886,596 (1959).
[339] A. Cohen and B. Heath-Brown, *J. Chem. Soc.*, **1965**, 7179.
[340] D. V. Nightingale and V. Tweedie, *J. Am. Chem. Soc.*, **66**, 1068 (1944).
[341] M. Senkus, U.S. Patent 2,399,068 (1946).
[342] N. L. Drake and A. A. Ross, *J. Org. Chem.*, **23**, 717 (1958).
[343] C. Grob and E. Jenny, U.S. Patent 3,118,946 (1964).
[344] D. Nightingale and J. R. Janes, *J. Am. Chem. Soc.*, **66**, 352 (1944).
[345] C. D. Hurd, U.S. Patent 2,483,201 (1949).
[346] A. Sonn and A. Schellenberg, *Ber.*, **50**, 1513 (1917).
[347] H. A. Smith and W. C. Bedoit, Jr., *J. Phys. Colloid Chem.*, **55**, 1089 (1951).
[348] J. B. Tindall, U.S. Patent 2,647,930 (1953).
[349] J. B. Tindall, U.S. Patent 2,636,901 (1953).
[350] a. K. Kindler, E. Brandt, and E. Gelhaar, *Ann.*, **511**, 209 (1934); b. G. Hahn and F. Rumpf, *Ber.*, **71**, 2141 (1938); c. J. L. Bills and C. R. Noller, *J. Am. Chem. Soc.*, **70**, 957 (1948); d. H. M. E. Cardwell and F. J. McQuillin, *J. Chem. Soc.*, **1949**, 708; e. J. Daly, L. Horner, and B. Wittkop, *J. Am. Chem. Soc.*, **83**, 4787 (1961); f. M. Green, U.S. Patent 3,062,884 (1962).
[351] B. Reichert, *Arch. Pharm.*, **274**, 505 (1936).
[352] a. W. D. McPhee, E. S. Erickson, Jr., and V. J. Salvador, *J. Am. Chem. Soc.*, **68**, 1866 (1946); b. K. Rorig, *J. Org. Chem.*, **15**, 391 (1950).
[353] a. G. Stochdorph and O. V. Schickl, Ger. Patent, 848,197 (1952); b. H. L. Curtis, U.S. Patent 3,187,046 (1965).
[354] G. A. Bonetti, C. B. De Savigny, and C. Michalski, U.S. Patent 3,226,442 (1965).
[355] a. B. Reichert and W. Koch, *Arch. Pharm.*, **273**, 265 (1935); b. B. Reichert and W. Hoffman, *Arch. Pharm.*, **274**, 153 (1936).
[356] W. K. Seifert and P. C. Condit, *J. Org. Chem.*, **28**, 265 (1963).
[357] O. W. Chandler, Fr. Patent 1,343,869 (1963).
[358] P. Lipp and H. Mettegang, *Ber.*, **76**, 1275 (1943).
[359] G. D. Buckley, R. L. Heath, and J. D. Rose, *J. Chem. Soc.*, **1947**, 1500.
[360] G. D. Buckley and T. J. Elliot, *J. Chem. Soc.*, **1947**, 1508.
[361] R. C. Elderfield and H. A. Hageman, *J. Org. Chem.*, **14**, 605 (1949).
[362] a. D. Lonchamp and P. Baumgartner, *Compt. Rend.*, **257**, 668 (1963); b. H. C. Godt, Jr. and J. F. Quinn, *J. Am. Chem. Soc.*, **78**, 1461 (1956).
[363] D. Jerchel, Ger. Patent 884,368 (1953).
[364] L. B. Clapp, J. F. Brown, Jr., and L. Zeftel, *J. Org. Chem.*, **15**, 1043 (1950).
[365] a. O. Moidenhauer, W. Irion, and R. Pfluger, Ger. Patent 906,697 (1954); b. W. Irion, D. Mastaglio, and H. Döser, Ger. Patent 928,529 (1955).
[366] D. T. Warner and O. A. Moe, *J. Am. Chem. Soc.*, **74**, 1064 (1952).
[367] a. B. Belleau and J. Burba, *J. Med. Chem.*, **6**, 755 (1963); b. V. B. Piskov, *J. Gen. Chem. USSR*, **34**, 3481 (1964) (English translation).
[368] V. V. Perekalin and A. S. Sopova, *J. Gen. Chem. USSR*, **28**, 656 (1958) (English translation).

[369] a. M. C. Kloetzel, *J. Am. Chem. Soc.*, **69**, 2271 (1947); b. M. C. Kloetzel, J. L. Pinkus, and R. M. Washburn, *J. Am. Chem. Soc.*, **79**, 4222 (1957); c. M. C. Kloetzel, F. L. Chubb, and J. L. Pinkus, *J. Am. Chem. Soc.*, **80**, 5773 (1958); d. M. L. Stein and A. Burger, *J. Am. Chem. Soc.*, **79**, 154 (1957).

[370] W. Davey and D. J. Tivey, *J. Chem. Soc.*, **1958**, 2276.

[371] a. W. E. Noland and R. A. Hovden, *J. Org. Chem.*, **24**, 894 (1959); b. R. V. Heinzelman, W. C. Anthony, D. A. Lyttle, and J. Szmuszkovicz, *J. Org. Chem.*, **25**, 1548 (1960).

[372] a. B. Reichert, *Arch. Pharm.*, **274**, 369 (1936); b. K. Kindler and W. Peschke, *Ann.*, **519**, 291 (1935).

[373] M. Senkus, *J. Am. Chem. Soc.*, **68**, 1611 (1946).

[374] E. B. Hodge, U.S. Patent 2,681,934 (1954).

[375] a. S. Malkiel and J. P. Mason, *J. Am. Chem. Soc.*, **64**, 2515 (1942); b. M. Compton, H. Higgins, L. MacBeth, J. Osburn, and H. Burkett, *J. Am. Chem. Soc.*, **71**, 3229 (1949).

[376] a. C. A. Price and C. A. Sears, *J. Am. Chem. Soc.*, **75**, 3275 (1953); b. J. A. Robertson, *J. Org. Chem.*, **13**, 395 (1948).

[377] E. Schmidt and A. Ascher, *Ber.*, **58**, 356 (1925).

[378] R. Wilkendorf and M. Tre'nel, *Ber.*, **56**, 611 (1923).

[379] W. D. Guither, D. G. Clark, and R. N. Castle, *J. Heterocyclic Chem.*, **2**, 67 (1965).

[380] a. M. Bergmann, L. Zervas, and H. Rinke, *Zeit. Physiol. Chem.*, **224**, 40 (1934); b. H. Gibian and E. Schroeder, *Ann.*, **642**, 145 (1961).

[381] C. Gros, M. Privat de Garilhe, A. Costanapagiotis, and R. Schwyzer, *Helv. Chim. Acta*, **44**, 2042 (1961).

XI

REDUCTION OF NITROSO, AZO, AND AZOXY GROUPS AND OTHER GROUPS RELATED TO THE AZO SYSTEM

NITROSO GROUP

In Aromatic Systems

Nitroso compounds, like nitro compounds, are readily reduced to the corresponding amines under low-pressure conditions with palladium, platinum, or nickel catalysts. Conditions described in the hydrogenations of nitrosobenzenes [1a, b] and nitrosopyrimidines [2a, b] to the corresponding amines should be applicable to reductions of most aromatic nitroso compounds. At times minor modifications may be required. When the sodium salt of the oxime form of 4-nitrosodiphenylamine was hydrogenated in water, it was necessary to do so in the presence of benzene to prevent precipitation of 4-aminodiphenylamine which, in the absence of benzene, caused cessation of hydrogen uptake before all of the nitroso compound was reduced [3].

References to reductions of nitroso groups in aromatic compounds which contain other reducible functions appear to be unavailable. When it is found necessary to carry out such hydrogenations, analogous reductions of nitro compounds should be used for comparison.

Nitrosamines

In general, attempts in the past to hydrogenate nitrosamines to unsymmetrical hydrazines with Raney nickel or palladium or platinum catalysts have resulted in cleavage and formation of amines [4a, b, c]. An attempt to prepare 1,1,2-trimethyltriazane, $(CH_3)_2NNH(CH_3)NH_2$, by palladium on carbon-catalyzed reduction of trimethylnitrosohydrazine yielded 95% of trimethylhydrazine [5]. The tendency toward cleavage of the nitroso groups during catalytic reduction was used to obtain a good yield of N-butyl p-anisidine from the N-nitroso compound [6a] in preference to chemical reduction which caused O-dealkylation. The nitroso group was used as a protecting group in order to obtain pure monoquaternary salts of piperazines. Removal of other protecting groups by hydrolytic means caused partial breakdown of quaternary salt whereas removal of the nitroso group was readily accomplished by catalytic hydrogenation [6b].

Reductions to hydrazines are complicated by a competing side reaction in which nitrosamine is converted to tetrazene which in turn is reduced to tetrazane. The latter compounds are unstable and decompose to secondary amines and nitrogen [4a]:

$$2R_2NNO \xrightarrow[-2H_2O]{2H_2} R_2NN\!=\!NNR_2 \xrightarrow{H_2} (R_2NNH)_2 \rightarrow 2R_2NH + N_2$$

More recently successful reductions without cleavage have been reported. Dimethylnitrosamine in aqueous solution was converted to 1,1-dimethylhydrazine with cobalt catalyst at 90° and 280 atm [7]. A variety of pressure and temperature conditions and the use of a number of catalysts have been described for the reduction of dialkylnitrosamines in aqueous or alcoholic solution to 1,1-dialkylhydrazines [8a–d]. The use of various salts to increase yields of unsymmetrical hydrazines has also been described [9a–d]. The additions of organic and inorganic salts to the reactions were beneficial in reductions with palladium and platinum catalysts and to a lesser degree with rhodium [9d] but did not prevent formation of amines in reductions with Raney nickel. Although it is not known how the presence of salts effect reduction of nitrosamines, it has been suggested that their presence might increase polarization of the substrate and favor hydrogenation of the nitrosamine to hydrazine.

The choice of conditions to produce hydrazine by reduction of nitrosamines may be critical as exemplified in a study of the hydrogenation of nitrosodimethylamine. For good yields it was necessary to reduce a dilute solution of nitrosamine with a low ratio of catalyst to substrate and to also control the reaction temperature [10]. The use of 10% palladium on carbon gave better results than reductions with palladium on calcium carbonate or barium sulfate.

In contrast to the reported successes in the hydrogenations of nitrosamines to unsymmetrical hydrazines [8cd, 9a, 10], failures resulted in attempted reductions of 3-methyl-4-nitroso-2-phenylmorpholine with platinum, palladium, rhodium, and ruthenium catalysts at low pressure or up to 125 atm [11]. Only 3-methyl-2-phenylmorpholine was obtained.

The reduction of N-nitrosodimethylamine in concentrated hydrochloric acid in the presence of platinum on carbon at 75 atm and 25° gave hydroxylamine and dimethylamine [12] according to the following equation:

$$(CH_3)_2NNO \xrightarrow{2H_2} NH_2OH + (CH_3)_2NH$$

Halonitroso Compounds

9-Chloro-10-nitrosodecalin was hydrogenated to 9-chloro-10-hydroxylaminodecalin with prereduced platinum in ethyl acetate [13]. In the reduction of 1-aryl-1-chloro-2-nitroethanes with Raney nickel at room temperature and atmospheric pressure the chlorine-free amines were obtained [14]. When 1-chloro-1-phenyl-2-nitroethane was reduced in the presence of cyclohexylamine, the product of reaction was 2-(cyclohexyl)-2-(phenyl)acetaldoxime. When N-nitroso-bis(2-chloroethyl)amine was catalytically hydrogenated, no single simple product could be isolated [15].

AZO AND AZOXY GROUPS

In general the purpose of reducing compounds containing these groups is to obtain aromatic-type amines which at times may be difficult to obtain by other procedures. Hydrogenations are readily carried out with palladium, platinum, and nickel catalysts under low-pressure conditions. Elevated temperature and pressure conditions have been employed with nickel but they are usually unnecessary.

The following examples illustrate the ease with which azo compounds are reduced to amines. Azoxy compounds are reduced just as readily. 2,6-Dihydroxy-4-phenylazobenzamide in acetic acid [16a], 1-(2-hydroxyethyl)-3,5-dimethyl-4-phenylazopyrazole in methyl alcohol [16b], and 1,2-diphenyl 4-phenylazopyrazolidin-3,5-dione in ethyl alcohol [16c] were converted to the corresponding amines and aniline by hydrogenation over palladium on carbon. 4-Hydroxy-2,6-dimethyl-5-phenylazopyrimidone was reduced to 5-amino-4-hydroxy-2,6-dimethylpyrimidine with palladium on barium sulfate [16d].

Platinum oxide was employed for the hydrogenation of methyl and ethyl 2,6-dihydroxy-3-phenylazoisonicotinic acid in ethyl acetate to the esters of 2,6-dihydroxy-3-aminoisonicotinic acid [17a] and for the preparation of 2,5-diamino-6-hydroxy-, -6-methylamino- and 6-dimethylamino-4-methylpyrimidines from the corresponding 5-phenylazopyrimidines [17b]. Platinum on carbon catalyzed the reduction of 2,6-diamino-5-phenylazopyrimidine in aqueous hydrochloric acid to 2,3,6-triaminopyrimidine [17c].

Other N=N Systems

Reduction of 2,6-dihydroxy-3-phenylazoisonicotinic acid with Raney nickel in alkaline solution gave 92% yield of 3-amino-2,6-dihydroxyisonicotinic acid [17a]. Other very good results in hydrogenations with nickel are seen in references [18a–d].

In general there may be little choice among the three catalysts. However, in the hydrogenation of 2,2′,5,5′-tetraphenyl-3,3′-azopyrrole to 3-amino-2,5-diphenylpyrrole only palladium on carbon gave good results [19].

The azoxy group can be reduced to azo and either can be reduced to a hydrazine linkage. 3-Methyl-4-(phenylazo)azoxybenzene was reduced in ethyl acetate with Raney nickel to 3-methyl-4-(phenylazo)azobenzene [20a] (Eq. 11.1). 4,4′-Diamidinoazobenzene, 4,4′-dicyanoazo-, and 4,4′-dicyano-

$$C_6H_5N=N-C_6H_3(CH_3)(N\to O)-N=NC_6H_5 \xrightarrow{H_2}_{Ni(R)} C_6H_5N=N-C_6H_3(CH_3)-N=NC_6H_5$$

(11.1)

azoxybenzenes were converted to the corresponding hydrazines by reduction with palladium on carbon [20b] and methyl 5-methoxy-8-phenylazoquinoline-3-carboxylate was reduced to the hydrazine with Raney nickel under moderate conditions [16c].

There has been little attack on aromatic ring systems during partial or complete hydrogenation of either azo or azoxy compounds. The azo linkage has been reduced to an amino group without affecting the carbonyl group of an aminoketone [18d] and without attack of a cyano group [20b]. Selectivity in the presence of halogen can probably be maintained by reduction on acidic media.

OTHER N=N SYSTEMS

Compounds such as ethyl diazoacetate and diazomethanes, $RCHN_2$ where R is hydrogen, alkyl, or aryl or where hydrogen is replaced with another aryl or alkyl group, usually lose nitrogen on catalytic reduction. When R is acyl, as in diazoketones, a variety of products is obtained depending on the substrate and the reaction medium. In most instances the —CN_2 ring system is ruptured to yield amine. Hydrazines and hydrazones have also been obtained [21a]. When the acyl group is aroyl, there is a tendency toward accompanying reduction of ketone to alcohol. It occurred during the hydrogenation of methyl benzoyldiazocetate in 70% acetic acid with palladium on carbon [21b]. Reduction was stereospecific to yield the methyl ester of DL-*erythro*-3-phenylserine (Eq. 11.2).

$$\underset{\text{Pd on C}}{\overset{\text{O}}{\text{C}_6\text{H}_5\overset{\|}{\text{C}}\text{CN}_2\text{COOCH}_3 \xrightarrow{4\text{H}_2}}} \text{C}_6\text{H}_5\overset{\text{OH}}{\underset{|}{\text{C}}}\text{H}\overset{\text{NH}_2}{\underset{|}{\text{C}}}\text{HCOOCH}_3 \qquad (11.2)$$

Reduction of ketone probably can be controlled as it was when 2-diazoacetophenone, $\text{C}_6\text{H}_5\text{COCHN}_2$, was converted to 2-aminoacetophenone by hydrogenation with palladium oxide in ethyl acetate containing acetic acid [21a]. When the reaction was carried out in the absence of acid 2,5-diphenylpyrazine was obtained from self-condensation of aminoketone and spontaneous dehydrogenation.

When the acyl group was alkanoyl the carbonyl group was not affected. Ethyl 2-diazo-4-bromo-3-oxooctadecanoate was hydrogenated in alcoholic hydrogen bromide over palladium on carbon. The carbonyl group was untouched but debromination took place to yield the hydrohalide salt of ethyl 2-amino-3-oxooctadecanoate [21c]. When the acyl group was phenylacetyl, reduction gave aminoalcohol or aminoketone; when it was phenylpropionyl, hydrogenation gave a hydrazone [21a].

In another N=N system the reduction of *N*-aryl-*N'*-fluorodiimide *N*-oxides with platinum oxide in alcohol yielded anilines [32]:

$$\text{ArN}(\rightarrow \text{O})\!\!=\!\!\text{NF} \xrightarrow{4\text{H}_2} \text{ArNH}_2$$

REFERENCES

[1] L. K. J. Tong and M. C. Glesmann, *J. Am. Chem. Soc.*, **78**, 5827 (1956); b. W. Pfleiderer, E. Liedik, R. Lohrmann, and M. Rukwied, *Ber.*, **93**, 2015 (1960).
[2] a. R. Mozingo and G. S. Fonken, U.S. Patent 2,447,523 (1948); b. A. H. Homeyer, U.S. Patent 2,646,432 (1953).
[3] T. H. Newby and B. A. Hunter, U.S. Patent 2,974,169 (1961).
[4] a. C. Paal and W. N. Yao, *Ber.*, **63**, 57 (1930); b. G. F. Grillot, *J. Am. Chem. Soc.*, **66**, 2124 (1944); c. H. R. Nace and E. P. Goldberg, *J. Am. Chem. Soc.*, **75**, 3646 (1953).
[5] A. F. Graefe, *J. Org. Chem.*, **23**, 1230 (1958).
[6] a. E. Lorz and R. Baltzly, *J. Am. Chem. Soc.*, **73**, 93 (1951); b. M. Harfenist and E. Magnien, *J. Am. Chem. Soc.*, **79**, 2215 (1957).
[7] H. Suessenguth, H. Meier, H. Corr, and W. Simon, U.S. Patent 3,255,248 (1966).
[8] a. G. V. Mock, U.S. Patent 3,187,051 (1965); b. J. B. Tindall, U.S. Patent 3,178,479 (1965); c. J. W. Churchill and J. F. Haller, Brit. Patent 801,534 (1958); d. D. R. Levering and L. G. Maury, Brit. Patent 797,483 (1958).
[9] a. W. B. Tuemmler and H. J. S. Winkler, U.S. Patent 2,979,505 (1961); b. D. A. Lima, U.S. Patent 3,154,538 (1964); c. D. N. Thatcher, U.S. Patent 3,102,887 (1963); d. G. W. Smith and D. N. Thatcher, *Ind. Eng. Chem. Prod. Res. Dev.*, **1**, 117 (1962).

[10] K. Klayer, E. M. Wilson, and G. K. Helmkamp, *Ind. Eng. Chem.*, **52**, 119 (1960).
[11] M. J. Kalm, *J. Med. Chem.*, **7**, 427 (1964).
[12] I. L. Mador and L. J. Rekers, U.S. Patent 2,950,954 (1960).
[13] J. Meinwald, Y. C. Meinwald, and T. N. Baker, III, *J. Am. Chem. Soc.*, **86**, 4074 (1964).
[14] R. Holbein and R. Perrot, *Ann. Sci. Univ. Besancon Chim.*, [2], **2**, 113 (1956); through *Chem. Abstr.*, **53**, 6120 (1959).
[15] M. Ishidate, Y. Sakurai, and Y. Kuwada, *Chem. Pharm. Bull. (Tokyo)*, **8**, 543 (1960).
[16] a. K. Tomino, *Chem. Pharm. Bull. (Tokyo)*, **6**, 648 (1958); b. M. Bianchi, *Ann. Chim.* (Rome), **56**, 151 (1966); through *Chem. Abstr.*, **64**, 17571 (1966); c. K. M. Hammond, N. Fisher, E. N. Morgan, E. M. Tanner, and C. S. Franklin, *J. Chem. Soc.*, **1957**, 1062; d. H. S. Forrest, R. Hull, H. J. Rodda, and A. R. Todd, *ibid*, **1951**, 1.
[17] a. Z. B. Papanastassiou, A. McMillan, V. J. Czebotar, and T. J. Bardos, *J. Am. Chem. Soc.*, **81**, 6056 (1959); b. P. B. Russell, G. B. Elion, and G. H. Hitchings, *ibid*, **71**, 474 (1949); c. J. R. Vaughan, Jr., J. Krapcho, and J. P. English, *ibid*, p. 1885.
[18] a. I. A. Pearl, *J. Org. Chem.*, **10**, 209 (1945); b. E. A. Goldsmith and H. G. Lindwall, *ibid*, **18**, 507 (1953); c. T. J. Bardos, D. B. Olsen, and T. Enkoji, *J. Am. Chem. Soc.*, **79**, 4704 (1957); d. E. F. Elslager, D. B. Capps, D. H. Kurtz, F. W. Short, L. M. Werbel, and D. F. Worth, *J. Med. Chem.*, **9**, 378 (1966).
[19] A. Kreutzberger and P. A. Kalter, *J. Org. Chem.*, **26**, 3790 (1961).
[20] a. P. Ruggli and G. Bartusch, *Helv. Chim. Acta*, **27**, 1371 (1944); b. J. N. Ashley and S. S. Berg, *J. Chem. Soc.*, **1957**, 3089; c. K. N. Campbell, J. F. Kerwin, A. H. Sommers, and B. K. Campbell, *J. Am. Chem. Soc.*, **68**, 1559 (1946).
[21] a. L. Birkhofer, *Ber.*, **80**, 83 (1947); b. J. H. Looker and D. N. Thatcher, *J. Org. Chem.*, **22**, 1233 (1957); c. I. Sallay, F. Dutka, and G. Fodor, *Helv. Chim. Acta*, **37**, 778 (1954).
[22] T. E. Stevens and J. P. Freeman, *J. Org. Chem.*, **29**, 2279 (1964).

XII

NITRILES

In general the hydrogenation of a nitrile $RC\equiv N$ to primary amine is readily carried out. The major side product is secondary amine formed through the series of reactions [1] in Eq. 12.1. A second mechanism which requires

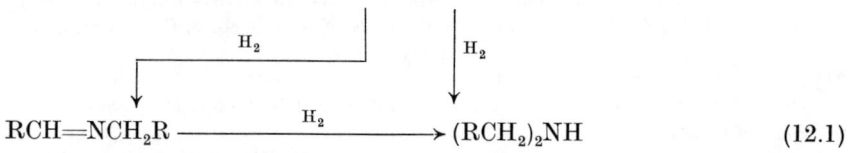

$$RC\equiv N \xrightarrow{H_2} RCH=NH \xrightarrow{H_2} RCH_2NH_2$$
$$RCH=NH + RCH_2NH_2 \rightarrow RCH(NH_2)NHCHR$$

$$RCH=NCH_2R \xrightarrow{H_2} (RCH_2)_2NH \tag{12.1}$$

the presence of water (or hydroxide ion) involves hydrolysis of aldimine to aldehyde and reaction of it with primary amine to form a Schiff base followed by hydrogenation [2].

CONTROL OF SECONDARY AMINE

Secondary amine formation can be minimized by carrying out reductions in the presence of ammonia. The rationale is that aldimine adds ammonia to form $RCH(NH_2)_2$ which gives primary amine upon hydrogenation [3].

The importance of excess ammonia, at least five molar equivalents, has been noted in reductions with Raney nickel [4] and with rhodium [5]. However, in the hydrogenation of 3,4-methylenedioxybenzyl cyanide in dioxane solution with Raney cobalt, equally high yields of primary amine were obtained in the absence or presence of ammonia [6]. There may be some dependence on nonhydroxylated solvent in cobalt reductions since reaction without ammonia in absolute alcohol gave much lower yield. On the other hand the use of ammonia in Raney cobalt hydrogenations of isophthalodinitrile in various solvents (dioxane was not included) caused increases in the yield of 1,4-*bis*-(aminomethyl)benzene [7]; the highest yield was obtained from hydrogenation in liquid ammonia.

Another cobalt catalyst, Urushibara cobalt, readily prepared from cobalt chloride and zinc dust, did not require ammonia to obtain 80–85% yields of primary amine from reductions of nitriles [8].

Hydrogenation in the absence of ammonia has yielded a predominance of primary amine when nickel was employed and sodium hydroxide was present as co-catalyst [9a, b, c]. Fluchaire and Chambret [9d] showed the effect of alkali by conducting a reduction of benzyl cyanide with Raney nickel washed free of it; 51.2% of 2-phenethylamine and 37.5% of *bis*-(2-phenethyl)amine were obtained. When 2% sodium hydroxide solution was added to similar reductions, yields of primary amine were raised to 92.5–95.5%. Some di- and triphenylacetonitriles were reduced to primary amines in good yield in alkaline solution with a large amount of platinized nickel catalyst [10a]. In one example methyl 1-cyano-1,1-diphenyl acetate was hydrogenated to primary amine with Raney nickel in the absence of ammonia in a slow reaction under 3 atm pressure [10b]. It is possible that occluded alkali in the nickel aided formation of primary amine but it is more likely that steric effects did and will prevent secondary amine formation in reduction of a cyano group attached to a tertiary carbon atom.

Secondary amine formation can be prevented or minimized by binding the primary amine as it is formed during the reduction process. Hydrogenations in acetic anhydride over platinum oxide led to good yields of acetylated primary amines [11a]. High yields were also obtained in reductions over nickel at 50° and 3.3 atm pressure in acetic anhydride in the presence of sodium acetate or sodium hydroxide as promoter or co-catalyst [11b]. The presence of sodium hydroxide made considerable reuse of the catalyst possible. The use of formamide or alkyl formates in nickel-catalyzed reductions of nitriles at elevated temperature and pressure gave high yields of formylated amines [12a, b].

Inhibition of secondary amine also results from hydrogenations of nitriles in acidic media with palladium or platinum catalysts. The method of Hartung [13], reduction in alcoholic hydrogen chloride with palladium on carbon catalyst, gives essentially primary amine [14a, b]. In one example using this medium, platinum oxide was substituted because reaction over palladium catalyst was too slow [14a]. Hydrogenation of benzonitrile in acetic acid over palladium on barium sulfate gave over 80% yield of benzylamine [15a]. In the reduction of phenylacetonitrile to 2-phenethylamine best results were obtained in acetic acid containing sulfuric acid or dry hydrogen chloride [15a]. Smooth hydrogenations of nitriles in acetic acid are reported in the presence of platinum oxide [15b, c]. The use of this catalyst in acetic acid containing sulfuric acid gave 90% yield of ethyl-3-amino-2-cyclohexylpropionate from the corresponding cyano ester [15d].

ALIPHATIC AND AROMATIC MONONITRILES

Hydrogenations of cyano compounds, RCN where R is an aliphatic or aromatic group or where an oxygen bridge exists between the cyano group and the R group, may be carried out by any of the previously described methods.

Reduction of these nitriles to primary amines have been carried out in good yield with Raney nickel and ammonia under elevated temperature and pressure conditions [16a, b]. There are instances where these hydrogenations are successful at low pressure. A specially prepared Raney nickel, W-7, which retains its activity for only a short period, catalyzed reduction at low pressure [17a] as did excessive amounts of normally active Raney nickel [17b, c]. Low-pressure nickel reductions of cyclohexyl, cyclohexylalkyl, and 3-phenoxypropyl cyanides were reported but the amounts of catalyst were not disclosed [17d]. Nitriles, pretreated in solution with Raney nickel and activated carbon to remove contaminants, were converted to primary amines in over 80% yield by hydrogenation with additional nickel [18]. Since many cyano compounds are prepared by reaction of organic halides with sodium cyanide, routine pretreatment of a solution of cyano compound with 5–10% weight of Raney may be advisable to prevent inhibition of the activity of any catalyst that is to be employed.

The addition of sodium hydroxide has co-catalyzed nickel reductions of nitriles in low-pressure reactions. However, its use has not been explored with too great a variety of cyano compounds. Nickel-catalyzed low-pressure hydrogenations in acetic anhydride in the presence of strong base or sodium acetate [11b, 19] may be a very promising method of obtaining primary amine even with the added step of hydrolysis of the acetylamine.

Rhodium-catalyzed hydrogenations of cyanoethyl ethers in ammoniacal solution under 3 atm pressure gave predominantly primary amines [5]. In the reduction of nitriles, RCN where R was phenyl or benzyl, it was less efficient. Hydrogenation of benzonitrile as a model compound gave too much dibenzylamine [20]; reduction of phenylacetonitrile under similar conditions led to equal amounts of primary and secondary amine [18]. Rhodium on carbon was found particularly effective in the conversions of ammonium 2-cyanoethylsulfonate to 3-aminopropylsulfonic acid, homotaurine [21a], and of sodium 2-aminopropylsulfonate in dilute aqueous ammonia [18].

Rhodium hydroxide and a mixed 70:30 rhodium-platinum oxide gave primary amines in good yields from reductions of adiponitrile and some aromatic nitriles at 85–100° and 80–100 atm [21b]. Lithium hydroxide promoted reductions with rhodium hydroxide and in some instances with the mixed oxides. Its use in the hydrogenation of benzonitrile led to substantial amounts of hexahydrobenzylamine. Sodium hydroxide promoted the same reduction but inhibited ring reduction to give 91% of benzylamine. However, only lithium hydroxide was useful for the rhodium hydroxide or rhodium-platinum oxide hydrogenations of isophthalonitrile and terephthalonitrile to the corresponding xylenediamines.

The method of Hartung [13], hydrogenation in alcoholic hydrogen chloride with palladium on carbon at 3 atm pressure, has been applied to nitriles RCN, where R was phenyl or substituted phenyl [13, 14a, b, 20, 21c], where it was aralkyl [22a, b], and where it was cyclobutyl [23]. Before 2-cyanobenzenesulfonamide was similarly hydrogenated it was first pretreated in solution with palladium on carbon [21c]. When R was an alkyl chain, platinum oxide was substituted because hydrogen uptake was too slow with palladium [14a]. Platinum oxide was also employed in the reduction of a substituted cyclohexyl cyanide [24].

The method of Rosenmund and Pfannkuch [15a], reduction in acetic acid or in acetic acid containing mineral acid with palladium on barium sulfate or with platinum oxide [15b, c, d], has resulted in the conversion of aliphatic and aromatic nitriles to primary amines [15b, c, d].

CYANO ACIDS AND ESTERS

Hydrogenations of α-cyano acids and esters as a means of obtaining the corresponding amines usually require elevated temperature and pressure when Raney nickel is employed. However, low-pressure reduction with nickel in acetic anhydride and sodium acetate has produced good yield of amino acid after hydrolysis [19]. The method of choice for this group of cyano compounds may be platinum-catalyzed hydrogenation in acetic acid containing sulfuric acid [15a, 25]. Palladium on carbon reduction in alcohol containing hydrochloric acid may also be useful [22b], although there is a report where an equal weight of catalyst was employed (26).

Reductions of β- and γ-cyano esters usually lead to the corresponding pyrrolidones and piperidones [27a, b]. Nickel-catalyzed hydrogenations of this group are carried out at 70–100° atm pressure in the absence of ammonia. Ammonia is not needed to prevent secondary amine formation because as primary amine is formed it reacts intramolecularly with the ester group to give the cyclic amide [27b]. Cobalt catalyst also has been employed [28b].

Reductions of γ-cyano esters over copper chromite at 200° and 175 atm yielded piperidines [28b]. 2-Piperidones were probably intermediates, but because of the reaction conditions were further converted to piperidines. A complicating side reaction was N-alkylation caused by the high temperature and release of alcohol from cyclization. Low-pressure hydrogenations of β- and γ-cyano esters also yielded cyclic amides. Ethyl 2-(2-pyridyl)-4-cyanobutyrate was converted to 3-(2-pyridyl)-2-piperidone in 85% yield in a reaction under 4 atm pressure [29a]; methyl 4-cyano-4-phenylbutyrate was reduced to 5-phenyl-2-piperdione at atmospheric pressure with an equal weight of catalyst [29b]. Hydrogenation of ethyl 2-cyano-4-phenylglutarate in acetic acid with nickel at 4 atm pressure gave 75% yield of 3-carbethoxy-5-phenyl-2-piperidone [29a].

Cyclization was prevented by carrying out the reaction in the presence of strong acid as in the platinum-catalyzed reductions of a β- and γ-cyano ester in acetic acid containing sulfuric acid [30a, b].

Another procedure, reduction in acetic anhydride, also prevented cyclization as in the platinum-catalyzed hydrogenation of ethyl 1-acetylamine-1-(3-cyanopropyl)malonate to ethyl 1-acetylamino-1-(3-acetylaminopropyl)-malonate [30c].

AMINONITRILES

α-Aminonitriles

There does not appear to be a particular catalyst or set of reaction conditions applicable to the hydrogenation of all α-aminonitriles to 1,2-diamines. Reduction is complicated by the liberation of hydrogen cyanide [31], a catalyst poison, or by cleavage, [27a];

$$(R)_2N \mid CH_2CN \xrightarrow{H_2} (R)_2NH + CH_3CN$$

The method of choice for the conversion of α-tertiaryaminonitriles, $(R)_2NCH_2CN$ where R is alkyl or where $(R)_2N$ represents a saturated heterocycle, to 1,2-diamines in good yield may be reduction in alcoholic ammonia with rhodium on alumina or rhodium on carbon at room temperature and 2–3 atm pressure [5]. The procedure was not successful with α-secondary aminonitriles, $RNHCH_2CN$. Hydrogen uptake was incomplete in the reduction of methylaminoacetonitrile; hydrogenation of cyclohexylaminoacetonitrile gave only cyclohexylamine [18].

Mixed results have been obtained in reductions of α-tertiaryaminonitriles in ammoniacal solution with Raney nickel under more vigorous reaction conditions. Hydrogenation of some α-piperidinonitriles resulted in rapid hydrogenolysis at the piperidino linkage [27a]; reduction of α-piperidinophenylacetonitrile at 85° and 50 atm pressure gave 1-benzylpiperidine as the major product [18]. Diethylaminoacetonitrile has been reported to undergo hydrogenolysis with nickel under a variety of conditions [32]. Meager yields of N,N-diethylenediamine from it have been reported [27a, 33a]. A procedure of cooling a mixture of nitrile and nickel catalyst to 30° before adding ammonia and solvent (ether) and subsequent rapid heating to 140° and 100 atm of hydrogen gave 61% yield of the 1,2-diamine [33b]. Despite its propensity toward hydrogenolysis, hydrogenation of the same nitrile with Raney cobalt in anhydrous cyclohexane and ammonia under elevated temperature and pressure gave 95% of diamine [34]; frequent reuse of the catalyst was noted and avoidance of water or alcohol in reduction was stressed. The use of methyl alcoholic ammonia in the nickel-catalyzed hydrogenation of morpholinoacetonitrile to N-(2-aminoethyl)morpholine [35a] or of ethyl alcoholic ammonia in the conversion of 1-cyanomethyl-4-(2-hydroxyethyl)-piperazine to 1-(2-aminoethyl)-4-(2-hydroxyethyl)piperazine [35b] resulted in good yields. In the reduction of other α-tertiaryaminonitriles, $RR'NCH_2CN$ where R and R' were dissimilar, with nickel in ether and ammonia at 125° and 80–95 atm, yields of diamines ranged from 41–88% [35c]. The effect of structure, catalyst, and reaction conditions has yet to be completely resolved in the hydrogenation of α-tertiaryaminonitriles.

At best reductions of α-secondary aminonitriles give only fair yields of 1,2-diamines, $RNH(CH_2)_2NH_2$. Methyleneaminoacetonitrile was hydrogenated with Raney nickel and ammonia to give 66% of N-methylethylenediamine [27a]. Reduction of nitriles, $(R)_2N(CH_2)_nNHCH_2CN$, in ammoniacal solution with nickel at 80–100° and 100–120 atm pressure yielded 35–51% of the corresponding diamines when n was 2 or 3 and R was methyl or $(R)_2N$ was piperidino; where n was 1 very low yields resulted [36a]. Reduction of α-anilinophenylacetonitrile in an acetic anhydride with platinum oxide under 2–3 atm pressure or of the N-acetyl compound gave 20% of (2-anilino-2-phenyl)ethylamine, 60% of aniline, and 10% N-benzylamine [36b].

Several methods of reducing α-primary aminonitriles to 1,2-diamines are available. One requires hydrogenation in acetic anhydride in the presence of platinum oxide under 2–3 atm pressure [31]. Reaction time is long but yields of N^1,N^2-diacetyl compounds are high. Reductions of acetylated aminonitriles in alcoholic ammonia over Raney nickel at 90° and 130 atm decreased the reaction time considerably [37a]. Hydrogenation under these conditions results in the formation of dihydroimidazoles which must be hydrolyzed to obtain the corresponding diamines (Eq. 12.2). A modification using Raney nickel (0.8 g per 1.0 g of acetylated or benzoylated aminonitrile) in alcohol containing 0.1 g of sodium hydroxide per 5.0 g of substrate gave very rapid reductions to yield products, ArCH(NRAcyl)CH$_2$NH$_2$, where R was H or alkyl [37].

$$RR'C(NHCOCH_3)CN \xrightarrow{2H_2} \begin{bmatrix} & CH_3 \\ & | \\ & NHC=O \\ RR'C & \\ & CH_2NH_2 \end{bmatrix}$$

$$\begin{array}{c} R' \\ R-\overset{|}{\underset{\underset{CH_3}{|}}{N}}\!\!\!\!-\!\!\!\!\underset{}{N} \end{array} \xrightarrow[\text{(2) Excess NaOH}]{\text{(1) Hydrolysis (HCl)}} RR'C(NH_2)CH_2NH_2 \qquad (12.2)$$

Primary 1,2-diamines have been prepared directly by hydrogenation of the aminonitrile in alcoholic hydrogen chloride in the presence of platinum oxide under 3 atm pressure [38]. No uptake of hydrogen occurred during similar reductions with palladium on carbon [18]. The method with platinum was not successful in attempted hydrogenations of α-tertiaryaminonitriles. 1-Cyclohexyl-2,5-dicyano-2,5-dimethylpyrrolidine underwent cleavage [38] as did piperidinoacetonitrile [18]. However, it may be worth investigating for reductions of α-secondaryaminonitriles since cyclohexylaminoacetonitrile was converted to N-(cyclohexyl)ethylenediamine in 84% yield [18].

β, γ, δ, and Other Aminonitriles

There are a number of useful procedures for the hydrogenation of aminonitriles, RR'N(CH$_2$)$_n$CN where $n = 2$ or more. Low-pressure rhodium-catalyzed reductions in alcoholic ammonia have given excellent results [5]. 3-(Cyclohexylamino)propionitriles were converted to the corresponding 1,3-propanediamines in high yield by means of this method [18]. When there is no danger of ring reduction, higher temperature and pressure may be applied in rhodium-ammonia reactions to achieve very rapid hydrogen uptake [18].

Hydrogenation in ammoniacal solution with Raney nickel under elevated temperature and pressure is a widely used procedure which generally gives good results [4, 39]. However, reduction of 1-(2-cyanoethyl)- and 1-(3-cyanopropyl)piperidines by this method resulted in hydrogenolysis at the piperidine nitrogen [26a], possibly because of reaction conditions. The conditions also appeared to have some effect on the hydrogenation of *tris*-(2-cyanoethyl)isocyanuric acid. Reduction at 80° and 97 atm yielded

57% of *bis*-3-aminopropylisocyanuric acid and propylene-1,3-diamine while reaction at 165° and 130 atm gave only 1-(3-aminopropyl)isocyanuric acid [40a] (Eq. 12.3).

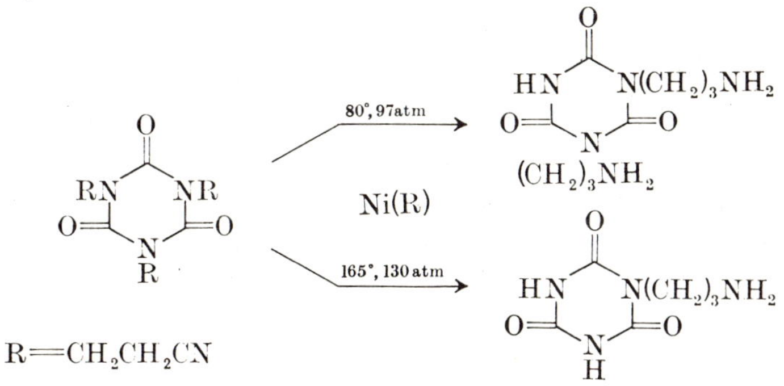

$$\text{(Eq. 12.3)}$$

R=CH$_2$CH$_2$CN

Although Raney cobalt and other cobalt catalysts have been employed for reduction of aminonitriles where $n = 2$ or more, there does not appear to be an advantage in replacing nickel with them in reactions at elevated temperature and pressure.

Low-pressure reductions with nickel and ammonia were employed for the reduction of some ε-tertiaryaminonitriles [41a, b]; hydrogenations in the absence of ammonia [42a, b] also produced reasonably good yields of the corresponding diamines.

Hydrogenations of similar type compounds in alcoholic hydrogen chloride with high catalyst ratios of platinum oxide under 2–3 atm pressure resulted in good yields of diamines [43a, b]. In most instances a high ratio was necessary when platinum oxide was employed [18].

Cyclization, through intramolecular condensation, can take place during hydrogenation of primary or secondary aminonitriles when three to four carbon atoms separate the amino and cyano groups. For example, 2-amino-4,5-dimethoxyphenylacetonitrile was converted to 6,7-dimethoxyindole by reduction over Raney nickel [44] (Eq. 12.4).

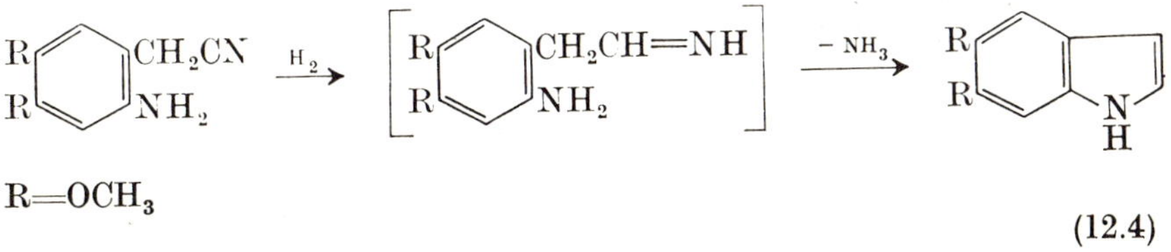

R=OCH$_3$

(12.4)

Hydrogenation of ethyl 4-benzylamino-2-cyano-2-phenylbutyrates and 5-benzylamino-2-cyano-2-phenylpentanoates,

$$C_6H_5CH_2NR(CH_2)_{2-3}C(Ar)(CN)COOC_2H_5$$

where R is hydrogen or methyl and Ar is phenyl or substituted phenyl, with palladium on carbon in alcoholic hydrogen chloride yielded pyrrolidines and piperidines [45a, b, c]. It was reported by another investigator that the method was unsatisfactory for the preparation of pyrrolidines because of poor yield [45d].

An attempt to cyclize N-(2-cyanoethyl)ethylenediamine to a 7-membered ring during reduction was generally unsatisfactory. Special conditions, continuous feeding of a dilute solution to Raney nickel or nickel on kieselguhr in a stirred autoclave at 130° and 45–100 atm, gave only 32% yield of 1,4-diazacycloheptane (homopiperazine) [46]. Other catalysts, Raney cobalt, copper chromite, and palladium on carbon were ineffective.

CYANOHYDRINS

The hydrogenation of mandelonitriles, RCHOHCN where R is phenyl or substituted phenyl, has been carried out in acidic medium with palladium or platinum catalyst [47a, b]. It is not surprising that hydrogenolysis of the hydroxyl group occurred in addition to hydrogenation to give phenethylamines since acidic conditions favor removal of benzylic-type hydroxyl groups (see Chapter XIX).

Aliphatic-type cyanohydrins tend to dissociate on standing to carbonyl compound and hydrogen cyanide. The effect of the latter can be responsible for the larger than normal amount of catalyst used for hydrogenations to these cyanhydrins to aminoalcohols. They should be freshly prepared and/or distilled immediately before reduction. If unstable, an alternate procedure is to pretreat a solution of substrate with Raney nickel and activated carbon before hydrogenation [18]. Reductions are carried out in acetic acid at room temperature and 2-3 atm pressure with platinum oxide [15c, 48a] or in acetic acid containing hydrochloric acid [48b, c]. In many instances although yields are good, reaction time is overlong. The use of a 3% ratio of platinum oxide, prereduced before use, as well as purity of substrate, may have been responsible for the short reaction period in the hydrogenation of tropinone cyanhydrin in glacial acetic acid to 3-aminomethyl-3-tropanol [49],

$$\underset{H}{\overset{H}{\diagdown}}C\boxed{NCH_3}C\underset{CN}{\overset{OH}{\diagup}} \xrightarrow{2\,H_2} \underset{H}{\overset{H}{\diagdown}}C\boxed{NCH_3}C\underset{CH_2NH_2}{\overset{OH}{\diagup}} \quad (12.5)$$

(Eq. 12.5). In the reduction of 1-cyanocycloalkanols the reactions were partially poisoned with carbon disulfide or hydrogen cyanide (probably to prevent hydrogenolysis of the OH group); 45–80% yields of 1-aminomethylcycloalkanols were obtained [50].

DINITRILES

Hydrogenation methods used for mononitriles can also be applied to dinitriles. It is possible to reduce one cyano group selectively but reaction usually leads to diamines, except in instances where intramolecular condensation takes place to form 5-, 6-, and sometimes 7-membered saturated heterocyclic rings.

In the nickel-catalyzed hydrogenation of succinonitirile in ammoniacal solution at 140° and 115 atm pressure, 45% of pyrrolidine and 20% of putrescine, $H_2N(CH_2)_4NH_2$, were obtained [51a]; reductions of ethyl 2,3-dicyano-2-phenylpropionate and 2,4-dicyano-2-phenylbutyrate gave the corresponding ethyl 3-phenylpyrrolidine- and piperidine-3-carboxylates [51b]. Hydrogenation of 3-hydroxyglutaronitrile with nickel gave a predominance of 3-hydroxypiperidine [52a], (Eq. 12.6). Rapid hydrogenation in the absence of ammonia yielded increased amounts of cyclized compound.

$$NCCH_2CHOHCH_2CN \xrightarrow[Ni, 50°, 50\,atm]{4H_2} \underset{\underset{H}{N}}{\bigcirc}^{OH} + H_2N(CH_2)_2CHOH(CH_2)_2NH_2$$

(12.6)

Reduction of 2-phenyl or 2-substituted phenylglutaronitriles in dioxane with nickel and ammonia at 120–140° and 110–120 atm pressure gave 3-phenylpiperidines [52b].

Formation of 7-membered rings from hydrogenation of dinitriles does not take place too readily. Reduction of phthalodinitrile in alcoholic ammonia with nickel at 75 atm initial pressure and 100° was reported to give 4,5-benzohexamethylenimine [53a] (Eq. 12.7).

$$\underset{CH_2CN}{\underset{CH_2CN}{\bigcirc}} \xrightarrow[Ni, 100°, 75\,atm]{4H_2} \bigcirc\!\!-\!\!NH$$

(12.7)

In similar reductions of phthalo- and substituted phthalodinitriles, bis-1,2-(2-(2-aminoethyl)benzenes were obtained in addition to much lower yields of benzohexamethylenimines [53b, c]. Hydrogenation of 1,4-dicyanobutene in tetrahydrofuran and ammonia with nickel at 100–120° and 50 atm pressure gave a 12:1 ratio of hexamethylenediamine to hexamethylenimine [53d].

Cyclization did not take place in a cobalt reduction of succinonitrile with a large excess of ammonia when reaction was carried out at 80° and 700 atm pressure; tetramethylenediamine was obtained in 68.5% yield [54a]. On the other hand the maximum yields of pentamethylenediamine from hydrogenation of glutaronitrile at 170° and 250 atm pressure was 41.5% with a cobalt catalyst [54b]. Use of high dilution may be a means of preventing cyclization during reduction [55a]. A more practical method, seen in the hydrogenation of 2-arylsuccinonitriles [43a, 55b], may be reaction in a strongly acidic medium to bind primary amine as it is formed. Reduction of 4-(2-cyanoethylamino)butyronitrile, $NC(CH_2)_2NH(CH_2)_3CN$, in alcoholic hydrogen chloride over platinum oxide gave 1,8-diamino-4-azaoctane (spermidine), $H_2N(CH_2)_4NH(CH_2)_3NH_2$ [55c].

Hydrogenations of α-aminodinitriles, $HN(CHR'CH)_2$ do not lead to cyclized products as readily as reductions of succino- and glutaronitriles. In most instances yields of piperazines from them are low. An 85% yield of piperazine was reported from reduction of N,N-bis-(cyanomethyl)methylamine in a flow system over Raney nickel at 90–105° and 300 atm pressure in the presence of ammonia [56]. Other methods have produced poor yields.

Reduction of a dinitrile may be interrupted after absorption of two molar equivalents of hydrogen to yield the corresponding amino nitrile as the major product. Nickel-catalyzed hydrogenation of adiponitrile in alcoholic ammonia gave primarily 6-aminocapronitrile [57a]. In another reaction equal weights of adiponitrile and aminocapronitrile were obtained [57b]. Hydrogenation of 1,4-dicyanocyclohexane with or without ammonia in the presence of Raney cobalt at 125° and 100 atm yielded 61.9% of 4-aminomethyl-1-cyanocyclohexane [57c]; nickel reduction in the presence of ammonia gave 70% of diamine and 25% of aminonitrile. In the aromatic series terephthalodinitrile was converted to 4-cyanobenzylamine in 74% yield by hydrogenation over palladium on carbon at 10–50° and 3–40 atm pressure [57d]; lower yield was obtained in the reduction of isophthalodinitrile. Reduction of the dinitrile in a flow system with palladium on platinum catalyst in the presence of ammonia gave very high yields of 3- and 4-cyanobenzylamines [57e].

ALDEHYDES FROM NITRILES

When nitriles are hydrogenated in the presence of water aldehydes are formed by rapid hydrolysis of aldimine [58a, b].

$$RCN \xrightarrow{H_2} RCH + NH \xrightarrow{H_2O} RCHO + NH_3$$

Reactions are often carried out in the presence of phenylhydrazines or semicarbazide to bind the aldehyde before it can be reduced to alcohol [59a–e]. In these reductions [59a–e] Raney nickel was employed at 25–70° and 60–100 atm pressure; yields of aldehyde derivatives were 50–70% except in the attempt to obtain 6-purinecarboxaldehyde semicarbazone from 6-cyanopurine where the yield was 24% [59c].

Hydrogenation in the absence of a carbonyl binding agent requires the presence of acid to neutralize ammonia to prevent it from reacting with the aldehyde to re-form aldimine. Reduction of benzonitrile and phthalodinitrile in dilute sulfuric acid with nickel at 45° and slightly above atmospheric pressure gave benzaldehyde and phthalaldehyde respectively; oxalic acid was employed in the reduction of 3-cyanopyridine to pyridine-3-carboxaldehyde in 70% yield [60a]. Carbon dioxide was present in a nickel-catalyzed reduction which gave 72% yield of pyridine-3-carboxaldehyde [60b].

Aminopentoses and hexoses have been obtained from low-pressure palladium-catalyzed hydrogenations of aminonitriles in dilute hydrochloric acid [61a, b, c]. Similar hydrogenations of α-cyano esters,

$$(R)_2C=C(CN)COOC_2H_5 \quad \text{and} \quad (R)_2CHCH(CN)COOC_2H_5,$$

gave amines or alcohols (through aldehydes) dependent on the structure of the substrate. It was not possible to apply the procedure to the formation of benzaldehyde or benzyl alcohol from benzonitrile [18]; only primary and/or secondary amine was obtained.

SECONDARY AND TERTIARY AMINES FROM REDUCTION OF NITRILES

It has been shown that a predominance of secondary amine, $(RCH_2)_2NH$, results from hydrogenation of nitriles in neutral media in the absence of ammonia. If other amines are substituted for ammonia in the reduction it is possible to produce secondary and tertiary amine, RCH_2NHY and RCH_2NYZ where RCH_2, Y, and Z may be dissimilar.

Hydrogenation of 3,4-dimethoxyphenylacetonitrile in pure benzene containing methylamine in the presence of palladium black under 2 atm pressure gave 78% yield of N-methyl-2-(3,4-dimethoxyphen)ethylamine [63a]; reduction in the presence of diethylamine gave the corresponding N,N-diethyl derivative. Reductions with palladium on barium sulfate in the presence of other secondary amines yielded tertiary amines [63b]. Hydrogenations of nitriles with Raney nickel and primary and secondary amines have been carried out under varying degrees of temperature and pressure [64a, b, c]. The use of salts of secondary amines were found necessary to obtain high yield of tertiary amines during reduction [64d]. In a Raney nickel reduction of phenylacetonitrile and excess dimethylamine and an equal amount of acetic acid yield of N,N-dimethyl-2-phenethylamine was 85% [64d]. When

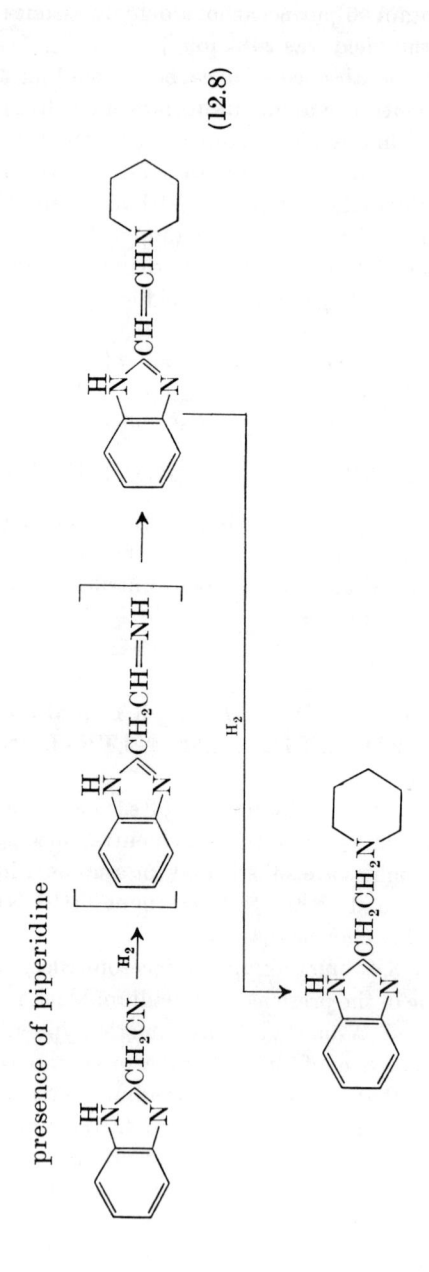

presence of piperidine

$$\text{benzimidazole-CH}_2\text{CN} \xrightarrow{H_2} [\text{benzimidazole-CH}_2\text{CH=NH}] \rightarrow \text{benzimidazole-CH=CHN(piperidine)}$$

$$\downarrow H_2$$

$$\text{benzimidazole-CH}_2\text{CH}_2\text{N(piperidine)}$$

(12.8)

the reaction was repeated in the absence of acid, only 37.5% of tertiary amine was obtained.

In the nickel-catalyzed hydrogenation of 2-cyanomethylbenzimidazole in the presence of piperidine, tertiary amine was not obtained directly [64c]. Instead 2-(1-piperidinovinyl)benzimidazole was isolated and later converted to tertiary amine through reduction with palladium (Eq. 12.8).

In many instances primary amine still results from reduction of nitrile even in the presence of other primary amines or secondary amines. Separation of it from the desired secondary or tertiary amine is sometimes difficult because of small differences in boiling points [18, 64e]. It may be necessary to corroborate results by chromatography to insure that the product of reduction is essentially one component [18].

SELECTIVE REDUCTIONS

Preferential hydrogenation of the cyano group should not be expected over that of a carbon to carbon triple bond or of a nitro group attached to an aromatic ring. In most instances olefinic bonds will also be reduced first (Chapter IX). There are exceptions, however, in the latter case. The cyano group has been selectively hydrogenated in certain unsaturated cyanoacetic esters, $RR'C=C(COOC_2H_5)CN$ where the double bond was tetrasubstituted [62]. The conversion of 2,3,3-triphenylpropenonitrile, also containing a tetrasubstituted bond to 2,3,3-triphenylpropylamine was accomplished with considerable difficulty; successful reduction in dioxane containing sodium hydroxide required twice as much catalyst as compound [65a]. It is likely that the difficulty in saturating the double bond was not only due to steric effects but also to the use of alkali as a promoter for the reduction of the cyano group which at the same time acted as an inhibitor for the hydrogenation of the double bond. A molecular model shows free access of the terminal cyano group to the catalyst and the hindering effect of the phenyl groups on the double bond. It suggests that the cyano group was reduced first.

Oleo- and eruconitriles, each of which contains a terminal cyano group and a symmetrically disubstituted double bond, were converted to unsaturated amines by hydrogenation in alkaline solution with Raney nickel [65b]. This appears to be another example favoring preferential reduction of a terminal cyano group due to promotion by alkali and poisoning by it of the catalyst sites for double bond absorption since neither bond is too hindered.

Preferential hydrogenation of a terminal nitrile function has also taken place in reductions of cyanocyclohexenes and cyclohexenylacetonitriles to the corresponding aminoalkylcyclohexenes with nickel and cobalt catalysts (usually in the presence of ammonia).

In contrast to the platinum-catalyzed reduction of 1-cyanocyclohexene in neutral medium to cyanocyclohexane and di(cyclohexylmethyl)amine [66a], 4-cyanocyclohexene with a less hindered double bond was converted to 4-aminomethylcyclohexene in 77–82% yields by hydrogenation at 95–100° and 100 atm in liquid ammonia with nickel on keiselguhr, Raney nickel, or Raney cobalt [66b].

The C≡N function in cyclohexenylacetonitriles, which has better contact with the catalyst than that group in cyanocyclohexenes, was also preferentially reduced. Cyclohexeneylacetonitrile gave 2-cyclohexenylethylamine in 90% yield by hydrogenation with Raney cobalt, which contains occluded alkali, at 60° and 90 atm [66c]; Raney nickel reduction with ammonia under somewhat similar conditions gave the same product containing 2–4% of cyclohexylethylamine [66d]. In another example cyclohexenyl-3-acetonitrile was hydrogenated to 2-(3-cyclohexenyl)ethylamine with nickel and ammonia [66a].

In the Presence of Aromatic Rings

A nuclear cyano group in a benzenoid ring should undergo reduction without affecting the ring. Saturation of the ring is more likely to take place in hydrogenations in the presence of some unsaturated heterocycles. Cyanopyridines and quinolines were hydrogenated to aminomethylpyridines and quinolines with palladium and platinum catalysts in acidic medium or with nickel and ammonia [18] but reduction of cyanopyridines at 70° and 60–70 atm with ruthenium and ammonia gave aminomethylpiperidines.

Hydrogenation of pyridines as quaternary salts facilitates saturation of the ring (Chapter XXIV). The combination of quaternization and removal of conjugation between the cyano group and the ring system by insertion of an alkylene bridge reversed the normal reaction and enabled ring reduction to take place preferentially in the rhodium on carbon or platinum oxide hydrogenations of 1-alkyl-2- or 3-cyanoalkylpyridinium halides to yield the corresponding alkyl cyanoalkylpiperidines [67]. However, the cyano group was reduced first in the hydrogenation of cyanoalkylpyridines as acid salts or when there was no alkylene bridge between the ring and cyano group as in alkyl 2-, 3-, or 4-cyanopyridinium halides.

When 4-substituted or 4,6-disubstituted 5-cyanopyrimidines were reduced, the expected 5-aminomethylpyrimidines were obtained [4, 68a, b] even though in the latter two instances as much or more freshly prepared W-7 Raney nickel than compound was employed. The hydrogenation of 5-cyano-2-methylpyrimidine with excess fresh W-7 nickel gave 70% yield of 5-amino-2-methyltetrahydropyrimidine [68a]. Reduction with an undisclosed amount of platinum oxide indicated no break in the curve during uptake of almost four equivalents. It is likely that the absence of steric effects allowed ring reduction to take place.

In the Presence of Nonaromatic Rings

There are very few examples where ring rupture takes place during the reduction of saturated cyanocarbocycles. Some ring opening took place in the hydrogenation of 3,3-dimethyl-2-dimethylaminocyclobutanecarbonitriles; 64% of N-methyl-3,3-dimethyl-2-dimethylaminocyclobutylamine, and 16% of 3,3-dimethylpiperidine were obtained [68c].

A few examples among saturated heterocycles indicated no ring cleavage as 1-(2-cyanoethyl)-2,2-dimethylethylenimine was reduced to 1-(3-aminopropyl)-2,2-dimethylethyleneimine with nickel and ammonia at 100° and 100 atm pressure [68d] and 1-(2-cyanoethyl)ethylenimine was converted to 1-(3-aminopropyl)ethylenimine in 80% yield with nickel at 40–50° and 100 atm pressure in the absence of ammonia [68e]. Attempted duplication of the latter reduction resulted in only 42% yield plus considerable secondary amine but no ring-opened compound was obtained [18].

Presence of Ketones

The selective hydrogenation of the cyano group can be accomplished although there is often a tendency toward self-condensation of the resulting aminoketone to yield cyclic compounds dependent on structure and reaction conditions.

3,4-Dimethoxybenzoyl cyanide, 3,4-$(CH_3O)_2C_6H_3COCN$, was reduced in glacial acetic acid over palladium black to 2-amino-3',4'-dimethoxyacetophenone when the reaction was stopped after absorption of two molar equivalents [69a]. Hydrogenation of 4-cyano-2,2,5,5-tetramethyltetrahydrofuran-3-one with nickel and ammonia gave predominantly the corresponding 4-aminomethyl-3-keto compound [69b]; 3-cyano-2-hydroxy-2,4,6-cycloheptatrienone was reduced to 3-aminomethyl-2-hydroxy-2,4,6-cycloheptatrienone over palladium on carbon in alcoholic hydrogen chloride [69c].

The hydrogenation of ketonitriles, RCOCH(R')CN where R' is alkyl, could be controlled to yield aminoketones with nickel catalyst at 120 atm when the reaction temperature was held below 35° [69d]. Reductions of the type, $RCO(CH_2)_{2-3}CN$, yielded saturated or partially saturated 5- and 6-membered heterocycles, no doubt through an intermediate aminoketone.

Presence of Reducible Nitrogen-Containing Groups

When a compound contains a cyano group and a nonaromatic nitro group, preferential hydrogenation may depend on reaction conditions, catalyst, and the relative positions of the two groups. Selective hydrogenation of a non-aromatic nitro group may be favored (Chapter X), but a thorough investigation is necessary before a categorical statement can be made. In contrast to expected conversion of the nitro group, nitroamines were obtained from reductions in acetic or propionic acid with palladium, of nitronitriles, RR'C(NO$_2$)(CH$_2$)$_2$CN where only R was hydrogen and R and R' were alkyl or substituted alkyl or RR'C was part of an alicyclic structure [70].

The problem of selective hydrogenation when oximino and cyano groups are present in the same molecule is incompletely resolved (see Chapter XIII). When ethyl 2-oximino-3-cyano-3-phenylpropionate was reduced in glacial acetic acid with platinum oxide or in acetic anhydride only the cyano group was affected [71a]. Cyclization took place in acetic acid to yield 3-oximo-4-phenyl-2-pyrrolidone, ethyl 4-acetylamino-2-oximino-3-phenylbutyrate was obtained from the reaction in acetic anhydride (Eq. 12.9). Reduction of the

$$C_6H_5CH(CN)C(COOC_2H_5)=NOH \xrightarrow[PtO_2]{AcOH} \begin{bmatrix} C_6H_5CHCH_2NH_2 \\ | \\ HON=CCOOC_2H_5 \end{bmatrix} \longrightarrow \underset{\underset{H}{N}}{C_6H_5}\!\!\!\diagdown\!\!\!=\!\!NOH \qquad (12.9)$$

$$\xrightarrow[PtO_2]{Ac_2O} CH_3CONHCH_2CH(C_6H_5)C(COOC_2H_5)=NOH$$

corresponding cyano oximino phenylpropionamide gave 4-amino-2-oximino-3-phenylbutyramide.

When the cyano and oximino groups were on the same carbon atom as in ethyl isonitrosocyanoacetate, NCC(=NOH)COOC$_2$H$_5$, the cyano group remained intact upon hydrogenation in acetic anhydride with palladium on

$$RHNH=C\!\!\begin{array}{c} COOC_2H_5 \\ \diagdown \\ (CH_2)_2CN \end{array} \xrightarrow{2H_2} \begin{bmatrix} RHNH=C\!\!\begin{array}{c} COOC_2H_5 \\ \diagdown \\ (CH_2)_3NH_2 \end{array} \end{bmatrix}$$

R = 4-methoxyphenyl

$$\longrightarrow \underset{\underset{H}{N}}{\bigcirc}\!\!=\!\!NNHR \qquad (12.10)$$

carbon, platinum on carbon, or platinum oxide [71b]. Reaction yielded ethyl acetylaminocyanoacetate.

There are very few references relating to selectivity of imino or substituted imino nitriles upon reductions. Hydrogenations of β-iminonitriles with nickel at 120–150° and 200–300 atm pressure gave 1,3-diamines or hydrogenolysis products [72a]. It is possible that the cyano group could be selectively reduced under milder conditions. The γ-nitrile function in the 4-methoxyphenylhydrazone of ethyl 4-cyano-2-oxobutyrate was selectively hydrogenated with nickel at 50–70° and 100 atm and underwent subsequent condensation with the ester group to yield 3-(4-methoxyphenylhydrazono)-2-piperidone [72b] (Eq. 12.10).

Presence of Hydrogenolyzable Groups

N- and O-Benzyl Groups. Raney nickel, the least likely of the commonly used catalysts to cause hydrogenolysis, may be the one of choice for the selective reduction of nitriles containing an *N*-benzyl group. It has been employed successfully with ammonia in the hydrogenations of *N*-benzyl-*N*-phenylaminoacetonitrile [73a], 1-benzyl-4-cyano-4-phenylpiperidine [73b], (*N*-benzyl-*N*-phenylacetyl)aminoacetonitrile [73c], and 5-(*N*-benzyl-*N*-methylamino)valeronitrile [73d] to the corresponding amines without debenzylation.

Of the noble metal catalysts only rhodium on a carrier has allowed reduction of the nitrile function to proceed in the presence of an *N*-benzyl group without causing its removal [5]. Examples with palladium on carbon and other palladium catalysts show that the protecting group will be cleaved [45b, d, 73b]. The use of platinum oxide caused partial debenzylation in the hydrogenation of 1-benzyl-3-cyanomethyl-2-oxindole [73e]. A method employing Raney nickel and ammonia may be best for the selective reduction of nitriles containing *O*-benzyl groups.

4-Benzyloxyphenylacetonitriles were hydrogenated to 2-(4-benzyloxyphenyl)ethylamines with nickel and ammonia [74a, b]; 5-benzyloxy-3-cyanomethyl-6-methoxyindole was converted to 5-benzyloxy-6-methoxytryptamine in a similar manner [74c]. Cobalt was substituted for nickel in the reduction of 4-benzyloxy-3,5-methoxyphenylacetonitrile to the corresponding benzyloxy phenethylamine [74d].

The nickel-ammonia method has not been applied to the reduction of mandelonitriles to 2-hydroxy-2-phenethylamines. Hydrogenation with palladium or platinum in acid solution has been shown to yield phenethylamines [47a, b] except when an *ortho* substituent is present and the benzylic hydroxyl group is not acylated [47b]. In that instance aminoalcohol is obtained. Reduction of the *O*-acyl compound in neutral solution indicates

256   Nitriles

that the first stage of reaction is elimination of the O-acyl group to give phenylacetonitrile [75].

Hydrogenolysis of the benzylic hydroxyl group is apparently activated by adjacence of the cyano group. When a methylene group separated them as in 3-cyano-4-chromanol, activation was decreased so that the cyano group was preferentially hydrogenated with platinum oxide in acetic anhydride to yield 3-acetylaminomethyl-4-chromanol [76].

Halogen. Nitriles containing aromatic halogen will undergo reduction to amines without dehalogenation with most catalysts other than palladium, particularly when the halogen is chlorine. Palladium has not been employed too often because of its tendency to induce dehalogenation. It caused dehalogenation in the hydrogenations of chlorocyanopyridines [70a–d] but it must be noted that an equal weight of catalyst to substrate was present.

Reductions with nickel and ammonia at 50–100° and 50–100 atm have resulted in high yields of chloro-containing amines from chlorocyano compounds [73d, 77a, b]. Hydrogenation of 4-butylamino-7-chloro-3-cyanoquinoline with nickel and ammonia at 0.5 atm gave 99% yield of 3-aminoethyl-4-butylamino-7-chloroquinoline [77c]. Nickel was also employed in reduction of 3-cyano-1-(2-cyanoethyl)-2-(4-chlorophenyl)indole in acetic anhydride and sodium acetate at 3.3 atm pressure to give 86% yield of 3-acetamidomethyl-1-(3-acetamidopropyl)-2-(4-chlorophenyl)indole [78].

Rhodium on a carrier has been successfully applied to reductions in alcoholic ammonia without causing loss of halogen [5, 18].

Platinum oxide catalyzed the reduction of 6-chloro-4-cyanoquinoline to 4-aminomethyl-6-chloroquinoline in good yield in alcoholic hydrogen chloride [79a]. Use of platinum oxide in acetic anhydride gave 80% yield of N-acetyl-2-(3-chloro-4'-methoxybiphenyl)ethylamine from the chlorobiphenylacetonitrile [79b].

Hydrogenations of nitriles containing nuclear bromine have been carried out with Raney nickel [81a, b, c]. Low yields [81b, c] suggest that loss of bromine may have occurred. In the latter reference neutral Raney nickel was used in alcoholic ammonia to convert 2-bromo-4,5-dimethoxyphenylacetonitrile at 100° and 70 atm to the bromodimethoxyphenethylamine. The addition of potassium hydroxide to a similar experiment caused complete loss of bromine. The use of nickel in acetic anhydride [78] might prevent hydrogenolysis.

Platinum oxide reduction of 4-cyanobenzenesulfon(3',5'-dibromo)anilide in alcoholic hydrogen chloride resulted in 78% yield of 4-aminomethylbenzenesulfon(3',5'-dibromo)anilide [82].

There appear to be no reported or attempted reductions in the presence of iodine.

Aliphatic and alicyclic type nitriles containing chlorine have been reduced with nickel [83a, b]. In the hydrogenation of 4-*trans*-cyano-5-(chloromethyl)-cyclohexene the cyclic product obtained could only have resulted from the intermediate 2-aminomethylcyclohexylmethyl chloride (Eq. 12.11).

$$\text{cyclohexene-CH}_2\text{Cl, CN} \xrightarrow{3H_2} \left[\text{cyclohexane-CH}_2\text{Cl, CH}_2\text{NH}_2 \right] \longrightarrow \text{bicyclic-NH} \quad (12.11)$$

REFERENCES

[1] J. v. Braun, G. Blessing, and F. Zobel, *Ber.*, **56**, 1988 (1923).
[2] K. Kindler, *Ann.*, **485**, 113 (1931).
[3] E. J. Schwoegler and H. Adkins, *J. Am. Chem. Soc.*, **61**, 3499 (1939).
[4] W. Huber, *J. Am. Chem. Soc.*, **66**, 876 (1944).
[5] M. Freifelder, *J. Am. Chem. Soc.*, **82**, 2386 (1960).
[6] W. Reeve and W. M. Eareckson, III, *J. Am. Chem. Soc.*, **72**, 3299 (1950).
[7] D. V. Sokol'skii and F. Bizhanov, *Chem. Abstr.*, **57**, 9703 (1962).
[8] S. Saito, *J. Pharm. Soc. Japan*, **76**, 351 (1956); through *Chem. Abstr.*, **50**, 8941 (1956).
[9] a. M. Grunfeld, U.S. Patent 2,449,036 (1948); b. M. Kawakatsu, M. Taniyama, and N. Sawa, Japanese Patent 8719 (1963); through *Chem. Abstr.*, **59**, 11331 (1963); c. V. Guth, G. Leitich, W. Specht, and F. Wessely, *Monatsh*, **94**, 1262 (1963); d. Fluchaire and F. Chambret, *Bull. Soc. Chim.* (France), **1944**, 22.
[10] a. J. Decombe, *Compt. Rend.*, **222**, 90 (1946); b. R. R. Burtner and J. W. Cusic, *J. Am. Chem. Soc.*, **65**, 262 (1943).
[11] a. W. H. Carothers and G. A. Jones, *J. Am. Chem. Soc.*, **47**, 3051 (1925); b. F. E. Gould, G. S. Johnson, and A. P. Ferris, *J. Org. Chem.*, **25**, 1658 (1960).
[12] a. M. Sekiya, *J. Pharm. Soc. Japan*, **70**, 524 (1950); b. p. 551; through *Chem. Abstr.*, **45**, 5639, 5640 (1951).
[13] W. Hartung, *J. Am. Chem. Soc.*, **50**, 3370 (1928).
[14] a. E. Miller, J. M. Sprague, L. W. Kissinger, and L. F. McBurney, *J. Am. Chem. Soc.*, **62**, 2099 (1940); b. H. Dahn, P. Zoller, and U. Solms, *Helv. Chim. Acta*, **37**, 565 (1954).
[15] a. K. Rosenmund and E. Pfannkuch, *Ber.*, **56**, 2258 (1923); b. A. M. Mattocks and O. S. Hutchinson, *J. Am. Chem. Soc.*, **70**, 3516 (1948); c. A. G. Anderson, Jr. and D. O. Barlow, *J. Am. Chem. Soc.*, **77**, 60 (1955); d. F. F. Blicke and W. A. Gould, *J. Org. Chem.*, **23**, 1102 (1958).
[16] a. H. Adkins, *Reactions of Hydrogen*, University of Wisconsin Press, Madison, Wisconsin, 1937, p. 53; b. W. P. Utermohlen, Jr., *J. Am. Chem. Soc.*, **67**, 1505 (1945).
[17] a. R. E. Bowman and J. F. Cavalla, *J. Chem. Soc.*, **1954**, 1171; b. R. H. F. Manske and W. R. Ashford, *J. Am. Chem. Soc.*, **73**, 5144 (1951); c. K. Harsanyi, K. Takacs, D. Korbonits, L. Tardos, G. Leszkovszky, and I. Erdelyi, Hung. Patent 151,256 (1964); through *Chem. Abstr.*, **60**, 14431 (1964); d. C. G. Skinner, P. D. Gardner, and W. Shive, *J. Am. Chem. Soc.*, **79**, 2843 (1957).
[18] M. Freifelder, unpublished results.
[19] Y. Kakimoto and M. D. Armstrong, *J. Biol. Chem.*, **236**, 3283 (1961).

[20] M. Freifelder and Y. H. Ng, *J. Pharm. Sci.*, **54**, 1204 (1965).
[21] a. M. Freifelder and D. Cota, U.S. Patent 3,218,352 (1965); b. Y. Takagi, S. Nishimura, K. Taya, and K. Hirota, *Sci. Papers, Inst. Phys. Chem. Res.*, (Tokyo), **61**, 114 (1967); c. G. Cignarella and U. Teotina, *J. Am. Chem. Soc.*, **82**, 1594 (1960).
[22] a. F. G. Bordwell and M. Knell, *J. Am. Chem. Soc.*, **73**, 2351 (1951); b. F. Leonard, A. Wajngurt, M. Klein, and C. M. Smith, *J. Org. Chem.*, **26**, 4062 (1961).
[23] E. R. Buchman and D. R. Howton, *J. Am. Chem. Soc.*, **70**, 2517 (1948).
[24] T. Oishi, *Yagugaku Zasshi*, **85**, 382 (1965); through *Chem. Abstr.*, **63**, 5604 (1965).
[25] K. H. Schaaf and F. D. Pickel, U.S. Patent, 2,365,295 (1944).
[26] E. Testa, F. Fava, and L. Fontanella, *Ann.*, **614**, 167 (1958).
[27] a. C. F. Winans and H. Adkins, *J. Am. Chem. Soc.*, **55**, 4167 (1933); b. C. F. Koelsch, *J. Am. Chem. Soc.*, **65**, 2093 (1943).
[28] a. H. Frick and A. H. Lutz, U.S. Patent 2,784,200 (1957); b. W. Barr and J. W. Cook, *J. Chem. Soc.*, **1945**, 438.
[29] a. R. K. Hill, C. E. Glassick, and L. J. Fliedner, *J. Am. Chem. Soc.*, **81**, 737 (1959); b. A. V. El'tsov, A. G. Chigarev, and N. T. Staryka, *Chem. Abstr.*, **63**, 16299 (1965).
[30] a. A. Yokoo and S. Akutagawa, *Bull. Chem. Soc. (Japan)*, **35**, 644 (1962); b. K. Babor, J. Ježo, V. Kalač, M. Karvaš, and K. Tihlarik, *Chem. Zvesti*, **14**, 679 (1960); through *Chem. Abstr.*, **55**, 17620 (1961); c. D. T. Warner and O. A. Moe, *J. Am. Chem. Soc.*, **70**, 3918 (1948).
[31] H. Reihlen, G. V. Hessling, W. Hühn, and E. Weinbrenner, *Ann.*, **493**, 20 (1932).
[32] R. Quelet and M. Paty, *Bull. Soc. Chim. France*, **1947**, 795.
[33] a. S. Kawahara and S. Kawakami, *Yakugaku Zasshi*, **81**, 149 (1961); through *Chem. Abstr.*, **55**, 12410 (1961); b. F. King and R. M. Acheson, *J. Chem. Soc.*, **1946**, 681.
[34] J. A. Lambrech, Brit. Patent 745,684 (1956).
[35] a. H. Baganz and H. Milster, *Arch. Pharm.*, **291**, 118 (1959); b. T. Y. Shen, E. F. Rogers, and L. H. Sarett, U.S. Patent 3,089,876 (1963); c. J. Corse, J. T. Bryant, and H. A. Shonle, *J. Am. Chem. Soc.*, **68**, 1905 (1946).
[36] a. V. G. Yashunskii, *J. Gen. Chem. USSR*, **33**, 185 (1963) (English translation); b. N. J. Leonard, G. W. Luebner, and E. H. Burk, Jr., *J. Org. Chem.*, **15**, 979 (1950).
[37] a. W. L. Hawkins and B. S. Biggs, *J. Am. Chem. Soc.*, **71**, 2530 (1949); b. A. Funke and P. Kornman, *Bull. Soc. Chim. France*, **1949**, 241.
[38] M. Freifelder and R. B. Hasbrouck, *J. Am. Chem. Soc.*, **82**, 696 (1960).
[39] F. C. Whitmore, H. S. Mosher, R. R. Adams, R. B. Taylor, E. C. Chapin, C. Weisel, and W. Yanko, *J. Am. Chem. Soc.*, **68**, 725 (1944).
[40] B. Lloyd, U.S. Patent 3,042,672 (1962).
[41] a. K. N. Campbell, J. F. Kerwin, A. H. Sommers, and B. K. Campbell, *J. Am. Chem. Soc.*, **68**, 1559 (1946); b. R. C. Jones, Jr., C. C. Price, and A. Y. Sen, *J. Org. Chem.*, **22**, 783 (1957).
[42] a. J. H. Burckhalter, E. M. Jones, W. F. Holcomb, and L. A. Sweet, *J. Am. Chem. Soc.*, **65**, 2012 (1943); b. J. H. Burckhalter, W. S. Brinigar, and P. E. Thompson, *J. Org. Chem.*, **26**, 4070 (1961).
[43] a. J. Harley-Mason and A. H. Jackson, *J. Chem. Soc.*, **1954**, 1165; b. L. N. Yakontov and M. V. Rubstov, *J. Gen. Chem. USSR*, **29**, 2307 (1959) (English translation).
[44] H. Plieninger, Ger. Patent 963,603 (1959).
[45] a. F. Bergel, N. C. Hindley, and A. L. Morrison, Brit. Patent 563,665 (1944); b. F. Bergel, N. C. Hindley, and A. L. Morrison, U.S. Patent 2,405,555 (1946); c. F. Bergel, A. L. Morrison, and H. Rinderknecht, U.S. Patent 2,446,804 (1958); d. J. F. Cavalla, *J. Chem. Soc.*, **1959**, 851.

[46] F. Poppelsdorf and R. C. Myerly, *J. Org. Chem.*, **26**, 131 (1961).

[47] a. W. H. Hartung, *J. Am. Chem. Soc.*, **50**, 3370 (1928); b. J. S. Buck, *J. Am. Chem. Soc.*, **55**, 2593 (1933); K. Kindler, H.-G. Helling, and E. Sussner, *Ann.*, **605**, 200 (1957).

[48] a. J. N. Ashley and M. Davis, *J. Chem. Soc.*, **1952**, 63; b. M. W. Goldberg and H. Kirchenstein, *Helv. Chim. Acta*, **26**, 288 (1943); c. Pl. A. Plattner, A. Furst, and A. Studer, *Helv. Chim. Acta*, **30**, 1091 (1947).

[49] A. C. Cope, H. R. Nace, and L. L. Estes, Jr., *J. Am. Chem. Soc.*, **72**, 1123 (1950).

[50] B. Tchoubar, *Bull. Soc. Chim. France*, **1949**, 160.

[51] a. H. P. Schultz, *J. Am. Chem. Soc.*, **70**, 2666 (1948); b. F. Bergel, A. L. Morrison, and H. Rinderknecht, U.S. Patent 2,446,803 (1948).

[52] a. K. Bowden and P. N. Green, *J. Chem. Soc.*, **1952**, 1164; b. N. Sugimoto and H. Kugita, Jap. Patent 3672 (1956); through *Chem. Abstr.*, **51**, 13939 (1957).

[53] a. P. Ruggli, B. B. Bussemaker, W. Muller, and A. Staub, *Helv. Chim. Acta*, **18**, 1388 (1935); b. J. H. Wood, M. A. Perry, and C. C. Tung, *J. Am. Chem. Soc.*, **73**, 4689 (1951); c. R. P. Mull, U.S. Patent 3,093,632 (1963); d. P. Kurtz and R. Schroter, Ger. Patent 850,615 (1952); through *Chem. Abstr.*, **47**, 5429 (1953).

[54] a. Brit. Patent 576,015 (1946); through *Chem. Abstr.* **42**, 591 (1948); b. A. Silverstone, U.S. Patent 2,790,804 (1957).

[55] a. E. Strack and H. Schwaneberg, *Ber.*, **67**, 39 (1934) b. J. Harley-Mason and A. H. Jackson, *J. Chem. Soc.*, **1954**, 3651; c. M. Danzig and H. P. Schultz, *J. Am. Chem. Soc.*, **74**, 1836 (1952).

[56] a. W. R. Miller, U.S. Patent 2,809,196 (1957).

[57] a. W. Reppe, *Ann.*, **596**, 1 (1955); b. J. Swerdloff, U.S. Patent 2,762,835 (1956); c. C. M. Hale, Jr., V. L. Brown, Jr., and T. E. Stanin, Brit. Patent 1,042,910 (1966); d. E. W. Pietrusza, R. E. Brown, and M. B. Mueller, U.S. Patent 3,050,544 (1962); e. D. C. Quin, Brit. Patent 814,631 (1959).

[58] a. H. Rupe and E. Hodel, *Helv. Chim. Acta*, **6**, 865 (1923); b. H. Rupe and F. Becherer, *Helv. Chim. Acta*, **6**, 880 (1923).

[59] a. A. Gaiffe, *Chim. Ind.* (Paris), **93**, 259 (1965); b. H. Plieninger and G. Werst, *Ber.*, **88**, 1956 (1955); c. A. Giner-Sorolla, I. Zimmerman, and A. Bendich, *J. Am. Chem. Soc.*, **81**, 2515 (1959); d. A. E. Rogers, U.S. Patent 3,078,274 (1963); e. F. Zymalkowski and P. Tinapp, *Ann.*, **699**, 98 (1966).

[60] a. G. Mignonac and P. J. H. Bourbon, U.S. Patent 2,945,862 (1960); b. D. Eilhauer, W. Hoefling, and G. Reckling, E. Ger. Patent 43,047 (1966); through *Chem. Abstr.*, **65**, 5446 (1966).

[61] a. R. Kuhn, W. Kirschenlohe, and W. Bister, *Ann.*, **600**, 115 (1956); b. R. Kuhn and W. Kirschenlohe, *Ann.*, **600**, 126 (1956); c. R. Kuhn and D. Weiser, *Ann.*, **602**, 208 (1957).

[62] S. Umio, *Yakugaku Zasshi*, **79**, 1133 (1959); through *Chem. Abstr.*, **54**, 3376 (1960).

[63] a. K. Kindler and F. Hesse, *Arch. Pharm.* **271**, 439 (1933); b. K. Kindler, K. Schrader, and B. Middlehoff, *Arch. Pharm.*, **283**, 184 (1950).

[64] a. J. G. Abrams and E. C. Chapin, U.S. Patent 3,209,029 (1965); b. M. Julia, P. Manoury, and C. Voillaume, *Bull. Soc. Chim.* France, **1965**, 1417; c. R. W. Balsiger, H. Hanni, and O. Schindler, *Helv. Chim. Acta*, **49**, 889 (1966); d. H. Henecka and R. Lorenz, Ger. Patent 1,029,380 (1958); e. R. Schlapfer and H. Spiegelberg, U.S. Patent 2,798,077 (1957).

[65] a. S. Wawzonek and E. M. Smolin, *J. Org. Chem.*, **16**, 746 (1951); b. G. Reutenauer and C. Paquot, *Compt. Rend.* **224**, 478 (1947).

[66] a. L. Bauer and J. Cymerman, *J. Chem. Soc.*, **1950**, 1826; b. L. C. Beegle, U.S. Patent 2,818,431 (1957); c. O. Schneider and J. Hellerbach, *Helv. Chim. Acta*, **33**,

1437 (1950); d. V. N. Sokolova and O. Yu. Magidson, USSR Patent 130,517 (1960); through *Chem. Abstr.*, **55**, 7321 (1961).

[67] M. Freifelder, *J. Pharm. Sci.*, **55**, 535 (1966).

[68] a. C. D. May and P. Sykes, *J. Chem. Soc.* (C), **1966**, 649; b. P. Nesbitt and P. Sykes, *J. Chem. Soc.*, **1954**, 3057; c. K. C. Brannock, A. Bell, R. D. Burpit, and C. A. Kelly, *J. Org. Chem.*, **29**, 801 (1964); d. D. S. Tarbell and D. F. Fukushima, *J. Am. Chem. Soc.*, **68**, 2499 (1946); e. C. H. Bestian, *Ann.*, **566**, 210 (1950).

[69] a. K. Kindler and W. Peschke, Ger. Patent 571,795 (1933); b. H. Richet, R. Dulou, and G. Dupont, *Bull. Soc. Chim. France*, **1947**, 693; c. J. W. Cook, J. D. Loudon, and D. K. V. Steel, *J. Chem. Soc.*, **1954**, 530; d. R. H. Wiley and H. Adkins, *J. Am. Chem. Soc.*, **60**, 914 (1938).

[70] V. V. Young, U.S. Patent 2,864,863 (1958).

[71] a. J. Stanek and J. Urban, *Coll. Czechoslov. Chem. Commum.*, **15**, 397 (1950); b. M. Fields, D. E. Walz, and S. Rothchild, *J. Am. Chem. Soc.*, **73**, 1000 (1951).

[72] a. H. Adkins and G. M. Whitman, *J. Am. Chem. Soc.*, **64**, 150 (1942); b. H. Henecka, H. Timmler, R. Lorenz, and W. Geiger, *Ber.*, **90**, 1060 (1957).

[73] a. Brit. Patent 605,784 (1948); b. C. E. Kwartler and P. Lucas, *J. Am. Chem. Soc.*, **69**, 2582 (1947); c. G. Erhart and F. Nolte, Ger. Patent 830,954 (1952); d. A. R. Surrey, A. J. Olivet, and J. O. Hoppe, *J. Am. Chem. Soc.*, **76**, 4920 (1954); e. J. Harley-Mason and R. F. Ingelby, *J. Chem. Soc.*, **1958**, 3639.

[74] a. I. T. Strukov, *J. Gen. Chem.* USSR., **31**, 2528 (1961) (English translation); b. T. Kametani, N. Wayatsuma, and F. Sasuki, *Yakugaku Zasshi*, **86**, 913 (1966); c. B. Heath-Brown and P. G. Philpott, *J. Chem. Soc.*, **1965**, 7165; d. A. Brossi, M. Baumann, and R. Borer, *Monatsh.*, **96**, 25 (1965).

[75] K. Kindler and K. Schrader, *Ann.*, **564**, 49 (1949).

[76] G. B. Bachmann and H. A. Levene, *J. Am. Chem. Soc.*, **70**, 599 (1948).

[77] a. C. E. Kwartler and P. Lucas, *J. Am. Chem. Soc.*, **68**, 2395 (1946); b. G. Erhart, *Arch. Pharm.*, **295**, 197 (1962); c. C. Price and V. Boekelheide, *J. Am. Chem. Soc.*, **68**, 1246 (1946).

[78] J. T. Suh and B. M. Puma, *J. Org. Chem.*, **30**, 2253 (1965).

[79] a. T. S. Work, *J. Chem. Soc.*, **1942**, 426; b. J. Sam and K. Aparajithan, personal communication.

[80] a. M. J. Reider and R. C. Elderfield, *J. Org. Chem.*, **7**, 286 (1942); b. L. A. Perez-Medina, R. P. Mariella, and S. M. McElvain, *J. Am. Chem. Soc.*, **69**, 2574 (1947); c. E. Testa and A. Vecchi, *Gazz. Chim. Ital.*, **87**, 467 (1957); d. J. L. Green, Jr. and J. A. Montgomery, *J. Med. Chem.*, **6**, 294 (1963).

[81] a. M. Schorr, H. Leditschke, D. Fussgaenger, and F. Bauer, Ger. Patent 1,103,936 (1961); b. M. Sayhun and J. A. Faust, U.S. Patent 3,089,826 (1963); c. C. Viel, *Ann. Chim. (Paris)*, **8**, 515 (1963).

[82] A. E. Senear, M. M. Rapport, J. F. Mead, J. T. Maynard, and J. B. Koepfli, *J. Org. Chem.*, **11**, 378 (1946).

[83] a. O. Cass, U.S. Patent 2,499,847 (1950); b. H. Christal, A. Donche, and F. Plenat, *Bull. Soc. Chim.* France, **1966**, 1315.

XIII

REDUCTION OF OXIMES

The hydrogenation of oximes to amines proceeds stepwise either through the hydroxylamine, reaction (a), or through the imine, reaction (b). The formation of imine and the release of water from reduction are responsible for the by-products found in these hydrogenations. In (c) the imine reacts with primary amine to form an addition product. It then loses ammonia to give an azomethine which is reduced to secondary amine (the formation of secondary amine follows the pattern seen in the hydrogenation of nitriles, Chapter XII). Water resulting from reduction causes hydrogenolysis of imine to aldehyde or ketone. Either reacts (d) with primary amine to form the corresponding aldimine or ketimine, which is hydrogenated to secondary amine or, as in reaction (e), the aldehyde or ketone is reduced to the corresponding alcohol or in some instances to a hydrocarbon (Eq. 13.1).

(a) $RR'C=NOH \xrightarrow{H_2} RR'CHNHOH \xrightarrow{H_2}$ amine $+ H_2O$

(b) $\xrightarrow{H_2} H_2O + RR'C=NH \xrightarrow{H_2}$ amine

(c) $RR'C=NH + RR'CHNH_2 \longrightarrow RR'C(NH_2)NHCHRR' \xrightarrow[H_2]{-NH_3}$

$(RR'CH)_2NH$ \hfill (13.1)

(d) $RR'C=NH \xrightarrow[-NH_3]{H_2O} RR'C=O \xrightarrow{amine} RR'C=NCHRR' \xrightarrow{H_2}$ sec. amine

(e) $\xrightarrow{H_2} RR'CHOH \xrightarrow{H_2} RR'CH_2$

REACTION CONDITIONS AND CATALYSTS

High yield of primary amine from hydrogenation of an oxime is dependent on speed of reaction and/or means of preventing or minimizing formation of secondary amine, the most predominant by-product. It has been shown by Mignonac [1a] that the presence of ammonia inhibits formation of secondary amine in the reduction of oximes. There are innumerable references which report reduction in acidic media as another means of preventing secondary amine formation. The use of acidic media is not necessarily restricted to hydrogenations with noble metal catalysts. Nickel has been used in the reduction of oximes in alcoholic acetic acid and in glacial acetic acid [1b, c]. Reduction of the O-acetyl derivative of the oxime has been employed as another means of decreasing secondary amine formation [1d].

Cobalt

Hydrogenations with cobalt catalyst are usually limited to elevated temperature and pressure conditions. Raney cobalt gave excellent yields of primary amines in reductions of oximes at 80–125° and 200–220 atm pressure [2]; secondary amine formation was extremely low in such reductions with or without ammonia.

Nickel

Most reductions of oximes with nickel are carried out at 25–100° and 20–130 atm. Some typical examples describe hydrogenations in the absence of ammonia [3a–e]. Often, better yields of amines are obtained from similar reductions in the presence of ammonia [3b, 4a–d]. The amount of catalyst used in either type reduction is usually greater than 20% of the weight of substrate. Notable exceptions are seen in the hydrogenations of some 1-substituted phenyl-2-propanone oximes at 80–120° and 200 atm with a 2.5% ratio [4b] and in the reduction of 3-phenylindanone oxime at room temperature and 100 atm pressure with freshly prepared catalyst.

Hydrogenations of oximes can be highly exothermic [3b, 5]. *Caution must be exercised in reactions with excessive amounts of Raney nickel at higher pressures, particularly if the catalyst is comparatively fresh and especially when more than a moderate amount of oxime is to be hydrogenated.* Reduction of 0.1–0.2 molar amounts of some cyclopentanone oximes, which had been carried out in alcohol at 50–75° and 100–125 atm pressure with about 25% by weight of Raney nickel, were mildly exothermic [5]. However, when 1.5 moles were similarly hydrogenated, it was necessary to stop the reaction completely because the rise in temperature was more rapid than the decrease in pressure. The same reduction was safely run with much less catalyst (5–10%).

Caution must also be exercised in reducing oximes with nickel under similar reaction conditions in the absence of solvent. In a small-scale run an exothermic reaction carried the temperature to 120° [5]. On larger scale, the temperature rose from 50° to 275°, despite a reduction in the amount of catalyst. However, since a rise in temperature was expected, the reaction did not get out of control because only 50% of the necessary amount of hydrogen was used initially; more hydrogen was added after the exothermic reaction abated.

Low-pressure hydrogenations of oximes have been carried out with Raney nickel. In these reductions in the absence of ammonia, yields varied from 55–80% [6a–c]. A series of aliphatic oximes was hydrogenated to the corresponding amines in 43–85% yield [7]. In this series secondary amines were not obtained, but ketones were found as by-products. Low-pressure reductions in the presence of ammonia have given good yields of amines [8a, b]. In one instance an active Raney nickel catalyst, W-7, was used [8c]. Oximoalcohols of the type, $RR^2C(OH)C(=NOH)R'$ where R, R', R^2 are aliphatic or aromatic groups, were reduced with nickel at atmospheric pressure to give aminoalcohols in 55–74% yields [9a, b]. Platinized nickel was reported to be useful in difficult reductions. It was employed in the presence of alkali to hydrogenate the oxime of 2,2-dimethyl-1-phenylhexanal to the corresponding amine at atmospheric pressure [10].

In spite of the references to the use of nickel at low pressure, it has been this author's experience that, when it is used for the reduction of oximes, best results are obtained at elevated temperature and pressure conditions unless the catalyst is freshly prepared or if it is a very active one [5]. The hydrogenation of 2,2-diphenylcyclohexanone oxime in alcoholic ammonia with Raney nickel promoted with chloroplatinic acid at 75° and 90 atm pressure was complete in 0.25 hr and gave 80% of 2,2-diphenylcyclohexylamine. Reduction at 3 atm pressure was too slow to be useful. Hydrogenation of 2-isopropyl-5-methylcyclohexanone oxime with Raney nickel in alcoholic ammonia at elevated temperature and pressure gave 78.5% yield of the amine in less than 2 hr. Low-pressure reduction was extremely slow. The hydrogenations of cycloheptanone oxime with nickel in the absence of ammonia provide a more striking example of poor yield and long reaction time in low-pressure reactions with nickel. At 90° and 90 atm pressure, reduction in alcohol with a 5% ratio of nickel to substrate gave 91% yield of cycloheptylamine in about 1 hr. In contrast, reduction with a 10% catalyst ratio at 55–60° and 3 atm pressure was complete in 24–27 hr; 62% yield of cycloheptylamine was obtained along with dicycloheptylamine.

Platinum

Hydrogenations of oximes with platinum catalysts are best carried out in acidic media [5]. When oximes are reduced in neutral solvent with platinum, poisoning often takes place because of formation of amine, unless an excessive amount of catalyst is used. In addition, if reduction does proceed in neutral media, very poor yields of primary amines and a predominance of secondary amines invariably result. This is illustrated in the low-pressure hydrogenations of (3-ethoxy-2-methoxy)- and (3,4,5-trimethoxy)phenylacetaldoximes which gave high yields of secondary amines [11a], in the reduction of 4-methyl-2-pentanone oxime which yielded 39% of primary amine and 56% of secondary amine [11b], in that of 3-carbethoxybutanone oxime [11c] and of tropinone-3-oxime [11d]. The reduction of the oxime of 17-yohimbone to 17-aminoyohimbane may be an exception [12], but it must be pointed out that 20% by weight of catalyst was used and yield of amine was not reported.

The presence of a free carboxyl group in an oxime compound may have sufficient neutralizing effect on the resulting amine to prevent poisoning of platinum catalyst during reduction in neutral solvent. This point has not been investigated except in the hydrogenation of the oxime of 3-(2-quinuclidinyl)-3-ketopropionic acid in alcohol where the yield of amino acid was 78% [13]. However, since the amount of catalyst used in the reaction was excessively large, 45% by weight of substrate, the effect of the presence of a carboxyl group on these reductions cannot be evaluated from this example.

A rate study emphasizing that the reduction of cyclohexanone oxime with platinum catalyst proceeded best in acidic media also indicated that it was more effective than other noble metal catalysts in the hydrogenation of this compound in acidic media.

This does not necessarily mean that it is the method of choice for the reduction of all oximes. In many instances good results have been obtained, at other times it was often necessary to employ high catalyst ratios although this may have been due to the structure and/or purity of the substrate.

The oxime of (4-butoxy-3-methoxyphenyl)acetone was hydrogenated in glacial acetic acid under 1.3 atm pressure with a 2% ratio of platinum oxide to yield 56% of the corresponding amine [5]. Less catalyst was used to reduce benzophenone oxime in acetic acid and acetic anhydride under 2 atm pressure of hydrogen to obtain 94–100% yields of *N*-acetylbenzhydrylamine [5]. The hydrogenation of 2-methoxy- and 2,4-dimethoxyphenylacetaldoxime with 4% by weight of platinum oxide gave 65–85% yields of the substituted phenethylamines [15]. The by-products in these reductions were *bis* substituted phenylethylhydroxylamines formed by reaction of the intermediate monohydroxylamine with the phenethylamine (Eq. 13.2).

$$\text{ArCH}_2\text{CH}=\text{NOH} \begin{cases} \xrightarrow{\text{H}_2} [\text{ArCH}_2\text{CH}_2\text{NHOH}] + \text{ArCH}_2\text{CH}_2\text{NH}_2 \xrightarrow{-\text{NH}_3} (\text{ArCH}_2\text{CH}_2)_2\text{NOH} \\ \\ \xrightarrow{2\text{H}_2} \text{ArCH}_2\text{CH}_2\text{NH}_2 \end{cases} \quad (13.2)$$

Hydrogenation of the oxime of 1,1-diphenylacetone in acetic acid with 3% by weight of platinum oxide was slow [5], probably because of steric effects. Hydrogen uptake was very slow, only 25% uptake in 24 hr, in alcoholic acetic acid with 20% by weight of 5% platinum on carbon, apparently because of the low metal to substrate ratio.

3-Hydroxy-5-hydroxymethyl-2-methylpyridine-4-aldoxime was reduced to the corresponding amine in almost quantitative yield in alcoholic hydrogen chloride with about 50% by weight of platinum oxide [16]. Steric effects, no doubt, made employment of a large catalyst ratio necessary. However, the amount used casts some doubts on the purity of the substrate. In such instances pretreatment of the oxime in neutral solvent with Raney nickel may remove impurities and allow the use of less catalyst [5].

Poor results were obtained in the hydrogenation of the oxime of quinuclidine-3-one in dilute hydrochloric acid with platinum oxide [17]; 3-hydroxyquinuclidine was obtained along with about 20% yield of 3-aminoquinuclidine. The yield of amine was increased to 46% by reduction with Raney nickel at 100° and 30 atm pressure. The platinum-catalyzed hydrogenation of 4,4-dimethyl-1-tetralone oxime in alcoholic hydrogen chloride gave 10% yield of amine and over 50% of 4,4-dimethyl-1-tetralol [5]. Reduction of the oxime in alcoholic ammonia with Raney nickel at 90° and 90 atm gave 4,4-dimethyl-1,2,3,4-tetrahydro-1-naphthylamine.

Palladium

In many instances it may also be necessary to employ acid medium for reductions of oximes with palladium. Hartung studied the effect of hydrochloric acid in low-pressure reductions of aldoximes with palladium on carbon [18]. He found that reduction of benzaldoxime in the presence of one equivalent of acid gave a mixture of benzylamine and dibenzylamine, while in the presence of three equivalents only the primary amine was obtained.

The hydrogenation of 2,3,4-trimethoxybenzaldoxime with 10% palladium on carbon in neutral solution yielded only 26% of primary amine [19].

On the other hand, a ketoxime, benzoin oxime, $C_6H_5CHOHC(C_6H_5)$=NOH, was readily hydrogenated in alcohol to DL-*erythro*-1,2-diphenylethanolamine in 85% yield with palladium on alumina at 40° and 4 atm pressure [20]. In this instance it is likely that steric effects prevented formation of secondary amine. The presence of a free carboxyl group may have not only allowed the reduction of the oxime of 4-keto-4-phenylbutyric acid to be carried out in neutral solvent with palladium on carbon at room temperature and 2 atm pressure [5], but may have also bound the amino group so it could not react further to form secondary amine. Yield of amino acid was 65%.

Hydrogenations of oximes with palladium catalysts in acidic media, when successful, give good yields of amines in not too lengthy reaction periods as exemplified by the reduction of the oxime of 3-(3,4,5-trimethoxybenzoyl)oxy-4-methoxybenzaldehyde in glacial acetic acid [21a], of the oxime of benzimidazol-4-aldehyde in methanolic hydrogen chloride [21b], and of 2,3-dimethoxyphenylacetaldoxime in glacial acetic-sulfuric acid [21c]. The latter utilizes a procedure in which the oxime in acetic acid is added dropwise to a stirred suspension of palladium oxide in acetic acid containing sulfuric acid.

It is of interest that the oximes of cyclopentanone and substituted cyclopentanones, cyclohexanone, and cycloheptanone could not be hydrogenated with palladium on carbon in acidic media [5]. Poor rate of uptake was reported in the reduction of cyclohexanone oxime in acidic, alkaline, and neutral media with palladium catalyst [14]. No uptake of hydrogen occurred in the reduction of 3,3-diphenyl-2-butanone oxime with palladium on carbon in alcoholic hydrogen chloride [22]. The failure in this instance may be due to steric effects. To achieve reduction to the corresponding amine it was necessary to carry out the reaction in alcoholic ammonia at 75° and 70 atm pressure of hydrogen for 8 hr in the presence of a 50% weight ratio of Raney nickel activated with chloroplatinic acid.

Ruthenium and Rhodium

Very little work on the hydrogenation of oximes with ruthenium catalyst is reported. Hydrogen uptake proceeded very slowly in the ruthenium on carbon reduction of cyclohexanone oxime in most of the solvents used [14]; there was a noticeable increase in rate during absorption of the second equivalent of hydrogen in reductions in water and particularly in sodium hydroxide solution. 5% Rhodium on a carrier appears particularly adaptable for the reduction of cycloheptanone oxime [23] and some cyclopentanone oximes [5]. Hydrogenation was carried out under low-pressure conditions in the absence of ammonia giving high yields of primary amines. Hydrogen uptake in the reduction of cyclohexanone oxime in alkaline solution with rhodium was rapid but the products of reaction were not examined [14].

OXIMINO ACIDS, ESTERS, AND AMIDES

Hydrogenations of α-oximino, esters, and amides appear to be successful under low-pressure conditions in acid, basic, and neutral media with palladium and with nickel. Reductions with platinum catalyst may be limited to acidic conditions. Hydrogenations of α-oximino acids with palladium may be less troublesome in alkaline solution.

Hydrogenations of a series of oximes of the type $C_6H_5C(=NOH)COR$, where R was OH, OC_2H_5, and various substituted amino groups, were carried out in aqueous ammonia with both palladium on calcium carbonate and with Raney nickel [24]. Reductions in the absence of ammonia were satisfactory but hydrogen uptake was slower. Nickel was particularly useful in reductions where R was a sulfur-containing amine. In those instances the palladium catalyst was poisoned.

Ethyl isonitrosomalonate was first reduced with Raney nickel at low pressure [25]. It was later shown that better results were obtained with a low ratio of palladium on carbon or palladium on alumina [26]. Comparative reductions showed that higher yields of ethyl aminomalonate were obtained in shorter reaction periods with 5% palladium on carbon than with an equal weight of Raney nickel [5].

Difficulties appear to be associated with hydrogenation of α-oximino acids. When some were reduced with palladium on carbon in the presence of hydrochloric acid, it was found that uptake of the first molar equivalent of hydrogen was rapid and that absorption of the second equivalent was considerably slower [28a,b]. It was reported that the resulting amino acid was the inhibitor [28b]. In the reductions of 2-(hydroxyimino)-3-(3,4-methylenedioxyphenyl)propionic acid [28c] and 2-(hydroxyimino)-3-(4-methoxyphenyl)propionic acid [28d] to the corresponding substituted phenylalanines under similar conditions, 40% and 75% by weight of catalyst were employed, apparently to overcome any inhibitory effect. 2-(Ethoxyimino)-3-phenylpropionic acid was reduced in neutral solution (alcohol) with 30% by weight of palladized carbon to give 93% of DL-phenylalanine [29]. The hydroxyimino compound gave only 43% yield of the DL-amino acid with W-6 Raney nickel, 40% by weight of substrate, in reduction in water at 2–3 atm pressure [30].

When the hydroxyimino group is further removed from the ester or acid group, cyclization usually takes place if the formation of 5- and 6-membered rings is possible, especially in reductions at elevated temperature and pressure. 5-Methyl-2-pyrrolidone was prepared from ethyl 4-(hydroxyimino)valerate with nickel in this manner [31a]. 3,6-Dimethylpiperazin-2-one was obtained from the hydrogenation of ethyl 2-(2-hydroxyiminopropyl)-aminopropionate with nickel under 100 atm pressure [32]. Reduction of oximino diesters with copper chromite at 260° and 350 atm was used as a general method of obtaining 1-azabicyclo compounds [33]. Under these reaction conditions the carbonyl groups of the bicyclic imide were converted to methylene groups to give an azabicyclo compound (Eq. 13.3). Reduction

$$\begin{array}{c} (CH_2)_n COOR \\ C=NOH \\ (CH_2)_n COOR \end{array} \xrightarrow{2H_2} \begin{array}{c} (CH_2)_n COOR \\ CHNH_2 \\ (CH_2)_n COOR \end{array} \longrightarrow \begin{array}{c} (CH_2)_n \quad N \quad (CH_2)_n \\ \| \quad \| \\ O \quad O \end{array} \Bigg\downarrow 2H_2 \\ \begin{array}{c} (CH_2)_n \quad N \quad (CH_2)_n \end{array}$$

$n = 2$ or 3 (13.3)

of ethyl 3-(2-hydroxyiminocyclohexyl)-3-phenylpropionate in alcohol under less strenuous conditions gave 1-ethyl-4-phenyldecahydroquinoline [34]. Ethylation of the ring nitrogen was caused by the reaction temperature (200°) (the mechanism of N-alkylation during reduction will be discussed in a later section). It is possible to carry out reductions of oximino acids without causing cyclization if the reaction is not allowed to become too warm. Some 4-(hydroxyimino)-4-substituted phenylbutyric acids were hydrogenated at room temperature and atmospheric pressure to give the corresponding amino acids; when the reductions were warmed, the corresponding 5-phenyl-2-pyrrolidones were obtained [35].

INDANONE OXIMES

There is no general method for the catalytic reduction of indanone oximes to aminoindanes. Ethyl 1-hydroxyiminoindane-6-oxyacetate was hydrogenated with platinum black catalyst in acetic anhydride to the corresponding 1-acetylamino compound [36], but reductions in alcohol with Raney nickel or platinum black or in alcoholic hydrogen chloride with platinum black did not produce a pure compound. The hydrogenation of 1-indanone oxime with Raney nickel at 70° and 50 atm in alcohol gave a mixture of 35% of 1-aminoindane and 60% of diindanylamine [37]; reduction of 1-hydroxyiminoindane-3-carboxylic acid under the same conditions in alcoholic ammonia gave 50% of primary amine plus other material which was not identified. High yields were reported in low-pressure reductions of 5-butoxy and 5-hexyloxy-1-hydroxyiminoindanes in methanol containing concentrated hydrochloric acid in the presence of palladium black [38]. Reaction time was long (25 hr) but the amount of catalyst used was not reported. When 5-isopropoxy-1-(isopropoxyimino)indane was hydrogenated under the same conditions, reaction time was almost halved (alkoxyimino groups appear to reduce more readily than the C=NOH function itself). Reduction of 5-hexyloxy-1-hydroxyiminoindane with Raney nickel in alcoholic ammonia at 70° and 130 atm pressure resulted in an excellent yield of primary amine in less than 2 hr. 1-Hydroxyimino-3-methylindane was reduced to the amine in 78% yield with 14% by weight of 5% palladium on carbon in alcoholic hydrogen chloride at 60–70° and 2 atm pressure [5]. Reduction of 1-hydroxyiminoindane under the same conditions gave only 45% yield of primary amine hydrochloride.

The reduction of 2-indanone oxime to 2-aminoindane is considered a difficult one [39 and references therein]. Successful conversion of the oxime to amine was reported for hydrogenation in alcoholic hydrogen chloride with a palladium on carbon catalyst especially activated by treatment with sodium acetate during its preparation [40]. Hydrogen uptake was not very rapid, 8–10 hr, despite the activity of the catalyst and the large amount used, 70% by weight. Hydrogenations of 2-indanone oxime to amine with palladium, platinum, rhodium, nickel, and cobalt catalysts under a variety of conditions were investigated [39]. Reductions with Raney nickel in basic solution (ammonia, potassium carbonate, methanolic sodium methylate, or sodium hydroxide) at low pressure gave up to 92–94% yield of 2-aminoindane. Generally as much catalyst as compound was required. Surprisingly Raney cobalt was also effective under similar conditions giving 85% yield of primary amine. Reductions in acidic medium (glacial acetic acid plus 2.2 moles of sulfuric acid/mole of oxime) gave high yields of 2-aminoindane when a 50% weight ratio of palladium on carbon was employed. Lowering the amount of catalyst increased the time of reaction. Decreasing the amount of sulfuric acid decreased the yield; use of acids other than sulfuric resulted in negligible yield of amine. Rhodium on carbon and platinum on carbon catalysts were found to be ineffective.

DIOXIMES

In general hydrogenations of dioximes are carried out in the same manner as monoximes. There have been no investigations concerning the effect of adjacent groups on the reductions of 1,2-dioximes. There may be an implication of steric effects in the reported failure to reduce 1-phenyl-1,2-*bis*-hydroxyiminopropane in dioxane with Raney nickel at atmospheric pressure since other dioximes, RC(=NOH)C(=NOH)R' where R and R' = H and where R = CH$_3$ and R' = CH$_3$, C$_4$H$_9$, and C$_5$H$_{11}$, were hydrogenated without difficulty [41]. No uptake of hydrogen occurred in the attempted reduction of 1-phenyl-1,2-*bis*-hydroxyiminopropane in alcoholic hydrogen chloride with palladium on carbon [5]. Nevertheless, a more hindered compound, benzil dioxime, was converted to 1,2-diphenylethylenediamine in 60% yield by hydrogenation in aqueous sodium hydroxide solution at room temperature and atmospheric pressure with Raney nickel [42].

Formation of small quantities of piperazines have been reported in the hydrogenation of 1,2-dioximes [41]. Reduction of 1,2-dioximes at elevated temperature and pressure may result in large amounts of piperazines.

In the hydrogenation of dioximes of the type, HON=CH(CH$_2$)$_n$CH=NOH where $n = 2$ to 4, it may be difficult to prevent cyclization. No uptake occurred in the reduction of succindialdehyde oxime with Raney nickel in alcoholic ammonia at room temperature and 100 atm [5]. On warming to 60° hydrogen absorption began; only pyrrolidine was obtained. Low-pressure reduction of 2,5-*bis*-hydroxyiminohexane in methyl alcoholic acetic acid with platinum oxide at room temperature and atmospheric pressure yielded 55% of 2,5-dimethylpyrrolidine; the hydrogenation of some steroidal dioximes under similar conditions resulted in formation of pyrrolidine rings attached to the D ring of the steroids [43].

Cyclization may be the preferred reaction in the hydrogenation of amino oximes where from four to six carbon atoms separate a primary or secondary amino group from the hydroxyimino group [36].

AMIDOXIMES TO AMIDINES

Since amidoximes can be prepared in good yield, catalytic reduction of them may provide a better method for the preparation of amidines than synthesis through iminoethers from nitriles. Hydrogenations of amidoximes have been carried out at 60–80° and 30 atm pressure with Raney nickel [44a, b]. There is a later report of a reduction of 4-aminoimidazole-5-amidoxime hydrochloride in water with nickel, presumably at low pressure [44c]. More recently amidoximes, R(CH$_2$)$_n$C(NH$_2$)=NOH, where R is a phenyl or a saturated heterocycle and $n = 1$ to 3, have been hydrogenated to amidines in absolute alcohol with 5% rhodium on alumina at 25° and 3 atm pressure [45]. Palladium on carbon was also employed but absorption of hydrogen proceeded at a slower rate.

PARTIAL REDUCTION

A sharp break in the rate of uptake was observed after absorption of one molar equivalent of hydrogen in the reduction of 2-hydroxyiminopropionic acid in alcoholic hydrogen chloride in the presence of palladium on carbon [28a]. None was noted in similar reductions of some 2-alkoxyimino acids [46]. Interruption of hydrogenations of both types at uptake of one equivalent yielded a mixture of starting material and the corresponding amino acid [46].

Oximes have been reduced catalytically to hydroxylamines under low-pressure conditions in alcoholic or aqueous alcoholic hydrogen chloride in the presence of platinum oxide but results were erratic due perhaps to differences in structure. Butyraldoxime was reduced to N-butylhydroxylamine but isobutyraldoxime gave N,N-*bis*-isobutylhydroxylamine; benzaldoxime was reduced to benzylamine and dibenzylamine, and acetophenone oxime and benzophenone oxime gave the corresponding primary amines [47]. The lack of selectivity in the latter two reductions may be due to proximity of the aromatic ring. When the benzene ring was separated from the hydroxyimino group by a methylene group, good yield of DL-1-(4-methoxyphenyl)-2-hydroxyaminopropane was obtained from reduction of the corresponding phenylpropanone oxime [48a]. 2-Hydroxyamino-4-(4-dimethylaminophenyl)-butane was obtained from the corresponding oxime but yield was not reported [48b].

Hydrogenation of cyclic ketoximes gave varying yields of hydroxylamines: N-cyclohexyl, 65% [49a]; N-cycloheptyl, 62% and N-cycloctyl, 31% [49b]; and DL-1-N-(1,2,3,4-tetrahydronaphthyl), 30% [48a]. In the reduction of cyclohexanone oxime, hydrogen absorption decreased sharply after uptake of one equivalent.

Platinum-catalyzed reductions of alkoxyimino compounds in the presence of hydrogen chloride also gave a variety of results. Very little O-alkyl-hydroxylamine was obtained from the hydrogenations of O-methyl and O-ethyl acetoximes and O-methyl acetaldoxime. Reversal apparently takes place to carbonyl compound and N-alkoxyamine; the latter compound was further reduced to ammonium chloride and an alcohol [50]. More stable, higher molecular weight O-methyl aliphatic oximes, 3-methoxyiminopentene, and 4-methoxyiminoheptane, were hydrogenated to the corresponding methoxyamines. The reduction of O-methyl benzaldoxime gave a mixture of benzylamine and dibenzylamine. The lack of selectivity in the reduction of an oxime when the hydroxyimino group is in conjugation with an aromatic ring was previously noted [47]. Removal of the activating effect by the presence of a methylene group between the alkoxyimino group and the ring was apparently responsible for the better than average yields of

1-phenyl- and 1-(3- and 4-pyridyl)-2-alkoxyaminopropanes from the platinum-catalyzed hydrogenations of the corresponding alkoxyimino compounds in alcoholic hydrogen chloride [51].

In another group of reductions 2-methoxyaminoheptane, 1-cyclopentyl-2-methoxyaminopropane, and 1-cyclohexyl-2-methoxyaminopropanes were obtained in 40–50% yields from the corresponding *O*-alkyl oximes [52].

SELECTIVITY IN THE PRESENCE OF OTHER FUNCTIONS

Cyano Groups

Most studies on the hydrogenation of cyano oximino acids and esters have been centered on the preparation of diamino acids, such as lysine, with little interest in preferential reduction. In addition, the amounts of catalyst used in these reactions would preclude selectivity. However, it was shown that uptake of hydrogen uptake almost came to a halt after two equivalents were absorbed during the reduction of carbon 14-acetate-labeled ethyl-2-hydroxyimino-2-cyanoacetate in acetic anhydride in the presence of either 5% platinum on carbon or platinum oxide at atmospheric pressure [53a]. Best results, 85–88% yields of the labeled ethyl 2-acetylaminocyanoacetate, were obtained with platinum on carbon. There was no change in rate in similar reductions with Raney nickel or with palladium on carbon, although it was possible to interrupt the palladium reduction after 2 molar equivalents to obtain 40–75% yields of ethyl 2-acetylaminocyanoacetate. Interruption of the nickel-catalyzed reaction gave unidentified products.

A molecular model of the hydroxyimino compound shows that both reducible groups are in good contact with the catalyst. Selective reduction in the cited example [53a] may indicate that the oximino group is preferentially adsorbed.

On the other hand, when ethyl 2-hydroxyimino-3-cyano-3-phenylpropionate was hydrogenated with platinum oxide in acetic acid, the cyano group was selectively reduced to form 3-hydroxyimino-4-phenyl-2-pyrrolidinone [53b]; in acetic anhydride the corresponding 1-acetylpyrrolidinone was formed. When the amide, instead of the ester, was similarly reduced, the product was 4-amino-3-phenyl-2-hydroxyiminobutyramide. Reductions of these compounds may be a contraindication of the preferential hydrogenation of the C=NOH function. However, molecular models appear to show that the hydroxyimino group in these compounds is somewhat hindered.

In another example, when 4-(1-cyano-1-phenyl)methenyl-1-hydroxyimino-2,5-cyclohexadiene was reduced in methyl alcohol with Raney nickel, the cyano group was untouched; 90% yield of 4-aminodiphenylacetonitrile was obtained [53c] (Eq. 13.4).

$$C_6H_5C(CN)=\langle\rangle=NOH \xrightarrow[25°, 75 \text{ atm}]{2 H_2, Ni(R)} C_6H_5CH(CN)\langle\rangle NH_2 \quad (13.4)$$

4-Aminodiphenylacetonitrile was obtained in similar reductions in the presence of potassium hydroxide, phenolate, acetate, and iodide or when weak bases such as pyridine and aniline were present [53d]. Hydrogenation of the oximino compound in the presence of ammonia and strong nitrogen bases as butylamine, diethylamine, triethylamine, and piperidine followed by treatment with water after concentration of the reduction solution gave 4-aminobenzophenone. A mechanism for this reaction, amine-catalyzed attack by water after partial reduction of the oxime and elimination of hydrogen cyanide, was proposed (Eq. 13.5). Unfortunately, the amount of hydrogen absorbed was not disclosed. This observation could have given added support to the mechanism.

$$RC(Y)=\langle\rangle=NOH \xrightarrow[-H_2O]{H_2} RC(Y)=\langle\rangle=NH$$

$$\downarrow H_2O \qquad\qquad\qquad\qquad (13.5)$$

$$RC(Y)(OH)C_6H_4NH_2 \xrightarrow{-HY} RC(=O)C_6H_4NH_2$$

$R = C_6H_5, Y = CN$

Although most of the examples suggest that an oximino group should be reduced in preference to a cyano group, further work is warranted on the effect of media, catalysts, and structure on selectivity when both groups are present in a substrate.

Ketones

In acid solution the hydroxyimino group is preferentially hydrogenated in the presence of a keto group [54a, b]. The method of Hartung and co-workers [55a, b], reduction of isonitrosoacetophenone and some isonitrosopropiophenones in alcohol containing at least 3 molar equivalents of hydrogen chloride in the presence of palladium on carbon catalyst, may be applied to other hydroxyiminoketones. In many such reductions uptake of hydrogen ceased or the reaction rate decreased markedly after absorption of two equivalents of hydrogen so that reactions could be interrupted and good yields of amino ketones obtained. Examples of hydrogenations of other 2-oximino ketones by the above method are seen in the preparation of

2-amino-1-indanone [56a] and 2-amino-4,6-dimethoxy-1-indanone [56b] from the 2-hydroxyimino compounds and in the conversion of 2-hydroxyiminocyclohexanone-6-acetic acid to *cis*-2-aminocyclohexanone-6-acetic acid [57]. In a modification of the method, acetic acid containing hydrochloric acid was used in the reduction of 4-methoxybenzil monooxime to the corresponding amino ketone [58].

Other modifications using acetic acid–acetic anhydride as the reaction medium [59a] or acetic anhydride alone [59b] gave acetylamino keto compounds from reductions of conjugated hydroxyimino ketones.

The conversion of 2-hydroxy-4-hydroxyiminotropone in alcoholic hydrogen chloride with palladium to 4-(1-aminoethyl)-2-hydroxytropone is an example of selective reduction among conjugated hydroxyimino ketones [60].

Platinum oxide has given good selectivity in hydrogenations of conjugated hydroxyimino ketones [61a–e] but there is a tendency toward continued uptake in its presence to form aminoalcohols.

Reductions of α-hydroxyimino ketones in neutral solvent often lead to self-condensation products. Amino ketones are formed but are converted to pyrazines [31a, b; 55a]. In certain palladium-catalyzed reductions of the type, RCOC(R′)=NOH where R = phenyl or ethyl, the oxime of the corresponding alcohol was isolated along with unidentified products [54b]. When hydrogenations are carried out at elevated temperature and pressure, there is not only greater opportunity for self-condensation, but often the carbonyl group will also be reduced. The lack of selectivity under such conditions can be seen in reductions of oximinoacetoacetanilides with nickel at 90–95° and 100 atm pressure [62a] and in the hydrogenation of sebacoin oxime with rhodium at 100° and 70 atm pressure [62b]. In each instance only amino alcohols were obtained.

When elevated temperature and pressure conditions are used during these reductions to obtain amino alcohols, the chemist must be aware of the danger of hydrogenolysis if the carbonyl group is adjacent to an aromatic ring. The reduction products in such instances would be benzylic-type alcohols and subject to cleavage.

Hydrogenolyzable Groups

O- and N-Benzyl Groups. Selective hydrogenation of the hydroxyimino group in compounds also containing an *O*- or *N*-benzyl group or one related to them will depend on the catalyst, at times on the amount of catalyst, and perhaps on the reaction medium.

Nickel, in general a poor catalyst for hydrogenolysis, may be useful in obtaining selective reduction of the C=NOH function in compounds also containing benzylic-type alcohols, esters, or ethers. Its use has been demonstrated in the hydrogenations of 3-hydroxyimino-2-phenyl-2-butanol [9a] and hydroxyimino compounds of the type, ArC(OH)(alkyl)C(alkyl)=NOH [9b].

A Mannich-type compound, 4-hydroxy-3-(piperidinomethyl)benzaldoxime was reduced with Raney nickel to 4-hydroxy-3-(piperidinomethyl)benzylamine without undergoing hydrogenolysis [63a]. In contrast only monoamines were obtained when the reductions of 4-benzyl-4-R-substituted-2-butanone oximes, $C_6H_5CHRC_2(CH_3)$=NOH where R = 1-piperidino or 4-morpholino, was carried out at 100 atm or 3 atm [63b].

More study is necessary to determine if selective reductions in the above instances [63a, b] depend on whether an aldoxime or ketoxime is to be reduced.

Platinum oxide gave poor results in the conversion of (3-hydroxymethyl-4-methoxy)phenylacetaldoxime to 2-(3-hydroxymethyl-4-methoxy)phenethylamine [64]. Loss of the benzylic hydroxyl group took place during the hydrogenation of 3-hydroxyimino-2-phenyl-2-butanol in glacial acetic acid with platinum oxide to yield 3-phenyl-2-butylamine [65]. However, in this reduction the use of a 10% ratio of catalyst suggests an attempt to remove the hydroxyl group.

The danger of noncomitant loss of an O-benzyl group in reduction with palladium is pointed out in the hydrogenation of 7-benzyloxy-2,3-dihydro-1-oxo-1H-pyrrolo[1,2-a]indole [66] (Eq. 13.6).

$$C_6H_5CH_2O\text{-indole-N-R-=NOH} \xrightarrow{3 H_2} HO\text{-indole-N-R-NH}_2 \quad (13.6)$$

R = H

Hydrogenolysis must be attributed to the catalyst itself, not necessarily to the amount used, since in a similar reduction with platinum oxide where R = OOCCH$_3$ the 7-benzyloxy compound was obtained, despite the large amount of catalyst. In the palladium-catalyzed hydrogenation of 2-hydroxyimino-3-(4,6-dibenzyloxy-2-pyrimidyl)propionic acid, only debenzylation took place yielding 2-hydroxyimino-3-(4,6-dihydroxy-2-pyrimidyl)propionic acid [67]. In the reduction of α-hydroxyiminobenzyl 4-benzyloxyphenyl ketone with palladium, it was reported that both debenzylation and conversion of the oxime took place before the carbonyl group was attacked [67]. These results suggest that palladium catalysts should not be considered when the chemist is concerned with reducing the C=NOH function in the presence of an O-benzyl grouping.

Halogens. Selective reduction of the oximino group in the presence of halogen is dependent on the reactivity of the halogen and on the catalyst. The medium can be a factor if reaction is to be carried out in the absence of acid since loss of halogen is facilitated by the presence of nitrogen bases as well as inorganic bases (Chapter XX).

In general, nuclear chlorine will be unattacked during reduction of a halogen-containing oxime. Retention of nuclear bromine depends on the catalyst employed and sometimes on the amount [5]. Alkyl chloride should be retained, as should alkyl bromide under acidic conditions. However, dehalogenation appears to be the preferred reaction in the reduction of α-chlorocycloalkanonoximes (Chapter XX).

Palladium catalysts, because they are generally so effective for removal of halogen, should be avoided although some selective reductions have been carried out with them. Halogen was retained during the hydrogenation of 3'- and 4'-chloro-2-hydroxyiminopropiophenones,

$$ClC_6H_4C(=O)C(CH_3)=NOH,$$

in alcoholic hydrogen chloride with palladium on carbon to the corresponding aminoalcohols [69a, b]. The 2'-chloro compound did not yield aminoalcohol in pure form [69b]. Loss of halogen occurred during the reduction of the 4'-chloro compound if water was present [69a]. However, it was shown that not only could dehalogenation be prevented but that aminoketone could be obtained by reduction of 3'-chloro-4'-methoxy-2-hydroximinopropiophenone in excess methanolic hydrogen chloride by using 5% by weight of 10% palladium on carbon [69c].

Complete debromination occurred in the hydrogenation of 3'-bromo-2-hydroximinopropiophenone with palladium on carbon in excess hydrogen chloride to yield 2-aminopropiophenone [69b]. The possibility of selective debromination in such cases is suggested by the reduction of the oxime of 4-bromophenoxyacetone under similar conditions [5]. At one equivalent absorption there was a very sharp break in the rate of uptake; after two equivalents analysis of the reduction solution indicated complete removal or bromine. The product of reaction at this point, 1-(4-phenoxy)isopropylamine was obtained in 52% yield.

Successful reduction without loss of bromine was achieved during hydrogenations with Raney nickel at 100° and 30–50 atm pressure in absolute alcohol containing 5 molar equivalents of dry ammonia [5]. Moderate yields, 52–67%, of this and other 1-(halophenoxy)isopropylamines were obtained employing this procedure. The major side reaction was hydrolysis of hydrogenolysis to halophenols.

Very high yield, 88.5%, unaccompanied by loss of bromine was reported when 4-bromobenzophenone oxime was hydrogenated in alcohol with 5%

by weight of Raney nickel at 50–100° and 100 atm pressure to 4-bromobenzhydrylamine [69d].

Some dehalogenation was encountered in a reduction of 1-(4-chlorophenyl)-acetone oxime with nickel in alcoholic ammonia at 3 atm pressure [5]. It was evident that it was not concurrent but occurred after reduction to amine was complete as shown by the absence of chloride ion when the reaction was interrupted at 90–95% of the required two equivalents of hydrogen. Apparently the ketone was not 100% pure.

It is of interest when the same oxime was reduced in alcoholic hydrogen chloride with platinum on carbon that some dehalogenation took place [5]. It also occurred during reduction in acetic acid–acetic anhydride with 20% and 10% by weight of 5% platinum on carbon but it was totally eliminated when reduction was carried out in the same medium with 5% by weight of catalyst.

A platinum catalyst was employed for the hydrogenation of 4-(*bis*-2-chloroethyl)aminobenzaldoxime in neutral solvent to give 58% yield of 4-(*bis*-2-chloroethyl)aminobenzylamine [70].

Unsaturated Heterocycles

In most instances when the substrate contains a hydroxyimino group and an unsaturated heterocyclic system, reduction of the hydroxyimino group should be expected without attack on the ring system. However, complete selectivity is governed by the choice of catalyst and its activity, as well as reaction conditions and susceptibility of the ring system to hydrogenation (Chapter XXIV).

The imidazole ring not readily reduced, was not affected during the hydrogenation of imidazole-4-aldoxime with palladium on carbon in methanolic hydrogen chloride [21b]. The pyrrole ring, generally resistant to nickel reduction except under high-pressure conditions, remained intact during the hydrogenation of oximes of acetylpyrroles with Raney nickel at 100–150° and 200 atm [71a]. The reduction of an oxime in the presence of a sulfur-containing ring took place when palladium on carbon, equal to the weight of substrate was employed [71b].

Among more readily reducible ring systems, saturation of the furan ring was controlled by the use of aged Raney nickel during the reduction of the oxime of 2-acetylfuran in acetic anhydride at 70–80° and 100 atm for 1 hr [72a]. When the hydrogenation was carried out with active W-7 Raney nickel in methyl alcohol for 16 hr, 1-(tetrahydro-2-furyl)-1-ethylamine was obtained although there is little doubt that the oxime function was reduced first. The hydroxyimino group in a side chain attached to a furan ring was also selectively hydrogenated with Raney nickel under low-pressure conditions [71b, 72b] and with palladium on carbon, in alcohol containing acetyl chloride, under 15 atm pressure [72a].

The oxime of 3-acetylpyridine was converted to 3-(1-aminoethyl)pyridine by high-pressure reduction with Raney nickel [71a]. 2-, 3- and 4-(4-Diethylamino-1-hydroxyiminobutyl)pyridines were hydrogenated to the corresponding 2-, 3-, and 4-pyridylbutanediamines by means of Raney nickel at 70° and 70 atm pressure [73a]. Ethyl 2-hydroxyimino-2-(2-pyridyl)acetate was easily reduced to ethyl 2-amino-2-(2-pyridyl)acetate in 75% yield with palladium on carbon in alcohol under 3 atm pressure [73b]. Pyridoxaldoxime was hydrogenated without accompanying ring reduction with platinum oxide in alcohol containing hydrochloric acid [16].

When the oxime of 2-phenacylpyridine was reduced in glacial acetic acid with 6% by weight of platinum oxide, 2-(2-amino-2-phenyl)ethylpiperidine was obtained in 75% yield [73c]. The reaction conditions and amount of catalyst used suggest that the ring-reduced compound was sought. There is also a report in the reduction of an oximino ester, $RCH_2C(=NOH)COOC_2H_5$ where R is 4-pyridyl-1-oxide, that concurrent attack of the pyridine ring occurred but the catalyst and reaction conditions were not described [73d].

Saturation of the nitrogen-containing ring accompanied conversion of oxime to amine during reduction of $RCH_2C(=NOH)COOC_2H_5$, where R = 2-quinolyl, with Raney nickel in ethyl acetate at 60° and 70 atm to give 35% of ethyl 2-amino-3-(1,2,3,4-tetrahydro-2-quinolyl)propionate and 20% of ethyl 2-amino-3-(2-quinolyl)propionate [74a]. Low-pressure reduction with palladium on carbon in alcohol seen in [73b] might be more effective in preventing ring reduction.

In contrast when R was 4-quinolyl, reduction was carried out as acid in dilute aqueous ammonia with Raney nickel at room temperature and atmospheric pressure; 68.5% yield of β-(4-quinolyl)alanine was obtained in a slow reaction which required additional catalyst [74b].

It is of interest that the same method could not be applied when R was 4-pyridyl. Reduction with Raney nickel was too slow. When palladium black was used under similar conditions, uptake of hydrogen was incomplete. There was no attempt to employ higher temperature and somewhat higher pressure in either reaction.

The pyrazine and pyridazine rings were unaffected during hydrogenation of 2-hydroxyimino-3-(2-pyrazinyl or 3- or 4-pyridazinyl)propionic acids in aqueous ammonia under 4 atm pressure with 50% by weight of 5% palladium on carbon [75]. The nitrogen-containing ring was untouched when the 2-quinoxalyl compound was reduced as ester with Raney nickel at 60° and 70 atm [74a].

When the 3-substitutuent was 4-pyrimidyl the ring was unattacked, but when it was 2-pyrimidyl ring reduction did occur [75]. The investigators suggested from their data that nuclear hydrogenation took place in preference to reduction of the oxime. They ascribe the difference in results to the greater susceptibility of 2-alkyl substituted pyrimidines to ring reduction in the presence of palladium.

Some 5,6-dihydropyrimidyl compound was observed in the hydrogenation of 2-hydroxyimino-3-(4,6-dibenzyloxy-2-pyrimidyl)propionic acid in methyl alcohol with palladium on carbon when the reaction was terminated at uptake of four molar equivalents [67]. However, none was obtained in the nickel-catalyzed reduction of 2-hydroxyimino-3-(4,6-dihydroxy-2-pyrimidyl)propionic acid in aqueous ammonia.

REFERENCES

[1] a. G. Mignonac, Brit. Patent 282,083 (1928); b. A. Stoll, E. Jucker, and A. Ebnöther, *Helv. Chim. Acta*, **38**, 559 (1955); c. H. Zinnes, R. A. Comes, and J. Shavel, Jr., *J. Org. Chem.*, **31**, 162 (1966); d. K. W. Rosenmund and E. Pfankuch, *Ber.*, **56**, 2258 (1923).

[2] W. Reeve and J. Christian, *J. Am. Chem. Soc.*, **78**, 860 (1956).

[3] a. H. Adkins, *Reactions of Hydrogen*, University of Wisconsin Press, Madison, 1937, p. 92; b. M. Carmack, O. H. Bullitt, Jr., G. R. Handrick, L. W. Kissinger, and I. Von, *J. Am. Chem. Soc.*, **68**, 1222 (1946); c. R. Jones, Jr., C. C. Price, and A. K. Sen, *J. Org. Chem.*, **22**, 783 (1957); d. E. Profft and A. Jamar, *Arch. Pharm.*, **289**, 90 (1956); e. F. W. Bollinger, F. N. Hayes, and S. Siegel, *J. Am. Chem. Soc.*, **75**, 1729 (1953).

[4] a. F. E. King, T. Henshall, and R. L. St. D. Whitehead, *J. Chem. Soc.*, **1948**, 1373; b. E. R. Shepard, J. F. Noth, H. D. Porter, and C. K. Simmans, *J. Am. Chem. Soc.*, **74**, 4611 (1952); c. M. K. Goo-on, L. H. Schwartzman, and G. F. Woods, *J. Org. Chem.*, **19**, 305 (1954); d. J. A. Barltrop, R. M. Acheson, P. G. Philpott, K. E. MacPhee, and J. S. Hunt, *J. Chem. Soc.*, **1956**, 2928.

[5] M. Freifelder, unpublished results.

[6] a. K. N. Campbell, A. H. Sommers, J. F. Kerwin, and B. K. Campbell, *J. Am. Chem. Soc.*, **68**, 1556 (1946); b. H. T. Huang, D. S. Tarbell, and H. R. V. Arnstein, *J. Am. Chem. Soc.*, **70**, 4181 (1948); c. W. F. Newhall, *J. Org. Chem.*, **24**, 1673 (1959).

[7] D. C. Iffland and T-F. Yen, *J. Am. Chem. Soc.*, **76**, 4180 (1954).

[8] a. J. Biel, E. G. Schwarz, E. P. Sprengler, H. A. Leiser, and H. L. Friedman, *J. Am. Chem. Soc.*, **76**, 3149 (1954); b. E. Szarvasi and L. Neuvy, *Bull. Soc. Chim. France*, **26**, 1393 (1959); c. D. E. Ames, D. Evans, T. F. Gray, P. J. Islip, and K. E. Richards, *J. Chem. Soc.*, **1965**, 2636.

[9] a. P. Fréon and S. Ser, *Compt. Rend.*, **225**, 1336 (1947); b. *ibid*, **226**, 1098 (1948).

[10] J. DeCombe, *Compt. Rend.*, **222**, 90 (1946).

[11] a. A. Dornow and G. Petsch, *Arch. Pharm.*, **284**, 153 (1951); b. A. Dornow and A. Frese, *Arch. Pharm.*, **285**, 463 (1952); c. M. Avram, C. D. Nenitzescu, and M. Maxim, *Ber.*, **90**, 1424 (1957); d. S. Archer, T. R. Lewis, and M. J. Unser, *J. Am. Chem. Soc.*, **79**, 4194 (1957).

[12] W. Voegtli, U.S. Patent 2,793,210 (1957).

[13] L. N. Yakhontov and N. V. Rubstov, *J. Gen. Chem. USSR*, **29**, 2307 (1959) (Eng. trans.).

[14] E. Breitner, E. Roginski, and P. N. Rylander, *J. Chem. Soc.*, **1959**, 2918.
[15] B. Reichert and W. Koch, *Arch. Pharm.*, **273**, 265 (1935).
[16] S. Kreisky, *Monatsh.*, **89**, 685 (1958).
[17] E. E. Mikhlina, V. Ya. Vorobéva, and M. V. Rubstov, *Chem. Abstr.*, **65**, 2219 (1966).
[18] W. H. Hartung, *J. Am. Chem. Soc.*, **50**, 3370 (1928).
[19] C. D. Gutsche and H. E. Johnson, *J. Am. Chem. Soc.*, **76**, 1776 (1954).
[20] A. LaManna and L. Fontanella, *Il Farmaco, Ed. Sci.*, **10**, 687 (1955).
[21] a. T. Nógrádi, *Monatsh.*, **88**, 1087 (1957); b. R. A. Turner, C. F. Huebner, and C. R. Scholz, *J. Am. Chem. Soc.*, **71**, 2801 (1949); c. A. Lindemann, *Helv. Chim. Acta.*, **32**, 69 (1949).
[22] H. E. Zaugg, M. Freifelder, and B. W. Horrom, *J. Org. Chem.*, **15**, 1911 (1950).
[23] M. Freifelder, W. D. Smart, and G. R. Stone, *J. Org. Chem.*, **27**, 2209 (1962).
[24] P. Moses, R. Dahlbom, and B. Sjöberg, *Arkiv. Kemi.*, **1964**, 451.
[25] P. A. Levene and A. Schormüller, *J. Biol. Chem.*, **106**, 595 (1934).
[26] E. Schipper and A. R. Day, *J. Am. Chem. Soc.*, **74**, 333 (1952).
[27] M. Vignau, *Bull. Soc. Chim. France*, **1952**, 638.
[28] a. K. E. Hamlin, Jr. and W. H. Hartung, *J. Biol. Chem.*, **145**, 349 (1942); b. K. L. Waters and W. H. Hartung, *J. Org. Chem.*, **10**, 524 (1945); c. R. H. Barry, A. M. Mattocks, and W. H. Hartung, *J. Am. Chem. Soc.*, **70**, 693 (1948); d. W. H. Hartung, A. M. Mattocks, and R. I. Ellin, *J. Pharm. Sci.*, **53**, 553 (1964).
[29] A. F. Ferris, *J. Org. Chem.*, **24**, 1726 (1959).
[30] R. Gaudry and R. A. McIvor, *Can. J. Chem.*, **29**, 427 (1951).
[31] a. C. F. Winans and H. Adkins, *J. Am. Chem. Soc.*, **55**, 4167 (1933); b. R. Lukeš and J. Palacek, *Coll. Czechoslov. Chem. Commun.*, **29**, 1582 (1964); c. E. D. Clair, F. H. Clarke, W. A. Edminton, and K. Wiesner, *Can. J. Research, B*, **28**, 745 (1950).
[32] K. Masuzawa, M. Masaki, and M. Ohta, *Bull. Chem. Soc. Japan*, **38**, 2078 (1965).
[33] N. J. Leonard and W. E. Goode, *J. Am. Chem. Soc.*, **72**, 5404 (1950).
[34] G. M. Badger, J. W. Cook, and T. Walker, *J. Chem. Soc.*, **1948**, 2011.
[35] K. W. Rosenmund and P. Engels, *Arch. Pharm.*, **284**, 209 (1951).
[36] C. F. Koelsch and R. A. Scheiderbauer, *J. Am. Chem. Soc.*, **65**, 2311 (1943).
[37] K. V. Levshina, A. I. Gavrilova, and S. I. Sergievskaya, *J. Gen. Chem. USSR*, **30**, 3601 (1960) (Eng. trans.).
[38] H. Richter and M. Schenck, U.S. Patent 2,832,804 (1958).
[39] W. E. Rosen and M. J. Green, *J. Org. Chem.*, **28**, 2797 (1963).
[40] N. Levin, B. E. Graham, and H. G. Kolloff, *J. Org. Chem.*, **9**, 380 (1944).
[41] S. Ser, L. Piaux, and P. Freon, *Compt. Rend.*, **229**, 376 (1949).
[42] N. K. Kochetkov and N. V. Dudykina, *J. Gen. Chem. USSR*, **29**, 4038 (1959) (Eng. trans.).
[43] F. C. Uhle and W. A. Jacobs, *J. Biol. Chem.*, **160**, 243 (1945).
[44] a. H. J. Barber and A. D. H. Self, U.S. Patent 2,375,611 (1945); b. H. C. Carrington, *J. Chem. Soc.*, **1955**, 2527; c. M. A. Stevens and G. B. Brown, *J. Am. Chem. Soc.*, **80**, 2762 (1958).
[45] R. P. Mull, R. H. Mizzoni, M. R. Dapero, and M. E. Egbert, *J. Med. Chem.*, **5**, 651 (1962).
[46] L. Neelankantan and W. H. Hartung, *J. Org. Chem.*, **23**, 964 (1958).
[47] G. Vavon and M. Krajcinovic, *Bull. Soc. Chim. France*, [4], **43**, 231 (1928).
[48] a. F. Benington, R. D. Morin, and L. C. Clark, Jr., *J. Med. Chem.*, **8**, 100 (1965); b. J. Finkelstein, J. A. Romano, E. Chiang, and J. Lee, *J. Med. Chem.*, **6**, 153 (1963).
[49] a. G. Vavon and A. L. Berton, *Bull. Soc. Chim. France*, [4], **37**, 296 (1925); b. E. Müller, D. Fries, and H. Metzger, *Ber.*, **88**, 1891 (1955).

[50] L. W. Jones and R. T. Major, *J. Am. Chem. Soc.*, **52,** 669 (1930).
[51] R. T. Major and K. W. Ohly, *J. Med. Chem.*, **4,** 51 (1961).
[52] J. Levy, U.S. Patent 3,207,787 (1965).
[53] a. M. Fields, D. E. Walz, and S. Rothchild, *J. Am. Chem. Soc.*, **73,** 1000 (1950); b. J. Staněk and J. Urban, *Coll. Czechoslov. Chem. Commun.*, **15,** 397 (1950); c. R. B. Davis and J. D. Benigni, *J. Chem. Eng. Data*, **8,** 578 (1963); d. R. B. Davis, D. D. Carlos, and G. S. Mattingly, *J. Org. Chem.*, **30,** 2607 (1965).
[54] a. W. H. Hartung and J. C. Munch, *J. Am. Chem. Soc.*, **51,** 2262 (1929); b. W. H. Hartung, *J. Am. Chem. Soc.*, **53,** 2248 (1931).
[55] a. W. H. Hartung, J. C. Munch, W. A. Deckert, and F. Crossley, *J. Am. Chem. Soc.*, **52,** 3317 (1930); b. W. H. Hartung, J. C. Munch, E. Miller, and F. Crossley, *J. Am. Chem. Soc.*, **53,** 4149 (1931).
[56] a. N. Levin, B. E. Graham, and H. G. Kolloff, *J. Org. Chem.*, **9,** 380 (1944); b. R. Heinzelman, U.S. Patent 2,549,685 (1951).
[57] F. Ramirez and J. W. Sargent, *J. Am. Chem. Soc.*, **77,** 6297 (1955).
[58] C. J. Collins, M. M. Staum, and B. M. Benjamin, *J. Org. Chem.*, **27,** 3525 (1962).
[59] a. N. F. Albertson, B. F. Tullar, J. A. King, and B. B. Fishburn, *J. Am. Chem. Soc.*, **70,** 1150 (1948); b. R. H. Wiley and O. H. Borum, *J. Am. Chem. Soc.*, **70,** 1666 (1948).
[60] T. Nozoe, K. Takase, and K. Suzuki, *Bull. Chem. Soc. Japan*, **38,** 362 (1965).
[61] a. E. T. Stiller, U.S. Patent 2,422,598 (1947); b. V. Bruckner, G. Fodor, J. Kiss, and J. Kovács, *J. Chem. Soc.*, **1948,** 855; c. J. Murphy, *J. Org. Chem.*, **26,** 3104 (1961); d. M. Regitz and B. Eistert, *Ber.*, **96,** 3120 (1963); e. R. Huisgen, G. Binsch, and L. Glosez, *Ber.*, **97,** 2628 (1964).
[62] a. G. Ehrhart and I. Hennig, *Ber.*, **89,** 1568 (1956); b. P. E. Fanta, L. J. Panya, W. R. Groskopf, and H.-J. Su, *J. Org. Chem.*, **28,** 413 (1963).
[63] a. N. H. Cromwell, *J. Am. Chem. Soc.*, **68,** 2634 (1946); b. N. H. Cromwell, Q. T. Wiles, and O. Schroeder, *ibid.*, **64,** 2432 (1942).
[64] B. Reichert and E. A. dem Kampe, *Arch. Pharm.*, **277,** 261 (1939).
[65] B. Witkop, *J. Am. Chem. Soc.*, **70,** 1424 (1948).
[66] W. A. Remers, R. H. Roth, and M. J. Weiss, *J. Org. Chem.*, **30,** 2910 (1965).
[67] R. H. Springer, W. J. Haggerty, Jr., and C. C. Cheng, *J. Heterocyclic Chem.*, **2,** 49 (1965).
[68] W. D. McPhee and E. S. Erickson, Jr., *J. Am. Chem. Soc.*, **68,** 624 (1946).
[69] a. W. Hartung, J. C. Munch, and F. S. Crossley, *J. Am. Chem. Soc.*, **57,** 1091 (1935); b. B. L. Zenitz and W. H. Hartung, *J. Org. Chem.*, **11,** 444 (1946); c. H. V. Daeniker, *Helv. Chim. Acta*, **49,** 1543 (1966); d. C. F. Winans, *J. Am. Chem. Soc.*, **61,** 3564 (1939).
[70] F. Bergel, J. L. Everett, J. J. Roberts, and W. C. J. Ross, *J. Chem. Soc.*, **1955,** 3835.
[71] a. H. Adkins, I. A. Wolff, A. Pavlic, and E. Hutchinson, *J. Am. Chem. Soc.*, **66,** 1293 (1944); b. W. Ried and W. Reitz, *Ber.*, **89,** 2570 (1956).
[72] a. N. Clauson-Kaas, N. Elming, and Z. Tyle, *Acta. Chem. Scand.*, **9,** 1 (1955); b. C. Lespagnol, J. C. Cazin, and M. Cazin, *Bull. Soc. Chim. France*, **1965,** 942.
[73] a. M. S. Bloom, D. S. Breslau, and C. R. Hauser, *J. Am. Chem. Soc.*, **67,** 2206 (1945); b. G. Van Zyl, L. De Vries, R. H. Decker, and E. T. Niles, *J. Org. Chem.*, **26,** 3373 (1961); c. R. F. Shuman, H. F. Hansen, and E. D. Amstutz, *ibid.*, **27,** 1970 (1962); d. R. L. Bixler and C. Niemann, *ibid.*, **23,** 575 (1958).
[74] a. W. Ried and H. Schiller, *Ber.*, **86,** 730 (1953); b. D. F. Elliott, A. T. Fuller, and C. R. Harrington, *J. Chem. Soc.*, **1948,** 85.
[75] W. J. Haggerty, Jr., R. H. Springer, and C. C. Cheng, *J. Heterocyclic Chem.*, **2,** 1 (1965).

XIV

CARBONYL GROUPS

Aldehydes are readily hydrogenated to alcohols; ketones, because they are more hindered, are somewhat more difficult to reduce. In either case, further reduction of the alcohol to hydrocarbon is dependent on structure of the compound, reaction conditions, the reaction medium, and the catalyst; this does not usually take place when aliphatic aldehydes or ketones are hydrogenated catalytically.

There does not appear to be a particular catalyst or set of reaction conditions applicable to hydrogenations of all aldehydes and ketones. In general, reductions with noble metal catalysts can be carried out at 1–3 atm pressure. Certain special freshly prepared nickel catalysts may be employed under similar conditions, but in most instances higher pressures are required for nickel reductions. Hydrogenations of carbonyl-containing compounds with copper chromite require more vigorous conditions than with nickel. However, it may be advantageous to employ it in the reduction of substrates which also contain an aromatic ring because saturation of the ring does not usually take place in its presence. In addition, unless reaction temperature is too high, concomitant hydrogenolysis of the resulting alcohol does not occur except in the reduction of certain phenolic aldehydes and ketones.

ALDEHYDES

When aldehydes are hydrogenated over platinum oxide or platinum or ruthenium on carbon, inactivation of the catalyst may take place (it does not occur in the presence of palladium or rhodium). Inactivation can be overcome by aeration of the reaction mixture [1a] or prevented by reduction in the presence of a small amount of oxygen [1b, c]. Reduction over platinum oxide may be promoted by addition of specific amounts of ferric chloride or stannous chloride [1a, 2] as it was in the conversion of valeraldehyde to 1-pentanol (valeryl alcohol). Improvement in these instances appears to be limited to reactions in alcohol [2].

When reductions of aldehydes are carried out with nickel catalyst promoters are frequently added. Sodium hydroxide, for example, was effective in the hydrogenation of furfuraldehyde to furfuryl alcohol [3a]; when it was added to the reduction of glucose with nickel on kieselguhr, the reaction temperature and pressure were greatly decreased [3h]. The use of other promoters for hydrogenations of aldehydes will be noted in ensuing sections of this chapter.

Aliphatic Aldehydes

Although ruthenium catalysts are not generally useful under low-pressure conditions, 5% ruthenium on carbon was particularly effective in the hydrogenation of heptaldehyde to heptyl alcohol in aqueous alcohol [1b]. It was also employed in the reduction of dextrose to sorbitol at 130° and 5–6 atm [1d]. Its use should be extended to other aldehydes for a better evaluation of its capabilities. It must be noted, however, that it is usually necessary to activate supported ruthenium catalysts for low-pressure reductions by prehydrogenation in solvent before addition of substrate.

Raney nickel-catalyzed reduction of 3-methoxybutyraldehyde at elevated temperature and pressure yielded 3-methoxybutanol [4a]. Quantitative yield of rhamnitol was obtained from nickel reduction of D-rhamnose [4b]. Hydrogenation of glucose and other sugars in aqueous solution with Raney nickel and magnesium metal at 95–95° and 20 atm pressure [4c] appears to be an excellent method worthy of modification (substitution of organic or hydrated organic solvent) and extension as a general one for the reduction of aldehydes and ketones. Reaction period was short, quantitative yields were obtained, and the catalyst could be reused indefinitely if a small amount of magnesium was added after every tenth hydrogenation.

Reductions of hexaldehyde under 1–3 atm pressure have been carried out with highly active W-6 Raney nickel [4d]. The addition of triethylamine decreased reaction time almost 200%. The addition of platinic chloride and triethylamine had a more profound effect. However, W-6 Raney nickel has certain disadvantages. It must be freshly prepared under special conditions and it retains its high activity for only a few weeks.

Aromatic Aldehydes

The hydrogenation of an aromatic aldehyde to the corresponding alcohol, $ArCH_2OH$, may be complicated by further hydrogenolysis to hydrocarbon. This side reaction, which is easily controlled, is dependent on the catalyst (sometimes on the amount), on the severity of the reaction conditions and, in certain instances, on the presence of a phenolic hydroxy group *ortho* or *para* to the aldehyde function. Examples of such hydrogenolyses are seen in the reductions of 2- and 4-hydroxybenzaldehydes with copper chromite at 200–250° and 220–240 atm to yield cresols [5]. In general, the presence of these hydroxyl groups does not induce hydrogenolysis in reductions of aromatic aldehydes with other catalysts under moderate conditions.

Although supported palladium catalysts are very active for most hydrogenolyses, they may be the ones of choice for the hydrogenation of aromatic aldehydes in neutral solvent to benzyl-type alcohols. Reductions of 3-hydroxy-4-methoxybenzaldehyde [6a], 2,4-dihydroxybenzaldehyde [6b], and 3,4-dihydroxy-3,4-dimethoxy-, and 4-isopropoxybenzaldehyde [7] were successful with palladium on carbon at 1–3 atm pressure. Hydrogen uptake is usually very rapid in these reactions. Any danger of hydrogenolysis can be controlled by employing a moderate amount of catalyst [7]. The presence of an inhibitor can also prevent hydrogenolysis, as illustrated by the addition of *N,N*-diethylnicotinamide to the conversions of substituted benzaldehydes to benzyl alcohols in high yields over palladium black [8].

Reductions of benzaldehydes with platinum are less prone to hydrogenolysis than with palladium but hydrogen uptake can be slow and a promoter may be required [7]. 3,5-Dimethoxybenzaldehyde was converted to the alcohol in 95% yield in the presence of platinum catalyst [9a] and 2,3-dimethoxybenzyl alcohol was obtained from a platinum-catalyzed reduction of redistilled aldehyde [9b]. Promoters were not used in the above reactions [9a, b] but it was necessary to add ferric chloride to the hydrogenation of 3,4-dimethoxybenzaldehyde to the corresponding alcohol to obtain uptake of one molar equivalent in 18 hr [7].

Hydrogenations of substituted benzaldehydes with commercial Raney nickel are usually carried out at pressures well above atmospheric [10a–c]. There is little danger of conversion of $ArCH_2OH$ to $ArCH_3$ unless reaction conditions are too vigorous or unless a 2- or 4-hydroxybenzaldehyde is to be reduced [5]. These can and have been hydrogenated with nickel to the corresponding benzyl alcohols by controlling reaction conditions. 2-Hydroxybenzaldehyde was converted to 2-hydroxybenzyl alcohol with Raney nickel at room temperature and atmospheric pressure [11a]; reduction at 50–60° gave the same product and a trace of cresol. At 55° and 15–59 atm *ortho* cresol was the only product. In another example, 4-hydroxy-3-methoxybenzaldehyde was reduced with nickel at 40–50° and 130 atm to yield 90% of 4-hydroxy-3-methoxybenzyl alcohol [11b].

Aralkyl Aldehydes

Any of the catalytic methods cited in the two preceding sections should be applicable for conversion of aralkylaldehydes to alcohols because the resulting products, $Ar(CH_2)_nOH$ where n is at least 2, are no longer susceptible to loss of alcoholic hydroxyl except under extreme conditions.

Amino Aldehydes

Reductions of aldehydes of Mannich bases, $R_2NCH_2C(CH_3)_2CHO$, to amino alcohols were successful when reactions were carried out as hydrochloride salts in aqueous solution at pH 3.0–6.0 with Raney nickel at 79–80° and 10 atm pressure [12a]. With noble metal catalysts some hydrogenolysis to form R_2NH resulted in every instance. Hydrogenation of free aminoaldehyde gave poor yield even with nickel. In other instances high yields of 2,2-dimethyl-3-dimethylaminopropanol and 2,2-dimethyl-3-(1-piperidyl)-propanol were obtained by reduction of the corresponding aminoaldehydes as bases with nickel in alcohol at 70–80° and 120 atm [12b].

Nickel was also employed for hydrogenations of 2-diethylamino- and 2-(1-piperidino)alkanals to aminoalcohols [12c]. The reduction of 2-diethylaminobutyraldehyde was carried out at room temperature and atmospheric pressure; with more hindered compounds the pressure was raised to 84 atm.

MONOKETONES

Promoters or promoted catalysts are used more frequently in the reduction of ketones than in the reduction of aldehydes.

The addition of metal salts was required in platinum-catalyzed hydrogenations [1a, 2, 13a]. Oddly enough sodium hydroxide, assumed to be a poison for platinum, decreased reaction time in the reduction of some α-ketols [13b]. It also activated hydrogenations of ketones at atmospheric pressure with ruthenium catalyst [13c].

The presence of alkali in Raney nickel or in reductions of ketone with it to which sodium or potassium hydroxide has been added is reported to greatly increase the reaction rate [4d, 13d, e]. Sodium hydroxide was a promoter in platinized-nickel reductions [13f]. The addition of sodium hydroxide and chlorplatinic acid improved W-6 Raney nickel hydrogenations of ketones [13g]. Lithium hydroxide was shown to increase the reaction rate in nickel reduction of aliphatic ketones to a greater degree than sodium hydroxide [14a].

Triethylamine was effective in promoting nickel-catalyzed hydrogenation of ketones [4d]. Reductions of aliphatic, alicyclic, and aromatic ketones at 1–3 atm with W-4 Raney nickel [14b] were promoted to the greatest degree by a combination of platinic chloride, sodium hydroxide, and triethylamine [14a]. This procedure may be useful since W-4 nickel is relatively easy to prepare and does not lose its activity as does W-6 nickel.

A combination of rhodium and platinum oxides may also be of value for the reduction of many ketones [14c].

Dialkyl Ketones

Hydrogenolysis does not usually occur during the hydrogenation of dialkyl ketones to secondary alcohols. Palladium, of all the commonly employed catalysts, appears to be the least adaptable for reductions of these ketones [13c, 15]. High-pressure methods using Raney nickel or copper chromite are applicable but not necessary unless very rapid reaction is desired or if the carbonyl group is highly hindered as in pinacolone, $(CH_3)_3CCOCH_3$ [16a, b].

Promoted nickel reductions at low pressure proceeded readily except with diisopropyl ketone [14a]. 5-Acetoxy-2-pentanone was converted to 5-acetoxy-2-pentanol with unpromoted Raney nickel under 2 atm pressure [16c].

Ruthenium-catalyzed low-pressure reductions of acetone and methyl ethyl ketone in water containing sodium hydroxide [13c] and of methyl isobutyl ketone in neutral or alkaline solution [15] suggest that these methods should be studied more extensively for possible wider application. Although success was obtained in the latter preference, the use of a large amount of catalyst may have been responsible for overcoming steric effects. In contrast, when ethyl 2,2-difluoro-3-ketosuccinate was reduced with a moderate amount of ruthenium dioxide, 1.5% by weight of substrate, reaction conditions were far more vigorous, 75–80° and 70–140 atm [17a].

In another hydrogenation at 3 atm pressure, that of ethyl 10-acetoxy-3-ketodecanoate, 90% yield of the corresponding 3-hydroxy compound was obtained but 10% by weight of ruthenium dioxide was employed and reaction time was 40 hr [17b].

Platinum-catalyzed reductions of dialkyl ketones have not been too widely studied. It was reported that methyl isobutyl ketone was more rapidly hydrogenated in acetic acid and in aqueous hydrochloric acid over platinum catalyst than in neutral solvent [15]. Acetic acid may be a useful medium for these hydrogenations but aqueous hydrochloric in most instances should prove impractical because of poor solubility of most ketones in it. The use of alcoholic hydrogen chloride in these reductions may lead to ether formation [18a].

High concentrations of substrate in acetic acid was reported necessary for rapid reduction of ketones [15b]. It was also suggested that trifluoracetic acid was the most useful medium for platinum reductions of aliphatic ketones. However, in many instances the resulting alcohols were esterified.

Despite the emphasis on acidic conditions the platinum-catalyzed reduction of levulinic acid in ether was more rapid and gave higher yield of hydroxy acid than reaction in acetic acid [18c].

Platinum-catalyzed reductions in neutral solvent may be particularly useful for the conversion of keto esters to the corresponding hydroxy esters. Ethyl 2-keto-3-methylsuccinate was reduced in 2–3 hr to ethyl 2-hydroxy-3-methylsuccinate in 88–92% yield [19a]. High yields (82–88%) of hydroxy esters were obtained from hydrogenations of keto esters,

$$CH_3CO(CH_2)_{1-5}COOC_2H_5 (19b);$$

the reduction rate decreased as the number of methylene groups increased. Ethyl acetoacetate was reduced in water with platinum oxide and in tetrahydrofuran with platinum on carbon to ethyl 3-hydroxybutyrate uncontaminated with ethyl butyrate [19c]. The use of ruthenium on carbon in water is also of value in the same reactions.

Cycloalkanones

Unsubstituted Cycloalkanones. Unsubstituted cycloalkanones have been reduced to hydroxycycloalkanes with a number of catalysts under a variety of conditions. Only palladium was found ineffective [15].

Rhodium-platinum oxide [14c] showed some promise for the low-pressure hydrogenation of cyclopentanone and cyclohexanone. Rapid hydrogen uptake occurred during reductions of cyclohexanone at 1–3 atm with ruthenium in aqueous solution containing sodium hydroxide as promoter [13c] or with ruthenium or rhodium on carbon in water or aqueous sodium hydroxide [15]. The latter procedure was also used for the hydrogenation of cyclopentanone to cyclopentanol [15]. Cyclohexanone was readily reduced with platinum on carbon in water, dilute aqueous sodium hydroxide or dilute hydrochloric acid [15] but the catalyst was poisoned in the hydrogenation of cyclopentanone in alkaline solution.

Hydrogenations of cyclopentanone at elevated temperature and pressure with nickel or copper chromite have given high yields of cyclopentanol [7, 20a, b]. Cycloheptanone was converted to cycloheptanol with nickel at low pressure but 50% by weight of catalyst was used and reaction period covered 14 hr [20c].

Substituted Cycloalkanones. The hydrogenation of a substituted cycloalkanone is more involved than that of an unsubstituted one. A substituent adjacent to the carbonyl function can have a profound effect on its reducibility. In addition reduction of a substituted cycloalkanone will yield a mixture of stereoisomeric alcohols which may be difficult to separate.

The reaction medium is an important factor in these reductions because of its effect on the stereochemistry of the resulting alcohols. The catalyst may also be a factor in the formation of the stereoisomeric alcohols although the effect may be due to occluded acid or alkali in the catalyst.

Hydrogenations in the presence of base are assumed to give equatorial alcohol while reactions in acidic media lead to axial alcohols [21]. However, in many instances the stereochemistry of the reduction products is not reported. There is a prime need for a thorough investigation in this area to determine the effect of structure, catalyst, and medium on the stereochemistry of the reaction products.

CYCLOBUTANONES AND CYCLOPENTANONES. Examples include the reductions of N-(2,2-dimethyl-3-oxocyclobutyl)-N-propylacetamide and N-(2,2-dimethyl-3-oxocyclobutyl)caprolactam to the corresponding alcohols in high yield with ruthenium on carbon or with Raney nickel at 100° and 210 atm [22].

Among cyclopentanones, 2,2-*bis*-hydroxymethylcyclopentanone was hydrogenated at low pressure with 50% by weight of freshly prepared nickel catalyst in a slow reaction which gave almost quantitative yield of triol. The application of heat and/or higher pressure should result in more rapid reduction of this hindered ketone and allow the use of less catalyst. 3-Oxocyclopentylacetic acid in aqueous alkali was converted to 3-hydroxycyclopentylacetic acid with Raney nickel under 3.3 atm hydrogen pressure [23b]. Reductions in alcohol with nickel, platinum oxide, or iron-promoted platinum oxide were unsatisfactory.

CYCLOHEXANONES AND LARGER CYCLOALKANONES. Some mono- and disubstituted cyclohexanones were readily hydrogenated at 1–3 atm with promoted W-4 Raney nickel [14a]. The reduction of 2,2,6-trimethylcyclohexanone was more difficult and 2,2,6,6-tetramethylcyclohexanone could not be completely hydrogenated to the alcohol under the same conditions.

3-Methylcyclohexanone was reduced in acetic acid with platinum oxide [24a]. The presence of a branched chain in 2-position made it necessary to employ a 10% ratio of platinum black in the hydrogenation of 2-isopropylcyclohexanone in acetic acid [24b]. A mixture consisting of 89% *cis* and 11% *trans* 2-isopropylcyclohexanol was obtained.

In other examples among substituted cyclohexanones, cyclohexanone-4-carboxylic acid gave good yield of *cis*-4-hydroxycyclohexanecarboxylic acid upon reduction in dilute alkali with Raney nickel [25a]. Uptake of hydrogen in the conversion of 5-(2-oxocyclohexyl)pentanoic acid to the corresponding alcohol under similar conditions was very slow [25b] due no doubt to steric effects. When methyl 6,6-dimethylcyclohexanone-2-acetate was reduced in alcohol with Raney nickel, the corresponding *cis*-hydroxy ester was obtained; hydrogenation in acetic acid with platinum gave the *trans* ester [25c].

Among cycloheptanones, the reduction of 4-*t*-butylcycloheptanone is of interest because of the effect of the *t*-butyl group [25d]. Hydrogenation in acetic acid over platinum gave a preponderance of *trans* alcohol, the presence of concentrated hydrochloric acid in a similar reaction yielded the *cis* alcohol as the major product.

Reductions of 2-hydroxycyclopentadecan- and 2-hydroxycyclohexadecanones were carried out with platinum oxide in ethyl alcohol to give presumably *cis* diols [26].

STEROIDAL KETONES. Hydrogenations of steroidal mono ketones to alcohols have generally been carried out in neutral solvent with Raney nickel or platinum oxide under mild conditions [27a–c]. Poor solubility may necessitate the use of acetic acid and can lead to the formation of a different epimer. The addition of strong acid to reductions in alcohol may induce ether formation [18a, 27d].

Estrone, containing a phenolic hydroxyl in the A ring, has been hydrogenated in alkaline solution [27a] (Eq. 14.1). It was also reduced as the

$$\text{Estrone} \xrightarrow{H_2} \text{Estradiol} \quad (14.1)$$

sodium salt of the sulfonic ester with platinum oxide [27a]. Estrone was reduced in alcohol to estradiol in good yield with copper chromite at 125° and 125 atm pressure but Raney nickel reduction at 100° and 100 atm caused saturation of the aromatic ring [7].

Alkyl Cycloalkyl Ketones

Cyclopropyl methyl ketone was hydrogenated to the corresponding carbinol with platinum oxide [28]. The yield of 1-cyclopropyl-1-ethanol in this reduction was greatly improved and reaction time was decreased by the addition of a drop of 40% sodium hydroxide solution to the reaction in alcohol with platinum [7]. Enhanced reaction rate due to sodium hydroxide was also encountered in the platinum-catalyzed hydrogenation of 1-acetylcyclohexanol to 1-(1-hydroxyethyl)cyclohexanol [13b].

Dicyclic Ketones

Catalytic hydrogenations of dicyclic ketones to the corresponding carbinols have not been reported. Molecular models suggest that the carbonyl group cannot make contact with the catalyst as it can in alkyl cycloalkyl ketones. Attack at the carbonyl group will only take place after ring rupture as exemplified in the hydrogenation of dicyclopropyl ketone. No uptake of hydrogen was observed during platinum-catalyzed reductions but hydrogenation over palladium on carbon gave unreacted ketone, 1-cyclopropyl-1-butanol, and 4-heptanol [7] (Eq. 14.2).

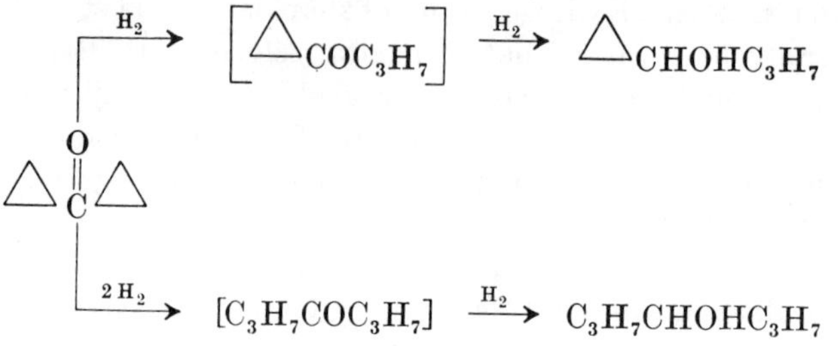

$$\triangle\overset{O}{\underset{C}{\|}}\triangle \quad \begin{array}{c} \xrightarrow{H_2} [\triangle COC_3H_7] \xrightarrow{H_2} \triangle CHOHC_3H_7 \\ \\ \xrightarrow{2H_2} [C_3H_7COC_3H_7] \xrightarrow{H_2} C_3H_7CHOHC_3H_7 \end{array} \qquad (14.2)$$

Ketones Containing Aromatic Rings

The hydrogenation of ketones, ArCOR where R is alkyl or substituted alkyl, leads to a benzyl-type secondary alcohol which may be subject to further hydrogenolysis. As a rule this seldom takes place in neutral medium. Indeed, palladium on carbon, the most efficient catalyst for O-debenzylations, may be the one of choice for the hydrogenation of acetophenone to the carbinol in alcohol under low-pressure conditions [29]. Platinum oxide also appears useful for the same reduction [14c, 29], as does Raney nickel when promoted with alkali [29].

Raney nickel catalyzed the hydrogenations of 2-(benzoylmethyl)- and 2-(4-methoxybenzoylmethyl)adipic acid in alkaline solution [30]. The resulting products were lactones formed by cyclization of the hydroxy acids (Eq. 14.3).

$$RC_6H_4COCH_2CH(COOH)(CH_2)_3COOH \xrightarrow{H_2}$$
$$[RC_6H_4CHOHCH_2CH(COOH)(CH_2)_3COOH] \longrightarrow$$

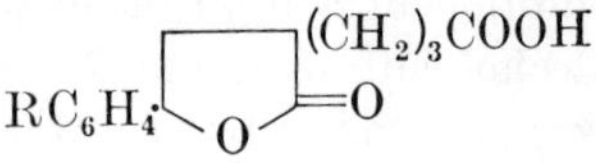

$$RC_6H_4\underset{O}{\overset{(CH_2)_3COOH}{\diagdown\diagup}}=O$$

(14.3)

R = H or 4-OCH$_3$

When rapid conversion of substituted aryl alkyl ketones to carbinols is desired, hydrogenation with Raney nickel or with copper chromite at elevated temperature and pressure [5] should be considered, except when a

2- or 4-hydroxy group is attached to the aromatic ring. In those instances reduction will yield alkylphenols [5]. However, it is likely that hydrogenolysis takes place because of the elevated temperature—hydrogenolysis in the reductions of alkyl 2- or 4-hydroxyphenyl ketones with nickel was reported to be dependent on temperature [31].

Successful low-pressure hydrogenations of 2- and 4-hydroxybenzaldehydes in neutral solvent with Raney nickel [11a, b] and with palladium on carbon [6a, b] without hydrogenolysis suggest that either procedure should also be applicable for hydrogenations of phenolic ketones. Platinum oxide gave good yield in a related reduction at 3 atm when amyl 2-hydroxyphenyl ketone was converted to 1-(2-hydroxyphenyl)-1-hexanol [7].

There is usually little danger of saturating the aromatic ring when reductions in neutral medium are carried out with palladium, platinum, or nickel under moderate conditions. Exceptions are noted in the hydrogenation of 9-benzoylanthracene with Raney nickel which yielded 9-benzoyl-1,2,3,4-tetrahydroanthracene as the major product [32a] and in the reductions of 9-acetyl- and 9-propionylanthracenes with platinum oxide which gave only the 9-acyl-1,2,3,4-tetrahydroanthracenes [32b].

Rhodium is contraindicated in the hydrogenation of aryl ketones because of the tendency toward ring reduction [15, 29]. The 70–30 rhodium-platinum combination of Nishimura [14c] has the same tendency although it may be possible to interrupt the reaction after uptake of one equivalent to obtain carbinol containing the intact aromatic ring. Ruthenium, ineffective for these reductions at low pressure, yielded 1-cyclohexyl-1-ethanol as well as 1-phenyl-1-ethanol upon hydrogenation of acetophenone at 70° and 70 atm [7].

Reductions of aryl ketones in acidic media can lead to hydrocarbon in the presence of palladium or platinum [32c]; saturation of the ring often takes place when similar reductions are carried out over platinum [32c] or rhodium-platinum oxide [14c].

When diaryl ketones, as benzophenone, are hydrogenated, palladium on carbon should not be employed unless inhibitors are present. Diphenylmethanes were always obtained in the reductions of mono- and disubstituted benzophenones even when a small amount of catalyst was employed unless it was inactivated with nicotinamide or N,N-diethylnicotinamide [33]. A small amount of pyridine or quinoline might also inhibit hydrogenolysis [7].

Examples of hydrogenations of benzophenones to benzhydrols without hydrogenolysis are seen in the platinum-catalyzed reductions of 4-(2-diethylaminoethoxy)benzophenone hydrochloride [7] and 2-amino-4'-methoxybenzophenone at 2–3 atm [34a] and in the low-pressure hydrogenation of 4-dimethylaminobenzophenone with commercial Raney nickel [7]. High-pressure methods with copper chromite and Raney nickel [34b] and with platinized nickel [34c] have also been successful giving high yields of benzhydrols.

When a carbon or a hetero atom separates the benzene ring and the carbonyl group, the tendency toward hydrogenolysis of the resulting hydroxyl group is almost completely removed, particularly in reductions in nonacid media. In such instances palladium catalysts are especially useful. Palladium on carbon catalyzed the very rapid hydrogenation of benzyl trifluoromethyl ketone to 1,1,1-trifluoro-3-phenyl-2-propanol in 98% yield [35a]. It was also employed in the reduction of diphenylpyruvic acid to 3,3-diphenyllactic acid [35b].

A nickel catalyst, nickel on kieselguhr, was employed to reduce phenoxyacetone and 2-naphthoxyacetone to the corresponding carbinols at 90–100 atm [35c].

Aminoketones

In general the best method for hydrogenating these compounds is as a neutral salt in alcohol, aqueous alcohol, or water. Reduction as base may cause inactivation of the catalyst. In the case of Mannich-type aminoketones, reduction as base can lead to extensive deamination.

Palladium, platinum and, to a lesser degree, nickel have been employed for reductions of the salts of aminoketones.

ω-Aminoacetophenones and Related Ketones.
For the reduction of ω-aminoacetophenones and related compounds, ArCOCNXY where X and Y are H or alkyl or NXY is dialkylamino or part of a cyclic system, the method of Mannich and Thiele [36a], hydrogenation as a salt over palladium on carbon, appears to be the one of choice.

The only precaution is to avoid an excess of strong acid which might cause hydrogenolysis of the benzylic alcohol [36b]. When the reduction solution is essentially neutral, there is little danger of hydrogenolysis even with the especially active catalyst of Iwamoto and Hartung [37a] because of a distinct break in the rate curve after the carbonyl group has been converted to carbinol [7].

The following references [37b–f] employed Mannich and Thiele's method. In one instance palladium on barium sulfate was used [37g].

Platinum oxide was employed in hydrogenations of similar aminoketones [38a–f] but reaction time was often long [38b] and a high catalyst ratio [38c] or additional catalyst was necessary to obtain complete uptake of hydrogen [38d].

Mannich-Type Aminoketones. When reduction of a Mannich-type aminoketone, RCOCCH$_2$NXY where R is alkyl or aryl and NXY is a tertiary nitrogen system, is carried out as the salt the major side reaction, deamination, can usually be avoided. It is difficult to suggest a catalyst of choice for these compounds. Palladium [39a–e] and platinum [40a–g] have been employed for hydrogenations of the salts as has Raney nickel [40e, 41a, b]. Nickel was used in the reduction of 4-dimethylamino-3-phenylbutanone hydrochloride when reaction failed with palladium on carbon [41c].

Occasionally hydrogenation as base is successful as in the reduction of methyl-2-(4-morpholino)-3-benzoylpropionate with platinum [39e] but more often deamination either accompanies formation of aminoalcohol or is the major reaction. Diethylamine resulted from palladium and platinum-catalyzed hydrogenations of 4-diethylamino-2-butanone [42a], but the aminoalcohol was obtained from reduction in acetic acid with platinum. 2-Dibutylaminoethyl and 2-diamylaminoethyl (4-methoxy-1-naphthyl) ketone underwent cleavage on reduction with Raney nickel or palladium [42b]. Hydrogenation of these particular compounds as hydrochloride salts was only satisfactory with platinum oxide.

The effect of factors, other than reduction as base, such as the catalyst and structure of the substrate on the occurrence of deamination in the hydrogenation of Mannich-type aminoketones has been studied only meagerly. There are a few examples of the effect of structure seen in nickel-catalyzed reductions of ω-dimethylaminopropiophenones as hydrochloride salts where deamination occurred when the carbon atom adjacent to the ketone was primary or secondary but not when it was tertiary [41b] (Eq. 14.4). In other

$$C_6H_5COC(CH_3)_2CH_2N(CH_3)_2 \cdot HCl \xrightarrow{H_2}$$
$$C_6H_5CHOHC(CH_3)_2CH_2N(CH_3)_2 \cdot HCl \quad (14.4)$$

$$C_6H_5COCHRC(R)_2N(CH_3)_2 \cdot HCl \xrightarrow{H_2}$$
$$\text{Aminoalcohol} + C_6H_5COCHRCH(R)_2 + (CH_3)_2NH$$

R = H or alkyl

hydrogenations deamination occurred as the dialkylamino group increased in size [40c].

Miscellaneous Linear Ketones. Hydrogenation of 6-dimethylamino-4,4-diphenyl-3-heptanone hydrochloride was slow but successful with platinum oxide [43a, b] but unsuccessful with palladium [43a] or Raney nickel [43b]. The more hindered 6-dimethylamino-4,4-diphenyl-5-methyl-2-hexanone could not be reduced catalytically [43a, c].

Hydrogenations of aminoketones of the type, RCH$_2$COCH$_2$NXY where R is aliphatic, aromatic, cyclic, or saturated heterocyclic and NXY is dialkylamino or part of a cyclic system, have been carried out as the salt in alcohol or water with platinum oxide [44b]. As should be expected, reduction as base was slow.

Saturated N-Heterocyclic Ketones. Hydrogenation of the carbonyl group in 1-substituted-3-pyrrolidones and 3- and 4-piperidones was readily carried out in neutral solvent with platinum oxide [45a–e]. When reduction of the substituted 3-pyrrolidone was carried out in methyl alcohol, the 1-substituted-3-hydroxypyrrolidine was obtained [45a]; hydrogenation in glacial acetic acid with an excessive amount of catalyst yielded only the 1-substituted pyrrolidine. Raney nickel catalyzed the hydrogenations of 1-alkyl and 1-cyclohexyl-4-piperidones to 1-substituted-4-hydroxypyridines [12b, 46a–c]. There appear to be no reports on the use of palladium catalysts in these reductions.

Palladium, platinum, rhodium, and ruthenium catalysts have been employed for the hydrogenation of azabicycloalkanones to the corresponding alcohols. Mixtures of epimers are usually obtained. The composition of the reduction products is dependent on structure, media, and catalyst. In the hydrogenation of azabicyclooctanone, nonanone, and decanones with platinum oxide, the predominance of epimers depended more on structure [47]. When oxopyrrolizidines, oxoindolizidines, and oxoquinolizidines were reduced with palladium, platinum, rhodium, and ruthenium in ethyl alcohol or aqueous hydrochloric acid, the mixtures of alcohols varied with the nature of the catalyst and the acidity of the medium [48a–c].

Miscellaneous Ketones

Ketones such as 1-indanones and 1-tetralones are related to aryl alkyl ketones and should be similarly reduced to secondary alcohols. Hydrogenations of 1-indanones have been carried out with platinum oxide [49a] and Raney nickel [49b] at low pressure or with nickel at 105° and 80 atm pressure [49c]. 1-Tetralones have been reduced at low pressure with Raney nickel [49a] or with alkali-promoted platinized nickel [49e]. Palladium on carbon was employed to convert 2-amino-1-indanone hydrochloride to 2-aminoindanol in very high yield [50a] and a number of 5- and 6-substituted-2-amino-1-indanones were hydrogenated at 60° and 2 atm to the corresponding indanols without dehydroxylation [50b].

There should be little concern about loss of hydroxyl function (unless reaction conditions are too vigorous) in reductions of 2-indanones and 2-tetralones or related compounds in neutral solvent, since the effect of the benzene ring on the hydrogenolysis of the resulting alcohols is almost completely removed.

Quinones require only one molar equivalent of hydrogen for conversion to diphenolic compounds. Best results were obtained by reduction in neutral solvent with palladium on carbon; platinum on carbon was less efficient [51a]. In one instance platinum oxide, prereduced before the addition of substrate, was used for a rapid reduction of 4-methylcarbostyril-5,6-quinone to 5,6-dihydroxy-4-methylcarbostyril [51b] (Eq. 14.5). Hydrogenation of

$$\text{(14.5)}$$

quinones in mineral acid with platinum may lead to saturation of the ring [51a]. Acetic acid (96%) was used in the platinum-catalyzed hydrogenation of duroquinone to 2,3,5,6-tetramethylhydroquinone [51c], but it must be noted that ring reduction of a completely substituted benzenoid compound does not take place readily.

Generally there is such a sharp break in the reaction rate after absorption of one equivalent [7, 51a] that most low-pressure hydrogenations of quinones, except those in mineral acid, could be interrupted to obtain diphenols. The difference in rate also suggests that reductions of poorly soluble quinones could be carried out at higher temperature and/or pressure with little danger of overhydrogenation if hydrogen uptake is monitored.

When hydrogen uptake is impaired because of poor solubility of the quinone, methyl cellosolve may prove to be extremely useful as it was in the rapid reduction of 2,6-dimethoxybenzoquinone to 2,6-dimethoxyhydroquinone with palladium on carbon at slightly above atmospheric pressure [7].

DIKETONES

Most hydrogenations of diketones to diols are readily carried out under mild conditions. In many reductions hydroxyketones can be obtained.

In the hydrogenation of diketones containing phenolic hydroxyl groups adjacent to the carbonyl function, palladium may be less preferred because of the possibility of hydrogenolysis. The amount of catalyst may also be a contributing factor. These points are illustrated in the hydrogenations of 3′,6′-dihydroxy-1,2-benzocycloheptene-3,7-dione with palladium on carbon or with 3% by weight of platinum oxide where 3′,6′-dihydroxy-1,2-benzocyclohepten-3-one was obtained [51d]. Hydrogenolysis was prevented by

296   *Carbonyl Groups*

carrying out the reaction at atmospheric pressure with less than 1% by weight of Adams' catalyst or by hydrogenation as 3',6'-diacetate (Eq. 14.6).

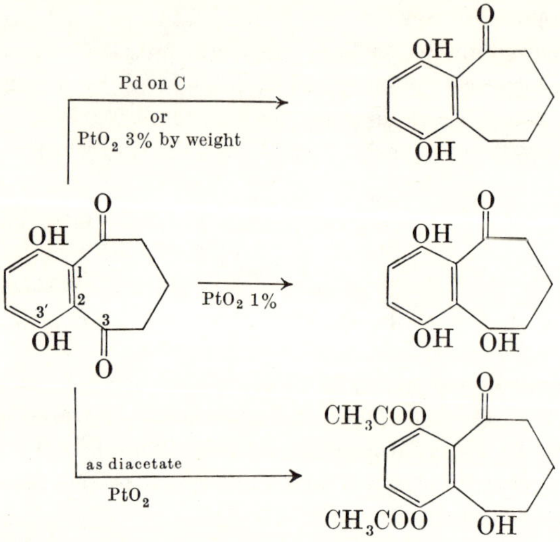

(14.6)

1,2-Diketones

Elevated temperature and pressure conditions have been employed for reductions of 1,2-diketones, as diacetyl with Raney nickel [52a] and substituted benzoins with copper chromite [52b]. In most instances low-pressure methods should not only be adequate but should result in far less hydrogenolysis. For example, Raney nickel at atmospheric pressure catalyzed the hydrogenation of 1,4,7,7-tetramethylbicyclo[2,2,1]heptane-2,3-dione [53a] (Eq. 14.7). Similar conditions were employed for the reduction of 1,4-

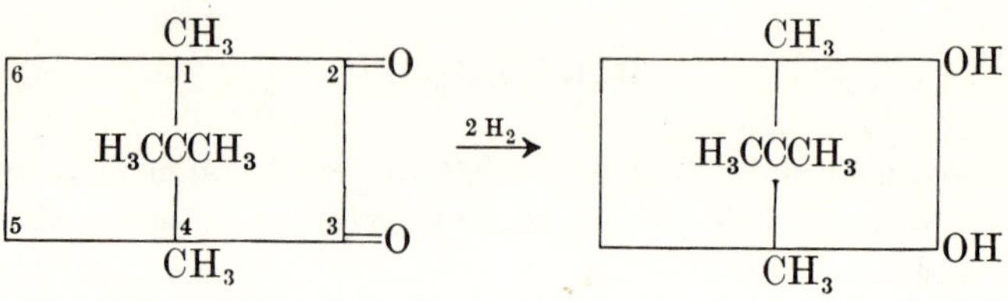

(14.7)

diphenylbutane-1,2-dione [53b] and of 1,3-diphenylpropane-1,2-dione [53c] to the corresponding diols. There was a marked difference in absorption after uptake of one equivalent in the latter reductions [53b, c] because of activation of the 1-keto group by the benzene ring. Interruption at this point gave the corresponding diphenylalkan-1-ol-2-ones.

Diketones 297

When reduction of the enol form of the propanedione,

$$C_6H_5C(=O)C(OH)=CHC_6H_5,$$

was interrupted the 1-alkanol-2-one was also obtained. Benzaldehyde was noted in hydrogenation of both the enol and keto form [53c] but other possible cleavage products were not identified.

1,3-Diketones

Hydrogenation of 1,3-diketones may be accompanied by cleavage on either side of the methylene group or by other side reactions if conditions are too vigorous. As a rule such conditions are unnecessary, but they were in an attempt to reduce a hindered ketone to a glycol. Optimum conditions for the reduction of 2,2,4,4-tetramethylcyclobutane-1,4-dione consisted of reaction in methyl alcohol with ruthenium on carbon at 125° and 75–100 atm pressure [54a]. A 98% yield of a mixture of *cis* and *trans* diols was obtained. Similar reductions with Raney nickel, nickel on kieselguhr, and copper chromite gave lower yields, 60–68%. Reduction in methyl alcohol over Raney was accompanied by formation of 1-hydroxy-2,2,4-trimethyl-3-pentanone and methyl 3-oxo-2,2,4-trimethylpentanoate through the following reactions (Eq. 14.8).

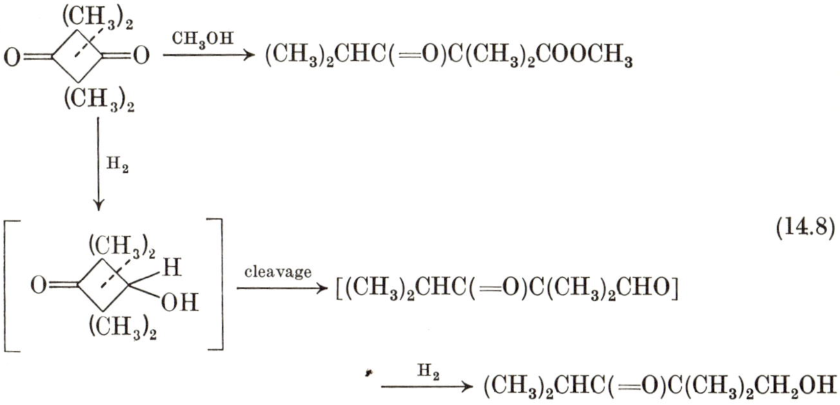

$$O=\underset{(CH_3)_2}{\overset{(CH_3)_2}{\square}}=O \xrightarrow{CH_3OH} (CH_3)_2CHC(=O)C(CH_3)_2COOCH_3$$

$$\downarrow H_2$$

$$\left[O=\underset{(CH_3)_2}{\overset{(CH_3)_2}{\square}}\underset{OH}{\overset{H}{\diagdown}} \right] \xrightarrow{cleavage} [(CH_3)_2CHC(=O)C(CH_3)_2CHO] \quad (14.8)$$

$$\xrightarrow{H_2} (CH_3)_2CHC(=O)C(CH_3)_2CH_2OH$$

Side reaction in the reduction with nickel was avoided by using non-alcoholic solvents. Hydrogenations in these solvents with Raney nickel at 150° and 100 atm gave 94% yield of *cis* and *trans* cyclobutanediols [54b]; similarly, a supported nickel catalyst [55] also gave high yield.

Hydrogenation of the tetramethylcyclobutanedione under low-pressure conditions yielded ketoalcohol with Raney nickel [54a] or with nickel promoted with triethylamine [56a].

In reductions where hydrogenolysis can occur, because the resulting product is a benzylic alcohol, platinum oxide and nickel gave good results. 2,2-Dimethylindane-1,3-dione was converted to either ketoalcohol or diol by hydrogenation over platinum oxide in acetic acid [56b]. The high yield of 3-hydroxy-2-(1-naphthyl)indan-1-one from reduction of the diketone with nickel [56c] suggests that little hydrogenolysis would be caused by its use in reductions of this type.

Hydrogenation of 1,3-diones which do not contain an activating aromatic ring may yield only ketoalcohol. 1,1,1-Trifluorohexane-2,4-dione in ether was reduced to 1,1,1-trifluoro-2-hydroxyhexan-4-one with palladium on carbon; loss of the hydroxy group accompanied reduction with platinum oxide or high-pressure hydrogenation over copper chromite [56d]. In an attempted partial hydrogenation of ethyl 2,4-dioxopentanoate with platinum oxide, equal parts of starting material, ethyl 2-oxopentanoate,

$$C_3HCOCO_2C_2H_5,$$

and ethyl 2-hydroxy-4-oxopentanoate, $CH_3COCH_2CHOHCO_2C_2H_5$, were obtained [56e].

In another example, uptake of hydrogen stopped abruptly at one equivalent in the reduction of 3-(1-methylcyclopentyl-2,5-dione) propionic acid with platinum oxide [57]. The free acid, 3-(2-hydroxy-1-methylcyclopentyl-5-one)propionic acid, 20%, was obtained. It can be assumed that the hydroxyl and carbonyl groups were generally *cis* to each other since an 80% yield of lactone was also obtained.

Other Polycarbonyl Compounds

Cleavage does not usually accompany hydrogenation of 1,4- or 1,5-dicarbonyl compounds. However, condensation may take place leading to a cyclic structure dependent on catalyst and conditions. 3,7-Dimethyl-5-oxooctanal, $(CH_3)_2CHCH_2C(=O)CH_2CH(CH_3)CH_2CHO$, was converted to a mixture of *cis* and *trans* 2-isobutyl-4-methyltetrahydropyran, 67%, and 3,7-dimethyloctane-1,5-diol, the minor product, by hydrogenation over a supported nickel catalyst at 125° and 135 atm [55], while reduction with copper chromite at 150° and 165 atm gave predominantly diol [58]. Except for palladium, which has little utility in the reduction of aliphatic carbonyl compounds, it is likely that low-pressure hydrogenation with nickel, platinum, or possibly ruthenium would also yield diol.

When 2-acetylcyclopentane-1,3-dione was hydrogenated with palladium on carbon, the carbonyl group in the side chain was attacked to the exclusion of the other carbonyl groups to give 2-ethylcyclopentane-1,3-dione [59].

CARBONYL TO METHYLENE

It is often desirable to convert the carbonyl function to —CH$_2$—, the ease of which depends on the structure of the substrate and activating groups. In most instances the presence of acid, especially strong acid, catalyzes the reaction.

Aromatic Aldehydes and Ketones, ArCOR

When the compound, ArCOR, to be reduced contains a carbonyl group conjugated with a benzenoid system, where R is H, aryl, alkyl, aminoalkyl, an alkanoic acid, or is part of a cyclic system as 1-tetralone, the method of choice is low-pressure hydrogenation in acidified solvent with palladium on carbon although other palladium catalysts may be employed.

Although there is a possibility of ring reduction in acidic media with palladium, it generally does not take place at moderate temperature; instead more rapid and more complete conversion of the carbonyl group to methylene results.

Reaction can be carried out in unacidified alcohols [60a–c] or in other oxygen-containing solvents which tend to aid complete reduction conversion [60d], but hydrogenation in neutral solvent usually requires excess catalyst to induce further hydrogenolysis of the intermediate secondary benzyl alcohol.

Reductions have been carried out in acetic acid [61a–c] but the best medium may be acetic acid containing either perchloric acid [61d] or sulfuric acid [61e]. Hydrogenation in acetic acid containing trifluoroacetic acid also appears to be a very useful procedure for the conversion of ArCOR to ArCH$_2$R [7]. The addition of two equivalents of methanesulfonic acid to the hydrogenation of 3-nitrobenzophenone in acetic acid resulted in 86% yield of 3-aminodiphenylmethane [61f].

It was usually necessary to use strong acid in palladium-catalyzed reductions of aroylalkanoic acids, ArCO(CH))$_n$COOH, in acetic acid to

$$Ar(CH_2)_{n+2}COOH.$$

Reaction was often rapid and yields were very high [62a–e].

1-Methyl-1-(1-piperidyl)propiophenone in acetic acid containing perchloric acid did not proceed beyond the aminoalcohol stage during reduction with palladium [63a] due, no doubt, to steric effects. In general among less hindered aminoketones the carbonyl group was readily converted to methylene in acid media with palladium [63b].

Examples of palladium-catalyzed reductions in other acidic media are those of propiophenone and substituted propiophenones to propylbenzenes in alcoholic hydrogen chloride [64a], 3,3-dimethyl-1-indanone to 3,3-dimethylindane in alcohol containing sulfuric acid [64b], and 6-(2-diethylaminoethoxy)-2-methylindanone to the indane in 7% hydrochloric acid [64c]. Perchloric acid-acidified ethyl acetate was also found to be a useful medium [64d].

The attempted conversions of 2-acetyl and 2-phenylacetylbenzoic acid in acid solution to the corresponding hydrocarbon gave lactones because reaction stopped at the hydroxy stage [65a]. In this instance reduction in alkaline solution prevented formation of lactone and enabled reaction to proceed to hydrocarbon [65b].

Although platinum catalysts have been employed in conversion of conjugated aryl ketones to the corresponding hydrocarbon, their use in acidic media can lead to ring reduction. Saturation of the ring with it was reported in neutral solvent in the hydrogenation of 4,4-dihydroxybenzophenone while a similar reaction in the presence of palladium on carbon gave 4,4-dihydroxydiphenylmethane [66].

Miscellaneous Carbonyl Compounds

Ethyl diphenylpyruvate, $(C_6H_5)_2CCOCO_2C_2H_5$, was converted to ethyl 3,3-diphenylpropionate with an undisclosed amount of Raney nickel in a slow reduction at 3 atm pressure in which a second portion of catalyst was added [67a].

The presence of acid is apparently necessary to carry the reduction of alicyclic and saturated heterocyclic ketones through the hydroxy stage. When 2,3- and 2,6-dihydroxycyclohexanones were hydrogenated with platinum oxide in the presence of sulfuric acid, cyclohexane-1,2- and 1,3-diols were obtained; triols resulted in the absence of acid [67b]. 1-Methyl-*trans*-decahydro-7-quinolone was converted to 1-methyl-*trans*-decahydroquinoline in aqueous hydrochloric acid in the presence of platinum oxide [67c]; hydrogenation in neutral solvent with palladium on carbon gave the 7-α-hydroxy compound. When some 2,3-dihydro-4-(1H)-isoquinolones were hydrogenated in acetic acid with palladium on carbon, reactions were carried out at 80–120° and 65–80 atm to yield 1,2,3,4-tetrahydroisoquinolines [67d, e]. Platinum-catalyzed reduction of ethyl 1-carbethoxy-2-methyl-4-pyrrolidone-3-acetate in acetic acid yielded ethyl 1-carbethoxy-2-methylpyrrolidine-3-acetate [67f].

SELECTIVE HYDROGENATION OF THE CARBONYL GROUP

In Aromatic Ring Systems

Benzenoid Rings. There is such a great difference in the rate of hydrogenation of a carbonyl function over that of a benzenoid ring that there is usually little danger of ring reduction in nonacidic media with nickel, palladium, or platinum. A few exceptions with platinum [32a, b] and palladium [66] have been noted. The use of rhodium is contraindicated [15, 29]. Saturation of the ring also took place in ruthenium-catalyzed hydrogenations of aryl alkyl ketones at elevated temperature and pressure [7].

The presence of excess strong acid may cause overhydrogenation in reductions with palladium or platinum catalysts [41a] if hydrogen absorption is not interrupted at the proper point.

Unsaturated N-Heterocycles.

PYRROLES AND INDOLES. A similar difference in rate exists between the carbonyl group and unsaturated heterocyclic systems. In instances where the ring system is not readily reduced, vigorous conditions can be used with some catalysts. For example, ethyl 3-acetyl-2,5-dimethylpyrrole-5-carboxylate was hydrogenated to ethyl 3-ethyl-2,5-dimethylpyrrole-5-carboxylate at 160° and 75 atm with Raney nickel [68a] and ethyl 3-(3-indolyl)-2-oxopropionate was converted to the corresponding hydroxy ester over Raney nickel at 145 atm pressure [68b].

Reductions of other indole ketonic compounds to the corresponding alcohols have been carried out at low pressure with palladium on carbon [69a], palladium oxide [69b], and platinum oxide [69c].

IMIDAZOLES AND BENZIMIDAZOLES. Examples of selective reduction in the presence of an imidazole ring are seen in the conversion of imidazole-4-(5)aldehyde to the hydroxymethylimidazole over palladium oxide promoted with ferric chloride [70a, b]. Examples of hydrogenations of imidazole carbonyl compounds in acetic acid indicate little attack on the ring. 2-Benzoyl-5-phenylimidazole was reduced to the carbinol with a rather large amount of palladium on carbon without affecting any of the rings or causing hydrogenolysis of the OH group [70c]. The carbonyl group in ethyl 6-oxo-6-[4(5)-methyl-2-imidazolone-5(4)] hexanoate was converted to methylene with platinum oxide without reducing the imidazole double bond [71a]; the corresponding imidazolidinone was obtained only after a long reaction period. The carbonyl function in the side chain of another imidazolone was similarly and selectively reduced to methylene [71b].

Some 1-phenacylbenzimidazoles and related compounds were converted to 1-(2-hydroxy-2-phenylethyl)- and 1-(3-hydroxy-3-phenylpropyl)benzimidazoles at 70–80° and atmospheric pressure by reduction in glacial acetic acid with palladium [72c].

PYRIMIDINES. The carbonyl group in pyrimidine compounds is also preferentially hydrogenated. 4-Methyl-2-phenyl-5-pyrimidyl 2-dialkylaminoalkyl ketones and 5-acetyl-4-dialkylamino-5-phenylpyrimidines were reduced to pyrimidyl carbinols with platinum oxide in alcohol [72a, b]. Extended reaction did not cause further hydrogen uptake.

PYRIDINES. Although the pyridine ring is the least difficult aromatic system to reduce, selective conversion of a carbonyl function to carbinol or methylene in its presence should generally be expected. Low-pressure reduction with palladium or platinum in neutral solvent or with palladium in the presence of an equivalent of acid is the method of choice for the conversion to carbinol without affecting the ring. At times low-pressure reactions with Raney nickel in neutral solvent are useful. Reductions with rhodium are not. It has caused saturation of the ring in hydrogenation of pyridyl carbonyl compounds [73].

Reductions with nickel or copper chromite at elevated temperature and pressure should be selective since neither reduces the pyridine ring with ease but hydrogenation with ruthenium at 70–90° and 70 atm usually leads to the formation of considerable amounts of piperidyl carbinol [7].

Selective conversion of the carbonyl group to methylene can be carried out with palladium in acetic acid containing perchloric acid or in aqueous mineral acid with platinum but hydrogen uptake must be watched to avoid ring reduction [7]. Examples of selectivity follow.

The hydrogenations of aminomethyl pyridyl ketones, $RCOCH_2NH_2$ where R = 2-, 3-, and 4-pyridyl, as salts gave the isomeric 2-hydroxy-2-(pyridyl)-ethylamines with palladium black [74a] and platinum black [74b]. Reduction of ethyl 2-nicotinylpropionate ($RCOCH(CH_3)COOC_2H_5$ (R = 3-pyridyl) in glacial acetic acid with platinum on barium sulfate or with platinum oxide in alcohol gave the corresponding carbinol, $RCHOHCH(CH_3)COOC_2H_5$ [74b]. In the hydrogenation of ethyl nicotinoylacetate, $RCOCH_2COOC_2H_5$, with palladium on barium sulfate in acetic acid containing perchloric acid the carbonyl group was converted to methylene to yield ethyl 3-(3-pyridyl)-propionate [74c].

It is known that quaternization of the pyridine nitrogen aids ring reduction but the conversion of (4-acetoxy-3-methoxyphenacyl)pyridinium bromide in water with platinum to 2-hydroxy-2-(4-acetoxy-3-methoxyphenyl)ethyl pyridinium bromide is another indication of attack at the carbonyl group in preference to ring reduction among pyridines [74d] (Eq. 14.9.)

$$\underset{\underset{Br}{(-)}}{\overset{+}{N}}-CH_2\overset{O}{\underset{\|}{C}}R \xrightarrow{H_2} \underset{\underset{Br}{(-)}}{\overset{+}{N}}-CH_2CHOHR \qquad (14.9)$$

Unexpected results were obtained in the hydrogenation of 3-acetylpyridine [73]. When it was hydrogenated over palladium on carbon or rhodium on alumina, the expected 3-(1-hydroxyethyl)pyridine was only a by-product. The major product was 3-acetyl-1,4,5,6-tetrahydropyridine formed in the following manner (Eq. 14.10).

$$\underset{N}{\bigcirc}COCH_3 \xrightarrow{H_2} \underset{\underset{H}{N}}{\overset{H\ H}{\bigcirc}}COCH_3 \xrightarrow{H_2} \underset{\underset{H}{N}}{\bigcirc}COCH_3 \qquad (14.10)$$

Reduction of 3-acetylpyridine over platinum oxide did yield the carbinol but it was contaminated with starting material and 3-acetyl-1,4,5,6-tetrahydropyridine; no uptake of hydrogen occurred in hydrogenation with nickel.

The reduction of 2-acetylpyridine was more or less straightforward. It was converted in alcohol to 2-(1-hydroxyethyl)pyridine in high yield by hydrogenation with palladium on carbon or in the presence of a 20% weight ratio of Raney nickel. Reaction over platinum or rhodium gave 60–70% yields plus 23–25% of starting material and varying amounts of piperidyl carbinol.

The hydrogenation of 4-acetylpyridine was only complicated by the formation of pinacol. Reduction in alcohol over platinum oxide gave good yield of 4-(1-hydroxyethyl)pyridine and some 2,3-di-(4-pyridine)butane-2,3-diol. Similar reductions with palladium on carbon or rhodium on a carrier gave large amounts of the pinacol. The pinacol could be obtained in very high yield from hydrogenation of the ketone in alcohol with Raney nickel or by reduction of a very concentrated aqueous solution with platinum oxide. In each instance only one mole of hydrogen was absorbed per two moles of substrate (Eq. 14.11).

$$2RCOCH_3 \xrightarrow{H_2} RC(CH_3)(OH)C(CH_3)(OH)R \qquad (14.11)$$

R = 4-pyridyl

QUINOLINES, ISOQUINOLINES, AND RELATED HETEROCYCLES. Palladium and platinum catalysts appear best for selective hydrogenation of the carbonyl group in quinolines and isoquinolines [75a–e].

Of the isomeric, 2-, 3-, and 4-acetylquinolines only the reduction of the 2-acetylquinoline is reported. Hydrogenation in aqueous or anhydrous alcohol with platinum oxide yielded 2-(1-hydroxyethyl)quinoline, while reduction in the presence of some hydrochloric acid gave the pinacol [76].

In quinoline compounds the carbonyl group can be selectively converted to methylene during reduction with palladium in acetic acid containing perchloric acid [74c].

In hydrogenations of carbonyl groups in heterocyclic systems related to quinolines and isoquinolines, the ring was not affected when reductions were carried out in neutral solvent with palladium on carbon [77a, b]. When the platinum-catalyzed reduction of a quaternary compound, 4-oxo-1,2,3,4-tetrahydrobenzo[c]quinolinium bromide, was not interrupted after conversion of the carbonyl function to carbinol, the nitrogen-containing ring became saturated [77b]. The hydrogenation of 1,3-dimethyl-9-oxo-2-azafluorene in alcohol with platinum yielded the corresponding 9-hydroxy compound [77c]; in reduction with Raney nickel under similar conditions, conversion of the carbonyl group to methylene occurred to give 1,3-dimethyl-2-azafluorene.

Unsaturated O-Heterocycles. Examples of hydrogenation of the carbonyl functions in furanoid compounds indicate reduction of that function in preference to attack of the ring. Alkali-promoted Raney nickel reductions of furfuraldehydes to furfuryl alcohols [3a] and low-pressure hydrogenations with platinum illustrate this point. It was also possible to maintain selectivity with copper chromite at higher pressure conditions shown in the reduction of 5-hydroxymethyl-2-furaldehyde to 2,5-dihydroxymethylfuran at 125° and 100 atm [7]. More vigorous conditions caused formation of tetrahydrofuran and other side reaction products.

Presence of Functional Groups

The carbonyl group is not reduced catalytically as readily as an acetylene linkage, an aromatic nitro or nitroso group, or an oxime. In most instances it will not reduce as readily as an azomethine, a carbon to carbon double bond, or a nitrile group.

Olefinic Bonds. The olefinic bond in α,β-unsaturated aldehydes and ketones is usually preferentially reduced but the effect can be reversed at least in aldehydes in platinum-catalyzed hydrogenations by a combination of zinc ion, which inhibits double bond saturation, and ferrous ion, which is a promoter for carbonyl reductions [78a–c]; in the absence of either one selective conversion of the carbonyl group is lost.

Examples in Chapter IX indicate that monosubstituted double bonds and symmetrically disubstituted ones not conjugated with the carbonyl function are also preferentially reduced. The hydrogenations of cyclohexene-4-aldehyde with copper chromite to 4-hydroxymethylcyclohexene [79a, b] may be an exception.

There are too few examples of the reduction of substrates containing a carbonyl function and a trisubstituted double bond to give a clear indication of selectivity. Reduction may be tipped in favor of the ketonic carbonyl by the use of an alkaline nickel catalyst and the presence of alkali as a promoter. The hydrogenation of 3-acetonyl-5-phenyl-2-pentenoic acid,

$$C_6H_5(CH_2)_2C(CH_2COCH_3)=CHCOOH,$$

in alkaline solution to 3-(2-hydroxypropyl)-5-phenyl-2-pentenoic acid with W-7 Raney nickel [79c] may be an example of this. It is possible in this case that the effect is essentially steric—a molecular model shows that the double bond is highly hindered while the carbonyl group makes good contact with the catalyst.

Other examples of preferential reduction of a ketone in the presence of a trisubstituted double bond are seen in the hydrogenations of 5,6-double bonded, unconjugated, steroidal ketones in neutral solvent with Raney nickel [80a–d]. Selectivity in these examples is due to the known poor reducibility of the 5,6-double bond in neutral solvent.

Cyano Groups. In most instances a cyano group will be preferentially hydrogenated in the presence of a ketone (Chapter XII), but there are very few examples of reductions in the presence of an aldehyde. When 3-cyanopropionaldehyde was hydrogenated with palladium on carbon at 25° and 120 atm pressure or with nickel on kieselguhr or with copper chromite at below 100°, 3-cyano-1-propanol was obtained [81a].

There is an example where an unhindered ketone carbonyl group was reduced to methylene in the presence of a cyano group but selectivity appears to be due to steric effects. 1-Cyano-1-(2-carboxyethyl)-7,8-dimethoxy-4-tetralone was hydrogenated in acetic acid containing sulfuric acid to the corresponding 1-cyanotetrahydronaphthalene over palladium on carbon [87b]. A molecular model clearly shows the carbonyl group in good contact with the catalyst while the cyano group is either out of plane or is prevented from reaching the catalyst surface by the carboxyethyl group in 1-position and the methoxy group in 8-position.

Presence of Hydrogenolyzable Groups

O-Benzyl Groups. When there is competition between the ketone function and a benzyl ether, it is likely that the carbonyl group can be preferentially reduced if palladium catalyst is avoided. 2-Amino-3'-benzyloxypropiophenone hydrochloride in alcohol was converted to 2-amino-1-(3-benzyloxyphenyl)-1-propanol in 86% yield over platinum oxide [82a]. Raney nickel might also be useful since it is generally ineffective in debenzylations except at elevated temperature and pressure. Copper chromite was employed with good results in the reduction of 4-benzyloxyacetophenone to 1-(4-benzyloxyphenyl)-1-ethanol [82b].

It might be assumed that an aldehyde group, because it is less hindered, could also be reduced selectively. However, the few examples show, at least among aromatic aldehydes, that this is difficult to accomplish except with copper chromite.

When 2-benzyloxy-3-methoxy- and 4-benzyloxy-3-methoxybenzaldehydes were hydrogenated over Raney nickel at low pressure, debenzylation appeared to be favored or was concurrent with conversion of alcohol to aldehyde; reduction of 3-benzyloxy-4-methoxybenzaldehyde under the same conditions gave only 12% yield of 3-benzyloxy-4-methoxybenzyl alcohol [7]. The use of copper chromite at 125–150° and 100 atm, however, did result in selectivity; high yield of the benzyloxy-methoxybenzyl alcohol was obtained [7].

N-Benzyl Groups. Palladium catalysts have been used with mixed success in the reduction of benzylaminoketones [83a, b], but there are other examples in which debenzylation appears to be the prime reaction (Chapter XIX). In another example ω-benzylaminoacetophenone hydrochloride was hydrogenated at 7 atm pressure [83c]; preferential debenzylation occurred. On the other hand, other attempts with as much as 50% by weight of a similar supported commercial catalyst resulted in selective reduction of the ketone to yield 2-benzylamino-1-phenyl-1-ethanol [7].

Low-pressure methods with platinum [84] or with Raney nickel may be more promising since neither catalyst is particularly active for *N*-debenzylation.

Halogen. In general the carbonyl group will be reduced in preference to a halogen if the halogen is not too reactive and if hydrogenation is carried out in neutral medium. Selectivity may also be aided by the presence of acid.

Nickel and platinum may be the preferred catalysts. Palladium has been used but it has the greatest potential for causing dehalogenation in low-pressure hydrogenations. High-pressure methods employing Raney nickel or copper chromite have given good selectivity.

HALOGEN-CONTAINING ALDEHYDES. Platinum oxide was extremely useful for the selective reduction of chlorine-containing aldehydes when an iron salt, as promoter for the aldehyde group, was present. 2-Chloro- and 4-chlorobenzaldehydes were converted to the chlorobenzyl alcohols in very high yield by low-pressure hydrogenation over platinum oxide promoted with ferrous chloride [85]. Reduction of 4-chlorobenzaldehyde with platinum and ferrous chloride carried out at 25° and 70 atm gave a 95% yield of the chloroalcohol [7].

Reduction of the same aldehyde was carried out with Raney nickel at 80° and 70 atm to give 86% of the corresponding alcohol [7]. Reaction was also carried out in benzene with palladium catalyst at low pressure with nicotinamide as an inhibitor to give about 70% of 4-chlorobenzyl alcohol [8]. The ascribed purpose of the inhibitor was to prevent hydrogenolysis of the resulting benzyl alcohol but it may have also prevented dehalogenation.

HALOGEN-CONTAINING KETONES. Platinum oxide was successfully employed in the hydrogenation of chloro-containing ketones seen in those of 4-(dialkylaminoacetyl)-6-chloro-5-methyl-2-phenylpyrimidines [86a] and 2-dialkylamino-3'-chloro-4'-hydroxyacetophenones [86b] to the corresponding alcohols without loss of chlorine.

When it was used to catalyze the reduction of (2-chlorophenoxy)acetone, considerable loss of chlorine resulted [7]. However, a Raney nickel reduction was successful although slow; 82.5% yield of 1-(2-chlorophenoxy)-2-propanol was obtained.

Palladium catalysts have been used with mixed success. 5-Chloro-4-methyl-1-tetralone was reduced to 5-chloro-4-methyl-1,2,3,4-tetrahydronaphthalene in alcohol with palladium on barium sulfate [87a]. Attempts to convert 2-acyl-4-chlorophenols to 2-alkyl-4-chlorophenols in alcohol by hydrogenation over palladium black resulted in considerable dehalogenation [87b]. Loss of chlorine was kept at a low level by reduction in acetic acid containing sulfuric acid, perchloric acid, or methanesulfonic acid. The hydrogenation of 2-amino-3'-chloro-4'-methoxypropiophenone hydrochloride in alcohol over palladium on carbon was interrupted after absorption of one equivalent to yield 65% of 1-(3-chloro-4-methoxyphenyl)-1-propanol hydrochloride [87c]. 4-Chlorobenzophenone was reduced in alcohol containing nicotinamide or N,N-diethylnicotinamide in the presence of palladium on carbon to 4-chlorobenzhydrol in 65% yield [35a].

In examples with copper chromite, 3'-chloroacetophenone was converted to 1-(3-chlorophenyl)-1-ethanol in 84% yield by reduction at 140° and 65 atm [87d] and 5-chloro-4-methyl-1-tetralone was hydrogenated to the chlorotetralol at 150° and 180 atm [87a].

It is likely that there will not be selective reduction of the carbonyl group in α-haloketones, at least not with platinum or palladium, as seen from the hydrogenations of 4-benzoyl-4-chloro or 4-bromo-1-methylpiperidine hydrohalides [88]. This will be discussed at greater length in the chapter on dehalogenation.

Selective reduction of the carbonyl group in the presence of bromine may possibly be achieved only with nickel catalyst. 4-Bromobenzophenone was converted to 4-bromobenzhydrol in 72% yield by reduction with Raney nickel under 100 atm pressure [89a]; unsuccessful reduction was reported with palladium in the presence of inhibitors [35a]. Catalytic hydrogenation of 1,4-*bis*-(4-bromobenzoyl)benzene resulted in loss of bromine [89b].

REFERENCES

[1] a. W. H. Carothers and R. Adams, *J. Am. Chem. Soc.*, **45**, 1071 (1923); b. P. N. Rylander and J. G. Kaplan, U.S. Patent 3,144,490 (1964); c. P. N. Rylander and J. G. Kaplan, U.S. Patent 3,150,192 (1964); d. G. G. Boyers, U.S. Patent 2,868,847 (1959).
[2] E. B. Maxted and S. Akhtar, *J. Chem. Soc.*, **1959**, 3130.
[3] a. R. Paul, *Bull. Soc. Chim. France*, **1946**, 208; b. S. Yamomoto, *J. Soc. Chem. Ind. Japan*, **46**, 901 (1943); through *Chem. Abstr.*, **42**, 6320 (1948).
[4] a. R. C. Elderfield, B. M. Pitt, and I. Wempen, *J. Am. Chem. Soc.*, **72**, 1334 (1950); b. W. T. Haskins, R. M. Hann, and C. S. Hudson, *ibid.*, **68**, 628 (1946); c. L. A. Flexser, U.S. Patent 2,421,416 (1947); d. H. Adkins and H. R. Billica, *J. Am. Chem. Soc.*, **70**, 695 (1948); e. D. R. Levering, F. L. Morritz, and E. Lieber, *ibid.*, **72**, 1190 (1950).
[5] D. Nightingale and H. D. Radford, *J. Org. Chem.*, **14**, 1089 (1949).
[6] a. H. W. Bersch, *Arch. Pharm.*, **277**, 279 (1939); b. J. H. Birkinshaw, A. Bracken, and H. Raistrick, *Biochem. J.*, **37**, 726 (1943).
[7] M. Freifelder, unpublished results.
[8] K. Kindler, H.-G. Helling, and E. Sussner, *Ann.*, **605**, 200 (1957).
[9] a. R. Adams, S. McKenzie, Jr., and S. Loewe, *J. Am. Chem. Soc.*, **70**, 664 (1948); b. H. S. Mason, *ibid.*, **67**, 1538 (1945).
[10] a. J. L. Bills and C. R. Noller, *J. Am. Chem. Soc.*, **70**, 957 (1948); b. E. Profft. *Arch. Pharm.*, **292**, 70 (1959); c. J. Jirkovsky and M. Protiva, *Coll. Czechoslov. Chem. Comm.*, **29**, 400 (1964).
[11] a. E. Kelecsenyi-Dumesnil, *Bull. Soc. Chim. France*, **1955**, 815.
[12] a. W. Wenner, *J. Org. Chem.*, **15**, 301 (1950); b. F. Leonard and L. Simet, *J. Am. Chem. Soc.*, **77**, 2855 (1955); c. R. Muths, Fr. Patent 1,122,148 (1956); through *Chem. Abstr.*, **53**, 14942 (1959).
[13] a. L. Long, Jr. and A. Burger, *J. Org. Chem.*, **6**, 852 (1941); b. G. P. Hennion and E. J. Watson, *ibid.*, **23**, 656 (1958); c. Y. Takagi, *Sci. Papers Inst. Phys. Chem. Res., Tokyo*, **57**, 105 (1963), d. M. Delepine and A. Horeau, *Bull. Soc. Chim. France*, **1937, 31**; e. J. Forrest and S. H. Tucker, *J. Chem. Soc.*, **1948**, 1137; f. M. Delepine and A. Horeau, *Bull. Soc. Chim. France*, **1937**, 431; g. E. Lieber and G. B. L. Smith, *J. Am. Chem. Soc.*, **58**, 1417 (1936).
[14] a. R. B. Blance and D. T. Gibson, *J. Chem. Soc.*, **1954**, 2487; b. A. A. Pavlic and H. Adkins, *J. Am. Chem. Soc.*, **68**, 1471 (1946); c. S. Nishimura, *Bull. Chem. Soc. Japan*, **34**, 32 (1961).
[15] E. Breitner, E. Roginski, and P. N. Rylander, *J. Org. Chem.*, **24**, 1855 (1959).
[16] a. H. Adkins and R. Connor, *J. Am. Chem. Soc.*, **53**, 1091 (1931); b. A. P. Howe and H. B. Hass, *Ind. Eng. Chem.*, **38**, 251 (1946); c. R. C. Elderfield, W. J. Gensler, F. Brody, J. D. Head, S. C. Dickerson, L. Wiederhold, C. B. Kremer, H. A. Hageman,

F. J. Kreysa, J. M. Griffing, S. M. Kupchan, B. Newman, and J. J. Maynard, *J. Am. Chem. Soc.*, **68**, 1579 (1946).

[17] a. M. S. Raasch, U.S. Patent 2,824,888 (1958); b. E. E. Smissman, J. F. Muren, and N. A. Dahle, *J. Org. Chem.*, **29**, 3517 (1964).

[18] a. M. Versele, M. Acke, and M. Anteunis, *J. Chem. Soc.*, **1963**, 5598; b. P. E. Peterson and C. Casey, *J. Org. Chem.*, **29**, 2325 (1964); c. H. A. Scheutte and R. W. Thomas, *J. Am. Chem. Soc.*, **52**, 3010 (1930).

[19] a. H. W. Schwerp, *J. Am. Chem. Soc.*, **68**, 912 (1946); b. E. J. Lease and S. M. McIlvain, *ibid.*, **55**, 806 (1933); c. P. N. Rylander and S. Starrick, *Engelhard Ind. Tech. Bull.*, **7**, 106 (1966).

[20] a. D. D. Hey and O. C. Musgrave, *J. Chem. Soc.*, **1949**, (3156); b. R. L. Jacobs, and G. L. Goerner, *J. Org. Chem.*, **21**, 836 (1956); c. L. Ruzicka, Pl. A. Plattner, and H. Wild, *Helv. Chim. Acta*, **28**, 395 (1945).

[21] a. R. J. Wicker, *J. Chem. Soc.*, **1956**, 2165; b. P. Anziani and R. Cornubert, *Bull. Soc. Chim. France*, **1945**, 359; c. R. Helg and H. Schinz, *Helv. Chim. Acta*, **35**, 2406 (1952).

[22] J. C. Martin, Fr. Patent 1,415,321 (1965); through *Chem. Abstr.*, **64**, 4966 (1966).

[23] a. H. Gault and J. Skoda, *Bull. Soc. Chim. France*, **1946**, 316; b. J. Meinwald and E. Frauenglass, *J. Am. Chem. Soc.*, **82**, 5235 (1960).

[24] a. A. K. Macbeth and J. A. Mills, *J. Chem. Soc.*, **1945**, 709; b. W. Huckel and R. Neidlein, *Ber.*, **91**, 1391 (1958).

[25] a. N. R. Campbell and J. Hunt, *J. Chem. Soc.*, **1950**, 1379; b. R. T. Conley and R. F. Czaja, *J. Org. Chem.*, **27**, 1643 (1962); c. G. Tschudi and H. Schinz, *Helv. Chim. Acta*, **35**, 1230 (1952); d. D. R. Roberts, *J. Org. Chem.*, **30**, 4375 (1965).

[26] M. Stoll and A. Commarmont, *Helv. Chim. Acta*, **31**, 1077 (1948).

[27] a. W. Dirscherl, U.S. Patent 2,045,702 (1936); b. P. L. Julian, J. W. Cole, A. Magnani, and E. W. Meyer, U.S. Patent 2,484,833 (1949); c. M. Harnik, *J. Org. Chem.*, **28**, 3386 (1963); d. J. C. Babcock and L. F. Fieser, *J. Am. Chem. Soc.*, **74**, 5472 (1952); e. G. A. Grant and C. V. Seemann, U.S. Patent 2,392,660 (1946).

[28] W. F. Bruce, G. Mueller, J. Seifter, and J. L. Szabo, U.S. Patent 2,494,084 (1950).

[29] M. Freifelder, T. Anderson, Y. H. Ng, and V. Papendick, *J. Pharm. Sci.*, **53**, 967 (1964).

[30] A. Horeau and J. Jacques, *Bull. Soc. Chim. France*, **1946**, 382.

[31] H. D. Redford, *Univ. Microfilms Pub.* 1373, Ann Arbor, Mich.

[32] a. A. Horeau and J. Jacques, *Bull. Soc. Chim. France*, **1946**, 71; b. E. L. May and E. Mosettig, *J. Am. Chem. Soc.*, **70**, 686 (1948); c. W. Theilacker and E. G. Drossler, *Ber.*, **87**, 1676 (1954).

[33] L. M. Werbel, E. F. Elslager, and W. M. Pearlman, *J. Org. Chem.*, **29**, 967 (1964).

[34] a. E. J. Engels, M. Lamchen, and A. J. Wicken, *J. Chem. Soc.*, **1959**, 2694; b. J. H. McCrackin, J. G. D. Schulz, and A. C. Whitaker, U.S. Patent 3,203,982 (1965); c. A. Horeau and J. Jacques, *Bull. Soc. Chim. France*, **1948**, 53.

[35] a. R. G. Jones, *J. Am. Chem. Soc.*, **70**, 143 (1948); b. F. F. Blicke and J. A. Faust. *ibid.*, **76**, 3156 (1954); c. C. D. Hurd and P. Perletz, *ibid.*, **68**, 38 (1946).

[36] a. C. Mannich and E. Thiele, *Arch. Pharm.*, **253**, 181 (1915); b. W. Kindler, B. Hedeman, and E. Scharfe, *Ann.*, **560**, 215 (1949).

[37] a. H. K. Iwamoto and W. H. Hartung, *J. Org. Chem.*, **9**, 513 (1944); b. J. R. Corrigan, M. J. Langerman, and M. L. Moore, *J. Am. Chem. Soc.*, **67**, 1894 (1945); c. H. Bretschneider, *Monatsh.*, **78**, 82 (1948); d. N. Rubin and A. R. Day, *J. Org. Chem.*, **5**, 54 (1940); e. H. S. King and T. S. Work, *J. Chem. Soc.*, **1940**, 1307; f. A. L. Allewelt and A. R. Day, *J. Org. Chem.*, **6**, 384 (1941); g. K. W. Rosenmund and E. Karg, *Ber.*, **75**, 1850 (1942).

[38] a. J. F. Hyde, H. Browning, and R. Adams, *J. Am. Chem. Soc.*, **50**, 2287 (1928); b. E. L. May and E. Mosettig, *J. Org. Chem.*, **11**, 1 (1946); c. D. W. Mathieson and G. Newberry, *J. Chem. Soc.*, **1949**, (1133); d. A. J. Castro, D. K. Brain, H. D. Fisher, and R. K. Fuller, *J. Org. Chem.*, **19**, 1444 (1954); e. E. D. Hornbaker and A. Burger, *J. Am. Chem. Soc.*, **77**, 5314 (1955); f. H. K. Muller and G. Rieck, *Arch. Pharm.*, **299**, 809 (1966).

[39] a. C. Mannich and G. Heilner, *Ber.*, **55**, 356 (1922); b. C. Mannich and D. Lammering, *ibid.*, **55**, 3510 (1922); c. J. B. Wright and E. H. Lincoln, *J. Am. Chem. Soc.*, **74**, 6301 (1952); d. J. M. Sprague, U.S. Patent 2,686,808 (1954); e. N. H. Cromwell, P. L. Creger, and K. E. Cook, *J. Am. Chem. Soc.*, **78**, 4412 (1956).

[40] a. R. F. Phillips, C. H. Shunk, and K. Folkers, *J. Am. Chem. Soc.*, **70**, 1661 (1948); b. R. E. Lutz, R. E. Jordan, and W. L. Truett, *ibid.*, **72**, 4085 (1950); c. E. L. May and E. Mosettig, *J. Org. Chem.*, **11**, 105 (1946); d. A. Burger and W. B. Bennett, *J. Am. Chem. Soc.*, **72**, 5414 (1950); e. J. H. Birckhalter and S. H. Johnson, Jr., *ibid.*, **73**, 4827 (1951); f. L. Katz and L. S. Karger, *ibid.*, **74**, 4085 (1952); g. L. Katz, L. S. Karger, and M. S. Cohen, *J. Org. Chem.*, **19**, 1225 (1954).

[41] a. I. N. Nazarov and E. M. Cherkasova, *J. Gen. Chem. USSR*, **28**, 122 (1958) (Eng. trans.); b. E. M. Cherkasova, *ibid.*, **30**, 2826 (1960) (Eng. trans.); c. C. F. Huebner and H. A. Troxell, *J. Org. Chem.*, **18**, 736 (1953).

[42] a. K. Tuda, S. Fukushima, and A. Oguri, *J. Pharm. Soc. Japan*, **61**, 69 (1941); through *Chem. Abstr.*, **36**, 3154 (1942); b. S. Winstein, T. L. Jacobs, D. Seymour, and G. B. Linden, *J. Org. Chem.*, **11**, 215 (1946).

[43] a. M. E. Speeter, W. M. Byrd, L. C. Cheney, and S. B. Binkley, *J. Am. Chem. Soc.*, **71**, 57 (1949); b. E. L. May and E. Mosettig, *J. Org. Chem.*, **13**, 459 (1948); c. E. L. May and E. Mosettig, *ibid.*, **13**, 663 (1948).

[44] a. R. E. Lutz and J. W. Wilson, III, *J. Org. Chem.*, **12**, 767 (1947); b. J. W. Wilson, III, R. E. Lutz, and R. H. Jordan, *ibid.*, **12**, 776 (1947).

[45] a. R. Kuhn and G. Osswald, *Ber.*, **89**, 1423 (1956); b. L. E. Craig and D. S. Tarbell, *J. Am. Chem. Soc.*, **71**, 465 (1949); c. F. F. Blicke and J. Krapcho, *ibid.*, **74**, 4001 (1952); d. E. A. Mailey and A. R. Day, *J. Org. Chem.*, **22**, 1061 (1957).

[46] a. R. Fankhouser, C. A. Grob, and V. Krasnobajew, *Helv. Chim. Acta*, **49**, 690 (1966); b. J. Levy and G. O. Bernotsky, U.S. Patent 2,767,190 (1956); d. J. Levy and G. O. Bernotsky, U.S. Patent 2,776,293 (1957).

[47] H. O. House, H. C. Müller, C. G. Pitt, and P. P. Wickham, *J. Org. Chem.*, **28**, 2407 (1963).

[48] a. C. P. Rader, G. E. Wicks, Jr., R. L. Young, Jr., and H. S. Aaron, *J. Org. Chem.* **29**, 2252 (1964); b. C. P. Rader, R. L. Young, Jr., and H. S. Aaron, *ibid.*, **30**, 1536 (1965); c. H. S. Aaron, C. P. Rader, and G. E. Wicks, Jr., *ibid.*, **31**, 3502 (1966).

[49] a. W. Huckel, M. Sachs, J. Yantschulewitsch, and F. Nerdel, *Ann.*, **518**, 155 (1935); b. J. Colonge and G. Weinstein, *Bull. Soc. Chim. France*, **1952**, 462; c. C. F. Koelsch and R. A. Scheiderbauer, *J. Am. Chem. Soc.*, **63**, 2311 (1943); d. E. R. Alexander and A. Mudrak, *ibid.*, **72**, 3194 (1950); e. J. Jacques and A. Horeau, *Bull. Soc. Chim. France*, **1950**, 512.

[50] a. N. Levin, B. E. Graham, and H. G. Kolloff, *J. Org. Chem.*, **9**, 380 (1944); b. R. V. Heintzelman, U.S. Patents 2,549,684 and 2,549,685 (1951).

[51] a. E. F. Rosenblatt, *J. Am. Chem. Soc.*, **62**, 1092 (1940); b. R. R. Holmes, J. Conrady, J. Guthrie, and R. McKay, *ibid.*, **76**, 2400 (1954); c. F. Jung, Can. Patent 392,259 (1940); through *Chem. Abstr.*, **35**, 1416 (1941); d. N. M. Brown, A. M. Khan, and G. R. Procter, *J. Chem. Soc. (C)*, **1966**, 1889.

[52] a. F. C. Foster and L. P. Hammett, *J. Am. Chem. Soc.*, **68**, 1737 (1946); b. R. C. Fuson and J. Corse, *ibid.*, **67**, 2054 (1945).

[53] a. H. Rupe and W. Thommen, *Helv. Chim. Acta*, **30**, 933 (1947); b. P. Ruggli, P. Weis, and H. Rupe, *ibid.*, **29**, 1788 (1946); c. P. Ruggli and A. H. Lutz, *ibid.*, **30**, 1070 (1947).

[54] a. R. E. Hasek, E. U. Elam, J. C. Martin, and R. G. Nations, *J. Org. Chem.*, **26**, 700 (1961); b. E. U. Elam and R. Hasek, U.S. Patent 3,190,928 (1965).

[55] Girdler G-49 nickel, Chemetron Corporation, Louisville, Kentucky.

[56] a. G. I. Poos and J. T. Suh, U.S. Patent 3,287,390 (1966); b. A. Aebi, E. Gyurech-Vago, E. Hofstetter, and P. Weser, *Pharm. Acta. Helv.*, **38**, 407 (1963); c. T. Dumpis and G. Venags, *Dokl. Akad. Nauk. SSSR*, **142**, 92 (1962); through *Chem. Abstr.*, **56**, 15433 (1962); d. F. Misani, J. Speers, and A. M. Lyon, *J. Am. Chem. Soc.*, **78**, 2801 (1956); e. R. W. Wynn and A. H. Corwin, *J. Org. Chem.* **15**, 203 (1950).

[57] R. E. Brown, D. M. Lustgarten, R. J. Stanaback, and R. I. Meltzer, *J. Org. Chem.*, **31**, 1489 (1966).

[58] K. C. Brannock and H. E. Davis, *J. Org. Chem.*, **31**, 980 (1966).

[59] H. Smith, Brit. Patent 1,051,836 (1966).

[60] a. D. A. Peck and T. I. Watkins, *J. Chem. Soc.*, **1950**, 445; b. W. J. Genzler, E. M. Healy, I. Onshuus, and A. L. Bluhm, *J. Am. Chem. Soc.*, **78**, 1713 (1956); c. J. S. Moffatt, *J. Chem. Soc. (C)*, **1966**, 734; d. K. Kindler and L. Blaas, *Ber.*, **77**, 584 (1944).

[61] a. G. N. Walker, *J. Am. Chem. Soc.*, **77**, 6699 (1955); b. G. Schill, *Ann.*, **691**, 79 (1966); c. E. C. Horning and D. B. Reisner, *J. Am. Chem. Soc.*, **71**, 1036 (1949); d. A. A. Khalaf and R. M. Roberts, *J. Org. Chem.*, **31**, 89 (1966); e. H. Oelschlager, *Arzneim*, **8**, 538 (1958); f. R. N. Baker and B-T. Ho, *J. Heterocyclic Chem.*, **2**, 72 (1965).

[62] a. E. R. Alexander and A. Mudrak, *J. Am. Chem. Soc.*, **72**, 3194 (1954); b. W. L. Mosby, *J. Org. Chem.*, **18**, 485 (1953); c. J. Novak and F. Sorm, *Coll. Czechoslov. Chem. Commun.*, **19**, 1264 (1954); d. D. J. Goldsmith, *J. Org. Chem.*, **26**, 2078 (1961); e. R. H. Baker and W. Jenkins, *J. Am. Chem. Soc.*, **68**, 2102 (1946).

[63] a. C. L. Stevens and C. H. Chang, *J. Org. Chem.*, **27**, 4392 (1962); b. N. H. Cromwell and K. E. Cook, *ibid.*, **23**, 1327 (1958).

[64] a. W. H. Hartung and F. S. Crossley, *J. Am. Chem. Soc.*, **56**, 158 (1934); b. J. W. Wilt and C. A. Schneider, *J. Org. Chem.*, **26**, 4196 (1961); c. J. Sam and J. N. Plampin, *J. Am. Chem. Soc.*, **82**, 5205 (1960); d. C. D. Gutsche, N. N. Saha, and H. E. Johnson, *ibid.*, **79**, 4441 (1957).

[65] a. R. Huisgen and E. Rauenbusch, *Ann.*, **641**, 51 (1961); b. J. Kollonitsch, H. E. Mertel, and V. I. Verdi, *J. Org. Chem.*, **27**, 3362 (1962).

[66] M. Levine and S. C. Temin, *J. Org. Chem.*, **22**, 85 (1957).

[67] F. F. Blicke and J. A. Faust, *J. Am. Chem. Soc.*, **76**, 3156 (1954); b. T. Posternak and F. Ravenna, *Helv. Chim. Acta*, **30**, 441 (1947); c. C. A. Grob and H. J. Wilkens, *ibid.*, **48**, 808 (1965); d. G. Grethe, H. L. Lee, M. Usković, and A. Brossi, *J. Org. Chem.*, **33**, 494 (1968); e. G. Grethe, V. Toome, H. L. Lee, M. Usković, and A. Brossi, *ibid.*, p. 504; f. F. Hamaguchi and S. Oki, *Yakugaku Zasshi*, **86**, 148 (1966); through *Chem. Abstr.*, **64**, 14158 (1966).

[68] a. H. Plieninger and U. Lerch, *Ann.*, **698**, 196 (1966); b. J. Ratusky and F. Sorm, *Coll. Czechoslov. Chem. Commun.*, **23**, 467 (1958).

[69] a. R. H. Roth, W. A. Remers, and M. J. Weiss, *J. Org. Chem.*, **31**, 1012 (1966); b. K. N. F. Shaw, A. McMillan, A. G. Gudmonson, and M. D. Armstrong, *ibid.*, **23**, 1171 (1958); c. J. Madinaveita, *J. Chem. Soc.*, **1937**, 1927.

[70] a. J. W. Cornforth and H. T. Huang, *J. Chem. Soc.*, **1948**, 731; b. R. G. Jones and K. C. McLaughlin, *J. Am. Chem. Soc.*, **71**, 2449 (1949); c. H. Behringer and U. Turck, *Ber.*, **99**, 1815 (1966).

[71] a. R. Duschinsky and L. A. Dolan, *J. Am. Chem. Soc.*, **78**, 2079 (1945); b. G. B. Brown and M. F. Farger, *ibid.*, **68**, 1507 (1946).

[72] a. B. Graham, A. M. Griffith, C. S. Pease, and B. E. Christensen, *J. Am. Chem. Soc.*, **67**, 1294 (1945); b. R. A. Clarke, B. Graham, and B. E. Christensen, *ibid.*, **70**, 1088 (1948); c. S. Herrling, H. Keller, and N. Mückter, Ger. Patent 1,021,850 (1958).

[73] M. Freifelder, *J. Org. Chem.*, **29**, 2895 (1964).

[74] a. H. O. Burrus and G. Powell, *J. Am. Chem. Soc.*, **67**, 1468 (1945); b. A. Burger and C. R. Walter, Jr., *ibid.*, **72**, 1988 (1950); c. E. Graef, J. M. Fredericksen, and A. Burger, *J. Org. Chem.*, **11**, 257 (1946); d. B. Riegel and H. Wittcoff, *J. Am. Chem. Soc.*, **68**, 1913 (1946).

[75] a. P. Rabe, R. Pasternack, and K. Kindler, *Ber.*, **50**, 144 (1917); b. H. King and T. S. Work, *J. Chem. Soc.*, **1940**, 1307; c. H. King and T. S. Work, *ibid.*, **1942**, 401; d. A. Brossi and O. Schnieder, *Helv. Chim. Acta*, **39**, 1376 (1956); e. K. N. Campbell and J. F. Kerwin, *J. Am. Chem. Soc.*, **68**, 1837 (1946).

[76] K. N. Campbell, C. H. Helbing, and J. F. Kerwin, *J. Am. Chem. Soc.*, **68**, 1840 (1946).

[77] a. A. Fozard and G. Jones, *J. Org. Chem.*, **30**, 1523 (1965); b. G. Jones and J. Wood, *Tetrahedron*, **21**, 2529 (1965); c. H. H. Fox, J. I. Lewis, and W. Wenner, *J. Org. Chem.*, **16**, 1259 (1951).

[78] a. W. F. Tuley and R. Adams, *J. Am. Chem. Soc.*, **47**, 3061 (1925); b. R. Adams and B. S. Garvey, *ibid.*, **48**, 477 (1926); e. R. H. Bray and R. Adams, *ibid.*, **49**, 2101 (1927).

[79] a. Brit. Patent 870,009 (1961); b. J. Falbe and F. Korte, *Ber.*, **98**, 1928 (1965); c. G. R. Ames and W. Davey, *J. Chem. Soc.*, **1956**, 3001.

[80] a. U. Westphal, Y. L. Wang, and H. Heilman, *Ber.*, **72**, 1233 (1939); b. St. Kauffman and G. Rosenkranz, *J. Am. Chem. Soc.*, **70**, 3052 (1948); c. P. Wieland and K. Miescher, *Helv. Chim. Acta*, **32**, 1764 (1949); d. H. Heusser, P. Th. Herzig, A. Furst, and Pl. A. Plattner, *ibid.*, **33**, 1093 (1950).

[81] a. T. Komatsu, R. Iwanaga, and J. Kato, U.S. Patent 3,141,895 (1964); b. E. C. Horning and R. U. Schoch, Jr., *J. Am. Chem. Soc.*, **70**, 2945 (1948).

[82] a. R. Pessina, A. Botto, F. Gatti, and G. Valenti, *Farmaco, Sci. Ed.*, **20**, 97 (1965); b. R. Simonoff, *J. Am. Chem. Soc.*, **69**, 2073 (1947).

[83] a. T. Immediata and A. R. Day, *J. Org. Chem.*, **5**, 512 (1940); b. B. R. Baker and F. I. McEvoy, *ibid.*, **20**, 127 (1955); c. R. Simonoff and W. R. Hartung, *J. Am. Pharm. Assoc.* **35**, 306 (1946).

[84] R. Howe, L. H. Smith, and J. S. Stephenson, U.S. Patent 3,255,249 (1966).

[85] W. H. Carothers and R. Adams, *J. Am. Chem. Soc.*, **46**, 1675 (1924).

[86] a. R. A. Clarke and B. E. Christensen, *J. Am. Chem. Soc.*, **70**, 1818 (1948); b. L. S. Fosdick, O. Fancher, and K. F. Urbach, *ibid.*, **68**, 840 (1946).

[87] a. J. Cason and D. D. Phillips, *J. Org. Chem.*, **17**, 298 (1952); b. K. Kindler, H. Oelschlager, and P. Heinrich, *Ber.*, **86**, 501 (1953); c. H. U. Daeniker, *Helv. Chim. Acta*, **49**, 1543 (1966); d. W. S. Emerson and V. E. Lucas, *J. Am. Chem. Soc.*, **70**, 1180 (1948).

[88] R. E. Lyle and H. J. Troscianiec, *J. Org. Chem.*, **24**, 333 (1959).

[89] a. C. F. Winans, *J. Am. Chem. Soc.*, **61**, 3564 (1939); b. A. Galun, A. Kaluszyner, and E. D. Bergmann, *J. Org. Chem.*, **27**, 2373 (1962).

XV

REDUCTION OF THE C=N BOND

The C=N bond in most compounds is so readily hydrogenated that high-pressure conditions are seldom necessary. When reducing an aldimine or a ketimine it is best, because of their relative instability, to carry out the reaction shortly after preparation of the imine to obtain the best yield of amine [1]. When the C=N bond is part of a nonaromatic heterocycle, such a precaution is unnecessary.

One must be aware of steric effects in the reduction of imines. They can change the ease of reducibility or the course of reaction. Hydrogenations of 2-(1-iminoethyl)-1,1,3-trimethylcyclohexane, in alcohol over platinum oxide

$$\text{cyclohexane ring with substituents } (CH_3)_2, C(CH_3)=NH, CH_3$$

[2a] and of N-(1-trifluoromethylbenzylidene)-1-phenylisopropylamine, $C_6H_5CH_2CH(CH_3)N=C(CF_3)C_6H_5$, with 5% platinum on carbon or with platinum oxide could not be effected [1]. In each case a molecular model showed that there was little or no contact between the C=N bond and the catalyst [1a]. Uptake was incomplete in the hydrogenation of N-(5-diethylamino-2-pentylidene)-4-methoxy-3-propylaminoaniline under 30 atm pressure in the presence of 10% by weight of platinum oxide [2b].

Steric effects changed the course of reaction in the reduction of N-benzylidene-2,2,6-trimethylcyclohexylamine in acidified alcohol with platinum oxide; saturation of the benzene ring took place without affecting the azomethenyl group [2a]. Molecular models show that the benzene ring is in position to be reduced while the C=N bond is either out of plane or is prevented from making contact with the catalyst because of the methyl groups in 2- and 6-positions [1a].

When a primary imine, RCH=NH, is hydrogenated, secondary amine may form through condensation of primary amine and imine [3] in the same manner as in reductions of cyanides (Chapter XII). Secondary amine formation seldom takes place when the imino nitrogen is substituted.

PRIMARY IMINES

Hydrogenations of imines, RCH=NH or RR′C=NH, are best carried out in anhydrous solvent since imines may be hydrolyzed in part to aldehydes or ketones which can lead to formation of secondary amines by reductive alkylation of primary amine.

Platinum oxide has been employed to give high yields of primary amines from reductions of imines in neutral solvent [4a–c]. It is likely that 5% platinum on carbon or 5% palladium on carbon could be substituted for platinum oxide. Hydrogenation in solvent containing an equivalent of acid can speed up reaction and allow use of lower catalyst ratios [1].

N-SUBSTITUTED IMINES

Ketimines

Hydrogenations of N-substituted ketimines to the corresponding secondary amines have been carried out at low pressure in neutral solvent with platinum oxide [4b, 5a] and with palladium on carbon [5b, c]. Reductions with nickel may require more rigorous conditions. 3-(2-Diethylaminoethylimino)-tropane was converted to 3-(2-diethylaminoethylamino)tropane over Raney nickel at 60° and 70 atm [5d]; hydrogenation over platinum under mild conditions gave 85% yield in a short reaction period.

Aldimines

N-Benzylidene and Related -ylideneamines. Aldimines of the type, ArCH=NR, undergo hydrogenation to N-substituted benzylamines with relative ease unless R is a highly substituted alkyl group or a ring system with a bulky group in 2-position or if the aromatic ring attached to the double-bonded C atom also contains a large 2-substituent. Hydrogenolysis of the resulting N-benzylamines, ArCH$_2$NHR, must be recognized as a competing side reaction.

Despite the fact that palladium catalysts are more likely to cause debenzylation than others employed for the hydrogenation of benzylidene-amines to benzylamines they are very useful for the reductions of aldimines, ArCH=NR. In general, the product of hydrogenation, the benzylamine, because of its basicity, inactivates the catalyst to some degree. Unless a deliberate attempt is made to cause hydrogenolysis by raising the temperature and pressure [6], secondary amines are obtained in high yields. Obviously hydrogen uptake must be controlled but there is usually a sharp enough break in the rate of uptake that reaction can be interrupted with little danger of debenzylation [1].

Employment of less catalyst in palladium reductions should substantially limit hydrogenolysis. Nuclear-substituted benzylidenemethylamines, in the absence of solvent, were hydrogenated with low ratios of 10% palladium on carbon to give high yields of N-methylbenzylamines [7a]. N-(Benzhydrylidene)aniline, $(C_6H_5)_2C=NC_6H_5$, was converted to N-phenylbenzhydrylamine in 87% yield by reduction with a modest amount of catalyst [7b]. Hydrogenations in alcohol of N-(2-, 3-, and 4-substituted and 2,4- and 3,4-disubstituted benzylidene)methylamines gave consistently high yields of N-methylbenzylamines in rapid reactions with 3–5% ratios of 5% palladium on carbon [1]. A very small amount of catalyst, 1.25% by weight, was employed in reductions of N^1-(4-methoxybenzylidene)-N^2N^2-dimethylethylenediamine and related compounds [7c].

Steric effects may necessitate an increase in the amount of catalyst. The hydrogenation of N-[2-(6-hexyloxy)benzylidene]methylamine,

$$2\text{-ROC}_6\text{H}_4\text{CH}=\text{NCH}_3, \ R=C_6H_{13},$$

was extremely slow unless a 10% weight ratio of catalyst was employed [1].

Platinum oxide and supported platinum catalysts find considerable use in conversions of benzylideneamines to benzylamines because they have less tendency to cause debenzylation than palladium catalysts. At times reduction of imine to secondary amine may not go to completion because the nitrogen base formed during reduction in neutral solvent inhibits the activity of platinum catalysts but this can be overcome by carrying out reactions in neutral solvent containing an equivalent of acid to neutralize the resulting base [1] or by the use of acidic conditions for reduction [8a–c].

Occasionally an excessive amount of platinum catalyst may be required in reductions in neutral solvent [9a], but more often 1–2% ratios of platinum oxide to substrate carry reactions to completion in a moderate amount of time giving high yields of secondary amines [8c, 9b–e]. The catalyst was reused several times in the conversion of 3-benzylideneamino-1,2-propanediol to 3-benzylamino-1,2-propanediol [9f].

Low ratios of 5% platinum on carbon (equivalent to 0.25–0.3% metal to substrate) gave high yields of N-benzylcyclobutylamine and N-benzylcyclohexylamine in fairly rapid reductions of the Schiff bases in alcohol [1].

Rhodium on a carrier is less useful for these reactions than palladium or platinum catalysts because its activity is more adversely affected by formation of secondary amine during reduction in neutral solvent. The inhibitory effect was overcome by the presence of an equivalent of acid, pKa 2.7 to 5.0 [10a]. Nevertheless, more rapid uptake and more complete conversions of Schiff bases to secondary amines took place with palladium on carbon in neutral solvent than with an equal amount of rhodium on carbon under the enhanced conditions [1].

Raney nickel does not compare favorably with palladium for speed of reaction in low-pressure reductions of benzylideneamines [1]. However, because it seldom causes debenzylation, it can prove advantageous in reactions which are not closely monitored [10b]. Debenzylation did take place in the reduction of methyl 3-(4-benzylideneaminophenyl)propionate with nickel at 3.3 atm [10c] but three times as much catalyst as substrate was employed.

It has been used very successfully in low-pressure hydrogenations of N-(1- and 2-naphthylidene)-1- and 2-menaphthylamines to the corresponding dimenaphthylamines [11a] and of N-benzylidenebenzhydrylamine to N-benzylbenzhydrylamine [11b]. These reactions are of particular interest because in each case the products of reduction contain readily hydrogenolyzable groups (see Chapter XIX).

The hydrogenations of N-benzylidenemethylamine to N-methylbenzylamine [11c], of nuclear-substituted benzylidene amino alcohols to the corresponding benzylaminoalkanols [11d], of N-benzylidenealkoxyanilines to N-benzylalkoxyaniles [11e] and conversion of some benzylideneaminocycloalkanes [11f, g] are other examples of the application of Raney nickel at low pressures.

The addition of an equivalent of acid such as was employed with rhodium [10a] should aid slow reactions without materially attacking nickel catalyst.

Raney nickel is of particular value for the reduction of benzylideneamines under more rigorous conditions. N-Benzylideneisopropylamine was converted to N-isopropylbenzylamine at 90° and 30 atm pressure [12]. Almost quantitative yield was obtained in the same reduction with commercially available Raney nickel at 90–100° and 100 atm [1]. It is a useful method for the preparation of N-alkyl nuclear-substituted benzylamines on large scale [1].

Bis Imines. Most reductions of *bis*-benzylidene alkanediamines and triamines have been carried out in neutral solvent with palladium or platinum catalysts at 1–3 atm pressure. N,N'-*bis*-(Benzylidene)ethylenediamine, $(C_6H_5HC=NCH_2)_2$, in ethyl acetate was hydrogenated over platinum oxide, 1% by weight, to N,N'-*bis*-benzylethylenediamine, $(C_6H_5CH_2NHCH_2)_2$, in high yield [13a]; almost comparable yield was obtained from reduction with a 1% ratio of 10% palladium on carbon. Palladium on alumina also was employed giving better results than platinum on alumina [13b]. In another comparison platinum oxide was superior to Raney nickel [13c]; Raney nickel was useful only at higher pressure. Very high yields of *bis*-benzylamino- and *bis*-benzhydrylaminodiethylenetriamines were obtained from palladium-catalyzed reductions of the *bis*-Schiff bases of diethylenetriamine, RCH=N(CH$_2$)$_2$]$_2$NH [13d]. Palladium on carbon was employed in hydrogenations of a series, (3,4-diCH$_3$C$_6$H$_3$CH=N)$_2$—(CH$_2$)$_n$-, n = 3–10, to give very high yields of the corresponding *bis*-(*N*-benzylamino)alkanes [1]. Platinum oxide was used to good advantage in reductions of *N,N'*-(nuclear-substituted benzylidene)-propane-1,3-diamines [13e, f].

Miscellaneous Aldimines. Poor results were obtained from Raney nickel reductions of aliphatic aldimines, RCH=NR' where R and R' were similar or different alkyl groups [14]. Good yield was obtained from the rapid platinum-catalyzed hydrogenation of freshly prepared *N*-(3-diethylamino-2,2-dimethyl)propylidene-3-diethylamino-2,2-dimethylpropylamine [15] (Eq. 15.1).

$$(C_2H_5)_2NCH_2C(CH_3)_2CH=NCH_2C(CH_3)_2CH_2N(C_2H_5)_2 \xrightarrow{H_2}$$
$$[(C_2H_5)_2NCH_2C(CH_3)_2CH_2]_2NH \quad (15.1)$$

Palladium catalysts may be best for reductions of other aldimines where there is no possibility of debenzylation. *N*-(3-Dimethylamino-2,2-dimethylpropylidene)-1-naphthylamine was hydrogenated to the corresponding *N*-propylnaphthylamine with palladium on carbon [16a]. The same catalyst was employed in reduction of some of 3,3-diphenylpropylideneamines [16b] and in the conversion of the aldimine, C$_6$H$_5$CH$_2$CH=NR where R was cyclopropyl, to *N*-cyclopropyl-2-phenylethylamine [1].

Poor results from palladium reductions of Schiff bases of 2-sulfamylaniline, 2-NH$_2$SO$_2$C$_6$H$_4$N=CHR, R = phenyl, pyridyl, benzyl, and cyclohexyl, caused the investigators to resort to chemical methods [16c]. In this instance the difficulty may be ascribed to a combination of the steric effects of the sulfamyl group in 2-position and the bulk of the R group. Hydrogenations of some methylideneamino compounds, C$_6$H$_5$CHRCH$_2$N=CH$_2$, with palladium on carbon were extremely slow [1]. Improvement was noted when the reactions were run in alcohol with platinum oxide or in alcohol containing an equivalent of acetic acid with rhodium on carbon. In each instance yields of *N*-methylamines were not exceptionally high. Hydrogenations of this type may require larger than normal amounts of catalyst for reaction. An alternate method-preparation of the *N*-benzylamine, methylation, followed by debenzylation over palladium on carbon should give good yield despite the number of steps involved.

THE UNCONJUGATED C=N BOND IN HETEROCYCLES

Hydrogenation of an unconjugated C=N bond in a heterocycle is relatively easy to accomplish unless it is in a sulfur-containing one as in methyl 4,4-dimethyl-5,6-dihydro-1,3(4H)-thiazine-2-acetate [17a]. Reductions with palladium, platinum, and rhodium catalyst failed apparently because of catalyst poisoning. In certain instances hydrogenation will not take place in neutral solution as in the reduction of 1-pyrrolines with platinum oxide [17b].

2-Methyl-4,5,6,7-tetrahydro-3[H]-azepine was converted to the hexahydroazepine with platinum oxide in aqueous acid [18] (Eq. 15.2). Platinum-

$$(15.2)$$

catalyzed hydrogenations of 3,4-dihydroisoquinolines in neutral solvent or in acidic media appear to be the favored method of obtaining 1,2,3,4-tetrahydroisoquinolines [19a–f]. Palladium on carbon was also employed [19f]. When 1-alkyl-6,7-dimethoxy-3,4-dihydroisoquinolines were reduced in alcohol with either platinum oxide or palladium on carbon, reaction was complete in 1 hr [19f]. No explanation was offered for the 4–6 hr reaction period for reduction of an aqueous or alcoholic solution of the hydrochloride with either catalyst.

Platinum oxide was employed in the hydrogenation of the C=N bond in a substituted 1,2,3,4,5,6,7,8-octahydroquinolizine [20a]; Raney nickel reduction of 1-methyl-1,2,3,4,5,6,7,8-octahydroquinolizine under 3 atm pressure yielded racemates of 1-methylquinolizidine [20b] (Eq. 15.3).

$$(15.3)$$

Reductions of the C=N bond in substituted 1,3-dihydro-2H-1,5-benzodiazepin-2-ones were carried out with Raney nickel [21a, b] (Eq. 15.4).

$$(15.4)$$

Conversions of 3-substituted 1,4-benzoxazine-2-ones to the corresponding morpholones were carried out in very high yield with Raney nickel or with supported palladium catalysts at 80° and 50 atm pressure of hydrogen in ethyl acetate or tetrahydrofuran [22]. When methyl alcohol was used in the Raney nickel reduction of the 3-phenyl compound, ring opening and methanolysis took place (Eq. 15.5).

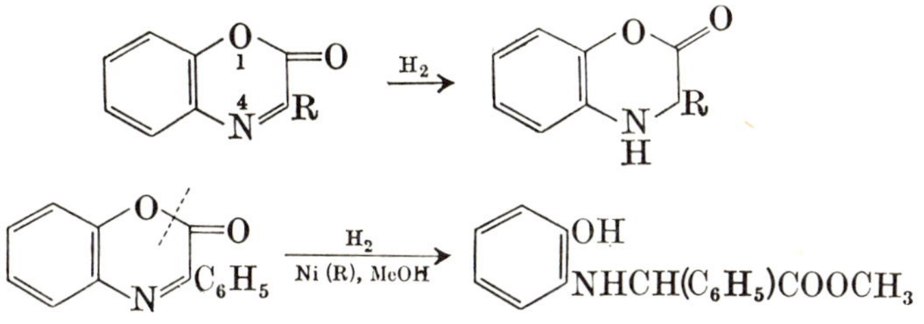

(15.5)

THE C=N BOND IN AZINES, HYDRAZONES AND SEMICARBAZONES

Azines

Hydrogenation of the C=N bond in aldazines and ketazines, $(RR'C=N)_2$, does not proceed as readily as reduction of a single C=N bond. Partial reversal to aldehyde or ketone or hydrogenolysis of the resulting hydrazines, —NH ⋮ NH—, to primary amines are factors which greatly affect yields.

Reductions of aliphatic aldazines and ketazines are generally slow [23a]. They have been carried out over platinum in acetic acid [23a] or in neutral solvent containing an equivalent of hydrogen chloride [23b]. In contrast to the latter report little or no hydrazines were obtained in other reductions of aliphatic ketazines with platinum oxide in alcohol or water containing hydrogen chloride [1].

Freshly prepared azine, distilled before use, and rapid reaction are essential to high yields of N,N'-dialkylhydrazines [1]. Reductions in neutral solvent over platinum oxide or preferably 5% platinum on carbon gave high yield of hydrazines in a number of instances. When the hydrogenation of 1,2-*bis*-(1-methylpropylidene)hydrazine, $[CH_3CH_2C(CH_3)=N]_2$, was carried out with platinum on carbon at 70 atm pressure, reaction was complete in 1 hr to give 82% of 1,2-*bis*-(1-methylpropyl)hydrazine [1]:

1,2-Dicyclobutyl- and dicyclopentylhydrazines were prepared from the ketazines by hydrogenation in acetic acid over platinum oxide [23c]. 1,2-Dibenzylhydrazine was obtained from reduction of 1,2-dibenzylidenehydrazine in methanolic hydrogen chloride with platinum oxide; by interrupting the reaction at one molar equivalent 1-benzyl-2-benzylidenehydrazine was obtained [23d]. 1,2-Dibenzylhydrazine was also obtained from the aldazine by reduction with palladium on calcium carbonate [23]. Palladium on carbon was employed in the hydrogenation of the azine of 3-methyl-3-(4-nitrophenyl)-2-butanone in dry tetrahydrofuran to 1,2-*bis*-[3-methyl-3-(4-acetaminophenyl)-2-butyl)hydrazine in 67% yield [23f] and in the conversion of C=N—N=C bonds in the cyclic compound, 3,8-diphenyl-1,2-diazacyclo-octa-2,8-diene [23g]. Platinum oxide was used as catalyst in the reduction of a similar compound in acetic acid [23h].

Hydrazones

In many instances the purpose of the reduction of hydrazones, RR'C=NNHR", is to obtain amines by cleavage of the resulting hydrazine. However, dependent on structure of the substrate, choice of catalyst, and mildness of reaction conditions, hydrazines can be obtained.

When R" is H, partial hydrolysis and condensation of the carbonyl compound with the hydrazone can take place yielding an azine and subsequently a disubstituted hydrazine on reduction (Eq. 15.6). Hydrogenation

$$RCH=NNH_2 \xrightarrow{H_2O} RCHO + NH_2NH_2$$

$$\left.\begin{array}{l} RCHO + RCH=NNH_2 \\ \text{or} \\ 2\ RCHO + NH_2\ NH_2 \end{array}\right\} \longrightarrow (RCH=N)_2 \xrightarrow{2\ H_2} (RCH_2NH)_2 \quad (15.6)$$

of the hydrazone of 1-phenyl-2-propanone, $C_6H_5CH_2C(CH_3)$=NNH$_2$, with Raney nickel yielded azine and 3-phenylisopropylamine [24]; reduction in neutral solvent with palladium or rhodium on a carrier or with ruthenium or platinum oxide yielded 1,2-*bis*-substituted hydrazine which was also obtained from slow reduction in aqueous acetic acid with platinum oxide. Hydrogenation in alcohol containing a molar equivalent of acetic acid with platinum oxide or platinum on carbon at 130 atm pressure gave 50–70% yields of 1-(3-phenylisopropyl)hydrazine [24]. In an attempted reduction in the same medium at 2 atm pressure with platinum on carbon, the 1,2-disubstituted hydrazine was the main identified product [1]. In the hydrogenation of the hydrazone of 3-hydroxy-5-hydroxymethyl-2-methylpyridine-4-aldehyde with platinum on carbon in methyl alcohol, only azine and 1,2-*bis*-(3-hydroxy-5-hydroxymethyl-2,4-dimethylpyridyl)hydrazine-4-methylhydrazine were obtained [1].

It is likely that success in the reduction of hydrazones, R"=H, to monosubstituted hydrazines is dependent on the stability of the hydrazone. When freshly prepared 1-benzylidenehydrazide was hydrogenated in anhydrous alcohol over palladium on carbon at 3 atm, uptake was complete in less than 30 min to give 82% yield of 1-benzylhydrazine [1]; when the hydrazone was allowed to stand for several days to a week before reduction, the yield dropped to 45–48%.

When R" is a substituent other than hydrogen, azine formation does not take place. When R" is phenyl or another aromatic system, conversion to the 1,2-disubstituted hydrazine, RR'CHNNHAr, may be accomplished, although cleavage of the resulting hydrazine to primary amine does occur even under moderate reaction conditions. Freshly distilled 1-isopropylidene-2-phenylhydrazine in aqueous alcoholic hydrogen chloride was hydrogenated over colloidal platinum to give 90% yield of 1-isopropyl-2-phenylhydrazine [25a]. It is likely that platinum oxide or platinum on carbon could be used under similar reaction conditions. In most instances employment of platinum oxide in reductions in neutral solvent gives fair to poor yield. 1-Isopropylidene-2-(4-carboxyphenyl)hydrazine was reduced to 1-isopropyl-2-(4-carboxyphenyl)hydrazine in ethyl alcohol over platinum oxide at 2.25 atm [25b]; the poor yield, less than 30%, could have been due to hydrogenolysis. When 1-isopropylidene-2-(2-pyridyl)hydrazine was hydrogenated over platinum oxide under similar conditions hydrogenolysis occurred; 2-aminopyridine was obtained and starting material was recovered [1].

When the system RR'NN=C is part of a cyclic structure, reduction to a cyclic hydrazine takes place in fairly good yield. Examples are seen in the hydrogenation of 3,4-dihydro-5*H*-benzo-2,3-diazepine in methyl alcohol with palladium on carbon to 1,2,3,4-tetrahydro-5-*H*-benzo-2,3-diazepine [26a] and in the reduction of 4,4a,5,6-tetrahydro-3*H*-pyridazo[2,3a]quinoline in acetic acid with platinum oxide to the corresponding 2,3,4,4a,5,6-hexahydro compound [26b].

When hydrogenations of hydrazones, RR'C=NNHacyl, are carried out high yields of hydrazines are usually obtained. In general 1-acylhydrazones are fairly stable. The tendency toward hydrolysis is greatly decreased and hydrogenolysis of the resulting hydrazine can be easily controlled [1]. Low-pressure reductions over platinum have been widely used among the group with good success. 1-Isonicotinoyl-2-alkyl- and cycloalkylhydrazines have been prepared in good yield by hydrogenation of the 2-alkylidene- or cycloalkylidene compounds in methyl alcohol over platinum oxide at room temperature and 2–3 atm pressure [27a]. Reduction of 1-isonicotinyl-2-benzylidenehydrazine required the presence of acid. Hydrogenations of the above group were unsuccessful with Raney nickel. In a similar series the C=N linkage of the acylated hydrazones was reduced with platinum in alcohol, water, or acetic acid at 60° and 3.3 atm [27b]. Very high yields of hydrazine, RCONHNHCH(CH$_3$)$_2$ where R was alkyl, cycloalkyl, benzoyl, and nuclear-substituted benzoyl, were obtained from platinum oxide-catalyzed reductions of the isopropylidenehydrazines in ethyl alcohol [1]. Platinum on carbon reductions in alcohol of 1-benzoyl, 1-phenylacetyl,

1-(2-naphthoyl and 1- and 2-naphthylacetyl)-2-isopropylidenehydrazines resulted in 74–83% yields of the corresponding 1-substituted-2-isopropylhydrazines [27c]. The same catalyst was also employed in a mixture of alcohol and acetic acid to convert 1-carbethoxy-2-(2-indanylidene)- and [2-2-(1,2,3,4-tetrahydro-naphthylidene)]hydrazines to 1-carbethoxy-2-(2-indanyl)- and 2-(2-(1,2,3,4-tetrahydronaphthyl)hydrazines [27d].

In the reduction of hydrazones, RCONHN=CHR' where R was 2- and 3-indolyl or where RCO was 3-indolylacetyl and R' was alkyl and phenyl or where CH(R') was cycloalkylidene, palladium on carbon and Raney nickel were employed as well as platinum oxide [28a]. 3-Methylideneamino- and 3-ethylideneamine-2-oxazolidinones, which have a C(=O)NHN=C structure, were converted to 3-methylamino- and 3-ethylamino-2-oxazolidones with palladium on carbon [28b] (Eq. 15.7).

R=H, CH$_3$

$$\begin{array}{c}\text{O}\\\text{NN=CHR}\end{array} \xrightarrow[\text{7 atm}]{\text{H}_2\ \text{Pd on C}} \begin{array}{c}\text{O}\\\text{NNHCH}_2\text{R}\end{array} \qquad (15.7)$$

Semicarbazones and Related Compounds

Raney nickel was employed for hydrogenations of semicarbazones of keto acids, $C_6H_5(CH_2)_{0-2}C(=O)COOH$ and $C_6H_5C(=O)(CH_2)_2COOH$ to the semicarbazides in aqueous or preferably aqueous alkaline solution [29a]. Reduction of the esters in alcohol gave unidentifiable products. The semicarbazide of pinonic acid, $C_6H_5C(=O)(CH_2)_3COOH$ could not be hydrogenated.

2-Heptanone semicarbazone in acetic acid and ethyl alcohol was converted to 2-heptylsemicarbazide, $C_5H_{11}CH(CH_3)NHNHCONH_2$, by reduction with platinum on carbon [29b]. In a reaction related to the hydrogenation of semicarbazones, phenylisopropylideneaminoguanidine,

$$C_6H_5CH_2C(CH_3)=NNHC(=N)NH_2,$$

was reduced with platinum oxide to N-guanidino-1-methyl-2-phenethylamine, $C_6H_5CH_2CH(CH_3)NHNHC(=NH)NH_2$ [29c].

There does not appear to be any reason why palladium on a carrier should not also be useful in these reactions.

IN AMIDINES

Hydrogenation of amidines results only in hydrogenolysis to amines. This is to be expected since 1,1-diamines are unstable. N,N'-Di-(3-tolyl)benzamidine was converted to toluene and *m*-toluidine by reduction in acetic acid with palladium black [30a]. The first step in the reaction produced *m*-toluidine and *N*-benzyl-*m*-toluidine, the latter undergoing further hydrogenolysis. The main product of the reduction of *N*,*N*-dimethyl-1-phenylcinnamamidine with platinum oxide in alcohol was *N*,*N*-dimethyl-2,3-diphenylpropylamine [30b]. A very small amount of 2,3-diphenylpropylamine was also obtained.

SELECTIVE HYDROGENATIONS

Since the C=N linkage is readily reduced, it is to be expected that it will be preferentially hydrogenated over an aromatic ring system. This is indicated from numerous examples in the literature. Even with rhodium, which reduces ring systems so readily at low pressure, there was a distinct difference in rate of hydrogen absorption in the reaction of furfurylidenecyclopropylamine to *N*-(tetrahydrofurylmethyl)cyclopropylamine [1] (Eq. 15.8).

in EtOH + AcOH

$$\text{furfuryl-CH=N-cyclopropyl} \xrightarrow[\text{Rh on C}]{H_2} [\text{furfuryl-CH}_2\text{NH-cyclopropyl}] \xrightarrow{2H_2} \text{tetrahydrofuryl-CH}_2\text{NH-cyclopropyl}$$

20 min   6 hr

(15.8)

The C=N function will not be reduced in preference to an acetylenic linkage or in general in preference to an aromatic nitro group. There is a reported exception. *N*-(3-Nitro-4,5-diphenyl-2-furfurylidene)aniline in tetrahydrofuran was hydrogenated over palladium on carbon to yield 81% of *N*-2-furylmethyl-3-nitro-4,5-diphenylaniline [30c].

Presence of an Olefinic Bond

It is generally assumed when a substrate contains an olefinic bond and C=N bond that the carbon to carbon double bond will be reduced first. It is an assumption which may be based on too few examples and one in which the structure of the substrate and the presence of substituents on the double-bonded carbon and nitrogen atoms is not taken into consideration.

The double bond in conjugated compounds, $C_2H_5CH(CH_2OH)N=CHC(H$ or $C_2H_5)=CHR$ (R = methyl, propyl or phenyl) was selectively reduced over Raney nickel at low pressure (Chapter IX, Refs. 185 and 186). It was shown that there was a great difference in rate between the two reducible systems.

324   *Reduction of the* C=N *Bond*

This difference may be due to the effects of the groups surrounding the nitrogen atom, specifically HC—N=CH with CH₂OH and CH₂CH₃ substituents, which interfere with the contact between that portion of the molecule and the catalyst. This appears to be substantiated by molecular models.

There was no break in the rate in the Raney nickel-catalyzed hydrogenation of a nearly symmetrically conjugated compound, 4-benzylideneaminostilbene, $C_6H_5CH=NC_6H_4CH=CHC_6H_5$ [30d]. From molecular models [1a] there appears to be no difference in the ability of either function to reach the surface of the catalyst.

In another conjugated system the exo double bond in 3-isopropylidene-2,5,5-trimethyl-1-pyrroline was hydrogenated in preference to the cyclic C=N bond in a reaction carried out in absolute methyl alcohol with platinum oxide [30e]. A framework molecular model of the starting material helps to explain selectivity [1a]. It shows good contact of the pi bonds of the exo double bond with the catalyst surface while the methyl groups in 2- and 5-positions prevent the bond representing the unshared pair of electrons of the nitrogen atom from getting to the catalyst [1a]. At the same time one of the side-chain methyl groups also interferes with contact of the pi bond of the carbon atom of the N=C linkage.

In the hydrogenation of other conjugated compounds where the cyclic C=N bond was less hindered, that bond was reduced and an exocyclic double bond remained intact. Examples are seen in the palladium-catalyzed hydrogenations of 4-anilinomethylene-3-benzyl- and 3-phenylisoxazolin-5-one [30f] (Eq. 15.9).

$$C_6H_5NHCH=\underset{R}{\overset{O-O}{\underset{\parallel}{C-N}}} \quad \xrightarrow[\text{Pd on C}]{H_2} \quad C_6H_5NHCH=\underset{}{\overset{O-O}{\underset{}{C-N}}} \quad (15.9)$$

R = C_6H_5 or $C_6H_5CH_2$

However, when 3,4-cyclopenteno-2,6,6-trimethyl-5,6-dihydropyridine, representing another conjugated C=C—C=N-system, was reduced in methyl alcohol with platinum oxide, reaction did not proceed by preferential attack at either function. Instead 1,4-addition followed by rearrangement appeared to have taken place to yield 3,4-cyclopentano-2,6,6-trimethyl-1,4,5,6-tetrahydropyridine [30g] (Eq. 15.10). It is of interest that when the reduction was carried out under acidic conditions, two equivalents of hydrogen were absorbed rapidly and that when the tetrahydro compound was reduced under the same conditions, conversion to piperidine required 50 hr. It is likely that reduction of the dihydropyridine in acid proceeded by a different

$$\text{(15.10)}$$

mechanism than in base and, as the authors assume, that the 1,4,5,6-tetrahydropyridine could not be the intermediate in the acidic hydrogenation.

In reductions where the two functions were not conjugated, the indolinene C=N bond in the alkaloid strictamine was preferentially reduced in neutral solvent in the presence of platinum oxide [31a]; N-(3,7-dimethyloct-6-enylidene)cyclohexylamine, $(CH_3)_2C=CH(CH_2)_2CH(CH_3)CH_2CH=NC_6H_{11}$, was converted to N-(3,7-dimethyloct-6-enyl)cyclohexylamine,

$$(CH_3)_2C=CH(CH_2)_2CH(CH_3)(CH_2)_2NHC_6H_{11},$$

with Raney nickel under mild conditions [31b], and N-benzylidene-2-(cyclohexenyl)ethylamine, $C_6H_5CH=N(CH_2)_2R$ (R = cyclohexenyl), was selectively reduced in ammoniacal alcohol to N-benzyl-2-(cyclohexenyl)-ethylamine, $C_6H_5CH_2NH(CH_2)_2R$, in almost quantitative yield over palladium on carbon [31c].

Presence of a Carbonyl Group

Although there are very few cited examples, hydrogenation of the C=N linkage should be preferred over that of the carbonyl function. The relationship of the reducibility of the C=N bond to the C=C bond and, by analogy, the selective reduction of the olefinic bond in the presence of most carbonyl groups are arguments for that assumption. The hydrogenation of the perchlorate salt of the methyl ether of 8-aza-8,9-dehydroestrone in alcohol over platinum oxide is one example. Hydrogen uptake, which was rapid, stopped at one molar equivalent to give a mixture of bases of 8-azaestrone [32] (Eq. 15.11).

Presence of a Cyano Group

Further study of reduction of a number of different type compounds in different media is necessary in order to determine specific selectivity in substrates containing both functions. When the 4-methoxyphenylhydrazone of ethyl 4-cyano-2-oxobutyrate was hydrogenated over Raney nickel at 50–70° and 100 atm, the cyano group was reduced while the phenylhydrazone portion of the molecule was unaffected (Chapter XII, Ref. 72b). When 7-cyano-5-phenyl-1,3-dihydro-2H-1,4-benzodiazepin-2-one was subjected to reduction with platinum at low pressure, the cyano group was untouched to give 7-cyano-5-phenyl-1,3,4,5-tetrahydro-2H-1,4-benzodiazepin-2-one [33]. In this instance the result is not entirely unexpected since cyano groups directly attached to an aromatic ring are not readily reduced under the described conditions.

Presence of Hydrogenolyzable Groups

N- and O-Benzyl and Related Groups. Reduction of the C=N linkage in compounds containing *O*- or *N*-benzyl groups is probably best carried out with Raney nickel or platinum, rather than palladium, to insure better selectivity. Raney nickel was employed for the conversion of *N*-(3,4-dioxmethylenebenzylidene)-2-(3,4-dibenzyloxyphen)ethylamine to *N*-(3,4-dioxymethylenebenzyl)-2(3,4-dibenzyloxyphen)ethylamine [34a] and in reductions of imines which produced high yields of amines containing hydrogenolyzable groups, RCH$_2$NHCH$_2$R′ where R = 2-menaphthyl and R′ = 1-menaphthyl [12a] and (R)$_2$CHNHCH$_2$R where R = phenyl [12b].

The internal C=N bond in 6-benzyloxy-7-methoxy-1-methyl-3,4-dihydroisoquinoline was reduced over platinum oxide to give almost quantitative yield of 6-benzyloxy-7-methoxy-1-methyl-1,2,3,4-tetrahydroisoquinoline [1].

The C=N bond in 4-(pyridoxylideneamino)isoxazolidin-3-one which contains a reducible ArCH$_2$OH group, was selectively hydrogenated over Raney nickel, platinum oxide, or palladium on carbon [34b]. Palladium may be used in reductions of this type because hydrogenolysis of the hydroxyl group in an ArCH$_2$OH system does not take place with the same ease as hydrogenolyses of *O*-benzyl-type groups, ArCH$_2$OR where R is not hydrogen.

Halogens. Since the C=N bond yields an amine on reduction, there is always danger of dehalogenation in the hydrogenation of substrates containing it and a halogen other than fluorine. Any accompanying dehalogenation will depend on the basicity of the amine resulting from reduction and in certain instances on the proximity of the amine and halogen; it will also depend on reducibility of the halogen and on the type of halide as well as the catalyst.

Loss of aromatic chlorine occurred during reduction of imino compounds with palladium on carbon [1, 35]. This may be expected since palladium is the catalyst of choice for dehalogenations (Chapter XX).

Palladium can be used in reductions in which the end product is not a strong base as in the hydrogenation of N-(4-chlorobenzylidene)aniline and N-benzylidene-4-chloroaniline with palladium on carbon [1].

Lead-poisoned palladium on calcium carbonate allowed selective conversion of 6-chloro-2-(N-methyliminoethylene)-4-phenylquinazoline to 6-chloro-2-(N-methylaminoethyl)-4-phenylquinazoline [36a, b].

In other selective hydrogenations, N-(4-chlorobenzylidene)-N',N'-dimethylethylenediamine was reduced to the N-4-chlorobenzyl compound with a small amount of catalyst [7]; 4-(4-chlorobenzylamino)-3-isoxazolidinone was obtained from the hydrogenation of the 4-chlorobenzylideneamino compound with 5% by weight of palladium on carbon when the reaction was held at 15° [37].

However, a quantitative determination of ionic chlorine resulting from reduction of N-(4-chlorobenzylidene)methylamine with 2% by weight of 5% palladium on carbon in alcohol, in acidified alcohol, and in alcohol containing pyridine as a poison [38] suggests that palladium should not be employed in the hydrogenation of halogen-containing imines (Table 15.1) unless the halogen is fluorine or if the resulting halogen-containing amine is a weak base.

Table 15.1 Loss of Halogen[a]

$$4\text{-ClC}_6\text{H}_4\text{CH}=\text{NCH}_3 \xrightarrow{\text{H}_2} 4\text{-ClC}_6\text{H}_4\text{CH}_2\text{NHCH}_3$$

Solvent	HCL (%)[b]
EtOH	31.2
EtOH + 0.1 mole HCl	20.0[c]
EtOH + 0.3 mole AcOH	21.0
EtOH + 0.125 mole pyridine	17.0

[a] Reaction was stopped after absorption of 1H$_2$.
[b] Determined by titration of ionic halogen. [c] By difference between ionic halogen before and after reduction.

The catalysts of choice for selective conversion of the C=N bond in halogen-containing imino compounds are platinum oxide and 5% platinum on carbon although it is likely that platinum on other carriers should also be useful. Reduction in acidic medium with a small amount of catalyst should be an additional safeguard in preventing or minimizing dehalogenation especially when the halogen is bromine or iodine [35].

5% Platinum on carbon gave high yields of 1-(2-, 3- and 4-chlorobenzoyl)-2-isopropylhydrazines from the corresponding isopropylidene compounds [35]. When the same catalyst ratio, 7.5% by weight, was used in the reductions of N-(2-, 3- and 4-chlorobenzylidene)methylamines which gave bases stronger than the aforementioned hydrazines some dehalogenation took place [1]; when the amount of catalyst was decreased to 5% ionic chlorine could not be detected. A study of the effect of position on dehalogenation in the latter reductions with palladium, platinum, and rhodium on carbon and with Raney nickel revealed that loss of chlorine was greatest in 2-position, less in 3-, and least in 4-position [38]. This order was borne out to a certain extent in the reductions of 1-(2- and 4-bromobenzoyl)-2-isopropylidenehydrazines where substantial loss of bromine was noted during reaction of the 2-bromo compound with the same amount of platinum on carbon used for 1-(4-bromobenzoyl)-2-isopropylidenehydrazine [1].

Examples of the utility of platinum oxide are seen in the hydrogenations of N,N-*bis*-(2-chloroethyl)-N'-alkylidenehydrazines [39a], of 6- and 7-chloroquinolyl-4-methyleneimines [39b], of N',N'-*bis*-(2- and 4-chlorobenzylidene)ethylenediamines [39c], and of N-(4-chlorobenzylidene)benzylamine [9d] in neutral solvent to the halogen-containing amines in good yield.

It also gave good results in selective reductions of cyclic C=N bonds in the presence of aryl chloride. Examples are seen in the conversions in acid medium of 7-chloro-5-phenyl-1,3-dihydro-2H-1,4-benzodiazepin-2-ones to the 7-chlorotetrahydrobenzodiazepin-2-ones [40a, b] and of 7-chloro-5-phenyl-2,3-dihydro-1,4-benzothiazepine-1,1-dioxide to the 7-chlorotetrahydro compound [40c]. When 2,8-dichloro-6,12-diphenyldibenzo[b,f][1,5]diazocine was reduced with platinum oxide in acetic acid containing hydrogen chloride, a transannular reaction took place giving an indoloindole with the halogens intact; in reduction in acetic acid alone, a mixture of dichloro-*cis*-tetrahydrodiazocine and the dichlorotetrahydroindoloindole was obtained [41a, b] (Eq. 15.12).

Platinum oxide catalyzed the hydrogenation of 6-chloro-7-sulfamyl-1,2,4-benzothiadiazine-1,1-dioxide in neutral solvent to 6-chloro-7-sulfamyl-3,4-dihydro-1,2,4-benzothiadiazine-1,1-dioxide in 88% yield [42]. When a benzyl group was in 3-position the C=N bond could not be reduced with palladium or platinum. At 100° and 100 atm hydrogenation over platinum oxide gave 50% yield of 6-chloro-3-cyclohexylmethyl-7-sulfamyl-1,2,4-benzothiadiazine-1,1-dioxide.

There was a suggestion of dehalogenation in the reduction of 1-benzoyl-2-(1-chloroisopropylidene)hydrazine in acetic acid with platinum oxide where a mixture of products was obtained [43a, b]. The mixture contained starting material, 1-benzoyl-2-isopropylhydrazine and an unknown when hydrogen uptake was interrupted at one molar equivalent.

(15.12)

+ Dichlorotetrahydroindoloindole

Raney nickel may find some use in selective reductions because it is a poor dehalogenative catalyst. It catalyzed hydrogenations of N-benzylidene-2,5-dichloroaniline and N-(2-chlorobenzylidene)aniline to the N-benzyl compounds in reactions at 100 atm pressure and temperature below 125° [44a] and gave good yield in the high-pressure reduction of N-(4-chlorobenzylidene)-2-aminoaniline to N-(4-chlorobenzyl)-2-aminoaniline [44b]. However, when it was employed for the hydrogenations of N-[4-(bis-2-chloroethyl)-aminobenzylidene]methylamine, $(ClCH_2CH_2)_2NC_6H_4CH=NCH_3$ [44c] and of 2-(5-bromo-2-furfurylideneamino)pyridine [44d] to the corresponding halobenzyl and furfuryl compounds yields were not especially high.

Reductions of the C=N bond in the presence of halogen with Raney nickel might be improved by the addition of excess acetic acid to neutralize both occluded alkali in the catalyst and the effects of the nitrogen bases resulting from reaction.

A comparison of 5% rhodium on carbon and platinum on carbon for the hydrogenations of N-(2-, 3-, and 4-chlorobenzylidene)methylamines showed that some dehalogenation occurred with rhodium [38].

REFERENCES

[1] M. Freifelder, unpublished results; 1a. Molecular models constructed by the author.
[2] a. H. L. Lochte, J. Horeczy, P. L. Pickard, and A. D. Barton, *J. Am. Chem. Soc.*, **70**, 2012 (1948); b. H. J. Barber, F. W. Major, and W. R. Wragg, *J. Chem. Soc.*, **1946**, 613.

[3] C. F. Winans and H. Adkins, *J. Am. Chem. Soc.*, **55**, 2051 (1933).
[4] a. P. L. Pickard and D. J. Vaughan, *J. Am. Chem. Soc.*, **72**, 876 (1950); b. C. L. Stevens, P. Blumberg, and M. Munk, *J. Org. Chem.*, **28**, 331 (1963); c. J. Krapcho, U.S. Patent 3,145,209 (1964).
[5] a. C. L. Stevens, M. E. Munk, C. H. Chang, K. G. Taylor, and A. L. Schy, *J. Org. Chem.*, **29**, 3146 (1964); b. R. N. Blomberg and W. F. Bruce, Brit. Patent 702,985 (1954); c. K. Harada and K. Matsumoto, *J. Org. Chem.*, **32**, 1794 (1967); d. S. Archer, T. R. Lewis, and M. J. Unser, *J. Am. Chem. Soc.*, **79**, 4194 (1957).
[6] P. Karrer and E. Schick, *Helv. Chim. Acta*, **26**, 800 (1943).
[7] a. W. Beck, I. A. Kaye, I. C. Kogon, H. C. Klein, and W. J. Burlant, *J. Org. Chem.*, **16**, 1434 (1951); b. I. A. Kaye, I. C. Kogan, and C. L. Parris, *J. Am. Chem. Soc.*, **74**, 403 (1952); c. C. W. Sondern and P. J. Breivogel, U.S. Patent 2,582,292 (1952).
[8] a. J. S. Buck and R. Baltzly, *J. Am. Chem. Soc.*, **63**, 1964 (1941); b. R. Baltzly and A. P. Phillips, *ibid.*, **71**, 3421 (1949); c. N. E. Levin, B. E. Graham, and H. G. Kolloff, *J. Org. Chem.*, **9**, 380 (1944).
[9] a. J. A. Barltrop, R. M. Acheson, P. G. Philpott, K. E. MacPhee, and J. S. Hunt, *J. Chem. Soc.*, **1956**, 2928; b. H. G. Kolloff and N. Levin, U.S. Patent 2,541,967 (1951); c. L. Bauer, J. Cymerman, and W. J. Sheldon, *J. Chem. Soc.*, **1951**, 2342; d. R. L. Hinman and K. L. Hamm, *J. Org. Chem.*, **23**, 529 (1958); e. D. J. Wadsworth, *ibid*, **32**, 1184 (1967); A. W. Frank and C. B. Purves, *Can. J. Chem.*, **33**, 365 (1955).
[10] a. M. Freifelder, *J. Org. Chem.*, **26**, 1835 (1961); b. I. Ziderman and E. Dimant, *ibid*, **31**, 223 (1966); c. K. A. Hyde, E. M. Acton, W. A. Skinner, L. Goodman, and B. R. Baker, *ibid*, **26**, 3551 (1961).
[11] a. H. Dahn, P. Zoller, and U. Solms, *Helv. Chim. Acta*, **37**, 565 (1954); b. H. Dahn and U. Solms, *ibid*, **35**, 1162 (1952); c. N. H. Cromwell, R. D. Babson, and C. E. Harris, *J. Am. Chem. Soc.*, **65**, 312 (1943); d. G. Y. Lesher and A. R. Surrey, *ibid*, **77**, 636 (1955); e. F. Benington, R. D. Morin, and L. C. Clark, Jr., *J. Org. Chem.*, **23**, 19 (1958); f. K. E. Hamlin and M. Freifelder, *J. Am. Chem. Soc.*, **75**, 369 (1953); g. J. Finkelstein, E. Chiang, F. M. Vane, and J. Lee, *J. Med. Chem.*, **9**, 319 (1966).
[12] E. A. Steck, L. L. Hallock, and C. M. Suter, *J. Am. Chem. Soc.*, **70**, 4063 (1948).
[13] a. M. A. Rebenstorf, U.S. Patent 2,773,098 (1956); b. G. Parker, Brit. Patent 753,500 (1956); c. J. L. Szabo and W. F. Bruce, U.S. Patent 2,868,833 (1959); d. F. A. Sowinski, W. A. Lott and J. Bernstein, U.S. Patent 2,951,092 (1960); e. J. H. Billman and J. L. Meisenheimer, *J. Med. Chem.*, **6**, 682 (1963); f. J. H. Billman and M. Sami Khan, *ibid*, **9**, 347 (1966).
[14] M. R. Tiollais, *Bull. Soc. Chim. France*, **1947**, 959.
[15] M. Freifelder and Y. H. Ng, *J. Med. Chem.*, **8**, 122 (1965).
[16] a. L. M. Werbel, D. B. Capps, E. F. Elslager, W. Pearlman, F. W. Short, E. A. Weinstein, and D. F. Worth, *J. Med. Chem.*, **6**, 637 (1963); b. Fr. Patent 1,334,040 (1963); through *Chem. Abstr.*, **60**, 1642 (1964); c. E. Magnien, W. Tom, and W. Oroshnik, *J. Med. Chem.*, **7**, 821 (1964).
[17] a. A. I. Meyers and J. M. Greene, *J. Org. Chem.*, **31**, 556 (1966); b. P. J. A. Demoen and P. A. J. Jannsen, *J. Am. Chem. Soc.*, **81**, 6281 (1959).
[18] J. H. Boyer and F. C. Canter, *J. Am. Chem. Soc.*, **77**, 3287 (1955).
[19] a. A. J. Hillard and G. E. Hall, *J. Am. Chem. Soc.*, **74**, 666 (1952); b. J. M. Osbond, *J. Chem. Soc.*, **1953**, 1648; c. M. F. Grundon and H. J. H. Perry, *ibid*, **1954**, 3531; d. R. A. Robinson, U.S. Patent 2,683,146 (1954); e. A. Brossi, O. Schnider, and M. Walter, U.S. Patent 2,830,992 (1958); f. P. N. Craig, F. P. Nabenhauer, P. M. Williams, E. Macks, and J. Toner, *J. Am. Chem. Soc.*, **74**, 1316 (1952).

[20] a. G. C. Morrison, W. A. Cetenko, and J. Shavel, Jr., *J. Org. Chem.*, **32**, 2768 (1967); b. N. J. Leonard, R. W. Fulmer, and A. S. Hay, *J. Am. Chem. Soc.*, **78**, 3457 (1956).
[21] a. W. Ried and P. Stahlhofer, *Ber.*, **90**, 825 (1957); b. p. 832.
[22] a. E. Biekert, D. Hoffman, and F. J. Meyer, *Ber.*, **94**, 1676 (1961).
[23] a. K. A. Taipele, *Ber.*, **56**, 954 (1923); b. H. J. Lochte, J. R. Bailey, and W. A. Noyes, *J. Am. Chem. Soc.*, **43**, 2597 (1921); c. A. Burger, R. T. Standridge, and, E. J. Ariens, *J. Med. Chem.*, **6**, 221 (1963); d. H. H. Fox and J. T. Gibas, *J. Org. Chem.*, **20**, 60 (1955); e. R. L. Huang, H. H. Lee, and M. S. Malhotra, *J. Chem. Soc.*, **1964**, 5951; f. C. G. Overberger and H. Gainer, *J. Am. Chem. Soc.*, **80**, 4556 (1958); g. C. G. Overberger and I. Tashlick, *ibid*, **81**, 217 (1959); h. J. M. van der Zanden and G. de Vries, *Rec. Trav. Chim.*, **76**, 519 (1957).
[24] J. A. Biel, A. E. Drucker, T. F. Mitchell, E. F. Sprengler, P. A. Nuhfer, A. C. Conway, and A. Horita, *J. Am. Chem. Soc.*, **81**, 2805 (1959).
[25] a. R. C. Goodwin and J. R. Bailey, *J. Am. Chem. Soc.*, **47**, 167 (1925); b. Neth. Appl. 6,506,201 (1965); through *Chem. Abstr.*, **64**, 12690 (1966).
[26] a. E. Schmitz and R. Ohme, *Ber.*, **95**, 2012 (1962); b. M. Nagata, *Yakugaku Zasshi*, **86**, 608 (1966); through *Chem. Abstr.*, **65**, 15356 (1966); c. M. J. Kalm, *J. Med. Chem.*, **7**, 427 (1964).
[27] a. H. H. Fox and J. T. Gibas, *J. Org. Chem.*, **18**, 994 (1953); b. H. L. Yale, K. Losee, J. Martins, M. Holsing, F. H. Perry, and J. Bernstein, *J. Am. Chem. Soc.*, **75**, 1933 (1953); c. M. H. Weinsig and E. B. Roche, *J. Pharm. Sci.*, **54**, 1216 (1965); d. P. G. Marshall, P. A. McCrea, and J. P. Revill, Brit. Patent 1,019,363 (1966); through *Chem. Abstr.* **64**, 12620 (1966).
[28] a. A. Alemany, M. Bernabe, C. Elloriaga, E. F. Alvarez, and M. Lora-Tamayo, *Bull. Soc. Chim. France*, **1966**, 2486; b. F. R. Cervi, M. Hauser, G. S. Sprague, and H. J. Trofkin, *J. Org. Chem.*, **31**, 631 (1966).
[29] a. P. Chabrier and A. Sekera, *Compt. Rend.*, **226**, 818 (1948); b. J. H. Barnes and P. G. Marshall, Brit. Patent 970,978 (1964); c. J. E. Robertson, J. H. Biel, and F. Di Pierro, *J. Med. Chem.*, **6**, 381 (1963).
[30] a. G. Kubiczek, *Monatsh.*, **74**, 100 (1942); b. S. Wawzonek and R. C. Nagler, *J. Am. Chem. Soc.*, **77**, 1796 (1955); c. D. S. James and P. E. Fanta, *J. Org. Chem.*, **28**, 390 (1963); d. G. Drefahl and J. Ulbricht, *Ann.*, **598**, 174 (1956); e. A. I. Meyers and J. J. Ritter, *J. Org. Chem.*, **23**, 1918 (1958); f. G. Shaw, *J. Chem. Soc.*, **1951**, 1017; g. A. I. Meyers, J. Schneller, and N. K. Ralhan, *J. Org. Chem.*, **28**, 2944 (1963).
[31] a. H. K. Schnoes, K. Biemann, J. Mokry, I. Kompis, A. Chatterjee, and G. Ganguli, *J. Org. Chem.*, **31**, 1641 (1966); b. A. G. Caldwell and E. R. H. Jones, *J. Chem. Soc.*, **1946**, 597; c. R. Grewe, R. Hamann, G. Jacobsen, E. Noltre, and K. Riecke, *Ann.*, **581**, 88 (1953).
[32] R. E. Brown, D. M. Lustgarten, R. J. Stanaback, and R. I. Meltzer, *J. Org. Chem.*, **31**, 1489 (1966).
[33] Neth. Appl. 6,414,904 (1965); through *Chem. Abstr.*, **64**, 2114 (1966).
[34] a. E. J. Forbes, *J. Chem. Soc.*, **1955**, 3926; b. K. Folkers, U.S. Patents 2,776,296 and 2,801,248 (1957).
[35] M. Freifelder, W. B. Martin, G. R. Stone, and E. L. Coffin, *J. Org. Chem.*, **26**, 383 (1961).
[36] a. L. H. Sternbach, E. Reeder, A. Stempel, and A. I. Rachlin, *J. Org. Chem.*, **29**, 332 (1964); b. L. H. Sternbach, personal communication.
[37] E. B. Hodge, U.S. Patent 2,967,866 (1961).
[38] M. Freifelder and D. A. Dunnigan, unpublished results.

[39] a. V. W. Schulze, G. Letsch, and H. Fritsche, *J. Prakt. Chem.*, **1966**, 96; b. K. N. Campbell, A. H. Sommers, J. F. Kerwin, and B. K. Campbell, *J. Am. Chem. Soc.*, **68**, 1851 (1946); c. J. H. Billman, J. C. Ho, and L. R. Caswell, *J. Org. Chem.*, **22**, 538 (1957).

[40] a. L. H. Sternbach and E. Reeder, *J. Org. Chem.*, **26**, 4936 (1961); b. L. H. Sternbach, G. A. Archer, J. V. Early, R. I. Fryer, E. Reeder, and N. Wasyliw, *J. Med. Chem.*, **8**, 815 (1965); c. L. H. Sternbach, H. Lehr, E. Reeder, T. Hayes, and N. Steiger, *J. Org. Chem.*, **30**, 2812 (1965).

[41] a. W. Metlesics, R. Tavares, and L. H. Sternbach, *J. Org. Chem.*, **31**, 3356 (1966); b. W. Metlesics and L. H. Sternbach, *J. Am. Chem. Soc.*, **88**, 1077 (1966).

[42] C. W. Whitehead, J. J. Traverso, F. J. Marshall, and D. E. Morrison, *J. Org. Chem.*, **26**, 2809 (1961).

[43] a. B. T. Gillis and R. E. Kadunce, *J. Org. Chem.*, **32**, 91 (1967); b. B. T. Gillis, personal communication.

[44] a. C. F. Winans, *J. Am. Chem. Soc.*, **61**, 3564 (1939); b. M. Schenck, H. Richter, and H. Vogel, Ger. Patent 901,649 (1954); through *Chem. Abstr.*, **52**, 12926 (1958); c. F. Bergel, J. L. Everett, J. J. Roberts, and W. C. J. Ross, *J. Chem. Soc.*, **1955**, 3835; d. K. Hayes, G. Gever, and J. Orcutt, *J. Am. Chem. Soc.*, **72**, 1205 (1950).

XVI

REDUCTIVE AMINATION

Reductive amination is a process by which aldehydes or ketones react with ammonia in the presence of hydrogen and a catalyst to yield primary amines. The first work in this area used nickel catalyst [1]. These reactions are related to the hydrogenation of imines (Chapter XV). Presumably aldimine or ketimine or an addition product which leads to imine by dehydration is the intermediate through which reaction takes place (Eq. 16.1).

$$RR'C{=}O + NH_3 \longrightarrow RR'C(OH)NH_2 \xrightarrow{-H_2O} RR'C{=}NH \xrightarrow{H_2} RR'CHNH_2$$
$$R' = H \text{ or } R \tag{16.1}$$

Since the imine is presumed to be an intermediate in reductive amination, secondary amine can form by reaction of primary amine and imine through the same mechanism as in the hydrogenation of nitriles (Chapter XII), or by further reaction of the primary amine and aldehyde or ketone. Tertiary amine can also form from secondary amine and aldehyde but this is less likely to take place with secondary amine and ketone.

Isolation of imine and subsequent hydrogenation is reported to be a better method of obtaining amines from ammonia and aldehydes or ketones than by reductive amination [2a]. The difficulty in obtaining imines, RCH=NH, in high yield and their instability make this a moot point. In many instances another two-step reaction, conversion of carbonyl compound to oxime and subsequent hydrogenation, is a more desirable method of obtaining primary amine. Nevertheless, under the proper conditions reductive amination can be a useful procedure.

Excess ammonia is required for good yield of primary amine in the reductive amination process [2b]. Reduction with equimolar amounts of ammonia and carbonyl compound led to as much secondary as primary amine [3a]. Excess aldehyde allowed formation of tertiary as well as secondary amine [3a]; a deficiency of ammonia gave difurfurylamine from furfuraldehyde [3b]. Reductive amination with 2 moles of (2-methoxyphenyl)acetone in alcohol over platinum oxide gave only secondary amine [3c]. It may be advisable to employ about 5 moles of ammonia [4].

The relationship of reductive amination to hydrogenations of imines might suggest that reactions should be carried out under anhydrous conditions to avoid hydrolysis and subsequent side reaction (see Chapter XV). Nevertheless, aqueous ammonia has been employed in these reactions. However, there are very few comparisons of yields of primary amines with those obtained from reductive amination under anhydrous conditions to state categorically that anhydrous conditions are preferred.

The presence of an equivalent of acid or of an ammonium salt of an acid in addition to excess ammonia during reductive amination can also lead to improved yield of primary amine.

ALDEHYDES

One of the complications in the reductive amination of aldehydes is the formation of polymers or complex intermediates when certain aldehydes and ammonia are mixed. These often lead to a mixture of primary and secondary amines upon hydrogenation. For example, the complex formed by combining furfuraldehyde and ammonia gave furfurylamine and difurfurylamine after reduction with nickel [2a] (Eq. 16.2).

$$\text{furfuryl-CHO} \xrightarrow{NH_3} [\text{furfuryl-CH=N}]_2 \text{CH-furfuryl} \xrightarrow{H_2} \text{furfuryl-CH}_2\text{NH}_2 + [\text{furfuryl-CH}_2]_2 \text{NH} \quad (16.2)$$

The manner of mixing aldehyde and ammonia may prevent formation of complexes and lead to high yield of primary amine. A very useful procedure consisted in cooling the aldehyde in alcoholic solution to -20 to $-25°$ before adding liquid ammonia. After addition of catalyst the bomb was sealed and warmed quickly to room temperature, hydrogen was added and reaction was carried out under the desired temperature and pressure conditions to usually give good yield of primary amine [4]. The effect of cooling reactants and its influence on yield may also be seen when propionaldehyde and ammonia were mixed in methyl alcohol at $-5°$ and hydrogenated over Raney nickel at 70 atm and 135°; yield of propylamine was 82% [5]. When ammonia was passed into an uncooled solution, subsequent reduction gave only 20% yield. In the successful example mixing in the cold apparently prevented aldol condensation.

Progressive introduction of aldehyde to a heated mixture of ammonia, solvent, and catalyst in a hydrogen atmosphere is another means of preventing either polymerization or aldol condensation. When the aldehyde is added at elevated temperature, it apparently reacts with ammonia and splits out water quickly to yield imine. In the large excess of ammonia and a very high ratio of catalyst to substrate, imine is rapidly reduced to amine as it is formed. High yields of propylamine and butylamine were obtained from propionaldehyde and butyraldehyde and ammonia in this manner [6a, b]. Gradual addition of acetaldol, $CH_3CHOHCH_2CHO$, to a suspension of Raney nickel in methyl alcohol and ammonia heated to 125° under 30 atm of hydrogen gave 61% yield of 3-hydroxybutylamine [6c]; slow addition of 2-hydroxyadipaldehyde to Raney nickel and ammonia at 80–90° and 70 atm gave 2-hydroxyhexane-1,6-diamine [6d].

Another means of aminating aldehydes that undergo polymerization is by reaction of them as diacetate. Hexahydrobenzylidenediacetate, $C_6H_{11}CH(OOCCH_3)_2$, was converted to hexahydrobenzylamine over Raney nickel and ammonia at 100° and 140 atm [6e].

In many instances the yield of primary amine may depend on the rapidity of reaction. This may be brought about by an increase in the amount of catalyst used for reduction. In a comparison of the reductive amination of furfuraldehyde with 10% and with 20% by weight of Raney nickel, 60% yield of primary amine was obtained at 150° and 150 atm for 1 hr with a 10% catalyst ratio. When a 20% catalyst ratio was employed, hydrogen uptake was complete at 100° and 100 atm pressure by the time the reaction temperature was reached; yield of furfurylamine was 70–75% [4].

Catalysts

Although platinum oxide was employed for hydrogenations of imines, RCH=NH (Chapter XV, Refs. 4a, b, c), it has found very little use in reductive aminations in the presence of aldehyde. The very slow conversion of platinum oxide in ammoniated solvent to its active catalytic form may be the reason for this [7]. Prereduction in neutral or acidic media might be helpful. Reductive aminations over platinum on carbon or palladium on carbon should also be studied.

5% Rhodium on alumina was studied in the reductive amination of 3,4-dimethoxybenzaldehyde under low-pressure conditions [4]. It appeared to have some promise when a slight excess of a molar equivalent of ammonium acetate or acetic acid was present along with excess ammonia. Yield of 3,4-dimethoxybenzylamine was about 60% when reaction was carried out in alcoholic ammonia containing ammonium acetate. In reduction with alcoholic ammonia in the absence of ammonium acetate or acetic acid only

336 Reductive Amination

41% of primary amine resulted. Ammonium chloride may also be employed but is less desirable because of solubility problems [4].

Raney nickel has found the widest application in reductive aminations of a variety of aldehydes at elevated temperature and pressure conditions. In general, low-pressure reductive aminations with nickel were unsuccessful or required excessive amounts of catalyst [4].

Successful higher pressure aminations include the conversion of galactose to galactamine [8a] and 3-hydroxy-2,2-dimethylpropionaldehyde to the corresponding propanolamine [8b]. Reductive amination of 5-hydroxypentanal (from 2,3-dihydropyran) under 100 atm pressure of hydrogen was found to be a convenient method for the preparation of the difficultly obtainable 5-amino-1-pentanol [8c]. The reaction of 4-ethyl-4-formylhexanoic acid, another aliphatic aldehyde, with ammonia under elevated pressure yielded 5,5-diethyl-2-piperidone through reductive amination and subsequent cyclization [8d] (Eq. 16.3). Difficulty was encountered in an

$$OHCC(C_2H_5)_2(CH_2)_2COOH + NH_3 \xrightarrow{H_2}_{Ni(R)}$$

$$\left[\begin{array}{c} \text{structure with } CH_2, H_2C, C(C_2H_5)_2, O=C, CH_2, HO, HN, H \end{array} \right] \xrightarrow{-H_2O} \text{5,5-diethyl-2-piperidone} \qquad (16.3)$$

attempt to obtain N,N-diethyl-2,2-dimethylpropylamine from the aminoaldehyde, $(C_2H_5)_2NCH_2C(CH_3)_2CHO(4)$. No uptake occurred at 25° and 50 atm pressure; at 100 atm pressure of hydrogen uptake was only 70%. A 43% yield of diamine was obtained along with a by-product, which was identified as N-(3-diethylamino-2,2-dimethylpropylidene)-N',N'-diethyl-2,2-dimethylpropane-1,3-diamine,

$$(C_2H_5)_2NCH_2C(CH_3)_2CH=NCH_2C(CH_3)_2CH_2N(C_2H_5)_2.$$

It is likely that difficulty in this amination was due to the inhibitory effect of the strongly basic amino group on the catalyst more so than steric effects since pivaldehyde, $(CH_3)_3CCHO$, was converted to the amine in 77% yield under the same conditions [4].

Other aldehydes, such as tetralin-1-aldehyde and related 1-formylbenzocyclanes, were converted to 1-aminomethylbenzocyclanes in 60–80% yield by high-pressure amination with nickel [8e]. Fairly good yields of benzylamines have been obtained by amination of benzaldehyde and substituted benzaldehydes under similar conditions [9a–e] as was 5-hydroxymethyl-2-furfurylamine from 5-hydroxymethylfurfuraldehyde [9f].

Reductive aminations of 1,5- and 1,6-dialdehyde over Raney nickel usually resulted in cyclized products unless the temperature was controlled. An example of cyclization of a 1,5-dialdehyde during reaction is seen in the amination of 2-(2-hydroxyethyl)pentane-1,5-dial (from the hydrolysis of 3,4-dioxabicyclo[4,2,0]nona-5-ene) to 3-(2-hydroxyethyl)piperidine [10a] (Eq. 16.4). When 2-hydroxy-1,6-hexanedial was reacted with excess ammonia

$$\text{(dioxabicyclic compound)} \xrightarrow[\text{HOH}]{\text{HCl}} \text{OCH(CH}_2\text{)}_2\text{CH(CH}_2\text{CH}_2\text{OH)CHO}$$

$$\xrightarrow[\text{NiR, 100°, 100 atm}]{2\text{H}_2 + \text{NH}_3}$$

$$\left[\begin{array}{cc} \text{CH}_2 & \\ \text{CH}_2 \diagup \diagdown \text{CHCH}_2\text{CH}_2\text{OH} \\ | & | \\ \text{CH}_2 & \text{CH}_2 \\ | & | \\ \text{NH}_2 & \text{NH}_2 \end{array} \right] \longrightarrow \text{(piperidine with CH}_2\text{CH}_2\text{OH)} \quad 68\% \quad (16.4)$$

over Raney nickel at temperature below 150°, 1,6-diamino-2-hexanol was obtained in 50% yield [6d, 10b]. The same result was obtained with Raney cobalt [10b]. A major impurity was assumed to be 2-hydroxymethylpiperidine. However, work by Whetstone [10c] suggests that the impurity is a saturated 7-membered ring compound, 3-hydroxyperhydroazepine.

In contrast to the tendency toward cyclization at high temperature, 2-hydroxy-1,6-hexanedial yielded 1,6-diaminohexane on reductive amination at 225° over Raney nickel [10d], hydrogenolysis of the secondary alcohol occurring because of the high temperature.

KETONES

When ketones and ammonia are mixed prior to reduction, polymeric materials or complex intermediates do not form as with aldehydes. However, steric effects play a greater part in reductive amination of ketones than they do in similar reactions with aldehydes. The poor yield of benzhydrylamine from benzophenone is an example [2b]. 3,3-Diphenyl-2-butanone and ethyl 3-keto-2,2,3-triphenylpropionate could not be reductively aminated [4]. If formation of imine was possible, it is likely that in these two instances steric effects made contact with the catalyst difficult or impossible.

Since steric effects are a factor in formation of the intermediate ketimine, it may be necessary to allow the mixture of ketone and ammonia to stand for several hours before hydrogenation or to heat the mixture in a sealed vessel before hydrogenation to obtain ketimine and subsequently good yield of tertiary amine [4]. As a result of slow formation of imine at times, primary amine may be contaminated with secondary alcohol from competing hydrogenation of unreacted ketone. Too often the alcohol is not readily removed by fractionation [4]. However, despite some failings, reductive amination of ketones can be a useful procedure. At times it has been found superior to the two-step reduction of ketone to oxime and oxime to amine.

Amination was not especially satisfactory in attempts to react ammonia and cyclopentanone and particularly 2-alkylcyclopentanones over Raney nickel at 75–80° and 70–80 atm [4]. Although high yield of cyclopentylamine was obtained, it was contaminated with cyclopentanol. In reductive aminations of 2-alkylcyclopentanones the alkyl group in 2-position affected formation of imine. The maximum yield of 2-methylcyclopentylamine was only 30%; in each instance nonbasic material was the major component. In this series, conversion of ketone to oxime and subsequent reduction lead to higher yields of cyclopentylamines. A similar procedure was used for cycloheptylamine when reductive amination gave cycloheptylamine containing cycloheptanol.

Phenol may replace cyclohexanone in reductive amination to form cyclohexylamine. The reaction of equimolar amounts of phenol and aqueous ammonia in the presence of rhodium on alumina at 80–100° under a pressure of 2–4 atm of hydrogen gave cyclohexylamine in 65–80% yield; with gaseous ammonia yields were from 80–95% [11a]. Small amounts of cyclohexanol, phenylcyclohexylamine, and dicyclohexylamine were obtained along with unreacted phenol. Similar reactions in the presence of rhodium on carbon gave 72% yield of cyclohexylamine [11b]. When palladium on carbon was employed, the major product was dicyclohexylamine, 67%; cyclohexylamine was obtained in only 21% yield. The yield of dicyclohexylamine could be raised to 95% by carrying the amination in an excess of phenol at 90° and 10 atm pressure.

Nickel-Catalyzed Reactions

Raney nickel has been widely used in reactions of different type ketones with excess ammonia under both moderate hydrogenation conditions and at elevated temperature and pressure. It is probably the catalyst of choice for these reactions.

In the aliphatic series 2- and 3-aminoalkanes were prepared by amination of the corresponding ketones at 50–60° and 90–100° and 6 atm [12a]; 2-aminoheptane was obtained from 2-heptanone at 90° and 30–40 atm [12b].

Reductive amination of diheptyl ketone at 150° and 72 atm gave much higher yield of 8-aminopentadecane than hydrogenation of the oxime; 55% against 5% [12c]. 3-Hydroxy-3-methyl-2-pentylamine and other 2-amino-3-hydroxyalkanes were prepared by reductive amination of the hydroxyketones at 70° and 90 atm [12d].

Although contamination of cycloalkanol has been noted by gas-liquid chromatography in reductive aminations of cyclopentanone and cycloheptanone over nickel, good results are reported by others. Cyclopentylamine was obtained in 84% yield from amination of cyclopentanone at 50° and 3 atm [13a], 87% of cyclohexylamine resulted from hydrogenation of cyclohexanone in ammonia at 80–90° and 100 atm [13b], 99% yield was reported in another reduction over nickel [13c], and 4-t-butylcyclohexanone was converted to 4-t-butylcyclohexylamine with nickel and ammonia in 81% yield [13d].

In other examples among aliphatic ketones, 1-cyclohexyl-3,3,3-trifluoroacetone was converted to 1-cyclohexyl-3,3,3-trifluoroisopropylamine [14a], 1-cyclohexylisopropylamine was obtained from cyclohexylacetone [14b]; 1-cyclopentylisopropylamine resulted from reductive amination of cyclopentylacetone in one instance [14c] and from 1-cyclopentylideneacetone in another [14d].

Haskelburg reported that reductive amination of phenylacetone, $C_6H_5CH_2COCH_3$, over nickel at atmospheric pressure was a better method of obtaining 1-phenylisopropylamine, $C_6H_5CH_2CHNH_2CH_3$, than a two-step synthesis of ketone to oxime and subsequent hydrogenation [15a]. In related reactions, RCOR' where R is an aromatic ring or aralkyl and R' is alkyl, good yields of the corresponding amines have been obtained from aminations over nickel at different temperature and pressure conditions; examples are given in Table 16.1.

Where R is aryloxymethyl, good yield is dependent on the reaction temperature. Hydrogenation of phenoxyacetone with nickel and ammonia at 120° and 100 atm gave 51.5% of 1-phenoxyisopropylamine [14b]; reduction at room temperature by another group resulted in 60% yield [16]. Hydrogenations involving phenoxy compounds often result in cleavage at the O—C bond, $C_6H_5O \mid CR$, to give phenols especially at elevated temperature [4].

Examples of reductive amination involving a saturated heterocyclic ketone may be seen in the reduction of tetrahydro-4-pyrone with nickel and ammonia at high pressure to give 75% yield of 4-aminotetrahydropyran [17a] and in reduction at room temperature and atmospheric pressure with 100% by weight of W-7 Raney nickel which gave 78% yield [17b].

Reductive aminations of keto acids and esters, $RCO(CH_2)_2COO(H)$ or C_2H_5, over nickel often result in formation of 2-pyrrolidones through

Table 16.1

RCOR′ + NH$_3$ → RCH(NH$_2$)—R′

R	R′[a]	Reference
Benzyl		14a
2-Furyl		15b
3,4-Dihydroxymethyl-2-furyl		15c
2-(2-Furyl)ethyl		15d
3-(2-Furyl)propyl		15d
2-(2-Furyl)butyl		15e
(5-Methyl-2-pyridyl)methyl		15f
5,6,7,8-Tetrahydro-1-naphthylmethyl		15g
5,6,7,8-Tetrahydro-2-naphthylmethyl		15g

[a] R′ = methyl except where R = benzyl; in that example R′ was trifluoromethyl.

cyclization of the amino acid or ester because of elevated temperature and pressure used in the reactions [18a, b, c]. Amination of 4-keto-4-phenylbutyric acid with nickel chromium oxide at 150° and 50 atm gave 4-amino-4-phenylbutyric acid; when the reaction was carried out at 190° and 150 atm, 5-phenyl-2-pyrrolidone was obtained [18b].

Cyclization may be avoided by carrying out amination of the keto acid in alkali as shown in the hydrogenation of 2-oxoglutaric acid with nickel and ammonia in potassium hydroxide solution at 105° and 140 atm which gave DL-glutamic acid [18d].

In most instances reductive aminations of aminoketones over Raney nickel give good yields of diamines. The conversion of 1-(4-oxopentyl)pyrrolidine to 1-(4-aminopentyl)pyrrolidine is an example [19a]. Other examples include the conversion of 5-diethylamino-2-pentanone to 5-diethylamino-2-pentylamine [15a, 19b] and dibutylaminoacetone to 1-dibutylaminoisopropylamine [15a] in very good yield under low-pressure conditions. The latter compound was also obtained in good yield by amination at 75° and 40 atm [19c]. Reductive amination of 4-(N-butyl- and N-ethyl-N-2-hydroxyethylamino)-2-butanones under 70 atm pressure of hydrogen resulted in very high yields of the corresponding diamines [19d].

Reductive amination of 1-substituted-4-piperidones with nickel at elevated temperature and pressure gave reasonably good yield of 1-substituted-4-aminopiperidines [17a, 20a, b], although in many instances 1-substituted-4-piperidinols were contaminants. When 1-phenyl-2,5-dimethyl-4-piperidone was aminated, cleavage of the starting material to aniline and aliphatic ketone and further reaction of that ketone with ammonia were complicating side reactions [20b] (Eq. 16.5).

Cleavage of the Mannich-type substituent in 3-position was the side reaction which took place during the amination of 3-dimethylaminomethyl-1,2,5-trimethyl-4-piperidone [20b]; 4-amino-1,2,3,5-tetramethylpiperidine

$$H_3C\text{-piperidone} \xrightarrow[H_2]{NH_3} \text{4-amino-1,2,5-trimethyl-4-phenylpiperidine (46.5\%)}$$

$$\searrow C_6H_5NH_2 + (CH_3)_2CHCOC_3H_7 \xrightarrow[H_2]{NH_3}$$

$$(CH_3)_2CHCH(NH_2)C_3H_7 \quad (16.5)$$

was obtained. It is of interest that hydrogenolysis did not take place when the 3-substituted piperidone was reduced to piperidinol.

Reductions over Noble Metals

Low-pressure reductive aminations of ketones in neutral media over palladium, platinum, or rhodium do not give good results. Such reactions with palladium require large amounts of catalyst. When heptadecane-7,11-dione was aminated in the presence of palladium on carbon and palladium on strontium carbonate, more catalyst than compound was used [21a]. Large quantities of active catalyst were also needed for reductive aminations of keto acids [21b, c, d]. Rhodium on a carrier is not useful for low-pressure reductive amination except when an acid or an ammonium salt is present along with excess ammonia [4]. In general this also applies to platinum.

Noble metal catalysts may be more useful under elevated temperature and pressure conditions often employed with Raney nickel but their applicability for reductive amination under such conditions has not been investigated except in isolated instances.

Effect of Acidic Agents

Reductive amination of ketones as a class may be greatly improved by the addition of an equivalent or a slight excess of acid or by the addition of an equivalent amount of an ammonium salt to an ammoniacal solution of substrate. A procedure employing acetic acid and ammonium acetate gave

95% yield of 1-(2,5-dimethoxyphenyl)isopropylamine from 2,5-dimethoxyphenylacetone with Raney nickel at 90° and 80 atm [22]. Similar reactions over nickel on kieselguhr or palladium on barium sulfate gave comparable results.

Low-pressure reductions of ketones with ammonia and ammonium chloride gave fair yields of primary amines in the presence of prereduced platinum [23]; hydrogenation in the absence of ammonium chloride gave lower yields. Reductive amination of cycloheptanone in excess ammonia and ammonium acetate under low-pressure conditions with rhodium on carbon or on alumina or with palladium on carbon gave 53.5–54% yields of cycloheptylamine [4].

SELECTIVE HYDROGENATIONS

When reductive amination of a carbonyl compound containing an aromatic ring is carried out, there is little danger of subsequent saturation of the ring unless reaction conditions are too vigorous.

Presence of Olefinic Bonds

The selective reductive amination of an unsaturated carbonyl compound is dependent on the geometry of the substrate and on substitution about the olefinic bond.

In the amination of 4-(2,6,6-trimethyl-1-cyclohexen-1-yl)-2-methyl-2-butanal the trisubstituted double bond in the side chain was reduced along with formation of amine but the tetrasubstituted internal bond remained intact [24a] (Eq. 16.6). Similar results were obtained in the nickel-catalyzed

$$\text{[cyclohexenyl]}\text{CH}_2\text{CH}=\text{C}(\text{CH}_3)\text{CHO} + \text{NH}_3 \xrightarrow[\text{Ni(R), 150°, 150 atm}]{2\text{H}_2}$$

$$\text{[cyclohexenyl]}\text{CH}_2\text{CH}_2\text{CH}(\text{CH}_3)\text{CH}_2\text{NH}_2 \quad (16.6)$$

amination of a related ketone, 4-(2,6,6-trimethyl-1-cyclohexen-1-yl)-3-buten-2-one [24b]. The tetrasubstituted cyclohexenyl bond remained unchanged; 4-(2,6,6-trimethyl-1-cyclohexen-1-yl)-2-butylamine was obtained in 95% yield [24b].

An example shows that a trisubstituted double bond in an unsaturated open-chain ketone was unaffected during amination; 6,10-dimethylundec-9-en-2-one, $(CH_3)_2C=CH(CH_2)_2CH(CH_3)(CH_2)_3COCH_3$, gave 6,10-dimethylundec-9-enyl-2-amine [24c]. In another example it is seen that, in the amination of 6,10-dimethylundeca-3,5,9-trien-2-one,

$$(CH_3)_2C=C(CH_2)_2C(CH_3)=CHCH=CHCOCH_3,$$

only the trisubstituted double bond was not reduced; saturation of the unsymmetrically disubstituted 5,6-bond and the symmetrically disubstituted 3,4-bond accompanied formation of amine [24c].

There is one example of selectivity where a symmetrically disubstituted bond in a ring system was not reduced during amination. 2-Formyl-1,3,4-dihydro-2H-pyran was converted to 2-aminomethyl-3,4-dihydro-2H-pyran by amination over cobalt catalyst at 130–135° and 150 atm [24d].

Presence of Halogen

Reductive aminations of chlorobenzaldehydes have been successfully carried out with Raney nickel at 40–70° and 90 atm [10a] and at 70° and 70 atm [10b] without loss of halogen. 3-Chlorobenzaldehyde was converted to 3-chlorobenzylamine in 60% yield by amination over 5% rhodium on alumina under 2 atm of hydrogen. There was no evidence of dehalogenation [4]. Uptake of hydrogen was slow but rate of uptake and yield could probably be improved by the addition of acetic acid or an equivalent of ammonium acetate.

1-(4-Chlorophenyl)-3-(2-ethyl-4,5-dimethoxyphenyl)-1-propanone was reductively aminated over Raney nickel at 130° and 150 atm [25a].

Although phenylacetone was aminated over Raney nickel under mild conditions, it was not possible to obtain more than 10% yield of 4-chlorophenylisopropylamine from reductive amination of 4-chlorophenylacetone [25b]. In this instance the problem appeared to be more of poor reaction to form the intermediate imine rather than dehalogenation. The major product of reaction was nonbasic indicating competing reduction of ketone to alcohol.

REFERENCES

[1] G. Mignonac, *Compt. Rend.*, **172**, 223 (1921).
[2] a. C. F. Winans and H. Adkins, *J. Am. Chem. Soc.*, **55**, 2051 (1933); b. E. J. Schwoegler and H. Adkins, *J. Am. Chem. Soc.*, **61**, 3499 (1939).
[3] a. B. M. Vanderbilt, U.S. Patent 2,219,879 (1940); b. C. F. Winans, U.S. Patent 2,217,630 (1940); c. R. V. Heinzelman, U.S. Patents 2,647,929 (1953) and 2,930,731 (1960).

[4] M. Freifelder, unpublished results.
[5] Brit. Patent 728,702 (1955).
[6] a. J. F. Olin and E. J. Schwoegler, U.S. Patent 2,373,705 (1945); b. Brit. Patent 615,715 (1949); c. C. N. Robinson, Jr. and J. F. Olin, U.S. Patent 2,477,943 (1949); d. R. R. Whetstone and S. A. Ballard, U.S. Patent 2,636,051 (1953); e. W. W. Prichard, U.S. Patent 2,456,315 (1948).
[7] R. Willes and W. B. Worrall, *J. Org. Chem.*, **26**, 5233 (1961).
[8] a. F. Kagan, M. A. Rebenstorf, and R. V. Heinzelman, *J. Am. Chem. Soc.*, **79**, 3541 (1957); b. J. R. Caldwell, U.S. Patent 2,618,658 (1952); c. G. F. Woods and H. Sanders, *J. Am. Chem. Soc.*, **68**, 2111 (1946); d. H. A. Bruson and W. D. Niederhauser, U.S. Patent 2,505,459 (1950); e. G. Seidl, R. Huisgen, and J. H. M. Hill, *Tetrahedron*, **20**, 633 (1964).
[9] a. C. F. Winans, *J. Am. Chem. Soc.*, **61**, 3564 (1939); b. A. R. Surrey and G. Y. Lesher, *J. Am. Chem. Soc.*, **78**, 2573 (1956); c. M. W. Goldberg and S. Teitel, U.S. Patent 2,879,293 (1959); d. O. Scherer and F. Schacher, Ger. Patent 865,455 (1953); through *Chem. Abstr.*, **52**, 17182 (1958); e. G. Grethe, H. L. Lee, M. Uskoković, and A. Brossi, *J. Org. Chem.*, **33**, 491 (1968); f. N. Elming and I. Clauson-Kass, *Acta Chem. Scand.*, **10**, 1603 (1956) and U.S. Patent 2,944,059 (1960).
[10] a. R. Paul and S. Tchelitcheff, *Bull. Soc. Chim. France*, **1954**, 1139; b. H. Schulz and H. Wagner, *Angew. Chem.*, **62**, 105 (1950); c. R. R. Whetstone in C. W. Smith, *Acrolein*, Wiley, New York, 1962, Chapter 12, Ref. 16; d. T. Saito, Jap. Patent 217/62; through *Chem. Abstr.*, **58**, 448 (1963).
[11] a. Brit. Patent 1,031,169 (1966); b. F. H. Van Munster, U.S. Patent 3,351,661 (1967).
[12] a. H. L. Bami, H. H. Iyer, and P. G. Guha, *J. Indian Inst. Sci.*, **29A**, 9 (1947); b. Brit. Patent 625,085 (1949); c. E. J. Borrows, B. M. C. Hargreaves, J. E. Page, J. C. L. Resuggan, and F. A. Robinson, *J. Chem. Soc.*, **1947**, 197; d. Brit. Patent 646,951 (1950).
[13] a. J. R. Corrigan, M-J. Sullivan, H. W. Bishop, and A. W. Ruddy, *J. Am. Chem. Soc.*, **75**, 6258 (1953); b. I. M. Nazarov and N. I. Shvetsov, *J. Gen. Chem. USSR* (Eng. trans.), **27**, 1300 (1958); c. B. Choffe and R. Stern, Brit. Patent 1,050,589 (1966); d. W. Huckel and K. Heyder, *Ber.*, **96**, 220 (1963).
[14] a. W. R. Nes and A. Burger, *J. Am. Chem. Soc.*, **72**, 5409 (1950); b. M. Polonovski, M. Pesson, and J. Bededev, *Compt. Rend.*, **233**, 1120 (1951); c. M. Mousseron, R. Jacquier, and H. Christol, *Bull. Soc. Chim. France*, **1957**, 600; d. F. B. Zienty, U.S. Patent 2,885,439 (1959).
[15] a. L. Haskelburg, *J. Am. Chem. Soc.*, **70**, 2811 (1948); b. Brit. Patent 801,169 (1958); c. W. B. McDowell and H. Moes, U.S. Patent 2,855,407 (1958); d. I. F. Bel'skii, *J. Gen. Chem. USSR* (Eng. trans.), **32**, 2860 (1962); e. A. A. Ponomarev, A. P. Kriven'ko, and M. V. Noritsina, *ibid*, **33**, 1732 (1963); f. A. Furst and A. Boller, U.S. Patent 3,133,077 (1964); g. R. Urban and O. Schnider, *Monatsh.*, **96**, 9 (1965).
[16] H. D. Moed and J. Van Dijk, *Rec. Trav. Chim.*, **75**, 1215 (1956).
[17] a. E. T. Golovin, V. M. Bystrov and Yu. I. Semenova, *J. Gen. Chem. USSR*, **34**, 1277 (1964) (Eng. trans.); b. C. A. Grob and V. Krasnobajew, *Helv. Chim. Acta*, **47**, 2145 (1964).
[18] a. A. P. Dunlop and E. Sherman, U.S. Patent 2,681,349 (1954); b. F. Michael and W. Flitsch, *Ber.*, **88**, 509 (1955); c. N. R. Easton and R. D. Dillard, *J. Org. Chem.*, **27**, 3602 (1962); d. D. E. Floyd, U.S. Patent 2,610,212 (1952).
[19] a. R. H. Reitsema and J. H. Hunter, *J. Am. Chem. Soc.*, **71**, 750 (1949); b. M. M. Neeman, *J. Chem. Soc.*, **1946**, 811; c. S. Archer and C. M. Suter, *J. Am. Chem. Soc.*,

74, 4296 (1952); d. A. R. Surrey and H. F. Hammer, *J. Am. Chem. Soc.*, **72,** 1814 (1950).
[20] a. R. C. Fuson, W. E. Parham, and L. J. Reed, *J. Am. Chem. Soc.*, **68,** 1239 (1946); b. I. N. Nazarov and E. T. Golovin, *J. Gen. Chem. USSR*, **26,** 1679 (1956) (Eng. trans.).
[21] a. D. E. Ames and R. E. Bowman, *J. Chem. Soc.*, **1952,** 1055; b. R. Schoenheimer and S. Ratner, *J. Biol. Chem.*, **127,** 301 (1939); c. F. Knoop and H. Oesterlin, *Z. Physiol. Chem.* **148,** 305 (1925); d. *ibid*, **177,** 186.
[22] M. Green, U.S. Patent 3,187,047 (1965).
[23] E. R. Alexander and A. L. Misegades, *J. Am. Chem. Soc.*, **70,** 1315 (1948).
[24] a. M. W. Goldberg and S. Teitel, U.S. Patent 2,736,747 (1956); b. T. Kralt, H. D. Moed, and J. Van Dijk, *Rec. Trav. Chim.*, **77,** 177 (1958); c. H. Moed, T. Kralt, and J. Van Dijk, *ibid.*, p. 196; d. Neth. Appl. 6,501,213 (1965); through *Chem. Abstr.*, **64,** 5048 (1965).
[25] a. Brit. Patent 969,977 (1964); through *Chem. Abstr.*, **62,** 1597 (1965); b. M. Freifelder, Y. H. Ng, and P. F. Helgren, *J. Med. Chem.*, **7,** 381 (1964).

XVII

REDUCTIVE ALKYLATION

In this book reductive alkylation is defined as a reaction in which a mixture of primary or secondary amine and carbonyl compound are catalytically hydrogenated to yield secondary and tertiary amine respectively (in the early work on this reaction nickel was employed as catalyst [1a]). Tertiary amine can also be obtained from reductive alkylation of primary amine with excess aldehyde; alkylation with excess ketone is much more difficult because of steric effects. Amides also have been reductively alkylated but it is not a general reaction.

Alkylations that give secondary amines are related to hydrogenations of imines (Chapter XV). Presumably aldimine or ketimine or an addition product, which leads to imine by hydrogenation, is the intermediate through which reaction takes place. Reductive alkylation involving secondary amine and carbonyl compound to form tertiary amine proceeds (a) through hydrogenolysis of an addition product or (b) through an enamine, if it can form, and subsequent reduction of the double bond (Eq. 17.1).

$$\text{RC}(\text{CH}_3)=\text{O} + (\text{R})_2\text{NH} \begin{array}{c} \text{(a)} \longrightarrow \text{RC}(\text{CH}_3)(\text{OH})\text{N}(\text{R})_2 \\ \\ \text{(b)} \xrightarrow{-\text{H}_2\text{O}} \text{RC}(=\text{CH}_2)\text{N}(\text{R})_2 \end{array} \xrightarrow{\text{H}_2} \text{RCH}(\text{CH}_3)\text{N}(\text{R})_2 \quad (17.1)$$

Reductive alkylation is dependent on the ease with which aldehyde or ketone reacts with amine to form a reducible intermediate. Because aldehydes are less hindered than ketones they react more readily. In instances where a ketone and aldehyde are present in the same molecule, reaction should take place between amine and aldehyde. Most of the examples in the literature are those of ketoaldehydes, ArCOCHO, which give aminoketones, ArCOCH$_2$NHR, after reduction with amines at low pressure with Raney nickel [1b, c, d].

The basicity of the amino group is also a factor. When two amino groups are present in the same molecule the more basic one should undergo preferential alkylation. In the reductive alkylation of 2-amino-5-aminomethyl-4-methylpyrimidine with 5-diethylamino-2-pentanone and platinum oxide at 75° and 20 atm, 2-amino-5-(5-diethylamino-2-pentyl)aminomethyl-4-pyrimidine was obtained [1e]. There was no reaction of the ketone with the weakly basic 2-amino group. When merimine, 2,3-dihydro-1H-pyrrolo[3,4-c]pyridine, was reacted with formaldehyde the more basic secondary amine was methylated in preference to the aromatic type primary amine at 3 atm pressure in the presence of palladium on carbon [1f]. Similar results were obtained from alkylation with cyclohexanone and a mixture of 5% palladium on carbon and 10% platinum on carbon (Eq. 17.2). Alkylation of the 7-amino group required a long reaction period.

$$\text{structure} \longrightarrow \text{structure} \quad (17.2)$$

R = methyl or cyclohexyl

PROCEDURES

In many reductive alkylations amine and carbonyl compound can be mixed in solvent and hydrogenated immediately. This is especially desirable if there is a tendency for further reaction of the Schiff base as exemplified in the formation of 1,2,3,4-tetrahydroisoquinolines resulting from isolation of the Schiff base from 2-phenethylamine and aldehydes [2a, b] (Eq. 17.3).

$$\text{PhCH}_2\text{CH}_2\text{NH}_2 + R\text{CHO} \xrightarrow{-\text{H}_2\text{O}} [\text{intermediate}] \longrightarrow \text{tetrahydroisoquinoline-NH-R} \quad (17.3)$$

When formation of the reducible intermediate is slow because of steric effects, isolation and reduction of the intermediate may be necessary.

It is often advantageous, though not always necessary, to isolate and hydrogenate the Schiff base from a primary amine and an aromatic aldehyde if the imine is stable and readily obtained [3, 4a, b]. It was especially useful in alkylations of alkanediamines with disubstituted benzaldehydes [3]. When alkanediamine and two equivalents of aldehyde were mixed in solvent

348   Reductive Alkylation

and hydrogenated at low pressure over palladium, N,N^1-dibenzylalkanediamine was obtained in good yield but the final product required considerable purification. When the Schiff bases were isolated after complete removal of water and similarly reduced, yields of pure products were very high. When water is removed there is less chance for hydrolysis of imine and possible further alkylation of secondary amine to tertiary amine.

Alkylation of 2-aminoalkanols with aldehydes or ketones may proceed through hydrogenation of Schiff base or by hydrogenolysis of an intermediate oxazolidine, dependent on the structure of carbonyl compound and amino alcohol (Eq. 17.4). Aldehydes or unhindered ketones and aliphatic

$$RC(R')=O + NH_2CH_2CH_2OH \xrightarrow{-H_2O} \begin{array}{c} RR'C=NCH_2CH_2OH \xrightarrow{H_2} \\ \\ \underset{R'}{\overset{R}{>}}C\underset{NH-CH_2}{\overset{O-CH_2}{<}} \xrightarrow{H_2} \end{array} RR'CHNHCH_2CH_2OH \quad (17.4)$$

R = H or R

aminoalcohols are assumed to react to give oxazolidines; highly branched ketones and aliphatic aminoalcohols form alkylideneaminoalcohols [5a, b]. Reductive alkylation of 1,3-aminoalkanols goes through a Schiff base or a tetrahydrooxazine, also dependent on structure of the reactants [5c]. When an arylalkanolamine is reductively alkylated, the imine is the suggested intermediate because of the effect of the aromatic ring [5d].

In reductive alkylations of 1,2- and 1,3-aminoalkanols the reactants in solution have been hydrogenated immediately over platinum at 20–60° and 2–3 atm pressure [5a–d]. However, a mixture of disobutyl ketone and 1-amino-2-propanol could not be reduced under these conditions but isolation of the intermediate followed by hydrogenation over platinum yielded 1-(2,6-dimethyl-4-heptyl)amino-2-propanol,

$$[(CH_3)_2CHCH_2]_2CHNHCH_2CHOHCH_3 \text{ [5b]}.$$

Similarly, a mixture of butyraldehyde and 1-phenyl-2-amino-1-propanol did not yield the N-butyl compound on reduction [5c]. It was obtained, however, on hydrogenation of the isolated intermediate. Reductive alkylations of nuclear-substituted 2-aminoindanols with benzaldehyde over platinum oxide or palladium on carbon gave good yields of 2-benzylaminoindanols [5e], but it was found advantageous at times to reduce the isolated oxazolidine. In the reductive alkylation of 2-aminoethanol, 3-aminopropanol, and 1-amino-2-propanol with freshly distilled benzaldehyde in alcoholic solution with 5% palladium on carbon at 3 atm, good yields of *N*-benzyl compounds were obtained when the solutions were allowed to stand prior to reduction [3]. Yields were increased by 10–15% when water was removed and the intermediates hydrogenated under the same conditions.

In contrast to the greater ease of reduction of the intermediate from an aldehyde and a 2-aminoalcohol [5a, b], reduction of a mixture of benzaldehyde and 1-[2-(2-dimethylaminoethoxy)]-2-amino-1-propanol over platinum took place at lower temperature and in a shorter time than hydrogenation of the isolated intermediate [5f].

At times it may be necessary to allow reactants to stand to insure complete interaction before attempting hydrogenation. Mixing of reactants in a water-immiscible solvent such as thiophene-free benzene, standing, and treatment with anhydrous magnesium sulfate to remove water is another procedure. It was effective in the reductive alkylations of cyclopropylamine with cycloalkanones over platinum oxide [6].

When reductive alkylation of a primary amine does not take place, it may be necessary to add a condensing agent. Anhydrous 5-aminotetrazole could not be alkylated unless a strong base, such as guanidine or triethylamine, was present [7a]. An acidic agent, zinc chloride, aided the alkylation of 4-amino-2,3-dimethyl-1-phenyl-5-pyrazolone with acetone over platinum on barium sulfate to yield 2,3-dimethyl-4-isopropylamino-1-phenyl-5-pyrazolone [7b]; ammonium chloride, hydrochloric acid, and acetic acid were also employed.

Acidic agents may be useful in carrying out reductive alkylations of secondary amines with aldehydes and particularly with ketones. Reaction in acidic solvent may also be helpful especially when attempting alkylations to produce tertiary amines.

Reductive alkylation of secondary amines with aldehydes may require the presence of an alkaline dehydrating agent such as anhydrous potassium carbonate so that enamine can form [7c] and undergo reduction. In the alkylation of pyrrolidine with desoxybenzoin, $C_6H_5CH_2C(=O)C_6H_5$, over platinum oxide, hydrogen uptake would not take place unless the reactants were first heated with calcium oxide to remove water before hydrogenation [7d]. In certain instances, when reactants are not volatile they may be heated at 50° under reduced pressure in the absence of solvent prior to reduction. This procedure gave high yields of 4-benzyl and 4-substituted benzyl-1-substituted piperazines when applied to alkylations of 1-substituted piperazines with benzaldehydes [3].

Reductive alkylation may be carried out with substrates which lead to amines by reduction as nitro, nitroso, azo, and hydrazo compounds, oximes, and hydroxylamines [4a]. The advantage of using precursors depends on individual cases. In most instances, starting with amine gives best results [3, 4a]. Acetals and ketals may be used for alkylation. Phenol can replace cyclohexanone in the preparation of N-substituted cyclohexylamines. (When intermediates are isolated, references to their reduction may be found in Chapter XV.)

CATALYSTS

Palladium- and platinum-catalyzed low-pressure procedures may be the ones of choice for reductive alkylations. Similar reductions over rhodium usually proceed at a much slower rate [3]. Rhodium can be used to advantage in alkylations in the presence of halogen [3]. Although platinum has been used more frequently than palladium in reductive alkylations, it is difficult to state which is preferred. Reductive alkylation of benzylamine with 3-dimethylaminopropionaldehyde over 5% platinum on carbon was very rapid and gave 90% yield of N^1-benzyl-N^3,N^3-diethyl-2,2-dimethylpropane-1,3-diamine [8]. A study of the same reaction with equal amounts of 5% palladium on carbon and platinum on carbon showed little difference in yield and purity of reduction product; reaction time with palladium was slightly longer [3]. In this instance platinum was preferred because of lesser danger of debenzylation of the product of reduction. In other alkylations there appeared to be little choice between platinum oxide and palladium on carbon [3].

It is reported that platinum oxide should be prereduced before use to reductive alkylation. This, too, is a moot point which may depend on the mixture to be reduced and on variations in different lots of catalyst [3].

More vigorous conditions have often been employed in reductive alkylations with Raney nickel. The use of such conditions may be a matter of convenience because of availability of pressure equipment, not necessarily due to lower activity of nickel. Indeed, when condensation between amine and carbonyl compound is slow to take place, reaction under elevated temperature and pressure may be desirable.

Low-pressure alkylations with nickel are aided by the use of a condensing agent such as sodium acetate. It was employed for the reaction of aromatic amines or precursors and aldehydes [9a, b]. The presence of an equivalent of weak acid, as acetic, to neutralize the effect of the base on the catalyst may also enable Raney nickel to perform adequately in low-pressure reductive alkylations [3]. Reaction at elevated temperature may also increase the usefulness of nickel for alkylations at low pressure. N,N^1-*bis*-(cyclohexylmethyl)ethylenediamine was obtained from ethylene diamine and excess 3-cyclohexenecarboxaldehyde over Raney nickel at 60° and 3 atm of hydrogen [9c].

Base metal and noble metal sulfides have catalyzed alkylations of aromatic amines with aliphatic ketones [10a, b]. Elevated temperature, 140–245°, and pressures from 80–160 atm were employed for reactions with the base metal sulfides. Of that group, rhenium sulfide was the most active. Less vigorous conditions were required for reductive alkylations over the more active noble metal sulfides [10b]. It was shown that rhodium sulfide could be prepared *in situ* by addition of hydrogen sulfide to a reductive alkylation at 140° and 80–90 atm with rhodium on carbon and gave as good results as rhodium sulfide on carbon under the same conditions [10c]. Rhodium sulfide was claimed to be the most active of the sulfided catalysts for reductive alkylations. Noble metal sulfides have not been employed in a wide enough variety of reductive alkylations to state a general applicability. They are, however, useful in reductive alkylations in the presence of nuclear chlorine and bromine and might be worth studying in alkylations in the presence of other reducible functions.

SECONDARY AMINE FROM PRIMARY AMINE

Alkylations with Aldehydes

Although equimolecular amounts of aldehyde and primary amine are usually employed in reductive alkylations, a slight excess of amine can be beneficial. An excess of amine is desirable in alkylations with the more reactive aldehydes; an excess of some of the lower aliphatic aldehydes can lead to tertiary amine. Fresh aldehydes or freshly distilled aldehydes should be employed.

Aliphatic Aldehydes. It is sometimes difficult to prevent dimethylation in reductive alkylation of the more basic primary amines with formaldehyde [5e, 11a]. At times dimethylation is not only associated with the reactivity of formaldehyde but also occurs because of the difficulty in getting an exact amount of aldehyde for monomethylation. Better control can be maintained by the use of paraformaldehyde, $(CH_2O)_x$ [3]. It is simpler to get a more exact amount of aldehyde than when employing aqueous formaldehyde [3].

Paraformaldehyde was the alkylating agent for reductive monomethylation of a series of aralkylamines where the alkylene chain was straight or branched [3]. Mixtures of equimolar amounts of amine and paraformaldehyde in alcohol were allowed to stand. The solutions were then added to alcohol containing an equivalent of acetic acid and hydrogenated over rhodium on carbon or platinum oxide under 2–3 atm pressure. High yields were obtained with either catalyst but more rapid uptake of hydrogen resulted from the use of platinum oxide.

Platinum oxide catalyzed the monomethylation of aminoacetaldehydediethylacetal with paraformaldehyde in anhydrous alcohol containing fused sodium sulfate [11b]; 75% yield was obtained.

An alternate method of obtaining monomethylamines by reaction of the appropriate aldehyde and methylamine is less complicated and preferable to reductive methylation of the primary amine with formaldehyde. N-Methylbutylamine was prepared by reductive alkylation of methylamine with butyraldehyde over Raney nickel [12a]. Alkylation of methylamine with glucose over nickel on kieselguhr at 85° and 70 atm gave N-methylglucamine, $HOCH_2(CHOH)_4CH_2NHCH_3$ [12b]; activity of the catalyst was improved by storage over methylamine. Hexahydrobenzaldehydediacetate,

$$C_6H_{11}CH(OCOCH_3)_2,$$

and methylamine were hydrogenated over Raney nickel at 75–100° and 70 atm to give 90% yield of N-methyl cyclohexylmethylamine,

$$C_6H_{11}CH_2NHCH_3 \text{ [12c]}.$$

Some of the problems associated with reductive monomethylation of basic primary amines can be overcome by mixing the amine and an excess of aqueous formaldehyde, allowing the mixture to stand until formation of the azomethine is complete, followed by extraction of it and hydrogenation as imine [3].

Reductive monomethylation of aromatic amines with formaldehyde is easily controlled even when a slight excess of formaldehyde is employed. There is usually a very significant difference in rate between the first and second equivalent of hydrogen during the reaction [3]. When some 3- and 4-(N,N-disubstituted aminoalkoxy)anilines, $(R)_2N(CH_2)_nOC_6H_4NH_2$, were alkylated with excess formaldehyde in the presence of a 10% ratio of 5% palladium on carbon, the first equivalent of hydrogen was adsorbed in 2 hr and the second in 6–15 hr [3]. Monomethylanilines of this type could be obtained in 85–90% yield by interrupting the reduction after one equivalent uptake. In the reductive alkylation of 4-aminobenzoic acid with excess formaldehyde over platinum oxide, uptake of hydrogen stopped after one molar equivalent was absorbed [3]. Similar difficulties were noted in attempted dimethylations of 3- and 4-(aminoethyl)phenylacetic acids over palladium on carbon [3]. Frequent additions of catalyst and formaldehyde were required before dimethylation was complete.

Excess formaldehyde is usually employed in monomethylation of aromatic amines in order to insure complete formation of azomethine. It may not be necessary. 4-Methylaminophenol was obtained by reductive alkylation of an alkaline solution of 4-aminophenol with a molar equivalent of formaldehyde in the presence of Raney nickel at 35–40° and 1–3 atm [13a, b, c].

When reductive alkylations of the more basic primary amines are carried out with C_2 to C_4 aldehydes, dialkylation is controlled by the use of equimolar amounts of reactants or with a slight excess of amine if it can be readily removed by fractionation. The danger of dialkylation decreases with branched or long-chain aldehydes.

Gradual addition of a reactive aldehyde to an amine during hydrogenation is a procedure which was found useful in the reductive alkylation of propylamine over Raney nickel at 160–195° and 50–100 atm to yield dipropylamine [14a]. In most instances, hydrogenations of mixtures of aldehyde and amine give good yields of monosubstituted amine. *N*-Butylcyclohexylamine was obtained by reductive alkylation of amine with butyraldehyde over Raney nickel at elevated temperature and pressure [14b]. *N*-Propylcyclohexylamine was similarly prepared [3]. Immediate hydrogenation of a mixture of cyclohexylamine and propionaldehyde in alcohol did not take place at room temperature and 3 atm over palladium on carbon [3]. Reductive alkylation was successful when the two components were mixed in thiophene benzene and allowed to stand over anhydrous magnesium sulfate [6] and then reduced; 75% yield of *N*-propylcyclohexylamine was obtained. When using this procedure with lower aldehydes and low molecular weight amines, cooling was maintained during addition of aldehyde. Reductions were carried out with platinum oxide, platinum on carbon, or palladium on carbon [3]. In other procedures aldehyde is added to a cooled solution of amine in alcohol and hydrogenated [15], or the mixture is allowed to stand in the cold for a considerable period to allow imine to form before subjecting the solution to reduction [3]. In certain instances imine may not form readily or if it does, steric effects may prevent reduction. A mixture of cyclopropanecarboxaldehyde and isopropylamine did not undergo reaction [3]. The desired amine was obtained by alkylation of cyclopropylamine with acetone [6]. No uptake of hydrogen occurred in the attempted alkylation of 2-amino-2-methyl-1-butanol with propionaldehyde under mild conditions [3]. More vigorous conditions are probably required.

There is usually little danger of tertiary amine formation in reductive alkylations of aromatic amines with aliphatic aldehydes (formaldehyde may be an exception) when reactions are carried out under neutral or alkaline conditions [16a]. Aniline and 4-methoxyaniline have been alkylated with acetaldehyde in the presence of sodium acetate and Raney nickel at 15–100° and 1–4 atm [16a]; *N*-heptylaniline and *N*-butylnaphthylamine have been similarly prepared from amine and aldehyde. Platinum oxide was also used in the reaction to obtain *N*-ethylaniline. When a nitro compound is the precursor, the presence of a hydroxy or amino group *ortho* or *para* to the nitro group may lead to tertiary amine [16a]. This may not necessarily apply to reductive alkylation of an aromatic amine containing those groups in similar positions. In the tetracycline series the amino group *ortho* to the phenolic OH group was only monoalkylated with an excess of formaldehyde or acetaldehyde over palladium on carbon [16b] (Eq. 17.5).

$$\text{tetracycline-NH}_2 \xrightarrow{\text{HCHO}, \text{H}_2} \text{CH}_3\text{NH-tetracycline} \quad (17.5)$$

Excess aldehyde was used in other alkylations without formation of tertiary amines. 4,5-Dimethoxy-2-methylaniline gave only *N*-monomethyl and *N*-monoethylanilines upon reductive alkylation with three equivalents of formaldehyde and acetaldehyde, respectively, in the presence of Raney nickel [16c]. Excess aldehyde was employed in reductive alkylations over Raney nickel at 60–70° and 4 atm pressure to give 4-acetylamino-4'-ethyl-, 4'-propyl-, and 4'-octylaminophenylsulfones [16d].

In most alkylations equivalent quantities are employed. This procedure was successful in conversion of 4-aminosalicylic acid to 4-butylaminosalicyclic acid over nickel at 1–20 atm or with platinum oxide at low pressure [17a]. Reductive alkylation of methyl 3-(4-aminophenyl)propionate with an equivalent of acetaldehyde and Raney nickel and sodium acetate as condensing agent gave 61% yield of the 4-ethylamino compound [17b]. Increasing the amount of aldehyde led to diethylamino compound.

Alkylations of aminopyridines with ethyl glyoxylate in aqueous hydrochloric acid over palladium on carbon at 1–3 atm gave *N*-substituted glycines or esters, RNHCH$_2$COOR' where R = 2-pyridyl [18a], 3-pyridyl [18b], and 3-(4-carbethoxy)pyridyl [18c]. In the presence of two equivalents the *bis* product, RN(CH$_2$COOR')$_2$, was obtained [18b]. Reaction time was quite long in the alkylation of 2-aminopyridine.

Alkylation of 5-amino-6-methylpyrimidine required an excess of freshly distilled acetaldehyde to obtain 5-ethylamino-6-methylpyrimidine from low-pressure reduction in alcohol with palladium on carbon [19]; the reaction was unsuccessful with Raney nickel.

Reductive alkylation of primary amines with aralkyl aldehydes resembles alkylation with aliphatic aldehydes except that the bulk of these aldehydes precludes tertiary amine formation. In these alkylations, unlike those with benzaldehyde and related aldehydes, there is no possibility of hydrogenolysis of the product of reduction. Therefore there is greater freedom in the choice of catalyst and reaction conditions. There is usually less difficulty from aldolization (caused by the effect of strong nitrogen base) than takes place with lower aliphatic aldehydes. Nevertheless, formation of aldol was apparently responsible for the poor yield in the reduction alkylation of 3,4-dimethoxyphenethylamine and of phenethylamine with 3,3-diphenyl-propionaldehyde [20a]. The components were heated at 100° for a short period, then hydrogenated in alcohol with palladium on carbon. In the latter reaction, the aldol, $(C_6H_5)_2CH_2CH_2CHOHCH[CH(C_6H_5)_2]CHO$, was isolated along with N-(3-diphenylpropyl)phenethylamine. In general it may not be necessary to heat the reactants. 3-Amino-1-phenyl-2-pyrrolidone and phenylacetaldehyde were hydrogenated in alcohol with prereduced platinum to give 71.5% yield of 3-(2-phenethylamino)-1-phenyl-2-pyrrolidone [20b]. A mixture of phenethylamine and phenylacetaldehyde in alcohol was allowed to stand and then hydrogenated over platinum oxide to give good yield of (diphenethyl)amine, $(C_6H_5CH_2CH_2)_2NH$ [3]. The procedure of mixing of components in thiophene-free benzene and treatment with drying agent [6] gave good yield of N-cyclopropylphenethylamine from cyclopropylamine and phenylacetaldehyde with 5% palladium on carbon [3].

In examples of alkylations with related aldehydes in the heterocyclic series, indolyl-3-acetaldehyde and hydrazine hydrate in methyl alcohol containing acetic acid yielded 3-(2-hydrazinoethyl)indole on reduction with platinum oxide [21a]. In alkylations of benzylamine and related pyridyl and furylmethylamines in absolute alcohol with 3-(2-acetylamino-4-hydroxy-6-methyl-5-pyrimidyl)propionaldehyde at 2–3 atm pressure in the presence of Raney nickel or platinum oxide, good yields of the substituted pyrimidylpropylamines were obtained [21b] (Eq. 17.6). Best results were obtained

$$CH_3CONH\underset{OH}{\overset{N}{\underset{N}{\bigcirc}}}\overset{CH_3}{\underset{CH_2CH_2CHO}{}} + RNH_2 \xrightarrow[2-3\,atm]{H_2 \atop PtO_2}$$

$$CH_3CONH\underset{OH}{\overset{N}{\underset{N}{\bigcirc}}}\overset{CH_3}{\underset{(CH_2)_3NHR}{}} \quad (17.6)$$

R = benzyl, 2-, 3-, and 4-pyridylmethyl, 2-furylmethyl, and 3-pyridyl

from reductions with a 3% ratio of platinum oxide based on total weight of amine and aldehyde. Reaction time was 1.5 hr compared with 18 hr for nickel. Reductive alkylation of 3-aminopyridine with the same aldehyde took 80 hr with a 50% weight ratio of 5% palladium on carbon. Equimolecular quantities of amine and aldehyde were employed except in the alkylation of benzylamine. There the use of excess amine caused some hydrolysis of the 2-acetylamino group.

Aromatic Aldehydes. Generally reductive alkylations of primary amines with aromatic aldehydes give high yield of secondary amines, $ArCH_2NHR$. Formation of intermediate is usually fairly rapid and the intermediate is reasonably stable. It is for these reasons that ingredients can be mixed and hydrogenated without isolation of the Schiff base although it has been noted that reduction of the isolated imine often gives better results [3, 4a, b]. Since most of the Schiff bases from these aldehydes are stable, the components also can be mixed and heated to remove water resulting from condensation and subsequently hydrogenated in solution, usually under mild conditions. When low-boiling amines are to be alkylated, it is advisable to allow the reactants to stand before reduction.

It is advisable to use an unopened bottle of aromatic aldehyde or to redistill it before use because these aldehydes are so readily oxidized. This is particularly important if palladium catalyst is employed for reductions since the resulting products are N-benzylamines or amines related to them and, therefore, subject to hydrogenolysis. When the calculated amount of hydrogen is absorbed in a reduction which contains less than 100% of intermediate low yield may be the result, especially if elevated temperature is employed during reaction.

Platinum oxide or platinum on a carrier is probably best for alkylations with aromatic aldehydes under low-pressure conditions because they do not cause debenzylation as readily as palladium.

Platinum oxide catalyzed the reductive alkylation of methylamine with benzaldehyde and substituted benzaldehydes [22a], of ethylamine with 3-hydroxybenzaldehyde [22b], and cyclopentylamine with benzaldehyde [22c]. It was the catalyst in the preparation of N-2-hydroxyethyl-4-dimethylaminobenzylamine [22d] and N-3-hydroxypropylbenzylamine [22e]. In the alkylation of alkanolamines with 1,2,3,4-tetrahydrophenanthrene-9-aldehyde, the materials were refluxed in alcohol and then reduced [22f]; reaction time varied from 4–24 hr. In slow reductions the addition of an equivalent of acetic acid should be considered. In Reference 22d reaction time was about 10 hr due to the low catalyst ratio, 0.25%.

Platinum on carbon can be substituted for platinum oxide and gives equally good results in reductive alkylations with aromatic aldehydes [3]. In the reaction of N^4-(4-aminobenzoyl)-1,4-phenylenediamine with benzaldehyde in methyl cellosolve in the presence of 5% platinum on carbon, it was necessary to raise the temperature to 60° because of poor solubility [3].

Palladium catalyst can be used in reductive alkylations with aromatic aldehydes. The possibility of debenzylation of the reduction product does exist but the catalyst loses some of its activity during the process and does not cause hydrogenolysis unless the reaction is allowed to run unguarded. Use of lower catalyst ratios can lessen the danger of further reaction by causing a very significant difference between the rate of alkylation and that of hydrogenolysis of the secondary amine [3].

Reductive alkylations with palladium on carbon are often rapid. 2-Aminomethylindole was alkylated with nuclear disubstituted benzaldehydes in its presence to give high yields of N-benzylated aminomethylindoles in a short reaction period [23a]. Rapid reduction was also reported in the alkylation of methylamine with 4-dimethylamino- and 3,4-methylenedioxybenzaldehydes which gave yields of 58–69% [23b]. In this study, however, hydrogenation of the isolated imines was preferred. Reduction of a mixture of phenethylamine and benzaldehyde which had been allowed to stand gave 86% yield of N-benzylphenethylamine in a rapid reaction in alcohol with palladium on carbon [3]. Similar treatment of 5-aminomethyl-5-methyl-2-pyrrolidone and benzaldehyde gave good yield of 5-benzylaminomethyl-5-methyl-2-pyrrolidine [3].

Palladium on carbon was employed in the alkylations of ethanolamine [23c], of isopropylamine [23d], and of 5-diethylaminopentylamine [23e] with benzaldehyde or substituted benzaldehydes to give benzylated amines in good yields. Reductive alkylation of norcamphane, 2-amino-2,3,3-trimethyl-[2,2,1]bicycloheptane, with benzaldehyde over palladium on carbon gave 95% yield of N-benzyl compound [24a]. It is of interest that high yields of 3- and 2-benzylaminomethylpyridines were obtained from hydrogenations of a mixture of benzaldehyde and 3-aminomethylpyridinine and benzylamine and pyridine-2-carboxaldehyde in alcohol with palladium on carbon at 60° and 17 atm [24b]. Such conditions could lead to debenzylation.

Reductive alkylations of amines with benzaldehyde and substituted benzaldehydes have been carried out with Raney nickel at 1–3 atm pressure [24a, b, c] and have resulted in high yields of the corresponding amines. In most instances the rate of reaction does not compare with reductions over platinum or palladium [3], but debenzylation usually will not take place in the presence of nickel even when reactions are carried out at elevated temperature and pressure unless the catalyst is a highly active one.

Reductive alkylations with aldehyde, ArCHO, where Ar is an unsaturated heterocycle, are carried out in the same manner as when Ar is a benzenoid system. As in the hydrogenation of benzylideneamines (Chapter XV) ring reduction will not take place with nickel, palladium, or platinum unless reaction conditions are vigorous. When a mixture of benzylamine and furfuraldehyde was hydrogenated with Raney nickel at 100° and 100 atm,

358   Reductive Alkylation

80% yield of N-benzyl-1,2,3,4-tetrahydro-2-furylmethylamine was obtained [3]. When furfuraldehyde and a large excess of ethylenediamine (to obtain monoalkylation) were reacted in alcohol over Raney nickel at 80° and 100 atm, the furan ring was not reduced; yield of N-furylethylenediamine was 68% [26a].

N-Furylphenethylamine was obtained from reductive alkylation of phenethylamine with furfuraldehyde at atmospheric pressure with Raney nickel [22b]; other furylphenethylamines were obtained by reductive alkylation with Raney nickel at low pressure [26b]. In other alkylations in the furan series, aminodiethylacetal was alkylated with 5-methyl-2-furaldehyde over platinum oxide to give 64% of N-(5-methyl-2-furyl)aminodiethylacetal [26c]. When methylamine was alkylated with the same aldehyde in the presence of palladium on carbon under 2 atm pressure, yield of 2-methylaminomethyl-5-methylfuran was only 23%. Other catalysts were not tried in the same reaction.

Palladium on carbon gave good results in alkylations with pyridine-carboxaldehydes [24b]. Ethyl N-3-picolylglycinate was obtained in 50–60% yield from reductive alkylation of the ethyl ester of glycine with pyridine-3-carboxaldehyde in the presence of palladium on carbon [27a]; later Raney nickel gave better results. Reductive alkylation of tyramine with 3-hydroxy-5-hydroxymethyl-2-methylpyridine-4-carboxaldehyde (pyridoxal) over platinum oxide gave the corresponding 4-pyridylamino acid [27b] (Eq. 17.7).

$$RCHO + \underset{OH}{\underset{|}{C_6H_4}}-CH_2CHNH_2COOH \xrightarrow{H_2, PtO_2} \underset{OH}{\underset{|}{C_6H_4}}-CH_2CH(COOH)NHCH_2R$$

$$R = \text{3-hydroxy-5-hydroxymethyl-2-methylpyridin-4-yl}$$

(17.7)

A mixture of pyridoxal and 3,4-dihydroxyphenethylamine and of pyridoxal and 1-hydroxy-2-(3,4-dihydroxyphenyl)ethylamine were reduced over platinum oxide to the corresponding N-(pyridylmethyl)phenethylamines [2b].

Hydrogenations of mixtures of quinoline-2- or 4-aldehyde with 4 to 5 equivalents of primary amines, NHR where R = methyl to amyl, 2-hydroxyethyl, and phenethyl, were carried out with an equal weight of 5% palladium on barium sulfate; yields of resulting amines were very high [28]. The reason for the use of large excesses of amine is not clear except to inactivate the catalyst so that further reduction would not take place.

Alkylations with Ketones

Reductive alkylations of amines with ketones do not proceed with the same ease as alkylations with aldehydes because of steric effects. Groups in 2-position in cyclic ketones and in ketones of the type, ArCOR, can make alkylations more difficult because of slow formation of the intermediate or from the blocking effect of those groups on the reduction of the intermediate.

Because of steric effects, dialkylations with ketones do not take place too readily; it is therefore safe to use excess ketone except under vigorous reaction conditions.

Aliphatic Ketones. The procedure of adding ketone to low-boiling amine in thiophene-free benzene during cooling, allowing the solution to stand over drying agent, followed by hydrogenation at 2–3 atm in the presence of platinum oxide gave good yields of *N*-isopropyl- and *N*-cycloalkylcyclopropylamines [6]. *N*-Isopropylcyclopropylamine was also obtained by reduction of an alcoholic solution of acetone and cyclopropylamine which had been allowed to stand prior to hydrogenation [3].

In examples of reductive alkylations with aliphatic ketones, *N*-butyl-2-hexylamine was prepared from butylamine and 2-hexanone in alcohol over platinum oxide at 3 atm pressure [29a]; *N*-methyl-2-heptylamine was obtained from hydrogenation of a mixture of methylamine and 2-heptanone over prereduced platinum [29b]. Reduction over platinum with 1,3-*bis*-(acetylamino)acetone and nitromethane as amine precursor gave 58% yield of *N*-methyl-1,3-*bis*(acetylamino)isopropylamine,

$$CH_3CONHCH_2CH(NHCH_3)CH_2NHCOCH_3 \text{ [29c]}.$$

Although platinum is used most frequently in these reactions, alkylation of N^1-cyclohexyl-N^1-methylpropane-1,3-diamine with an equivalent amount of cyclohexanone in alcohol in the presence of 10% by weight of 5% palladium on carbon (based on total weight of components) was complete in 20 min to give 85% yield of N^1,N^3-dicyclohexyl-N^1-methylpropane-1,3-diamine, $C_6H_{11}N(CH_3)CH_2CH_2NHC_6H_{11}$ [3]. *N*-Methylcyclohexylamine was obtained from palladium on carbon reduction of a benzene solution of cyclohexanone and anhydrous methylamine previously treated with drying agent [3].

The alkylation of primary amines with phenol as a precursor for cyclohexanone is a very useful and economical method for obtaining *N*-monosubstituted cyclohexylamines, $C_6H_{11}NHR$ [29d]. Amine and phenol are heated under 2–3 atm hydrogen pressure to about 80° in the presence of palladium on carbon or rhodium on carbon to give good yield of *N*-alkylcyclohexylamine. Other monohydric phenols react in the same manner. Platinum catalyst causes conversion of phenols to cyclohexanols.

Dicyclohexylamine was prepared in very high yield by reductive alkylation of aniline with phenol and at least 9% of 5% palladium on carbon (based on total weight of components) at about 4 atm pressure and below 100° [29e]. It is difficult to say whether aniline was first reduced to cyclohexylamine followed by reductive alkylation or whether N-phenylcyclohexylamine, $C_6H_5NHC_6H_{11}$, was formed and reduced to dicyclohexylamine; small amounts of each were obtained.

Palladium on carbon was also specified in a similar procedure which gave over 75% yield of dicyclohexylamine by reduction at 100–200° and 7–13 atm [29f]. The first reaction [29e] can be carried out with safety in a Parr shaker.

Nickel catalyst has been employed in reductive alkylations with aliphatic ketones but often under elevated temperatures and pressures (without indication whether such conditions were necessary). The alkylation of methylamine with (bicyclo[2,2,1]hept-2-yl)acetone was carried out with Raney nickel at 3 atm and 70–90° [30a] (Eq. 17.8). When methylamine was

$$C_6H_{11}CH_2COCH_3 + CH_3NH_2 \xrightarrow[\text{Ni(R)}]{H_2} C_6H_{11}CH_2CH(CH_3)NHCH_3 \quad (17.8)$$
$$75\%$$

alkylated with 2-phenylcyclohexanone over nickel at 95° and 100 atm [30b], such conditions were probably necessary because of the steric effects of the phenyl ring; 82.3% yield of N-methyl-2-phenylcyclohexylamine was obtained.

Copper chromite was employed in the alkylation of methylamine with a hindered ketone, 2,6-dimethyl-4-heptanone (diisobutyl ketone), at 160–180° and 100 atm pressure [31a]. Oddly enough the reaction did not work with Raney nickel. A mixture of methylamine and methyl isobutyl ketone was also hydrogenated over copper chromite at 180° and 170 atm to give 91% yield of N-,4-dimethyl-2-pentylamine [31b]. It is likely in the latter reduction that other catalysts would be effective under milder conditions. In this instance high pressure conditions were probably used as a matter of convenience because of the amount of material (4.2 moles) to be reduced.

Alkylation of aralkylamines and related amines with aliphatic-type ketones over platinum at low pressure generally give high yields of secondary amines. Mixtures of benzylamine and cyclopentanone [32a], of 4-(2-aminoethyl)-1,2,3-triazole and acetone [12b], of 3-(2-aminoethyl)-1,2,4-triazole and acetone [32c], of 3-(2-aminoethyl)indazole and acetone [32d], and of 1-(2-methoxyphenyl)isopropylamine and acetone [32e] were hydrogenated in this manner to give the corresponding secondary amines.

Palladium on carbon was employed for the alkylation of 2-ethyl-2-(3-methoxyphenyl)butylamine with acetone, 2-butanone, and other alkanones and with cyclohexanone at 50° and pressures up to 50 atm of hydrogen [33]. Very high yields of N-substituted butylamines were obtained. It is likely that more moderate conditions could be employed as in the palladium-catalyzed alkylation of 3,3-diphenylpropylamine with cyclohexanone at 5–12 atm which yielded 91.5% of N-cyclohexyl-3,3-diphenylpropylamine [20a].

When diamines such as ethylenediamine are alkylated with ketones, platinum catalysts are generally favored. Excess ketone may be used to insure alkylation at each nitrogen. There is little chance of obtaining more than one substituent on each nitrogen. When the ketone is cyclic, dialkylation at one of the nitrogens is almost impossible [3]. When excess diamine is used alkylation takes place at only one of the amino groups [34a].

Prereduced catalyst was used for the preparation of N-cyclohexylalkanediamines, $C_6H_{11}NH(CH_2)_nNH_2$ where n was 2–6 and where it was 10 [34b]. Yields ranged from 63–87%; reaction time varied from 2 to 18 hr. Prereduction of platinum oxide was found unnecessary in reruns of these alkylations; hydrogen uptake was just as rapid, sometimes more rapid, than when the catalyst was prereduced [3].

Platinum oxide was employed in the alkylation of ethylenediamine and propane-1,3-diamine with cyclohexanone to yield N^1,N^2-bis(cyclohexyl)-ethylenediamine and N^1,N^3-bis(cyclohexyl)propane-1,3-diamine respectively [34c]. Platinum oxide and platinum on carbon were employed for reductive alkylations of ethylenediamine with two equivalents of aliphatic ketone at moderate temperatures and low to high pressure conditions [34d]. Low-pressure reactions did not work efficiently unless water was removed or unless the Schiff bases were isolated. In the alkylation of ethylenediamine with excess 4-methyl-2-pentanone over platinum on carbon, the reaction was carried out at 32–70° and 40–55 atm to yield N^1,N^2-bis(1,3-dimethylbutyl)ethylenediamine [34c]. Other bis-alkylated ethylenediamines were similarly prepared.

Reductive alkylation of hydrazine with excess cyclopentanone proceeded readily in aqueous solution in the presence of platinum oxide at 2.8 atm to give 74% yield of 1,2-dicyclopentylhydrazine as the hydrochloride salt [35a]. Monocyclopentylhydrazine was prepared from equimolar amounts of hydrazine hydrate and cyclopentanone by reduction in dilute hydrochloric acid over platinum oxide at 100 atm [35b].

Reductive alkylation of hydrazine with excess cyclobutanone did not yield 1,2-dicyclobutylhydrazine. Instead, ammonium chloride and dicyclobutylamine as hydrochloride were obtained [35a]. According to the proposed scheme, cyclobutanone, because of its reactivity, condensed further with 1,2-dicyclobutylhydrazine in acid solution to undergo conversion to a tetrasubstituted hydrazine which was then hydrogenolyzed (Eq. 17.9).

It was only possible to obtain 1,2-dicyclobutylhydrazine by reduction of an intermediate, assumed to be 1,2-cyclobutylidenehydrazine, $(C_4H_6{=}N)_2$, in acetic acid.

$$2\,\square{=}O + NH_2\cdot NH_2\cdot HCl \xrightarrow[Pt]{H_2} [\square{-}NH{-}]_2 + 2\,\square{=}O$$

$$(\square{-}N{-}N{-}\square)_2 \longrightarrow 2(\square{-}NH)_2 \qquad (17.9)$$

When phenylhydrazine was alkylated with cyclobutanone and cyclopentanone, the intermediate was isolated and hydrogenated in acetic acid over platinum oxide. Uptake of hydrogen was interrupted after one equivalent to prevent hydrogenolysis of the hydrazine.

Although reduction of the isolated 1-isonicotinyl-2-alkylidenehydrazines,

$$RCONHN{=}C\underset{R'}{\overset{R'}{\diagup}}$$

, may be the preferred method of obtaining the corresponding isonicotinylhydrazines (Chapter XV), reductive alkylations of 1-isonicotinylhydrazine with aliphatic ketones have been carried out in the presence of platinum oxide at room temperature and 3 atm [36a], at 60–70° and 3 atm [36b], and at 30–40° and 2–30 atm [36c]. In the last reference the less reactive longer chain ketones and 1-isonicotinylhydrazine were heated, then dissolved in alcohol and hydrogenated to yield 1-isonicotinyl-2-(5-hydroxy-2-pentyl)-, 2-(4-hydroxy-4-methyl-2-pentyl)-, and 2-(1,4-dimethoxy-2-butyl)hydrazines.

There should be little concern about hydrogenolysis of the $N\!-\!N$ bond in reductive alkylations of 1-acylhydrazines. Acylation greatly decreases the possibility of cleavage of the nitrogen to nitrogen bond [3].

Very high yields of N-substituted 2-aminoalcohols were obtained by low-pressure reductive alkylation of the aminoalcohol with aliphatic ketones in the presence of prereduced platinum [5a]; reaction time was reasonably rapid (3–10 hr) except for alkylations with diethyl ketone which was slower and with dipropyl and dibutyl ketones and l-menthone which required 20–30 hr reaction periods while heating at 50–60°. Palladium on carbon was found to be a less effective catalyst requiring frequent additions of fresh catalyst. Raney nickel and copper chromite were also employed, but at 150–160° and 70–140 atm pressure.

In other alkylations of aminoalcohols, an excess of 1,3-diamine-2-propanol was reacted with cyclohexanone over prereduced platinum at 60° and 2 atm to yield N-cyclohexyl-1,3-diamino-2-propanol [34a]; 1,3-diphenyl-3-amino-1-propanol was treated in methyl alcohol with acetone and with cyclohexanone and reduced over 10% platinum on barium sulfate at room temperature and 3.4 atm [36d]. 3-Isopropylamine- and 3-cyclohexylamino-1,3-diphenyl-1-propanols were obtained but the amount of catalyst was over 50% of the total weight of amine and ketone.

Palladium on carbon (5%) catalyzed the alkylation of 3-aminopropanol with cyclohexanone [3]. Uptake of hydrogen was complete in 1 hr at 20° and 30 atm with a 12.5% ratio of catalyst based on total weight of components; yield of 3-cyclohexylamino-1-propanol was 75% after distillation. Loss was probably mechanical since the yield of solid product before distillation was 95.5%. In the reductive alkylation of 1-hydroxy-1-(3,4-diethoxyphenyl)isopropylamine with excess acetone and with 2-nonanone at 1 atm, a large amount of 10% palladium on carbon was employed [36e], although this could be necessary because of the very small size run.

The occurrence of dialkylation in reductive alkylation of aromatic amines with aliphatic and indeed with most ketones is a minor reaction unless vigorous conditions are employed. The simpler 2-alkanones may cause dialkylation under such conditions when used in excess, but dialkylation usually does not take place with more complicated ketones or with cycloalkanones. 4-Aminosalicylic acid was alkylated with acetone at 1–20 atm pressure in the presence of Raney nickel or platinum oxide to yield 4-isopropylaminosalicyclic acid [25b]; 4-aminophenol and precursors, 4-nitro and 4-nitrosophenol, were reductively alkylated over platinum oxide [37a]. The use of acetic acid or sodium acetate may aid condensation and subsequent reductive alkylation of aromatic amines with aliphatic ketones over platinum or Raney nickel [9a, 16a]. In many instances high-pressure methods have been employed using platinum on a carrier [36b, c], with palladium in acetic acid containing sulfuric acid [37d], and with copper chromite [37e]. Commercial nickel on kieselguhr [38a] catalyzed the alkylation of 4-aminodiphenylamine with 4-methyl-2-pentanone at 165° and 52–70 atm to give 58% of 4-(4-methyl-2-pentylamino)diphenylamine [38b]; 26% of 4-methyl-2-pentanol was also obtained. The addition of sulfur-containing compounds (those containing sulfur atoms with unshared electron pairs) to the reductions increased the yield of alkylation product and minimized reduction of ketone. The additives apparently inactivate the catalyst for competing reduction of the carbonyl function and allow more complete condensation of amine and ketone.

A very useful method for reductive alkylation of aromatic amines with aliphatic ketones when elevated temperatures and pressure are required is hydrogenation over noble metal sulfides on a support [10b, c]. Alkylation of 4-aminodiphenylamine with 4-methyl-2-pentanone over 5% platinum sulfide on carbon gave 99% yield of 4-(4-methyl-2-pentylamino)diphenylamine [10b]; alkylation of 1,5-naphthalenediamine with acetone, with 2-butanone, and with cyclohexanone over 5% rhodium sulfide on carbon (or prepared *in situ*) gave 1,5-diisopropylamino-, 1,5-di-(2-butylamino)- and 1,5-dicyclohexylaminonaphthalenes [10c]. When a sulfided catalyst is employed, the alkylation process can be more economical because less ketone is necessary [38c]. When comparative alkylations with unsulfided platinum on alumina and sulfided catalyst (prepared *in situ*) were carried out, five times more carbinol resulted from alkylation with unsulfided catalyst [38c]. N^1,N^4-*bis*(2-Octyl)phenylenediamine was obtained from alkylation of 1,4-phenylenediamine or 4-nitroaniline with 2-octanone over the sulfided catalyst at 130° and 55 atm. The inability of sulfided catalysts to reduce ketones under the reaction conditions necessary for alkylation favors condensation of amine and ketone and leads to good yield. In addition, sulfided catalysts reputedly do not lose sulfur, nor do they poison or attack the reaction vessel [39].

Aromatic Ketones. Reductive alkylation of amines with diaryl ketones may be difficult because of steric effects. Often it is expedient to isolate the intermediate or to employ elevated temperature and pressure and an acidic agent to force condensation to take place. In the attempted alkylation of aniline with benzophenone it was necessary to isolate the anil before hydrogenation [40a]. N-Phenylbenzhydrylamine, $(C_6H_5)_2CHNHC_6H_5$, was obtained in 78% yield from the palladium on carbon reduction of the imine in 2 hr at 4 atm pressure. The alkylation of methylamine with 4-methoxybenzophenone in alcohol containing acetic acid was carried out at 150° and 140 atm over Raney nickel (15% by weight) to yield 71% of N-methyl-4-methoxybenzhydrylamine [40b].

Alkylations with acetophenone and other related aromatic ketones, ArCOR where R is alkyl, are less difficult than alkylations with diaryl ketones because of lesser steric effects. Reactions with them at low pressure can be slow [5a]. Alkylation of 2-aminoethanol with acetophenone in alcohol in the presence of platinum oxide or prereduced platinum was so slow that it was necessary to isolate the intermediate and subject it to hydrogenation over platinum oxide at 100 atm pressure [3]; reaction time was 1.5 hr under those conditions, yield of 2-(1-phenyl-1-ethyl)aminoethanol, $C_6H_5CH(CH_3)NHCH_2CH_2OH$, was 80%. 2-Acetyldibenzofuran and 2-aminoethanol were hydrogenated in isopropyl alcohol over Raney nickel at 140° and 100–130 atm to give the N-substituted aminoalcohol [40c]. Alkylations of 3,3-diphenylpropylamine with acetophenone and with propiophenone were carried out in methyl alcohol with 5% palladium on carbon at

Reductive alkylation of primary amines with ketones, $Ar(CH_2)_nCOCH_3$, where Ar is an unsaturated heterocyclic system, are also carried out in the same manner as reactions with phenylacetone. Amine, $ArCH_2CH(CH_3)NHR$, where Ar was 5-methyl-2-pyridyl, was prepared by high-pressure reduction of amine, RNH_2, and ketone, $ArCH_2COCH_3$, with Raney nickel [44a]; where Ar was 3-indolyl, nickel at 100° and 50 atm [44b] or palladium on carbon at low pressure [44c] catalyzed the reaction. In a related reaction, a solution of 4-(3-indolyl)-2-butanone and 4-hydroxyphenethylamine was hydrogenated over platinum oxide at low pressure to give secondary amine, but reaction time was quite long, 40 hr [44d].

Aryloxyalkanones. Reductive alkylation of primary amines with aryloxyalkanones are similar to those with aralkanones, except that hydrolysis or hydrogenolysis, $ArO|(CH_2)_nCOR$, may take place, particularly if reaction conditions are too vigorous. The difficulty in preventing cleavage probably accounts for the very modest yields usually obtained from alkylations with these ketones.

Alkylations of a number of amines with phenoxyacetone over Raney nickel at 120° and 100 atm gave a maximum of yield of 60% with 2-aminoethanol [45a] and 22.5–40% yields with methyl, ethyl, and isopropylamines. Alkylation of 2-aminoethanol with phenoxyacetone and nickel at 3 atm gave only 25% of secondary amine [45b]. Raney nickel catalyzed alkylations of alkylamines with phenoxyacetone at room temperature and 100 atm [45c]; yields ranged from 20 to 65%. When a mixture of phenoxyacetone and methylamine was hydrogenated with a 2% weight ratio of platinum oxide to components, at 25° and 2–3 atm, about 30% yield of N-methyl-1-phenoxyisopropylamine was obtained [3]. When a 4% weight ratio of 5% platinum on carbon was substituted (equal to 0.2% ratio of metal to substrate), less cleavage occurred and the yield rose to 50%. Alkylation of other amines with nuclear-substituted phenoxyacetones are described but yields are not reported [45d, e]. High yield was obtained from reductive alkylation of benzylamine with phenoxyacetone in the presence of platinum oxide at 3 atm pressure; N-benzyl-1-phenoxyisopropylamine was not isolated but debenzylated to give 1-phenoxyisopropylamine; overall yield from the two reactions was 90% [45b].

Keto Acids and Esters. Alkylations of primary amines with keto acids, $RCH_2CO(CH_2)_nCOOH$, proceed readily to give secondary amines when $n = 0 - 2$. When $n = 3$ or 4, formation of 5- or 6-membered saturated nitrogen-containing rings often takes place.

Reductive alkylations of 5-diethylamino-2-pentylamine,

$$(C_2H_5)_2N(CH_2)_3CH(CH_3)NH_2,$$

with phenylpyruvic acid, $C_6H_5CH_2COCOOH$, and 4-dimethylaminophenyl-pyruvic acid over platinum oxide under 2 atm pressure gave 60 and 70% yields, respectively, of 2-(5-diethylamino-2-pentylamino)-3-phenyl- and 3-(4-dimethylaminophenyl)propionic acids [46a].

Reductive alkylation of optically active benzylamines, $C_6H_5CHRNH_2$ where R is not H, with 2-oxo acids followed by hydrogenolysis has been used as a method to obtain optically active amino acids [46b] (Eq. 17.11).

$$C_6H_5\overset{*}{C}H(CH_3)NH_2 + RCOCOOH \xrightarrow[Pd(OH)_2]{H_2}$$
$$[C_6H_5\overset{*}{C}H(CH_3)N=C(R)COOH] \longrightarrow C_6H_5\overset{*}{C}H(CH_3)NH\overset{*}{C}HRCOOH$$
$$\overset{*}{R}CHNH_2COOH \xleftarrow{Pd(OH)_2} \qquad\qquad\qquad\qquad (17.11)$$

R = alkyl or benzyl

Yield of amino acid varied with the structure of the keto acid. Alkylation of 3-methyl-2-oxobutyric acid with D-(—)α-methylbenzylamine and further reduction gave only 15.7% yield of D-(—)-valine; similar reaction of amine with 3,3-dimethyl-2-oxobutyric acid gave no amino acid. The magnitude of induced symmetry was dependent on the substrate and the catalyst. Reductive alkylation of optically active phenylglycine with 2-ketobutyric acid with palladium on carbon or palladium hydroxide on carbon and continuing hydrogenolysis gave optically active 2-aminobutyric acid [46c]. Optically active alanine and glutamic acid were prepared in a similar manner. Palladium hydroxide was the preferred catalyst. Palladium on carbon induced more rapid hydrogenation but caused some change in rotation of the amino acid.

Platinum oxide or palladium on carbon are equally useful in catalyzing alkylations with keto acids and esters which result in cyclic products. Reduction of ethyl-4-oxopentanoate and 2-aminoethanol in alcohol with prereduced platinum (3 atm pressure) gave quantitative yield of 1-(2-hydroxyethyl)-5-methyl-2-pyrrolidone [47a]. Hydrogenations of phenethylamine with methyl 4-oxopentanoate and with methyl 4-oxo-4-phenylbutyrate at 60° and 1 atm with 10% palladium on carbon gave 5-methyl and 5-phenyl-1-(2-phenethyl)-2-pyrrolidones [47b] in long drawn-out reactions due, probably, to the very low catalyst ratio, 1% (0.1% based on metal contact), and a desire to insure cyclization.

Alkylations with keto acids in the presence of Raney nickel were carried out at higher temperature, 50–100°, and greater pressures, 35–80 atm [48a, b, c]. In the alkylation of aniline with (2-oxocyclopentyl)acetic acid over Raney nickel there was little hydrogen uptake until the temperature was raised to 135° [48a].

Aminoketones. Reductive alkylation of an amine with an aminoketone may require more catalyst than usual because of the basicity of the aminoketone or the resulting substituted diamine. Yields may also be affected by competing reduction of the aminoketone because of slow formation of imine. Alkylations with Mannich-type ketones, $(R)_2NCH_2CH_2COR'$, may give poor yield because of competing hydrogenolysis. No product was obtained from reduction of methylamine and 4-diethylamino-2-butanone,

$$(C_2H_5)NCH_2CH_2COCH_3,$$

over nickel at 25° and 2 atm [3]. When a mixture of 3-dimethylaminoethyl-1,2,5-trimethyl-4-piperidone and methylamine was hydrogenated over Raney nickel at an initial pressure of 100 atm and 140° for 2 hr, none of the desired product was obtained [49]. Hydrogenolysis accompanied alkylation to yield 1,2,3,5-tetramethyl-4-methylaminopiperidine (33%); the second product, 3-dimethylaminomethyl-4-hydroxy-1,2,5-trimethylpiperidine (21% yield), arose from reduction of the starting ketone (Eq. 17.12). On the other hand, almost quantitative yield of N-(4-diethylamino-2-butyl)-4-methoxyaniline resulted from alkylation of 4-methoxynitrobenzene, as amine precursor, and excess 4-diethylamino-2-butanone, a Mannich aminoketone, with Raney nickel in the presence of tertiary amine hydrochloride as condensing agent [50a]. Successful alkylations of 3,3-diphenylpropylamine with Mannich-type aminoketones were carried out at 50–60° and 5–12 atm [20a]. Unfortunately the effect of structure, condensing agents, and catalysts on alkylation with these ketones have not been well studied.

There should be little concern about cleavage during reductive alkylation with aminoalkanones when there is only one and at least three carbon atoms between the tertiary amino group and the carbonyl function. 1-(2-Oxopropyl)piperidine, $C_5H_{10}NCH_2COCH_3$, and methylamine were mixed in alcohol and hydrogenated over platinum oxide at 2–3 atm to give 60% yield of 1(2-aminopropyl)piperidine; diethylaminoacetone and methylamine were reduced over Raney nickel at 70° and 40 atm to yield 71% of N^1N^1-diethyl-N^2-methyl-1,2-propanediamine, $(C_2H_5)_2NCH_2CH(CH_3)NHCH_3$ [3]. Alkylations of some aromatic amines with 5-diethylamine-2-pentanone were carried out at 25–60° and 3–5 atm with Raney nickel or platinum oxide or with palladium on a carrier in the presence of a condensing agent [50a, b]. In most cases reaction time was long.

When a mixture of 6-methoxy-8-nitroquinoline, as amine precursor, and 5-diethylamino-2-pentanone was reduced with palladium on barium sulfate at 60° and 3 atm in the presence of a tertiary amine hydrochloride or alcoholic hydrogen chloride as condensing agent, 8-(5-diethylamino-2-pentylamino)-6-methoxyquinoline was obtained after 48 hr [50a]. When the results could not be reproduced [50c], the Schiff base was isolated and hydrogenated in the

$$\begin{array}{c}\text{structure with } H_3CN, O, CH_3, CH_2N(CH_3)_2, CH_3\end{array} + CH_3NH_2 \xrightarrow{H_2} \text{products} \quad (17.12)$$

presence of platinum oxide at room temperature and 15 atm pressure to give 80% yield of desired product. In another procedure when the same ketone was pumped into a solution of the aminoquinoline in methyl alcohol at 130° and 40 atm pressure of hydrogen in the presence of Raney nickel, only 26–29% yield was obtained [50d]. The major product was 2-methyl-8-methoxy-5,6-dihydro-4-imidazo[i,j]quinoline, obtained in 50% yield. It may have resulted from continuing reduction of the 8-substituted aminoquinoline to the corresponding 1,2,3,4-tetrahydroquinoline followed by condensation and cleavage to form the 3-ringed system. It is also possible, and perhaps more reasonable to assume, that, before the addition of aminoketone, heating the starting material in the presence of catalyst and hydrogen caused conversion in part to 8-amino-6-methoxy-1,2,3,4-tetrahydroquinoline. It then underwent reductive alkylation and condensation (Eq. 17.13).

The ease of alkylation of amines with cyclic aminoketones can depend on the structure of the amine or ketone and how readily condensation between the two takes place. The reaction of 3-tropinone and methylamine was carried out with platinum oxide at 50° and 3 atm in 90 min to give very high yield of 3-methylaminotropane [51a]. Reaction with benzylamine took 3 hr. When 3-tropinone was used to alkylate N-ethyl-N-phenylethylenediamine, the materials were allowed to stand overnight to allow completion of condensation before hydrogenation with Adams' catalyst [51b]. Alkylation of a hindered amine, benzhydrylamine, with 3-tropinone was very slow [51c].

Benzylamine was alkylated with 1-ethyl- and 1-methyl-3-piperidone and N,N-diethylethylenediamine was alkylated with 1-methyl-3-piperidone to give 61–67% yields of the corresponding 1-alkyl-3-substituted aminopiperidines [52a]. The materials were mixed with cooling and subsequently hydrogenated over platinum oxide at 2–3 atm pressure. Prereduced platinum was employed for the low-pressure alkylation of N^1,N^1-dimethyl-1,3-propanediamine with 1-methyl-3-piperidone; 87% yield of 1-methyl-3-(3-dimethylaminopropylamino)piperidine was obtained [52b].

1-Ethyl-4-piperidone and amines yielded 1-ethyl-4-substituted aminopiperidines in low-pressure reductions over platinum oxide [53a–d]. 2-Methyltetrahydro-4-pyrone and methylamine were hydrogenated over Raney nickel at 75–80° and 80 atm to form 2-methyl-4-methylaminotetrahydropyran in 49% yield [54].

A group adjacent to the carbonyl function in cyclic aminoketones appears to have a hindering effect on condensation with amine when attempting alkylations with them. In the reduction of a mixture of methylamine and 1,2,5-trimethyl-4-piperidone with Raney nickel at an initial pressure of 149 atm and 65–69°, 46% yield of 1,2,5-trimethyl-4-methylaminopiperidine was obtained along with 21% of 4-hydroxy-1,2,5-trimethylpiperidine [49].

(17.13)

Alkylations of higher aliphatic amines and other larger amines with 1-alkyl-2,5-dimethyl-4-piperidones gave yields of 4-hydroxypiperidines as high as 90% because reducible intermediates were not formed.

It is apparent that the major difficulty in reductive alkylation of amines with aminoketones lies in effecting complete condensation between the two reactants. Until this can be accomplished it is difficult to suggest a catalyst of choice. It is also possible that, when formation of intermediate imine can be brought about, cleavage will not occur during alkylation with Mannich ketones.

Much of the work on reductive alkylation of primary amines with diketones, ArCOCOR, has been directed toward formation of aminoalcohols. Usually one carbonyl group undergoes condensation with the amine to form an imine, presumably the less hindered one. Reduction may also involve an intermediate enol which by ketonization is converted to aminoketone [55a] (Eq. 17.14).

$$C_6H_5COCOC_6H_5 + RNH_2 \xrightarrow{-H_2O} \begin{bmatrix} C_6H_5CCC_6H_5 \\ \parallel \ \parallel \\ O \ \ NR \end{bmatrix} \xrightarrow{H_2} C_6H_5C-CH \atop \underset{O \ \ NHR}{\parallel \ \ |}$$

$$\xrightarrow{H_2} \begin{bmatrix} C_6H_5C=CC_6H_5 \\ | \ \ \ | \\ HO \ \ NHR \end{bmatrix}$$

(17.14)

High yield of 2-methylamino-1,2-diphenylethanone,

$$C_6H_5CH(NHCH_3)C(=O)C_6H_5,$$

resulted from reduction of a mixture of benzil and methylamine over Raney nickel [55a]. When the intermediate imine was isolated, hydrogenation of it with nickel or with platinum gave aminoalcohol; the aminoketone was obtained only when palladium catalyst was employed. The aminoketone was also obtained when a mixture of benzil and excess methylamine was reduced in the presence of palladium on carbon but the yield was only 32% [55b]. When ethylamine was substituted for methylamine, reductive alkylation with benzil gave only 6% yield of the corresponding (N-ethyl)aminoketone.

It was possible to get both carbonyl groups in 1-phenylpropane-1,2-dione to react with methylamine to obtain N^1,N^2-dimethyl-1-phenylpropane-1,2-diamine, $C_6H_5CH(NHCH_3)CH(NHCH_3)CH_3$, by hydrogenation of the ketone and excess methylamine with a mixture of equal amounts of 5% platinum and 5% palladium on carbon [55c]. Equal amounts of diamine and ephedrine,

$C_6H_5CHOHCH(NHCH_3)CH_3$, were obtained. If either catalyst was used separately or if platinum oxide was employed, ephedrine predominated in a 4:1 ratio.

Alkylation of amines with steroidal diketones is dependent on the reactivity of the carbonyl function. In the hydrogenation of pyrrolidine and 5α- or 5β-androstane-3,17-dione over palladium on carbon, only the 3-keto group underwent reaction to yield the corresponding 3α- and 3β-pyrrolidin-1-yl-5α- or 5β-androstane-17-ones [55d].

FORMATION OF TERTIARY AMINE

Dialkylation of Primary Amines

Dialkylation of primary amines with ketones to form tertiary amines is extremely difficult. An attempt to obtain N,N-diisopropylaniline from reduction of aniline and excess acetone gave low yield [56a]. When excess ketal and acetone and aniline were hydrogenated in acetic acid containing sulfuric acid at 50° and 120 atm over 5% palladium on carbon, good yield was obtained. Other dialkylated anilines were obtained from alkylations with an excess of the appropriate ketal over platinum on alumina or platinum, palladium, or ruthenium on carbon at 135–155° and 100 atm. Palladium on carbon was considered the best catalyst.

Reductive dialkylations with aldehydes are less difficult than similar reactions with ketones, but formation of tertiary amine is dependent on the reactivity of the aldehyde and its structure.

Reductive alkylations of basic amines with excess formaldehyde usually give N,N-dimethylamines in high yield. Hydrogenation of a mixture of primary amine and excess formaldehyde in alcohol over platinum oxide or platinum on carbon in the presence of an equivalent of acetic acid is a useful method [3]. The presence of acid appeared to be beneficial (mineral acid can be substituted but may cause precipitation of the salt of the amine). When 5% palladium on carbon was employed in similar reductive dimethylations it was necessary, at times, to filter and rehydrogenate with fresh catalyst [3].

Good results were obtained with a 15% weight ratio of 10% palladium on carbon in the conversions of 5-methoxy-1,2,3,4-tetrahydro- and 5-methoxy-1,1-dimethyl-1,2,3,4-tetrahydro-2-naphthylamines to the corresponding 2-dimethylamino compounds at 50° and atmospheric pressure [56b]. The steric effects of the methyl groups in 1-position were not noted.

Steric effects were probably responsible for the difficulty in the dimethylation of 2-butyl-2-ethylhexylamine, $(C_4H_9)_2C(C_2H_5)CH_2NH_2$, over prereduced platinum [56c]; formaldehyde and catalyst were added several times before reaction was complete.

Reductive alkylations of aralkylamines with excess aldehyde in the presence of sodium acetate and Raney nickel at 3 atm pressure gave dimethylated products with formaldehyde in 51–85% yields [11a], but larger aldehydes gave mixtures of mono- and dialkylated amines. As the chain length increased, only monoalkylated amines were obtained. The presence of an equivalent of acetic acid used for dimethylation [3] might aid dialkylation with some of the lower aldehydes.

Dialkylations of aromatic amines or their precursors with excess aliphatic aldehyde are often carried out under acidic conditions at 1–4 atm with platinum or nickel [16a]. 1-Benzoyl-5-nitro-7-azaindole was converted to 1-benzoyl-5-dimethylamino-7-azaindole with excess formaldehyde and Raney nickel in the presence of acetic acid [57a]. N,N-Dibutylaniline was prepared in very high yield when excess butyraldehyde was pumped into a mixture of aniline in alcohol containing a catalytic amount of acetic acid and a 10% weight ratio of 5% palladium on alumina at room temperature under 200 atm pressure of hydrogen [57b]. When the reactants were mixed at once and hydrogenated, poor yield resulted.

Alkylation of aliphatic amino acids in aqueous solution with excess formaldehyde in the presence of palladium on carbon gave high yields of dimethylated aminoacids [58a]; as much catalyst as aminoacid was used. In most instances optically active acids were alkylated without racemization. Reductive alkylation of glycine with straight-chain aldehydes gave dialkylaminoacids up to di-n-heptyl; dialkylation of alanine was more difficult and required higher reaction temperature. N,N-Dialkyl derivatives of glycine were difficult to obtain by reduction with branched chain aldehydes; dialkylation with simpler aldehydes became more difficult as the distance between the amino and carboxyl groups increased. However, 3-aminobutyric acid was converted to 3-diethylaminobutyric acid by reduction in aqueous solution with excess acetaldehyde and palladium on carbon but reaction took 24 hr [58b]. 4-Amino-3-hydroxybutyric acid was rapidly converted to 3-hydroxy-4-dimethylaminobutyric acid in 80% yield upon reduction with excess formaldehyde and 10% palladium on carbon [58c]. 6-Dimethylaminohexanoic acid was similarly prepared from the amino acid and excess formaldehyde [58d]. High catalyst ratios were used in these dialkylations.

2-Dimethylaminobenzoic acid was prepared from anthranilic acid and excess formaldehyde by hydrogenation in acetic acid with about 35% by weight of 10% palladium on carbon [58a]. 4-Dimethylaminobenzoic acid was obtained by reduction and subsequent dialkylation of its amine precursor, 4-nitrobenzoic acid, in aqueous alcohol with an active palladium catalyst prepared *in situ*. Sulfanilic acid in aqueous solution was also dialkylated with formaldehyde.

An alkylation of 4-aminobenzoic acid in alcohol containing an equivalent of acetic acid with excess formaldehyde and platinum oxide stopped after one equivalent was absorbed. When an equivalent of hydrochloric acid was added to the reaction of the aminoacid and excess formaldehyde in alcohol at 2.5 atm pressure in the presence of the same amount of catalyst used in the first reduction (2% by weight), hydrogen uptake was complete in 1 hr to give 85–88% yield of 4-dimethylaminobenzoic acid [3].

Alkylation of Secondary Amines

With Ketones. The difference in alkylating primary amines to obtain secondary amines with ketones and with aldehydes is magnified in the corresponding reactions with secondary amines. Steric effects are a greater factor in alkylation of secondary amines with ketones than they are in alkylations with aldehydes. Formation of a reducible intermediate from the reaction of secondary amine and ketone is often slow except with the most reactive ketones; reduction of ketone to carbinol is a strongly competing reaction. This is illustrated in the alkylation of dimethylamine with phenylacetone over Raney nickel at 100° and 70 atm when 26% of N,N-dimethyl-1-phenylisopropylamine was obtained along with 60–70% of 1-phenyl-2-propanol, and in the reaction of 4(4-methoxyphenyl)-2-butanone and 1-phenylpiperazine under the same conditions [3]. In the latter reaction the basic material consisted of 75% of recovered 1-phenylpiperazine and 20% of 1-(4-methoxyphenyl-2-butyl)-4-phenylpiperazine. Sulfided noble metal catalysts may be useful in these reactions because of their comparative unreactivity toward reduction of the carbonyl group. A mixture of quinoline and acetone hydrogenated at 160° and 100 atm over 4% sulfided platinum on alumina yielded 61% of 1-isopropyl-1,2,3-tetrahydroquinoline [59]. 1-Cyclohexyl- and 1-(2-butyl)-1,2,3,4-tetrahydroquinolines were similarly prepared from quinoline and cyclohexanone and from quinoline and 2-butanone, respectively. In these reactions partial or complete reduction of the heterocyclic portion of the quinoline ring was the first step followed by reductive alkylation.

Elevated temperature and pressure conditions were used in the preparation of 1-isopropyl-2-methyl-5,6-diphenylpyrrolidine from the pyrrolidine and acetone over Raney nickel [60a], in the alkylation of diethylamine with cyclopentanone over nickel to N,N-diethylcyclopentylamine [60b] and in the platinum-catalyzed alkylation of piperidine with 1-phenyl-2-pentanone to form 1-(1-phenyl-2-phentyl)piperidine [60c].

High-pressure methods are useful for alkylations with inexpensive ketones where a large excess can be used. Under such conditions there is sufficient ketone for reaction with secondary amine despite reduction of part of it to carbinol. When alkylations are to be carried out with ketones which are expensive or difficult to prepare, it is advisable to attempt isolation of the reducible intermediate and to hydrogenate it under the proper conditions.

Alkylations of secondary amines with reactive ketones as phenylacetone and cyclic ketones give fairly good yields of tertiary amine upon reduction with platinum or palladium at 2–3 atm pressure. In most instances equimolar amounts of reactants are employed. A mixture of pyrrolidine and 2-methoxyphenylacetone was hydrogenated under 3 atm pressure in the presence of platinum oxide to give 90% yield of 1-[1-(2-methoxyphenyl)-2-propyl]pyrrolidine [7d]. 3-(5-Methylaminovaleryl)indole in alcohol underwent cyclization on reduction with platinum oxide to form 3-(1-methyl-2-piperidyl)indole (Eq. 17.15). Dialkylamines, dimethyl to dibutyl, were

$$\text{indole-C(CH}_2)_4\text{NHCH}_3 \xrightarrow{\text{H}_2, \text{Pt}} \text{3-(1-methyl-2-piperidyl)indole} \quad (17.15)$$

alkylated with 5- and 8-alkoxy-2-tetralones over platinum oxide [56b]; yields were not reported except for N,N-dibutyl-5-methoxy-1,2,3,4-tetrahydro-2-naphthylamine (20%). Platinum oxide catalyzed the alkylation of some large size secondary amines, RR'NH, where R was benzyl and R' was 2-pyridyl, 2-dimethylaminoethyl, or 2-(1-pyrrolidino)ethyl, with 1-methyl and 1-ethyl-4-piperidones [61b]. The amine and ketone were mixed with cooling before reduction; yields of tertiary aminos ranged from 21–54%.

N,N-Dimethyl and N,N-diethylcyclohexylamines were prepared by reductive alkylations of dimethylamine and diethylamine with cyclohexanone in the presence of either platinum oxide or palladium on carbon [3]. Alkylation with phenol as precursor for cyclohexanone to produce N-alkylcyclohexylamine [29d] was applied to the preparation of N,N-dialkylcyclohexylamines with good results. 1-Cyclohexylpiperidine and 4-cyclohexylmorpholine were obtained from reductions of equimolar amounts of phenol and appropriate secondary amine over palladium on carbon at 105° and 4 atm in the absence of solvent [62]. Hydrogenation of a mixture of pyridine and phenol was carried out under similar conditions to also give 1-cyclohexylpiperidine.

With Aldehydes. Reductive alkylation of a secondary amine with formaldehyde to produce the corresponding tertiary amine is a relatively simple process unless steric effects are a factor. Palladium, platinum, and nickel catalysts have been employed. A useful procedure for methylation of unhindered secondary amines consists of the addition of excess formaldehyde to a cooled solution of amine followed by immediate hydrogenation with platinum oxide. This method gave 86–99% yields of tertiary amine compared to 71% when the same components were allowed to stand before reduction [3]. Some di(2,3- and 3,4-dialkoxyphenethyl)amines were converted to tertiary amines in 90% yields by reduction with a 3 mole excess of formaldehyde in the presence of platinum oxide [63a].

Another worthwhile method is alkylation with formaldehyde in the presence of acetic acid and palladium on carbon. It gave good yields of ethyl 1-methylpiperidine-2-, 3- and 4-carboxylates [63b]. The reactions were carried out at 80 atm presumably because of the amount of material to be reduced. The same reductions at 2–3 atm pressure gave comparable yields [3]. Palladium-catalyzed methylation of 2-benzhydryl-3-hydroxypiperidine was carried out at 3 atm in the presence of acetic acid [63c]. Reaction of 4-ethoxy-4-phenylpiperidine as hydrochloride salt with formaldehyde over palladium gave the N-methyl compound [63d].

Many methylations with formaldehyde and palladium have been run in neutral solvent with good results. 2-Hydroxymethylpiperidine was converted to 2-hydroxymethyl-1-methylpiperidine by reduction with formaldehyde and palladium on carbon in alcohol containing sodium acetate [63e]. Methylation of 2-(2-hydroxyethyl)piperidine under the same conditions was a very slow reaction and was eventually run at 80° and 80 atm to give 84% of 2(2-hydroxyethyl)-1-methylpiperidine in 1–2 hr [3]. A more hindered compound, dimethyl 4-methoxypiperidine-2,6-dicarboxylate was methylated at 60° and 100 atm to give 97% yield of tertiary amino compound [63f]. It was surprising that 2(2-hydroxy-2,2-diphenylethyl)piperidine could be methylated so readily, 1 hr at 2–3 atm with palladium on carbon [3]. Success may have been due to standing with a large excess of formaldehyde for 12 hr which allowed opportunity for formation of addition product or enamine.

Alkylation of secondary amines with formaldehyde and Raney nickel at low pressure also results in high yields of N-methylamines, CH_3NRR'. 1(2-Hydroxyethyl)-4-methylpiperazine was prepared from 1-(2-hydroxyethyl)piperazine in 87.5% yield by this method [64a]. In other instances a mixture of formaldehyde and secondary amine was refluxed [64b] or was allowed to stand for varying periods before hydrogenation with Raney nickel [64c–f]. In the alkylation of 2-benzylimidazoline, paraformaldehyde was used as alkylating agent and a 60% weight of nickel was employed; fairly good yield of 1-methyl-2-benzylimazoline was obtained but uptake of hydrogen was slow [64g].

Among higher aliphatic aldehydes, butyraldehyde was used to obtain 4-(N-butyl-N-methylamino)-3,5-xylenol from 4-methylamino-3,5-xylenol by reduction with palladium on carbon in the presence of sodium acetate at

3 atm [65a]. The corresponding N-isobutyl and N-amyl derivatives were also obtained. It is likely that low-pressure methods should be applicable unless aldehydes are somewhat hindered. In the alkylations of dimethylamine and diethylamine with 2,4,4-trimethylpentanal, $(CH_3)_3CCH_2CH(CH_3)CHO$, reductions with Raney nickel at 100° and 50 atm gave 27% and 16% yields respectively of dimethylamino and diethylamino compounds [65b]. Elevated temperature and pressure were also used for the conversions of quinoline and butyraldehyde and quinoline and isobutyraldehyde over sulfided platinum on alumina to 1-butyl and 1-isobutyl-1,2,3,4-tetrahydroquinolines [59], but such conditions are necessary for sulfided noble metal catalysts.

Reductive alkylations with aralkylaldehydes are effected very readily. Diphenylacetaldehyde and dimethylamine were mixed in alcohol at 0° and then warmed slightly. The mixture was hydrogenated under 2-3 atm pressure with palladium or platinum catalyst to yield N,N-dimethyl-2,2-diphenylethylamine [66]; N,N-diethyl-2,2-diphenylethylamine and 4-(2,2-diphenylethyl)morpholine were prepared from diphenylacetaldehyde and diethylamine and morpholine respectively. Palladium on carbon catalyzed the reduction of phenylacetaldehyde and 2,2-dimethyl-3-ketopiperazine in alcohol to yield 1-(2-phenethyl)-2,2-dimethyl-3-ketopiperazine and also catalyzed the reductive alkylation of 1-(2-hydroxy-2,2-diphenylethyl)piperazine with the same aldehyde [3]. In the latter reaction the temperature was held at 50-60°.

Secondary amines can be readily alkylated with aromatic aldehydes to yield tertiary amines. Raney nickel alkylations of piperazines have been carried out at 150° and 100 atm pressure after benzaldehyde and piperazine were refluxed in butyl alcohol [67]. A method which gave 90-95% yield of 4-benzyl-1-substituted piperazines consisted of heating the substituted piperazine and excess redistilled benzaldehyde without solvent under reduced pressure for several hours followed by hydrogenation in alcohol over Raney nickel at 100° and 70 atm pressure [3]. Elevated temperature favors formation of reducible intermediate and Raney nickel is very useful under these conditions because it does not cause hydrogenolysis of the resulting product when an aromatic aldehyde is used for alkylation.

High yield (75%) was also obtained from palladium-catalyzed reductions of benzaldehyde and substituted benzaldehydes and 1-substituted piperazines with palladium on carbon at 3 atm and 60° [3]. Obviously there is always a possibility of debenzylation in these reductions with palladium but it can be controlled by limiting the amount of catalyst. Overhydrogenation would occur in the alkylation of dimethylamine with 2-hydroxybenzaldehyde over 10% palladium on carbon despite a low catalyst ratio if the reaction was not watched closely [68a]. Good yield, 86%, of 2-dimethylaminomethylphenol was obtained when the reaction was interrupted at the proper point.

380 *Reductive Alkylation*

Hydrogenolysis from the use of palladium catalyst may have been responsible for the modest yield, 53.5% of 4-dimethylaminomethylphenol from 4-hydroxybenzaldehyde and dimethylamine [68b]. The possibility of hydrogenolysis may have been the reason that the Lindlar catalyst, a lead-poisoned palladium catalyst, was employed for the reductive alkylation of dimethylamine with methyl 2-furfaldehyde-4-carboxylate [68c]. It may also have been used to prevent ring reduction.

Platinum oxide was used for the low-pressure reduction of a mixture of 4-dimethylaminobenzaldehyde and pyrrolidine [69a]. When it was employed in the alkylation of 2-ethylaminoethanol with 9-anthraldehyde the intermediate, 2-anthranyl-3-ethyloxazolidine, was first isolated [69b]. When the alkylation of 1,2,3,4-tetrahydroisoquinoline with indole-3-aldehyde was carried out, the materials were heated in benzene. After distillation of solvent, which azeotropically removed water, the resultant enamine was hydrogenated over platinum oxide under 2–3 atm pressure [3].

SELECTIVE HYDROGENATION

Acetylenic bonds and nitro groups will be reduced before reductive alkylation takes place. In general aromatic ring systems will not be affected during reductive alkylation. As noted in the hydrogenation of imines, selectivity in the presence of certain reducible groups is a point to be resolved.

In the Presence of Olefinic Bonds

As noted in the related hydrogenation of imines in the presence of an olefinic double bond (Chapter XV), whether a carbon to carbon double bond will be affected during reductive alkylations will also depend on the type of double bond and on whether it is di-, tri-, or tetrasubstituted. In general cyclohexenyl double bonds will remain intact. In reductions where the system 2,6,6-trimethylcyclohexen-1-yl was R, 4-(2-diethylaminoethoxy)-benzaldehyde and the amine, $RCH_2CH_2(CH_3)NH_2$, were hydrogenated with nickel at 60° and 100 atm [70a], the double bond remained intact. In alkylations of methylamine [70b] and N,N-diethylenediamine [70c] with a cyclohexenyl aldehyde also containing a trisubstituted double bond in the side chain, $RCH_2CH=C(CH_3)CHO$, the side chain bond was saturated during alkylation but the tetrasubstituted cyclohexenyl bond was untouched. In alkylations of amines with ionones containing a related carbon skeleton that was aliphatic and not cyclic, the trisubstituted double bond remained intact on hydrogenation with platinum oxide [70d] (Eq. 17.16).

$$\begin{array}{c}\mathrm{CH_3}\mathrm{CH_3}\\\diagdown\diagup\\\mathrm{C}\mathrm{O}\\\mathrm{HC}\diagup\|\\\phantom{\mathrm{HC}}(\mathrm{CH_2})_3\mathrm{CCH_3} + \mathrm{R'NH_2}\\\mathrm{H_2C}\mathrm{CHCH_3}\\\diagdown\diagup\\\mathrm{C}\\\mathrm{H_2}\end{array} \xrightarrow[\substack{\mathrm{PtO_2}\\ 2\text{–}3\,\mathrm{atm}}]{\mathrm{H_2}} \begin{array}{c}\mathrm{H_3C}\mathrm{CH_3}\\\diagdown\diagup\\\mathrm{C}\\\mathrm{HC}\diagup\\\phantom{\mathrm{HC}}(\mathrm{CH_2})_3\mathrm{CH(CH_3)NHR'}\\\mathrm{H_2C}\mathrm{CHCH_3}\\\diagdown\diagup\\\mathrm{C}\\\mathrm{H_2}\end{array} \quad (17.16)$$

R′ = CH$_3$, C$_2$H$_5$, CH(CH$_3$)$_2$, (CH$_2$)$_2$CH(CH$_3$)$_2$, (CH$_2$)$_3$N(CH$_3$)$_2$, CH(CH$_3$)CH$_2$C$_6$H$_5$

The trisubstituted double bond in 3,4-dihydro-2H-pyran-2-carboxaldehyde also remained intact after reaction with primary and secondary amines under vigorous hydrogenation conditions with cobalt catalyst [70e] (Eq. 17.17).

$$\text{(CH}_3\text{)H}\underset{}{\overset{}{\bigcirc}}\underset{\text{CHO}}{\overset{\text{H(CH}_3\text{)}}{\diagup}} + \text{R'NH}_2 \text{ or } (\text{R'})_2\text{NH} \xrightarrow[\substack{\text{Co, 130–135°,}\\ \text{150–230 atm}}]{\text{H}_2}$$

$$\underset{}{\overset{\text{O}}{\bigcirc}}\text{CH}_2\text{N}\diagdown \qquad (17.17)$$

In general alkylation of an amino compound also containing a carbonyl function should take place without reducing that group to carbinol. When the amino group in a tetracycline was methylated with formaldehyde in the presence of palladium on carbon, neither the double bond present in the molecule nor the carbonyl groups were reduced [16b]. This was not entirely unexpected in the case of the double bonds since each was tetrasubstituted. 4-Aminoacetophenone was converted to 4-dimethylaminoacetophenone by reductive alkylation with formaldehyde and prereduced platinum under 3 atm pressure [71a]. The carbonyl group in 4-nitrophenylacetone remained intact when the compound as amine precursor was reacted with formaldehyde and Raney nickel at 14 atm pressure of hydrogen [71b]. 10-Methylamino-*cis*-1,2,3,4a,10a-hexahydro-9-(10H)phenanthrone was obtained in 61% yield from methylation of the corresponding 10-aminoketone with formaldehyde and platinum oxide [71c].

In the Presence of Hydrogenolyzable Groups

The C=N bond was shown to be reduced selectively in the presence of hydrogenolyzable groups (Chapter XV). Selective reductive alkylation of primary amines with aldehydes or ketones should also be expected in the presence of hydrogenolyzable groups, although not always attained.

Hydrazines. Indole-3-acetaldehyde and hydrazine or 1-butylhydrazine and indole-3-acetone and hydrazine were hydrogenated in methyl alcohol containing acetic acid in the presence of platinum oxide to give the corresponding substituted hydrazines [21a]. 1-(2-Methoxyphenylisopropyl)-hydrazine and acetone were reduced with platinum oxide to give the corresponding 1,2-substituted hydrazine [72a]. Yields in alkylations of 4-amino 2,3-disubstituted morpholines with various carbonyl compounds at 4 atm pressure over platinum oxide may indicate that some cleavage of the nitrogen to nitrogen bond did take place [72b]. The nitrogen to nitrogen bond in *N*-aminoheterocycles seems to be less stable than the same bond in a linear 1,2-hydrazine.

Hydroxylamines. Reductive alkylation of *O*-methylhydroxylamine with cyclohexylacetone over platinum oxide at 25° and 4 atm of hydrogen pressure gave only 40% yield of *N*-methoxy cyclohexylisopropylamine, $C_6H_{11}CH_2CH(CH_3)NHOCH_3$ [73]. A maximum yield of 35% was obtained on attempted alkylation of hydroxylamine with cyclohexanone at 55–60° and 2 atm in the presence of platinum oxide [3].

O- and N-Benzyl and Related Groups. Palladium catalysts should not be employed in reductive alkylations in the presence of *O*-benzyl or related hydrogenolyzable groups. It may be possible to carry out palladium-catalyzed alkylations in the presence of certain *N*-benzyl protecting groups. However, the use of Raney nickel or platinum oxide should allow alkylation to take place without accompanying hydrogenolysis of either *O*- or *N*-benzyl type groupings.

3-(2-Aminoethyl)-5-benzyloxyindazole was treated with acetone and hydrogenated in alcohol with platinum oxide; no explanation was given for the low yield (32%) of 3-(2-isopropylaminoethyl)-5-benzyloxyindazole [54a]. 1-Carbobenzyloxypiperazine and propionaldehyde were reduced over platinum oxide to 1-carbobenzyloxy-4-propylpiperazine but yield was not reported [74b].

Excellent yields are reported in reductive alkylations when an *N*-benzyl group was present. Raney nickel was the catalyst for the methylations of *N*-(3,4-dimethoxybenzyl)-3,4-dimethoxyphenethylamine with formaldehyde which gave an 80% yield of *N*-(3,4-dimethoxybenzyl)-*N*-methyl-3,4-dimethoxyphenethylamine [64b]. It was also employed in methylation of *N*-benzyl-2-cyclohexyloxycyclopropylamine [75]; almost quantitative yield of *N*-benzyl-*N*-methyl-2-cyclohexyloxycyclopropylamine was obtained. Platinum oxide or platinum on carbon was used in the alkylation of 2-benzylaminoethanol with propionaldehyde to give 80–88% yields of 2-(*N*-benzyl-*N*-propylamino)ethanol [3]. A series of *N*-(substituted benzyl)-alkanediamines, $ArCH_2NH(CH_2)_nNHCH_2Ar$, was subjected to reductive alkylation with formaldehyde and platinum oxide to give the corresponding N-benzyl N-methyl compounds [3]. There was no evidence of debenzylation. A benzylaminoketone was reacted with methylamine and platinum oxide to give 90% yield of diamine [3] (Eq. 17.18).

$$\text{C}_6\text{H}_5\text{CH}_2(\triangle)\text{NCH}_2\text{COCH}_3 + \text{CH}_3\text{NH}_2 \xrightarrow{\text{H}_2} \text{C}_6\text{H}_5\text{CH}_2(\triangle)\text{NCH}_2\text{CH}(\text{CH}_3)\text{NHCH}_3 \quad (17.18)$$

Halogens. Loss of halogen during reductive alkylations depends on the halogen and the catalyst. Palladium catalyst has been used in reductive alkylations in the presence of aromatic chlorine [76a, b] but it is more advisable to use other catalysts.

Platinum as oxide or platinum on carbon generally gives good results without causing loss of aromatic chlorine if a moderate amount of catalyst is employed.

However, loss of chlorine can occur particularly when the product of alkylation is a fairly strong base. 2-(N-Benzylamino)-1-(4-chlorophenyl)-propane was prepared from alkylation of benzylamine with 1-(4-chlorophenyl)-2-propanone with an amount of 5% platinum on carbon equal to about 5% of the total weight of the two components [77a]. When the reaction was not interrupted at the proper point, extensive dehalogenation took place. This was controlled by using a 3% ratio of catalyst which extended the reaction time from 30 min with a 5% ratio to 3 hr [3]. Platinum oxide 0.5% of the combined weight of reactants also catalyzed the same alkylation without causing loss of chlorine. Procedures employing small amounts of platinum as oxide or on a support gave good yields of racemic N-methyl-chlorophenyl and N-methyl-dichlorophenylisopropyl-amines from the corresponding 1-(chlorophenyl)-2-propanones and methylamine [3]. The reductive alkylation of methylamine with 1-(4-chlorophenyl)-1-phenyl-2-propanone over platinum on carbon at 2–3 atm pressure gave only 33% yield of secondary amine [3]. Dehalogenation did not take place but formation of 1-(4-chlorophenyl)-1-phenyl-2-propanol, 50%, from competing hydrogenation of the ketone and mechanical loss from distillation account for a material balance of almost 100%. Low yield 14%, was obtained from alkylation of methylamine with 1-(2-chlorophenoxy)-2-propanone with platinum oxide [3]. In this instance the poor yield came from hydrolysis or hydrogenolysis of the ketone to 2-chlorophenol. The yield of N-methyl-1-(2-chlorophenoxyl)isopropylamine was raised to 36.5% by reduction of methylamine and ketone over Raney nickel. There was no evidence of chloride ion but 2-chlorophenol was recovered.

A related reduction with 1-(2,4-dichlorophenoxy)-2-propanone and Raney nickel gave 47% yield of methylated amine [77b]. Raney nickel should prove useful in reductive alkylations of chlorine-containing compounds despite the mediocre yields seen with chlorophenoxyacetones. These compounds cannot give a good assessment of the value of nickel in reductive alkylations of chlorine-containing compounds because they are subject to hydrolysis or hydrogenolysis at the ArO:C bond.

Low-pressure methylations of chlorobenzylalkanediamines,

$$(Cl-C_6H_5CH_2NH)_2(CH_2)_n,$$

were carried out with excess formaldehyde and 15% by weight of 5% rhodium on carbon at 2–3 atm pressure to give N-methylamines, $[Cl-C_6H_4CH_2N(CH_3)]_2(CH_2)_n$, in 68–75% yields [3]. There was no evidence of loss of chlorine but uptake of hydrogen was slow, 8–24 hr.

N,N-Diisopropyl-2-chloroaniline was obtained by reductive alkylation of 2-chloroaniline with excess acetone dimethyl ketal over 5% ruthenium on carbon at 50° and 120 atm [56a]; loss of chlorine was negligible.

Reductive alkylation of chloroanilines, through chloronitrobenzenes as precursors, with aldehydes or ketones and noble metal sulfides may be an excellent means of carrying out reactions without losing chlorine [78]. Palladium sulfide could not be employed in alkylation of bromine-containing amines but it is likely that 5% platinum sulfide on carbon could catalyze the reaction without causing loss of bromine. It was used in the hydrogenation of 4-bromonitrobenzene at 100–130° and 42 atm initial pressure to give 99.5% of 4-bromoaniline [78].

Since loss of bromine was minimized or prevented in the reductive of bromine-containing imines by the use of low ratios of 5% platinum on carbon and the presence of acid (Chapter XV) it is possible the same procedure could be applied to reductive alkylations without debromination.

In an alkylation in the presence of aliphatic halide platinum oxide catalyzed the dimethylation of N,N-di(2-chloroethyl)-1,4-phenylenediamine with formaldehyde [79a]. The Schiff base was isolated and hydrogenated with Raney nickel when methylamine was reacted with 4-di(2-chloroethyl)-aminobenzaldehyde. Palladium on carbon catalyzed the methylations of 1-amino-3-bromoadamantane with an equivalent amount of formaldehyde and with excess formaldehyde to give 80% yield of 3-bromo-1-methylaminoadamantane and 88% yield of 3-bromo-1-dimethylaminoadamantane.

REFERENCES

[1] a. G. Mignonac, *Compt. Rend.*, **172**, 223 (1921); b. O. Kovacs and G. Fodor, *Ber.*, **84**, 795 (1951); c. E. Boehringer, J. Liebrecht, W. Mayer-List, W. Boehringer, and H. A. Boehringer, Brit. Patent 920,623 (1963); d. P. Pratesi, E. Grava, L. Lilla, A. La Manna, and L. Villa, *Farmaco (Pavia)*, *Ed. Sci.*, **18**, 932 (1963); e. C. C. Price, N. J. Leonard, D. Y. Curtin, and R. H. Reitsema, *J. Org. Chem.*, **12**, 497 (1947); f. W. B. Wright, Jr., *ibid.*, **24**, 1016 (1959).
[2] a. C. Schopf and W. Salzer, *Ann.*, **544**, 1 (1940); b. D. Heyl, E. Luz, S. A. Harris, and K. Folkers, *J. Am. Chem. Soc.*, **74**, 414 (1952).
[3] M. Freifelder, unpublished results.
[4] W. S. Emerson, *Organic Reactions*, Vol. IV, Wiley, New York, 1949, Chapter III; b. E. H. Woodruff, J. P. Lambooy, and W. E. Burt, *J. Am. Chem. Soc.*, **62**, 922 (1940).
[5] a. A. C. Cope and E. M. Hancock, *J. Am. Chem. Soc.*, **64**, 1503 (1942); b. *ibid.*, **66**, 1453 (1944); c. E. M. Hancock, E. M. Hardy, D. Heyl, M. E. Wright, and A. C. Cope, *ibid.*, **66**, 1747 (1944); d. E. L. Engelhardt, F. S. Crossley, and J. M. Sprague, *ibid.*, **72**, 2718 (1950); e. R. V. Heinzelman, H. G. Kolloff, and J. H. Hunter, *ibid.*, **70**, 1386 (1948); f. R. I. Meltzer and A. D. Lewis, *J. Org. Chem.*, **22**, 612 (1957).
[6] M. Freifelder and B. W. Horrom, *J. Pharm. Sci.*, **52**, 1191 (1963).
[7] a. R. A. Henry and W. G. Finnegan, *J. Am. Chem. Soc.*, **76**, 926 (1954); b. A. Skita and W. Stühmer, Ger. Pat. 932,677 (1955); through *Chem. Abstr.*, **52**, 20200 (1958); c. P. L. deBenneville, U.S. Patent 2,578,787 (1951); d. R. V. Heinzelman and B. D. Aspergren, *J. Am. Chem. Soc.*, **75**, 3409 (1953).
[8] M. Freifelder and Y. H. Ng, *J. Med. Chem.*, **8**, 122 (1965).
[9] a. W. S. Emerson and P. M. Walters, *J. Am. Chem. Soc.*, **60**, 2023 (1938); b. W. S. Emerson, U.S. Patent 2,298,284, (1940); c. L. C. Cheney, U.S. Patent 2,767,161 (1956).
[10] a. F. S. Dovell and H. Greenfield, *J. Org. Chem.*, **29**, 1265 (1964); b. F. S. Dovell and H. Greenfield, *J. Am. Chem. Soc.*, **87**, 2767 (1965); c. H. Greenfield and F. S. Dovell, *J. Org. Chem.*, **31**, 3053 (1966).
[11] a. E. H. Woodruff, J. P. Lambooy, and W. E. Burr, *J. Am. Chem. Soc.*, **62**, 922 (1940); b. Von F. Fischer and H. Riese, *J. Prakt. Chem.*, **294**, 177 (1961).
[12] a. H. A. Shonle and J. W. Corse, U.S. Patent 2,424,061 (1947); b. H. R. Arnold, U.S. Patent 2,205,552 (1940); c. W. W. Prichard, U.S. Patent 2,456,315 (1948).
[13] a. Brit. Patent 600,426 (1948); through *Chem. Abstr.*, **42**, 7793 (1948); b. G. S. Myers, U.S. Patent 2,571,053 (1951); c. O. Hora, G. S. Scholz, and J. Franke, Ger. Pat. 1,138,793 (1962); through *Chem. Abstr.*, **58**, 6749 (1963).
[14] a. J. F. Olin and E. J. Schwoegler, U.S. Patent 2,373,705 (1945); b. C. F. Winans and H. Adkins, *J. Am. Chem. Soc.*, **54**, 306 (1932).
[15] A. C. Cope, N. A. Lebel, H-H. Lee, and W. R. Moore, *J. Am. Chem. Soc.*, **79**, 4720 (1957).
[16] a. W. S. Emerson, U.S. Patent 2,380,420 (1945); b. J. H. Boothe and J. Petisi, U.S. Patent 3,148,212 (1964); c. P. Rayet, M. Prost, and M. Urbain, *Helv. Chim. Acta*, **39**, 87 (1956); d. J. K. Tandan, S. K. Chatterjee, and N. Anand, *J. Sci. Ind. Research* (India), **15B**, 419 (1956); through *Chem. Abstr.*, **51**, 4300 (1957).
[17] a. W. Grimme, H. Emde, H. Schmitz, and J. Wallner, U.S. Patent 2,766,278 (1956); b. W. A. Skinner, H. F. Gram, and B. R. Baker, *J. Org. Chem.*, **25**, 777 (1960).

[18] a. M. Augustin and H. Dehne, *J. Prakt. Chem.*, **285**, 119 (1965); b. J. M. Tien and I. M. Hunsberger, *Chem. Ind.*, **1955**, 119; c. W. Luttke and O. Hundiecker, *Ber.*, **99**, 2146 (1966).

[19] C. G. Overberger, I. C. Kogon, and W. J. Einstman, *J. Am. Chem. Soc.*, **76**, 1953 (1954).

[20] a. K. Harsányi, D. Korbonits, and P. Kiss, *J. Med. Chem.*, **7**, 623 (1964); b. K. Okumura and K. Inoue, *Chem. Pharm. Bull.*, **12**, 718 (1964).

[21] a. R. A. Robinson, U.S. Patent 2,947,758 (1960); b. B. R. Baker and J. Novotny, *J. Heterocyclic Chem.*, **4**, 23 (1967).

[22] a. C. R. Noller and P. D. Kneeland, *J. Am. Chem. Soc.*, **68**, 201 (1946); b. T. Kralt, W. J. Asma, and H. D. Moed, *Rec. Trav. Chim.*, **80**, 330 (1961); c. J. R. Corrigan, M-J Sullivan, H. W. Bishop, and A. W. Ruddy, *J. Am. Chem. Soc.*, **75**, 6258 (1953); d. I. M. Kaye, *ibid.*, **73**, 5003 (1951); e. Z. Horii, T. Inoi, S. Kim. Y. Tamura, A. Suzuki, and H. Matsumoto, *Chem. Pharm. Bull.*, **13**, 1151 (1965); f. E. L. May, *J. Org. Chem.*, **11**, 353 (1946).

[23] a. T. Nogradi, *Monatsh.*, **88**, 1087 (1957); b. W. Beck, I. A. Kaye, I. C. Kogon, H. C. Klein, and W. J. Burlant, *J. Org. Chem.*, **16**, 1434 (1951); c. A. R. Surrey, *J. Am. Chem. Soc.*, **76**, 2214 (1954); d. A. R. Surrey and M. K. Rukwid, *ibid.*, **77**, 3798 (1955); e. T. J. Work, *J. Chem. Soc.*, **1942**, 426.

[24] a. C. A. Stone and co-workers, *J. Med. Chem.*, **5**, 665 (1962); b. T. S. Gardner, E. Wenis, and J. Lee, *ibid.*, **3**, 461 (1961).

[25] a. D. M. Balcom and C. R. Noller, *Organic Synthesis*, Vol. 30, Wiley, New York, 1950, p. 59; b. Brit. Patent 735,047 (1955); c. J. M. Bobbitt, J. M. Kiely, K. L. Khanna, and R. Ebermann, *J. Org. Chem.*, **30**, 2247 (1965).

[26] a. A. A. Ponomarev, I. M. Skvortsov, and N. P. Maslennika, *J. Gen. Chem. USSR*, **33**, 1113 (1963) (Eng. trans.); b. M. Spillman and K. Hoffman, *Helv. Chim. Acta*, **37**, 1699 (1954); c. M. P. Mertes, R. F. Borne, and L. E. Hare, *J. Org. Chem.*, **33**, 133 (1968).

[27] a. H. N. Wingfield, Jr., *J. Org. Chem.*, **25**, 1671 (1960); b. D. Heyl, E. Luz, S. A. Harris, and K. Folkers, *J. Am. Chem. Soc.*, **70**, 3669 (1948).

[28] F. Zymalkowski, *Arch. Pharm.*, **292**, 682 (1959).

[29] a. R. C. Elderfield and H. A. Hageman, *J. Org. Chem.*, **14**, 605 (1949); b. L. P. Friz and D. Della Bella, *Bull. Chim. Farm.*, **101**, 527 (1962); through *Chem. Abstr.*, **59**, 1465 (1963); c. T. A. Geissman, M. J. Schlatter, and I. D. Wenn, *J. Org. Chem.*, **11**, 736 (1946); d. F. Van Munster, U.S. Patent 3,355,940 (1967); e. R. M. Robinson, Brit. Patent 956,116 (1964); f. L. J. Dankert and D. A. Permoda, U.S. Patent 2,571,016 (1951).

[30] a. W. Klavehn, Ger. Patent 945,391 (1956); b. Belg. Patent 631,005 (1963); through *Chem. Abstr.*, **61**, 611 (1964).

[31] a. J. B. Tindall, U.S. Patent 2,640,855 (1953); b. W. J. Bailey and C. N. Bird, *J. Org. Chem.*, **23**, 996 (1958).

[32] a. J. R. Corrigan, M-J Sullivan, H. W. Bishop, and A. W. Ruddy, *J. Am. Chem. Soc.*, **75**, 6258 (1953); b. J. C. Sheehan and C. A. Robinson, *ibid.*, **71**, 1436 (1949); c. C. Ainsworth and R. G. Jones, *ibid.*, **75**, 4915 (1953); d. C. Ainsworth, *ibid.*, **79**, 5242 (1957); e. R. V. Heinzelman, U.S. Patent 2,625,566 (1953).

[33] Brit. Patent 971,752 (1964).

[34] a. D. E. Pearson, W. H. Jones, and A. C. Cope, *J. Am. Chem. Soc.*, **68**, 1225 (1946); b. J. H. Short, W. L. Chan, M. Freifelder, D. G. Mikolasek, J. L. Schmidt, H. G. Schoepke, C. Shannon, and G. R. Stone, *J. Med. Chem.*, **6**, 596 (1963); c. J. A. Harpham, R. J. J. Simkins, and G. F. Wright, *J. Am. Chem. Soc.*, **72**, 341 (1950); d. R. G. Shepherd and R. G. Wilkinson, *J. Med. Chem.*, **5**, 823 (1962); e. J. P.

Thomas, R. G. Wilkinson, G. S. Redin, and R. G. Shepherd, U.S. Patent 3,192,113 (1965).
[35] a. A. Burger, R. T. Standridge, and E. J. Ariens, *J. Med. Chem.*, **6,** 221 (1963); b. J. Drúey, P. Schmidt, K. Eichenberger, and M. Wilhelm, Ger. Patent 1,082,258 (1960); through *Chem. Abstr.*, **55,** 23384 (1961).
[36] a. H. H. Fox and J. T. Gibas, *J. Org. Chem.*, **18,** 994 (1953); b. H. L. Yale, K. Losee, J. Martins, M. Holsing, F. M. Perry, and J. Bernstein, *J. Am. Chem. Soc.*, **75,** 1933 (1953); c. W. Wenner, *J. Org. Chem.*, **18,** 1333 (1953); d. A. Skita, W. Stühmer, and W. Heinrich, Ger. Patent 874,916 (1953); through *Chem. Abstr.*, **52,** 12918 (1958); e. B. J. McLoughlin, Brit. Patent 975,291 (1964); through *Chem. Abstr.*, **62,** 2737 (1965).
[37] a. R. T. Major, U.S. Patent 1,978,433 (1934); b. R. H. Rosenwald and J. R. Hoatson, Brit. Patent 753,740 (1956); through *Chem. Abstr.*, **51,** 5827 (1957); c. Brit. Patent 860,923 (1961); through *Chem. Abstr.*, **55,** 16480 (1961); d. A. Gaydasch and J. T. Arrigo, U.S. Patent 3,234,281 (1966); e. A. C. Ruggles, U.S. Patent 2,494,059 (1950).
[38] a. Nickel G-49A, Chemetron Corp., Louisville, Kentucky, U.S.A.; b. W. Budd. Fr. Patent 1,463,529 (1966); through *Chem. Abstr.*, **67,** 90517 (1967); c. E. J. Bicek, U.S. Patent 3,209,030 (1965).
[39] Personal communication from Dr. H. Greenfield, Uniroyal, Naugatuck, Conn., U.S.A.
[40] a. I. A. Kaye, I. C. Kogon, and C. L. Parris, *J. Am. Chem. Soc.*, **74,** 403 (1952); b. H. G. O. Becker and E. Fanghänel, *J. Prakt. Chem.*, **298,** 58 (1964); c. Brit. Patent 687,892 (1953); through *Chem. Abstr.*, **48,** 4594 (1954); d. G. L. Collier, A. H. Jackson, and G. W. Kenner, *J. Chem. Soc. (C)*, **1967,** 66; e. A. R. Brown and F. C. Copp, *ibid.*, **1954,** 879; f. F. S. Dovell and H. Greenfield, Neth. Appl. 6,402,424 (1964); through *Chem. Abstr.*, **62,** 11733 (1965).
[41] a. R. A. Robinson, U.S. Patent 2,908,691 (1965); b. W. Wenner, U.S. Patent 2,382,686 (1945).
[42] a. M. Freifelder, *J. Med. Chem.*, **6,** 813 (1963); b. M. Freifelder and G. R. Stone, *J. Am. Chem. Soc.*, **80,** 5270 (1958).
[43] a. R. V. Heinzelman, U.S. Patent 2,930,731 (1960); b. Belg. Patent 659,425 (1965) through *Chem. Abstr.*, **64,** 2003 (1965); c. J. F. Kerwin, T. F. Herdegen, R. Y. Heisler, and G. E. Ullyott, *J. Am. Chem. Soc.*, **72,** 3983 (1950); d. *ibid.*, p. 940; e. I. Scriabine, U.S. Patent 2,509,309 (1950); f. K. N. Campbell, J. R. Corrigan, and B. K. Campbell, *J. Org. Chem.*, **16,** 1712 (1951).
[44] a. A. Furst and A. Boller, U.S. Patent 3,133,077 (1964); b. B. Heath-Brown and P. G. Philpott, *J. Chem. Soc.*, **1965,** 7165; c. A. Kalir and S. Szara, *J. Med. Chem.*, **9,** 341 (1966); d. Brit. Patent 861,428 (1961); through *Chem. Abstr.*, **58,** 12516 (1963).
[45] a. M. Polonovski, M. Pesson, and J. Bededeu, *Compt. Rend.*, **233,** 1120 (1951); b. J. F. Kerwin, G. C. Hall, F. J. Milnes, I. H. Witt, R. A. McLean, E. Macko, E. J. Fellows, and G. E. Ullyot, *J. Am. Chem. Soc.*, **73,** 4162 (1951); c. H. Suter and H. Zutter, *Ann.*, **576,** 215 (1952); d. J. F. Kerwin and G. E. Ullyot, U.S. Patent 2,683,719 (1954); e. J. Mills and R. W. Kattau, U.S. Patent 3,235,597 (1966).
[46] a. L. E. Hellerman, C. C. Porter, H. J. Lowe, and F. K. Koster, *J. Am. Chem. Soc.*, **68,** 1890 (1946); b. R. G. Hiskey and R. C. Northrop, *ibid.*, **83,** 4798 (1961); c. K. Harada, *J. Org. Chem.*, **32,** 1790 (1967).
[47] a. W. B. Reid, U.S. Patent 2,719,851 (1955); b. V. Boekelheide and J. C. Godfrey, *J. Am. Chem. Soc.*, **75,** 3679 (1953).

[48] a. A. Bertho and G. Rodl, *Ber.*, **92**, 2218 (1959); b. E. R. Shepard and D. E. Morrison, U.S. Patent 2,956,058 (1960); c. H. McKennis, Jr., L. B. Turnbull, E. R. Bowman, and E. Tamaki, *J. Org. Chem.*, **28**, 383 (1963).

[49] I. N. Nazarov and E. T. Golovin, *J. Gen. Chem.*, *USSR*, **26**, 1679 (1956) (Eng. trans.).

[50] a. Brit. Patent 547,302 (1942); b. Brit. Patent 547,301 (1942); c. H. J. Barber, D. H. O. John, and W. R. Wragg, *J. Am. Chem. Soc.*, **70**, 2282 (1948); d. R. C. Elderfield, F. J. Kreysa, J. H. Dunn, and D. D. Humphreys, *ibid.*, **70**, 40 (1948).

[51] a. S. Archer, T. R. Lewis, and M. J. Unser, *J. Am. Chem. Soc.*, **79**, 4194 (1957); b. S. Archer, T. R. Lewis, M. J. Unser, J. O. Hoppe, and H. Lape, *ibid.*, p. 5783; c. G. B. Payne and K. Pfister, U.S. Patent 2,678,317 (1954).

[52] a. R. H. Reitsema and J. H. Hunter, *J. Am. Chem. Soc.*, **71**, 1680 (1949); b. J. H. Biel, W. K. Hoya, and H. A. Leiser, *ibid.*, **81**, 2527 (1959).

[53] a. R. H. Reitsema, U.S. Patent 2,476,912 (1949); b. 2,496,955; c. 2,496,956; d. 2,496,957 (1950).

[54] E. T. Golovin, V. M. Bystrov, and V. I. Chumenko, *J. Gen. Chem. USSR*, **34**, 1283 (1964) (Eng. trans.).

[55] a. W. B. Wheatley, W. E. Fitzgibbon, and L. C. Cheney, *J. Org. Chem.*, **18**, 1564 (1953); b. R. Hinderling, B. Prijs, and H. Erlenmeyer, *Helv. Chim. Acta*, **38**, 1415 (1955); c. S. Hyden, S. Emmanuele, H. Wetstein, and G. Wilbert, *J. Med. Chem.*, **10**, 953 (1967); d. M. Davis, E. W. Parnell, and J. Rosenbaum, *J. Chem. Soc.*, **1966**, 1983.

[56] a. A. Gaydasch and J. T. Arrigo, U.S. Patent 3,234,281 (1966); b. D. E. Ames, D. Evans, T. F. Grey, P. J. Islip, and K. E. Richards, *J. Chem. Soc.*, **1965**, 2636; c. E. A. Zuech, R. F. Kleinschmidt, and J. E. Mahan, *J. Org. Chem.*, **31**, 3713 (1966).

[57] a. Neth. Appl. 6,510,648 (1966); through *Chem. Abstr.*, **65**, 13711 (1966); b. J. C. Martin and R. H. Hasek, U.S. Patent 2,947,784 (1960).

[58] a. R. E. Bowman and H. H. Stroud, *J. Chem. Soc.*, **1950**, 1342; b. A. H. Beckett and A. F. Casy, *ibid.*, **1955**, 900; c. F. Keller, R. R. Engel, and M. W. Klohs, *J. Med. Chem.*, **6**, 202 (1963); d. W. Barbieri, L. Bernardi, and S. Coda, *Tetrahedron*, **21**, 2453 (1966).

[59] L. Schmerling, U.S. Patent 3,247,209 (1966).

[60] a. F. J. Villani and N. Sperber, U.S. Patent 2,852,526 (1958); b. M. A. Shonle and J. W. Corse, U.S. Patent 2,424,063 (1947); c. E. Seeger and A. Kottler, Ger. Patent 1,225,646 (1966); through *Chem. Abstr.*, **65**, 18564 (1966).

[61] a. J. C. Powers, *J. Org. Chem.*, **30**, 2534 (1965); b. R. H. Reitsema and J. M. Hunter, *J. Am. Chem. Soc.*, **70**, 4009 (1948).

[62] R. M. Robinson, U.S. Patent 3,314,952 (1967); see also Fr. Patent 1,358,073 (1964).

[63] a. A. Dornow and G. Petsch, *Arch. Pharm.*, **284**, 153 (1951); b. R. F. Feldkamp, J. A. Faust, and A. J. Cushman, *J. Am. Chem. Soc.*, **74**, 3831 (1952); c. L. A. Walter and N. Sperber, U.S. Patent 2,997,478 (1961); d. A. F. Casy and N. A. Armstrong, *J. Med. Chem.*, **8**, 57 (1965).; e J. A. Moore and E. C. Capaldi, *J. Org. Chem.*, **29**, 2860 (1964); f. E. A. Steck, L. T. Fletcher, and R. P. Brundage, *ibid.*, **28**, 2233 (1963).

[64] a. J. B. Wright, H. G. Kolloff, and J. H. Hunter, *J. Am. Chem. Soc.*, **70**, 3098 (1948); b. E. J. Forbes, *J. Chem. Soc.*, **1955**, 3926; c. O. Schnider, A. Brossi, and K. Vogler, *Helv. Chim. Acta*, **37**, 710 (1954); d. C. A. Grob and E. Renk, *ibid.*, **37**, 1672 (1954); e. A. Brossi, L. H. Chopard-dit-Jean, J. Wursch, and O. Schnider, *ibid.*, **43**, 583 (1960); f. R. Urban and O. Schnider, *Monatsh.*, **96**, 9 (1965); g. A. Marxer, *J. Am. Chem. Soc.*, **79**, 467 (1957).

References

[65] a. W. W. Kaeding, U.S. Patent 3,053,895 (1962); b. E. J. Gasson, A. R. Graham, A. F. Millidge, K. M. Robson, W. Webster, A. H. Wild, and D. P. Young, *J. Chem. Soc.*, **1954**, 2170.

[66] E. Eidebenz, Ger. Patent 725,844 (1942).

[67] H. G. Morren, Belg. Patent 506,695 (1951).

[68] a. S. B. Cavitt, H. R. Sarrafizadeh, and P. D. Gardner, *J. Org. Chem.*, **27**, 1211 (1962); b. W. S. Saari, *ibid.*, **32**, 4074 (1967); c. G. Zwicky, P. G. Waser, and C. H. Eugster, *Helv. Chim. Acta*, **42**, 1177 (1959).

[69] a. F. Haglid and I. Wellings, *Acta Chem. Scand.*, **17**, 1743 (1963); b. W. T. Hunter, J. S. Buck, F. W. Gubitz, and C. H. Bolen, *J. Org. Chem.*, **21**, 1512 (1956).

[70] a. M. W. Goldberg and S. Teitel, U.S. Patent 2,774,766 (1956); b. *ibid.*, U.S. Patent 2,736,746 (1956); c. *ibid.*, U.S. Patent 3,064,052 (1962); d. T. Kralt, H. D. Moed, and J. Van Dijk, *Rec. Trav. Chim.*, **77**, 196 (1958); e. Neth. Appl. 6,501,213 (1965); through *Chem. Abstr.*, **64**, 5048 (1965).

[71] a. D. E. Pearson and J. D. Bruton, *J. Am. Chem. Soc.*, **73**, 864 (1951); b. J. Finkelstein, J. A. Romans, E. Chiang, and J. Lee, *J. Med. Chem.*, **6**, 153 (1963); c. J. G. Murphy, *J. Org. Chem.*, **26**, 3104 (1961).

[72] a. E. I. Schumann, M. E. Grieg, R. V. Heinzelman, and P. H. Seay, *J. Med. Chem.*, **3**, 567 (1961); b. M. J. Kalm, *ibid.*, **7**, 427 (1964).

[73] J. Levy, U.S. Patent 3,207,787 (1965).

[74] a. C. Ainsworth, *J. Am. Chem. Soc.*, **80**, 965 (1958); b. L. Goldman and R. P. Williams, U.S. Patent 2,724,713 (1955).

[75] J. Finkelstein, E. Chiang, F. M. Vane, and J. Lee, *J. Med. Chem.*, **9**, 319 (1966).

[76] a. F. Leonard, A. Wajngurt, M. Klein, and C. M. Smith, *J. Org. Chem.*, **26**, 4062 (1961); b. H. Muxfeldt and W. Rogalski, *J. Chem. Soc.*, **87**, 933 (1965).

[77] a. M. Freifelder, Y. H. Ng, and P. F. Helgren, *J. Med. Chem.*, **7**, 381 (1964); b. Belg. Patent 626,725 (1963) through *Chem. Abstr.*, **60**, 10602 (1964).

[78] Neth. Appl., 6,409,250 (1965); through *Chem. Abstr.*, **63**, 11428 (1965).

[79] a. F. Bergel, J. L. Everett, J. J. Roberts, and W. C. J. Ross, *J. Chem. Soc.*, **1955**, 3835; b. C. A. Grob and W. Schwarz, *Helv. Chim. Acta*, **47**, 1870 (1964).

XVIII

HYDROGENOLYSIS OF ALLYLIC O- AND N-SYSTEMS

Both the allylic oxygen and nitrogen systems, C=C—C—O—X or —N—X are known to be susceptible to hydrogenolysis. Reductions where the double bond is part of an aromatic system will be discussed in Chapter XIX, Debenzylation.

ALLYLIC OXYGEN COMPOUNDS

When the C=C linkage is not part of an aromatic system cleavage at the C—O bond is dependent on a number of factors which include the nature of X, the accessibility of the C—O—X portion of the molecule to the catalyst compared with that of the double bond, the acidity of the medium used as solvent, and sometimes the catalyst. There are other complications that make this reaction not too straightforward and sometimes unpredictable.

When X is hydrogen, alkyl, aryl, or acyl, hydrogenolysis, if it takes place, leads to hydrocarbon. When O—X is part of a lactone ring the product of hydrogenolysis is an acid [1a, b, c].

Hydrogenolysis precedes double-bond reduction and saturation of the double bond usually follows rapidly. In certain instances it has been possible to interrupt reaction after absorption of one equivalent of hydrogen to obtain hydrogenolysis product with an intact double bond. If the double bond is reduced first, hydrogenolysis generally will not take place except under extreme conditions.

Cleavage between the carbon oxygen linkage should occur when the double bond in the system is tetrasubstituted. Reduction of 2-hydroxycyclohexenylacetic acid lactone with platinum in alcohol gave the expected cyclohexylacetic acid [2a]. On the other hand structural differences may produce

Allilyc Oxygen Compounds 391

different results. In the hydrogenation of 4-(1-acetoxy-2-propylidene)-2-phenyl-2-oxazoline-5-one and related oxazolinones in dioxane with palladium on carbon, saturation of the double bond occurred with little or no hydrogenolysis of the allyl acetate group [2b]. In reductions of substituted acrylic acids and esters under the same conditions, hydrogenolysis was the only reaction (Eq. 18.1).

$$R = C_6H_5 \tag{18.1}$$

A molecular model suggests that the exocyclic double bond in the oxazoline compound cannot be prevented from making contact with the catalyst because of its more or less fixed position, due to the strong bonding effect of the highly conjugated system, while the CH_2O-acetyl portion of the molecule with much weaker bonding power may be out of contact with the catalyst most of the time.

In contrast the tetrasubstituted bond in the acrylic acids and esters can be completely surrounded because of greater freedom of movement by all the substituents.

Saturation of the trisubstituted double bond was the principal reaction in the hydrogenolysis of the allylic lactone, 2-hydroxycyclohexylideneacetic, acid, in alcohol with platinum oxide [2a] (Eq. 18.2).

$$\tag{18.2}$$

85%　　　　　11.3%

On the other hand, when 2-hydroxycyclohex-3-enylacetic acid lactone, which contains a symmetrically disubstituted double bond, was hydrogenated under similar conditions the major reaction was formation of cyclohexylacetic acid, 71% [1a]; only a small amount of saturated lactone was obtained (Eq. 18.3). When the reduction was carried out in acetic acid

[1c], the yield of saturated lactone rose to 25% and that of cyclohexylacetic acid dropped to 55%.

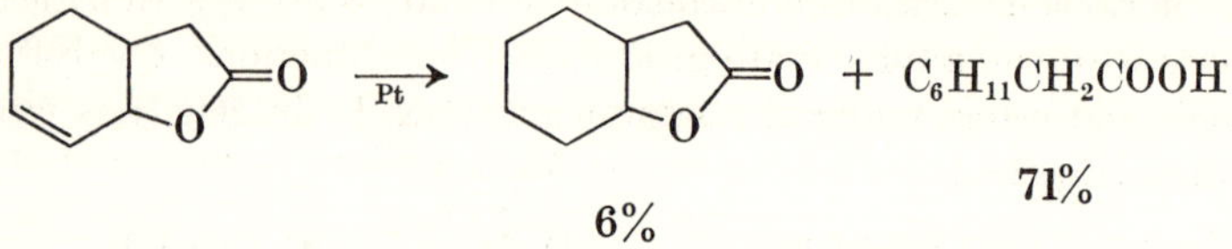

$$\text{(structure)} \xrightarrow{\text{Pt}} \text{(structure)} + C_6H_{11}CH_2COOH \quad (18.3)$$

6% 71%

The acidity of the solvent is a very important factor in the hydrogenolysis of the allylic oxygen system. Weak acid usually has little effect. If reductions of allylic-type alcohols or ethers to hydrocarbons are to be carried out, the presence of strong acid is necessary. When 3α- and 3β-hydroxy, acetoxy, or methoxycholest-4-enes were hydrogenated in ethyl acetate over platinum oxide, the major products were the corresponding coprostanols and cholestanols [3a]. Yields of hydrocarbons increased slightly in an ethyl acetate-acetic acid mixture. The amount of hydrocarbons rose to 65–75% in ethyl acetate containing sulfuric acid and were as high as 96% in solvent containing perchloric acid. When cinnamyl alcohol, $C_6H_5CH{=}CHCH_2OH$, was reduced in ethyl alcohol with noble metal catalysts only saturated alcohol was obtained [3b]. The best conditions for this hydrogenolysis was reaction in acetic acid containing hydrochloric acid with platinum oxide as catalyst. Platinum oxide in alcohol containing hydrogen chloride catalyzed the hydrogenolysis of cholest-4-ene-3β, 6β-diol to 5α-cholestane (3c); the investigators showed that hydrogenolysis of each allylic hydroxyl group preceded reduction of the double bond.

Whether an allylic oxygen grouping is axial or equatorial can also affect hydrogenolysis. In the palladium-catalyzed reduction of trans-(—)-α-santonin (with an equatorial oxygen) in acetic acid the lactone remained intact, while the cis-isomer santonin-C (with an axial oxygen) underwent cleavage and double-bond reduction [4] (Eq. 18.4).

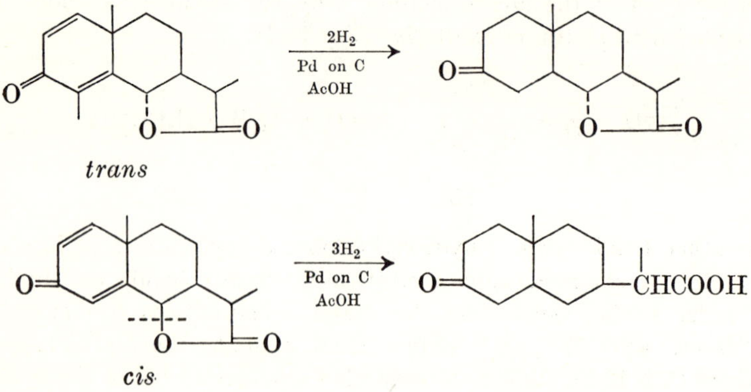

(18.4)

Double-bond migration may occur in the hydrogenolysis of nonaromatic allylic systems. It may precede or accompany reduction. If it precedes hydrogenation, the position of the double bond will determine whether or not hydrogenolysis will take place. Migration of the double bond accompanied cleavage of the lactone ring in the reduction of ψ-santonin with palladium on carbon in acetic acid [5a] (Eq. 18.5). It was reported that cleavage of the

$$\psi\text{-Santonin} \xrightarrow[\text{Pd}]{\text{H}_2} \quad (18.5)$$

lactone always took place in hydrogenations in the ψ-santonin series [5b], but it might be accompanied by reduction of the double bond and also of the carbonyl function. Accompanying saturation of the double bond and conversion of the carbonyl group to alcohol did result from platinum-catalyzed reactions [5c]. In the palladium on strontium carbonate-catalyzed reduction of 8-desoxy-ψ-santonin (see Eq. 18.5) only one equivalent of hydrogen was absorbed [5b]. Cleavage of the lactone took place as did migration of the double bond, presumably after hydrogenolysis, to give 64% yield of 2-(1-oxo-7-hydroxy-9-methyl-1,2,3,4,6,7,8,9-octahydro-6-naphthyl)-propionic acid.

A complication in the hydrogenolysis of cyclic allylic-type alcohols is the formation of ketone. The hydrogenation of the unsaturated axial alcohol, bicyclo[3,2,1]oct-3-en-2-ol, with 30% palladium on carbon resulted in only 50–60% uptake of an equivalent of hydrogen to yield saturated alcohol and bicyclo[3,2,1]octan-2-one, the latter presumably by intramolecular transfer of hydrogen on the catalyst surface [6]. When palladium on barium sulfate catalyzed the reduction, one molar equivalent was absorbed to give the saturated axial alcohol. In each instance the saturated ketone may have been the intermediate leading to saturated alcohol. In neither case was there evidence of hydrogenolysis.

Occasionally migration of the hydroxyl group, allylic rearrangement, may occur during reduction of allylic-type alcohols but this is a minor problem [7].

Catalysts

The effects of different catalysts in the hydrogenolysis of various allylic oxygen systems (alcohols, ethers, esters, and lactones) has not been systematically studied. It is difficult to suggest one of choice. Based on analogy with the reduction of related O-benzyl compounds where palladium is the catalyst of choice, it should be equally valuable for hydrogenolyses in the aforementioned systems. On the other hand, platinum catalysts may be more useful because they cause the least amount of double-bond isomerization (see Chapter IX, Olefins). When reductions in this series are to be carried out with platinum oxide in neutral solvent, the chemist must be aware of the inhibiting effect of sodium impurity (inherent in the preparation of the catalyst) on hydrogenolysis [7a].

It is likely that the choice of catalyst lies between palladium and platinum depending on the type of compound to be reduced and on the reaction medium. There is a report which states that platinum oxide is better for the hydrogenolysis of allyl esters [76]. There is little work to indicate that nickel may be useful for hydrogenolyses of O-allyl systems.

ALLYLIC NITROGEN COMPOUNDS

Most of the studies on hydrogenolyses of the C=C—C—N—X system where the double bond is nonaromatic have been confined to those reactions where the nitrogen atom was quaternary. Allyl trimethylammonium iodide was hydrogenated in the presence of platinum oxide or platinum black in acetic acid containing sodium acetate to give 63% of trimethylamine hydriodide [8a]. No uptake of hydrogen was reported in the presence of palladium on barium sulfate. The reduction of cinnamyltrimethylammonium chloride, $C_6H_5CH=CHCH_2\overset{+}{N}(CH_3)_3Cl^-$, with palladium in acetic acid containing sodium acetate yielded trimethylamine hydrochloride, propylbenzene, and 1-phenyl-1-propene.

In certain instances, dependent on structure, saturation of the double bond is a competing or even the preferred reaction. In one case 1,1-dimethyl-1,2,5,6-tetrahydropyridinium iodide, hydrogenated in alcohol over platinum oxide, gave 80% yield of 1,1-dimethylpiperidinium iodide [8b]. In a similar reduction equal amounts of saturated quaternary compound and 1-dimethylaminopentane, the product of hydrogenolysis, were obtained [8c] (Eq. 18.6).

$$\underset{\underset{CH_3\ \ CH_3\ \ I^{(-)}}{}}{\overset{+}{N}} \longrightarrow \underset{\underset{CH_3\ \ CH_3\ \ I^{(-)}}{}}{\overset{+}{N}} + C_5H_{11}N(CH_3)_2 \cdot HI \qquad (18.6)$$

In contrast to the above examples of reduction of the allylic entity in a cyclic system, some indole alkaloids underwent cleavage during hydrogenation with platinum in alcohol [8c]. The exclusive course of reaction was considered to be due to the affinity of the indole portion of the molecule for the catalyst.

SELECTIVE REDUCTIONS

O- and *N*-Allyl Compounds

There are too few examples of selective reductions between the *O*- and *N*-allylic systems. If the nitrogen atom is not quaternized it is likely that hydrogenolysis of the *O*-allyl system will take place preferentially. Elymoclavine, an alkaloid containing both hydrogenolyzible groupings, was hydrogenated with platinum oxide in acetic acid to give dihydroagroclavine

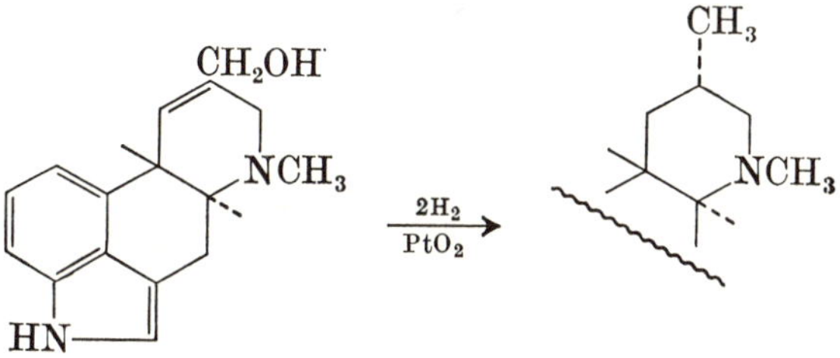

(18.7)

Elymoclavine

[9a] (Eq. 18.7). The allylic alcohol system was completely reduced; there was no cleavage at the C=CCH$_2$—N(CH$_3$)-system, which was part of the heterocyclic ring.

When some 3-alkoxy-6-dialkylaminomethyl-3,5-dienyl steroids were hydrogenated over prereduced palladium on carbon, hydrogenolysis of the allylic nitrogen system took place without cleavage of the enol ether when the nitrogen atom was quaternary [9b]. A fairly large amount of catalyst was necessary to achieve rapid reduction in order to counteract gradual decomposition of the quaternary salt. The presence of an equivalent of sodium acetate and, in some instances, addition of acetic acid to maintain a pH at 7.0–7.5 prevented decomposition and led to improved yield. Termination of reaction at one equivalent of hydrogen gave the 6-methyl enol ethers (Eq. 18.8). Reduction of the amino oxide under similar conditions was also

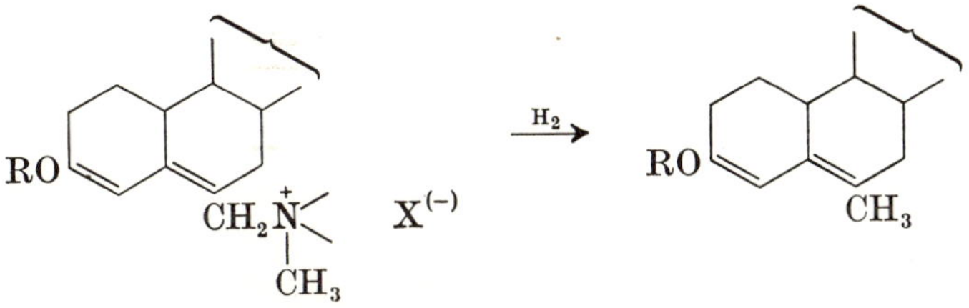

(18.8)

satisfactory, but hydrogenations of the free 6-dialkylamino enol ethers gave mixtures of products.

Experiments by the same group also indicated that when two allylic oxygen functions were present one could not be hydrogenolyzed selectively. Attempts to reduce 6-formyl- or 6-hydroxymethyl-3-alkoxy-3,5-dienes without concomitant attack of the enol ether system could not be accomplished.

In the Presence of Functional Groups

There are too few examples to make any generalizations about selectivity in the presence of other reducible groups. However, because of the relationship of the O- and N-allyl systems to O- and N-benzyl groups, application of examples found in the section on debenzylation in Chapter XIX should be helpful.

It has been noted that hydrogenolysis of the carbon-oxygen bond in an allylic oxygen system is usually accompanied by reduction of the double bond, although in certain instances hydrogenolysis may be selective. It was shown in the allylic nitrogen system that selective hydrogenolysis was possible when the nitrogen atom was quaternary [8a, 9b].

Reduction of cis-1,2-dibenzoyl-3-(4-morpholino)-1-propene,

$$C_6H_5COCH=C(COC_6H_5)CN_2\diagup\diagdown O,$$

in alcohol over palladium on barium sulfate or as hydrochloride in the presence of platinum oxide gave 1,2-dibenzoylpropane, $C_6H_5COCH_2CH(CH_3)COC_6H_5$ [9c]. On the other hand reduction in acetic acid in the presence of platinum oxide yielded 1,2-dibenzoyl-3-(4-morpholino)propane [9d]. A mechanism for the different results is suggested [9c] but the material balance in these hydrogenations (a maximum of 40% of material is accounted for) leaves much to be desired. Reductions followed by gas-liquid chromatography of the neutralized solutions should give a better insight into these reactions.

Hydrogenolysis of the carboallyloxy group was accomplished without cleaving the N—N bond of phenylhydrazine in the hydrogenation of 1-(N-carboallyloxy-L-leucyl)-2-phenylhydrazine,

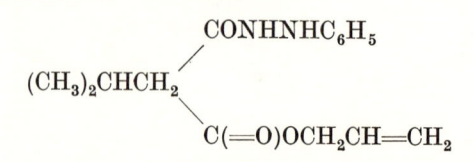

$$(CH_3)_2CHCH_2\diagdown\begin{matrix}CONHNHC_6H_5\\ \\ C(=O)OCH_2CH=CH_2\end{matrix}$$

to 1-leucyl-2-phenylhydrazine [9e]. In this instance the result is not unexpected. The carboallyloxy group is very readily removed; acylation makes the N—N bond fairly resistant to cleavage [10].

REFERENCES

[1] a. J. Meinwald, M. C. Seidel, and B. C. Cadoff, *J. Am. Chem. Soc.*, **80,** 6303 (1958); b. W. E. Noland, J. H. Cooley, and P. A. McVeigh, *ibid.*, **81,** 1209 (1959); c. J. Meinwald and E. Frauenglass, *ibid.*, **82,** 5235 (1960).
[2] a. M. S. Newman and C. A. VanderWerf, *J. Am. Chem. Soc.*, **67,** 233 (1945); b. E. Galantay, A. Szabo, and J. Fried, *J. Org. Chem.*, **28,** 98 (1963).
[3] a. C. W. Shoppee, B. D. Agashe, and G. H. R. Summers, *J. Chem. Soc.*, **1957,** 3107; b. S. Nishimura, T. Onoda, and A. Nakamura, *Bull. Chem. Soc. Jap.*, **33,** 1356 (1960); c. S. Nishimura and K. Mori, *ibid.*, **32,** 102 (1959).
[4] W. G. Dauben, W. K. Hayes, J. S. P. Schwarz, and J. W. McFarland, *J. Am. Chem. Soc.*, **82,** 2232 (1960).
[5] a. W. G. Dauben and P. D. Hance, *J. Am. Chem. Soc.*, **77,** 2451 (1955); b. W. G. Dauben, J. S. P. Schwarz, W. K. Hayes, and P. D. Hance, *ibid.*, **82,** 2239 (1960); G. R. Clemo and W. Cocker, *J. Chem. Soc.*, **1946,** 30.
[6] H. L. Goering, R. W. Greiner, and M. F. Sloan, *J. Am. Chem. Soc.*, **83,** 1391 (1961).
[7] a. M. C. Dart and H. B. Henbest, *J. Chem. Soc.*, **1960,** 3563; b. Koczka, G. Bernáth, and A. Molnár, *Acta. Chim. Acad. Sci. Hung.*, **53,** 205 (1967).
[8] a. H. Emde and H. Kull, *Arch. Pharm.*, **274,** 173 (1936); b. R. R. Renshaw and R. C. Conn, *J. Am. Chem. Soc.*, **60,** 744 (1938); c. J. P. Dickinson, J. Harley-Mason, and J. H. New, *J. Chem. Soc.*, **1964,** 1858.
[9] a. S. M. Olin, U.S. Patent 3,029,243 (1962); b. D. Burns, G. Cooley, M. T. Davies, A. K. Hiscock, D. N. Kirk, V. Petrow, and D. M. Williamson, *Tetrahedron*, **21,** 569 (1965); c. P. S. Bailey and R. E. Lutz, *J. Am. Chem. Soc.*, **67,** 2232 (1945); d. R. E. Lutz and P. S. Bailey, *ibid.*, p. 2229; e. H. B. Milne, J. E. Halver, D. S. Ho, and M. S. Mason, *ibid.*, **79,** 637 (1957).
[10] M. Freifelder, unpublished results.

XIX

DEBENZYLATION AND RELATED REACTIONS

Debenzylations are related to hydrogenolyses of allylic oxygen and nitrogen systems, C=C—C—OR or —NR, which are known to undergo cleavage under the proper conditions (Chapter XVIII). However, debenzylations, hydrogenolyses at ArC—O—R and ArC—N—R, are more readily accomplished and far more predictable because these reactions are not subject to double-bond migration and other vagaries associated with hydrogenolyses of nonaromatic allylic oxygen and nitrogen systems (Chapter XVIII). Furthermore, because double bonds in aromatic systems are not readily reduced, the allylic-type moiety is maintained during the process of hydrogenolysis.

Cleavage between the carbon and oxygen and carbon and nitrogen bonds is not necessarily confined to benzenoid compounds where Ar is mono-, di- or triphenyl, 1- or 2-naphthyl, and other related types but also includes systems where Ar is an unsaturated heterocyclic ring.

CATALYSTS AND CONDITIONS

A number of catalysts have been employed for the hydrogenolyses of O- and N-benzyl type groups [1]. There is little doubt that palladium catalysts are the ones of choice for these reactions. They are not only the most active under the mild conditions used for removal of the blocking group, but they generally will not cause ring reduction during reaction in an acidic medium (if the aromatic ring is hydrogenated, debenzylation will not take place).

This author's preference is 5% palladium on activated carbon. It is the most active of the supported palladium catalysts for debenzylations and on a metal-weight basis is far superior to palladium oxide [2].

It is most unusual when removal of benzyl-type blocking groups does not take place in the presence of a palladium catalyst. They were not completely removed in the reduction of D-benzyloxypantatheine-4-dibenzylphosphate over palladium on barium sulfate [3]. A large amount of platinum oxide was employed to accomplish this. The use of palladium on carbon instead of palladium on barium sulfate might have led to better results. In related reactions in which the substrate contained benzyl phosphate esters as well as a benzyl ether, 5% palladium on carbon, 20% by weight of the substrate, caused complete removal of the benzyl groups in hydrogenations at 40–50° and 3 atm pressure [2].

Hydrogenolysis did not take place in the palladium-catalyzed reduction of 1-β-phenyl-1,5-D-anhydroglucitol tetraacetate [4]; the conditions, which palladium catalyst was employed, and the catalyst ratio were not disclosed. When reaction was carried out in acetic acid containing perchloric acid in the presence of 6% by weight of platinum oxide to substrate, saturation of the benzene ring was the only reaction (Eq. 19.1).

$$
\begin{array}{c}
\text{AcO-CH}_2\text{-O-AcO-C}_6\text{H}_5 \xrightarrow{\text{Pd}} \text{C}_6\text{H}_5\text{CH}_2\text{CHCHCHCHCH}_2\text{OH (with OH groups)} \\
\downarrow \text{PtO}_2 \text{ acidic conditions} \\
\text{AcO-CH}_2\text{OAc-O-AcO-C}_6\text{H}_{11}
\end{array}
$$
(19.1)

At times seemingly poor results may be due to impurities in the substrate. Uptake of hydrogen was incomplete in the hydrogenation of N-4-benzyloxyphenylacetylurea in methyl cellosolve with 5% palladium on carbon [2]; no uptake occurred during reduction of the corresponding 4-carbobenzyloxy compound, $C_6H_5CH_2OC(=O)C_6H_5CH_2CONHCONH_2$. However, pretreatment of a solution of each compound with Raney nickel followed by hydrogenation over palladium on carbon gave N-(4-hydroxyphenylacetyl)urea, 4-$HOC_6H_4CH_2CONHCONH_2$, in good yield.

The use of palladium catalyst may be contraindicated in the hydrogenolyses of optically active, unsymmetrically substituted benzyl alcohols, esters, and ethers if retention of configuration is desired. Palladium causes inversion; hydrogenation with nickel, copper, and cobalt catalysts favor retention of configuration [5a, b]. These catalysts should be used in such instances if reduction can be effected in their presence.

Removal of *O*-benzyl and related protecting groups is enhanced by carrying out reactions in acidic media. When benzyl alcohols are reduced in acetic acid or in acetic acid containing strong acid, deoxygenation is assumed to proceed through the acetate [5b]. Since it has been shown that the rate of displacement in reductions of compounds, $ArCH_2OR$ (where R = hydrogen, acetyl, or trifluoroacetyl), is $CF_3CO > CH_3CO > H$ [5b], it might be advisable to add trifluoroacetic acid to a hydrogenolysis when difficulty is encountered.

When deoxygenation of benzyl alcohols is carried out in neutral solvent containing mineral acid, reaction may proceed through an ester. It may also go through a benzyl-type halide if hydrogen halide is used to catalyze hydrogenolysis. It is also possible in mineral acid-catalyzed hydrogenolyses of alcohols, $ArCHOHCH_2R$, that dehydration takes place, $ArCH=CHR$, followed by reduction of the double bond.

The advantage of the use of palladium in acid-catalyzed hydrogenolyses of *O*-benzyl groups may be seen in reactions among carbohydrates containing these groups. Such debenzylations gave quantitative yields of sugars and toluene with palladium, whereas in similar reactions in the presence of platinum there was competition between hydrogenolysis and ring reduction [6]; or, as in one example, only saturation of the ring took place [4].

The removal of *N*-benzyl and related groups with palladium catalyst is also enhanced by reaction in acidic media. However, the effect is probably not catalytic but is more likely due to neutralization of the potential inhibiting effect of the amino nitrogen on the activity of the catalyst. Reduction may be carried out in acetic acid or in solvent containing an equivalent or a slight excess of an equivalent of most acids [2].

When *O*-debenzylation of a compound containing a basic nitrogen atom is to be carried out, it is advisable to use an equivalent of acid or to reduce the compound as an acid salt. The hydrogenolysis of 6-benzyloxy-7-methoxy-1-methylisoquinoline in alcohol with 5% palladium on carbon required a 15–18 hr reaction period at 70° and 3 atm, while reaction as hydrochloride salt was complete in 2 hr to yield 6-hydroxy-7-methoxy-1-methylisoquinoline [2].

O-DEBENZYLATION AND RELATED REACTIONS

Arylmethanols

Benzyl alcohols, $ArCH_2OH$, and related compounds generally are reduced to hydrocarbons so readily in neutral solution or in acidic media with palladium catalysts under mild conditions, that there does not seem to be any reason to consider other catalysts for this purpose.

Nuclear substitution appears to have little effect on hydrogenolysis of benzyl alcohols. 2,3-Dihydroxy-4-methylbenzyl alcohol in ethyl alcohol was converted to 3,6-dimethylcatechol with palladium on carbon [7a], and 7,8-methylenedioxy-1-(3,4-methylenedioxyphenyl)naphthalene-2,3-dimethanol was reduced to the corresponding 2,3-dimethylnaphthalene in ethyl acetate with the same catalyst [7b]. The effect of an amino group on the activity of the catalyst can be overcome easily by the addition of an equivalent of acid [2]. The presence of a bulky group adjacent to the —CH$_2$OH group may impede hydrogen uptake, but reduction in acetic acid or in acetic acid containing some strong acid should aid deoxygenation.

It might be anticipated that hydrogenolysis should also take place when Ar is an unsaturated heterocycle, but examples appear to be limited to only a few systems.

Palladium on carbon was employed in the hydrogenolyses of some 2-alkyl-3-hydroxy-4,5-dihydroxymethylpyridine hydrochloride salts in dilute hydrochloric acid [7c]. Deoxygenation occurred exclusively in 4-position to yield 2-alkyl-3-hydroxy-5-hydroxymethyl-4-methylpyridine, as it did in the hydrogenation of the 2-methyl-4,5-dihydroxymethyl compound with platinum oxide under the same conditions, yielding 3-hydroxy-5-hydroxymethyl-2,4-dimethylpyridine [7d].

In the pyrimidine series 5-hydroxymethyluracil, which may be viewed as 2,4-dihydroxy-5-hydroxymethylpyrimidine, was converted to thymine in 50% acetic acid with platinum oxide or rhodium on alumina [7e] (Eq. 19.2).

Thymine

(19.2)

Platinum oxide was preferred in this reaction because it did not cause reduction of the 5,6-double bond even with an equal weight of catalyst. Reaction with rhodium in 50% acetic acid also yielded thymine but saturation of the 5,6-bond occurred if uptake of hydrogen was allowed to continue. Hydrogenation in water or 5% acetic acid with rhodium led to reduction of the double bond, not hydrogenolysis. Palladium catalyst was not tried.

α-Substituted Arylmethanols

The removal of oxygen in α-substituted arylmethanols can be influenced by the substituent on the side chain. If the substituent is a straight-chain alkyl group, little difficulty should be encountered in reductions with palladium catalyst, particularly in acetic acid or in the presence of an acid catalyst. For example, 1-phenylethanol, $C_6H_5CHOHCH_3$, was rapidly converted to ethylbenzene over palladium catalysts (from palladium chloride or palladium oxide) in acidic media, particularly when the catalyst was not prereduced [8a].

Hydrogenolyses may be carried out in neutral solvent. N-Cyclohexyl-4-hydroxy-4-phenylbutyramide, $C_6H_5CHOH(CH_2)_2CONHC_6H_{11}$, was reduced in methyl alcohol to N-cyclohexyl-4-phenylbutyramide with 10% palladium on carbon under 3 atm pressure [8b], and 3-(1-hydroxypentadecyl)veratrole in ethyl alcohol was converted to 3-pentadecylveratrole in 90% yield by a similar procedure [8c]. More catalyst than compound was used in the latter hydrogenolysis. However, a more normal amount of catalyst, 10% by weight, was employed in the reduction of 3-(1-hydroxypentadecyl)-4,5-dimethylveratrole in ethyl acetate containing some sulfuric acid at 60° and 4 atm pressure to yield 4,5-dimethyl-3-pentadecylveratrole [8d].

Acid-catalyzed hydrogenolysis with palladium on carbon was employed for the conversion of 1-hydroxy-7,9-dimethoxy-1,2,3,4-tetrahydrophenanthrene, a benzyl-type alcohol, to 7,9-dimethoxy-1,2,3,4-tetrahydrophenanthrene [8e] (Eq. 19.3).

$$\underset{\text{CH}_3\text{O}}{\overset{\text{CH}_3\text{O}}{\text{OH}}} \xrightarrow[\text{3 hr}]{\text{H}_2 \atop \text{Pd on C}} \underset{\text{CH}_3\text{O}}{\overset{\text{CH}_3\text{O}}{}} \qquad (19.3)$$

83%

Acid catalysis was also employed to aid removal of the benzylic hydroxyl group attached to a tertiary carbon atom in 1-alkyl-1-indanols and 1-phenyl-1,2,3,4-tetrahydro-1-naphthol [8f] (Eq. 19.4).

$$\text{in AcOH + HClO}_4$$

$$\underset{(\text{CH}_2)_n}{\overset{\text{HO} \quad \text{R}}{}} \xrightarrow[\text{Pd on C}]{\text{H}_2} \underset{(\text{CH}_2)_n}{\overset{\text{H} \quad \text{R}}{}} \qquad (19.4)$$

$n = 2$ or 3

A small amount of perchloric acid caused a significant increase in reaction rate in the hydrogenolyses of 1,2-diethyl-1,2-di(4-anisyl)- and di(substituted-4-anisyl)ethanols in acetic acid [9]. It also caused a decrease in yield, which led to the suggestion that the acid catalyst should be employed only when hydrogen uptake was too slow to be practical. It was also stated that reduction in acetic acid was slower with palladium on carbon than with palladium oxide. This may be attributed, in this instance, to the higher ratio of metal to substrate when palladium oxide was used. The highest yield [75%], however, was obtained by reaction in acetic acid with palladium on carbon. Removal of the hydroxyl group did not take place in alcohol.

In the reduction of arylmethanols, ArCHOHR, where R is an aromatic or an unsaturated heterocyclic ring, any expected activation by the second ring should be counteracted by steric effects. Hydrogenolysis of benzhydrol, $(C_6H_5)_2CHOH$, in alcohol with palladium on carbon proceeded at a moderate rate [10a]. 2,3,4,5,6-Pentafluorobenzhydrol in alcohol was reduced very slowly to 2,3,4,5,6-pentafluorodiphenylmethane with palladium on carbon [10b], due no doubt to steric effects. Acid catalysis might have aided reaction as it did in the hydrogenolysis of 4-(1-hydroxy-2-methylpropyl)-triphenylmethanol. It was converted to 4-(1-hydroxy-2-methylpropyl)-triphenylmethane within 24 hr in alcohol containing hydrogen chloride by hydrogenation with a small amount of palladium on carbon under 2 atm pressure [10c].

In hydrogenations of arylmethanols where R was pyridine 1-(5-indanyl)-(5-methyl-2-pyridyl)-1-ethanol was reduced with palladium on carbon in alcohol containing an excess of one equivalent of perchloric acid to yield 68% of 1-(5-indanyl)-1-(5-methyl-2-pyridyl)ethane, in a 60-hr reaction period [11a] (Eq. 19.5). When the reaction was carried out in concentrated

$$\text{structure} \xrightarrow[\text{Pd on C}]{H_2} \text{structure} \qquad (19.5)$$

hydrochloric acid with an equal weight of palladium on carbon, uptake of hydrogen was complete in 2 hr to give 73% yield of the same product. These examples illustrate the advantage of the use of palladium catalysts in debenzylations in acidic media without causing saturation of reducible ring systems.

404   *Debenzylation and Related Reactions*

When platinum oxide was used as catalyst in the reduction of some pyridyl phenylmethanols in alcohol containing excess hydrochloric acid, ring reduction accompanied hydrogenolysis when R was 3-pyridyl [11b] (Eq. 19.6). Loss of the hydroxy group did not take place with the 2- or 4-pyridyl compounds. It should be noted that in these reactions saturation of the pyridine ring was the prime purpose, not hydrogenolysis.

in EtOH + HCl

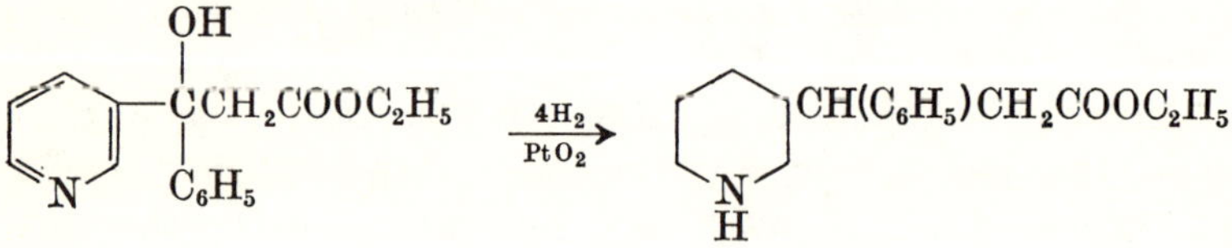

$$\text{pyridyl-C(OH)(C}_6\text{H}_5\text{)CH}_2\text{COOC}_2\text{H}_5 \xrightarrow[\text{PtO}_2]{4\text{H}_2} \text{piperidyl-CH(C}_6\text{H}_5\text{)CH}_2\text{COOC}_2\text{H}_5 \quad (19.6)$$

Hydrogenolyses of benzyl alcohols, ArCHOHCHRY, where R is H or alkyl and Y is amino or alkylamino or a structure related to it, should be carried out in acidic medium with palladium catalyst. 2-Amino-1-phenyl-1-butanol, $C_6H_5CHOHCH(NH_2)C_2H_5$, was reduced to 1-phenyl-2-butylamine in acetic acid containing perchloric acid in the presence of palladium on barium sulfate [12a]. Ephedrine, $C_6H_5CHOHCH(CH_3)NHCH_3$, was converted to N-methyl-1-phenylisopropylamine, $C_6H_5CH_2CH(CH_3)NCHH_3$, by hydrogenation in acetic acid containing sulfuric acid in the presence of palladium black [12b]. Similar conditions were employed to obtain 2-amino-6-methoxyindane from 2-amino-6-methoxyindanol [12c]. Palladium on carbon catalyzed the deoxygenation of 10-dimethylaminomethyl-9-hydroxy-*cis*-1,2,3,4,4a,9,10,10a-octahydrophenanthrene in acetic acid containing hydrochloric acid to the corresponding 10-dimethylaminomethyloctahydrophenanthrene [12d] (Eq. 19.7).

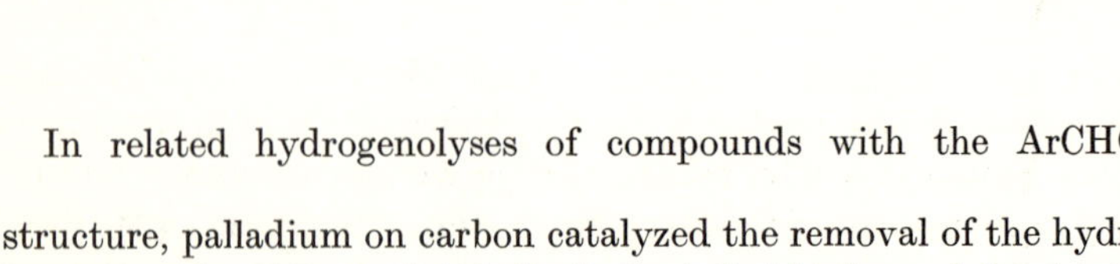

$$\text{9-OH-10-CH}_2\text{N(CH}_3\text{)}_2\cdot\text{HCl octahydrophenanthrene} \xrightarrow[\text{Pd on C}]{\text{H}_2} \text{10-CH}_2\text{N(CH}_3\text{)}_2\cdot\text{HCl octahydrophenanthrene}$$

72%

(19.7)

In related hydrogenolyses of compounds with the ArCHOHCH$_2$N< structure, palladium on carbon catalyzed the removal of the hydroxyl group in some 5,8-, 6,7-, and 7,8-disubstituted 4-hydroxy-1,2,3,4-tetrahydroisoquinolines [12e] (Eq. 19.8).

Hydrogenolysis of 10-hydroxy-2-methyldecahydroisoquinolyl-1-(α-phenyl)methanol in acetic acid containing sulfuric acid with an equal weight

in 20% HCl

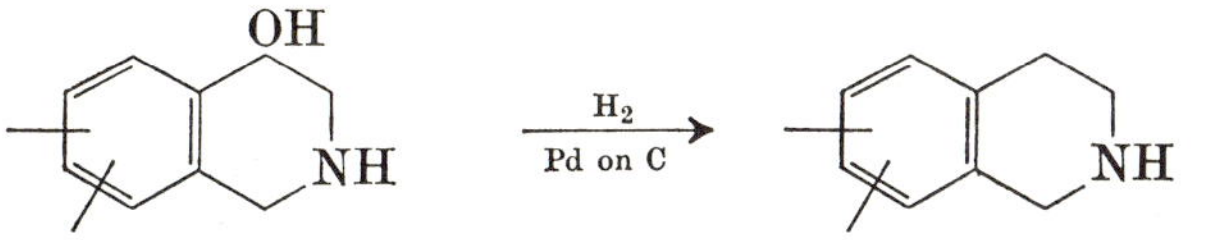

$$\text{(structure)} \xrightarrow[\text{Pd on C}]{\text{H}_2} \text{(structure)} \quad (19.8)$$

of palladium on barium sulfate did not yield 1-benzyl-10-hydroxy-2-methyl-decahydroisoquinoline [13]. Instead an epoxide and 1-benzyl-2-methyl-1,2,3,4,5,6,7,8-octahydroisoquinoline were obtained. It is possible that reaction may have proceeded through the epoxide, obtained in 66% yield, followed by cleavage of the epoxide ring to 1-benzyl-10-hydroxy-2-methyl-isoquinoline and subsequent dehydration to yield 1-benzyl-2-methyl-1,2,3,4,5,6,7,8-tetrahydroisoquinoline (Eq. 19.9).

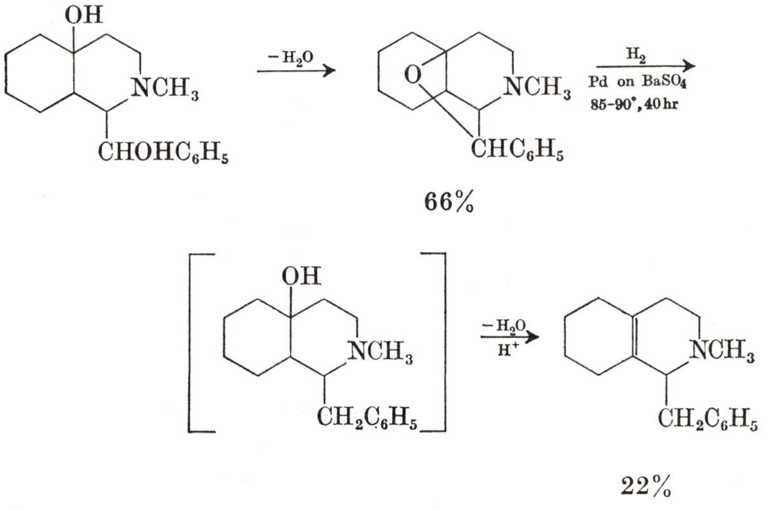

$$\text{(structure with OH, NCH}_3\text{, CHOHC}_6\text{H}_5\text{)} \xrightarrow{-\text{H}_2\text{O}} \text{(epoxide structure, NCH}_3\text{, CHC}_6\text{H}_5\text{)} \xrightarrow[\substack{\text{Pd on BaSO}_4 \\ 85-90°, 40\text{ hr}}]{\text{H}_2}$$

66%

$$\left[\text{(structure with OH, NCH}_3\text{, CH}_2\text{C}_6\text{H}_5\text{)}\right] \xrightarrow[\text{H}^+]{-\text{H}_2\text{O}} \text{(structure with NCH}_3\text{, CH}_2\text{C}_6\text{H}_5\text{)} \quad (19.9)$$

22%

Removal of the hydroxyl group in compounds like N-methyl-4-hydroxy-4-phenylpiperidine may be difficult because normally the strongest points of attachment of the molecule to the surface of the catalyst are through the pi bonds of the benzene ring and the unshared pair of electrons of the basic nitrogen atom. On this basis a molecular model shows the hydroxyl group out of plane.

It may be necessary to apply more energy to cause a change in the bonding pattern, as was done in the hydrogenation of the above compound with Raney nickel at 150° and 125 atm [14a].

Benzyl Esters of Organic Acids

In general, the hydroxyl group of a benzylic alcohol is hydrogenolyzed more readily as an ester than as the parent alcohol [5b]. Reduction leads to high yield of hydrocarbon and the acid from which the ester was formed [1]. Esterification may not be necessary for the hydrogenolysis of a primary benzyl alcohol, $ArCH_2OH$, to hydrocarbon. However, it was very useful in the conversions of 3-hydroxymethylpyridine and 2- and 4-hydroxymethylquinoline as acetate or benzoate to the corresponding methylpyridine and methylquinolines in very high yield [14b]. Reduction as ester should be attempted when deoxygenation of an α-substituted arylmethanol is found difficult.

Reductions are best carried out with palladium on carbon [2], although other palladium catalysts have been employed [1]. Difficult hydrogenolyses should be run in acidic media.

The presence of acid is not always necessary. Reaction can be carried out in alcohol, as in the hydrogenolysis of α-(1-dimethylamino-2-dodecyl)benzyl acetate as hydrochloride salt to N,N-dimethyl-2-benzyldodecylamine, $C_{10}H_{21}CH(CH_2C_6H_5)CH_2N(CH_3)_2$, over active palladium on carbon catalyst [15]. Ethyl acetate may be a useful solvent for hydrogenolysis of benzyl-type acetates because of the maintenance of acetate ion during reduction. It is possible that its usefulness may also be due to the presence of acetic acid in it—acetic acid was found in ethyl acetate which had been standing for a period and was not redistilled prior to use in hydrogenation [2]. The presence of acetic anhydride may also be advantageous in the hydrogenolysis of benzyl acetates.

Hydrogenolysis of mandelic acid, $C_6H_5CHOHCOOH$, and related compounds is reported to be difficult [10a]. However, reaction of methyl 4-methoxymandelate as the acetate gave methyl 4-methoxyphenylacetate in 90% yield after hydrogenation in acetic acid with palladium black [16a]. More rapid reaction was obtained when sulfuric acid was present. Hydrogenolysis of related pyridyl and quinolyl compounds, $RCH(OOCC_6H_5)CONH_2$ where R = 2-, 3-, or 4-pyridyl, or 2- or 4-quinolyl, over palladium on barium sulfate in alcohol gave high yields of the corresponding pyridyl- and quinolylacetamides [16b].

Esterification by means of benzyl alcohol is used to protect an acid function so that the resulting compound can take part in further synthesis. The acid function can then be retrieved after hydrogenation of the benzyl ester. Examples of such procedures can be seen in the preparation of 1,1,2,2-ethanetetracarboxylic acid from the hydrogenolysis of the corresponding tetrabenzyl ester in alcohol with 5% palladium on carbon at 1 atm overpressure (15 psig), and that of 1,2,3,4-dibenzo-1,3-cyclooctadien-6,7-dicarboxylic acid from the corresponding ester [17a] (Eq. 19.10).

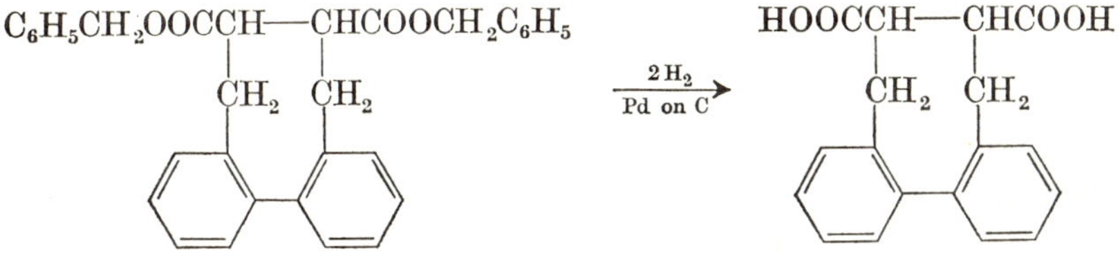

$$C_6H_5CH_2OOCCH\text{---}CHCOOCH_2C_6H_5 \quad \xrightarrow[\text{Pd on C}]{2H_2} \quad HOOCCH\text{---}CHCOOH$$

(19.10)

Removal of the protecting group by hydrogenation of benzyl 2-methyl-4-oxo-4,5-dihydropyrrole-3-carboxylate with palladium on strontium carbonate was the last step in the synthesis of the corresponding acid [17b]. 5% Platinum on carbon was employed for the removal of the benzyl groups in the synthesis of a complex L-lysine ester [17c] (Eq. 19.11).

$$HBr \cdot H_2N(CH_2)_4CHNH(CHCOOCH_2C_6H_5)\Big]_2\text{---}C\!=\!O \xrightarrow[\text{5\% Pt on C}]{H_2}$$

$$H_2N(CH_2)_4CHNHCHNH(CHCOOH)\Big]_2 C\!=\!O \quad (19.11)$$

Hydrogenolysis of benzyl esters is especially useful in compounds containing other ester groups, all of which would be subject to cleavage if hydrolysis was employed. Methyl-2-carboxypyrrolidine-1-acetate was obtained in 82% yield by reduction of the benzyl ester of the 2-carboxy compound in alcohol with a very small amount of 5% palladium on carbon [18a]. α-Acetylamino-α-(4-carbomethoxyindole-3-methyl)malonic acid was obtained in 75% yield from the dibenzyl malonate by hydrogenation with palladium on carbon [18b]. O-Carbamyl-D-, L-, and D,L-serine resulted from a cleavage of the O-carbamyl-N-carbobenzyloxyserines as benzyl esters over palladium black in aqueous dioxane [18c] (Eq. 19.12).

$$\underset{\underset{\underset{O}{\|}}{HNCOCH_2C_6H_5}}{H_2NCOCH_2CHCOCH_2C_6H_5} \xrightarrow[\text{Pd}]{2H_2} H_2NCOCH_2CH(NH_2)COOH \quad (19.12)$$

94%

Hydrogenolyses of benzyl esters, $ROOC(CH_2)_nCOOCH_2C_6H_5$ where $n = 2$ to 7 and R was dialkylaminoalkyl, in alcohol in the presence of 5% palladium on carbon gave high yields of the corresponding free acids, $ROOC(CH_2)_nCOOH$ [2]. Hydrogen uptake under 2–3 atm pressure was generally so rapid that it was not necessary to add an acid to neutralize the effect of the basic nitrogen atom on the activity of the catalyst. Release of the free carboxyl group resulting from hydrogenolysis appeared sufficient for this purpose.

A benzhydryl ester may also be useful for protection of the carboxyl group. In most instances hydrogenolysis of a number of benzhydryl esters in methyl alcohol or ethyl acetate with palladium black gave quantitative yields of the free acids [19].

Benzyl Phosphates

The catalyst of choice for hydrogenolysis of benzyl phosphates is palladium. Although platinum has been employed in the hydrogenolysis of benzyl esters of carboxylic acids [1], it is generally ineffective for the conversion of benzyl phosphates to the free acid. Platinum catalysts are useful only for the cleavage of phenyl esters.

Palladium on carbon was employed for the hydrogenolysis of dibenzyl ethyl phosphate in alcohol to ethyl phosphate [20a]. A mixture of palladium oxide and palladium on carbon gave glucose-6-phosphate from reduction of 1,2-isopropylidene-D-glucose dibenzyl phosphate in alcohol and subsequent removal of the isopropylidene group [20b]. When the sodium salt of 4-nitrobenzyl isopropyl phosphate was hydrogenated in water and methyl alcohol with palladium black, p-toluidine and isopropyl phosphate were obtained [20c]. A mixture of palladium oxide and palladium on carbon was used for the preparation of adenosine triphosphate from the corresponding tetrabenzyl phosphate [20d].

N-Phosphorylamino acids and peptides have resulted from low-pressure hydrogenolysis of benzyl and substituted-benzyl phosphate esters of these compounds with palladium black [21a, b]. A phosphoryl urea, N-(butylcarbamoyl) phenylphosphonamide as sodium salt, $C_6H_5P(OH)(=O)$-$NHCONHC_4H_9$, was obtained by hydrogenation of the benzyl ester with more than an equal weight of 10% palladium on carbon [21c].

Benzyl Ethers

Benzyl ethers, $C_6H_5CH_2OR$, are very readily cleaved by catalytic reduction in neutral or acidic solution. Based on rapidity of reaction and high yields of products of reduction the use of palladium on carbon gives the best results [1, 2].

Benzyl Aryl Ethers. There are many examples of successful hydrogenolyses of these compounds to the corresponding phenolic structures [1]. There does not appear to be any group attached to the aryl portion of benzyl aryl ethers that has a strongly deterring effect on cleavage between the benzyl group and the oxygen atom. The inability to convert 4-benzyloxynaphthoic acid to 4-hydroxynaphthoic acid in alcohol, ethyl acetate, or acetic acid with palladium on carbon or with Raney nickel might be attributed to the effect of the carboxyl group, since reduction as methyl ester with Raney nickel at room temperature and atmospheric pressure was successful [22a]. On the other hand, the lack of success may be an isolated instance. Hydrogenation as free acid in other aromatic systems generally gave good results. 2,5-Dibenzyloxybenzoic acid was converted to 2,5-dihydroxybenzoic acid, gentisic acid, in 84% yield by reduction with palladium on carbon [22b]. 3-Alkoxy-4-benzyloxybenzoic acids, where the groups in 3-position were isopropoxy, *n*-butyloxy, isobutyloxy, and 4-methylbutyloxy, were hydrogenolyzed to the corresponding 3-alkoxy-4-hydroxybenzoic acids with 5% palladium on carbon [22c]. No difficulty was noted when 3-benzyloxypicolinic acid hydrochloride was hydrogenated in 50% alcohol over palladized carbon; yield of 3-hydroxypicolinic acid hydrochloride was 77% [22d].

Reductions as esters gave 79–90% yield of alkyl 3,4,5-trihydroxybenzoates from alkyl 3,4,5-tribenzyloxybenzoates after reaction in hot alcohol with palladium on carbon under 2–3 atm of hydrogen [23]. In some instances the ratio of supported catalyst to substrate was 50% but the catalyst was reused thirty times. There was no comparison to determine whether debenzylations would be as successful when the free acid was employed.

Cleavage of the benzyl group in the presence of a strong acidic function, the sulfonamide group, took place very rapidly in the palladium reduction of 2-benzyloxybenzenesulfonamide [24]. It is difficult to say whether the ease of conversion to 2-hydroxybenesulfonamide was due to the effect of the sulfonamide group, the reaction medium (acetic acid), or to the large amount of catalyst employed, 50% by weight, since no other experiments were reported.

Groups adjacent to the benzyl ether do not have too great an effect on hydrogenolysis except perhaps to hinder reaction somewhat if the adjacent group is large. 3-Methyl-6,7-dimethoxy-1-(2-benzyloxy-4-methoxyphenyl)-isoquinoline was reduced to the corresponding 1-(2-hydroxy-4-methoxyphenyl)isoquinoline in 96% yield with palladium on carbon [25a]. A more hindered compound, the corresponding 1-(2-benzyloxy-3,4-dimethoxyphenyl)isoquinoline was rapidly debenzylated by using a 35% ratio of palladium on carbon [25b]. 8-Benzyloxy-6,7-dimethoxy-2-methyl-1-(4-methoxyphenyl)-1,2,3,4-tetrahydroisoquinoline in alcoholic hydrogen chloride was hydrogenolyzed to 8-hydroxy-6,7-dimethoxy-2-methyl-1-(4-methoxyphenyl)1,2,3,4-tetrahydroisoquinoline in 80% yield in less than 1 hr in the presence of 10% palladium on carbon [25c]. 2,4-Dibenzyloxy-5-substituted pyrimidines, in which the 5-substituent was a sugar moiety, were readily reduced to the corresponding 5-substituted uracils with palladium on carbon [25d]. A completely substituted furan, 3,4-dibenzyloxyfuran-2,5-dimethanol dibenzoate, was reduced without incident over palladium on carbon in tetrahydrofuran containing powdered potassium carbonate [25e]. Potassium carbonate was used to neutralize benzoic acid resulting from hydrogenolysis of the O-benzoylfuranmethanol portion of the molecule. Uptake of hydrogen amounted to four equivalents to yield 4-hydroxy-2,5-dimethyl-3(2H)-furanone (Eq. 19.13).

$$C_6H_5COOH_2C \underset{C_6H_5CH_2O}{\overset{O}{\diagup\diagdown}} \underset{OCH_2C_6H_5}{CH_2OOCC_6H_5} \xrightarrow{4H_2}$$

$$\left[CH_3 \underset{HO}{\overset{O}{\diagup\diagdown}} CH_3 \atop OH \right] \rightleftharpoons H_3C \underset{HO}{\overset{O}{\diagup\diagdown}} \underset{O}{\overset{H}{\diagdown}} CH_3 \quad (19.13)$$

When methyl 2,3-dibenzyloxybenzoate in alcohol was hydrogenated over palladized carbon, the reaction was stopped after absorption of one molar equivalent; the more hindered 2-substituent was cleaved to yield 88% of methyl 3-benzyloxy-2-hydroxybenzoate [25f]. This may suggest that a bulky group adjacent to a benzyl ether aids hydrogenolysis.

Hydrogenolyses of benzyl aryl ethers, $C_6H_5CH_2OAr$ where Ar is an unsaturated nitrogen-containing heterocycle, to the corresponding free hydroxy compounds are carried out in the same manner as when Ar is a benzenoid system, except that acidic media may be necessary to counteract the effect of the nitrogen atom on activity of the catalyst. Palladium on carbon was employed in the following examples. Its use in the pyridine series has been described [22d]. 4-Amino-5-benzyloxypyrimidine was converted to 4-amino-5-hydroxypyrimidine in 72% yield [26a], and 2-amino-4,5-dihydroxypyrimidine was obtained in 76% yield from the 5-benzyloxy compound [26b]. 2-Benzyloxy-5-methoxypyrazine was reduced to 2-hydroxy-5-methoxypyrazine and 2,5-dibenzyloxypyrazine was hydrogenated to 2-hydroxy-5-benzyloxypyrazine or to piperazine-2,5-dione when hydrogen uptake was allowed to continue [26c] (Eq. 19.14).

6-Benzyloxy-2-fluoro- and 2-methoxypurines were hydrogenolyzed in alcohol to yield the corresponding 2-substituted 6-hydroxypurines [26d], and 5-benzyloxyimidazo[1,2-a]pyridine gave the tautomeric imidazo[1,2-a]pyridine-5-(1H)-one, not the 5-hydroxy compound [26e] (Eq. 19.15).

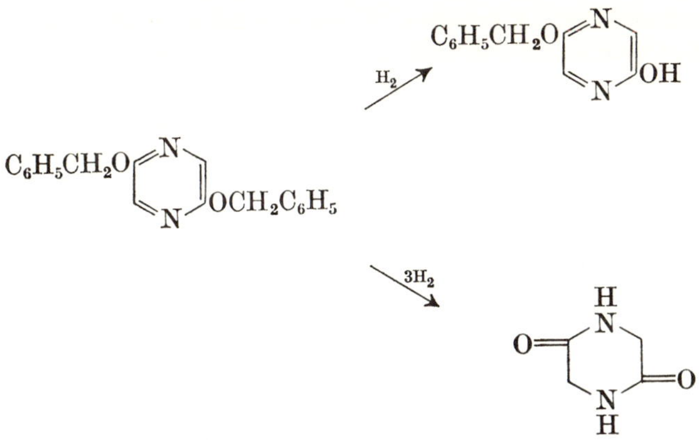

O-Debenzylation and Related Reactions

(19.14)

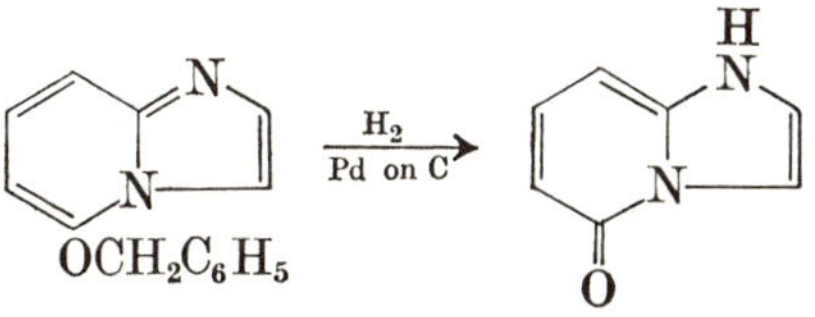

(19.15)

Palladium oxide catalyzed the removal of the benzyl group of 2-benzyloxy-8-(3-diethylaminopropyl)amino-6-methoxyquinoline in alcohol containing acetic acid to give 8-(3-diethylaminopropyl)amino-6-methoxycarbostyril in 91% yield [26f]. Hydrogen uptake was slow in alcohol. Debenzylation would not take place when reduction of the dihydriodide salt was attempted with various palladium catalysts or with Raney nickel. This is not very surprising since hydrogen iodide and iodide salts often inhibit the activity of hydrogenation catalysts (iodide ion is a fairly strong catalyst poison).

No difficulty should be anticipated in the cleavage of benzyl ethers when the ether group is attached to the benzene ring of heterocycles such as quinoline, isoquinoline, indole, and others. Palladium on carbon is the favored catalyst and usually gives the best results [2, 27a, b, c, d, e, f].

Benzyl Ethers of Aliphatic Compounds. Palladium on carbon was used to cleave the benzyl group of DL-2-acetylamino-4-benzyloxy-2-methylbutyric acid [28a]; the hydroxy acid was not isolated but was converted to lactone. The same catalyst was used to remove the benzyl group of the N-ethylcarbamate of 1-benzyloxy-2-propanol, $C_6H_5CH_2OCH_2CH(CH_3)$-$OC(=O)NHC_2H_5$ [28b]. Palladium on carbon from palladium nitrate catalyzed the hydrogenation of N-(1-benzyloxy-2-propyl)-3-(N-methyl-1,2,3,4-tetrahydro-2-naphthyl)propionamide to the N-(1-hydroxy-2-propyl) amide [28c]. The dimethylacetal of 2-benzyloxy-D-glyceraldehyde-3-diphenylphosphate as cyclohexylamine salt was reduced over palladium on carbon prepared from palladium chloride. The catalyst was prehydrogenated and washed to remove any acid present before it was used for the hydrogenolysis. The corresponding 2-hydroxy compound was obtained [28d]. A similarly prepared palladium catalyst was employed to remove the benzyl protecting groups from furanoses and pyranoses [28e, f, g]. Palladium hydroxide or palladium on a carrier prepared from palladium hydroxide may be useful in reductions of these compounds, because neither contains occluded mineral acid which could have an adverse effect on the substrate.

Miscellaneous Benzyl Ethers. Benzyl 4-(4-pyridyl)butyl ether, an aralkyl benzyl ether, was hydrogenated in acetic acid over 5% palladium on carbon to give 78% of 4-(4-pyridyl)butanol [29a]. 2-Phenyltetrahydropyran, which may be considered as a benzyl cyclic ether, was cleaved in acetic acid containing perchloric acid to yield 5-phenylpentanol upon hydrogenation with palladium on carbon [29b] (Eq. 19.16). 2-Phenyl-1,3-dioxane in acetic acid

$$\text{(tetrahydropyran-}C_6H_5) \xrightarrow{\text{Pd on C}} C_6H_5(CH_2)_5OH \qquad (19.16)$$
$$72\%$$

containing sulfuric acid also underwent ring opening to give 75% yield of 2-hydroxyethyl 2-phenethyl ether, $C_6H_5(CH_2)_2O(CH_2)_2OH$. When 2,3-diphenyl-1,4-dioxane was hydrogenated in acetic acid containing perchloric acid with palladium on carbon, complete cleavage occurred to ethylene glycol and 1,2-diphenylethane [29b]. Reduction in acetic acid at 90° with 10% palladium on barium sulfate gave the same result [29c]. Hydrogenation of 2,5-diphenyl-1,4-dioxane under the same conditions yielded only 2-phenylethanol.

Benzyl Acetals

Palladium catalysts are favored for hydrogenolysis of benzyl acetals, although platinum catalysts have been used in neutral solvent [1]. Reduction in acidic medium with platinum will also lead to hydrogenolysis but it may be accompanied by ring reduction [6] so that cleavage will be incomplete. In the platinum-catalyzed reduction of 1,2-benzylidene-D-glucofuranose in acetic acid, 1,2-hexahydrobenzylidene-D-glucofuranose was obtained in 88% yield [30]. Reaction in alcohol with palladium gave D-glucose.

Arylmethyl Carbanilates

Hydrogenations of the phenylurethans of arylmethanols, $ArCH_2OC(=O)$-NHC_6H_5 where Ar was 2-, 3-, and 4-pyridyl, were carried out with palladium oxide in methyl alcohol, and with palladium oxide in dioxane when Ar was

2-furyl [31]. Very high yields of the corresponding methyl compounds were obtained. There was no attempt to determine whether the procedure was superior to direct hydrogenolysis of the parent alcohol or to hydrogenolysis of the acetate of the arylmethanols.

N-DEBENZYLATIONS

The removal of N-benzyl and protecting groups does not take place with the same ease as the hydrogenolysis of O-benzyl groups, except when the protecting group is carbobenzyloxy. This group is usually removed so readily that platinum or nickel may also be used in addition to the preferred catalyst, palladium.

Palladium catalysts are not only the most frequently used for all other type N-debenzylations [1], but are also the ones of choice [1, 2]. Platinum catalysts have been employed but are much less efficient [2]. Raney nickel usually requires far more vigorous conditions than the mild ones used in palladium-catalyzed reductions. Rhodium has a greater tendency to cause saturation of the aromatic ring in preference to debenzylation [2, 32]. Of the palladium catalysts, those on carriers are preferred over palladium oxide because much less actual metal is required for reductions with the supported catalysts. Because of the large amount of metal involved in palladium oxide debenzylations, close to 10% by weight of substrate, ring reduction may take place. This occurred in the hydrogenation of benzyl-dimethylanilinium chloride, $C_6H_5\overset{+}{N}(CH_3)_2CH_2C_6H_5\ Cl^{\ominus}$ with it. The N-benzyl group was removed but the product of reduction was N,N-dimethylcyclohexylamine [33].

5% or 10% palladium on carbon appears to be the most useful or even preferred palladium catalyst [1, 2, 10a, 34a, b]. Palladium hydroxide on carbon containing 20% metal [34c] was substituted in some N-debenzylations which had been unsuccessful with 10% palladium on carbon prepared from palladium chloride [34d]. In this instance difficulty with 10% palladium on carbon may have been due to lack of purification of the substrate, or success due to the higher ratio of metal to substrate in the palladium hydroxide-catalyzed hydrogenolyses. More useful information about the superiority of this catalyst would have resulted from a comparison of its reducibility with 20% palladium on carbon prepared from palladium chloride.

Aqueous media may be employed for debenzylations; 75–80% aqueous alcohol and 75–80% acetic acid are both useful. The amount of water should be kept low enough so that the catalyst remains free-flowing during

414   Debenzylation and Related Reactions

the release of toluene (or other hydrocarbons depending on the protecting agent). Otherwise the catalyst may become a smeary mass coated either with hydrocarbon, insoluble in the aqueous solvent, or with water, insoluble in the hydrocarbon-containing solvent. In either case hydrogen uptake will be impaired. When this occurs the addition of alcohol should improve the reaction rate [2].

A study confined to N-benzylamines showed that nuclear substitution decreased the ease of hydrogenolysis [10a, 34a] but never to the extent of preventing removal of the protecting group [2]. A methyl group in α-position also lessened the ease of hydrogenolysis [34b]. α-Alkyl substitution caused a far greater degree of difficulty than nuclear substitution [2].

A comparison of hydrogenolyses of benzyl-like protecting groups revealed the following order:

1-menaphthyl > 2-menaphthyl > 9-fluorenyl

> benzhydryl > biphenylmethyl > benzyl [35a, b].

Debenzylation of quaternary amines was shown to take place more readily than debenzylation of tertiary amines [34b]. Removal of the protecting group from secondary amines takes place less readily. Deamination of benzylamine was reported to be extremely difficult [1, 33].

The presence of acid is not always necessary in N-debenzylations, though it will prevent inactivation of the catalyst when the amine is a strong base. Reduction of quaternary compounds usually does not require the addition of acid because of the release of HX after debenzylation. The presence of acid may also be unnecessary in the reduction of N-benzylanilines and related benzylated aromatic amines, because they and the anilines resulting from hydrogenolysis are comparatively weak bases and should have little inhibiting effect on the activity of the catalyst. If a mineral acid is to be added a great excess will prevent hydrogenolysis [2, 34a].

Carbobenzyloxy Groups

Cleavage of the carbobenzyloxy group is usually accomplished so readily that there is little danger of reducing other functions except the most reactive ones. In much of the earlier work palladium black was employed as catalyst [1]. Most of the work after 1950 has been carried out with palladium on a support. In general, on a weight basis of metal to substrate, much less supported catalyst is necessary [2]. Hydrogenolysis may be carried out in neutral solvent or under acid conditions [1]. Reduction gives toluene and carbon dioxide, the latter due to spontaneous cleavage of an intermediate carbamate:

$$C_6H_5CH_2OC(=O)NHR \xrightarrow{H_2} \text{toluene} + [HOC(=O)NHR] \longrightarrow RNH_2 + CO_2$$

The amount of carbon dioxide released is equal to the molar amount of hydrogen absorbed. In analytical procedures an equivalent amount of sodium hydroxide should be present to trap carbon dioxide in order to get the exact hydrogen absorption. At other times, depending on the solvent, unless a very concentrated solution of substrate is reduced, the pressure of hydrogen above gravity is usually sufficient to keep the carbon dioxide in solution so that it will not substantially affect a reading of the drop in pressure [2].

There are many examples of the hydrogenolytic removal of the protecting carbobenzyloxy group among amino acids and peptides. Glycyl-L-histidyl-L-tyrosine and L-seryl-L-tyrosine were obtained from the carbobenzyloxy compounds by reduction with palladium on barium sulfate [36a]. Palladium on carbon catalyzed the hydrogenolysis of the carbobenzyloxy group of the ester of a tripeptide in alcohol containing hydrogen chloride [36b]. Palladium black in 75% acetic acid was used for the reduction of 2,6-(dicarbobenzyloxyamino)pimelic acid to give a 60% yield of 2,6-diaminopimelic acid, $HOOCCH(NH_2)(CH_2)_3CH(NH_2)COOH$ [36c]; hydrogenation of the corresponding pimelic acid diamide gave 2,6-diaminopimelic acid diamide. L-Leucyl-4-methoxy-2-naphthylamide was obtained by hydrogenolysis of the carbobenzyloxyleucyl derivative in methyl alcohol over 10% palladium on carbon [36d]. The hydrogenolysis of 3-(4-carbobenzyloxyglycylaminophenyl)propionic acid, $4\text{-}C_6H_5CH_2OC(=O)NHCH_2CONHC_6H_4(CH_2)_2COOH$, in 50% acetic acid with a 40% ratio of 5% palladium on carbon was fairly rapid; reaction yielded 94% of 3-(4-glycylaminophenyl)propionic acid [36e].

When esters of N-carbobenzyloxyaminoacids are reduced, removal of the protecting group may be less difficult than when hydrogenations of the free acids are attempted. Difficulty may also be due to impurities. It may be advisable to pretreat the substrate with Raney nickel before hydrogenolysis [2].

The presence of a strong acidic group did appear to hamper cleavage in the hydrogenation of 4-carbobenzyloxyaminobenzenesulfonamide to sulfanilamide, $4\text{-}NH_2C_6H_4SO_2NH_2$, with palladium on carbon in water [36f]. Hydrogen uptake was slow as it was in the reduction of N-carbobenzyloxysulfanilic acid as sodium salt. It is possible that the difficulty in both instances may have been due to coating of the catalyst because of reaction in water. Better results might have been obtained in 50% alcohol.

416   Debenzylation and Related Reactions

Hydrogenolysis of a series of diethyl carbobenzyloxyglycylaminomalonates with palladium on carbon gave 90–97% yields of 3-carbethoxypiperazine-2,5-diones [37a] (Eq. 19.17). The reduction of ethyl α-carbobenzyloxyamino-

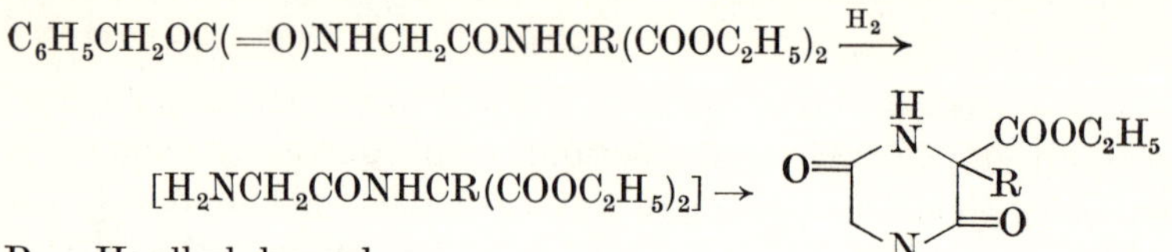

$$C_6H_5CH_2OC(=O)NHCH_2CONHCR(COOC_2H_5)_2 \xrightarrow{H_2}$$

$$[H_2NCH_2CONHCR(COOC_2H_5)_2] \rightarrow$$

R = H, alkyl, benzyl

(19.17)

α-2-carbethoxyethylmalonate with palladium on carbon yielded 5,5-dicarbethoxy-2-pyrrolidine by removal of the carbobenoxy group and subsequent intramolecular condensation [37b] (Eq. 19.18). Palladium on

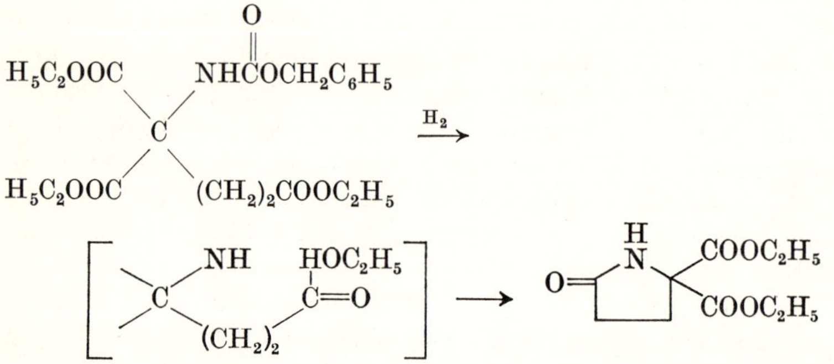

(19.18)

carbon also catalyzed the hydrogenolysis of dipotassium α-carbobenzyloxyamino-α-5-pyrimidylmethyl malonate in water under 4 atm pressure [37c]; 57% of the monopotassium salt of α-amino-α-(5-pyrimidylmethyl)malonic acid was obtained.

The removal of the carbobenzyloxy group from compounds which yield strongly basic amines may be carried out in neutral solution or in the presence of acid. In general the expected product is obtained. 7-Amino-2,3,7,8-tetrahydro-6-methyl-1*H*-cyclopenta[7,8]naphtho[2,3-*b*]furan was obtained by hydrogenolysis of the 7-carbobenzyloxyamino compound with palladium on carbon in alcohol [38a]. 5-Amino-5-phenyl-2-piperidone was obtained from reduction of the 5-carbobenzyloxyamine in acetic acid with palladium on carbon [38b]. 1,2-*trans*-Diaminocyclopropane resulted from cleavage of the *N*-carbobenzyloxy groups in methyl alcohol containing hydrochloric acid in the presence of hydrogen and palladium on barium sulfate [38c]. The hydrogenation of 4-carbobenzyloxyamino-1-phenyl-2-pyrrolidone with palladium on carbon in a dioxane-acetone mixture, which contained sodium hydroxide solution, yielded 78% of 4-amino-1-phenyl-2-pyrrolidone [38d].

When the hydrogenolysis of 8-carbobenzyloxy-3,8-diazabicyclo[3,2,1]-octane-2,4-dione was carried out with palladium on carbon in 80% methyl alcohol, the expected product was obtained in low yield [39a]. The major product, ethyl 5-carbamylpyrrolidine-2-carboxylate, apparently resulted from reaction of the hydrogenolysis product with alcohol (Eq. 19.19). The

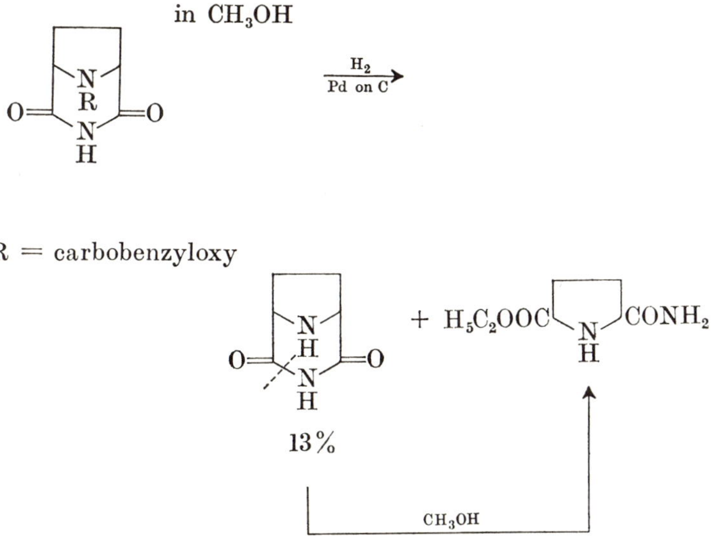

in CH$_3$OH

$\xrightarrow{\text{H}_2 \atop \text{Pd on C}}$

R = carbobenzyloxy

+ H$_5$C$_2$OOC―N(H)―CONH$_2$

13%

CH$_3$OH

(19.19)

expected product, 4-aminooxazolidin-2-one, was not obtained from the removal of the protecting group when 4-carbobenzyloxyaminooxazolidin-2-one was hydrogenated in alcohol over palladium on carbon [39b]. The expected primary amine, apparently due to its instability, reacted with itself to split out ammonia and yield secondary amine (Eq. 19.20). When

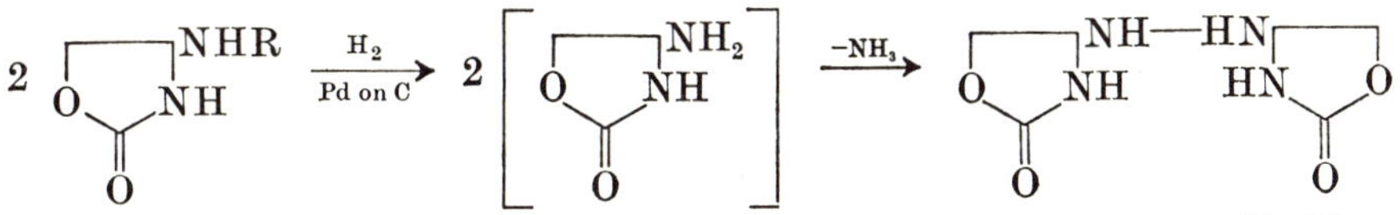

$2\;\begin{array}{c}\text{NHR}\\ \text{O}\diagup\text{NH}\\ \|\\ \text{O}\end{array} \xrightarrow{\text{H}_2 \atop \text{Pd on C}} 2\left[\begin{array}{c}\text{NH}_2\\ \text{O}\diagup\text{NH}\\ \|\\ \text{O}\end{array}\right] \xrightarrow{-\text{NH}_3} \begin{array}{c}\text{NH—HN}\\ \text{O}\diagup\text{NH}\quad\text{HN}\diagdown\text{O}\\ \|\qquad\qquad\|\\ \text{O}\qquad\qquad\text{O}\end{array}$

R = carbobenzyloxy (19.20)

N-carbobenzyloxy-O-acyl-2-aminophenols were hydrogenated over palladium on carbon the carbobenzyloxy group was removed, but the expected migration of the acyl group took place yielding 2-acylaminophenols [39c]. An unusual effect of pressure was noted in the palladium-catalyzed reduction of N-carbobenzyloxy-N-(2-fluoroethyl)aniline [39d]. Reaction under 1 atm pressure of hydrogen yielded N-(2-fluoroethyl)aniline; reduction under 4 atm pressure gave N-(2-fluoroethyl)cyclohexylamine.

N-Benzylamines

Quaternary Amines. The ease of removal of an N-benzyl group from a benzyl quaternary ammonium compound with palladium on carbon catalyst has been noted [34b]. Additional examples are seen in the hydrogenations of *bis* benzyl quaternary salts of N,N,N,N-tetramethyl and other N-tetrasubstituted alkyl and aralkylalkanediamines [40a]; removal of the benzyl groups was very rapid in reactions at 3 atm pressure; yields of tertiary amines were high. 4-Methyl-, 2,4-dimethyl-, and 4-phenethyl-2-oxo-1-phenyl-4-benzylpiperazinium chlorides in methyl alcohol were reduced to the corresponding 4-alkyl- or 4-phenethyl-1-phenylpiperazin-2-ones in over 90% yield in 10 min under 3 atm pressure of hydrogen [40b]. The methosulfates or methobromides of 1-benzyl-1-methyl-4-substituted piperazines were hydrogenated in ethyl alcohol in a similar manner [2]. The benzyl group in 10-aza-γ-carboline methobromide was also easily removed [40c] (Eq. 19.21).

$$\underset{\text{Br}^-}{\text{[benzimidazo-piperazinium]}} \xrightarrow[\text{Pd on C}]{\text{H}_2} \text{[benzimidazo-piperazine]}\cdot\text{NCH}_3 \cdot \text{HBr} \qquad (19.21)$$

Palladium-catalyzed cleavage of the N-benzyl group from benzylpyridinium and benzylquinolinium halides is of particular interest because hydrogenolysis occurs without saturation of the nitrogen-containing ring. Benzyl 4,5-dicarbomethoxy-3-hydroxy-2-methylpyridinium chloride was converted to dimethyl 3-hydroxy-2-methylpyridine-4,5-dicarboxylate by hydrogenation over palladium on carbon [41a], and 4-hydroxy- and 4-hydroxy-6,7-dimethoxyisoquinolines were obtained after hydrogenation of the benzyl quaternary chlorides in acetic acid with palladium on carbon [41b]. In contrast, reduction of related quaternary salts with platinum and rhodium yielded saturated nitrogen-containing rings with the N-benzyl group intact [2, 41c].

Tertiary Amines. Removal of the benzyl group or related groups from tertiary amines, $RR'NCH_2C_6H_5$ where R and R' are dissimilar groups which are alkyl, aralkyl, cycloalkyl, heteroalkyl, or saturated heteroalkyl, is readily accomplished. In a reduction related to this type, cleavage of the benzylamino bridge of the morphanthridine ring took place to give a substituted diphenylmethane when the amount of catalyst used was equal to the amount of substrate [41d].

Many examples of hydrogenolyses of N-benzylated tertiary amines are available [1]. In the earlier literature, platinum as well as palladium catalysts were employed. The more recent references overwhelmingly advocate palladium.

3-β-Methylaminocholestane was obtained from reduction of 3-β-(N-benzyl-N-methylamino)cholestane with palladium on carbon in alcohol [42a]. *trans*-N-Benzyl-N-methyl-2-cyclohexyloxycyclopropylamine was reduced to *trans*-N-methyl-2-cyclohexyloxycyclopropylamine with palladium on carbon in alcoholic hydrogen chloride [42b]. 3-(N-Benzyl-N-methylamino)-5-methyl-1-phenyl-2-pyrrolidone was hydrogenolyzed with palladium on carbon prepared *in situ* to yield 65% of 5-methyl-3-methylamino-1-phenyl-2-pyrrolidone [42c]. N-Benzyl-N,N-*bis*-3-(*cis*- and *trans*-3- and 4-methylcyclohexyl)propylamines were converted to the corresponding N,N-*bis*-(methylcyclohexyl)propylamines [42d]. 4-Methylamino-1,2-diphenylpyrazolidine-3,5-dione was obtained in 80% yield from hydrogenolysis of the corresponding 4-N-benzyl-4-methylamino compound with palladium on carbon [42e]; 4-butylamino-1,2-diphenylpyrazolidine-3,5-dione was prepared in a similar manner. 5-Ethylamino- and 5-methylamino-1,2,3,4-tetrazole resulted from palladium-catalyzed hydrogenations of 5-N-benzyl-N-ethyl and methylaminotetrazoles [42f].

Palladium on carbon catalyzed the hydrogenolysis of 4-(N-benzyl-N-cyclohexylamino)-2-hydroxyphenyl)-2-phenyl-N-cyclohexylbutyramide [42f]. Reduction gave 1-cyclohexyl-3-(2-hydroxyphenyl)-3-phenyl-2-pyrrolidone (Eq. 19.22). In the reaction where R was butyl, cyclohexylamine was split out to give 1-butyl-3-(2-hydroxyphenyl)-3-phenyl-2-pyrrolidone.

$$R = C_6H_{11} \text{ or } C_4H_9, \quad R' = C_6H_{11}$$

(19.22)

Palladium on carbon was employed for the removal of the benzyl group of some substituted alkanediamines, RR'NYN(CH$_3$)CH$_2$C$_6$H$_5$ where Y is an alkylene chain, R = phenyl, and R' = propionyl [43d], where R = phenyl and R' = acetyl [43b], and where R and R' = H [43c]. Yields were 71–75%. It was also used for the preparation of N^1,N^4-*bis*-(2-hydroxyethyl)butane-1,4-diamine [43d]. Reactions were carried out with the salts in neutral solvent or with the bases in acidic solution.

Debenzylations among tertiary amines, $\diagdown$NCH$_2$C$_6$H$_5$ where the tertiary nitrogen atom is part of a saturated heterocyclic ring, are readily performed over palladium catalyst. Since these compounds are strong bases, reductions are best carried out in acid solution or as salts in neutral or acid solution. When the base is reduced in neutral solvent, it may be advisable to use 15–20% by weight of supported palladium and/or to warm the reaction to 50–60° to avoid slow uptake of hydrogen [2].

Typical examples are seen in the reduction of 1-benzyl-3-hydroxyazetidine hydrochloride with palladium on carbon to 3-hydroxyazetidine [44]. The 1-benzyl group was readily hydrogenolyzed from a number of 1-benzyl substituted pyrrolidines with palladium on carbon to give the corresponding pyrrolidines [45a, b, c, d]. Dimethyl 1-benzylpyrrolidine-2,5-dicarboxylate as hydrochloride was converted to dimethyl pyrrolidine-2,5-dicarboxylate in 99% yield by reduction with palladium on carbon at 3.3 atm pressure [45e]. When the corresponding diethyl ester underwent debenzylation as base, the reaction was carried out at 40° and 30 atm pressure [45f]; yield was 90%. High yield was also obtained in the same reaction at room temperature and 3 atm pressure with a 20% weight ratio of 5% palladium on carbon; uptake of hydrogen was fairly rapid but was rather slow when the amount of catalyst was reduced [2]. In a large run, reduction as base in concentrated solution (190 g in 500 cc of alcohol) was carried out under 60 atm pressure to avoid refilling with hydrogen. Reaction was exothermic, hydrogen uptake was complete in less than 0.5 hr. However, the yield of diethyl pyrrolidine-2,5-dicarboxylate was only 45%. Self-condensation also took place to yield an imide, apparently due to the exothermic reaction (Eq. 19.23).

References 12f, and 46a, b, c, d are a few of the many available examples of hydrogenolyses of 1-benzyl substituted piperidines. Reductions of 1-benzyl-4-substituted piperidines, where the 4-substituent was dimethylamino, 1-pyrrolidino, or 1-piperidino, were carried out as dihydrochloride salts in 50% aqueous isopropyl alcohol at room temperature and atmospheric pressure to give 80–83% yields of the 4-substituted piperidines [46e].

Hydrogenolyses of 1-benzyl-4-(substituted aryl)piperazines were carried out as bases in 80% acetic acid or as hydrochloride salts in aqueous alcohol with 10–20% by weight of 5% palladium on carbon at 60° and 1–2 atm [2]. When substrates were pure, hydrogen uptake was usually complete in about

$$\text{ROOC}\underset{\underset{\text{CH}_2\text{C}_6\text{H}_5}{|}}{\text{N}}\text{COOR} \xrightarrow[\substack{5\% \text{ Pd on C} \\ (15\% \text{ by weight}) \\ 25°, 60 \text{ atm} \\ \text{exothermic}-60-70°}]{\text{H}_2}$$

$$\text{ROOC}\underset{\underset{\text{H}}{|}}{\text{N}}\text{COOR} + \left[\text{HN}\begin{matrix}\text{COOR}\\ \\ \text{COOR}\end{matrix}\right] \longrightarrow \text{ROOC}\underset{\underset{\text{H}}{|}}{\text{N}}\text{CON}\begin{matrix}\text{COOR}\\ \\ \text{COOR}\end{matrix}$$

45%

(19.23)

$R = C_2H_5$

1 hr. Some difficulty was noted in the hydrogenolyses of 4-aryl-1-benzyl-2,6-dimethylpiperazine. A larger amount of catalyst, such as was employed for the preparation of cis-2,6-dimethyl-piperazine from 1-benzyl-cis-2,6-dimethylpiperazine [47a], probably would have increased the reaction rate. The presence of a substituent in 2-position did not appear to have too strong a hindering effect on removal of the 1-benzyl group. 1-Benzyl-3,4-dimethyl-2-phenylpiperazine was reduced in acidified aqueous alcohol to 1,2-dimethyl-3-phenylpiperazine in 94% yield in about 5 hr with a 6% ratio of 10% palladium on carbon [47b]. There is also little danger of ring reduction with supported palladium catalysts during debenzylation in acidified solvent. 1-Benzyl-4-(4-pyridyl)piperazine was converted to 1-(4-pyridyl)piperazine in 85% yield by hydrogenation in methanolic hydrogen chloride with 40% by weight of palladized carbon [47c].

Examples of N-debenzylations in other heterocyclic systems are seen in the preparation of isoquinuclidine [48a], 4-amino- and 4-hydroxy-1-azacycloheptane [48b], 3,4-diazabicyclo[3,3,1]nonane [48c], 2-tosyl-2,5-diazabicyclo-[2,2,1]heptane [48d], 3-methyl-1,2,4,5-tetrahydro-1,3-benzodiazepine [48e], and 7-methoxy-1,2,3,4-tetrahydro-9H-pyrido[3,4-b]indole [48f]. All were obtained by reduction of the corresponding N-benzyl compounds with palladium on a support.

Difficulties encountered in the N-debenzylation of complex saturated nitrogen heterocycles may be entirely due to structure. In such cases molecular models may show why debenzylation will take place in one instance and not in another. The benzyl group of N-benzylpiperidino[4,3,2-a,b]-1,2,3,4-tetrahydroisoquinoline was readily removed by hydrogenation in alcohol with 20% by weight of 10% palladium on carbon [49a]. Debenzylation did not take place during attempts to reduce *N*-benzylpiperidino-[5,4,3-a,b]-1,2,3,4-tetrahydroisoquinoline with platinum oxide, Raney nickel, palladium oxide, or palladium on carbon [49b]. *N*-Benzylpiperidino[5,4,3-b,c]-4-phenylchromane was debenzylated rapidly with palladium on carbon [2]. The three reactions are seen in Eq. 19.24.

$$R = H \text{ or } C_6H_5 \tag{19.24}$$

The difference in reducibility may be seen if molecular models are made. One of benzylpiperidino[5,4,3-a,b]tetrahydroquinoline shows that the benzylated nitrogen atom in ring C is out of plane. Flipping over the model does not produce contact between the catalyst and the *N*-benzyl portion of the molecule [49c].

Models of the two compounds which were debenzylated suggest that in each case the *N*-benzyl group is relatively unobstructed in its approach to the catalyst [49c].

Difficulty has been experienced in some hydrogenolyses when the *N*-benzyl group is attached to a weakly basic nitrogen atom which is part of an unsaturated heterocyclic system. Catalytic reduction of 1-benzylindole gave poor results [50a]; hydrogenolysis of 1-benzylimidazole with palladium oxide was unsuccessful [33]. The benzyl group could not be removed from some 1-benzylpyrazoles in neutral solvent or in solvent containing acetic acid by hydrogenation with 40% by weight of 5% palladium on carbon at 3 atm pressure and temperatures up to 70° [2]. 1-Benzyl-1,2,3-triazole-4,5-dicarboxylic acid could not be reduced to 1,2,3-triazole-4,5-dicarboxylic acid with palladium on carbon at 2–3 atm pressure [50b]. In most of these reactions there was little or no uptake of hydrogen. Rapid uptake was observed when the last reduction was carried out at 150–200° and 70 atm but the yield was poor. Hydrogenation of 1-benzyl-1,2,3-triazole in dioxane to 1,2,3-triazole was quite successful at 100° and 90–100 atm in the presence of palladium on alumina; uptake of hydrogen was complete in 2 hr; yield was 93% [50c]. 1,2,3-Triazole was also obtained from the 1-benzyl compound by reduction with palladium on carbon at 175° and 80 atm [49b]. The yield was much lower, 47% along with 32% of recovered starting material, and reaction time was far longer, 84 hr, in this reduction. It is not possible to evaluate the effect of the solvent or catalyst support in the two reactions because in Ref. 49b much less catalyst was employed than in Ref. 49c.

In contrast to the poor results in the previously noted low-pressure hydrogenolyses, removal of the benzyl group from the tetrazole ring system took place with comparative ease with palladium oxide [50d, e, f] and with palladium on carbon [50g]. The N-benzyl group of a purine nucleoside was also removed by hydrogenolysis with palladium on carbon in an alcohol-water mixture at 80° and 3 atm [51]; the resulting product, 7-β-D-ribofuranosyladenine, was obtained in 31% yield. There was, however, no explanation of the low yield of mention of whether unchanged starting material was recovered.

MANNICH BASES. The hydrogenolysis of Mannich bases, $ArCH_2N(R_2)$ where $N(R)_2$ is dialkylamino or when it is a part of a saturated heterocycle, is related to removal of the benzyl group from N-benzyl tertiary amines. Reduction does not proceed with the same ease, however. Any conclusions to be drawn about optimum conditions and the best catalysts for their hydrogenolyses are complicated by their relative instability. In certain instances the presence of water or hydroxylated solvent may cause reversal of the base to its original components [2].

Hydrogenolyses of Mannich bases, where Ar is a phenol, have been carried out at elevated temperature and pressure with Raney nickel [52a] and with copper chromite [52b]. When Mannich bases of naphthols are reacted under similar conditions, hydrogenolysis is often accompanied by formation of methyl tetrahydronaphthols [52c]. The reduction of 1-dimethylamino-2-naphthol with W-7 Raney nickel, a very active catalyst, at 50–60° and 100 atm initial pressure yielded 1-methyl-2-naphthol when the reaction was interrupted at one molar equivalent [52c]. A low-pressure reaction could be carried out with the same catalyst when a minimum pressure of 2.4 atm was maintained; no uptake of hydrogen occurred at lower pressure. Unfortunately W-7 Raney nickel must be freshly prepared and does not maintain its high activity for a very long period. It is possible that an active nickel catalyst prepared by simpler means might be successful for low-pressure hydrogenolysis of this and related Mannich bases.

Palladium catalysts give the best results in the hydrogenolysis of Mannich bases of phenols and usually do not affect the aromatic ring. 5-Acetylamino-1-(1-piperidinomethyl)-2-naphthol was hydrogenated to 5-acetylamino-1-methyl-2-naphthol in 90% yield when the reaction was carried out in 50% aqueous methyl alcohol containing oxalic acid with 66% by weight of 1% palladium on strontium carbonate at 40° and 1 atm pressure [52c]. Reductions of other Mannich bases in the series with the same catalyst in neutral solution gave the expected products, but hydrogen uptake was never complete and yields were low. It is likely that palladium on carbon would have given better results [2].

The hydrogenolysis of 2-dimethylaminomethylphenol was reported to be controlled by the presence of groups *meta* to the basic side chain when reductions were carried out over 10% palladium on barium sulfate in 95% alcohol containing less than one equivalent of mineral acid. 2-Dimethylaminomethyl-4-nitrophenol and 2-dimethylaminomethyl-6-methoxy-4-nitrophenol were converted to the corresponding aminocresols by reduction under those conditions with 30% by weight of catalyst [52d, e]. Excess acid prevented debenzylation. Hydrogenolysis apparently took place before the nitro group was completely reduced; no reaction occurred when the amino group was substituted for nitro. Hydrogenolysis appeared to be dependent on the presence of mineral acid; reduction of 2-dimethylaminomethyl-4-nitrophenol in alcohol with the same catalyst yielded only 4-amino-2-(diethylaminomethyl)phenol. No observable uptake of hydrogen was noted in reductions under the best conditions when a methoxy or a phenol group was *meta* to the basic side chain.

Some of the observations noted in Refs. 52d and 52e may not apply to reductions of similarly substituted Mannich bases with different basic side chains or to hydrogenations employing other palladium catalysts. In contrast to the unsuccessful hydrogenolysis of 2-dimethylaminomethyl-4-phenylphenol [52d], 2-(4-morpholinomethyl)-4-phenylphenol in alcohol was reduced to 2-methyl-4-phenylphenol in 91% yield within 16 min at 80° and 5 atm initial pressure in the presence of 37% by weight of 10% palladium on carbon [52f]. 3-Methyl-5-phenylcatechol was obtained in a similar manner from 3-(morpholinomethyl)-5-phenylcatechol.

The following are examples of the reduction of Mannich bases where Ar is an unsaturated heterocycle. 5-(4-Morpholinomethyl)uracil was hydrogenated in 85% alcohol with palladium on carbon at 3 atm to give 32% yield of thymine [53a]. This example is included because of the aromatic character of the 5-position of the pyrimidine ring. 2-Dimethylaminomethyl-1,3-dimethylindole as quaternary salt was reduced in methyl alcohol with 40%

by weight of 10% palladium on carbon at room temperature and atmospheric pressure [53b]. Reaction was slow (10 hr) but 90% yield of 1,2,3-trimethylindole was obtained. When Raney nickel was employed, the reaction time was 50 hr. Raney nickel catalyzed the hydrogenolysis of 3-diethylaminomethyl-2,4,5-trisubstituted pyrroles. High yields were obtained but reaction time varied in reductions carried out at 100 atm and different temperatures [53c]. In a reaction in which Ar was furyl, the methobromide of methyl 3-dimethylaminomethylfuran-2-carboxylate was converted to methyl 3-methylfuran-2-carboxylate in 31% yield by reduction in alcohol with palladium oxide for 5 days under 2 atm pressure [53d]. 3-Hydroxy-1-phenyl-4-(1-piperidylmethyl)pyrazole was hydrogenated in alcohol with palladium on carbon to give a 75% yield of 3-hydroxy-4-methyl-1-phenylpyrazole [53e].

Although poor results are generally obtained from low-pressure hydrogenolyses of Mannich compounds with Raney nickel, successful reductions were carried out with 1-chloro-4-hydroxy-3-(diethylaminoethyl, morpholinomethyl, and piperidinomethyl)isoquinolines in dilute sodium hydroxide solution. Dehalogenation accompanied deamination to yield 4-hydroxy-3-methylisoquinoline [53f].

Secondary Amines. Certain N-alkylbenzylamines have been reported resistant to hydrogenolyses under mild conditions [1, 10a, 33]. There are, however, so many examples of successful reductions of related N-(substituted alkyl)benzylamines that suggest that the unsuccessful ones should be reexamined. It is likely that better results could be achieved by the use of more palladium catalyst and by increasing the reaction temperature and perhaps the pressure. The concentration of the substrate in the solvent may also be a factor. Concentrated solutions are debenzylated with greater ease than dilute solutions [2].

D-1-Benzylamino-2-propanol hydrochloride was hydrogenolyzed over palladium on carbon to yield 1-amino-2-propanol in a slow reaction under 3 atm pressure [54a]. Hydrogen uptake was also slow in the hydrogenation of 2-benzylaminomethyl-1-ethylpyrrolidine in alcohol with an equal weight of 5% palladium on carbon [54b]. This reaction might have proceeded more rapidly if the basic nitrogen atom was neutralized. 2,5-Dibenzylaminomethyl-1-methylpyrrolidine as dihydrochloride underwent debenzylation with 20% by weight of 5% palladium on carbon when the reduction was carried out at 60° and 1–2 atm [2]. Glucosamine was obtained from benzylaminoglucose within 2 hr when a large amount of 5% palladium on carbon was employed [54c]. Galactosamine was obtained in 77% yield from the N-benzyl compound with a more modest amount of 10% palladium on carbon at 50° and 2–3.3 atm in less than 2 hr [54d]. The hydrogenolysis of

1-benzylamino-1-deoxy-2,4-O-ethylidene-D-erythritol to the 1-amino compound with palladium on carbon was successful but slow, due probably to the low catalyst ratio [54e].

A very significant example of the ease of removal of the benzyl group from a N-(substituted monoalkyl)benzylamine may be seen in the hydrogenolysis of a concentrated solution of N^1-benzyl-N^3,N^3-diethyl-2,2-dimethylpropane-1,3-diamine in alcohol under 2 atm pressure of hydrogen in the presence of about 13% by weight of 5% palladium on carbon [54f]. Hydrogen uptake was complete in 1 hr, the yield of N,N-diethyl-2,2-dimethyl-propane-1,3-diamine was 78.5%.

3-(1-Isopropylamino)ethylindole may be considered to be related to a N-monoalkylbenzylamine. Hydrogenolysis in alcohol with 30% palladium on carbon at 60–80° and 3 atm pressure was carried out for 48 hr to obtain 3-ethylindole [54g] (Eq. 19.25).

$$\text{indole-CH(CH}_3\text{)NHCH(CH}_3\text{)}_2 \xrightarrow{H_2} \text{indole-C}_2\text{H}_5 + (CH_3)_2CHNH_2 \quad (19.25)$$

The reason for the extended reaction period was not disclosed but could have been due to the absence of acid and to the effect of the α-methyl group.

In general, palladium-catalyzed hydrogenolyses of N-benzyl and related N-protected amino acids take place more or less readily. As a rule yields of primary amino acids are high. At times results appear to be dependent on the reaction medium and perhaps on the catalyst itself.

2,3-Dibenzylaminosuccinic acid was reduced to 2,3-diaminosuccinic acid in 10% hydrobromic acid or in either alcohol or glacial acetic acid containing hydrogen chloride with palladium on carbon at 3 atm pressure [55a]. The free acid or alkali salts in neutral solvent could not be debenzylated. 2-Benzylaminosuccinamic acid, $HOOCCH_2CH(NHCH_2C_6H_5)CONH_2$, was converted to 2-aminosuccinamic acid by reduction with 30% palladium on carbon in 50% acetic acid at 60–70° [55b]. 3-Aminobutyric acid was obtained in 95% yield by hydrogenolysis of the N-benzyl compound in glacial acetic acid with the same catalyst [55c]. Reaction time was longer than in the previous reaction, presumably because of a lower catalyst ratio. Debenzylation in acetic acid with palladium on carbon was also employed in the preparation of 3-aminopropionic acid and other amino acids [55d]. Ethyl 2-(benzylaminomethyl)butyrate was hydrogenated in alcohol to the corresponding amino ester in 87% yield with 10% by weight of 10%

palladium on carbon at 60–65° and atmospheric pressure [55e]; reaction time was 1 hr. In one instance a protecting trityl group was removed with palladium black to give glycine [55f]:

$$(C_6H_5)_3CNHCH_2COOH \xrightarrow{H_2} H_2NCH_2COOH + (C_6H_5)_3CH$$

Slow uptake of hydrogen was reported in the hydrogenolysis of a secondary N-benzylamine. The removal of the benzyl group from 4-benzylamino-5-(N-methylformylamino)uracil with palladium catalyst at 45° required 85–95 hr [55g]. This may have been due to reduction in 5% aqueous acetic acid in which toluene was not completely soluble and because of this the activity of the catalyst was impaired.

Palladium on carbon and palladium hydroxide on carbon were used in the hydrogenolyses of optically active N-benzyl or α-substituted benzyl aminoacids [56a, b]. A faster rate of reduction was observed with palladium hydroxide on carbon but optical purity was reported to be higher with palladium on carbon [56a]; in reductions in aqueous sodium hydroxide, optical purity and yield were dependent on the amount of water used in the reactions.

Primary Amines. Although deamination of benzylamine is reported to be difficult [1, 33] it was readily accomplished by first purifying the base by redistillation and subjecting it to hydrogenation in 90–95% alcohol containing an equivalent of hydrochloric acid in the presence of over 20% by weight of 5% palladium on carbon at 25–50° and 2 atm pressure [2]. Phenylglycine, $C_6H_5CH(NH_2)COOH$, which is a hindered benzylamine, on the other hand, required a large amount of 5% palladium on carbon or palladium hydroxide on carbon, over 130% by weight, and a 24-hr reaction period for conversion to phenylacetic acid during reduction as sodium salt in aqueous solution [56a].

N-Benzylamides

Removal of the N-benzyl group from an amide is rather difficult [2]. When it does take place, elevated temperature and pressure conditions are usually required. Nevertheless an N-benzyl group was removed from a cyclic amide under low-pressure conditions. 1-Benzyl-3H-1,4-benzodiazepine-2,5-dione was converted to 3H-1,4-benzodiazepine-2,5-dione by reduction in acetic acid with 10% by weight of 10% palladium on carbon at 60° and 4 atm for 48 hr, [57a, b]. On the other hand, the benzyl group in a related cyclic amide, 4-benzyl-1-methyl-3H-1,4-benzodiazepine-2,5-dione, could not be removed by reduction in acetic acid in the presence of 35% by weight of 5% palladium on carbon under the same conditions [2].

As a general rule it may be expected that hydrogenation of most reducible groups will take place preferentially in the presence of an N-benzylamide [2].

An example may be seen in the reduction of 1-benzyl-5-benzylamino-3,3,5-trimethyl-2-pyrrolidone with Raney cobalt at 150° and 100 atm to 5-amino-1-benzyl-3,3,5-trimethyl-2-pyrrolidone [57c]. A surprising exception is the reported removal of a carbobenzyloxy protecting group from a cyclic amide in preference to one on a more basic amino group in the hydrogenolysis of L-1-carbobenzyloxy-3-carbobenzyloxyamino-2-pyrrolidone in alcohol with 30% by weight of platinum oxide. In this reaction hydrogen uptake stopped after absorption of one molar equivalent to yield a product which was identified as L-3-carbobenzyloxyamino-2-pyrrolidone [57d]. When the reaction was carried out in acetic acid both protecting groups were removed.

SELECTIVE REDUCTIONS

Selectivity among N-Benzyl-type Groupings

An order of selectivity for the removal of N-benzyl-like protecting groups has been shown [35a, b], but whether 100% cleavage of the most active one, 1- or 2-menaphthyl, could be achieved in the presence of the least active one, the N-benzyl group, was not established.

The order of reducibility of N-benzylamines, quaternary > tertiary > secondary > primary, is well known. Tribenzylamine can be selectively hydrogenated to dibenzylamine, and dibenzylamine can be converted to benzylamine in high yield. In certain instances an unsubstituted benzyl group may be completely cleaved in the presence of a *para*-substituted benzyl group. The hydrogenolysis of a quaternary compound, $R\overset{+}{N}(CH_3)_2R'Cl^{(-)}$ where R was *para*-substituted benzyl and R' was benzyl, yielded $RCH_2N(CH_3)_2$ exclusively [34a]. In the palladium-catalyzed reduction of a tertiary amine, $RCH_2N(CH_3)R'$ where R was 3-hydroxy-6-methyl-2-pyridyl and R' was benzyl, 3-hydroxy-6-methyl-2-methylaminopyridine was obtained [58]. However, there has been too little work on the effect of a nuclear substituent and its ring position on selectivity in the hydrogenolyses of secondary, tertiary, or quaternary dibenzylamines, where one N-benzyl group is substituted and the other unsubstituted, to state categorically that the unsubstituted benzyl group will be removed to the exclusion of the other. This requires further work on an analytical basis to determine the products of reaction after absorption of one molar equivalent of hydrogen.

Selectivity among O-Benzyl-type Groups

There are very few comparisons to note the effect of substituents on the removal of O-benzyl type groups. In certain instances it has been shown that one *O*-protecting group can be removed more readily than another but complete selectivity does not always result. In the palladium on carbon-catalyzed cleavage of benzyl-type ethers, it was shown that one protecting group was removed to a greater extent than another but 100% selectivity could not be accomplished [59a]. In the hydrogenolysis of benzyl-type esters, 1-naphthylmethyl acetate was completely cleaved in competition with benzyl acetate [59a].

In most instances esters of benzyl alcohols are more readily hydrogenolyzed than the parent alcohol [5b]. This may not apply when attempting to reduce the ester of a secondary benzylic alcohol in the presence of a primary benzylic type alcohol.

The geometry of the substrate can play a role in determining selectivity. Such an example is seen in the hydrogenation of ethyl 1-(2-hydroxy-2-phenethyl)-4-phenylpiperidine-4-acetate as hydrochloride salt in alcohol with an equal weight of palladium on carbon to which palladium chloride was added [59b]. The secondary hydroxyl group in the side chain, not the ester in 4-position, was removed to give ethyl 1-(2-phenethyl)-4-phenylpiperidine-4-acetate. This result is not unexpected if one compares the favored position of a related substrate, *N*-methyl-4-hydroxy-4-phenylpiperidine, to the surface of the catalyst, as shown earlier in this chapter.

Where steric effects do not play a very great part it may be advisable, in attempting to reduce a benzyl ester in the presence of a benzylic alcohol, to avoid the presence of strong acid, which might induce esterification and lead to partial hydrogenolysis of the benzylic alcohol.

Benzyl ethers should also be cleaved more readily than benzyl alcohols. The conversion of 1-(3,4-dibenzyloxyphenyl)-1-hydroxyisopropylamine to 1-(3,4-dihydroxyphenyl)-1-hydroxyisopropylamine [60a] and the preparation of 1-(4-hydroxy-3-methoxyphenyl)-1-hydroxypropylamine [60b] and 1-(3-hydroxyphenyl)-1-isopropylamine [60c] from the corresponding nuclear benzyloxy compounds with palladium on carbon at low pressure are examples of selective hydrogenations, as is the palladium black-catalyzed reduction of 2-amino-3-hydroxy-3-(3,4-dibenzyloxyphenylphenyl)propionic acid to 2-amino-3-hydroxy-3-(3,4-dihydroxyphenyl)propionic acid [60d]. It should be pointed out that selectivity in these examples may have been due to the difficulty in hydrogenolyzing secondary benzylic alcohols because of steric effects.

Competitive reductions of *O*-benzyl-type esters and ethers indicated that benzyl esters were usually preferentially cleaved unless a very easily removed group protected the ether function [59]. When benzyl benzoate and benzyl cyclohexyl ether were simultaneously hydrogenated in ether over palladium on carbon, cleavage of the ester was complete without attack on the benzyl ether; 1-naphthylmethyl acetate was selectively hydrogenolyzed without affecting benzyl cyclohexyl ether, and benzyl benzoate underwent complete cleavage in the presence of 1-naphthylmethyl cyclohexyl ether. In one instance, however, when the least easily cleaved ester in the series, benzyl acetate, was hydrogenated together with the most readily hydrogenolyzed ether, 4-phenylbenzyl cyclohexyl ether, a significant amount of 4-methylbiphenyl was obtained.

Although the examples suggest that benzyl-type esters are hydrogenolyzed more readily than benzyl ethers, there are no examples to substantiate this assumption when two such groups are contained in a single substrate. Whether complete selectivity will result in such cases is a matter for further study. It must be borne in mind that competitive hydrogenolysis with two substrates may not provide the same answer as when two such reducible groups are present in the same molecule because of the geometry of the substrate.

N-Benzyl in the Presence of O-Benzyl

It is generally assumed that O-debenzylation takes place with greater ease than removal of an N-benzyl group. This assumption should not be the basis for further assumption that an O-benzyl or a related group will always be selectively cleaved in the presence of an N-benzyl-type protecting group. There are too few examples of hydrogenations of benzylic alcohols, ethers, or esters in the presence of various N-benzyl groups to be very specific. It may be possible to draw some conclusions by dividing these hydrogenolyses into subgroups and comparing results. However, in certain instances where there appears to be a tendency toward O-debenzylation, the chemist must realize that selectivity may change if the N-protecting group is a readily reducible one as carbobenzyloxy.

N-Benzyl Groups in the Presence of Benzyl Alcohols. There do not appear to be examples of the reduction of substrates containing both a hydroxymethyl group and an aminomethyl group attached to an aromatic nucleus, where the hydroxyl group is primary, secondary, or tertiary and where the amino nitrogen is primary or secondary. This is a matter for further study.

There are examples of reductions of substrates which contain both an aryl tertiary carbinol and a tertiary N-benzyl group. The protecting group of the amine was removed in each instance by hydrogenolysis over palladium on carbon, usually under a few atmospheres of hydrogen. N-Benzyl-3-piperidyl benzilate was converted to 3-piperidyl benzilate in high yield by reduction in acetic acid [61a]. N-Benzyl-N-ethylaminomethyl benzilate hydrochloride was selectively hydrogenated in aqueous alcohol at 1.3 atm to N-ethylaminomethyl benzilate [2] (Eq. 19.26). In these examples steric effects make hydrogenolysis of the tertiary benzyl-type alcohol very difficult.

It is likely that a secondary N-benzyl group would also be removed selectively in similar reductions. In view of the slow deoxygenation of

$$(C_6H_5)_2\overset{\overset{\displaystyle OH}{|}}{C}COOCH_2N\overset{\displaystyle C_2H_5}{\underset{\displaystyle CH_2C_6H_5}{\diagdown}} \cdot HCl \xrightarrow[\text{Pd on C}]{H_2} \\ 25°, 1.3 \text{ atm}, <1\text{hr}$$

$$(C_6H_5)_2\overset{\overset{\displaystyle OH}{|}}{C}COOCH_2NHC_2H_5 + C_6H_5CH_3 \quad (19.26) \\ 100\%$$

benzhydrol, it is also likely that hydrogenolysis of a benzyl group attached to a tertiary nitrogen atom would also take place in preference to removal of oxygen in substrates containing both hydrogenolyzable groups.

Preferential hydrogenolysis of the N-benzyl group should be expected in compounds of the type

$$C_6H_5CH_2N\overset{\diagup (CH_2)_n \diagdown}{\underset{\diagdown (CH_2)_n \diagup}{}}C\overset{\diagup OH}{\diagdown C_6H_5}$$

because under normal conditions the benzylic hydroxy group is out of contact with the catalyst. A model of the related 4-hydroxy-1-methyl-4-phenylpiperidine emphasizes this point [49c]. Examples of the expected selectivity are shown in the palladium-catalyzed hydrogenations of 1-benzyl-4-hydroxy-4-phenylpiperidine [61b], 1-benzyl-4-hydroxy-3-methyl-4-phenylpiperidine [61c], 1-benzyl-4-hydroxy-3,5-dimethyl-4-phenylpiperidine [61d], and 1-benzyl-4-phenyl-1-azacycloheptan-4-ol [61e], which gave the corresponding cyclic secondary amino-4-phenyl-4-ols.

N-Benzyl Groups in the Presence of Benzyl Ethers. The few available examples of reductions of substrates containing a benzyl ether and a primary or secondary N-benzylamine suggest that the O-benzyl group will be removed preferentially. 1-Phenyl-2-(4-benzyloxy-3-methoxyphenyl)ethylamine, 4-$C_6H_5CH_2O$-3-$CH_3OC_6H_4CH_2CH(C_6H_5)NH_2$, was reduced in methyl alcohol with palladium on carbon at 55° and 1.3–4 atm to give a 99% yield of 1-phenyl-2-(4-hydroxy-3-methoxyphenyl)ethylamine [62a]. The O-benzyl groups in N-(3,4-dioxymethylenebenzyl)-2-(3,4-dibenzyloxyphenyl)ethylamine were selectively removed by reduction over 5% palladium on barium sulfate [62b]. N-(4-benzyloxy-3-methoxybenzyl)-2-(4-benzyloxy-3-methoxyphenyl)ethylamine on hydrogenation with palladium on carbon gave

N-(4-hydroxy-3-methoxybenzyl)-2-(4-hydroxy-3-methoxyphenyl)ethylamine [62c].

It is likely that unless the ether portion of the molecule cannot make contact with the catalyst, most benzyl ethers will also be selectively hydrogenolyzed in the presence of a tertiary *N*-benzylamine. A successful example is the palladium-catalyzed reduction of 1-(4-benzyloxybenzyl)-4-methylpiperidine to 1-(4-hydroxybenzyl)-4-methylpiperidine [62d]. When 1-benzyl-4-ethoxy-4-piperidine was hydrogenated with 15% by weight of 10% palladium on carbon, 4-ethoxy-4-phenylpiperidine was obtained [62e]. This result is not unexpected. The major points of contact between the catalyst and this substrate are the aromatic rings and the basic nitrogen atom with the OC_2H_5 portion of the molecule out of plane in the same manner as the hydroxyl group in the reduction of the related alcohol, 1-benzyl-4-hydroxy-4-phenylpiperidine, which was discussed earlier in this chapter.

A Mannich-type amine, 3-(dimethylaminomethyl)indole, containing a benzyloxy group in 4-, 5-, 6-, or 7-position was hydrogenated over palladium on carbon [62f]. The intent of the investigators was to obtain the corresponding hydroxyskatoles, so unfortunately no observation was made to determine if there was any difference in selectivity between the *O*-benzyl- and the *N*-benzyl-like groups. There is, however, an example of the hydrogenation of the 5-benzyloxy compound as hydrochloride salt with a specially prepared 1% palladium on carbon catalyst. The *O*-benzyl group was cleaved to yield 53% of 3-dimethylaminomethyl-5-hydroxyindole, 5-hydroxygramine [62g].

There is a description of the preferential removal of an *N*-benzyl group from a quaternary amine which also contained a benzyl ester. 3-Benzyl-8-benzyloxy-3-methyl-3-azabicyclo[3,2,1]octane tosylate upon hydrogenation with palladium gave 8-benzyloxy-3-methyl-3-azabicyclo[3,2,1]octane [62h].

N-Benzyl Groups in the Presence of Benzyl Esters. The palladium-catalyzed hydrogenations of 4-(1-benzyl-4-phenyl)piperidyl propionate gave conflicting and surprising results [61b]. Based on analogy to reductions of related 1-benzyl-4-phenylpiperidines, where the 4-hydroxy group [61b, c, d] and the 4-ethoxy group [62e] were not affected, it might be expected that the ester group in 4-position also would not be reduced. At uptake of one molar equivalent of hydrogen the major products were 4-phenylpiperidine, 38%, and unchanged starting material, 32%. 1-Benzyl-4-phenylpiperidine, 4-hydroxy-4-phenylpiperidine propionate, and 4-hydroxy-4-phenyl-1-propionylpiperidine were also obtained, the latter through *N*-debenzylation and migration of the propionyl group (Eq. 19.27). A repeated reduction yielded 29% starting material, 50.5% of 4-hydroxy-4-phenylpiperidine and only 9% of 4-phenylpiperidine. There was no explanation for the results

$$\underset{R'}{\overset{R}{\bigcirc}} \overset{OOCC_2H_5}{\underset{N}{\bigcirc}} \xrightarrow[Pd\ on\ C]{H_2} \underset{H}{\overset{R}{\bigcirc}}\overset{H}{\underset{N}{\bigcirc}} +$$

38%

(19.27)

$$\underset{R'}{\overset{R}{\bigcirc}}\overset{H}{\underset{N}{\bigcirc}} + \underset{H \cdot C_2H_5COOH}{\overset{R}{\bigcirc}}\overset{OH}{\underset{N}{\bigcirc}} + \underset{O=CC_2H_5}{\overset{R}{\bigcirc}}\overset{OH}{\underset{N}{\bigcirc}} + \text{starting material}$$

32%

R = phenyl, R' = benzyl

which showed O-hydrogenolysis to be the major reaction in one reduction and N-debenzylation the principal reaction in another.

Selective Reductions in the Presence of Other Reducible Groups

Selective reduction of an N- or O-benzyl group should not be expected in the presence of an acetylenic bond, a nitro group, or an azomethine. Reduction in the presence of a double bond may depend on whether the double bond is di-, tri-, or tetrasubstituted and whether an O- or an N-benzyl group is to be hydrogenolyzed. It is doubtful that selective hydrogenolysis of an O- or N-benzyl group will take place in the presence of a double bond unless the double bond is tetrasubstituted. The hydrogenation of 7-benzyloxy-3-(4-methoxyphenyl)-Δ^3-isoflavene with palladium on carbon in acetic acid gave both the corresponding 7-hydroxy-Δ^3-isoflavene and the 7-benzyloxy-isoflavone [63]. Reduction of the double bond and removal of the benzyl ether appeared to be concurrent in the attempted hydrogenolyses of the sodium salts of 4-benzyloxy-3-methoxy- and 3-benzyloxy-4-methoxy-cinnamic acids in aqueous solution with Raney nickel at slightly above 1 atm [2]. Selectivity appeared to have been obtained in the reduction of 2-benzyloxy-3-methoxycinnamic acid as sodium salt in water with Raney nickel or with palladium on carbon to 2-hydroxy-3-methoxycinnamic acid based on the known melting point, but elemental analysis could not confirm that the desired product was obtained [2].

O-Benzyl in the Presence of Carbonyl. Benzyl esters of keto acids have been selectively reduced to keto acids with palladium catalyst. Benzyl

434   Debenzylation and Related Reactions

(3-methoxy-5-phthalimidopentanoyl)malonate was reduced to the corresponding keto malonic acid in ethyl acetate and acetic acid with palladium on carbon [64a]. It was not isolated as the malonic acid but obtained as 4-methoxy-6-phthalimido-2-hexanone in 93% yield by decarboxylation. *N*-Benzyl pyruvylglycine, $CH_3C(=O)CONHCH_2C_6H_5$, was converted to pyruvylglycine in 80% yield by reduction in methyl alcohol with palladium black or with palladium oxide on barium sulfate [64b]. *trans-anti*-2-Carbomethoxy-5-methyl-3-(4-methoxyphenyl)cyclopentan-1-one-2-acetic acid was obtained in 96% yield by reduction of the corresponding benzyl ester with 5% palladium on calcium carbonate previously saturated with hydrogen in methyl alcohol [64c].

The *cis* and *trans* forms of 4-benzyloxycyclohexanone-2-acetic acid were each readily reduced to *cis*- (57%) and *trans*- (87%) 4-hydroxycyclohexanone-2-acetic acids [65a]. Although there are no available references on selectivity among reductions of compounds of the type $C_6H_5CH_2OArC(R)=O$ where R = alkyl, aralkyl, or aryl, it might be assumed that debenzylation will occur preferentially in the presence of palladium, because the carbonyl group is hindered.

Hydrogenation of some benzyloxy(benzaldehydes and phenylacetaldehydes) in absolute methyl alcohol in the presence of 10% by weight of palladium oxide or prereduced palladium from the oxide was reported to give the phenolic aldehydes [65b], but yields were not noted. On the other hand, reductions of benzyloxybenzaldehydes in the presence of palladium on carbon [2] and with Raney nickel indicated concurrent attack of both groups (Chapter XIV).

It is possible that selective debenzylation in the presence of aldehyde might be obtained with an inactivated palladium catalyst. Protection of the aldehyde group may be more advisable.

A benzyl ether was cleaved in the presence of an α,β-unsaturated ketone. 3-(2-Hydroxybenzoylmethyl)piperidine-2-one was obtained from the 2-benzyloxy compound by reduction in acetic acid with 2.5% palladium on barium sulfate [66] (Eq. 19.28).

$$\text{(19.28)}$$

There are numerous examples of the palladium-catalyzed reduction of a benzyl ether to a phenolic hydroxyl in compounds also containing an aminoketone. 2-Amino-2'-hydroxy-5'-methylacetophenone was obtained from 2-amino-2'-benzyloxy-5'-methylacetophenone by reduction with palladium on carbon in alcohol [67a]; 2-isopropylamino-3',4'-dihydroxypropiophenone was obtained in a similar manner from the corresponding 3',4'-dibenzyloxy compound [67b], and 2-(1-phenoxy-2-propylamino)-4'-benzyloxypropiophenone hydrochloride yielded 2-(1-phenoxy-2-propylamino)-4'-hydroxypropiophenone upon reduction in 80% alcohol with palladium on carbon [67c].

N-Benzyl in the Presence of Carbonyl. The N-protecting carbobenzyloxy group was removed from methyl α-(5-carbobenzyloxyamino-3-methoxypentanoyl)malonate and from methyl α-(4-carbobenzyloxyamino-3-methylbutyryl)malonate with palladium on carbon in acetic acid without attack on the carbonyl group [68a]. The same protecting group was selectively removed from the basic amino groups of 4-ketolysineamide with platinum [68b]. Hydrogenolysis of 2-(N-carbobenzyloxy-N-methylamino)-3',4'-diacetoxyacetophenone in acetic acid was selective in the presence of more than 10% by weight of palladium black, but when palladium on carbon was used in the same ratio of metal to substrate, the carbonyl group was converted in part to carbinol [68c].

Generally hydrogenolyses of N-benzyl phenacylamines with palladium have resulted in high yields of the corresponding aminoketone. 2-Benzylaminoacetophenone hydrochloride was reduced in alcohol over palladium on carbon to 2-aminoacetophenone hydrochloride in 88% yield [69a]. Reduction in water was reported to lead to 2-amino-1-phenylethan-1-ol. The effect of water may have been duplicated in hydrogenations of 5-(N-benzyl-N-methylglycyl)-2-methylbenzimidazole and the similarly substituted benzimidazolone [69b]. 2-(N-Benzyl-N-methylamino)-2'-methoxy-5'-methylpropiophenone was debenzylated selectively over palladium on carbon [67a]. The N-benzhydryl group as well as benzyloxy groups were cleaved in the hydrogenation of 2-benzhydrylamino-3',4'-dibenzyloxybutyrophenone with palladium to 2-amino-3',4'-dihydroxybutyrophenone [69c].

In reductions among other benzylaminoketones, 4-acetyl-1-benzylpiperidine was converted to 4-acetylpiperidine in 84% yield [69d] and 2-benzyl-7-methoxy-2,3-dihydro-4(1H)-isoquinolone was hydrogenated in acetic acid over palladium to 7-methoxy-2,3-dihydro-4(1H)-isoquinolone in 94% yield [69e]. Other related compounds were similarly reduced. Uptake of hydrogen continued in these reductions if the reaction was not interrupted after absorption of one molar equivalent. Some 1-benzyl-3-ketoquinuclidinium bromides were hydrogenolyzed in aqueous alcohol with palladium on carbon to 3-ketoquinuclidines in very high yield [69f, g]. 1-Benzyl-1-ethyl-4-(3,4-dimethoxyphenyl)-4-propionylpiperidinium chloride was converted in a similar manner to 1-methyl-4-(3,4-dimethoxyphenyl)-4-propionylpiperidine [69h].

In other hydrogenolyses of an *N*-benzyl group in the presence of a ketone, 3-(4-dibenzylaminophenyl)-2-butanone was reduced to 3-(4-aminophenyl)-2-butanone in 81% yield and 1,1-*bis*-(4-dibenzylaminophenyl)-2-propanone gave 1,1-*bis*-(4-aminophenyl)-2-propanone [70].

O- and N-Benzyl Groups in the Presence of Nitriles. Examples show that *O*- or *N*-benzyl groups will be removed by hydrogenation over palladium without affecting the nitrile group. In the hydrogenolysis of mandelonitriles, $C_6H_5CHORCH$ where R = H or acyl, reduction in neutral solution gave benzyl cyanide [71]. In the presence of an equivalent of strong acid phenethylamine was obtained, but in all experiments the first stage of reduction was conversion of the CHOH group to methylene. In a related reaction, $ArCH(OOCCH_3)CN$ where Ar was 4-quinolyl, hydrogenation in alcohol in the presence of palladium on barium sulfate yielded 4-cyanomethylquinoline [16b].

5-Benzyloxyindole-3-acetonitrile was hydrogenated in alcohol in the presence of 5% palladium on carbon at less than 2 atm pressure to give 86.5% of 5-hydroxyindole-3-acetonitrile [2].

Reduction of 4-(*N*-benzyl-*N*-methylamino)-3-methyl-2,2-diphenylbutyronitrile in a mixture of alcohol, acetic acid, and a small amount of water over palladium on carbon prepared *in situ* gave 94% yield of 4-methylamino-3-methyl-2,2-diphenylbutyronitrile [72a]. In another example of selective hydrogenolysis of an *N*-benzyl-type amine in the presence of a nitrile group, *N*,*N*-dimethyl-3,5-dicyanobenzylamine was converted to 3,5-dicyanotoluene [72b].

Benzyl Groups in the Presence of an Oximino Group. There do not appear to be any references on selective reduction of an *N*- or *O*-benzyl group in the presence of an aldoxime. Benzyl groups have been cleaved in the presence of α-oximino acids and amides. 3-(4,6-Dibenzyloxy-2-pyrimidyl)-2-hydroxyiminopropionic acid was reduced in absolute methyl alcohol with about 2% by weight of the 10% palladium on carbon to 3-(4,6-dihydroxy-2-pyrimidyl)-2-hydroxyiminopropionic acid [73a]. 2-Benzylamino-2-hydroxyiminoacetamide in methyl alcohol gave 2-hydroxyiminoglycinamide in quantitative yield by reduction with palladium black [73b]. When reduction was carried out under acidic conditions, hydrogen uptake stopped before one molar equivalent was absorbed, giving only 40% yield along with 2-iminoglycinamide and *N*-benzyloxamide, which apparently resulted from hydrolysis of the starting material, Eq. 19.29.

Hydrogenolysis has taken place selectively in the reduction of α-benzyloximino amides of the type $RC(CONYZ)=NOCH_2C_6H_5$ where R = benzyl or isobutyl, Y was hydrogen or methyl, and Z was phenyl, substituted phenyl, or aralkyl, in aqueous ammonia in the presence of palladium on carbon [74]. No uptake of hydrogen occurred during reduction at room temperature and low pressure when Z was aliphatic. Other reductions have

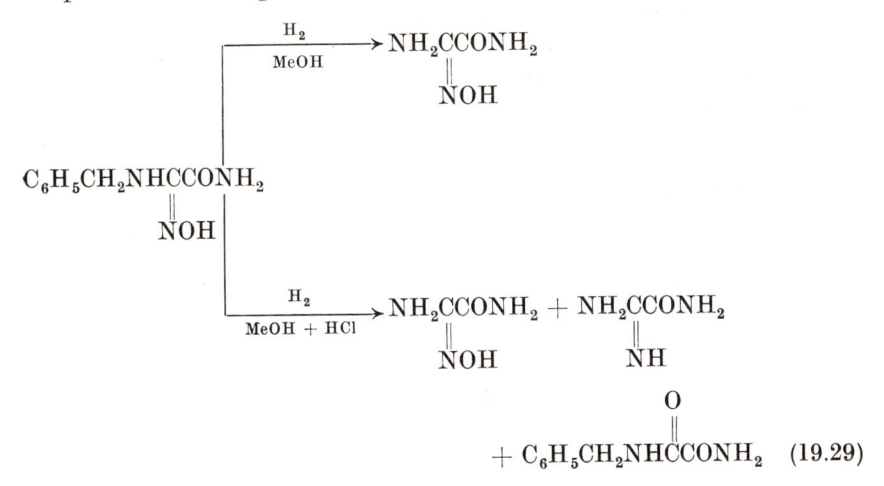

$$C_6H_5CH_2NHCCONH_2 \atop \| \atop NOH \begin{cases} \xrightarrow[\text{MeOH}]{H_2} NH_2CCONH_2 \atop \| \atop NOH \\ \\ \xrightarrow[\text{MeOH + HCl}]{H_2} NH_2CCONH_2 + NH_2CCONH_2 \atop \| \quad\quad\quad \| \atop NOH \quad\quad\quad NH \\ \quad\quad\quad\quad\quad\quad\quad\quad\quad\quad\quad\quad\quad O \\ \quad\quad\quad\quad\quad\quad\quad\quad\quad\quad\quad\quad\quad \| \\ + C_6H_5CH_2NHCCONH_2 \end{cases} \quad (19.29)$$

been described which involve the removal of the benzyl group from the oximino function, but the aim in these reactions was to obtain aminoacids or peptides.

Benzyl Groups in the Presence of Hydroxylamine. Most references are those which describe reductions of *O*-benzylhydroxylamines, YZNOCH$_2$-C$_6$H$_5$, where YZN represents a straight- or branched-chain amide or a cyclic amide or imide. Reaction, which could lead to conversion to YZNH by absorption of two molar equivalents of hydrogen, is usually very selective to yield a hydroxamic acid, YZNOH, upon hydrogenation in the presence of palladium.

N-Hydroxysuccinamic and *N*-hydroxyglutaramic acids were obtained from hydrogenations of the *N*-benzyloxy acids in ethyl methyl ketone with 5% palladium on strontium carbonate [75a]; *N,N*-diacetylhydroxylamine resulted in a similar manner from the *O*-benzyl compound. *N*-Hydroxysuccinimide and other cyclic imides were obtained from the corresponding benzyloxyimides. 3-Benzyloxy-2-methyl-4-quinazolone was rapidly and selectively reduced in ethyl acetate with 10% palladium on carbon to 3-hydroxy-2-methyl-4-quinazolone [75b]. *O*-Benzylleucylhydroxylamine, (CH$_3$)$_2$CHCH$_2$CH(NH$_2$)CONHOCH$_2$C$_6$H$_5$, was hydrogenated in alcohol with 5% palladium on carbon at room temperature and 40 atm pressure to yield leucinehydroxamic acid exclusively [75c]. In similar reductions of related compounds, (CH$_3$)$_2$CHCH$_2$CH[NHCH$_2$C(=O)alkyl]CONHOCH$_2$C$_6$H$_5$, the corresponding hydroxamic acids were obtained [75c, d]. In these reductions the ketone function, also, was not affected. When compounds of this type,

(CH₃)₂CHCH₂CHXCONHOCH₂C₆H₅ where X = H or OH, were hydrogenated with palladium at 40 atm pressure the amide was obtained exclusively [75e]. The corresponding hydroxamic acids were obtained only if pyridine was added to the reaction mixture.

The carbobenzyloxy group of an α-amino-*O*-substituted hydroxyamic acid could be hydrogenolyzed with palladium on carbon without cleaving the amido-oxy bond [76]:

C₆H₅CH₂OCONHCH(CH₃)CONHOCH₂COOR $\xrightarrow{H_2}$

H₂NCH(CH₃)CONHOCH₂COOR

However, the protecting group could not be removed from the nitrogen atom of a hydroxylamino group of a closely related compound without accompanying reduction of the amido-oxy end group:

C₆H₅CH₂OCONHOCH₂CONHOCH₂COOC₂H₅ ⟶

H₂NOCH₂CONHOCH₂COOC₂H₅

There was no mention of cleavage of the terminal hydroxylamine.

Debenzylation in the Presence of a Hydrazine. 1-Amino-5-benzyloxy-2-hydroxymethyl-4-pyridone was hydrogenated in alcohol with 2% palladium on carbon to give 85% yield of 1-amino-5-hydroxy-2-hydroxymethyl-4-pyridone [77]. Until more examples become available it is difficult to predict selectivity when either an *O*- or *N*-benzyl group is present in a substrate also containing a hydrazine-like nitrogen to nitrogen bond.

Debenzylation in the Presence of Halogen. When a benzyl group and a halogen are both present in a compound to be reduced, selectivity will depend on the reactivity of the halogen and, when an *N*-benzyl group is present, on the basicity of the compound.

Since aliphatic halides do not lose halogen readily, it is expected that an *O*-benzyl group should undergo hydrogenolysis selectively in the presence of an alkyl-type chloride. An example may be seen in the reduction of benzyl phenyl *N*-*bis*(2-chloroethyl)phosphoramidate to phenyl *N*-*bis*(2-chloroethyl)-phosphoramidate with palladium on carbon [78a]:

(ClCH₂CH₂)₂NP(=O)(OC₆H₅)OCH₂C₆H₅ $\xrightarrow{H_2}$

(ClCH₂CH₂)₂NP(=O)(OC₆H₅)OH

Benzyl 5-*bis*(2-chloroethyl)aminoindole-3-carboxylate was converted to 5-*bis*(2-chloroethyl)aminoindole-3-carboxylic acid by reduction with 15% by weight of platinum oxide in tetrahydrofuran containing hydrochloric acid [78b]. The halogen of the 2-chloroethyl groups attached to much more basic nitrogen atoms remained intact when N^1,N^4-bis(2-chloroethyl)-N^1,N^4-bis-(benzyl)-1,4-butanediamine as dihydrochloride salt was hydrogenated with 5% palladium on carbon in 95% acetic acid [78c]; high yield of N^1,N^4-bis(2-chloroethyl)-1,4-butanediamine was obtained.

An aromatic chloride should be retained during O-debenzylation if the substrate is nonnitrogenous. If the substrate also contains nitrogen, aromatic chloride will be retained if the nitrogen atom is nonbasic or weakly basic. Aryl chloride will probably be retained during hydrogenolysis of an O-benzyl group in a compound containing a basic nitrogen atom if the reaction is run in acidic medium or if neutralization takes place during the course of reduction as it did in the following reaction [79a] (Eq. 19.30).

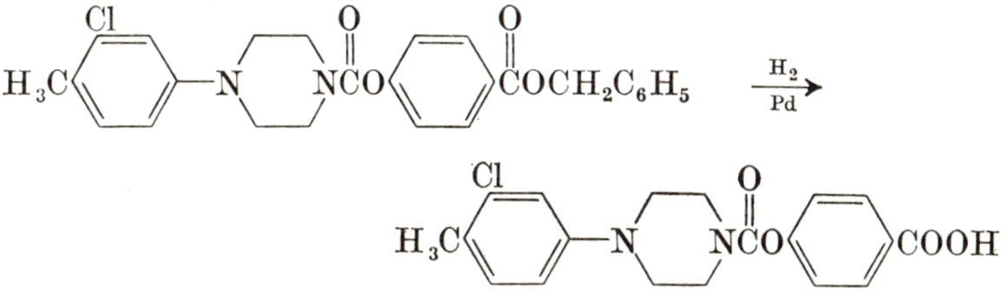

$$H_3C\text{-}C_6H_3(Cl)\text{-}N\text{-}\overset{\|}{NCO}\text{-}C_6H_4\text{-}\overset{O}{\overset{\|}{C}}OCH_2C_6H_5 \xrightarrow{H_2}{Pd}$$

$$H_3C\text{-}C_6H_3(Cl)\text{-}N\text{-}\overset{\|}{NCO}\text{-}C_6H_4\text{-}COOH \quad (19.30)$$

Loss of nuclear chlorine should not occur during N-debenzylation of weak bases such as aromatic amines. It did not take place when 4-benzylamino- and the structurally related 4-(2-furylmethylamino)-5-carboxy-2-chlorobenzenesulfonamides were reduced in methyl alcohol with palladium black, or when the latter compound was reduced in acetic acid with palladium on carbon [79b]. The 4-dibenzylamino compound was also reduced to 4-amino-5-carboxy-2-chlorobenzenesulfonamide. Some amino chlorobenzenedisulfonamides were obtained in over 90% yields from the corresponding benzylamino chlorobenzenedisulfonamides with palladium on carbon [2]. These amines are very weak bases. The danger of dehalogenation appears to increase with increased basicity [2]. Attempts to remove the N-benzyl group from a series of N-benzyl chlorophenethyl and chlorophenylisopropylamines resulted only in dehalogenation [80]. These secondary amines are far stronger bases than N-benzyl chloroanilines.

Loss of halogen accompanied debenzylation of some chlorine-containing benzylated tertiary amines. Some dehalogenation occurred in the reduction of 1-benzyl-5-(2,4-dichloro- and 2,4,6-trichlorophenoxymethyl)tetrazoles [50d]. Dehalogenation also accompanied debenzylation in the conversion of 10-benzyl-8-chloro-3-nitrophenoxazine to 3-amino-8-chlorophenoxazine in acetic acid during reduction with palladium on carbon [81]. Attempted debenzylations of 1-benzyl-4-(2- and 3-chlorophenyl)piperazines with palladium on carbon in aqueous hydrochloric acid were never completely successful [2]; analyses always indicated dehalogenation.

Debenzylation of a quaternary benzylamine containing nuclear chlorine was effected without dehalogenation in the reduction of N,N-dimethyl-N-4-chlorobenzyl-N-4-methoxybenzylammonium chloride to N,N-dimethyl-4-chlorobenzylamine with palladium on carbon [34a]. Although there are insufficient examples of selective debenzylation of quaternary benzylamines containing chlorine, it is possible that selectivity should be expected because the nitrogen atom is completely neutralized and cannot catalyze dehalogenation.

On the other hand, selectivity in the latter instance, where failure resulted among most secondary and tertiary benzylamines, may be due to greater ease of debenzylating quaternary amines than other benzylamines.

Although the carbobenzyloxy group is relatively easy to remove by hydrogenolysis with palladium catalyst, extensive dehalogenation accompanied the removal of that group in α-amino (2-, 3-, and 4-chloro)benzylpenicillins [82]. It may have been caused by the large amount of palladium on barium sulfate employed (to help overcome the poisoning effect of the sulfur atom in these penicillins).

Miscellaneous. N-Benzyl and N-carbobenzyloxy groups have been selectively hydrogenolyzed in the presence of a lactone ring. 3-Dibenzylaminotetrahydrofuran-2-one was hydrogenated in alcohol containing hydrogen chloride in the presence of palladium on carbon to give 3-aminotetrahydrofuran-2-one (2-amino-γ-butyrolactone) in 85% yield [83a]. Reduction with palladium black gave 30% yield of 3-benzylaminotetrahydrofuran-2-one plus 54% of recovered starting material. Reduction in neutral solvent was also carried out without ring opening. Hydrogenation in the presence of base gave 2-amino-4-hydroxybutyric acid.

3-Carbobenzyloxyaminotetrahydrofuran-2-one was converted to aminotetrahydrofuranone by hydrogenation in alcohol with palladium on carbon [83b] and some 3-(carbobenzyloxyaminoacetylamino)tetrahydrofuran-2-ones were reduced in alcohol containing hydrochloric acid with palladium black to give 3-(aminoacetylamino)tetrahydrofuran-2-one in 40% yield when the 3-substituent was CH$_2$(NH$_2$)CONH—, and 55% yield when it was CH$_3$CH(NH$_2$)CONH— [83c].

REFERENCES

[1] W. H. Hartung and R. Simonoff, *Organic Reactions*, Vol. VII, Wiley, New York, 1953, p. 263.
[2] M. Freifelder, unpublished results.
[3] T. L. Miller, G. L. Rowley, and C. J. Stewart, *J. Am. Chem. Soc.*, **88**, 2299 (1966).
[4] C. D. Hurd and H. Jenkins, *J. Org. Chem.*, **31**, 2045 (1966).

[5] a. S. Mitsui and Y. Kudo, *Chem. Ind.* (London), **1965**, 381; b. A. M. Khan, F. J. McQuillin, and I. Jardine, *J. Chem. Soc.* (C), **1967**, 136 and references therein.
[6] N. K. Richtmeyer, *J. Am. Chem. Soc.*, **56**, 1633, (1934).
[7] a. M. W. Baines, D. B. Cobb, R. J. Eden, R. Fielden, J. N. Gardner, A. M. Roe, W. Tertiuk, and G. L. Willey, *J. Med. Chem.*, **8**, 81 (1965); b. D. Brown and R. Stevenson, *J. Org. Chem.*, **30**, 1759 (1965); c. D. Heyl, E. Luz, S. A. Harris, and K. Folkers, *J. Am. Chem. Soc.*, **75**, 4080 (1953); d. S. A. Harris, *ibid.*, **62**, 3203 (1940); e. R. E. Cline, R. M. Fink, and K. Fink, *ibid.*, **81**, 2521 (1959).
[8] a. K. Kindler, H. G. Helling, and E. Sussner, *Ann.*, **605**, 200 (1957); b. N. H. Cromwell and K. E. Cook, *J. Am. Chem. Soc.*, **80**, 4573 (1958); c. H. E. Mason, *ibid.*, **67**, 1538 (1945); d. J. S. Byck and C. R. Dawson, *J. Org. Chem.*, **32**, 1084 (1967); e. R. A. Barnes and W. M. Bush, *J. Am. Chem. Soc.*, **80**, 4714 (1958); f. A. A. Khalaf and R. M. Roberts, *J. Org. Chem.*, **31**, 89 (1966).
[9] R. J. Pratt and E. V. Jensen, *J. Am. Chem. Soc.*, **78**, 4430 (1956).
[10] a. R. Baltzly and J. S. Buck, *J. Am. Chem. Soc.*, **65**, 1984 (1943); b. P. L. Coe, A. E. Jukes, and J. C. Tatlow, *J. Chem. Soc.*, (C), **1966**, 2020; c. R. Heck and S. Winstein, *J. Am. Chem. Soc.*, **79**, 3432 (1957).
[11] a. F. C. Uhle, J. E. Krueger, and A. E. Rogers, *J. Am. Chem. Soc.*, **78**, 1932 (1956); b. F. J. Villani, M. S. King, and F(lorence) J. Villani, *J. Med. Chem.*, **6**, 142 (1963).
[12] a. K. W. Rosenmund and E. Karg, *Ber.*, **75**, 1850 (1942); b. K. Kindler, B. Hedeman, and E. Schärfe, *Ann.*, **560**, 215 (1948); c. H. Richter and M. Schenck, Ger. Patent 952,441 (1956); through *Chem. Abstr.*, **53**, 2190 (1959); d. J. G. Murphy, *J. Org. Chem.*, **26**, 3104 (1961); e. J. M. Babbitt and J. C. Sih, *ibid.*, **33**, 856 (1968).
[13] R. Grewe, H. Köpnick and P. Roder, *Ann.*, **605**, 15 (1957).
[14] a. C. J. Schmidle and R. C. Mansfield, *J. Am. Chem. Soc.*, **77**, 5698 (1955); b. F. Zymalkowski, *Arch. Pharm.*, **287**, 505 (1954).
[15] J. M. Sprague, U.S. Patent 2,686,808 (1954).
[16] a. K. Kindler, E. Brandt, and E. Gehlhaar, *Ann.*, **511**, 209 (1934); b. F. Zymalkowski and W. Schauer, *Arch. Pharm.*, **290**, 218 (1957).
[17] a. L. V. Dvorken, R. B. Smyth, and K. Mislow, *J. Am. Chem. Soc.*, **80**, 486 (1958); b. J. Davoll, *J. Chem. Soc.*, **1953**, 3802; c. M. Sela and A. Berger, *J. Am. Chem. Soc.*, **77**, 1893 (1955).
[18] a. R. Adams and D. Fleš, *J. Am. Chem. Soc.*, **81**, 5803 (1959); b. F. C. Uhle, C. M. McEwen, Jr., H. Schröter, C. Yuan, and B. W. Baker, *ibid.*, **82**, 1200 (1960); c. C. G. Skinner, T. J. McCord, J. M. Ravel, and W. Shive, *ibid.*, **78**, 2412 (1956).
[19] E. Hardegger, Z. El Heweihi, and F. G. Robinet, *Helv. Chim. Acta*, **31**, 439 (1948).
[20] a. F. R. Atherton, H. T. Openshaw, and A. R. Todd, *J. Chem. Soc.*, **1945**, 382; b. F. R. Atherton, H. T. Howard, and A. R. Todd, *ibid.*, **1948**, 1106; c. L. Zervas and I. Dilaris, *J. Am. Chem. Soc.*, **77**, 5354 (1955); d. A. R. Todd and A. M. Michelson, U.S. Patent 2,645,637 (1953).
[21] a. L. Zervas, *Naturwissenschaften*, **27**, 317 (1939); b. L. Zervas and P. G. Katsoyannis, *J. Am. Chem. Soc.*, **77**, 5351 (1955); c. W. J. Fanshawe, V. J. Bauer, and S. R. Safir, *J. Med. Chem.*, **10**, 116 (1967).
[22] a. G. W. K. Cavill and J. R. Tetaz, *J. Chem. Soc.*, **1952**, 3634; b. S. G. Morris, *J. Am. Chem. Soc.*, **71**, 2056 (1949); c. L. Canonica, A. Bonati, and C. Tedeschi, *Ann. Chim.* (Rome), **46**, 465 (1956); d. J. T. Sheehan, *J. Org. Chem.*, **31**, 636 (1966).
[23] S. G. Morris and R. W. Riemenschneider, *J. Am. Chem. Soc.*, **68**, 500 (1946).
[24] C. A. Bartram, P. Oxley, D. A. Peak, and J. S. Nicholson, *J. Chem. Soc.*, **1958**, 2903.

[25] a. V. Bruckner, G. Fodor, J. Kovács, and J. Kiss, *J. Am. Chem. Soc.*, **70**, 2697 (1948); b. G. Fodor, V. Bruckner, J. Kiss, and J. Kovács, *ibid.*, **71**, 3694 (1949); c. A. Brossi and S. Teitel, *Helv. Chim. Acta*, **49**, 1757 (1966); d. W. Asbun and S. B. Binkley, *J. Org. Chem.*, **33**, 140 (1968); e. D. W. Henry and R. M. Silverstein, *ibid.*, **31**, 2391 (1966); f. S. A. Telang and C. K. Bradsher, *ibid.*, **30**, 752 (1965).
[26] a. J. F. W. McOmie and A. B. Turner, *J. Chem. Soc.*, **1963**, 5590; b. J. Davoll and D. H. Laney, *ibid.*, **1956**, 2124; c. G. W. H. Cheeseman and E. S. G. Törzs, *ibid.*, **1965**, 6681; d. J. F. Gerster and R. K. Robins, *J. Org. Chem.*, **31**, 3258 (1966); e. J. P. Paolini and R. K. Robins, *J. Heterocyclic Chem.*, **2**, 53 (1965); f. K. Mislow and J. B. Koepfli, *J. Am. Chem. Soc.*, **68**, 1553 (1946).
[27] a. H. Bruderer and A. Brossi, *Helv. Chim. Acta*, **48**, 1945 (1965); b. J. D. Benigni and R. L. Minnis, *J. Heterocyclic Chem.*, **2**, 387 (1965); c. R. G. Taborsky and P. Delvigs, *J. Med. Chem.*, **9**, 251 (1966); d. M. Julia, P. Manoury, and C. Voillaume, *Bull. Soc. Chim.*, France, **1965**, 1417; e. E. Shaw, *J. Am. Chem. Soc.*, **77**, 4319 (1955); f. A. S. F. Ash and W. R. Wragg, *J. Chem. Soc.*, **1958**, 3887.
[28] a. S. Terashima, K. Achiwa, and S. Yamada, *Chem. Pharm. Bull.*, **14**, 572 (1966); b. M. M. Baizer, J. R. Clark, and E. Smith, *J. Org. Chem.*, **22**, 1706 (1957); c. V. Burckhardt, W. Kündig, and P. Sieber, *Helv. Chim. Acta*, **35**, 1437 (1952); d. C. E. Ballou and H. O. L. Fischer, *J. Am. Chem. Soc.*, **77**, 3329 (1955); e. T. Y. Shen, H. M. Lewis, and W. V. Royle, *J. Org. Chem.*, **30**, 835 (1965); f. T. D. Inch and H. G. Fletcher, Jr., *ibid.*, **31**, 1815 (1966); g. S. J. Angyal and M. E. Tate *J. Chem. Soc.*, **1965**, 6949.
[29] a. G. M. Singerman, R. Kimura, J. L. Riebsomer, and R. N. Castle, *J. Heterocyclic Chem.*, **3**, 74 (1966); b. R. H. Baker, K. H. Cornell, and M. J. Cron, *J. Am. Chem. Soc.*, **70**, 1490 (1948); c. W. Stumpf, *Z. Electrochem.*, **57**, 690 (1953).
[30] J. C. Sowden and D. J. Kuenne, *J. Am. Chem. Soc.*, **74**, 686 (1952).
[31] T. Kametani, K. Fukumoto, and Y. Nomura, *Chem. Pharm. Bull.*, **6**, 467 (1958).
[32] M. Freifelder, *J. Am. Chem. Soc.*, **82**, 2386 (1960).
[33] L. Birkhofer, *Ber.* **75**, 429 (1942).
[34] a. R. B. Baltzly and P. B. Russell, *J. Am. Chem. Soc.*, **72**, 3410 (1950); b. *ibid.*, **75**, 5598 (1953); c. W. M. Pearlman, *Tetrahedron Letters*, **1967**, 1663; d. R. G. Hiskey and R. C. Northrop, *J. Am. Chem. Soc.*, **83**, 4798 (1961).
[35] a. H. Dahn, P. Zoller, and U. Solms, *Helv. Chim. Acta*, **37**, 565 (1954); b. H. Dahn, U. Solms, and P. Zoller, *ibid.*, **35**, 2117 (1952).
[36] a. R. F. Fischer and R. R. Whetstone, *J. Am. Chem. Soc.*, **76**, 5076 (1954); b. C. Ressler and V. DuVigneaud, *ibid.*, **76**, 3107 (1954); c. R. Wade, S. M. Birnbaum, M. Winitz, R. J. Koegel, and J. P. Greenstein, *ibid.*, **79**, 648 (1957); d. D. H. Rosenblatt, M. M. Nachlos, and A. M. Seligman, *ibid.*, **80**, 2463 (1958); e. W. A. Skinner, H. F. Gram, C. W. Mosher, and B. R. Baker, *ibid.*, **81**, 4639 (1959); f. H. Gregory, W. G. Overend, and L. F. Higgins, *J. Chem. Soc.*, **1949**, 2067.
[37] a. H. E. Zaugg, M. Freifelder, H. J. Glenn, B. W. Horrom, G. R. Stone, and M. R. Vernsten, *J. Am. Chem. Soc.*, **78**, 2626 (1956); b. G. H. Cocolas and W. H. Hartung, *ibid.*, **79**, 5203 (1957); c. W. J. Haggerty, Jr., R. H. Springer, and C. C. Cheng, *J. Heterocyclic Chem.*, **2**, 1 (1965).
[38] a. J. S. Moffett, *J. Chem. Soc.*, (C), **1966**, 725; b. D. C. Bishop and J. F. Cavalla, *ibid.*, p. 802; c. H. A. Staub and F. Vögtle, *Ber.*, **98**, 2691 (1965); d. K. Okumara, I. Inoue, M. Ikezaki, G. Hayashi, S. Nurimoto, and K. Shintomi, *J. Med. Chem.*, **9**, 315 (1966).
[39] a. G. Cignarella, G. Nathanson, and E. Occelli, *J. Org. Chem.*, **26**, 2747 (1961); b. E. D. Nicolaides, *ibid.*, **32**, 1251 (1967); c. L. Amundsen and C. Ambrosio,

ibid., **31**, 731 (1966); d. A. P. Martinez, W. L. Lee, and L. Goodman, *J. Med. Chem.*, **8**, 741 (1965).
[40] a. W. A. Lott and J. Krapcho, U.S. Patent 2,813,904 (1957); b. O. E. Fancher and S. Hayao, U.S. Patent 3,072,658 (1963); c. J. Schmutz and F. Künzle, *Helv. Chim. Acta*, **39**, 1144 (1956).
[41] a. F. Bergel and A. Cohen, U.S. Patent 2,493,520 (1950); b. G. Grethe, H. L. Lee, M. Uskoković, and A. Brossi, *J. Org. Chem.*, **33**, 494 (1968); c. M. Freifelder, *J. Pharm. Sci.*, **55**, 535 (1966); d. A. E. Drukker and C. I. Judd, *J. Heterocyclic Chem.*, **2**, 283 (1965).
[42] a. R. R. Sauers, *J. Am. Chem. Soc.*, **80**, 4721 (1958); b. J. Finkelstein, E. Chiang, F. M. Vane, and J. Lee, *J. Med. Chem.*, **9**, 319 (1966); c. K. Okumura and I. Inoue, *Chem. Pharm. Bull.*, **12**, 718 (1964); d. F. P. Lauduena, U.S. Patent 2,795,612 (1957); e. K. M. Hammond, N. Fisher, E. N. Morgan, E. M. Tanner, and C. S. Franklin, *J. Chem. Soc.*, **1957**, 1062; f. W. L. Garbrecht and R. M. Herbst, *J. Org. Chem.*, **18**, 1022 (1953); f. H. E. Zaugg and R. W. Denet, *ibid.*, **29**, 2769 (1964).
[43] a. W. B. Wright, Jr., H. J. Brabander, and R. A. Hardy, Jr., *J. Org. Chem.*, **26**, 485 (1961); b. *ibid.*, p. 2120; c. J. A. Faust, A. Mori, and M. Sahyun, *J. Am. Chem. Soc.*, **81**, 2214 (1959); d. W. W. Lee, B. L. Berridge, Jr., L. O. Ross, and L. Goodman, *J. Med. Chem.*, **6**, 567 (1963).
[44] S. S. Chatterjee and D. J. Triggle, *Chem. Commun.*, **1968**, 93.
[45] a. J. F. Cavalla, R. Jones, M. Welford, J. Wax, and C. V. Winder, *J. Med. Chem.*, **7**, 412 (1964); b. J. F. Cavalla, *J. Chem. Soc.*, **1959**, 851; c. B. R. Baker, R. E. Schaub, and J. H. Williams, *J. Org. Chem.*, **17**, 116 (1952); d. E. Jaeger and J. Biel, *ibid.*, **30**, 740 (1965); e. E. Schipper and W. R. Boehme, *ibid.*, **26**, 3599 (1961); f. G. Cignarella and G. Nathanson, *ibid.*, p. 1500.
[46] a. C. van de Westeringh, P. Van Daele, B. Hermans, C. Van der Eycken, J. Boet, and P. A. J. Janssen, *J. Med. Chem.*, **7**, 619 (1964); b. J. M. McManus, J. W. McFarland, C. F. Gerber, W. M. McLemore, and G. D. Laubach, *ibid.*, **8**, 766 (1965); c. C. A. Grob and V. Krasnobajew, *Helv. Chim. Acta*, **47**, 2145 (1964); d. H. Stetter and K. Zoller, *Ber.*, **98**, 1446 (1965); e. B. Hermans, P. Van Daele, C. van de Westeringh, C. Van der Eycken, J. Boey, and P. A. J. Janssen, *J. Med. Chem.*, **9**, 49 (1966).
[47] a. G. Cignarella, *J. Med. Chem.*, **7**, 241 (1964); b. R. Haberl, *Monatsh.*, **89**, 798 (1958); c. R. Ratovis, J. R. Boissier, and C. Dumont, *J. Med. Chem.*, **8**, 104 (1965).
[48] a. L. H. Werner and S. Ricca, Jr., *J. Am. Chem. Soc.*, **80**, 2733 (1958); b. S. Morosawa, *Bull. Chem. Soc. Japan*, **31**, 418 (1958); c. E. S. Nikitskaya, V. S. Usovskaya, and M. V. Rubstov, *J. Gen. Chem., USSR*, **32**, 2842 (1962) (Eng. trans.), d. P. S. Portoghese and A. A. Mikhail, *J. Org. Chem.*, **31**, 1059 (1966); e. G. DeStevens and M. Dughi, *J. Am. Chem. Soc.*, **83**, 3087 (1961); f. L. N. Iakontov and M. V. Rubstov, *J. Gen. Chem., USSR*, **28**, 3139 (1958) (Eng. trans.).
[49] a. T. Kametani, K. Kigesawa, H. Hiiragi, and H. Ishimara, *Chem. Pharm. Bull.*, **13**, 295 (1965); b. T. Kametami, K. Kigesawa, and T. Hayasaka, *ibid.*, p. 1225. c. From framework molecular orbital models constructed by the author.
[50] a. M. Julia, P. Manoury, and J. Igolan, *Compt. Rend.*, **251**, 394 (1960); b. R. H. Wiley, K. F. Hussing, and J. Moffatt, *J. Org. Chem.*, **21**, 190 (1956); c. H. Gold, *Ann.*, **688**, 205 (1965); d. R. M. Herbst and D. F. Percival, *J. Org. Chem.*, **19**, 439 (1954); e. D. F. Percival and R. M. Herbst, *ibid.*, **23**, 825 (1957); f. R. A. Henry, W. G. Finnegan, and E. Lieber, *J. Am. Chem. Soc.*, **76**, 2894 (1954); g. J. M. McManus and R. M. Herbst, *J. Org. Chem.*, **24**, 1643 (1959).

[51] J. A. Montgomery and H. J. Thomas, *J. Am. Chem. Soc.*, **85**, 2672 (1963).
[52] a. R. M. Acheson, *J. Chem. Soc.*, **1956**, 4232; b. W. J. Burke, J. A. Warburton, J. L. Bishop, and J. L. Bills, *J. Org. Chem.*, **26**, 4669 (1961); c. J. W. Cornforth, O. Kauder, J. E. Pike, and R. Robinson, *J. Chem. Soc.*, **1955**, 3348; d. W. E. Solodar and M. Green, *J. Org. Chem.*, **27**, 1077 (1962); e. M. Green and W. E. Solodar, U.S. Patent 3,187,049 (1965); f. D. L. Fields, J. B. Miller, and D. D. Reynolds, *J. Org. Chem.*, **29**, 2640 (1964).
[53] a. J. H. Burckhalter, R. J. Siewald, and H. C. Scarborough, *J. Am. Chem. Soc.*, **82**, 991 (1960); b. J. Thesing and P. Binger, *Ber.*, **90**, 1419 (1957); c. A. Treibs and R. Zinsmeister, *ibid.*, p. 87; d. J. D. Prugh, A. C. Huitric, and W. C. McCarthy, *J. Org. Chem.*, **29**, 1991 (1964); e. D. F. O'Brien and J. W. Gates, Jr., *ibid.*, **31**, 1538 (1966); f. M. Pesson and D. Richer, *Compt. Rend.*, **261**, 1339 (1965).
[54] a. R. L. Clark, W. H. Jones, W. J. Raich, and K. Folkers, *J. Am. Chem. Soc.*, **76**, 3995 (1954); b. R. H. Reitsema, *ibid.*, **71**, 2141 (1949); c. J. F. Carson, *ibid.*, **78**, 3728 (1956); d. F. Kagan, M. A. Rebenstorf, and R. V. Heinzelman, *ibid.*, **79**, 3541 (1957); e. I. Ziderman and E. Dimant, *J. Org. Chem.*, **31**, 223 (1966); f. M. Freifelder and Y. H. Ng, *J. Med. Chem.*, **8**, 122 (1965); g. H. R. Snyder and D. S. Matteson, *J. Am. Chem. Soc.*, **79**, 2217 (1957).
[55] a. W. Wenner, U.S. Patent 2,389,099 (1945); b. Y. Liwschitz and A. Zilkha, *J. Am. Chem. Soc.*, **76**, 3698 (1954); c. A. Zilkha and J. Rivlin, *J. Org. Chem.*, **23**, 94 (1958); d. A. Zilkha, E. S. Rachman, and J. Rivlin, *ibid.*, **26**, 376 (1961); e. T. Fujii, *Chem. Pharm. Bull. Tokyo*, **6**, 591 (1958); f. L. Zervas and D. M. Theodoropoulos, *J. Am. Chem. Soc.*, **78**, 1359 (1956); g. W. Pfleiderer and F. Sagi, *Ann.*, **673**, 78 (1964).
[56] a. K. Harada, *J. Org. Chem.*, **32**, 1790 (1967); b. K. Harada and K. Matsumoto, *ibid.*, p. 1794.
[57] a. M. Uskoković, J. Iacobelli, and W. Wenner, *J. Org. Chem.*, **27**, 3606 (1962); b. M. Uskoković and W. Wenner, U.S. Patent 3,244,698 (1966); c. N. M. Bortnick and M. F. Fegley, U.S. Patent 3,065,237 (1962); d. S. Wilkinson, *J. Chem. Soc.*, **1957**, 104.
[58] A. Stempel and E. C. Buzzi, *J. Am. Chem. Soc.*, **71**, 2969 (1949).
[59] a. W. E. Conrad and S. M. Dee, *J. Org. Chem.*, **23**, 1700 (1958); b. T. D. Perrine and N. B. Eddy, *ibid.*, **21**, 125 (1956).
[60] a. V. Bruckner and G. von Fodor, *Ber.*, **76**, 466 (1943); b. G. von Fodor, *ibid.*, p. 1216; c. R. Pessino, A. Botto, F. Gatti, and G. Valenti, *Farmaco Sci. Ed.*, **20**, 97 (1965); d. G. Ehrhart and H. Ott, U.S. Patent 2,737,526 (1956).
[61] a. J. H. Biel, U.S. Patent 2,955,114 (1960); b. B. G. Boggiano, V. Petrow, O. Stephenson, and A. M. Wild, *J. Chem. Soc.*, **1959**, 1143; c. P. M. Carabateas and L. Grumbach, *J. Med. Chem.*, **5**, 913 (1962); d. N. J. Harper, C. F. Chignell, and G. Kirk, *ibid.*, **7**, 726 (1964); e. A. F. Casy and H. Birnbaum, *J. Chem. Soc.*, **1964**, 5130.
[62] a. W. D. McPhee and E. S. Erickson, Jr., *J. Am. Chem. Soc.*, **68**, 624 (1946); b. E. J. Forbes, *J. Chem. Soc.*, **1955**, 3926; c. G. W. Kirby and H. P. Tiwari, *J. Chem. Soc.*, (C) **1966**, 676; d. J. Sam and D. M. Noble, *J. Pharm. Sci.*, **56**, 729 (1967); e. A. F. Casy and N. A. Armstrong, *J. Med. Chem.*, **8**, 57 (1965); f. R. A. Heacock and O. Hutzinger, *Can. J. Chem.*, **42**, 514 (1964); g. B. Marchand, *Ber.*, **95**, 577 (1962); h. H. O. House, P. P. Wickham, and H. G. Muller, *J. Am. Chem. Soc.*, **84**, 3139 (1962).
[63] K. H. Dudley, H. W. Miller, R. C. Corley, and M. E. Wall, *J. Org. Chem.*, **32**, 2317 (1967).

[64] a. B. R. Baker, R. E. Schaub, M. V. Querry, and J. H. Williams, *J. Org. Chem.*, **17**, 77 (1952); b. T. Wieland, K. H. Shin, and B. Heinke, *Ber.*, **91**, 483 (1958); c. G. S. Grinenko, V. I. Bayunova, S. D. Padobrazhnykh, and V. I. Maksimov, *J. Org. Chem. USSR*, **1**, 2182 (1965); Eng. trans.

[65] a. E. Fry, *J. Am. Chem. Soc.*, **76**, 284 (1954); b. C. Schöpf, E. Brass, E. Jacobi, W. Jorde, W. Mocnik, L. Neuroth, and W. Salzer, *Ann.*, **544**, 30 (1940).

[66] V. A. Zagorevskii, D. A. Zykov, and E. K. Orlova, *J. Gen. Chem.*, *USSR*, **34**, 541 (1964) (Eng. trans.).

[67] a. A. E. Ardis, R. Baltzly, and W. Schoen, *J. Am. Chem. Soc.*, **68**, 591 (1946); b. E. L. Engelhardt, U.S. Patent 2,582,587 (1952); c. H. D. Moed and J. van Dijk, *Rec. Trav. Chim.*, **75**, 1214 (1956).

[68] a. B. R. Baker, R. E. Schaub, and J. H. Williams, *J. Org. Chem.*, **17**, 97 (1952); b. N. Izumiya, Y. Fujita, F. Irreverre, and B. Witkop, *Biochemistry*, **4**, 2501 (1965); c. H. Bretschneider, *Monatsh.*, **78**, 71 (1948).

[69] a. R. Simonoff and W. H. Hartung, *J. Am. Pharm. Assoc., Sci. Ed.*, **35**, 307 (1946); b. J. R. Vaughan, Jr. and J. Blodinger, *J. Am. Chem. Soc.*, **77**, 5757 (1955); c. C. M. Suter and A. W. Ruddy, U.S. Patent 2,431,285 (1947); d. A. T. Nielson, D. W. Moore, J. H. Mazur, and K. H. Berry, *J. Org. Chem.*, **29**, 2898 (1964); e. G. Grethe, H. L. Lee, M. Uskoković, and A. Brossi, *ibid.*, **31**, 491 (1968); f. A. T. Nielsen, *ibid.*, **31**, 1053 (1966); g. F. I. Carroll, A. M. Ferguson, and J. B. Lewis, *ibid.*, p. 2957; h. K. Mieschler and H. Kaegi, U.S. Patent 2,714,108 (1955).

[70] J. B. Wright and E. S. Gutsell, *J. Am. Chem. Soc.*, **81**, 5193 (1959).

[71] K. Kindler and K. Schrader, *Ann.*, **564**, 49 (1949).

[72] a. B. Elpern, W. Wetterau, P. Carabateas, and L. Grumbach, *J. Am. Chem. Soc.*, **80**, 4916 (1958); b. K. Schenker and J. Druey, *Helv. Chim. Acta*, **45**, 1344 (1962).

[73] a. R. H. Springer, W. J. Haggerty, Jr., and C. C. Cheng, *J. Heterocyclic Chem.*, **2**, 49 (1965); b. H. U. Daeniker, *Helv. Chim. Acta*, **47**, 33 (1964).

[74] J. W. Martin, Jr., and W. H. Hartung, *J. Org. Chem.*, **19**, 338 (1954).

[75] D. E. Ames and T. F. Gray, *J. Chem. Soc.*, **1955**, 631; b. P. Mamalis, M. J. Rix and A. A. Sarsfield, *ibid.*, **1965**, 6278; c. M. Masaki and M. Ohta, *J. Org. Chem.*, **29**, 3165 (1964); d. M. Masaki, Y. Chigira, and M. Ohta, *ibid.*, **31**, 4143 (1966); e. M. Masaki, J. Ohtake, M. Sugiyama, and M. Ohta, *Bull. Chem. Soc. Japan*, **38**, 1802 (1965).

[76] M. Frankel, G. Zvilichofsky, and Y. Knobler, *J. Chem. Soc.*, **1964**, 3931.

[77] A. F. Thomas and A. Marxer, *Helv. Chim. Acta*, **43**, 469 (1960).

[78] a. O. M. Friedman and A. M. Seligman, *J. Am. Chem. Soc.*, **76**, 655 (1954); b. J. DeGraw and L. Goodman, *J. Med. Chem.*, **7**, 213 (1964); c. W. W. Lee, B. J. Berridge, Jr., O. L. Ross, and L. Goodman, *ibid.*, **6**, 567 (1963).

[79] a. D. Schmidt-Barbo, F. Hampe, M. Schorr, and G. Lammler, U.S. Patent 2,945,860 (1960); b. W. Siedel, K. Sturm, and W. Scheurich, *Ber.*, **99**, 345 (1966).

[80] M. Freifelder, *J. Org. Chem.*, **31**, 3875 (1966).

[81] B. Boothroyd and E. L. Clark, *J. Chem. Soc.*, **1953**, 1499,

[82] F. P. Doyle, G. R. Fosker, J. H. C. Nayler, and H. Smith, *J. Chem. Soc.*, **1962**, 1440.

[83] a. T. Sheradsky, Y. Knobler, and M. Frankel, *J. Org. Chem.*, **26**, 1485 (1961); b. Y. Knobler and M. Frankel, *J. Chem. Soc.*, **1958**, 1629; c. T. Sheradsky, Y. Knobler, and M. Frankel, *J. Org. Chem.*, **26**, 2710 (1961).

XX

DEHALOGENATION

INTRODUCTION

The removal of halogen from an organic compound by catalytic hydrogenation is usually accomplished without difficulty under mild conditions unless the halogen is fluorine. For dehalogenation, RX to RH, the order is RI > RBr > RCl; RF cannot be converted to RH by catalytic means.

As a rule, rupture of the carbon-fluorine bond does not take place except in certain highly activated systems or during reduction of an aromatic ring containing nuclear fluorine. Indeed an aromatic ring containing any nuclear halogen has not been saturated without accompanying dehalogenation.

It is known that allylic- and vinylic-type halides are dehalogenated more readily than saturated aliphatic-type halides [1a, b], but the ease of removal of a vinylic halide does not usually apply to fluorine. The carbon-fluorine bond is not sufficiently activated in the system ArCH=CHF to undergo hydrogenolysis, at least not under mild hydrogenation conditions. However, in the palladium-catalyzed hydrogenation of 5-fluorouracil, where the vinylic fluoride was activated by the carbonyl function of the cyclic amide, removal of fluorine took place in alkaline solution to give an 80% yield of uracil [1c].

It is also known that activation of one halogen occurs when two halogen atoms are attached to a single carbon atom [1b]. This activation apparently aided the reduction of $\beta,\beta,4$-trifluorostyrene with platinum oxide to 4-(2-fluoroethyl)fluorobenzene [1d]. However, neither the aryl nor vinyl fluorides were affected (Eq. 20.1).

$$F\text{-}C_6H_4\text{-}CH=CF_2 \xrightarrow{H_2} F\text{-}C_6H_4\text{-}CH=CHF + HF \qquad (20.1)$$

Loss of fluorine did occur during ring reduction of some fluorobenzenes with platinum [2a] and with nickel catalyst [2b, c]. It was shown, however, that defluorination did not take place before or after 4-fluorophenylacetic acid was converted to cyclohexylacetic acid [2c]; neither the starting material nor 4-fluorocyclohexylacetic acid could be hydrogenolyzed to the corresponding defluoro compounds. It might be assumed that loss of fluorine occurred during partial saturation of the ring, perhaps by 1,4-addition of hydrogen which would lead to an allylic fluoride capable of undergoing hydrogenolysis (Eq. 20.2).

$$R\text{-}C_6H_4\text{-}F \xrightarrow{H_2} \left[\begin{array}{c} R \\ H \end{array} \diagup \diagdown \begin{array}{c} F \\ H \end{array} \right] \xrightarrow{3H_2} R\text{-}C_6H_{11} \qquad (20.2)$$

Catalysts and Reaction Conditions

While the hydrogenative removal of halogen is aided by the adjacence of certain activating groups or functions, the most important factors in the process are the catalysts and the reaction medium.

Dehalogenations have been carried out in neutral media. This method is not preferred, although it may be useful in certain instances to insure the selective removal of one halogen in the presence of another.

If dehalogenation in neutral medium is to be employed, palladium catalysts generally give the best results because they are the least affected by the substrate or the resulting hydrogen halide. When nickel is used under similar conditions, larger amounts of catalyst are necessary to overcome the reputed poisoning effect of organic halide [3], although it is more likely that more nickel is required because it is dissolved by the hydrogen halide that is released. As a rule platinum is an inferior catalyst for dehalogenation; indeed, platinum and rhodium are often employed to minimize hydrogenolysis in selective hydrogenations of other reducible functions in the presence of halogen. If platinum is used for dehalogenation of an aryl halide, ring reduction may precede loss of halogen or can result because of acid catalysis from the release of HX. For example, after one molar equivalent of hydrogen was absorbed in the low-pressure reduction of 2-chlorophenol in ethyl alcohol with platinum oxide, a mixture was obtained which consisted of recovered starting material, cyclohexanol, and cyclohexane [4a]. When 3-benzyl-6-chloro-7-sulfamoyl-1,2,4-benzothiadiazine-1,1-dioxide in ethyl alcohol was hydrogenated with 10% by weight of platinum oxide at 100° and 70 atm, halogen was not removed but the 3-benzyl group was converted to 3-cyclohexylmethyl [4b].

Although platinum oxide itself is not recommended for dehalogenations, a trace amount added during the preparation of palladium on carbon promoted the hydrogenation of 2,6-dichloro-3-cyclopentyl-5-methylpyridine in methyl alcohol to 3-cyclopentyl-5-methylpyridine so strongly that reaction was complete in minutes, as compared to 11 hr with unpromoted palladium on carbon [4c]. On the other hand, the synergistic effect was negligible in the comparison of the reduction of 2,3,6-trichloropyridine with promoted and unpromoted catalyst. There also did not appear to be any particular activation when a combination of 10% by weight of 10% palladium on carbon and 1% by weight of platinum oxide was employed for the reduction of ethyl 4-(2-chloro-4-pyrimidyl)piperazine-1-carboxylate in aqueous sodium hydroxide [4d]; 48% of starting material was recovered and only 36% of ethyl 4-(4-pyrimidyl)piperazine-1-carboxylate was obtained.

Dehalogenations have been carried out in acidic medium but there are many disadvantages to such a procedure. Raney nickel will be attacked and platinum, as previously noted, may cause saturation of the ring of aromatic compounds. Dehalogenations with palladium in acetic acid as solvent are often slow [4a], although the presence of an acid acceptor may increase the speed of hydrogen uptake. The use of an acid stronger than acetic acid has an even more pronounced retarding effect in dehalogenations with palladium [1a, 4a].

According to a study of the literature and personal experience, the best method of removing chlorine, bromine, and iodine from an organic substrate is by hydrogenation in neutral solvent with palladium on a support in the presence of an acid acceptor. Busch and Stove [5a] showed that the halogen in organic halides could be cleaved quantitatively by reduction in alcohol containing excess potassium hydroxide in the presence of palladium on calcium carbonate. Sodium hydroxide, sodium acetate, and potassium carbonate as acid acceptors, as well as tertiary amines such as triethylamine, have been found satisfactory in dehalogenations in the presence of 5% palladium on carbon [4a]. Alkali hydroxides, ammonium hydroxide, and magnesium oxide are among other acid acceptors found useful in catalytic dehalogenations. Magnesium oxide is most often employed in dehalogenations in the pyrimidine series to prevent reduction of the ring. It has also been highly recommended as an acid acceptor in the hydrogenation of chloropyrazines and related haloheterocycles [5b].

The choice of base often is not critical. There are times when strong inorganic bases are contraindicated in the hydrogenolysis of very reactive halides. There are also examples in the removal of nuclear halogen from nitrogen-containing heterocycles where sodium acetate and ammonium hydroxide were not recommended, because the solutions after reduction were acidic enough to induce ring reduction. However, in these instances the amount of catalyst employed may have also been a factor.

As a rule when acid acceptors are used, the amount should be at least equal to the molar amount of HX released.

When the substrate contains a basic nitrogen, reduction with palladium may be carried out in the absence of a base since the substrate or reduction product acts as acid acceptor. 1-(2-Chlorophenyl)piperidine was easily converted to 1-phenylpiperidine by hydrogenation in alcohol with 5% palladium on carbon [4a]; reduction in 80% acetic acid was very slow. Weak bases as the haloanilines were rapidly dehalogenated with palladium on carbon in alcohol without an added acid acceptor [1a].

Nickel-catalyzed dehalogenations should also be carried out in the presence of an acid acceptor. In general nickel reductions, even under basic conditions, appear to require high catalyst ratios. Ion-exchange resins have been used as bases in dehalogenations with nickel [6a, b]; these reactions have also been carried out with a large amount of catalyst.

Condensation

One of the side reactions that can accompany dehalogenation is condensation, $2RX \rightarrow R\text{—}R$. As a rule this unusual reaction takes place in the presence of palladium catalysts, but conditions are generally so specific [6c] that it should not deter the chemist from employing palladium catalysts for dehalogenations.

ALIPHATIC HALIDES

Alkyl and Substituted Alkyl Halides

Halogen was not removed from primary and secondary alkyl chlorides and bromides upon reduction in neutral medium with palladium on carbon [1a, 7a] or with platinum or rhodium on carriers [7a]. The more reactive halogen in allyl chloride was removed during hydrogenation with palladium in the absence of base [1a].

The bromine atom in ethyl bromoacetate, activated by adjacence of the carboxyl group, was readily cleaved during reduction in alcohol with palladium [1a] but the activating effect was lost when the carboxyl group was beta to the bromine; no reaction occurred during a similar reduction of ethyl 3-bromopropionate. The presence of an acid acceptor probably would have aided debromination.

Monochloroacetic acid did not undergo loss of chlorine during palladium reduction in alcohol [1a], but did in the presence of base [5a]. On the other hand the halogen of N-(chloroacetyl)benzhydrylamine was cleaved under neutral conditions with palladium to give a 70% yield of N-acetylbenzhydrylamine [7b]. Apparently when a halogen is adjacent to a carboxyl group, COR, the least activation is obtained when R is OH.

450 *Dehalogenation*

A substituted alkyl chloride, 4-chloromethylquinolizidine hydrochloride, was dehalogenated in water containing excess potassium bicarbonate to 4-methylquinolizidine in 96% yield by hydrogenation with palladium on barium sulfate [7c]. The large amount of catalyst used, over 300% by weight, may be an indication of the unreactivity of alkyl and substituted alkyl chlorides toward hydrogenation.

It is difficult to evaluate the worth of Raney nickel in the dehalogenation of alkyl halides in the presence of base because of the very large amount of catalyst used in some of these reactions. A study employing nickel in methyl alcohol containing potassium hydroxide emphasized the inertness of primary and secondary alkyl monochlorides and showed that the corresponding bromides and iodides lost halogen, as did activated chlorides such as allyl and vinyl chloride [1b]. However the high catalyst ratio, 300% by weight of substrate, makes the practicality of the successful reactions questionable.

In other examples 60% by weight of Raney nickel catalyzed the reduction of ethyl 5-bromo-2-carbethoxyhexanoate to ethyl 2-carbethoxyhexanoate in alcohol and excess potassium hydroxide [7d] (Eq. 20.3). Sodium hydroxide was used with an unspecified amount of nickel in the hydrogenation of

in EtOH·KOH

$$CH_3CHBr(CH_2)_2CH(COOC_2H_5)_2 \xrightarrow[Ni(R)]{H_2} CH_3(CH_2)_3CH(COOC_2H_5)_2$$

(20.3)

1-chloromethyl-3-hydroxy-2-oxaadamantane to the corresponding 1-methyl compound [7e] (Eq. 20.4). Ethyl 2-iodomethylcyclohexanecarboxylate was

pres. NaOH

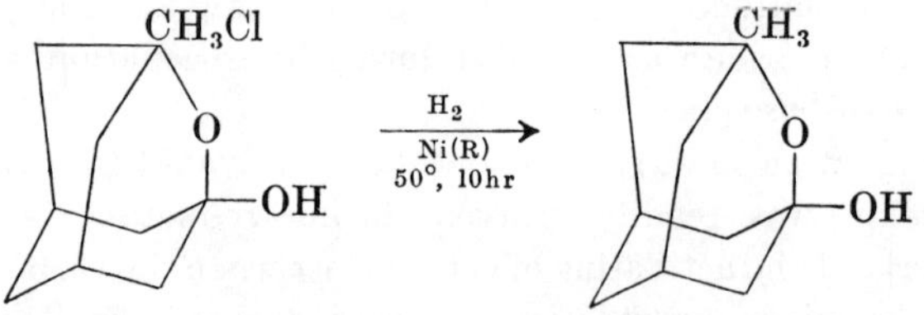

(20.4)

converted to ethyl 2-methylcyclohexanecarboxylate in alcohol containing pyridine [7f]. In this reaction the excessive amount of catalyst, 160% by weight of substrate, may have been necessary because of the inhibitory effect of pyridine.

Alicyclic Halides

There are not very many references on hydrogenations of alicyclic monohalides where a modest amount of catalyst is employed. 3-Bromobicyclo[3,

2,1]oct-2-en-6-yl formate, which contains a vinyl-type bromide, was reduced with a 25% weight ratio of 10% palladium on carbon in aqueous alkaline tetrahydrofuran [8a] (Eq. 20.5).

$$\text{HCOO-[bicyclic with Br]} \xrightarrow{2H_2} \text{HCOO-[bicyclic]} \qquad (20.5)$$

1-Chlorocyclohexane was converted to cyclohexane with 200% by weight of Raney nickel in methanolic potassium hydroxide [2b]. Whether such a large amount of catalyst was necessary to remove the vinyl-type chloride was not determined. About 50% by weight of Raney nickel was employed to dehalogenate cycloheptyl, cyclooctyl, and cyclononyl iodides to the corresponding cycloalkanes [8b]. It is likely that less catalyst would be required if an acid acceptor was present. Very high ratios of Raney nickel to substrate, 300–600%, were used in reductive dehalogenations of some substituted cyclopentyl bromides carried out in water or aqueous alcohol employing the base form of ion-exchange resins as acid binders [6a, b, 9a, b].

Halogen was not removed from the alicyclic compound 2-chloro-1,3-hydroxycyclopentanecarboxylic acid by reduction with palladium on carbon, with platinum catalyst in alcohol or acetic acid, or with palladium on carbon in methyl alcohol and potassium carbonate [10]. It is difficult to state that this was due to the unreactivity of the alicyclic halide. There are no examples available on the catalytic hydrogenolysis of unsubstituted alicyclic chlorides with base and a moderate amount of palladium or nickel. In this particular case construction of a molecular model indicates that steric effects can be responsible. It suggests that, because of the unshared pairs of electrons of the oxygen atom of the carboxyl and the 3-hydroxy groups, this particular part of the molecule makes the strongest contact with the catalyst, leaving the chlorine atom out of contact with the catalyst.

ARYL HALIDES, ArX

Halogens in Benzenoid Systems

Aryl halides are examples of vinyl halides. With the exception of fluorine, nuclear halogens should be readily removed without attack on the ring. Dehalogenation of these compounds is best carried out with palladium on a carrier in the presence of an acid acceptor.

Many aryl chlorides are not attacked by hydrogenation in the absence of base [1a, 4a]. The more reactive bromides and iodides may be removed without an acid acceptor [1a, 11] but the presence of one greatly increases the reaction rate [4a]. When the substrate contains a basic side chain, an acid acceptor is usually unnecessary. When the substrate is a weak base, such as a haloaniline, an acid acceptor may not be required [1a]. In this instance, however, the addition of one not only increases the reaction rate but also can lead to the use of less catalyst for dehalogenation [4a]. If the substrate contains several nitrogen atoms, one of which is very basic, it may be necessary to add one equivalent of acid or to hydrogenate the compound as a monoacid salt to prevent inactivation of the catalyst. An example is seen in the dehalogenation of 5-chloro-2-(3-dimethylaminopropylamino)aniline [4a]. Reduction with palladium on carbon alcohol at 25° and 2 atm was rather slow (15 hr), but was complete in several hours in alcohol containing an equivalent of hydrogen chloride. The hydrogenation of 2-chloro-9-(5-diethylamino-4-pentyl)amino-6-methoxyacridine was carried out as the dihydrochloride salt [4a]. In this instance, sodium acetate was also added to neutralize the release of hydrogen chloride. Reaction was complete in less than 2 hr to yield 86% of the 6-methoxy-6-substituted acridine.

The method preferred by this author consists of hydrogenation of the substrate under low-pressure conditions in alcohol (in some instances in water) containing sodium or potassium hydroxide sufficient to neutralize the release of hydrogen halide, in the presence of 10–20% by weight of 5% palladium on carbon. When a halogen is to be removed from a strongly acid compound such as a halobenzenesulfonamide, the reaction is best carried out in water (or aqueous alcohol) containing excess sodium hydroxide to act as acid acceptor and to maintain a solution of the sulfonamide [4a]. This procedure gave good results in the formation of 7-sulfamyl-3,4-dihydro-1,2,4-benzothiadiazine-1,1-dioxide from the 6-chloro derivative [12a] and in the dehalogenation of aminodichlorobenzenesulfonamides [12b].

Raney nickel has also been employed for removal of aryl halides. A moderate amount catalyzed the reductive dehalogenation of 2-amino-4-chloro-5-methylbenzenesulfonic acid in aqueous potassium hydroxide to give 75% yield of 2-amino-5-methylbenzenesulfonic acid [13a]. It was also used in the dehalogenation of tetrachlorobenzofurans in methanolic potassium hydroxide, but the weight of catalyst was not reported [13b, c].

Halogens in Unsaturated Heterocycles

Catalysts and reaction conditions used for the dehalogenation of aryl halides should also be applicable for the removal of nuclear halogen from unsaturated heterocycles; in general, hydrogenolysis will take place with greater facility.

In many instances the heterocyclic ring will not be attacked during or after removal of halogen from it. In other cases this can depend on whether an acid acceptor is present, on the amount of catalyst, and on reaction conditions in the absence of an acid acceptor, as illustrated in the following examples.

6-Chloro-1-imidazo[b]pyridine was dehalogenated at room temperature and 3 atm pressure with an equal weight of 5% palladium hydroxide on calcium carbonate [14a]; 29% yield of 1-imidazo[b]pyridine was reported with no explanation concerning the material balance. When 6-chloropyrido[2,3-d]v-triazole was hydrogenated with *three times the weight of catalyst at 80°*, saturation of the pyridine ring accompanied dehalogenation. Platinum on a support had the same effect under similar conditions. Success attained at room temperature was ascribed to reduction with a mixture of supported palladium and platinum catalyst (100% by weight of substrate) in the presence of a trace of sodium hydroxide. However, calculation of the amount of base employed showed that three equivalents were present. It seems more reasonable to suggest that the excess of strong base caused inactivation of the catalyst and was probably the most important factor in preventing ring reduction.

Furans. 4-Bromo-3-dimethylaminomethyl- and 4-bromo-3-(4-morpholinomethyl)-2,5-*bis*(4-bromophenyl)furans were dehalogenated in alcohol over palladium on barium sulfate to yield 3-dimethylaminomethyl- and 3-(4-morpholinomethyl)-2,5-diphenylfurans [14b]. The combination of the ready reducibility of the bromine atom and the presence of the basic side chain made an acid acceptor unnecessary.

Pyrroles. 3-(5-Ethoxycarbonyl-2-iodo-4-methylpyrrole)propionic acid was hydrogenated with 5% palladium on carbon on alcohol containing sodium acetate to yield 83% of the deiodinated pyrrolepropionic acid [15a], and ethyl 2-bromo-5-indo-4-methyl-3-pyrrolecarboxylate was reduced to ethyl 4-methyl-3-pyrrolecarboxylate with 10% palladium on carbon and magnesium oxide as acid acceptor [15b]. Another group of investigators found the latter method ineffective and carried out reactions with Raney nickel and magnesium oxide at 130–140° and 80 atm [15c]. Raney nickel in amount equal to the weight of substrate catalyzed the dehalogenation of diethyl 5-bromo-3-ethylpyrrole-2,4-dicarboxylate in water containing sodium hydroxide [15d], with no explanation of the need for the high catalyst ratio.

Pyridines, Quinolines, Isoquinolines, and Related Ring Systems. Nickel and palladium catalysts are used for the removal of nuclear halogen from the pyridine ring. Nickel-catalyzed reductions are run in alkaline media and usually require more vigorous conditions and/or higher catalyst ratios. See Table 20.1 for examples.

Dehalogenations with palladium may be carried out in neutral medium, but usually a large amount of catalyst is needed and reaction time is often

Table 20.1 Pyridines

Substituents	Catalyst	Ratio (%)	Temperature (°C)	Pressure	Medium	Reference
2,6-diCl-4-COOH 2,6-diBr-4-COOH	Ni from Ni(CO$_3$)$_2$	100	50	4 atm	H$_2$O, NaOH	16a
2,6-diCl-4-COOH	Ni(R)	10	86–120	33 atm	H$_2$O, NaOH	16b
4,6-diCl-2-CH$_3$- 3-C$_2$H$_5$O(CH$_2$)$_2$-	Ni(R) freshly prepared	30	25	atmospheric	CH$_3$OH, NaOH	16c

overlong. 2,3,6-Trichloro-4-cyclopentylpyridine containing some 2,6-dichloro-4-cyclopentylpyridine was hydrogenated in alcohol with 20% by weight of 5% palladium on carbon; after 36 hr 79% of starting material was recovered [17a]. Some 3,5-dialkyl-2,6-dichloropyridines were dehalogenated with 50% by weight of 15% palladium on carbon by hydrogenation for 24–36 hr [17b]. The removal of chlorine from 3-(4-chloro-3-hydroxy- and 4-chloro-3-methoxy-2-pyridylamino)propionic acids in aqueous solution was complete in 4 hr at 3 atm pressure [17c], but with 40% by weight of palladium black.

Removal of halogen from chloropyridines has been carried out with palladium catalysts in acid solution [18a, b, c]; an analysis of the weight ratio of palladium metal to substrate shows it to be rather high, whereas in alkaline media that ratio is about 1% or less.

In general the best results are obtained with palladium on a carrier in solvent containing an acid acceptor. Methyl 2,6-dichloroisonicotinate was dehalogenated in methyl alcohol containing excess triethylamine with 5% palladium on carbon within 1 hr at 50–60° and less than 3 atm pressure of hydrogen [19a]; ammonia, pyridine, morpholine, and other bases were also employed as acid acceptors. The combination of palladium on carbon and sodium acetate led to almost quantitative yield of 3-ethyl-4-methyl-6-isopropoxypyridine from reductive dehalogenation of the 2-chloro compound [19b]. 2,3-Diamino-5-bromopyridine was converted to 2,3-diaminopyridine in 78–86% yield by hydrogenation in dilute sodium hydroxide with a low ratio of 5% palladium on strontium carbonate [19c]. Successful dehalogenation with a small amount of catalyst should not be too unexpected in such reductions since bromine is more readily removed than chlorine.

It is likely that the bromodiaminopyridine could have been reduced to diaminopyridine without an acid acceptor since the amino groups would bind hydrogen bromide as it was formed. For example, 2-chloro-4-ethyl-5-nitropyridine, as amine precursor, was easily dehalogenated in alcohol to 3-amino-4-ethylpyridine in 77% yield with palladium on calcium carbonate [20a].

Some difficulty was experienced in the dehalogenation of 6-amino-2-chloro-4-methyl-5-phenylnicotinamide with palladium on carbon and sodium acetate as acid acceptor [20b]. The yield of 6-amino-4-methyl-5-phenylnicotinamide was high (96%) but it was necessary to add catalyst twice to complete the reaction. However, this was probably due to the reported presence of nitrile impurity in the starting material.

Impurity of the substrate in the removal of chlorine from the pyridine ring of 2,4-dichloro-5H-6,7-dihydro-1-pyridene with palladium on carbon in alcohol containing potassium hydroxide necessitated filtration of the reaction mixture and rehydrogenation with fresh catalyst to obtain 5H-6,7-dihydro-1-pyridene [20c] (Eq. 20.6). The technique employed, use of too small an

$$(H_2C)_3 \text{—pyridine ring with Cl, Cl} \xrightarrow[\text{Pd on C}]{2H_2} (H_2C)_3 \text{—pyridine ring} \quad 86\% \quad (20.6)$$

amount of catalyst for adequate hydrogenation and rehydrogenation with fresh catalyst, is a very useful one when there is a question of the purity of the substrate. It can often result in the use of less catalyst for reaction [4a].

The removal of nuclear chlorine, bromine, and iodine from the nitrogen-containing ring of quinolines and isoquinolines is carried out in the same manner as the reduction of halopyridines. Dehalogenations of 5-iodo-8-hydroxyisoquinoline [21a] and some 8-chloro-4-hydroxyquinolines in strongly alkaline solution gave good yields of the hydroxyisoquinoline and hydroxyquinolines on reduction with nickel catalyst, but ratios were high. Rapid reaction was obtained in the hydrogenation of 2-chloroquinoline-4-carboxylic acid in aqueous potassium hydroxide with a 15% ratio of Raney nickel at 50–60° and 3 atm pressure [21c]. In general, however, reduction with nickel does not proceed so rapidly with a moderate amount of catalyst. Raney nickel promoted with chloroplatinic acid catalyzed the conversion of 5-amino-1-chloroisoquinoline in alcoholic sodium hydroxide to 5-aminoisoquinoline within 4 hr but with 55% by weight of catalyst [21d]. Raney nickel and triethylamine as acid acceptor gave good yield in the hydrogenation of 1-chloro-4-ethoxyisoquinoline to 4-ethoxyisoquinoline but the amount of catalyst and time of reaction were not reported [21e].

Good results were obtained in dehalogenation with palladium in neutral solvent. 5% Palladium on carbon, 12% by weight, catalyzed the reductions of bis(4-chloro-2-methylquinoline-6-oxy)alkanes at room temperature and 2 atm pressure of hydrogen. Yields were above 90% and reaction time was about 2 hr [4a] (Eq. 20.7).

Dehalogenations of haloquinolines and isoquinolines have also been successful with supported palladium catalysts in acetic acid and sodium

456 Dehalogenation

acetate as acid acceptor. 6-Benzyl-4-chloroquinoline was reduced to 6-benzylquinoline in 83% yield by reaction in acetic acid-sodium acetate at

in MeOH

$$\left[\begin{array}{c}\text{Cl}\\ \text{H}_3\text{C}\diagdown\text{N}\diagup\diagdown\diagup\text{O}\end{array}\right]_2 (\text{CH}_2)_n \xrightarrow[\text{5\% Pd on C}]{2\text{H}_2} \left[\text{H}_3\text{C}\diagdown\text{N}\diagup\diagdown\diagup\text{O}\right]_2 (\text{CH}_2)_n$$

$n = 5\text{–}7$ (20.7)

room temperature with palladium on carbon [22a]. 2-Chlorolepidine was similarly converted to lepidine, but at 70° [22b] when reduction did not proceed smoothly at room temperature. It must be noted that the possibility of ring reduction exists in such reductions at temperatures above 35–40° [4a].

Examples of palladium-catalyzed dehalogenations of haloquinolines and isoquinolines in the presence of an acid acceptor (the best method) are given in Table 20.2.

Table 20.2 Dehalogenation with Palladium

Substrate	Support	Base	Reference
Quinolines			
2-Cl-6-OCH$_3$-4-CH$_3$	SrCO$_3$	NaOH	23a
3-Cl-6,8-diOCH$_3$-4-CH$_3$	carbon	KOH	23b
4-Cl-8-(1-piperidyl)-2-CH$_3$	carbon	MgO	23c
Isoquinolines			
3-Cl-5,6,7,8-tetrahydro		CH$_3$COONa	23d
3-Cl-4-OH		CH$_3$COONa	23e

There may be some contraindication to the use of sodium acetate as acid acceptor in dehalogenation of haloquinolines because the release of acetic acid may catalyze ring reduction, but this requires further substantiation. When ethyl 2,4-dichloroquinoline-3-carboxylate was hydrogenated with 30% by weight of 10% palladium on carbon in the presence of sodium acetate, the product of reduction was mainly ethyl 1,2-dihydroquinoline-3-carboxylate [23f]. Overhydrogenation may have been due to the use of the large amount of catalyst. Another factor to be considered in this instance is the possibility that removal of halogen *para* to the quinoline nitrogen may lead to ring reduction. A related example is the hydrogenolysis of 9-chloro-4,5-dimethylacridine, where the halogen is *para* to the ring nitrogen, with Raney nickel, which is notoriously ineffective for reducing a quinoline ring under moderate conditions. The product of reduction was the corresponding acridan (9,10-dihydroacridine) in the absence [23g] or presence [23h] of strong base.

Some examples of dehalogenations among phenanthrolines, which are related to quinolines or isoquinolines depending on the position of the ring nitrogens, are available. 4-Chloro-2-methyl-1,8-phenanthroline was hydrogenated in alcoholic potassium hydroxide with palladium on carbon at room temperature and 2.2 atm pressure [24a] (Eq. 20.8).

$$\text{structure} \xrightarrow{H_2} \text{structure} \tag{20.8}$$

When 1,3-dichloro-2-substituted-4,7-phenanthrolines in alcohol containing sodium ethylate were hydrogenated over Raney nickel at 3.6 atm pressure the reaction, if not interrupted, proceeded to three molar equivalents to yield 2-substituted-5,6-dihydro- and 2-substituted-7,8,9,10-tetrahydro-4,7-phenanthrolines [24b]. When hydrogen uptake was interrupted at two equivalents, 4,7-phenanthrolines were obtained if the 2-substituent was benzyl or phenyl. When it was methyl or ethyl, complex mixtures resulted. There was no attempt to substitute other acid acceptors or to employ palladium as catalyst.

When Raney nickel and sodium hydroxide were employed in the hydrogenation of 4,7-dibromo-3,8-diphenyl-1,10-phenanthroline, 59% yield of 3,8-diphenyl-1,10-phenthroline was obtained [24c]. The hydrogenation of 4-bromo-3-phenyl-1,10-phenanthroline was carried out in the same manner, but the identity of the reduction product was in question.

Naphthpyridines may be considered as quinolines or isoquinolines containing a nitrogen atom in the second ring. Removal of nuclear halogen from either ring is readily accomplished by hydrogenation over palladium. Reaction, however, is complicated by attendant alkoxylation of reactive halogens during reduction in methyl and ethyl alcohols containing alkali hydroxide, or by ring reduction in the presence of palladium and acid acceptors such as acetates and carbonates that give buffered solutions.

2-Methyl-5-chloro-1,8-naphthpyridine was reduced to 2-methyl-1,8-naphthpyridine in 87% yield in methanolic alcoholic potassium hydroxide with palladium on calcium carbonate and a trace of palladium on carbon [24d]. The hydrogenation of 6-amino-8-bromo-1,7-naphthpyridine in absolute alcoholic potassium hydroxide with palladium on carbon gave 81% yield of 6-amino-1,7-naphthpyridine [24e], but the dehalogenation of 3-amino-1-bromo-2,6-naphthpyridine under similar conditions gave 3-amino-2,6-naphthpyridine and some 3-amino-1-ethoxy-2,6-naphthpyridine [24f].

458   Dehalogenation

A number of products were obtained from the hydrogenation of 1-chloro-3-phenyl-2,7-naphthpyridine with palladium catalyst in the presence of potassium acetate [24g]. Dehalogenation of 1,3,6,8-tetrachloro-2,7-naphthpyridine in methyl alcohol with palladium chloride and potassium acetate gave only 1,2,3,4-tetrahydro-2,7-naphthpyridine [24h]; when an excess of stronger base, potassium carbonate, was used, 1,8-dimethoxy-2,7-naphthpyridine was obtained along with the tetrahydro compound. It was also reported that 8-chloro-1,7-naphthpyridine could not be dehalogenated without concurrent ring reduction [24c].

Since the presence of alkali hydroxide appears to suppress ring reduction among naphthpyridines, it may be worthwhile to employ methyl cellosolve (ethyleneglycol monomethyl ether) as solvent in these dehalogenations. Alkoxylation of active halogens did not take place in its presence during other dehalogenations with palladium and excess alkali hydroxide [4a].

Azepines and Benzazepines.   There are no references on the dehalogenation of haloazepines, per se, but removal of halogen from the nitrogen-containing ring of benzazepines has been explored to some degree. 2-Amino-4-bromo-1H-3-benzazepine was hydrogenated in alcohol containing potassium carbonate with 10% palladium on carbon to obtain 63% yield of 2-amino-1H-3-benzazepine [25a]. When the reaction was carried out in alcohol containing acetic acid, partial reduction of the ring took place to give 92% of 2-amino-1H-4,5-dihydro-3-benzazepine. In contrast, only one equivalent was absorbed in the hydrogenation of the 2-amino-4-bromo compound in acetic acid with 20% by weight of 5% palladium on carbon [25b]. When the reduction of 2-amino-4-bromo-6,7,8,9-tetrahydro-1H-3-benzazepine was attempted in aqueous dimethylformamide and potassium carbonate, or in dioxane and triethylamine, the desired product was not obtained [25c]. The use of sodium bicarbonate suspended in alcohol did not provide sufficiently rapid neutralization to prevent initial formation of the hydrobromide

(20.9)

of the dehalogenated base, which appeared to undergo very fast reduction of the azepine nucleus [25d]. Hydrogenation in alcohol without an acceptor gave the saturated cyclic amidine [25c] (Eq. 20.9).

Pyrazoles. A fairly large amount of active Raney nickel was employed for the reduction of 5-chloro-1,3-dimethylpyrazole in alcoholic sodium hydroxide to 1,3-dimethylpyrazole [26a]. A more normal amount of nickel catalyzed the hydrogenation of ethyl α-acetylamino-α-(4-bromo-3-oxo-1-methyl-2-phenyl-5-pyrazolylmethyl)malonate in 80% aqueous alcohol containing sodium bicarbonate [26b] (Eq. 20.10).

$$\text{Br}-\overset{\text{CH}_2\text{C}}{\underset{\text{N}-\text{NCH}_3}{\bigcirc}}\overset{\text{NHCOCH}_3}{(\text{COOC}_2\text{H}_5)_2} \xrightarrow[\text{Ni(R)}]{\text{H}_2} \text{O}=\overset{\text{CH}_2\text{C}}{\underset{\text{N}-\text{NCH}_3}{\bigcirc}}\overset{\text{NHCOCH}_3}{(\text{COOC}_2\text{H}_5)_2}$$
$$\underset{\text{C}_6\text{H}_5}{} \qquad\qquad \underset{\text{C}_6\text{H}_5}{}$$

(20.10)

It is likely that palladium on a support would also be useful in these hydrogenolyses.

Imidazoles and Benzimidazoles. References to the removal of halogen from the imidazole ring are almost nonexistent. However, since a halogen in 2-, 4-, or 5- position is reasonably reactive, it is likely that hydrogenolysis can be accomplished without difficulty. There should be no attack on the ring since imidazoles are not readily hydrogenated.

An example of the removal of halogen from the nitrogen-containing ring of a benzimidazole is seen in the reduction of 2-chloro-1-(2-isopropenyl)-benzimidazole to 1-isopropylbenzimidazole in alcohol with palladium oxide [26c].

1,2-Diazines. PYRIDAZINES. The favored method of removing nuclear halogen from the pyridazine ring appears to be low-pressure reduction of the substrate with supported palladium catalyst in alcoholic or aqueous solution containing a base. Usually only a small amount of catalyst, 3–5% by weight, is required, reaction is very rapid, and yield of pyridazine is high. Typical examples employing 5% palladium on carbon are seen in Table 20.3.

Table 20.3 Pyridazines

Substituent	Medium	Product	Reference
6-Cl-3-CH$_3$	EtOH, NH$_4$OH	3-CH$_3$	27a
3-OC$_4$H$_9$-6-Cl	EtOH, NH$_4$OH	3-OC$_4$H$_9$	27b
3-OCH$_2$COOH-6-Cl	aq. NaOH	3-OCH$_2$COOH	27c
3-NH$_2$-6-Cl(Br)	aq. NaOH	3-NH$_2$	27d

460 Dehalogenation

Equal parts of 10% platinum on carbon and 10% palladium on carbon, amounting to 20% of the weight of 6-chloro-3-sulfanilyamino-pyridazine, catalyzed the reduction in aqueous alcohol to give the dehalogenated product in 53% yield in one instance [28a] and 86% yield in another [28b]. The reason for the combination of catalysts was not disclosed.

The combination of Raney nickel and alkaline medium gave good yields of pyridazines from chloropyridazines but a very large amount of catalyst was usually necessary to insure rapid reaction [29a, b].

A number of mono- and dichloropyridazines and pyridazones were dehalogenated in neutral solvent. Good yields were obtained but in most instances larger than normal amounts of 10% palladium on carbon were employed [30a, b, c]; reaction time was not reported.

In contrast to the above results 3,6-dichloropyridazine, when hydrogenated in alcohol with 50% by weight of 10% palladium on carbon, gave only 27.8% of pyridazine [30d]. The same dehalogenation carried out in the presence of ammonium hydroxide as acid acceptor required only 7% by weight of catalyst; 60.7% yield of pyridazine was obtained.

CINNOLINES. References on catalytic dehalogenation of halocinnolines are very limited. Attempted removal of chlorine from 4-chlorocinnoline with palladium hydroxide on calcium carbonate in methyl alcohol yielded very little cinnoline [31a]; instead condensation took place (Eq. 20.11). It is

$$\underset{\substack{\text{Pd(OH)}_2\text{on CaCO}_3 \\ \text{4-5 atm, 0.25 hr}}}{\xrightarrow{\text{H}_2}} \qquad (20.11)$$

possible that different results might have been obtained if the reaction had been carried out in the presence of a soluble strong base instead of using the catalyst support as acid acceptor. For example, some 1-substituted-3-bromo-4-cinnolones were hydrogenated over palladium on carbon in alcohol containing triethylamine to give 66–84% yields of 1-substituted-4-cinnolones [31b].

PHTHALAZINES. 1-Chlorophthalazine was reduced to phthalazine in 58% yield by hydrogenation in alcoholic sodium hydroxide with 5% palladium on carbon [32].

1,3-*Diazines*. PYRIMIDINES. Halogen in 5-position is not as readily removed as those adjacent to the ring nitrogens. However, since 5-halopyrimidines resemble halobenzenes in reactivity they should undergo hydrogenolysis at low pressure with palladium in the presence of an acid acceptor in the same manner. Nevertheless moderate conditions failed in the dehalogenation of 5-chloro-2,4,6-triaminopyrimidine [33a]. Hydrogenation was carried out in methyl alcohol containing excess sodium acetate at 75° and

55 atm with palladium on carbon promoted with platinum oxide. Steric effects may have necessitated the use of elevated temperature and pressure, although it is possible that reaction at low pressure failed because a strong inorganic base was not used as acid acceptor.

The preceding example may be an exception but in general the reductive removal of nuclear halogen from the pyrimidine ring can be carried out under mild conditions with a supported palladium catalyst in the presence of an equivalent or an excess of base. Neutralization of the release of HX is necessary to prevent ring reduction [33b].

Although acidic conditions are not recommended, N,N'-bis(2-chloro-5-methyl-4-pyrimidyl)ethylenediamine was reduced to the corresponding dehalo compound in 72% yield in acetic acid with 60% by weight of 5% palladium on carbon [33c], and hydrogen uptake did not go beyond one equivalent when 2-chloro-4,6-dimethylpyrimidine was reduced at 70° and 1 atm with a 10% ratio of 5% palladium on carbon in acetic acid containing sodium acetate [33d].

The choice of base for dehalogenation in this series may be of considerable importance. Alkali carbonates were reported to be contraindicated because the release of the weak acid, carbonic acid, caused ring reduction during the hydrogenolysis of 2,5-dichloropyrimidine with palladium on barium carbonate in water [33b]. Instead greater responsibility should be placed on the amount of catalyst employed for reduction. Overhydrogenation took place in the dehalogenation of 2,4,6-trichloropyrimidine with palladium and sodium acetate [34a]. It was attributed to the slightly acidic buffered solution but here, too, the catalyst ratio appears high. In contrast to the last example, 2,4-dichloro-5-methoxypyrimidine was converted to 5-methoxypyrimidine by hydrogenation in alcohol with sodium acetate and 3% by weight of 10% palladium on carbon [34b], and ethyl 4-chloro-5-pyrimidylacetate was reduced to ethyl 5-pyrimidylacetate in 72% yield under similar conditions [34c].

When a solution of 4-chloro-2-methylpyrimidine in alcohol and a slight excess of ammonia was reduced with a moderate amount of palladium on carbon it was necessary to interrupt the reaction at one equivalent to maintain high yield [27a]. On the other hand there were no reported difficulties in the hydrogenations of 4-amino-6-chloro-2-alkoxy- or 4-amino-2-chloro-6-alkoxypyrimidines with palladium on carbon in alcoholic ammonia [35a], or in that of 4-amino-2-chloro-5-nitro-6-methylpyrimidine to 4,5-diamino-6-methylpyrimidine in alcohol and ammonium hydroxide with 40% by weight of 5% palladium on carbon [35b].

Since the pyrimidine ring is stable to hydrogenation under basic conditions [35c], strong bases as sodium and potassium hydroxide should be useful acid acceptors. For example, 2,4-diamino-6-chloro-5-phenylpyrimidine was reduced in alcoholic potassium hydroxide with palladium on strontium carbonate to give 91% of 2,4-diamino-5-phenylpyrimidine [36a], and 4-amino-6-chloro-2-methoxypyrimidine yielded 88% of 4-amino-2-methoxy-pyrimidine after palladium reduction in alcoholic sodium hydroxide [36b]. There is, however, some limitation to the use of strongly basic inorganic acceptors in the hydrogenation of substrates containing a reactive halogen. Hydrolysis may accompany hydrogenation in aqueous alcoholic solution and alkoxylation may be a complicating reaction in alcohol. The latter side reaction may be eliminated by the use of methyl cellosolve, *t*-butyl alcohol, or dioxane if a solution can be maintained. Another technique, the use of a two-phase mixture of ether and a slight excess of 20% aqueous sodium hydroxide, generally prevented hydrolysis in the hydrogenolysis of very active halopyrimidines [35c].

The strongly basic tertiary amine, triethylamine, was employed as acid acceptor in the preparation of 2-amino-5-[2-(1,3-dioxolan-2-yl]-6-methyl-pyrimidine from the reductive dehalogenation of the corresponding 4-chloropyrimidine with palladium on carbon [37].

By far the most widely used acceptor in the series is magnesium oxide, which is reported to prevent reduction of the ring. A few of the many successful applications with it as acceptor include the conversion of 5-amino-2-chloro-6-methylpyrimidine to 5-amino-6-methylpyrimidine [38a] and the dehalogenations of 2,6-dichloro-4-methylpyrimidine [38b] and 2,6-diamino-4-chloropyrimidine and related halodiaminopyrimidines [38c]. In conflicting examples 2-chloropyrimidine was reduced to pyrimidine in good yield in the presence of magnesium oxide and supported palladium catalyst, but it was necessary to stop the reaction after absorption of one equivalent of hydrogen [39a]; uptake also did not stop after one equivalent in the hydrogenation of 4,5-diamino-6-chloropyrimidine with palladium on carbon when magnesium oxide was present, but ceased spontaneously in reaction in water in its absence [39b]. In almost direct contrast, absorption did not proceed beyond the required point in the hydrogenation of the closely related 5-amino-6-chloro-4-methylaminopyrimidine in water in the presence of magnesium oxide with a large amount of catalyst, 50% by weight of 2.5% palladium on carbon [39c]; 71% yield of 5-amino-4-methylaminopyrimidine was obtained.

Another instance of overhydrogenation occurred when the reduction of 2-amino-4-chloro-5,6-trimethylenepyrimidine, in alcohol with a 10% ratio of 5% palladium on carbon and magnesium oxide as acid acceptor, was scaled up. On a very small scale 63% yield of 2-amino-5,6-trimethylene-pyrimidine was obtained [39d], but when the batch size was increased (the amount of increase was not mentioned), large amounts of product in which the pyrimidine ring was hydrogenated were reported.

Excess calcium oxide was also employed as acid acceptor, as seen in the reduction of ethyl 2,4-dichloropyrimidine-5-carboxylate in isopropyl alcohol to ethyl pyrimidine-5-carboxylate with a low ratio (4%) of 5% palladium on carbon at slightly above 1 atm pressure [40].

The removal of halogen from aminohalopyrimidines may be carried out without an acid acceptor. A number of 2- and 5-alkyl- and 2,5-dialkyl-4-amino-6-chloropyrimidines were dehalogenated with palladium on carbon [41a] or with palladium on barium sulfate [41b] in this manner without affecting the ring, the aminopyrimidine acting as acceptor.

It may not be necessary to employ an acid acceptor for the removal of nuclear halogen from pyrimidones or pyrimidinediones. 6-Chloro-2-isopropyl- and 5-alkyl-6-chloro-2-isopropyl-4-pyrimidones [41b] and 5-bromo-1,3-dimethyluracil [42a] were dehalogenated without ring reduction with palladium in neutral solvent. However, an acceptor may improve yield. 6-Iodouracil was converted to uracil in 90% yield with palladium on carbon in aqueous alcohol containing magnesium oxide [42b].

Reductions of halopyrimidones in acid solution usually give 5,6-dihydropyrimidones [1c, 42c].

QUINAZOLINES. Removal of halogen from the nitrogen-containing ring of quinazolines by reduction in alcoholic alkali leads to alkoxyquinazolines with palladium on carbon or Raney nickel [43]. Reduction of 4-chloroquinazoline in dioxane containing potassium hydroxide or sodium acetate continued beyond one equivalent but the reaction could be interrupted to obtain good yield of quinazoline.

The solvent appeared to be an important factor in the extent of hydrogen uptake when reduction was carried out in the absence of base. Hydrogenation in hydroxylated solvents, including methyl cellosolve and secondary and tertiary alcohols, with palladium on calcium carbonate yielded 3,4-dihydroquinazolines. Reduction in dioxane stopped at one equivalent.

The amount of catalyst used could also have contributed to the formation of 3,4-dihydroquinazoline, as it may have when 4-chloro-6-nitro- and 4-chloro-8-nitroquinazolines were hydrogenated in methyl alcohol with an equal weight of the same catalyst to yield 6- and 8-amino-3,4-dihydroquinazolines. Unfortunately the amount was not disclosed in the reduction of 4-chloroquinazoline in the absence of base.

1,4-*Diazines* PYRAZINES. Very good yields of pyrazines were obtained from dehalogenations with an acid acceptor and 40–100% by weight of palladium on carbon, Table 20.4. There was no indication why such catalyst ratios were necessary.

QUINOXALINES. Among haloquinoxalines where the halogen was on the nitrogen-containing ring, 3-chloro-2-(2-pyridyl)quinoxaline was hydrogenated

Table 20.4 Pyrazines

Substituent	Base	Dehalopyrazine (% Yield)	Reference
6-Cl-2-COOC$_2$H$_5$-3-NH$_2$-5-R, R = NH$_2$, OH, or OCH$_3$	MgO, 30% excess	65–85[a]	44a
6-Cl-2-NH$_2$-3-OCH$_3$ 3-Cl-2-NH$_2$-6-OCH$_3$	KOH	[b]	44b
3,5-diCl-2,6-diNH$_2$ 5,6-diCl-2,3-diNH$_2$ 5,6-diCl-2-NH$_2$-3-OCH$_3$	KOH	87	44c
5-Br-2,3-diNH$_2$ 5-Br-2-NH$_2$-3-OCH$_3$	KOH	90–98	44d
3-Cl-2-COOCH$_3$-5-C$_6$H$_5$	(C$_2$H$_5$)$_3$N	65	44e

[a] R = N(CH$_3$)$_2$, 38% yield; R = NHCH$_3$, no reaction. [b] Not reported.

with an equal weight of 5% palladium on carbon in tetrahydrofuran containing potassium hydroxide solution [45a]. Only 24% of 2-(2-pyridyl)quinoxaline was obtained after chromatography; unfortunately the remainder of the material was not identified; it is possible that the amount of catalyst employed caused extensive ring reduction.

Ring reduction did accompany dehalogenation when the hydrogenation of 3-chloro-2-methylquinoxaline was run in acetic acid containing sodium acetate with palladium on carbon at 60° and 2 atm, but it was an obvious attempt to obtain 2-methyl-1,2,3,4-tetrahydroquinoxaline [45b].

Triazines. In most instances the removal of halogen from the triazine nucleus is carried out under moderate conditions with palladium catalyst and an acid acceptor. In alcoholic alkali hydroxide, however, alkoxylation may compete with dehalogenation [46a].

2,4,6-Trichloro-1,3,5-triazine (cyanuric chloride) could not be dehalogenated under a variety of conditions [46b, c], ascribed to the poisoning effect of s-triazine on noble metal catalysts [46d]. 3,5-Dichloro-1,2,4-triazine also resisted dehalogenation [46a]. 3-Dimethylamino-5-chloro-1,2,4-triazine was dehalogenated in benzene containing triethylamine with 250% by weight of 10% palladium on carbon [46a]. The substrate was reported to be a mild poison which could account for the large amount of catalyst. If the benzene employed as solvent was not thiophene-free, it, too, could have an adverse effect and necessitate the use of higher than normal catalyst ratios.

Mono- and dichloro-1,3,5-triazines were reduced without incident when other substituents were attached to the ring carbons, examples are seen in Table 20.5.

Table 20.5 1,3,5-Triazines[a]

Substituents	Solvent	Base	Reference
2-NH$_2$-4,6-diCl	dioxane	(C$_2$H$_5$)$_3$N	46e
2-NHR-4,6-diCl, R = alkyl	dioxane or (CH$_3$)$_2$CHOH	CaO	46e
2-OC$_6$H$_5$-4,6-diCl			
2,4-di-OR-6-Cl[b], R = alkyl or phenyl	aq. MeOH	NaHCO$_3$	46f
2,4-diOCH$_3$-6-Cl[c]	abs. ether	(C$_2$H$_5$)$_3$N	46g

[a] Reactions carried out with 5% palladium on carbon at 1–3 atm. [b] At 35 atm. [c] Very high yield of dehalotriazine obtained. Reaction might be less hazardous if carried out in thiophene-free benzene.

The chlorine atom was also removed from a number of 2,4-diamino, substituted amino, and diamino-6-chloro-1,3,5-triazines by passing hydrogen into the stirred solutions at 70–78° containing an equal weight of 10% palladium on carbon and also containing an acid acceptor [46h].

Miscellaneous Unsaturated Fused Heterocyclic Rings. The removal of halogen from unsaturated fused-ring heterocycles is carried out in the same manner as dehalogenations of simple related systems, that is, with a supported palladium catalyst and an acid acceptor (see Table 20.6).

In most instances yield of dehalo compound was high and reaction was straightforward, even among the pyridazine compounds where there was a possibility of hydrogenolysis of the nitrogen–nitrogen bond.

Table 20.6

Substrate	Support	Acceptor	Reference
6-Br-5-CH$_3$-1H-imidazo[4,5-b]pyridine	C[a]	NaOH	47a
5-Cl-1-CH$_3$-imidazo[4,5-d]pyridine	C	CH$_3$COONa	47b
7-Br-3-OH-2-CH$_3$-pyrido[2,3-b]pyrazine	SrCO$_3$	NaOH	47c
7-Cl-8-NH$_2$-3-phenyl-s-triazolo[4,3-b]pyridazine	C	NaOH	48a
7-Cl-v-triazolo[4,5-c]pyridazine	C	NaOH	48b
4,7-diCl-2-phenylmidazo[4,5-d]pyridazine	C	NaOH	48c
6-Cl-2-phenylimidazo[1,2-b]pyridazine	C	—	48d
4-Cl-7H-pyrrolo[2,3-d]pyrimidine	C	—	49a
4-Cl-2,6-diCH$_3$-5H-pyrrolo[3,2-d]pyrimidine	C	MgO	49b
7-Cl-2,3-diCH$_3$- \} pyrazolo[1,5-a]pyrimidine 7-Cl-2,3,5-tri CH$_3$-	C	CH$_3$COONa	49c

[a] Promoted with platinum oxide.

(20.12)

An unexpected reaction took place in the reduction of 2-chloro-10H-pyridino[3,2-b]quinazolin-10-one in methyl alcohol containing ammonium hydroxide with 80% by weight of 15% palladium on carbon [49d] (Eq. 20.12). Rupture of the N—N bond was probably caused by the large amount of catalyst employed. The formation of dihydro and tetrahydro compounds was probably due to a combination of the amount of catalyst and the use of ammonium hydroxide, which gave an acid-buffered solution that promoted ring reduction or did not retard it. Hydrogenation under more basic conditions with a moderate amount of catalyst might give the expected reaction and eliminate further hydrogenation products.

Examples of dehalogenation among other heterocycles such as halopurines and haloheterocycles that are isomeric with them are given in Table 20.7.

Table 20.7 Dehalogenation of Halopurines and Isomeric Heterocycles

Substituents	Acceptor	Reference
Purines		
6-Cl-9-butyl	MgO	50a
6-Cl-9-(2-cyclohexenyl)	MgO	50b
6-Cl-9-(2-diethylaminoethyl)	MgO	50c
6-Cl-9-vinyl	MgO	50d
2,6-diCl-9-vinyl		
1H-Pyrazolo[3,4-d]pyrimidines		
4-Cl	NH$_4$OH	51a
4-Cl-1-CH$_3$	NH$_4$OH	51b
7-Cl-3-CH$_3$	NH$_4$OH	51c
s-Triazolo[1,5-a]pyrimidines		
7-Cl-2,5-diCH$_3$	NH$_4$OH	52a
7-Cl-5,6-diCH$_3$	CH$_3$COONa	52b
5,7-diCl-6-CH$_3$		
7-Cl-6-CH$_3$	—	52c

Palladium on carbon was used exclusively. Good yields were obtained except in Refs. 51b, 51c, and 52c.

Dehalogenations among pteridines have been studied only meagerly. 2-Chloro-4,6-dimethyl-7,8-dihydropteridine was reduced with palladium on carbon and magnesium oxide to 4,6-dimethyl-7,8-dihydropteridine in good yield [53a]. 2,4-Dichloro-6-hydroxy-7,8-dihydropteridine was dehalogenated with an equal weight of the same catalyst in the absence of base [53b]. Reaction time was long and yield was poor, 29%. The hydrogenation of 2,4-dichloro-5,6,7,8-tetrahydropteridine was also carried out without an acceptor. In this reduction, however, the nitrogen atoms in the piperazine portion of the molecule are basic enough to neutralize the release of hydrogen chloride; the yield of 5,6,7,8-tetrahydropteridine was 60%.

ARALKYL HALIDES

Halides, ArCH$_2$X

Aryl methyl halides, ArCH$_2$X where Ar is an unsaturated carbocycle or an unsaturated heterocycle, undergo dehalogenation so readily that, despite the general ineffectiveness of platinum oxide and Raney nickel for dehalogenation, each has been employed with some success.

Platinum oxide catalyzed the reduction of 2-bromomethyl-3-hydroxy-6-methylpyridine in alcohol to 3-hydroxy-2,6-dimethylpyridine [54a]. Ring reduction accompanied dehalogenation of the corresponding 2-chloromethyl compound but this was due to the high catalyst ratio, 40% by weight of substrate.

The combination of Raney nickel and sodium hydroxide gave a 90% yield of 3,4-dimethyl-2,5-diphenylfuran from the reduction of 3,4-*bis*(bromomethyl)-2,5-diphenylfuran in alcohol [54c]. Raney nickel catalyzed the removal of chlorine from 2-chloromethyl-5-methoxy-4-pyrone in alcoholic potassium hydroxide [54d] and was also used in alcohol containing sodium acetate in the reduction of 2-chloromethyl-5-hydroxy-4-pyrone to 2-methyl-5-hydroxy-4-pyrone [54e].

Palladium black was also used in the removal of an arylmethyl-type halide in the reduction of 3,6-*bis*(chloromethyl)pyridazin-4-one in alcohol-acetic acid containing sodium acetate [55]. In general it is less efficient than a supported catalyst for dehalogenation usually requiring more catalyst on a metal-to-substrate basis than a supported one [4a].

Palladium on a support is favored for removal of halogen from arylmethyl halides. Reaction is usually fairly rapid under mild conditions and a large amount of catalyst is seldom necessary. Often a very small amount is adequate, as in the dehalogenations of 5(6)-chloromethyl-1-isopropylindane [56a] and 6-chloromethyl-3-ethyl-1,1,3,5-tetramethylindane and other 6-chloromethyl polyalkylindanes [56b]. In the latter reductions very high yields were recorded.

Where Ar is benzenoid or furanoid, reactions can be carried out in the absence of an acid acceptor. 4-Methoxy-2,6-xylenol was obtained by dehalogenation of 2,6-*bis*(bromomethyl)-4-methoxyphenol in alcohol with 10% palladium on carbon [57a]. Reaction time was about 2 hr with 16% by weight of catalyst. 3-Chloromethyl-2,5-diphenylfuran was hydrogenated in alcohol with an equal weight of 5% palladium on barium sulfate to yield

2,5-diphenylfuran [57b]. An abrupt decrease in hydrogen uptake occurred after dehalogenation.

Although there is little danger of ring reduction under mild conditions in the dehalogenation of like compounds with palladium in the absence of base, the presence of an acceptor can often result in increased reaction rate and a decrease in the amount of catalyst used. This was evident in the dehalogenation of 2,6-*bis*(chloromethyl)-4-methoxyphenol to which sodium acetate was added. Hydrogen uptake was complete within minutes with a lower than normal catalyst ratio [4a].

It may be advisable to employ an acid acceptor in dehalogenations of ArCH$_2$X to ArCH$_3$, where Ar is an unsaturated nitrogen heterocycle, to prevent attack of the ring. However, none was used in the palladium on carbon hydrogenations of 5-chloromethyl-8-hydroxyquinoline [57c], 2-bromomethyl-3-hydroxypyridine, 2,6-*bis*(bromomethyl)-3-hydroxypyridine [57d], 2-alkyl-3-amino-5-aminomethyl-4-bromomethylpyridine [57e], 5-chloromethyluracil [57f], and 4-chloromethylimidazole [57g].

Some difficulty was experienced in duplicating the hydrogenation of 4-chloromethylimidazole but it was apparently due to the presence of impurities. The chloromethyl compound was routinely pretreated in alcohol with recovered catalyst from hydrogenation of the previous batch [4a]. Subsequent reductions were very rapid and yields were high. The same procedure was followed for the preparation of 2-methylimidazole from 2-chloromethylimidazole [4a].

Palladium on carbon was employed for the hydrogenation of 2,4-*bis*-(chloromethyl)-6-hydroxy-1,3,5-triazine chloroacetamidine salt in methyl alcohol containing triethylamine. The resultant product, the acetamidine salt of 6-hydroxy-2,4-dimethyl-1,3,5-triazine, was obtained in 87.5% yield [58a] (Eq. 20.13). The reduction of the chloroacetoamidine salt to the acetamidine

$$\text{ClH}_2\text{C} \underset{\text{N}}{\overset{\text{N}}{\diagdown}} \overset{\text{OH} \cdot \text{NH}_2 \overset{\text{NH}}{\overset{\|}{\text{C}}} \text{CH}_2\text{Cl}}{\underset{\text{N}}{\diagup}} \text{CH}_2\text{Cl} \quad \xrightarrow[\text{Pd on C}]{3\text{H}_2} \quad \text{H}_3\text{C} \underset{\text{N}}{\overset{\text{N}}{\diagdown}} \overset{\text{OH} \cdot \text{NH}_2 \overset{\text{NH}}{\overset{\|}{\text{C}}} \text{CH}_3}{\underset{\text{N}}{\diagup}} \text{CH}_3 \quad (20.13)$$

salt is not surprising in view of the relationship of the halogen in it to that in an N-heterocyclic methyl chloride, —N=CCH$_2$Cl.

Palladium on calcium carbonate was used to convert 2-amino-4-anilino-6-chloromethyl-1,3,5-triazine [58b] and 2,4-diamino-6-chloromethyl-1,3,5-triazine [58c] to the corresponding 6-methyltriazines, the dehalogenated aminotriazines acting as acid acceptors.

Halides, ArCHRX

When hydrogenolysis of a halide, ArCHRX where R is any substituent except H or X, is attempted, the removal of halogen may not proceed as readily as in the reduction of ArCH$_2$X to ArCH$_3$. The dehalogenation of 1-chloro-2-methyl-1-phenylisopropylamine, C$_6$H$_5$CHCl(CH$_3$)$_2$NH$_2$, as hydrochloride salt was carried out in alcohol at 50–80° and 3.3 atm pressure of hydrogen with 33% by weight of palladium on carbon [59a]. Substitution of Raney nickel or platinum oxide gave substantial amounts of 2-methyl-1-cyclohexylisopropylamine as well as the desired 2-methyl-1-phenylisopropylamine. However, 2-chloro-2-(substituted phenyl) ethylamines were hydrogenated as hydrochloride salts in water at 30° with 15–20% ratios of 10% palladium on carbon for several hours to give good yields of the corresponding phenethylamines [59b], and 1-(2-chloro-2-phenylethyl)pyrrolidine was reduced very rapidly to 1-(2-phenylethyl)pyrrolidine with an undisclosed amount of palladium on calcium carbonate [59c].

Halides, ArC-C$_n$X

The reactivity of halogen in aralkyl halides is substantially decreased as the aromatic ring moves further away from it. Since the halogen in such compounds is far less reactive than in a benzyl halide, dehalogenation should be carried out in the presence of an acid acceptor. Palladium is the preferred catalyst [4a], although others have been employed.

1-(3-Chloro-2-phenylbutyl)piperidine was hydrogenated with Raney nickel under 3 atm pressure for 18 hr [60a]. In this instance the compound itself, a fairly strong base, acted as acceptor. 2-(2-Chloroethyl)tetrazole was hydrogenated to 2-ethyltetrazole in 18 hr in methyl alcohol containing potassium carbonate with 5% platinum on carbon promoted with palladium chloride [60b]. cis-5-Bromo-4-phenyl-1,3-dioxane, which may be viewed as a substituted phenethyl bromide, was converted to 4-phenyl-1,3-dioxane in 90% yield by reduction in alcohol and triethylamine with 30% by weight of 5% palladium on carbon [60c] (Eq. 20.14). The trans-5-bromo and *cis*-5-chloro compounds were similarly reduced. 3-Chloro-3-(4-pyridyl)-2-indolinone, also an arylethyl halide, was dehalogenated with palladium on carbon, but without an acid acceptor [60d] (Eq. 20.15).

$$\underset{\substack{\text{R = 4-pyridyl}}}{\text{(indolinone-Cl)}} \xrightarrow[\text{Pd on C}]{H_2} \underset{73\%}{\text{(indolinone-H)}} \quad (20.15)$$

Removal of bromine from 4-(2-bromoethyl)-3,4-dihydrocoumarin was carried out in glacial acetic acid (hydrogenation in the presence of alkali would have caused ring opening). It is of interest that, despite the large amount of catalyst, 33% by weight of platinum oxide, high yield of 4-ethyl-3,4-dihydrocoumarin was obtained unaccompanied by attack on the benzene ring [60e] (Eq. 20.16).

$$\underset{\text{in AcOH}}{\text{(CH}_2\text{CH}_2\text{Br-coumarin)}} \xrightarrow[\text{PtO}_2]{H_2} \underset{94\%}{\text{(C}_2\text{H}_5\text{-coumarin)}} \quad (20.16)$$

HALOGEN ATTACHED TO A SATURATED HETEROCYCLE

References to the removal of halogen attached to a saturated heterocycle are limited. The dehalogenation of tetraacetyl-α-D-glucopyranosyl bromide at room temperature and atmospheric pressure with palladium black in the presence of triethylamine gave high yield of the corresponding debromo compound [61a]. The hydrogenolyses of related acetylated pyranosyl bromides in dry ethyl acetate with platinum oxide at room temperature and 3.3 atm pressure in the presence of diethylamine were reported to give better results than reductions with palladium [61b]. Platinum oxide, 10% by weight of substrate, was employed for the reduction of 5-bromo-5-nitro-2,2-diphenyl-1,3-dioxane in alcohol and ethyl acetate to 5-amino-2,2-diphenyl-1,3-dioxane, the reduction product acting as acid acceptor [61c]. The dehalogenation of 14-chloro-1-azatetradecan-2-one, carried out under 10 atm pressure in acetic acid with palladium on carbon and sodium acetate

472 *Dehalogenation*

as acid acceptor, gave 93% yield of the corresponding lactam [61d]. Palladium on carbon reduction in alcohol with magnesium oxide as base was employed for the removal of halogen from diethyl 3-bromo-1-phenylpyrrolidin-2-one-5,5-dicarboxylate [61e]; palladium on carbon and ammonium hydroxide catalyzed the conversion of 3-chloro-1-methyl-5-(3-pyridyl)-pyrrolidin-2-one to 1-methyl-5-(3-pyridyl)pyrrolidin-2-one [61f]. In one instance Raney nickel, almost equal to the weight of substrate, was used to obtain 5-amyl-2(3H)-furanone from the hydrogenation of 5-amyl-3,3-dichloro-2(3H)-furanone in potassium hydroxide solution [61g].

ACYL HALIDES

Hydrogenolysis of an acid chloride, the Rosenmund reaction [62], is used to prepare the corresponding aldehyde:

$$RCOCl \xrightarrow{H_2} RCHO + HCl$$

Unlike most catalytic hydrogenations it does not take place in a closed system but is carried out in dry hydrocarbon solvent, usually xylene, while passing hydrogen through the solution at reflux temperature, generally in the presence of palladium on barium sulfate, The end of the reaction is determined by the cessation of evolution of hydrogen chloride. There are many examples of the Rosenmund reduction available [63a, b, c, d, e]. There are, however, many problems associated with it. Even when the reaction is closely watched and is interrupted immediately after evolution of hydrogen chloride stops, there is no assurance that reduction to alcohol or to hydrocarbon will not take place as aldehyde is formed. There may be some control by using lower-boiling hydrocarbons so that these undesirable reactions will be limited.

In many instances, however, moderators or poisons are employed to inactivate the catalyst and prevent further reaction. The quinoline and sulfur combination is a common one [63a]. The addition of thiourea to the hydrogenolysis of benzoyl chloride in refluxing toluene in the presence of platinum oxide gave 96% yield of benzaldehyde [64a]. Tetramethylthiourea was found to be very effective in controlling the dehalogenation of pure benzoyl chloride to benzaldehyde in the presence of palladium on barium sulfate [64b]. Occasionally the compound itself is a catalyst inhibitor. 2,5-Dimethyl-3-thenoyl chloride was hydrogenolyzed in dry xylene with palladium on carbon in a vigorous stream of hydrogen to yield 55–65% of 2,5-dimethylthiophene-3-aldehyde [64c] (Eq. 20.17).

$$H_3C\underset{S}{\overset{COCl}{\bigcirc}}CH_3 \xrightarrow{\text{Pd on C}} H_3C\underset{S}{\overset{CHO}{\bigcirc}}CH_3 \quad (20.17)$$

At times the use of an inhibitor can prevent reaction from taking place. Glutarimide-4-acetyl chloride was converted to glutarimide-4-acetaldehyde by bubbling hydrogen through refluxing dry toluene in the presence of 10% by weight of 10% palladium on barium sulfate [65a]. The yield of aldehyde was 64%, but when the catalyst was poisoned with quinoline-sulfur no aldehyde could be obtained.

It may be advantageous not to purify an acid chloride thoroughly but to use it in its unpurified state. Such a procedure may inactivate the catalyst somewhat and prevent further reduction of the resultant aldehyde [4a].

Certain groups—aromatic nitro, aryl chloride, carbon–carbon double bonds—have remained intact during the hydrogenolysis of acid chloride to aldehyde when inhibitors were present [63a]. 4-Acetylbenzoyl chloride was hydrogenated to 4-acetylbenzaldehyde over palladium on barium sulfate without an inhibitor [65b], but a dimeric substance was reported when the reaction was carried out in the presence of an inhibitor.

Good results have been obtained from the hydrogenolyses of acid chlorides with or without catalyst inhibitors. However, the vagaries or inconsistencies associated with the Rosenmund reduction do not make it a completely reliable reaction. The substitution of the more labile acid bromides for acid chlorides might allow reaction to be carried out at a lower temperature than is generally employed, and lead to less side reaction and greater reliability of the reaction.

SELECTIVE REDUCTIONS

Among Dissimilar Halogens

On the basis of the known order of susceptibility of halogens to hydrogenation, it can be assumed that the more reactive one usually will be reduced when dissimilar halogens in similar-type halides are present in a substrate. For example, 4-bromo-2,6-dichlorophenol was rapidly and selectively reduced to 2,6-dichlorophenol in 86% yield with palladium on carbon in a mixture of benzene and cyclohexane or in an alcohol–water mixture containing sodium acetate [66a], 5-bromo-3-chloro-2,4-dihydroxypyridine was selectively hydrogenated to 3-chloro-2,4-dihydroxypyridine in alcoholic sodium hydroxide with palladium on carbon [66b], and 2-amino-4-bromo-3-chloropyridine was reduced in the same way to 2-amino-3-chloropyridine [66c].

However, in the pyridine series and in certain other unsaturated heterocycles, selectivity can be affected by the positions of the nuclear halogens. For example the halogen in 3(5)-position in pyridines is less susceptible to hydrogenolysis than the others. This would help to explain the lack of

474   *Dehalogenation*

selectivity in the hydrogenation of 3-amino-5-bromo-6-chloro-2-picoline with palladium on carbon; interruption of the reaction after absorption of one molar equivalent gave equal parts of 3-amino-2-picoline and starting material [66d].

There are too few examples of selectivity in compounds containing dissimilar halogens of different types. An aliphatic-type iodide was preferentially removed in the presence of aryl chloride when 5,7-dichloro-2-iodomethyl-3,4-dimethyl-2,3-dihydrobenzofuran was hydrogenated with palladium on carbon in methyl alcohol containing potassium acetate (67a) (Eq. 20.18).

$$\text{(structure with Cl, CH}_3\text{, CH}_2\text{I)} \xrightarrow[\text{Pd on C}]{\text{H}_2} \text{(structure with Cl, CH}_3\text{, CH}_3\text{)} \qquad (20.18)$$

The selective removal of bromine in the hydrogenations of 6-(bromoacetyl)-3-chlorophenanthrene [67b], 9-(bromoacetyl)-3-chlorophenanthrene [67c], and 10-(bromoacetyl)-3-chlorophenanthrene [67d] with palladium on carbon in alcohol took place as expected. It is likely in view of the reactivity of phenacyl halides that a phenacyl chloride would be preferentially dehalogenated in the presence of an aryl bromide.

Among Similar Halogens

In Similar-Type Halides. ON THE SAME CARBON ATOM. When several similar halogens are on the same carbon atom they can be completely removed or can be removed stepwise. The amount of acid acceptor can in many instances determine the extent of reduction. When 6-trichloromethylpurine was hydrogenated in alcohol with platinum on carbon the reaction could be interrupted to obtain 6-dichloromethylpurine in 75% yield [68a]. 6-Monochloromethylpurine was not obtained when the same reduction was interrupted after two equivalents were absorbed, but the use of excess sodium acetate made it possible to obtain it. 6-Tribromomethylpurine could be reduced to 6-dibromomethyl- or 6-bromomethylpurine, depending on the amount of sodium acetate employed.

1-Acetyl-3,3-dichlorohexahydro-2*H*-azepine-2-one was hydrogenated in acetic acid containing one equivalent of sodium acetate in the presence of palladium on carbon to yield 98% of the corresponding 3-monochloro compound [68b]. In the reduction of 3,3-dichlorohexahydro-2*H*-azepine-2-one two equivalents of sodium acetate were added, but when the reaction was interrupted at one equivalent of hydrogen monochloro compound was still obtained; yield was 88%. The related 7,7-dichlorohexahydroazepinone was hydrogenated to the 7-monochloro product by means of Raney nickel and 90% of an equivalent of triethanolamine [68c]. In other examples of partial reductions of RCX$_3$ and RCHX$_2$ systems, N-(trichloroacetyl)-2,5-diethoxy-4-nitrophenethylamine was hydrogenated in methyl alcohol in the presence of platinum oxide. The resulting amino group acted as acid acceptor and only one chlorine atom was removed [68d]. Reaction of the corresponding dichloroacetyl nitrodiethoxyphenethylamine under similar conditions yielded a monochloroacetamide (Eq. 20.19).

$$\underset{O_2N}{}\overset{OC_2H_5}{\underset{OC_2H_5}{\bigcirc}}(CH_2)_2NHCOCCl_3 \xrightarrow[\substack{PtO_2 \\ 3\ atm}]{4H_2} \underset{H_2N}{}\overset{OC_2H_5}{\underset{OC_2H_5}{\bigcirc}}(CH_2)_2NHCOCHCl_2$$

$$\underset{O_2N}{}\overset{OC_2H_5}{\underset{OC_2H_5}{\bigcirc}}(CH_2)_2NHCOCHCl_2 \xrightarrow{4H_2} \underset{H_2N}{}\overset{OC_2H_5}{\underset{OC_2H_5}{\bigcirc}}(CH_2)_2NHCOCH_2Cl$$

(20.19)

When 3,3-dibromocyclopropane-1,2-*cis*-diacetic acid was reduced in alcohol containing excess potassium hydroxide in the presence of Raney nickel, platinum oxide, or palladium on carbon or on barium sulfate, monobromo compound was obtained in 51% yield when the reaction was interrupted [68e]. Both bromines were removed after absorption of two equivalents upon hydrogenation in the presence of Raney nickel, but about 10% of pimelic acid was obtained in addition to 84% of cyclopropane-*cis*-1,2-diacetic acid, indicating that some cleavage of the cyclopropane ring occurred.

ON DIFFERENT CARBON ATOMS. When similar halogens are on different carbon atoms the adjacence of one may activate the other, but there are too few examples to state that one will be completely reduced in preference to the other. Other activating groups may have some influence on selectivity in similar instances but there is little or no study on their effect.

In Halobenzenes. An attempt was made to obtain 2-amino-5-chlorobenzenesulfonamide from 2-amino-4,5-dichlorobenzenesulfonamide [4a] based on the rationale that the sulfonamide group *para* to chlorine would activate that chlorine in the same manner as a nitro group increases the chemical reactivity of a halogen *para* to it. Reduction in aqueous alcohol containing sodium acetate gave a mixture of products, of which the major product was starting material, when the reaction in the presence of palladium on carbon was interrupted after one equivalent of hydrogen was absorbed. Hydrogenation in the absence of acid acceptor gave a product which melted reasonably close to that of the desired monochloro compound, but complete elemental analysis indicated a mixture which contained much dichloro compound.

When two similar halogens are nonadjacent, prediction of selectivity is doubtful unless one is highly activated.

In Unsaturated Heterocycles. Where similar halogens are attached to the nucleus in an unsaturated heterocyclic system, selective dehalogenation in many instances depends on the position of the halogen in relation to the hetero atom. In nitrogen-containing unsaturated heterocycles, halogen *meta* to the nitrogen atom is usually least readily removed.

The order of dehalogenation in the pyridine series is 4- > 2(6)- > 3(5)-halo. When 2-amino-4,6-dibromopyridine was hydrogenated with palladium on carbon in the presence of sodium hydroxide, 2-amino-6-bromopyridine was obtained; the reduction of 2-amino-5,6-dibromopyridine yielded 2-amino-5-bromopyridine [69].

When two similar halogens are in equivalent positions on the pyridine ring the less-hindered one will probably be removed, as illustrated in the palladium reduction of 3,5-dibromo-2,4-dihydroxypyridine to 3-bromo-2,4-dihydroxypyridine [66b].

Halogen in 4-position is more easily removed from the pyrimidine ring than a similar one in 2- or 5-position. There do not appear to be any comparisons of selectivity in reductions where a pyrimidine ring contains similar 4- and 6-halogens except in the hydrogenation of 5,7-dichloro-*s*-triazolo[1,5-a]pyrimidine with palladium on carbon in alcohol containing sodium bicarbonate [70]. Both halogens could be removed, but retention of chlorine in 5-position in this system suggests that when similar halogens are present in 4- and 6-position in a pyrimidine, the 4-halogen can be preferentially removed (Eq. 20.20).

$$\text{(20.20)}$$

The presence of 34% of 5-amino-2-chloro-6-methylpyrimidine in the attempt to reduce 5-amino-2,4-chloro-6-methylpyrimidine to 5-amino-6-methylpyrimidine in the presence of palladium on carbon and excess magnesium oxide [38a] is indicative of the preferential removal of a 4-halogen from the pyrimidine ring in the presence of a like halogen in 2-position. The hydrogenation of 2,6-dichloro-4-methylpurine to 2-chloro-6-methylpurine in

87% yield with 10% palladium on carbon and sodium acetate as acid acceptor is another example [71].

An example of selective reduction when chlorines are in 4- and 5-positions in the pyrimidine ring is seen in the hydrogenation of 6,7-dichloro-5-methyl-*s*-triazolo[1,5-a]pyrimidine to 6-chloro-5-methyl-*s*-triazolo[1,5-a]pyrimidine with 5% palladium on carbon in the absence of an acid acceptor [72].

Examples are not available on the comparative ease of the reducibility of like halogens in 6- and 2-positions, 6- and 5-positions, and 2- and 5-positions. Since 6-position is related to 4-position by a similar relationship to the ring nitrogens, it might be assumed that a halogen in 6-position would be removed more easily than a similar 2- or 5-halogen. The relationship of halogen in 2- and 5-position to hydrogenolysis is a matter for further study.

In Dissimilar-Type Halides. When selective dehalogenations of like halogens in dissimilar-type halides are attempted, those in the most reactive systems should be removed preferentially. An allylic chloride was selectively reduced in the presence of a vinylic chloride in the platinum-catalyzed hydrogenations of 1,2,3,4,5,5-hexachloro- and 1,2,3,4,5-pentachlorocyclopenta-1,3-dienes to 1,2,3,4-tetrachlorocyclopenta-1,3-diene [73a] (Eq. 20.21).

(20.21)

The presence of two halogens on a single carbon atom causes some activation of one of the halogens, but it is apparently not sufficient to compete with the removal of a like halogen in a vinyl halide. An example is seen in the palladium on carbon reduction of 3,5,5-trichloro-1,4-dihydroxycyclopent-2-enecarboxylic acid, where the vinyl chloride was removed and the double bond was saturated to yield 2,2-dichloro-1,3-dihydroxycyclopentanecarboxylic acid [10].

The activation of one halogen by a similar halogen on the same carbon atom is probably only sufficient to allow it to be reduced in the presence of a similar aliphatic halide, as seen in the hydrogenation of 1,1,1,5-tetrachloropentane in methyl alcohol and pyridine over Raney nickel under 100 atm pressure to 1,1,5-trichloropentane in 80% yield [73b].

478 Dehalogenation

Obviously a vinyl-type halide should be preferentially cleaved in the presence of any saturated system, $\diagdown\!\!\!\diagup$ CRX, containing the same halogen. The reduction of 3-chloro-2-(2-chloroethoxy)-2,5,6,7-tetrahydrooxepin with Raney nickel in alcoholic sodium hydroxide at 100 atm pressure to 2-(2-chloroethoxy)oxepan in 75% yield and that of 3,5-dichloro-1,4-dihydroxy-cyclopent-2-enecarboxylic acid to 2-chloro-1,3-dihydroxycyclopentane-carboxylic acid with palladium on carbon under low-pressure conditions [10] are typical examples (Eq. 20.22). In other examples, chlorines in 2- and

$$\text{(structures: 7-membered ring with Cl and OCH}_2\text{CH}_2\text{Cl} \xrightarrow{2H_2,\ \text{Ni R}} \text{reduced ring with OCH}_2\text{CH}_2\text{Cl)}$$

(20.22)

$$\text{(cyclopentene with HO, COOH, Cl, HO, Cl)} \xrightarrow{2H_2,\ \text{Pd on C}} \text{(cyclopentane with HO, COOH, Cl, HO)}$$

6-positions in the pyridine ring (these may be considered as vinyl chlorides) were removed in preference to a phenethyl chloride in the reduction of 2,6-dichloro-3-(2-chloroethyl)-4-picoline to 3-(2-chloroethyl)-4-picoline with palladium on carbon [74b], and in that of 4,6-dichloro-3 (2 chloroethyl)-2-picoline to 3-(2-chloroethyl)-2-picoline with the same catalyst [4a]. In the latter example hydrogen uptake was extremely rapid and did not proceed beyond two equivalents to give quantitative yield. Reduction of the same trichloro compound with 75% by weight of palladium on barium sulfate also stopped spontaneously to yield 3-(2-chloroethyl)-2-picoline [74c]. The investigators considered the results surprising, expecting the β-chloroethyl group to be more readily attacked. It is doubtful whether the chlorine in the side chain would be attacked unless excess base were present and the hydrogenation were allowed to continue.

In instances where halogen is attached to an unsaturated carbocycle and the same halogen is attached to an unsaturated heterocycle, the halogen in the heterocyclic ring will be selectively removed. In these reductions selectivity is probably due to the influence of the ring nitrogen, which makes one vinyl halide more susceptible to hydrogenation than the other. Numerous examples support the selective removal of halogen from the nitrogen-containing ring.

4,7-Dichloroquinoline was reduced to 7-chloroquinoline with Raney nickel in alcohol containing one equivalent of sodium hydroxide [75a]. The yield was not reported. The same reaction carried out in alcohol with 40% by weight of 10% palladium on carbon and excess potassium hydroxide as acid acceptor gave about 60% yield [75b]. Almost quantitative yield resulted from hydrogenation in alcohol containing one equivalent of sodium acetate or sodium hydroxide with a 10% ratio of 5% palladium on carbon [4a]. The chlorine in 4-position was selectively removed even when chlorine in 7-position was activated by adjacence of an amino group as illustrated in the palladium on carbon reduction of 8-amino-4,7-dichloroquinoline in alcohol and sodium acetate to 8-amino-7-chloroquinoline [75a]. The chlorine in the nitrogen-containing ring was also selectively removed in the hydrogenation of 4,7-dichlorocarbostyril [75d]. In this reaction, where it was found necessary to employ 1.25 molar equivalents of potassium hydroxide to keep the substrate or reduction product in solution in 95% alcohol, Raney nickel was highly selective, giving 93% yield of 7-chlorocarbostyril. In contrast, the use of palladium on barium sulfate gave a mixture of 7-chlorocarbostyril and carbostyril when hydrogen uptake was interrupted after one equivalent of hydrogen was absorbed. The main disadvantage in the nickel-catalyzed reaction was the long reduction time (45 hr). The use of a more suitable solvent for the substrate and the employment of a molar equivalent of acid acceptor as sodium acetate or triethylamine might aid the reduction. Under these conditions it should be possible to use a moderate amount of supported palladium catalyst for reduction without danger of overhydrogenation.

Other examples of preferential removal of halogen in nitrogen-containing rings can be seen in the hydrogenation of 4,7-dichlorocinnoline to 7-chlorocinnoline with palladium on calcium carbonate in methyl alcohol [31b], and that of 2,6-dichloro- and 2,7-dichloro-3-[3-(1-piperidinopropyl)amino]-quinoxaline to the corresponding 6- and 7-monochloro compounds in alcohol with Raney nickel [75e]. When dehalogenation of the 2,6-dichloroquinoxaline was attempted in alcohol in the presence of sodium hydroxide, alkoxylation occurred in 2-position.

The selective hydrogenation of 6-chloro-2-(4-chlorophenyl)imidazo[1,2,-b]-pyridazine in alcohol with palladium on carbon yielded 2-(4-chlorophenyl)-imidazo[1,2-b]pyridazine [48d]. In view of the greater ease of removal of halogen from the nitrogen-containing ring in these systems, it was not surprising to find that the removal of chlorine occurred preferentially over an aryl bromide when the corresponding 6-chloro-2-(4-bromophenyl)imidazo-pyridazine was reduced over palladium on carbon.

Prediction of selectivity when similar halogens are attached to different unsaturated heterocyclic systems is difficult because of the dearth of examples. In some instances analogy may be drawn from the relative reactivity of such halogens toward nucleophilic reagents but there are not enough examples to place complete reliance upon the reactivity toward nucleophiles as equivalent to susceptibility to reductive dehalogenation.

480 Dehalogenation

Some conclusions about preferential dehalogenation of a haloimidazole versus that of a similar halopyrimidine may be drawn from the hydrogenations of 2,6,8-trichloropurine to 2,8-dichloropurine and to 2-chloropurine with palladium on carbon in aqueous sodium hydroxide solution [76a] (Eq. 20.23), and the reduction to 2-chloropurine in the absence of an acid acceptor [76b]. The results suggest that halogen in 2-position in the imidazole ring should be removed in preference to a similar one in 2-position, but not in 4-position, in the pyrimidine ring.

(20.23)

The selective removal of chlorine in 8-position in 2,8-chloro-9-glycosylpurines [76a, d] further emphasizes the greater susceptibility of a 2-haloimidazole to hydrogenative removal over that of a similar halogen in 2-position in the pyrimidine ring.

When 2,4,6,7-tetrachloropteridine was hydrogenated in anhydrous benzene over palladium on carbon, 82% yield of 2,4-dichloro-6-hydroxy-7,8-dihydropteridine was obtained [53b]. The manner in which reaction took place was not clarified, nor was the molar amount of absorbed hydrogen indicated. It might be assumed that one of the chlorine atoms was hydrogenolytically removed. If this is so, then there is basis for the assumption that a halogen attached to a pyrazine ring would be removed in preference to a similar one attached to the pyrimidine ring in 2-, 4-, 5-, or 6-position.

In the Presence of Other Functions

The selective removal of halogen in the presence of an acetylenic linkage or an aromatic-type nitro or nitroso group should not be expected.

Olefinic Bonds. Removal of halogen in the presence of an olefinic bond depends on the catalyst and reaction medium, on the halogen itself and the susceptibility of the particular type of halide to hydrogenation, and on whether the double bond is di-, tri-, or tetrasubstituted.

Palladium catalysts and basic conditions favor hydrogenolysis of chlorine, bromine, or iodine from compounds also containing an olefinic bond. Other catalysts and neutral or acidic conditions generally lead to preferential reduction of the double bond; compare with the section on haloolefins in Chapter IX.

ALLYLIC HALIDES. Removal of an allylic chlorine, bromine, or iodine by hydrogen usually takes place before double-bond reduction, but it may not be possible to obtain complete selectivity unless an acid acceptor is present. The effect is seen in the reduction 4-chlorobut-2-enoic acid,

$$ClCH_2CH=CHCOOH,$$

to crotonic acid, $CH_3CH=CHCOOH$, in alkaline solution with palladium on barium sulfate. When the same reduction was carried out in the absence of base, butyric acid was obtained along with crotonic acid [77a]. A contribution to selectivity may also be the effect of base in retarding double-bond reduction. Selective removal of an allylic bromine in the presence of an α,β-unsaturated double-bonded system is seen in the hydrogenation of 14-bromocodeinone in chloroform containing 5–10% of methyl alcohol with 10% palladium on carbon [77b]. In this reduction the basic nitrogen-containing compound acted as acid acceptor. Bromine was selectively removed but migration of the double bond occurred, possibly from the release of hydrogen chloride from partial hydrogenolysis of chloroform; 81% of neopinone was obtained (Eq. 20.24).

(20.24)

VINYL HALIDES. Selective removal of halogen in a true vinyl halide is dependent on the number of substituents on the double bond. There are no examples of preferential dehalogenation among vinyl halides where the double bond is mono- or disubstituted. Examples of selective dehalogenation when the double bond is trisubstituted are seen in the reductions of 4-chloro-3-methylbut-3-enoic acid and amide, $ClCH=C(CH_3)CH_2COR$, in aqueous sodium hydroxide with palladium on carbon to 3-methylbut-3-enoic acid and amide, respectively [77a]. The conversions of *cis*- and *trans*-1,2-dibromo-1,2-bis-(2-quinolyl)ethylenes to the corresponding 1,2-dis-(2-quinolylethylenes by hydrogenation with prereduced palladium on strontium carbonate in alcohol containing potassium hydroxide [77d] are examples of selective dehalogenation when the double bond is tetrasubstituted.

When there is competition between a vinyl-type halide, such as an aryl halide, and a double bond other than the one in the halovinyl system, it appears likely that selective dehalogenation will not take place if the double bond is disubstituted. This is illustrated in the reduction of 2-chlorocinnamic acid in alkaline solution with palladium catalyst, which yielded only hydrocinnamic acid [77a], and the hydrogenations of nuclear-substituted bromocinnamic acids with palladium on carbon under alkaline conditions, which were never selective [4a].

Carbonyl Groups

ALDEHYDES. Examples of hydrogenations of compounds containing an aldehyde function and a halogen are limited. Despite the lack of information it is possible to suggest which will be reduced selectively. When the halogen is not particularly reactive, hydrogenation of aldehyde should take place preferentially in neutral or acidic medium (see Chapter XIV).

In general the use of palladium catalyst and an acid acceptor should favor removal of an activated chlorine. 4,6-Dichloropyrimidine-5-aldehyde was reduced to pyrimidine-5-aldehyde with palladium on carbon in methyl alcohol containing suspended excess magnesium oxide [78a]. The reactive nuclear chlorines may have been further activated for hydrogenolysis because of adjacence of the aldehyde group. For further knowledge about selectivity, it would be of interest to hydrogenate 2-chloropyrimidine-5-aldehyde, where the halogen is in a less reactive position, under the same conditions.

The reactive bromine atoms in methyl mucobromate,

$$OHCC(Br)=C(Br)COOCH_3,$$

were selectively hydrogenolyzed and the double bond was saturated during reduction in methyl alcohol with Raney nickel or nickel on kieselguhr at 120–140° and 40–80 atm initial pressure of hydrogen, to yield methyl 3-formylpropionate, $OHC(CH_2)_2COOCH_3$ [78b]. Higher yield was obtained by reducing the ethyl ester in ethyl alcohol. The removal of chlorines in methyl or ethyl mucochlorate would likely also be selective, since each halogen is considerably activated, one by adjacence to the aldehyde function and the other by proximity to the carboxyl group. Selective dehalogenation might have been accomplished by hydrogenation with palladium and an acid acceptor under much milder conditions.

The use of inhibitors prevented reduction of the aldehyde function in the hydrogenolysis of trichloroacetaldehyde to dichloro-, monochloro-, or acetaldehyde itself during hydrogenations with Raney nickel, cobalt, copper, and iron and with copper and nickel chromites at 50–150° and 120–150 atm [78c]. The most effective inhibitors were triphenylphosphine, thiophene, thiophenol, and thioures.

KETONES. The removal of halogen in the presence of a ketone is easier to accomplish and more predictable than selective dehalogenation in the presence of an aldehyde, particularly when palladium catalysts are employed.

Dehalogenation should be the preferred reaction in the reduction of substrates containing a ketonic function and an activated system as an allylic or benzylic halide and in the hydrogenation of α-haloketones. An acid acceptor may not be necessary in palladium-catalyzed reductions but probably should be present when platinum catalyst is employed. It may be possible to get good results with Raney nickel in similar reactions with acid acceptors if alkali hydroxide, a promoter for nickel-catalyzed ketone reductions, is avoided.

Allylic or Benzylic Type Haloketones. The hydrogenation of 5-bromo-5-phenylvalerophenone, $C_6H_5CHBr(CH_2)_3COC_6H_5$, in alcohol with palladium on calcium carbonate to 5-phenylvalerophenone is an example of the selective removal of an allylic- or benzylic-type halide in the presence of an activated ketone function [79a]. It is likely that the corresponding chloride could also be dehalogenated, especially if an acid acceptor was present.

α-Haloketones. The hydrogenation of 2-bromo-3-phenyl-3-(1-piperidyl)-propiophenone with platinum oxide [79b] is an example of selective dehalogenation in this group. In this instance the basic substrate acted as acid acceptor. 3-Phenylpropiophenone was obtained, debenzylation resulting from the large amount of catalyst that was employed (Eq. 20.25).

$$\underset{\text{piperidyl}}{\diagdown N}\overset{C_6H_5}{\underset{|}{C}}(Br)CH_2COC_6H_5 \xrightarrow{PtO_2} C_6H_5CH_2CH_2COC_6H_5 \qquad (20.25)$$

The reduction of 10-acetyl-6-bromoacetyl-1,1a,2,3,4,4a,9,10-octahydroacridine in alcohol with 5% palladium on carbon and sodium acetate to the corresponding 6,10-diacetyloctahydroacridine is another example [79c]. Similar examples of selectivity are seen in the hydrogenations of bromoacetyl-3-chlorophenylphenanthrenes to the corresponding acetyl 3-chlorophenanthrenes with palladium on carbon in the absence of an acid acceptor [67b]. In this reduction the absence of an acceptor prevented loss of nuclear chlorine.

Examples of selective reduction of halogen in α-haloketones among steroids are seen in the hydrogenations of *2β-bromo-7-methyl-5α-androstane-1α,17β-diol-3-one* and *2β-bromo-7α,17α-dimethyl-5-α-androstane-1α,17β-diol-3-one* to the debromo ketones in 1:1 methyl alcohol-tetrahydrofuran solution

484 *Dehalogenation*

containing sodium acetate and a small amount of acetic acid with 10% palladium on calcium carbonate [79d].

Further examples include the dehalogenations of 4-benzoyl-4-chloro-1-methylpiperidine hydrochloride and the corresponding 4-bromo compound as hydrobromide with palladium on carbon in methyl alcohol to form 4-benzoyl-1-methylpiperidine [79e]. When reduction was carried out in methyl alcohol with platinum oxide in the presence or absence of sodium carbonate, 4-(1-hydroxybenzyl)-1-methylpiperidine was obtained. The authors suggested that loss of hydrogen halide occurred and enol was formed. This did not take place in the palladium-catalyzed hydrogenation in alcohol. Loss of HX was prevented by the substitution of chloroform as solvent in the reduction with platinum oxide, which then gave 4-benzoyl-1-methylpiperidine (Eq. 20.26). The conversion of 2,4,6-tribromobicyclo[3,3,0]oct-7-en-1-

$$\underset{\text{X = Cl or Br}}{\underset{\text{R = CH}_3}{\text{RN}\diagup\diagdown\text{X}\diagup\diagdown\text{COC}_6\text{H}_5}} \xrightarrow[\text{or PtO}_2, \text{CHCl}_3]{\text{H}_2, \text{Pd on C, MeOH}} \text{RN}\diagup\diagdown\text{H}\diagup\diagdown\text{COC}_6\text{H}_5$$

$$\xrightarrow[\text{PtO}_2, \text{MeOH}]{\text{H}_2} \left[\text{RN}\bigcirc\text{COC}_6\text{H}_5 + \text{HX} \right] \longrightarrow$$

$$\text{RN}\bigcirc=C(OH)C_6H_5 \xrightarrow{H_2} \text{RN}\bigcirc CHOHC_6H_5 \qquad (20.26)$$

one to bicyclo[3,3,0]oct-7-en-1-one and the selective debromination of 6-acetoxy-2,4-dibromo[3,3,0]oct-7-en-1-one with palladium on carbon and sodium bicarbonate [79f] not only are examples of the selective removal of halogen in an α-haloketone but also illustrate the expected selective removal of an allylic bromide in the presence of an α,β-unsaturated ketone. Debromination without affecting the double bond may be expected since each bromine is a reactive one and the double bond is tetrasubstituted (Eq. 20.27).

$$\underset{\text{Br}}{\overset{\text{Br}}{\bigodot}}\text{Br} \xrightarrow[\text{NaHCO}_3]{3\text{H}_2, \text{Pd on C}} \bigodot \qquad (20.27)$$

Vinyl Halides. Examples appear to be limited to substrates that contain a ketone and a vinyl-type halide such as nuclear halogen in unsaturated carbocycles and heterocycles. From the examples, selective removal of halogen should take place when a combination of palladium catalyst and an acid acceptor is employed for hydrogenation.

2-Chloro-7-phenyltropone was converted to 2-phenyltropone by reduction in methyl alcohol containing sodium acetate [80a] (Eq. 20.28). The hydro-

$$\underset{C_6H_5}{\overset{Cl}{\bigodot}}=O \xrightarrow[CH_3COONa]{H_2 \atop Pd \text{ on } C} \underset{C_6H_5}{\bigodot}=O \qquad (20.28)$$

genation of 3'-chloro-2'-hydroxyacetophenone to 2'-hydroxyacetophenone with a small amount of palladium on carbon in alcohol containing sodium acetate [80b] is another example of selectivity when an aryl halide and a ketonic function are present in a substrate.

Obviously, removal of halogen should be expected when the ketone function is hindered as in benzophenone or any diaryl ketone. 2-Hydroxybenzophenone was obtained from chlorohydroxybenzophenone [80b]. 3-Indolyl 4-pyridyl ketone was obtained from 3-indolyl 4-(2-chloropyridyl) ketone with palladium on barium carbonate [80c]. The marked difference between the removal of halogen and attack of the carbonyl function in such cases is seen in the reduction of 2-amino-5-chlorobenzophenone with palladium on carbon or on calcium carbonate in alcohol with a variety of acid acceptors, where high pressure was employed (70 atm) and the ketone function remained intact [80d]. Very high yields of 2-aminobenzophenone were obtained.

The removal of aryl halide in the presence of a cyclic ketone should also be expected, especially when palladium catalyst is employed, since aliphatic ketones are not readily reduced with that catalyst. 4-Bromo-7-hydroxy-1-indanone was converted to 7-hydroxy-1-indanone with palladium on barium sulfate in alcohol [81a]. The reaction time (20 hr) would probably have been greatly decreased if an acid acceptor were present. In another example 7-chloro-4 6-dimethoxy-6-methylgrisan-3,2',4'-trione was dehalogenated to yield the dehalo triketone in 96% yield by reduction in aqueous sodium carbonate or in aqueous triethylamine with palladium on carbon [81b] (Eq. 20.29).

(20.29)

A combination of Raney nickel and potassium hydroxide proved useful in the reduction of 8-bromo-5-hydroxy-3-methyl-1-tetralone to 5-hydroxy-3-methyl-1-tetralone [81c].

Saturated Aliphatic Halides. There is a lack of examples of hydrogenations of compounds containing a ketone function and a saturated aliphatic-type halide, so that a general statement about selectivity cannot be made.

It can probably be assumed that when a compound contains this type of halide and an aliphatic ketone, the halogen, excepting fluorine, can be removed during hydrogenation with palladium catalyst under the proper conditions because aliphatic ketones are seldom reduced with that catalyst. This point is illustrated in the reduction of 4-trichloromethyl-4-methylcyclohexanone in ethyl alcohol containing potassium hydroxide to 4-dichloromethyl-4-methylcyclohexanone with palladium on carbon [82a]. 4-Dichloromethyl-4-methylcyclohexanone, however, could not be reduced under similar conditions.

In contrast, 4-dichloromethyl-4-methylcyclohexane was reduced in the presence of palladium by another investigator but results were dependent on which of two bases was employed [82b]. Hydrogenation was carried out at atmospheric pressure in methyl alcohol containing a tenfold excess of base in the presence of 10% palladium on carbon equal to 50% of the weight of substrate. When potassium hydroxide was used, 4,4-dimethylcyclohexanone was obtained after reduction. Dehalogenation of the dichloroketone in the presence of triethylamine gave 4,4-dimethylcyclohexanone, some partially dechlorinated material, and a mixture of the stereoisomeric 4-dichloromethyl-4-methylcyclohexanols.

In the same study, when triethylamine was used as base in the palladium-catalyzed reduction of 2-dichloromethyl-2-methylcyclohexanone, there was little attack at the carbonyl group; 2,2-dimethylcyclohexanone was obtained. Dehalogenation in the presence of excess potassium hydroxide gave varying results. The yield of dimethylcyclohexanone decreased considerably and differences in the rate and amount of hydrogen absorption were observed. 2,2-Dimethylcyclohexanol was obtained in certain reductions. One of the products also obtained was 6-methylheptanoic acid, which resulted from the hydrogenation of 7-chloro-6-methylhept-6-enoic acid formed by the attack of hydroxide ion on 2-dichloromethyl-2-methylcyclohexanone [82c]. There is little doubt that the anomalous results obtained in the reductions containing potassium hydroxide were due to reaction of the starting material with it. When the substrate, solvent, catalyst, and potassium hydroxide were mixed immediately and hydrogenated without delay under 4 atm pressure the formation of 6-methylheptanoic acid was eliminated. Two equivalents were rapidly absorbed; 2,2-dimethylcyclohexanone containing a small amount of high-boiling impurity was obtained in 70% yield.

Cyano Groups. Palladium catalysts are generally ineffective for the reduction of cyano groups except in strongly acidic medium (see Chapter XII). Most of the examples of dehalogenations in the presence of a cyano group are those in which nuclear halogen is selectively removed from unsaturated nuclear cyano-containing heterocycles by low-pressure hydrogenation with palladium in the presence of an acid acceptor.

Halocyanopyridines have been converted to the corresponding cyanopyridines in very high yield by reduction with palladium catalysts in the presence of magnesium oxide [83a], triethylamine [83b], and sodium acetate [83c]. 4-Chloro-5-cyanoquinoline was hydrogenated to 5-cyanoquinoline in 85% yield with palladium on carbon and potassium hydroxide in alcohol [84a] and 4-dimethylamino-6-chloro-5-cyano-2-methylpyrimidine was converted to 4-dimethylamino-5-cyano-2-methylpyrimidine in 86% yield by reduction with palladium hydroxide on calcium carbonate with calcium hydroxide as base [84b].

Dehalogenation with palladium was carried out in the absence of base when ethyl 2-chloro-3-cyano-6-methylisonicotinate was reduced with 5% palladium on barium sulfate, but twice as much catalyst as compound was used [85a]. In another instance 1-chloro-4-cyano-6,7-dimethoxyisoquinoline was selectively reduced to 4-cyano-6,7-dimethoxyisoquinoline by passing hydrogen into a refluxing cymene solution of the chloro compound containing 50% by weight of 10% palladium on carbon [85b].

An example of the removal of an aliphatic iodide in the presence of alicyclic cyanide is seen in the reduction of 2-cyanocyclohexylmethyl iodide to 2-cyano-1-methylcyclohexane at atmospheric pressure with 200% by weight of Raney nickel and pyridine as base [7f]. When a related chloro compound, 4-cyano-5-chloromethylcyclohexene, was hydrogenated with nickel at 80° and 65 atm pressure in the absence of pyridine, the halogen was not removed but reacted with the aminomethyl group formed by reduction of the cyano group, to give *trans*-hexahydroisoindoline (Eq. 20.30).

$$\text{cyclohexene-CH}_2\text{Cl, CN} \xrightarrow[\text{NiR}]{3H_2} \left[\text{cyclohexane-CH}_2\text{Cl, CH}_2\text{NH}_2 \right] \longrightarrow \text{bicyclic-NH}$$

(20.30)

Imino Groups. In general the imino function, $>\!\!C\!\!=\!\!N$, will be selectively reduced in the presence of halogen, Chapter XII. When the substrate contains a reactive halogen and a reducible imine function which is a part of a cyclic system, there may be a possibility of the selective removal of halogen. This is suggested by the dehalogenation of 9-chloro-6-methyl-3,4-dihydro-1*H*-azepino[5,4,3-cd]indole to the corresponding 3,4-dihydro compound with palladium on carbon in alcohol without added base [86] (Eq. 20.31). The yield of reduction product was low (11%) after several recrystallizations; there was no mention of other reaction products. The yield might have been increased by the use of an acid acceptor.

$$\text{(20.31)}$$

N-Oxides. Examples of selective dehalogenation in the presence of an *N*-oxide are confined to the removal of nuclear halogen from substituted pyridazine *N*-oxides and related compounds. The use of palladium on carbon and an acid acceptor, usually ammonium hydroxide or sodium hydroxide, gave high yields of the dehalo *N*-oxides [87a, b, c, d]. In one instance about 5.5% of 3,4-dimethylpyridazine was obtained (isolated by chromatography), along with about 80% of the desired dimethylpyridazine-1-oxide when 6-chloro-3,4-dimethylpyridazine-1-oxide was hydrogenated in water and ammonium hydroxide with 10% by weight of palladium on carbon [87e].

Oximes. In general, reduction of an oxime will take place in preference to removal of halogen (Chapter XIII) except when α-halooximes are hydrogenated with palladium catalyst.

A number of α-halooximes, such as 1-chloro-2-butanone oxime and 2-chlorocycloalkanone oximes, were rapidly reduced in methyl alcohol to the corresponding oximes in high yield with palladium on carbon [88a]. 2-Chloro-2-methylcyclohexanone oxime was also selectively reduced to the oxime but because of steric effects uptake of hydrogen was slow (10–20 hr).

Selective reductions of α-halocycloalkanone oximes have also been carried out with palladium in acidic medium to give very high yields of cycloalkanone oximes. These include reaction in methylene chloride containing sulfuric acid [88b], in acetic acid containing hydrochloric acid [88c], and reduction as the sulfate in acetic acid [88d]. In one instance 2-chlorocyclohexanone oxime was converted to cyclohexanone oxime by bubbling hydrogen through a solution in ethyl acetate containing palladium on carbon and sodium acetate as acid acceptor [88e].

Hydrogenation of α-halooximes with platinum catalysts usually result in reduction of the oxime to hydroxylamine or amine as well as removal of halogen [88a, f]. Side reaction in reductions with palladium on carbon has been reported only in Ref. 88f.

Hydrazine Linkage. Although rupture of the nitrogen–nitrogen bond can occur in the removal of halogen from compounds also containing a hydrazino linkage, examples show that, at least, it has not taken place during hydrogenolyses of aromatic-type halides.

Examples of selective dehalogenation are seen in the hydrogenation of 5-amino-4-chloro-6-hydrazinopyrimidine in alcohol with palladium on carbon to yield 86% of 5-amino-6-hydrazinopyrimidine as hydrochloride salt, and in the palladium reduction in the presence of magnesium oxide to yield the free base [89]. 5-Amino-6-(1-methylhydrazino)pyrimidine and 5-amino-4-(1-benzylhydrazino)pyrimidine were obtained in a similar manner from the corresponding 4- and 6-chloro compounds, respectively. The substituted hydrazino compounds were also obtained from the corresponding chloro compounds by refluxing buffered solutions of them (pH 7.0) with 4–10 times the weight of Raney nickel.

Nuclear halogen was selectively removed from 1-benzyl-4-hydrazino-7-chloroimidazo[4,5-d]pyridazine by hydrogenation with palladium on carbon and sodium hydroxide [4a]. It was necessary to carry out the reaction at 65–70° in methyl cellosolve because of poor solubility. Nevertheless, even at this elevated temperature there did not appear to be any cleavage of the hydrazino linkage.

These examples suggest that hydrogenolyses of halogen in reasonably active systems can be carried out without concomitant attack of the nitrogen to nitrogen bond. If it is found that cleavage does occur during reduction, it probably can be eliminated by acylation of the hydrazino linkage [4a]. Indeed this procedure should allow selective removal of halogen from less-active halides, except fluorides, in the presence of an N—N bond, particularly if an acid acceptor is employed to aid dehalogenation.

Benzyl Groups

O-BENZYL. There are not enough examples in every category to suggest that dehalogenation with palladium catalyst will be selective in the presence of an *O*-benzyl grouping, except when an allyl-type or benzyl-type chloride, bromide, or iodide is involved. The removal of such a chloride in the presence of a comparatively readily reducible benzyl ester is shown in the hydrogenation of 3-hydroxy-5-benzoyloxymethyl-4-chloromethyl-2-picoline with palladium on carbon to the corresponding ester of 3-hydroxy-2,4-dimethylpyridine-5-methanol in 85% yield [90].

It should also be possible to selectively remove nuclear halogen from benzylic alcohols with palladium in the presence of a strong base since benzyl alcohols are usually more difficult to hydrogenolyze than benzyl ethers and esters. However, there are no examples to substantiate this.

Since Raney nickel is not especially reactive for debenzylation, a combination of it and alkali should be especially useful in the hydrogenolysis of

490   *Dehalogenation*

halogen from halobenzyl alcohols. 3,5-Dibromo-4-hydroxy-2,6-dimethoxybenzyl alcohol was hydrogenated to 4-hydroxy-2,6-dimethoxybenzyl alcohol with Raney nickel and sodium hydroxide in methyl alcohol [91a]. Chlorohydroxybenzyl alcohols were similarly reduced to hydroxybenzyl alcohols without apparent debenzylation despite the large amount of catalyst employed [91b]. Reduction with nickel in aqueous alkaline solution gave lower yield than hydrogenations in methanolic sodium or potassium hydroxide [91c].

A large amount of Raney nickel, ten times the weight of substrate, was employed to remove an alkyl iodide from a cyclic ether also containing a benzyl ether in a reaction which made use of the hydrogen held by freshly prepared Raney nickel. The compound in acetic acid and catalyst in 50% aqueous acetic acid were stirred for 12 hr at 35° [92] (Eq. 20.32).

$$\text{structure with } CH_2I, OCH_2C_6H_5, OAc, NHCOC_6H_5, AcO \xrightarrow{H_2, NiR} \text{structure with } CH_3, OCH_2C_6H_5, OAc, NHCOC_6H_5, AcO \qquad (20.32)$$

N-BENZYL. The palladium-catalyzed hydrogenations of some *N*-benzyl chloroanilines containing several acidic functions resulted in debenzylation without loss of halogen (Chapter XIX). It was pointed out that the danger of dehalogenation rises with the increase in basicity of the substrate (Chapter XIX, Ref. 2a).

When compounds more basic than *N*-benzylchloroanilines, such as chlorodibenzylamine, and other *N*-benzylaralkylamines containing nuclear chlorine were reduced under acidic conditions with palladium on carbon (in one instance with platinum oxide), dehalogenation was the principal, and in most cases the only, reaction [93a]. Similar results were obtained during the hydrogenations of 2-chloro-4-benzylaminopyrimidine and the corresponding 5-methyl compound in acetic acid with more than 70% by weight of 7.5% palladium on carbon [33c]. In another example where a tertiary *N*-benzyl group was involved, *N*-benzyl-2-(2-dimethylaminoethylamino)-5-chloropyrimidine underwent dehalogenation during reduction in methyl alcohol with palladium on calcium carbonate [93b]. In this instance dehalogenation was surely aided by the basicity of the substrate or reduction product, either of which acted as acid acceptor. Selective removal of halogen also took place in the hydrogenation of another tertiary *N*-benzyl aryl-type chloride, when 8-benzyl-2-chloro-4-methyl-5,6,7,8-tetrahydropteridine was reduced with palladium on calcium carbonate in alcohol or with 50% by weight of palladium on carbon in acetic acid [94].

Miscellaneous Ring Systems. Since the cyclopropane ring is not readily cleaved, one might assume that successful removal of reactive halogen could be accomplished without ring opening. There are, however, too few references for substantiation. There is an example of selectivity where bromine was attached to the cyclopropane ring.

3,3-Dibromocyclopropane-*cis*-1,2-diacetic acid in excess methanolic potassium hydroxide was partially dehalogenated by hydrogenation over platinum oxide or with palladium on carbon or on barium sulfate to yield the 3-bromo compound [68e]. It is likely that the great excess of alkali inactivated the catalysts, since reaction took 24 hr. No ring rupture was reported. When hydrogenation of the dibromo or monobromo compound was carried out in a similar manner with Raney nickel, 85% yield of cyclopropane-*cis*-1,2-diacetic acid was obtained along with some pimelic acid, the product of dehalogenation and ring rupture.

In most instances halogen can also be removed selectively without affecting a lactone ring if reduction is carried out with palladium catalyst and an acid acceptor. However, in the attempted dehalogenation of some α-bromo lactones, ring opening did occur in certain instances with potassium acetate as acid acceptor [95]. No hydrogen uptake was observed in the absence of base. Selective debromination occurred in the hydrogenations of the α-bromo derivative of 5-hydroxybicyclo[2,2,1]heptane-2,3-dicarboxylic acid 3 → 5 lactone, of the α-bromo-lactone of dihydro-ψ-santonin, and of 12-bromooleanic acid lactone when potassium acetate was present. When 5α-bromo-4β-hydroxycyclohexane-1β,2β-dicarboxylic acid 2 → 4-lactone and the related 3β-methyl compound were similarly reduced, debromination took place, but in each reaction more cyclohexanedicarboxylic acid was obtained than debromolactone (Eq. 20.33). No change occurred when dioxan was

$$\underset{\text{Br}}{\overset{\text{O}\text{—}\text{C}=\text{O}}{\bigotimes_{R\ H}\text{COOR}}} \xrightarrow[\text{CH}_3\text{COOK}]{\text{I}+\text{H}^2 \atop \text{Pd on C}} \underset{}{\overset{\text{O}\text{—}\text{C}=\text{O}}{\bigotimes_{R\ H}\text{COOR}}} + \underset{}{\overset{R\ H}{\bigotimes\begin{array}{l}\text{—COOH}\\\text{—COOR}\end{array}}}$$

(20.33)

employed as solvent instead of aqueous alcohol, or when nickel was substituted for palladium. The authors reported that the debrominated lactones were resistant to hydrogenolysis under the experimental conditions. They suggested that ring opening was apparently contingent on dehalogenation and ultimately on structural factors.

Recently a very distinct improvement in the Rosemond reaction has been reported in which high yield aldehyde is obtained [96]. The reduction is carried out in toluene or xylene in a closed system at 65 psig and 35° in the presence of palladium on barium sulfate and three equivalents of sodium acetate-in most instances without a quinoline-sulfur regulator. Ingredients must be rigorously dried and significantly the acid chloride must be very pure.

REFERENCES

[1] a. R. Baltzly and A. P. Phillips, *J. Am. Chem. Soc.*, **68**, 261 (1946); b. L. Horner, L. Schläfer, and H. Kammerer, *Ber.*, **92**, 1700 (1959); c. R. Duschinsky, E. Pleven, and C. Heidelberger, *J. Am. Chem. Soc.*, **79**, 4559 (1957); d. S. A. Fuqua, R. M. Parkhurst, and R. M. Silverstein, *Tetrahedron*, **20**, 1625 (1964).

[2] a. F. Swartz, *Bull. Acad. Roy. Belg.*, **1920**, 399; b. M. W. Renoll, *J. Am. Chem. Soc.*, **68**, 1159 (1946); c. F. L. M. Pattison and B. C. Saunders, *J. Chem. Soc.*, **1949**, 2745.

[3] J. N. Pattison and E. F. Degering, *J. Am. Chem. Soc.*, **73**, 611 (1951).

[4] a. M. Freifelder, unpublished results; b. C. W. Whitehead, J. J. Traverso, F. J. Marshall, and D. E. Morrison, *J. Org. Chem.*, **26**, 2809 (1961); c. P. L. Pickard and H. L. Lochte, *J. Am. Chem. Soc.*, **69**, 14 (1947); d. K. L. Howard, H. W. Stewart, E. A. Conroy, and J. J. Denton, *J. Org. Chem.*, **18**, 1484 (1953).

[5] a. M. Busch and H. Stöve, *Ber.*, **49**, 1063 (1916); b. Personal communication, Dr. E. J. Cragoe, Jr., Merck, Sharp, & Dohme Res. Lab., West Point, Pennsylvania.

[6] a. G. E. McCasland and E. C. Horswill, *J. Am. Chem. Soc.*, **75**, 4020 (1953); b. R. A. B. Bannard and L. R. Hawkins, *Can. J. Chem.*, **39**, 1530 (1961); c. F. R. Mayo and M. D. Hurwitz, *J. Am. Chem. Soc.*, **71**, 776 (1949).

[7] a. G. E. Ham and W. P. Coker, *J. Org. Chem.*, **29**, 194 (1964); b. J. W. Schulenberg and S. Archer, *ibid.*, **30**, 1279 (1965); c. S. Pitha and I. Ernest, *Coll. Czechoslov. Chem. Commun.*, **24**, 2632 (1959); d. A. T. Blomquist and J. Wolinsky, *J. Org. Chem.*, **21**, 1371 (1956); e. H. Stetter, J. Gärtner, and P. Tacke, *Ber.*, **99**, 1435 (1966); f. H. Christol, A. Donche, and F. Plénat, *Bull. Soc. Chim. France*, **1966**, 1315.

[8] a. K. B. Wiberg and B. A. Hess, Jr., *J. Org. Chem.*, **31**, 2250 (1966); b. L. Ruzicka, Pl. A. Plattner, and H. Wild, *Helv. Chim. Acta*, **28**, 395 (1945).

[9] a. M. Nakajima, A. Hasegawa, and F. W. Lichtenthaler, *Ann.*, **680**, 21 (1964); b. A. Hasegawa and H. Z. Sable, *J. Org. Chem.*, **31**, 4154 (1966).

[10] A. W. Burgstahler, T. B. Lewis, and M. O. Abdel-Rahman, *J. Org. Chem.*, **31**, 3516 (1966).

[11] R. A. Lutz and R. B. Rowlett, Jr., *J. Am. Chem. Soc.*, **70**, 1359 (1948).

[12] a. W. J. Close, L. R. Swett, L. E. Brady, J. H. Short, and M. Vernsten, *J. Am. Chem. Soc.*, **82**, 1132 (1960); b. J. H. Short and U. Biermacher, *ibid.*, p. 1135.

[13] a. B. Loev and M. Kormendy, *J. Org. Chem.*, **27**, 1703 (1962); b. R. Huisgen, G. Binsch, and H. König, *Ber.*, **97**, 2884 (1964); c. G. Binsch, R. Huisgen, and H. König, *ibid.*, p. 2893.

[14] a. J. R. Vaughan, Jr., J. Krapcho, and J. P. English, *J. Am. Chem. Soc.*, **71**, 1885 (1949); b. R. E. Lutz and P. S. Bailey, *ibid.*, **67**, 2229 (1945).

[15] a. G. G. Kleinspehn and A. H. Corwin, *J. Am. Chem. Soc.*, **76**, 5641 (1954); b. *ibid.*, **75**, 5295 (1953); c. A. Treibs, R. Schmidt, and R. Zinsmeister, *Ber.*, **90**, 79 (1957); d. R. A. Nicolaus and R. Nicoletti, *Ann. Chim.* (Rome), **47**, 167 (1957).

[16] a. J. P. Wibaut, *Rec. Trav. Chim.*, **63**, 141 (1944); b. A. Bavley, M. Harfenist, W. A. Lazier, and W. M. McLamore, U.S. Patent 2,742,480 (1956); c. R. F. Raffauf, *Helv. Chim. Acta*, **33**, 102 (1950).

[17] a. H. L. Lochte and E. N. Wheeler, *J. Am. Chem. Soc.*, **76**, 5548 (1954); b. S. Heřmanék, *Coll. Czechoslov. Chem. Commun.*, **24**, 2748 (1959); c. S. J. Norton, C. G. Skinner, and W. Shive, *J. Org. Chem.*, **26** 1495 (1961).

[18] a. R. P. Mariella, *J. Am. Chem. Soc.*, **69**, 2670 (1948); b. R. P. Mariella and V. Kvinge, *ibid.*, **70**, 3125 (1949); c. L. N. Yakhontov and M. V. Rubstov, *J. Gen. Chem. USSR*, **32**, 425 (1962) (Eng. trans.).

[19] a. A. Bavley, M. G. Gollaher, and M. W. McLamore, U.S. Patent 2,745,838 (1956); b. M. Barash, J. M. Osbond, and J. C. Wickens, *J. Chem. Soc.*, **1959**, 3530; c. B. A. Fox and T. L. Threlfall, *Organic Synthesis*, Vol. 44, Wiley, New York, 1964, p. 34.

[20] a. C. Hansch, W. Carpenter, and J. Todd, *J. Org. Chem.*, **23**, 1924 (1958); b. J. A. Moore and H. H. Püschner, *J. Am. Chem. Soc.*, **81**, 6041 (1959); c. M. M. Robison, *ibid.*, **80**, 6254 (1958).

[21] a. F. Schenker, R. A. Schmidt, W. Leimgruber, and A. Brossi, *J. Med. Chem.*, **9**, 46 (1966); b. A. K. Mallams and S. S. Israelstam, *J. Org. Chem.*, **29**, 3549 (1964); c. K. N. Campbell and J. F. Kerwin, *J. Am. Chem. Soc.*, **68**, 1837 (1946); d. R. A. Robinson, *ibid.*, **69**, 1939 (1947); e. M. Pesson and D. Richer, *Compt. Rend.*, Ser. C., **262**, 1719 (1966).

[22] a. C. E. Kaslow and E. Aronoff, *J. Org. Chem.*, **19**, 857 (1954); b. F. W. Newmann, N. B. Sommer, C. E. Kaslow, and R. L. Shriner, *Organic Synthesis*, vol. 26, Wiley, New York, 1947, p. 45.

[23] a. J. Walker, *J. Chem. Soc.*, **1947**, 1687; b. J. D. White and D. S. Straus, *J. Org. Chem.*, **32**, 2689 (1967); c. R. Garner and H. Suchitzky, *J. Chem. Soc.*, (C), **1966**, 186; d. E. Schlittler and R. Merian, *Helv. Chim. Acta*, **30**, 1339 (1947); e. M. M. Robison and B. L. Robison, *J. Am. Chem. Soc.*, **80**, 3443 (1958); f. W. Barbieri and L. Bernardi, *Tetrahedron*, **21**, 2453 (1966); g. M. S. Newman and W. H. Powell, *J. Org. Chem.*, **26**, 812 (1961); h. A. Albert and J. P. Willis, *J. Soc. Chem. Ind.*, **65**, 26 (1946).

[24] a. F. Misani and M. T. Bogert, *J. Org. Chem.*, **10**, 347 (1945); b. A. L. Searles and R. M. Warren, *ibid.*, **18**, 1317 (1953); c. F. H. Case and R. Sasin, *ibid.*, **20**, 1330 (1955); d. E. V. Brown, *ibid.*, **30**, 1607 (1965); e. R. Tan and A. Taurins, *Tetrahedron Letters*, **1966**, 1233; f. *ibid.*, **1965**, 2737; g. J. M. Bobbitt and R. E. Doolittle, *J. Org. Chem.*, **29**, 2298 (1964); h. B. M. Ferrier and N. Campbell, *J. Chem. Soc.*, **1960**, 3513; i. H. Rapoport and A. D. Batcho, *J. Org. Chem.*, **28**, 1753 (1963).

[25] a. F. Johnson and W. A. Nasutavicus, *J. Heterocyclic Chem.*, **2**, 26 (1965); b. J. Gardent and G. Hazebroucq, *Bull. Soc. Chim. (France)*, **1968**, 600; c. W. A. Nasutavicus and F. Johnson, *J. Org. Chem.*, **32**, 2367 (1967); d. Personal communication from Dr. Johnson.

[26] a. C. L. Habraken and J. A. Moore, *J. Org. Chem.*, **30**, 1892 (1965); b. I. Ito and N. Oda, *Chem. Pharm. Bull.* (Tokyo), **14**, 297 (1966); c. J. Davoll, *J. Chem. Soc.*, **1960**, 308.

[27] a. R. G. Jones, E. C. Kornfeld, and K. C. McLaughlin, *J. Am. Chem. Soc.*, **72**, 3539 (1950); b. P. Coad, R. A. Coad, and J. Hyepock, *J. Org. Chem.*, **29**, 1751 (1964); c. R. Schönbeck, *Monatsh.*, **90**, 284 (1959); d. E. A. Steck, R. P. Brundage, and L. T. Fletcher, *J. Am. Chem. Soc.*, **76**, 3225 (1954).

[28] a. M. M. Rogers and J. P. English, U.S. Patent 2,712,011 (1955); b. M. M. Lester and J. P. English, U.S. Patent 2,790,798 (1957).

[29] a. W. J. Leanza, H. J. Becker, and E. F. Rogers, *J. Am. Chem. Soc.*, **75**, 4086 (1953); b. S. Linholter, A. Kristensen, R. Rosenorn, S. E. Nielsen, and H. Kaaber, *Acta Chem. Scand.*, **15**, 1660 (1961).
[30] a. J. Druey, K. Meier, and K. Eichenberger, *Helv. Chim. Acta*, **37**, 121 (1954); b. J. Druey, A. Hüni, K. Meier, B. H. Ringner, and S. A. Staehelin, *ibid.*, p. 510; c. K. Eichenberger, R. Rometsch, and J. Druey, *ibid.*, **39**, 1755 (1956); d. R. H. Mizzoni and E. Spoerri, *J. Am. Chem. Soc.*, **73**, 1873 (1951).
[31] a. J. S. Morley, *J. Chem. Soc.*, **1951** (1971); b. D. E. Ames, R. F. Chapman, H. Z. Kucharska, and D. Waite, *ibid.*, **1965**, 5391; c. E. J. Alford and K. Schofield, *ibid.*, **1953**, 1811.
[32] E. F. M. Stephenson, *Chem. Ind.*, **1957**, 174.
[33] a. S. J. Childress and R. L. McKee, *J. Am. Chem. Soc.*, **72**, 4271 (1950); b. B Lythgoe and L. S. Rayner, *J. Chem. Soc.*, **1951**, 2323; c. H. Ballweg, *Ann.*, **673**, 153 (1964); d. C. R. Hauser and R. M. Manyik, *J. Org. Chem.*, **18**, 588 (1953).
[34] a. N. Whittaker, *J. Chem. Soc.*, **1951**, 1565; b. W. J. Haggerty, Jr., R. H. Springer, and C. C. Cheng, *J. Heterocyclic Chem.*, **2**, 1 (1965); c. G. G. Massaroli and G. Signorelli, *Chem. Abstr.*, **65**, 8903 (1966).
[35] a. Y. Nitta, K. Okui, and K. Ito, *Chem. Pharm. Bull.* (Tokyo), **13**, 557 (1965); b. R. N. Prasad, C. W. Noell, and R. K. Robins, *J. Am. Chem. Soc.*, **81**, 193 (1959); c. V. H. Smith and B. E. Christensen, *J. Org. Chem.*, **20**, 829 (1955).
[36] a. B. H. Chase, J. P. Thurston, and J. Walker, *J. Chem. Soc.*, **1951**, 3439; b. W. Klötzer and J. Schantl, *Monatsh.*, **94**, 1178 (1963).
[37] a. B. R. Baker, B-T. Ho, and T. Neilson, *J. Heterocyclic Chem.*, **1**, 79 (1964).
[38] a. C. Overberger, I. C. Kogon, and W. J. Einstman, *J. Am. Chem. Soc.*, **76**, 1953 (1954); b. W. Pfleiderer and H. Mosthaf, *Ber.*, **90**, 728 (1957); c. D. J. Brown and T. Teitei, *J. Chem. Soc.*, **1965**, 755.
[39] a. M. P. V. Boorland, J. F. W. McOmie, and R. N. Timms, *J. Chem. Soc.*, **1952**, 4691; b. A. Bendich, P. J. Russell, Jr., and J. J. Fox, *J. Am. Chem. Soc.*, **76**, 6073 (1954); c. D. J. Brown, *J. Appl. Chem.*, **4**, 72 (1954); d. L. O. Ross, L. Goodman, and B. R. Baker, *J. Am. Chem. Soc.*, **81**, 3108 (1959).
[40] E. F. Godefroi, *J. Org. Chem.*, **27**, 2264 (1962).
[41] a. H. R. Henze, W. J. Clegg, and C. W. Smart, *J. Org. Chem.*, **17**, 1320 (1952); b. H. P. Henze and S. O. Winthrop, *J. Am. Chem. Soc.*, **79**, 2230 (1957).
[42] a. S. I. Wang, *J. Am. Chem. Soc.*, **80**, 6197 (1959); b. P. K. Chang and A. D. Welch, *J. Med. Chem.*, **6**, 428 (1963); c. H. Kny and B. Witkop, *J. Am. Chem. Soc.*, **81**, 6245 (1959).
[43] R. C. Elderfield, T. A. Williamson, W. J. Gensler and C. B. Kremer, *J. Org. Chem.*, **12**, 405 (1947).
[44] a. E. J. Cragoe, Jr., O. W. Woltersdorf, Jr., J. B. Bicking, S. F. Kwong, and J. H. Jones, *J. Med. Chem.*, **10**, 66 (1967); b. G. Palamidessi, L. Bernardi, and A. Leone, *Farmaco, Ed. Sci.*, **21**, 805 (1966); c. G. Palamidessi and F. Luini, *ibid.*, p. 811; d. B. Camerino and G. Palamidessi, *Gazz. Chim. Ital.*, **90**, 1807 (1960); e. E. Felder, D. Pitre, S. Boveri, and E. B. Grabitz, *Ber.*, **100**, 555 (1967).
[45] F. R. Pfeifer and F. H. Case, *J. Org. Chem.*, **31**, 3384 (1966); b. M. Munk and H. P. Schultz, *J. Am. Chem. Soc.*, **74**, 3434 (1952).
[46] a. C. Grundmann, H. Schroeder, and R. Ratz, *J. Org. Chem.*, **23**, 1522 (1958); b. C. Grundmann, H. Ulrich, and A. Kreutzberger, *Ber.*, **86**, 181 (1953); c. A. Burger and E. D. Hornbaker, *J. Am. Chem. Soc.*, **75**, 4579 (1954); d. C. Grundmann and A. Kreutzberger, *ibid.*, **77**, 44 (1955); e. R. Hort, H. N. Decker, and R. Berchtold, *Helv. Chim. Acta*, **33**, 1365 (1950); f. R. A. Cutler, U.S. Patent 3,097,205

(1963); g. I. Flament, R. Promel and R. H. Martin, *Helv. Chim. Acta*, **42**, 485 (1959); h. W. O. Foye and M. H. Weinswig, *J. Am. Pharm. Assoc. Sci. Ed.*, **48**, 327 (1959).

[47] a. H. Graboyes and A. R. Day, *J. Am. Chem. Soc.*, **79**, 6421 (1957); b. G. B. Barlin, *J. Chem. Soc.* (B), **1966**, 285; c. C. L. Leese and H. N. Rydon, *ibid.*, **1955**, 303.

[48] a. T. Kuraishi and R. N. Castle, *J. Heterocyclic Chem.*, **3**, 218 (1966); b. G. A. Gerhardt and R. N. Castle, *ibid.*, **1**, 247 (1964); c. M. Malm and R. N. Castle, *ibid.* **1**, 182 (1964); d. F. Yoneda, T. Otaka, and Y. Nitta, *Chem. Pharm. Bull.*, **12**, 1351 (1964).

[49] a. J. Davoll, *J. Chem. Soc.*, **1960**, 131; b. K. Imai, *Chem. Pharm. Bull.*, **12**, 1030 (1964); c. A. Takamizawa and Y. Hamashima, *ibid*, **13**, 1207 (1965); d. M. Yanai, T. Kinoshita, S. Nakashima, and M. Nakamura, *Yakugaku Zasshi*, **85**, 339 (1965); through *Chem. Abstr.*, **63**, 5638 (1965).

[50] a. J. A. Montgomery and C. Temple, Jr., *J. Am. Chem. Soc.*, **80**, 409 (1958); b. H. J. Schaeffer and R. D. Weimar, Jr., *ibid.*, **81**, 197 (1959); c. H. H. Lin and C. C. Price, *J. Org. Chem.*, **26**, 108 (1961); d. J. Pitha and P. O. P. Ts'O, *ibid.*, **33**, 1341(1968); e. R. Prasad, C. W. Noell, and R. K. Robins, *J. Am. Chem. Soc.*, **81**, 193 (1959); f. A. G. Beaman, W. Tautz, R. Duschinsky, and E. Grunberg, *J. Med. Chem.*, **9**, 373 (1966).

[51] a. R. K. Robins, *J. Am. Chem. Soc.*, **78**, 7 (1956); b. C. C. Cheng and R. K. Robins, *J. Org. Chem.*, **21**, 1240 (1956); c. R. K. Robins, L. B. Holum, and F. W. Furcht, *ibid.*, p. 833.

[52] a. Y. Makisumi, *Chem. Pharm. Bull.*, **9**, 883 (1961); b. Y. Makisumi, H. Watanabe, and K. Tori, *ibid.*, **12**, 204 (1964); c. Y. Makisumi and H. Kano, *ibid.*, **8**, 907 (1959).

[53] a. A. Albert and S. Matsuura, *J. Chem. Soc.*, **1962**, 2162; b. E. C. Taylor and W. R. Sherman, *J. Am. Chem. Soc.*, **81**, 2464 (1959).

[54] a. H. M. Wuest and E. H. Sakal, *J. Am. Chem. Soc.*, **73**, 1210 (1951); b. T. R. Govindachari, N. S. Narasimhan, and S. Rajadurai, *J. Chem. Soc.*, **1957**, 560; c. P. S. Bailey, S. S. Bath, W. F. Thomsen, H. H. Nelson, and E. E. Kawas, *J. Org. Chem.*, **21**, 297 (1956); d. K. N. Campbell, J. F. Ackerman, and B. K. Campbell, *ibid.*, **15**, 221 (1950); e. M. G. Brown, *J. Chem. Soc.*, **1956**, 2558.

[55] A. F. Thomas and A. Marxer, *Helv. Chim. Acta*, **41**, 1898 (1958).

[56] a. H. Arnold, *Ber.*, **80**, 172 (1947); b. S. H. Weber, R. J. Kleipool, and D. B. Spoelstra, *Rec. Trav. Chim.*, **76**, 193 (1957).

[57] a. W. J. Moran, E. C. Schreiber, E. Engel, D. C. Behn, and J. L. Yamins, *J. Am. Chem. Soc.*, **74**, 127 (1952); b. P. S. Bailey and R. E. Lutz, *ibid.*, **67**, 2232 (1945); c. J. H. Burckhalter and R. I. Leib, *J. Org. Chem.*, **26**, 4078 (1961); d. A. Stempel and E. C. Buzzi, *J. Am. Chem. Soc.*, **71**, 2969 (1949); e. D. Heyl, E. Luz, S. A. Harris, and K. Folkers, *ibid.*, **75**, 4080 (1953); f. J. H. Burckhalter, R. J. Seiwald, and H. C. Scarborough, *ibid.*, **82**, 991 (1960); g. R. A. Turner, C. F. Huebner, and C. R. Scholz, *ibid.*, **71**, 2801 (1949).

[58] a. H. Schroeder and C. Grundmann, *J. Am. Chem. Soc.*, **78**, 2447 (1956); b. S. L. Shapiro and C. G. Overberger, *ibid.*, **76**, 97 (1954); c. C. G. Overberger, F. W. Michelotti, and D. M. Carabateas, *ibid.*, **79**, 941 (1957).

[59] a. R. S. Shelton and M. G. VanCampen, Jr., U.S. Patent 2,408,345 (1946); b. H. Bretschneider, *Monatsh.*, **78**, 82 (1948); c. S. L. Shapiro, H. Soloway, and L. Freedman, *J. Am. Chem. Soc.*, **80**, 6060 (1958).

[60] a. C. F. Huebner and H. A. Troxell, *J. Org. Chem.*, **18**, 736 (1953); b. W. G. Finnegan and R. A. Henry, *ibid.*, **24**, 1565 (1959); c. L. L. Dolby, C. Wilkins, and

T. G. Frey, *ibid.*, **31**, 1110 (1966); d. G. Tacconi, S. Pietra, and M. Zaglio, *Farmaco, Sci. Ed.*, **20**, 470 (1965); e. J. A. Vida and M. Gut, *J. Org. Chem.*, **33**, 1202 (1968).

[61] a. L. Zervas and C. Zioudrou, *J. Chem. Soc.*, **1956**, 214; b. G. R. Gray and R. Barker, *J. Org. Chem.*, **32**, 2764 (1967); c. F. F. Blicke and E. L. Schumann, *J. Am. Chem. Soc.*, **76**, 3153 (1954); d. Fr. Patent 1,442,416 (1966); through *Chem. Abstr.*, **67**, 108258 (1967); e. A. K. Bose and M. S. Manhas, *J. Org. Chem.*, **27**, 1244 (1962); f. H. McKennis, Jr., L. B. Turnbull, E. R. Bowman, and E. Tamaki, *ibid*, **28**, 383 (1963); g. G. Dupont, R. Dulov, and C. Pigerol, *Bull. Soc. Chim., France*, **1955**, 1101.

[62] K. Rosenmund, *Ber.*, **51**, 585 (1918).

[63] a. E. Mosettig and R. Mozingo, *Org. Reactions*, **4**, 362 (1948); b. D. C. Ayres, B. G. Carpenter, and R. C. Denny, *J. Chem. Soc.*, **1965**, 3578; c. M. S. Newman and N. Gill, *J. Org. Chem.*, **31**, 3860 (1966); d. M. Carissimi, A. Cattaneo, R. D'Ambrosio, E. Grumelli, E. Milla, and F. Ravebna, *Farmaco, Ed. Sci.*, **20**, 106 (1965); e. R. Lukeš and J. Kovář, *Coll. Czechoslov. Chem. Commun.*, **21**, 1317 (1956).

[64] a. C. Weygand and W. Meusel, *Ber.*, **76**, 503 (1943); b. S. Affrossman and S. J. Thomson, *J. Chem. Soc.*, **1962**, 2024; c. E. V. Brown and J. A. Blanchette, *J. Am. Chem. Soc.*, **72**, 3414 (1950).

[65] a. D. D. Phillips, M. A. Acitelli, and J. Meinwald, *J. Am. Chem. Soc.*, **79**, 3517 (1957); b. W. K. Detweiler and E. D. Amstutz, *ibid.*, **72**, 2882 (1950).

[66] a. E. C. Britton and T. R. Keil, U.S. Patent 2,725,402; b. H. J. den Hertog and J. C. M. Schogt, *Rec. Trav. Chim.*, **70**, 353 (1951); c. H. J. den Hertog, J. C. M. Schogt, J. de Bruyn, and A. de Klerk, *ibid.*, **69**, 673 (1950); d. E. D. Parker and W. Shive, *J. Am. Chem. Soc.*, **69**, 63 (1947).

[67] a. D. P. Brust, D. S. Tarbell, S. M. Hecht, E. C. Hayward, and D. L. Colebrook, *J. Org. Chem.*, **31**, 2192 (1968); b. E. L. May and E. Mosettig, *ibid.*, **11**, 429 (1946); c. *ibid.*, p. 435; d. *ibid.*, p. 441.

[68] a. S. Cohen, E. Thom, and A. Bendich, *J. Org. Chem.*, **27**, 3545 (1962); b. R. J. Wineman, E-P. T. Hsu, and C. E. Anagnostopoulos, *J. Am. Chem. Soc.*, **80**, 6233 (1958); c. C. M. Brenner, and H. R. Rickenbacher, U.S. Patent 2,955,109 (1960); d. A. D. Phillips, *J. Am. Chem. Soc.*, **75**, 3621 (1953); e. K. Hoffman, S. F. Orochena, S. M. Sax, and G. A. Jeffrey, *J. Am. Chem. Soc.*, **81**, 992 (1959).

[69] H. J. den Hertog, *Rec. Trav. Chim.*, **65**, 129 (1946).

[70] Y. Makisumi, *Chem. Pharm. Bull.*, **9**, 801 (1961).

[71] A. G. Beaman, W. Tautz, R. Duschinsky, and E. Grunberg, *J. Med. Chem.*, **9**, 373 (1966).

[72] H. Kano, Y. Makisumi, S. Takahashi, and M. Ogata, *Chem. Pharm. Bull.*, **7**, 903 (1959).

[73] a. E. T. McBee and D. K. Smith, *J. Am. Chem. Soc.*, **77**, 389 (1955); b. A. N. Nesmeyanov, L. I. Zakharkin, and T. A. Kost, *Chem. Abstr.*, **50**, 7061 (1956).

[74] a. F. Nerdel, J. Buddrus, W. Brodowski, and P. Weyerstahl, *Tetrahedron Letters*, **1966**, 5385; b. J. R. Stephens, R. H. Beutel, and E. Chamberlin, *J. Am. Chem. Soc.*, **64**, 1093 (1942); c. A. W. Wilson and S. A. Harris, *ibid.*, **73**, 2388 (1951).

[75] a. A. R. Surrey and H. F. Hammer, *J. Am. Chem. Soc.*, **68**, 113 (1946); b. C. C. Price, N. J. Leonard, and R. H. Reitsema, *ibid.*, p. 1256; c. A. R. Surrey and H. F. Hammer, *ibid.*, p. 1244; d. R. E. Lutz, G. Ashborn, and R. J. Rowlett, Jr., *ibid.*, p. 1322; e. A. F. Crowther, F. H. S. Curd, D. C. Davey, and G. J. Stacey, *J. Chem. Soc.*, **1949**, 1260.

[76] a. H. Ballweg, *Ann.*, **649**, 114 (1961); b. H. Bredereck, H. Herlinger, and I. Graudums, *Ber.*, **95**, 54 (1962); c. J. Davoll, B. Lythgoe, and A. R. Todd, *J. Chem. Soc.*, **1948**, 1685; d. J. Davoll and B. A. Lowy, *J. Am. Chem. Soc.*, **74**, 1563 (1952).

[77] a. K. W. Rosenmund and F. Zetzsche, *Ber.*, **51**, 578 (1918); b. H. Conroy *J. Am. Chem. Soc.*, **77**, 5960 (1955); c. A. Mooradian and J. B. Cloke, *ibid.*, **68**, 785 (1946); d. I. M. Barfoot, D. L. Hammick, E. D. Morgan, and A. M. Roe, *J. Chem. Soc.*, **1957**, 1533.

[78] a. H. Bredereck, G. Simchen, A. Santos, and H. Wagner, *Angew. Chem. Internat. Edit.*, **5**, 671 (1966); b. Y. Hachihama and T. Shono, *Technol. Repts. Osaka Univ.*, **7**, 177 (1957); through *Chem. Abstr.*, **52**, 10046 (1958); c. B. Teichmann, *J. Prakt. Chem.*, **301**, 51 (1965).

[79] a. C. G. Overberger and J. J. Monagle, *J. Am. Chem. Soc.*, **78**, 4470 (1956); b. N. H. Cromwell and D. J. Cram, *ibid.*, **65**, 301 (1945); c. L. J. Sargent and J. H. Ager, *J. Org. Chem.*, **23**, 1938 (1958); d. Neth. appl. 6,511,073 (1966); through *Chem. Abstr.*, **65**, 775 (1966); e. R. E. Lyle and H. J. Troscianiec, *J. Org. Chem.*, **24**, 333 (1959); f. E. Ghera, R. Szpigielman, and E. Wenkert, *J. Chem. Soc.*, (C), **1966**, 1479.

[80] a. T. Mukai, *Bull. Chem. Soc. Japan*, **32**, 272 (1959); b. E. C. Britton and J. D. Head, U.S. Patent 2,590,813 (1952); c. J. C. Powers, *J. Org. Chem.*, **30**, 2534 (1965); d. G. Chase, L. A. Dolan, and D. Wagner, U.S. Patent 3,213,139 (1965).

[81] a. R. A. Barnes, E. R. Kraft, and L. Gordon, *J. Am. Chem. Soc.*, **71**, 3523 (1949); b. V. Arkley, G. I. Gregory, and T. Walker, *J. Chem. Soc.*, **1963**, 1603; c. Z. Horii, M. Hanaoka, S. Kim, and Y. Tamura, *ibid.*, p. 3940.

[82] a. K. Isogai, *Nippon Kagaku Zasshi*, **81**, 1854 (1960); through *Chem. Abstr.*, **56**, 2420 (1962); b. M. G. Reinecke, *J. Org. Chem.*, **29**, 299 (1964); c. M. G. Reinecke, *ibid.*, **28**, 3574 (1963).

[83] a. N. Sperber, M. Sherlock, D. Papa, and D. Kender, *J. Am. Chem. Soc.*, **81**, 705 (1959); b. D. M. Mulvey, S. G. Cottis, and H. Tieckelmann, *J. Org. Chem.*, **29**, 2903 (1964); c. J. M. Babbitt and D. A. Scola, *ibid.*, **25**, 560 (1960).

[84] a. C. C. Price, H. R. Snyder, O. H Bullitt, Jr., and P. Kovacic, *J. Am. Chem. Soc.*, **69**, 374 (1947); b. C. D. May and P. Sykes, *J. Chem. Soc.*, (C), **1966**, 649.

[85] a. M. J. Reider and R. C. Elderfield, *J. Org. Chem.*, **7**, 286 (1942); b. E. Wenkert and R. D. Haugwitz, *Can. J. Chem.*, **46**, 1160 (1968).

[86] M. von Strandtmann, M. P. Cohen, and J. Shavel, Jr., *J. Med. Chem.*, **8**, 200 (1965).

[87] a. G. Okusa and S. Kamiya, *Chem. Pharm. Bull.*, **16**, 142 (1968); b. M. Ogata, H. Kano, and K. Tori, *ibid.*, **11**, 1527 (1963); c. M. Ogata and H. Kano, *ibid.*, p. 29; d. S. Sako, *ibid.*, **14**, 303 (1966); e. T. Nakagome, *ibid.*, **11**, 721 (1963).

[88] a. von R. Biela, I. Hahnemann, H. Panovsky, and W. Pritzkow, *J. Prakt. Chem.*, **33**, 282 (1966); b. P. Ciatoni, L. Rivolta, and C. Divo, *Chim. Ind.* (Milan), **46**, 875 (1964); c. Belg. Patent 630,941 (1963); through *Chem. Abstr.*, **61**, 15991 (1964); d. G. Ribaldone, C. Brichta, and A. Nenz, Ital. Patent 685,597 (1965); through *Chem. Abstr.*, **66**, 28440 (1967); e. Belg. Patent 626,384 (1963), through *Chem. Abstr.*, **60**, 10568 (1964); f. H. Metzger, *Angew. Chem.*, **75**, 980 (1963).

[89] J. A. Montgomery and C. Temple, Jr., *J. Am. Chem. Soc.*, **82**, 4592 (1960).

[90] R. P. Singh and W. Kortynyk, *J. Med. Chem.*, **8**, 116 (1965).

[91] a. B. Eistert, H. Fink, and A. Müller, *Ber.*, **95**, 2403 (1962); b. H. Kammerer and M. Grossman, *ibid.*, **86**, 1492 (1953); c. H. Kammerer and M. Grossman, German Patent 1,010,528 (1957).

[92] K. Brendel, P. H. Gross, and H. K. Zimmerman, Jr., *Ann.*, **691**, 192 (1966).

[93] a. M. Freifelder, *J. Org. Chem.*, **31**, 3875 (1966); b. T. Naito and O. Nagase, Japan. Patent 831 (1953); through *Chem. Abstr.*, **48**, 2123 (1954).

[94] P. R. Brook and G. R. Ramage, *J. Chem. Soc.*, **1955**, 896.

[95] D. A. Denton, F. J. McQuillin, and P. L. Simpson, *J. Chem. Soc.*, **1964**, 5535.

[96] D. P. Wagner, H. Gurien and A. I. Rachlin, U.S. Patent 3,517,066 (1970).

XXI

HYDROGENOLYSIS OF CARBOXYL-CONTAINING GROUPS

ACIDS AND ESTERS

Hydrogenolysis of carboxylic acids and esters leads to the corresponding alcohol, but conditions for conversion generally are rather drastic, especially for acids.

Reductions of aliphatic acids have been carried out in water at about 150° with ruthenium dioxide or with ruthenium on a carrier, but pressures were extremely high, 500–950 atm [1a, b, c]. Hydrogenations of perfluoroalkanoic acids were carried out in alcohol with 5% ruthenium on carbon under less vigorous conditions, 175° and 330 atm [1d], probably because of esterification of the acid during reaction.

Rhenium oxides have also been employed in the hydrogenolyses of carboxylic acids at 150–200° and 250 atm in various solvents [2a, b, c]. The one advantage of rhenium appears to be in hydrogenation of aryl- and aralkylcarboxylic acids to alcohols without accompanying ring reduction [2b, c].

Copper chromite or copper oxide catalyzed the hydrogenolyses of aliphatic acids, C_{4-12}, at 300° and 250 atm to give high yields of alcohols with the C_{7-12} acids [3a]. Lower acids gave a high percentage of esters from condensation of the alcohol with unreduced acid.

$$CH_3(CH_2)_n COOH \xrightarrow{2H_2} CH_3(CH_2)_n CH_2OH \rightarrow CH_3(CH_2)_n COOCH_2(CH_2)_n CH_3$$

(21.1)

Dicarboxylic acids were hydrogenolyzed to diols at 300° and 270 atm with barium-promoted copper chromite in dioxane solution [3b]. Higher temperature led to monoalcohols and hydrocarbons.

It is less difficult to reduce the ester than the parent carboxylic acid. The difference is seen in the hydrogenation of monoethyl adipate with copper chromite, where the ester was converted to alcohol and the carboxylic acid was unaffected to give a quantitative yield of a mixture of 6-hydroxyhexanoic acid and its lactone [4] (Eq. 21.2).

$$\begin{array}{c}\text{COOH}\\|\\(\text{CH}_2)_4\\|\\\text{COOC}_2\text{H}_5\end{array}\xrightarrow[\substack{\text{CuCr}_2\text{O}_4\,225-250°,\\135\text{ atm init.}\\\text{pressure}}]{2\text{H}_2}\text{HO}(\text{CH}_2)_5\text{COOH}+\begin{array}{c}(\text{CH}_2)_2-\text{C}=\text{O}\\|\qquad\quad|\\(\text{CH}_2)_2-\text{O}\end{array}\quad(21.2)$$

Although an ester is more readily reduced than the parent acid, elevated temperature and pressure conditions are still required. Because of the more or less severe conditions, side reaction often occurs. The alcohol formed may undergo further hydrogenolysis to hydrocarbon; the resultant alcohol may react with starting ester to produce a larger and more difficulty hydrogenolyzable ester as seen in Eq. 21.1 [5a]. In one instance in an attempted hydrogenolysis of ethyl piperidine-2-carboxylate at 200–250° and 230 atm with copper chromite in the absence of solvent the required amount of hydrogen was absorbed, but instead of the desired 2-hydroxymethylpiperidine, high-boiling components were obtained [5b]. One of them resulted from self-condensation of the starting ester and accompanying decarboxylation because of the high temperature (Eq. 21.3). Other components, difficult

(21.3)

to identify, may have been products of self-condensation of 2-hydroxymethylpiperidine or products of reaction of it with starting ester.

N-Alkylation of primary and secondary amino groups by the product alcohol in the hydrogenolysis of amino esters is another complication resulting from the reaction conditions [5a, b], as is N-alkylation when primary alcohols are used as solvent in the hydrogenolysis of such esters. The latter can be avoided by the use of tertiary alcohols in these reactions [5b] or by the employment of suitable nonhydroxylated solvents.

Other complications arising from the reaction conditions include cleavage of 1,3-diols resulting from hydrogenolyses of malonic esters, β-hydroxy and β-keto esters [5a]. Copper chromite, which does not usually cause reduction of benzenoid rings, did so to a certain extent in the hydrogenolysis of ethyl 3-hydroxy-4-isopropylbenzoate and ethyl 3-amino-4-isopropylbenzoate [5c] and in that of ethyl 4-(1- and 2-naphthyl)butyrates which gave the corresponding 5,6,7,8-tetrahydronaphthylbutanols [5d].

The catalyst most commonly used for the hydrogenolyses of esters is barium-promoted copper chromite, about 10–20% by weight [5a]. However, reaction temperature can be lowered appreciably by the use of an equal weight of catalyst. This technique prevented many of the side reactions that occur in the hydrogenolyses of esters [6].

The technique of using more catalyst than compound has been applied in nickel-catalyzed hydrogenolyses of α-amino esters at 50–100° and 150–200 atm [7a] and in other nickel reductions at 330 atm [6, 7b]. There is usually little side reaction except saturation of any aromatic ring present with nickel under these conditions.

Good yields of diols were obtained from diesters, $(CH_2)_{2-17}(COOR)_2$, at 260° and 200 atm with a catalyst obtained from an alloy of aluminum, copper, and zinc treated with sodium hydroxide [8].

In an unusual reaction 3-hydroxymethyl-1,2,3,4-tetrahydroisoquinoline resulted from an attempt to form ethyl decahydroisoquinoline-3-carboxylate by reduction of ethyl 1,2,3,4-tetrahydroisoquinoline-3-carboxylate with 5% rhodium on alumina at 50° and 50 atm [9]. There is no other record of the use of rhodium for hydrogenolysis of an ester. It is of interest that reduction of the acid gave the desired decahydroisoquinoline-3-carboxylic acid.

Selective Reductions

In general, most reducible groups except amides will be hydrogenated in preference to a carboxylic acid or ester. When copper chromite is used, an *N*-benzyl group can be retained since copper chromite is generally ineffective for the hydrogenolysis of that group. Such an example is seen in the reduction of ethyl 2-benzylaminopropionate to 2-benzylamino-1-propanol at 180° and 200 atm initial pressure [10]. It is doubtful whether success could be achieved with nickel catalyst in a similar instance.

Ethylenic Bonds. Zinc chromium oxide, much less active than copper chromite, has been employed for the hydrogenolysis of unsaturated esters to unsaturated alcohols [11a]. However, it does not always lead to unsaturated alcohol [5b]. The presence of 3–13% of alumina in the zinc-chromium catalyst was reported to avoid reduction of the double bond in the hydrogenolysis [11b]. Normally reductions of unsaturated esters with copper chromite are not selective, but oleic acid was converted to oleyl alcohol by partially inactivating it with quinoline-sulfur [11c]. Other selective procedures involve hydrogenation of heavy metal salts of unsaturated acids under high-pressure conditions [11d]. Unsaturated alcohols were also obtained by reduction of unsaturated acids at 300° and 300 atm with 3:1 or 4:1 copper-cadmium catalysts prepared from a mixture of the carbonates [11e].

CARBOXAMIDES

The hydrogenolysis of carboxamides to amines is the most difficult of all hydrogenations to effect. Reaction conditions are more vigorous than is required for hydrogenolysis of acids. Reaction of primary amides generally leads to a mixture of primary and secondary amines during hydrogenation with copper chromite at 250° and 200–300 atm [12a]. The yield of primary amine from hydrogenolysis of an aliphatic primary carboxamide was increased by reduction in the presence of a large excess of ammonia at 350° and 350–400 atm [12b]. Hydrogenolysis with Raney nickel or with Raney cobalt in the presence of ammonia also increased the yield of primary amine, but not to the same degree; reaction temperature was appreciably lower, however.

Dioxane appeared to be the most useful solvent for hydrogenolysis; reduction in alcohol caused alcoholysis of amide to ester and subsequent hydrogenolysis to RCH_2OH instead of to RCH_2NH_2 [12a].

Hydrogenolyses of secondary and tertiary amides give fair-to-good yield of secondary and tertiary amines upon reduction with copper chromite [12a]. In certain instances there is some attendant cleavage. For example, the hydrogenolysis of 1-benzoylpiperidine gave toluene and benzyl alcohol, and that of 1-(2-phenylbutyryl)piperidine gave 32% of 2-phenyl-1-butanol as well as 65% of 1-(2-phenylbutyl)piperidine [12a]. Side reaction may depend on the acyl group since 1-heptanoyl, 1-nonanoyl, and 1-lauroylpiperidines were converted to the corresponding 1-heptyl-, 1-nonyl-, and 1-dodecyl-, piperidines in very high yield with little attending side reaction under the same vigorous conditions.

It is of interest that aliphatic tertiary carboxamides, $C_{11}H_{23}CON(alkyl)_2$, were hydrogenolyzed to tertiary amines in high yield under very mild conditions for copper chromite, 260° and 15–50 atm [13].

The hydrogenolysis of diamides, $(CH_2)_n(CONHR)_2$ yields diamines except when $n = 2$ or 3. In those instances ring closure took place to yield N-substituted 2-pyrrolidones and 2-piperidones [14]. In one reduction when $n = 3$ and the β-carbon was disubstituted, ring closure did not take place; instead a substituted cadaverine was obtained (Eq. 21.4). When $n = 4$

$$RC(=O)CH_2C(CH_3)_2CH_2C(=O)R \xrightarrow{4H_2} R(CH_2)_2C(CH_3)_2(CH_2)_2R$$
R = 1-piperidyl
(21.4)

the product obtained was not the lactam but a 1-substituted hexahydroazepine (Eq. 21.5).

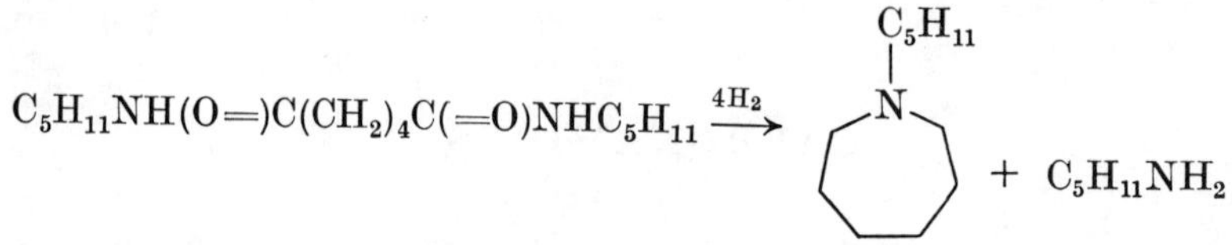

$$C_5H_{11}NH(O=)C(CH_2)_4C(=O)NHC_5H_{11} \xrightarrow{4H_2} \underset{\underset{C_5H_{11}}{|}}{N}\text{-cycloheptane} + C_5H_{11}NH_2$$

(21.5)

LACTAMS

There is a reported reduction of a lactam carried out at low pressure. This is seen in the conversion of α-norlupinone, octahydro-4*H*-quinolizine-4-one, in dilute hydrochloric acid to norlupinan, octahydro-4*H*-quinolizine, with 50% by weight of platinum oxide [15a]. Reduction with less catalyst or in acetic acid was very slow.

For the most part, hydrogenolyses of lactams to saturated heterocycles are carried out under vigorous conditions with copper chromite in nonhydroxylated solvent. 3,3,6,6-Tetramethylpiperazine-2-one and 2,5-dione were reduced to 2,2,5,5-tetramethylpiperazine in high yield in this way [15b]. The results from hydrogenolysis of the dione were dependent on its purity; up to 36% of tetramethylpiperazine-2-one was obtained when the dione contained impurities.

Elevated temperature and pressure were also employed in the reduction of 4-(3,4-dihydro-1*H*-2-pyridon-6-yl)butyramide to quinolizidine [15c] (Eq. 21.6).

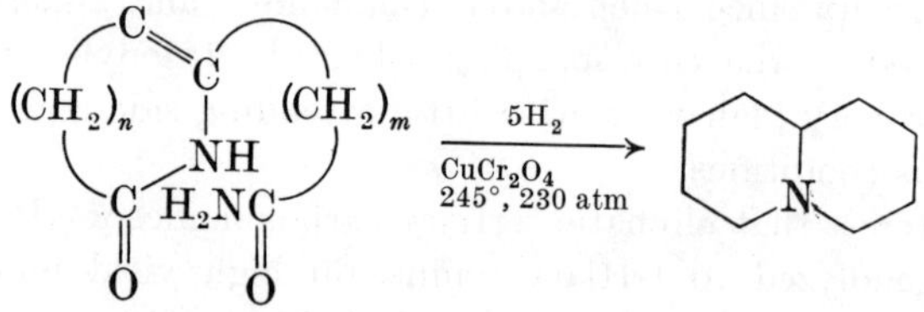

$$\xrightarrow[\text{CuCr}_2\text{O}_4]{5H_2}$$ (21.6)

$n = 2, m = 3$

Quinolizidine, pyrrolizidine, and indolizidine were prepared in a similar manner depending on whether $n = 1$ or 2 and $m = 2$ or 3 [15d]. However, pressure was much lower, 130 atm, probably because as much catalyst as compound was used for hydrogenolysis.

Since vigorous conditions are employed for hydrogenolyses of lactams hydroxylated solvents, except tertiary alcohols, should be avoided to prevent *N*-alkylation. The hydrogenation of 5,5-dimethyl-2-pyrrolidone in qutyl alcohol at 250° and 180 atm gave only 1-butyl-2,2-dimethylpyrrolidine and starting material [15e].

Although esters are generally reduced in preference to amides, it appears likely in the hydrogenolysis of ethyl 4-phenyl-2-piperidone-5-carboxylate over copper chromite in methyl alcohol that the cyclic amide function was reduced first [16a]; N-methylation also took place because of the reaction conditions to yield ethyl 1-methyl-4-phenylpiperidine-3-carboxylate. The low yield was attributed to further conversion of the above ester to carbinol which was obtained in 50.5% yield (Eq. 21.7).

$$\underset{\substack{\text{H}}}{\overset{\substack{C_6H_5}}{\bigcirc}}\!\!\text{COOC}_2\text{H}_5 \quad \xrightarrow[160-200°,\ 200\ \text{atm}]{\text{CuCr}_2\text{O}_4} \quad \underset{\substack{\text{CH}_3}}{\overset{\substack{C_6H_5}}{\bigcirc}}\!\!\text{COOC}_2\text{H}_5 \;+\; \underset{\substack{\text{CH}_3}}{\overset{\substack{C_6H_5}}{\bigcirc}}\!\!\text{CH}_2\text{OH}$$

in CH$_3$OH $\qquad\qquad\qquad\qquad\qquad\qquad\qquad\qquad\qquad$ 50.5% $\qquad$ (21.7)

In contrast, only the ester function in methyl 5-methyl-2-azabicyclo[3,3,1]-nonan-3-one-1-acetate was hydrogenolyzed in the presence of 300% by weight of copper chromite at 210° and 150 atm pressure of hydrogen to yield the corresponding ethanol [16b] (Eq. 21.8).

$$\xrightarrow[\text{CuCr}_2\text{O}_4]{2\text{H}_2} \qquad 65\% \qquad (21.8)$$

CYCLIC IMIDES

Since lactams can be reduced to saturated heterocycles with copper chromite, it is likely that cyclic imides could also be reduced to them. However, the only reported reductions are those of cyclic imides to lactams with catalysts other than copper chromite.

Succinimide resisted reduction with nickel [14] but was readily converted to 2-pyrrolidone by hydrogenation in tetrahydrofuran with Raney cobalt and ammonia at 250° and 90–120 atm, or with ruthenium on carbon in aqueous ammonia at 200° and 80–125 atm [17]. N-Alkyl succinimides and glutarimides were hydrogenolyzed with Raney nickel in dioxane at 200–220° and 200–400 atm to yield 1-substituted 2-pyrrolidones and piperidones [14].

Phthalimide has been reduced to phthalimidine at elevated temperature and pressure with Raney nickel [18a]; prepared *in situ* from phthalic anhydride in xylene and a large amount of ammonia, it underwent conversion without ring saturation to give phthalimidine in 75–80% yields by reduction with Raney cobalt and other cobalt catalysts, or with Raney nickel at 185° and 150–200 atm [18b]. However, when the benzene ring was out of conjugation with the carboximide, as in 1-phenethylsuccinimide and 1-phenethylglutarimide, it was converted to cyclohexyl [14].

ANHYDRIDES

Catalytic hydrogenation of carboxylic acid anhydrides yields a number of products dependent on the catalyst, on reaction conditions, and at times, on the reaction medium. The major product is lactone from reduction of one carbonyl group of the anhydride to methylene. The intermediate hydroxylactone has been isolated, but only in a few specific instances. Conversion of one of the anhydride carbonyl groups to methyl also takes place along with ring opening (Eq. 21.9). Other side reactions have been observed including formation of hydrocarbon.

(21.9)

Aromatic Acid Anhydrides

When aromatic 1,2-carboxylic acid anhydrides are hydrogenated, saturation of the aromatic ring may also occur, as seen in the reduction of phthalic anhydride in acetic acid with platinum black, which gave hexahydrophthalide and 2-methylhexahydrobenzoic acid [19a]; 2-hydroxymethylhexahydrobenzoic acid was also reported as one of the side-reaction products. Phthalic anhydride was converted to phthalide in 90% yield by reduction in dioxane with Raney nickel at essentially 30° and 100 atm [19b]. However, under more severe conditions with nickel on kieselguhr at 150° and 100–170 atm, reduction yielded 2-methylbenzoic acid and 2-methylhexahydrobenzoic acid along with phthalide [19c].

Nonaromatic Anhydrides

Succinic anhydride was reduced to butyrolactone with copper chromite [20a]. Because of the severity of the conditions only 29% yield was obtained, along with butyric acid resulting from further hydrogenolysis. Succinic acid, also found, apparently was a product of hydrolysis, not hydrogenolysis, since the investigators noted the presence of water during distillation of the reduction solution.

Reductions of succinic anhydride with palladium on carbon or on alumina, carried out in ethyl acetate under much milder conditions, 35–100° and 16–75 atm, gave 80–94% yields of butyrolactone [20b]. Hydrogenation with platinum oxide at atmospheric pressure also yielded butyrolactone as the major product [20c] and some butyric acid due possibly to the high catalyst ratio.

A limited number of Diels-Alder adducts (from maleic anhydride) have been hydrogenated at atmospheric pressure with fairly high ratios of platinum oxide [20, 21]. Results appear to be dependent on the reaction medium and to some degree on how long the reaction is allowed to run. The amount of catalyst may also be a factor.

4-Acetoxy-7-oxabicyclo[2,2,1]hept-5-ene-2,3-dicarboxylic acid anhydride when hydrogenated in ethyl acetate over platinum oxide (13% by weight) yielded mainly hemiacylal (hydroxylactone) [20c]; only the less-hindered carbonyl group of the anhydride was attacked. Extended reduction of the adduct led to lactone. Hydrogenation of the adduct in acetic acid yielded lactone and a mixture of hydroxylactone and methyl acid (Eq. 21.10).

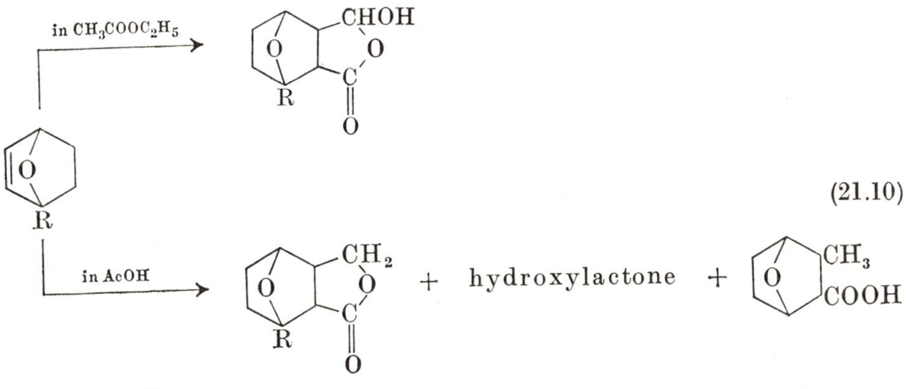

(21.10)

R = OOCCH₃

Hydroxylactones were also obtained from platinum reductions in ethyl acetate, of adducts where R was hydrogen and where R was hydrogen and a methylene group replaced the epoxide oxygen. However, the method did not apply to the hydrogenation of a related saturated anhydride or other saturated alicyclic 1,2-acid anhydrides; lactones, 2-methyl, and 2-hydroxymethyl acids were obtained, but no hydroxylactone [20c].

LACTONES

The hydrogenation of a lactone may lead to diol, monoalcohol, acid, or cyclic ether depending on the catalyst and reaction conditions [22].

The lactone of 4-hydroxyvaleric acid, γ-valerolactone, yielded 78.5% of 1,4-pentanediol and 8% of 1-pentanol after reduction with copper chromite at 250° and 200–300 atm [23a]. Hydrogenation in dioxane at 240–260° and 200 atm initial pressure gave 83% of 1,4-pentanediol; at 270–290° the yield of diol was low [23b]. At this temperature, 2-methyltetrahydrofuran was also obtained. Reduction of a butyrolactone, 4-hydroxy-7-phenylbutyric acid lactone, at 220° and 250 atm yielded equal parts of 7-phenylheptan-1-ol and 7-phenylheptane-1,4-diol [23c]. 3-Phenylphthalide was reduced in alcohol in 0.5 hr over Raney nickel at 80° and 90 atm to 2-benzylbenzoic acid in 75% yield [5a].

δ-Lactones have been hydrogenated with platinum oxide under low-pressure conditions. 1-Oxa-2-oxodecalin and steroids with related structures were reduced to cyclic ethers in 80–90% yields with platinum in acetic acid [24] (Eq. 21.11). Reaction time was drastically reduced by the addition of a

$$\underset{\text{PtO}_2}{\xrightarrow{2\text{H}_2}} \qquad (21.11)$$

R = H or CH$_3$

precise small amount of perchloric acid during hydrogenation with an almost equal weight of catalyst. No measurable uptake occurred in 1 hr during hydrogenations of γ- and ε-lactones under similar conditions.

Reduction of other γ-lactones with platinum in acetic acid has been attempted unsuccessfully [20c]. In one instance, however, a γ-hydroxy-γ-lactone did undergo reduction. The hemiacylal seen in Eq. 21.10 was hydrogenated to give some 1-acetoxy-3-methyl-7-oxabicyclo[2,2,1]heptane-2-carboxylic acid. However, the predominant reaction, hydrogenolysis of the hydroxyl group to yield the γ-unsubstituted lactone, suggests that cleavage proceeded through the hydroxylactone and further substantiates the unreactivity of γ-lactones to hydrogenation with platinum under mild conditions.

Some δ-sugar lactones were reduced to aldoses over platinum under low-pressure conditions [25a, b].

Rhodium on alumina, 150% by weight, was employed in the hydrogenation of ethyl 5,6-benzocoumarin-3-carboxylate at room temperature and 3.6 atm [26]. The purpose of the experiment was to cause saturation of the aromatic ring. The large amount of catalyst necessary for reduction does not indicate any facility of rhodium for hydrogenolyses of lactones. The result, however, does show that a lactone is hydrogenolyzed in preference to an ester group (Eq. 21.12).

$$\text{benzocoumarin-COOC}_2\text{H}_5 \xrightarrow[5\% \text{ Rh on Al}_2\text{O}_3]{10\text{H}_2} \text{decalin-CH}_2\text{CH(CH}_3\text{)COOC}_2\text{H}_5 + 2\text{H}_2\text{O} \quad (21.12)$$

REFERENCES

[1] a. W. F. Gresham, U.S. Patent 2,607,805 (1952); b. T. A. Ford, U.S. Patent 2,607,807 (1952); c. J. E. Carnahan, T. A. Ford, W. F. Gresham, W. E. Grigsby, and G. F. Hager, *J. Am. Chem. Soc.*, **77**, 3766 (1955); d. R. C. Schreyer, U.S. Patent 2,862,977 (1958).

[2] a. H. S. Broadbent, G. C. Campbell, W. J. Bartley, and J. H. Johnson, *J. Org. Chem.*, **24**, 1847 (1959); b. H. S. Broadbent and T. G. Selin, *ibid.*, **28**, 2343 (1963); c. H. S. Broadbent and W. J. Bartley, *ibid.*, p. 2345.

[3] a. A. Guyer, A. Bieler, and K. Jaberg, *Helv. Chim. Acta*, **30**, 39 (1947); b. A. Guyer, A. Bieler, and M. Sommaruga, *ibid.*, **38**, 976 (1955).

[4] D. C. Sayles and E. F. Degering, *J. Am. Chem. Soc.*, **71**, 3161 (1949).

[5] a. H. Adkins, *Organic Reactions*, Vol. VIII, Wiley, New York, 1959, p. 1; b. M. Freifelder, unpublished results; c. F. K. Signaigo and C. Sly, U.S. Patent 2,419,093 (1947); d. H. Adkins and E. E. Burgoyne, *J. Am. Chem. Soc.*, **71**, 3528 (1949).

[6] H. Adkins and H. R. Billica, *J. Am. Chem. Soc.*, **70**, 3121 (1948).

[7] a. H. Adkins and A. A. Pavlic, *J. Am. Chem. Soc.*, **69**, 3039 (1947); b. H. Adkins and H. R. Billica, *ibid.*, **70**, 3118 (1948).

[8] H. Indest, U.S. Patent 2,863,928 (1958).

[9] R. T. Rapala, E. R. Lavignino, E. R. Shepard, and E. Farkas, *J. Am. Chem. Soc.*, **79**, 3770 (1957).

[10] A. Stoll, J. Peyer and A. Hoffman, *Helv. Chim. Acta*, **26**, 929 (1943).

[11] a. J. Sauer and H. Adkins, *J. Am. Chem. Soc.*, **59**, 1 (1937); b. Belg. Patent

540,285 (1956); through *Chem. Abstr.*, **54**, 11991 (1960); c. W. Normann and G. V. Schuckmann, U.S. Patent 2,127,367 (1938); d. A. S. Richardson and J. E. Taylor, U.S. Patents 2,340,687, 2,340,689 and 2,340,691 (1944); e. H. Bertsch, E. Koenig, and H. Reinheckel, *Chem. Abstr.*, **68**, 12371 (1968).

[12] a. B. Wojcik and H. Adkins, *J. Am. Chem. Soc.*, **56**, 2419 (1934); b. A. Guyer, A. Bieler and G. Gerliczy, *Helv. Chim. Acta*, **38**, 1649 (1955).

[13] N. M. LeBard, L. R. Vertnik, R. Fisher, and K. E. McCaleb, U.S. Patent 3,190,922 (1965).

[14] J. H. Paden and H. Adkins, *J. Am. Chem. Soc.*, **58**, 2487 (1936).

[15] a. F. Galinovsky and E. Stern, *Ber.*, **76**, 1034 (1943); b. S. M. McElvain and E. H. Pryde, *J. Am. Chem. Soc.*, **71**, 326 (1949); c. K. Tsuda, S. Saeki, S-I. Imura, S. Okuda, Y. Sato, and H. Mishima, *J. Org. Chem.*, **21**, 1481 (1956); d. K. Tsuda and S. Saeki, *ibid.*, **23**, 91 (1958); e. R. C. Elderfield and H. A. Hageman, *ibid.*, **14**, 605 (1949).

[16] a. J. T. Plati, A. K. Ingberman, and W. Wenner, *J. Org. Chem.*, **22**, 261 (1957); b. M. W. Cronyn and G. H. Riesner, *J. Am. Chem. Soc.*, **75**, 1664 (1953).

[17] H. P. Liao and W. B. Tuemmler, U.S. Patent 3,092,638 (1963).

[18] a. H. Adkins and H. I. Cramer, *J. Am. Chem. Soc.*, **52**, 4349 (1930); b. A. Schütz and O. Stichnoth, Ger. Patent 1,015,435 (1957); through *Chem. Abstr.*, **53**, 22018 (1959).

[19] a. R. Willstatter and D. Jacquet, *Ber.*, **51**, 767 (1918); b. W. Theilacker and H. Kalenda, *Ann.*, **584**, 87 (1953); c. H. Adkins, B. Wojcik, and L. W. Covert, *J. Am. Chem. Soc.*, **55**, 1669 (1933).

[20] a. B. Wojcik and H. Adkins, *J. Am. Chem. Soc.*, **55**, 4939 (1933); b. B. R. Franko-Filipasic, J. M. Kolyer, and R. E. Borks, Jr., U.S. Patent 3,113,138 (1963); c. R. McCrindle, K. H. Overton, and R. A. Raphael, *J. Chem. Soc.*, **1962**, 4798.

[21] R. McCrindle, K. H. Overton, and R. A. Raphael, *Proc. Chem. Soc.*, **1961**, 313.

[22] H. Adkins, *Reactions of Hydrogen*, 3rd ed., University of Wisconsin Press, Madison, Wisc., 1944, p. 78.

[23] a. K. Folkers and H. Adkins, *J. Am. Chem. Soc.*, **54**, 1145 (1932); b. R. V. Christian, Jr., H. D. Brown, and R. M. Hixon, *ibid.*, **69**, 1961 (1947); c. K. Thewalt and W. Rudolph, *Ber.*, **96**, 2256 (1963).

[24] J. T. Edward and J. M. Ferland, *Chem. Ind. (London)*, **1964**, 975.

[25] a. J. W. E. Glattfield and G. W. Schimpff, *J. Am. Chem. Soc.*, **57**, 2204 (1935); b. O. Th. Schmidt and H. Müller, *Ber.*, **76**, 344 (1943).

[26] K. J. Liska and L. Salerni, *J. Org. Chem.*, **25**, 124 (1960).

XXII

HYDROGENOLYSES OF ALCOHOLS, ETHERS, ACETALS, AND KETALS

ALCOHOLS

Hydrogenolysis of an alcohol to hydrocarbon, excluding allylic and benzylic alcohols, which are covered in previous chapters, usually requires elevated temperature and pressure conditions [1a], particularly when reactions are carried out in a neutral medium.

Primary alcohols undergo hydrogenolysis more readily than secondary alcohols in neutral solvent. The reduction of glycerin to 1,2-propanediol is an example [1a, p. 72]. The preparation of 4-propylcyclohexanol from the copper chromite hydrogenation of 4-(3-hydroxypropyl)cyclohexanol at 250° and 340 atm is another example [1b].

In other hydrogenolyses removal of the hydroxyl group in aralkanols becomes more difficult with increasing distance between the ring and the alcohol function, and 1,3-diols generally are converted to monoalcohols [1a].

The catalyst most commonly used for the hydrogenolysis of alcohols at elevated temperature and pressure is copper chromite. Nickel catalysts have also been used, as seen in the removal of the hydroxyl group from 5-hydroxy-3,3,5-trimethyl-2-pyrrolidinones and related hydroxy-2-piperidinones at 140–185° and 80–130 atm [2a]. When the substrate contains an aromatic system, however, ring reduction may also take place in the presence of nickel. Cobalt on alumina was employed for the hydrogenolysis of 2,2,3-trimethyl-1-butanol at 300° and 740–965 atm for 18 hr 2[b]. Despite the vigorous conditions only 36% of hydrocarbon was obtained; much unreacted starting material was recovered.

Palladium and platinum catalysts have some utility in the hydrogenolyses of alcohols in strongly acidic solution. Reduction is dependent on the structure of the compound and in most instances on its ability to undergo dehydration to olefin. Attempts to convert 1-hydroxymethylcyclohexyl-amine in acetic acid containing perchloric acid to 1-methylcyclohexylamine with an equal weight of 20% palladium on carbon at 80–90° and 2 atm gave only recovered starting material [3a]. Reduction with 20% by weight of platinum oxide also failed (Eq. 22.1). However, 2-methyl-6-(2-hydroxy-

$$\text{in AcOH} + \text{HClO}_4$$

$$\underset{NH_2}{\overset{CH_2OH}{\diagdown}}\!\!\!\bigcirc\!\!\!\diagup \quad \xrightarrow{H_2}\!\!\!\!/\!\!\!/\!\!\!/\!\!\!\rightarrow \quad \underset{NH_2}{\overset{CH_3}{\diagdown}}\!\!\!\bigcirc\!\!\!\diagup \qquad (22.1)$$

propyl)pyridine was reduced to 2-methyl-6-propylpyridine in acetic acid containing perchloric acid with 5% palladium on barium sulfate [3b] because it underwent dehydration (Eq. 22.2). Ethyl 3-hydroxy-3-(3-isatylidene)-

$$\text{in AcOH} + \text{HClO}_4$$

$$H_3C\!\diagdown\!\underset{N}{\bigcirc}\!\!\diagup\!CH_2CHOHCH_3 \quad \xrightarrow{-H_2O}$$

$$\left[H_3C\!\diagdown\!\underset{N}{\bigcirc}\!\!\diagup\!CH\!\!=\!\!CHCH_3 \right] \quad \xrightarrow{H_2} \quad H_3C\!\diagdown\!\underset{N}{\bigcirc}\!\!\diagup\!C_3H_7 \qquad (22.2)$$

propionate and ethyl 2-hydroxy-2-(3-isatylidene)acetate were hydrogenated with palladium on carbon to yield ethyl 3-(3-oxindolyl)propionate [3c] and ethyl 2-(3-oxindolyl)acetate [3d], respectively, through a similar dehydration-hydrogenation procedure.

Platinum oxide, 10% by weight, catalyzed the removal of the hydroxyl group of *t*-butyl alcohol in trifluoroacetic acid [4]. In this instance it has been assumed that the alcohol was converted to trifluoroacetate by a rapid carbonium ion reaction prior to hydrogenation (Eq. 22.3).

$$(H_3C)_3COH + H_2 \xrightarrow{CF_3COOH} (H_3C)_3CH + H_2O \qquad (22.3)$$

There is also an example of successful dehydroxylation with a noble metal catalyst in neutral medium, shown in the hydrogenation of 3-hydroxy-4-methyl-2,2-diphenylmorpholine to 4-methyl-2,2-diphenylmorpholine in alcohol with only 15% by weight of 5% palladium on carbon [5].

ETHERS

Linear Ethers

Dialkyl and aryl alkyl ethers are fairly stable toward hydrogenolysis and usually require high temperature and pressure for cleavage with nickel or with copper chromite. In many instances when aryl alkyl ethers were hydrogenated over nickel at 150–200° and 150–250 atm, ring reduction was the preferred reaction [6]. Similar results were obtained from the hydrogenation of diphenyl ethers under the same conditions. Cleavage, when it occurred, yielded cyclohexanols and cyclohexanes. Alkoxy groups attached to saturated cyclic ketones have been removed by hydrogenation with nickel at 120–150° and 100 atm [2a] and at 90° and 50 atm [7a].

When aryl alkyl ethers or diaryl ethers are hydrogenated in the presence of noble metal catalysts, the primary reaction is saturation of the ring.

Cyclic Oxides

1,2-Epoxides usually undergo hydrogenolyses to alcohols with palladium and platinum catalysts under mild conditions. Reductions with nickel as a rule require elevated temperature and high pressure. As the epoxide ring increases in size cleavage becomes more difficult. The difference in the ease of reducibility between a 1,2- and a 1,4-epoxide may be seen in Eq. 22.4 [7b], where the 1,2-epoxide underwent cleavage and the 1,4-epoxide ring remained intact.

$$\begin{array}{c}\text{structure} \end{array} \xrightarrow[\substack{\text{NiR} \\ 60-65°,\,100\text{ atm}}]{\text{H}_2} \begin{array}{c}\text{structure}\end{array} \qquad (22.4)$$

Ethylene Oxides. Rupture of the ring can take place on either side of the oxygen atom. The direction appears to be dependent on substitution, on the reaction medium, and perhaps on the catalyst.

When styrene oxide, C₆H₅CH—CH₂ (with O bridging), was hydrogenated under acidic, basic, or neutral conditions only 2-phenylethanol, C₆H₅(CH₂)₂OH, was obtained [8a]. Reduction of substrates which contained an ArC—C- structure also yielded 1-aryl-2-hydroxy compounds. 1-Methyl-4-phenyl-3,4-epoxypiperidine was hydrogenated in methyl alcohol containing perchloric acid with platinum oxide to form 1-methyl-4-phenyl-3-piperidinol [8b] (Eq. 22.5).

in CH₃OH + HClO₄

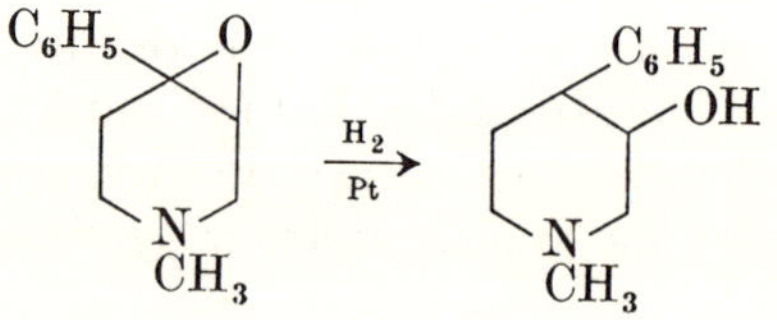

$$\text{[epoxypiperidine]} \xrightarrow[\text{Pt}]{\text{H}_2} \text{[piperidinol]} \qquad (22.5)$$

Sodium glycidate, C₆H₅CH—CHCOONa (with O bridging), was reduced in water with 5% by weight of 2% palladium on carbon to give 3-phenyllactic acid,

C₆H₅CH₂CHOHCOOH,

in high yield [8c], and 1-phenyl-1,2-epoxy-3-butene yielded 1-phenyl-2-butanol after hydrogenolysis with palladium on carbon in ether [8d].

When *cis*- and *trans*-2,3-dimethyl-2,3-diphenylethylene oxides were hydrogenated in alcohol, the products obtained were dependent on the catalyst. Hydrogenolysis of the epoxides gave 2,3-diphenyl-2-butanols, with retention of configuration over Raney nickel and with inversion over palladium on carbon [8e] (Eq. 22.6). The addition of alkali to nickel reductions caused inversion but did not produce any change in the palladium-catalyzed hydrogenolyses.

There appear to be too few examples of hydrogenolyses of epoxides, XCH—CHY (with O bridging) where X is H or a nonaromatic substituent and Y is not aryl, to generalize about ring opening. Hydrogenation of a terminal epoxide CH₂—CH(CH₂)ₙCOOC₂H₅ (with O bridging), in neutral solvent over palladium black yielded only the secondary alcohol [9]. In contrast, reduction of 1,2-epoxydecane with nickel at 150° and 60–75 atm in neutral sodium gave primary alcohol while reaction in the presence of sodium hydroxide gave secondary alcohol

[8a]. Reduction in the presence of phosphoric acid yielded a 1:1 mixture of primary and secondary alcohols.

The products of hydrogenolysis of 1,2-aliphatic disubstituted 1,2-epoxides were in some cases dependent on X and Y. When *cis*-6,7-epoxyoctadecanoic acid was hydrogenated in alcohol over palladium on carbon, equal amounts

$$
\begin{array}{c}
\text{C}_6\text{H}_5 \\
| \\
\text{HOCCH}_3 \\
| \\
\text{HCCH}_3 \\
| \\
\text{C}_6\text{H}_5 \\
\textit{erythro}
\end{array}
\qquad \text{(22.6)}
$$

(reaction scheme: *cis* epoxide → via NiR → erythro diol; *cis* epoxide → via Pd on C → threo diol; *trans* epoxide → via Pd on C → erythro; *trans* epoxide → via NiR → threo)

of 6- and 7-hydroxyoctadecanoic acids were obtained [10a]. It is of interest that low-pressure reduction with Raney nickel, with platinum oxide, and with palladium on calcium carbonate failed. *cis*-9,10-Epoxyoctadecanol and the acetate also gave a 1:1 mixture of 9- and 10-hydroxy compounds after hydrogenation with palladium on carbon [10b]. In contrast, the hydrogenolysis of methyl *cis*-9,10-epoxyoctadecanoate gave only the 10-hydroxy compound [10c]. This was attributed to the effect of the carboxyl group.

In a study of a series of cyclohexene and related cyclic oxides, hydrogenolysis over platinum catalyst was found to be acid-catalyzed [11] (the addition of strong acid to reductions greatly increased the reaction rate). Steroidal 1,2-epoxides as a rule gave axial alcohols after reduction in acidic medium; usually little or no hydrogen was absorbed during reaction in alcohol or dioxane. 1-Substituted-1,2-epoxides, as illustrated by the hydrogenation of 1-methylcyclohexene oxide, gave only 2-substituted-1-ols on reduction in acidic medium; no 1-methylcyclohexanol was obtained.

The following hydrogenolysis of a diepoxy compound is of interest. When *cis*-1,2,4,5-diepoxycyclohexane was hydrogenated in ethyl acetate with 40%

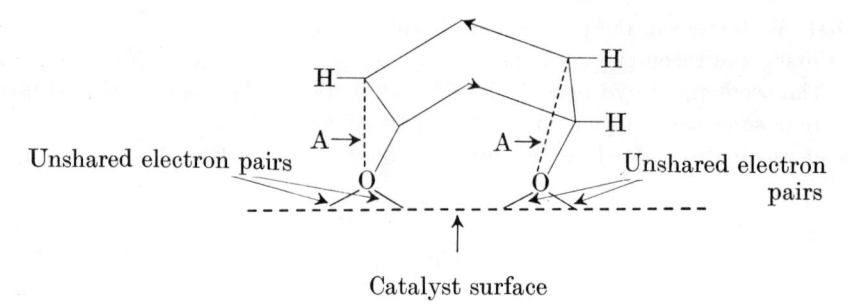

514 Hydrogenolyses of Alcohols, Ethers, Acetals, and Ketals

Fig. 22.1

by weight of 5% palladium on carbon (prereduced before use) at 95 atm for 36 hr, a mixture of about 63% of *cis*-cyclohexane-1,3-diol and 37% of *cis*-cyclohexane-1,4-diol was obtained [12]. A drawing of a framework molecular orbital model appears to explain why the 1,3-diol was the major product, Fig. 22.1. The model suggests that the favored position of the molecule is one in which it is bound to the surface of the catalyst by the unshared electron pairs of the oxygen atoms. As a result the carbon to oxygen bonds (A, A) are in the best position to undergo scission. The *trans* isomer did not undergo ring opening at 80° and 95 atm initial pressure. Hydrogenolysis of the isomers in acidic medium was not studied.

Some interesting chemistry is seen in the hydrogenation of some 2,2-dimethyl and 2,2-diphenyl-3-(3-oxindolyl)ethylene oxides with palladium on carbon in alcohol [13]. The epoxides yielded the isomeric alcohols but each was unstable. One alcohol underwent hydrogenolyses to yield a 3-substituted oxindole; the other underwent dealdolization to give the oxindole and ketone (Eq. 22.7).

SELECTIVE REDUCTIONS. In most instances when a substrate contains a readily reducible group and a 1,2-epoxide ring, hydrogenation should take place without ring opening, although this may be dependent to a great degree on the reaction medium.

Presence of Olefinic Bonds. Although reduction of a carbon–carbon double bond should be expected over opening of a 1,2-epoxide ring, the presence of an equivalent of silver nitrate was found to offer complete protection to the ethenoid linkage in hydrogenations of partially epoxidized unsaturated esters and vegetable oils with palladium on carbon in alcohol, probably by π-complex formation, according to the authors [14].

Presence of Carbonyl Groups. All of the described work is on the reduction of α-keto-1,2-epoxides. Among steroids, when such compounds are hydrogenated in neutral media the carbonyl group is reduced without affecting the epoxide ring [11]. Except for one reference on a hydrogenation in acetic

$$R = H \text{ or } CH_3, \ X \text{ and } Y = CH_3 \text{ or } C_6H_5 \tag{22.7}$$

acid with platinum oxide, in which the carbonyl group was preferentially reduced [15a], and one in which 4-phenyl-3,4-epoxy-2-butanone was selectively hydrogenated to 4-phenyl-3,4-epoxy-2-butanol in alcohol or ether with a platinized nickel catalyst [15b], most reports indicate that the epoxide ring is attacked to give a 1,2-ketol, although reduction may continue to yield 1,2-diol. When benzalacetophenone oxide (2,3-epoxy-1,3-diphenylpropan-1-one) was hydrogenated over platinum oxide, W-2 Raney nickel or 5% palladium on carbon in alcohol, ether, ethyl acetate, or acetic acid, ring opening was the preferred reaction, although continued reaction did yield 1,3-diphenylpropane-1,2-diol [16a]. Hydrogen uptake did stop upon hydrogenation with platinum oxide in ether to yield 2-hydroxy-1,3-diphenylpropan-1-one. When reduction of the same ketoepoxide was carried out in alcohol with palladium on carbon containing a trace of acid or a trace of base the same 1,2-ketol was obtained [16b]. Other reductions of α-ketoepoxides with Raney nickel [16c], with palladium [16d], and with palladized nickel [15b, 16e] were also selective, yielding 1,2-ketols.

Presence of Halogen. There are very few available examples of reductions involving substrates containing an epoxide ring and a reactive halogen. In one series, those of the hydrogenations of 1-oxa-3,4,6,7-tetrahalo-5-oxo-[5,2,0]bicycloocta-3,6-dienes with platinum oxide in neutral solvent, 2,3,5,6-tetrahalo-4-hydroxybenzyl alcohols were obtained [17] (Eq. 22.8). The

$$O=\underset{X\ X}{\overset{X\ X}{\bigoplus}}\underset{CH_2}{\overset{O}{\diagdown}} \xrightarrow[\text{prereduced Pt}]{H_2} HO\underset{X\ X}{\overset{X\ X}{\bigoplus}}CH_2OH \qquad (22.8)$$

4X's = tetrachloro, tetrabromo, or tetraiodo

results are not entirely unexpected. Platinum catalysts generally are not efficient for dehalogenations. In addition, aryl halides are not very readily removed in the absence of base (see Chapter XX).

POLYMETHYLENE OXIDES. Ring opening of tetrahydrofurans when carried out with copper chromite usually require high temperature and pressure. When 2-substituted tetrahydrofurans undergo hydrogenolysis, cleavage takes place between the oxygen and the carbon atom in 2-position. Examples are seen in the copper chromite reduction of 2-hydroxymethyltetrahydrofuran at 225° and 200 atm initial pressure [18a] and that of 5-alkyl- and 5-phenyl-2-hydroxytetrahydrofuran with Raney nickel [18b] to yield 1,5-pentanediol and various 1,4-alkanediols, respectively.

The hydrogenolysis of benzo-2,3-dihydrofurans is generally carried out under mild conditions with palladium or platinum catalysts to yield phenols.

Acetals and Ketals 517

The reductions of 5-hydroxy-2R- and 3-alkyl or phenyl-2,3-dihydrobenzofurans, where R is a saturated *N*-heterocyclic system, with palladium or platinum on carbon or on alumina to give the corresponding 2-substituted hydroquinones [19a] are a good illustration. The formation of 2-[2-(1-piperidyl)cyclopentyl]hydroquinone from the reduction of 5-hydroxy-2-(1-piperidyl)-2,3-trimethylene-2,3-dihydrobenzofuran in acetic acid is another example [19b] (Eq. 22.9).

Cleavage of a benzo-2,5-dihydrofuran also took place with relative ease. 2,3-Trimethylene-1,4-diphenyl-1,4-epoxy-1,4-dihydronaphthalene was hydrogenated in ethyl acetate with palladium on carbon [20a]. The alcohol was

not isolated per se but was dehydrated to yield 2,3-trimethylene-1,4-diphenylnaphthalene (Eq. 22.10).

ACETALS AND KETALS

The catalytic hydrogenolysis of acetals and ketals usually requires elevated temperature and pressure conditions [1a, p. 75]. However, cleavage of some ketals was found to take place under mild conditions over rhodium on alumina when some acid was present [21]. Reaction is assumed to proceed through formation of an enol ether by acid catalysis and subsequent reduction to a saturated ether (Eq. 22.11). Reductions were unsuccessful with

$$\underset{H_3C}{\overset{H_3C}{\diagdown}}C\underset{OR}{\overset{OR}{\diagup}} \underset{-ROH}{\overset{H^+}{\rightleftarrows}} \underset{H_2C}{\overset{H_3C}{\diagdown}}C{-}OR \xrightarrow[2\text{-}4\ \text{atm}]{H_2,\ Rh\ \text{on}\ Al_2O_3} (H_3C)_2CHOR \quad (22.11)$$

platinum and ruthenium under similar conditions; palladium was about half as active as rhodium. Hydrogenations in neutral or alkaline solution with rhodium failed even when the temperature was raised to 100°.

REFERENCES

[1] a. H. Adkins, *Reactions of Hydrogen*, 3rd Ed., University of Wisconsin Press, Madison, 1944, p. 69; b. L. M. Cooke, J. L. McCarthy, and H. Hibbert, *J. Am. Chem. Soc.*, **60**, 3052 (1941).

[2] a. N. M. Bortnick and M. F. Fegley, U.S. Patent 3,065,237 (1962); b. T. A. Ford, H. W. Jacobson, and F. C. McGrew, *J. Am. Chem. Soc.*, **70**, 3793 (1948).

[3] a. M. Freifelder, unpublished results; b. W. H. Tallent and E. C. Horning, *J. Am. Chem. Soc.*, **78**, 4467 (1956); c. P. L. Julian and H. C. Printy, *ibid.*, **75**, 5301 (1953). d. P. L. Julian, H. C. Printy, R. Ketcham, and R. Doone, *ibid.*, p. 5305.

[4] P. E. Peterson and C. Casey, *J. Org. Chem.*, **29**, 2325 (1964).

[5] A. L. Morrison, R. F. Long, and M. Königstein, *J. Chem. Soc.*, **1951**, 952.

[6] E. M. Van Duzee and H. Adkins, *J. Am. Chem. Soc.*, **57**, 147 (1935).

[7] a. C. H. Vasey, U.S. Patent 2,676,176 (1954); b. E. Vischer and T. Reichstein, *Helv. Chim. Acta*, **27**, 1332 (1944).

[8] a. M. S. Newman, G. Underwood, and M. Renoll, *J. Am. Chem. Soc.*, **71**, 3362 (1949); b. R. E. Lyle and W. E. Krueger, *J. Org. Chem.*, **30**, 394 (1965); c. K. Munakata and I. Moriyama, Jap. Patent 12, 785 (1965); through *Chem. Abstr.*, **64**, 11133 (1966); d. O. Grummitt and R. M. Vance, *J. Am. Chem. Soc.*, **72**, 2669 (1950); e. S. Mitsui and Y. Nagahisa, *Chem. Ind.* (London), **1965**, 1975.

[9] G. V. Pigulevsky and Z. Ya. Rubashko, *J. Gen. Chem. USSR*, **25**, 2191 (1955) (Eng. trans.).

[10] a. S. P. Fore and W. G. Bickford, *J. Org. Chem.*, **26**, 2104 (1961); b. S. P. Fore and W. G. Bickford, *ibid.*, **24**, 620 (1959); c. C. H. Mack and W. G. Bickford, *ibid.*, **18**, 686 (1963).

[11] F. J. McQuillin and W. O. Ord, *J. Chem. Soc.*, **1959**, 3169.

[12] T. W. Craig, G. R. Harvey and G. A. Berchtold, *J. Org. Chem.*, **32**, 3743 (1967).

[13] W. C. Anthony, *J. Org. Chem.*, **31**, 77 (1966).

[14] P. Subbarao, G. V. Rao, and K. T. Achaya, *Tetrahedron Letters*, **1966**, 379.

[15] a. J. Reese, *Ber.*, **75**, 384 (1942); b. T. I. Temnikova and V. A. Kropachev, *Zhur. Obshchei Khim.*, **19**, 2069 (1949); through *Chem. Abstr.*, **44**, 3941 (1950).

[16] a. W. Herz, *J. Am. Chem. Soc.*, **74**, 2928 (1952); b. S. Mitsui, Y. Senda, T. Shimodaira, and H. Ichikawa, *Bull. Chem. Soc.* (Japan), **38**, 1897 (1965); c. E. D. Bergmann, *J. Appl. Chem.*, **1**, 380 (1951); d. O. Dann and H. Hofmann, *Ber.*, **96**, 320 (1963); e. T. I. Temnikova and V. A. Kropachev, *Zhur. Obshchei Khim.*, **21**, 501 (1951); through *Chem. Abstr.*, **45**, 8447 (1951).

[17] B. Eistert, H. Fink, and A. Müller, *Ber.*, **95,** 2403 (1962).
[18] a. C. D. Nenitzescu and I. Necsiou, *J. Am. Chem. Soc.*, **72,** 3483 (1950); b. C. Glacet, *Ann. Chim.* (Paris), **1947,** 317.
[19] a. L. L. Skaletzky, U.S. Patent 3,291,798 (1966); b. Neth. Appl. 6,415,270 (1965); through *Chem. Abstr.*, **64,** 3511 (1966).
[20] G. Wittig, J. Weinlich, and E. R. Wilson, *Ber.*, **98,** 458 (1965).
[21] W. L. Howard and J. H. Brown, Jr., *J. Org. Chem.*, **26,** 1026 (1961).

XXIII

MISCELLANEOUS HYDROGENOLYSES

AZIDES

Azides are very readily hydrogenated to amines under mild conditions in neutral or acidic solvent with any of the commonly used catalysts. Reactions are fairly rapid [1] but they are often allowed to run for a longer period than necessary because no drop in pressure is observed—as much nitrogen is released as hydrogen is absorbed:

$$RN_3 \xrightarrow{H_2} RNH_2 + N_2$$

Excess hydrogen should be present. Some experiments have shown when a limited amount of hydrogen is present there appears to be a blanketing effect on the reaction [1] caused by the released nitrogen competing with hydrogen for adsorption on the catalyst

Raney nickel [2a, b], palladium [3a, b], and platinum catalysts [4a, b, c] have been employed with good effect for the reduction of azides. Few conclusions can be drawn about hydrogenations of azides in the presence of other reducible functions because of the scarcity of examples.

In the palladium on carbon reduction of 2-azidocyclooctanone oxime in moist alcohol containing hydrochloric acid the product obtained was assumed to be 2-aminocyclooctanone [5a]. Hydrolysis of oxime to ketone also took place during the reduction of 2-azido-5,9-cyclododecadienone oxime under similar conditions. Saturation of the double bonds also took place to yield 2-aminocyclododecanone (Eq. 23.1).

(23.1)

In another example the C=N bond remained intact during the reduction of 2-(2-azidobenzylideneamino)indazole with an equal amount of 5% palladium on carbon to yield 2-(2-aminobenzylideneamino)indazole [5b] (Eq. 23.2). The generality of these reactions is a matter for further study.

$$\text{[indazole]}-N-N=CH-\text{[C}_6H_4\text{-}N_3\text{]} \xrightarrow{\text{Pd on C}} \text{[indazole]}-N-N=CH-\text{[C}_6H_4\text{-}NH_2\text{]}$$

(23.2)

HYDRAZINES AND HYDRAZONES

The catalytic hydrogenolysis of hydrazines and hydrazones is often a means of preparing difficultly obtainable amines. The method finds considerable use in the sugar series where ozazones and phenylhydrazones are converted to amines.

A considerable number of amines was prepared from phenylhydrazones by hydrogenation with Raney nickel at 3 atm for 17–18 hr [6a]. 1-Amino-1-deoxy-2,4-O-ethylidene-D-erythritol was prepared in 46% yield from the hydrogenation of 2,4-O-ethylidene-D-erythrose phenylhydrazone with 10% by weight of palladium on carbon and in 74% yield with an undisclosed amount of Raney nickel [6b].

Raney nickel may be the catalyst of choice for the hydrogenolysis of phenylhydrazones to amines. In another series hydrogenation of the *syn* form of a phenylhydrazone with Raney nickel gave amine and some *anti*-phenylhydrazone while reduction with palladium or platinum only converted the *syn* form to *anti* [7]. Reduction of the *anti* form required high-pressure conditions, 125° and 80 atm pressure, with nickel.

Hydrogenolyses of cyclic hydrazones such as 2-pyrazolines have been carried out with Raney nickel under high-pressure conditions [8a, b]. These conditions may be unnecessary as evidenced by the rapid reduction of 4-amyl-5-hexyl-1-methyl-2-pyrazoline with Raney nickel in alcohol at room temperature and 2–3 atm pressure which gave 75% yield of N-methyl 2-amyl-1-hexylpropylene-1,3-diamine [8c].

The hydrogenolysis of another cyclic hydrazone, a tetrahydropyridazoquinoline, was carried out with platinum oxide to give very good yield of 2-(3-aminopropyl)-1,2,3,4-tetrahydroquinoline [8d] but a large amount of catalyst 15% by weight was employed (Eq. 23.3).

The hydrogenolysis of linear unacylated hydrazine can be carried out under mild conditions in neutral or acidic solution. Phenylhydrazines and related aromatic-type hydrazines are especially easy to reduce to amines. In fact, it is often difficult to retain the hydrazino group in such compounds which also contain other reducible groups [1].

$$\text{(structure)} \xrightarrow{2H_2}_{PtO_2} \text{(structure)}(CH_2)_3NH_2 \quad (23.3)$$

Raney nickel appears to be useful for hydrogenolysis of hydrazines, ArNHNHR where R is hydrogen, alkyl, or aryl. It was employed for the reduction of 4-ethoxy-1-hydrazinoisoquinoline to 1-amino-4-ethoxyisoquinoline at atmospheric pressure [9a] and for the preparation of 4,5-diaminopyridazine from 4-amino-5-hydrazinopyridazine [9b]. It was also the catalyst in the preparation of 2-aminoethanol, 3-amino-1-methylpiperidine, and 2-methylamino-1-methylpyrrolidine from the corresponding hydrazines [10a] and in the hydrogenolysis of 2,2'-biindazole to indazole at 40–50° [5b].

Prereduced platinum catalyzed the hydrogenolysis of 2-hydrazino-2-phenylbutane in acetic acid to 2-phenyl-2-butylamine [10b].

1,1-Disubstituted hydrazines such as N-amino heterocycles are also particularly easy to reduce to amines. For example, 1-amino-2,5-dimethylpyrrolidine and 1-amino-2,6-dimethylpiperidine were hydrogenated in methyl alcohol containing acetic acid with 5% rhodium on alumina and with platinum oxide to yield the dimethylpyrrolidine and the dimethylpiperidine [10c]. Reductions among 1-aminopiperazines with palladium on carbon or with Raney nickel also resulted in deamination, yielding piperazines [1]. Indeed, it is seldom possible to catalytically hydrogenate related N-nitroso amines to the corresponding hydrazines. The usual result is the parent saturated N-heterocycle.

In some instances cyclic hydrazines may be difficult to cleave. Hydrogenolyses of 2-pyrazolines were carried out at elevated temperature and pressure because it was reasoned that the N—N bond in the intermediate pyrazolidine would be difficult to cleave [8a]. In contrast the bond in a related compound, a substituted 1,5-diazabicyclo[3,3,0]octane was readily broken by reduction in water with Raney nickel at atmospheric pressure to yield 1,5-diazacyclooctane [11] (Eq. 23.4).

$$\text{(structure)} \xrightarrow{H_2} \text{(structure)} \quad (23.4)$$

R = H or CH_3

The linkage in a 6-membered ring was ruptured but not with the same ease. Phthalazine was reduced in two stages to o-xylene-α,α'-diamine through 1,2,3,4-tetrahydrophthalazine with Raney nickel or palladium on carbon for the first stage and with 30% by weight of Raney nickel at 50° and 3.5 atm for the second stage [12].

Hydrogenolysis of acylhydrazides usually requires more vigorous reaction conditions than those used for the reduction of most hydrazines. Acylation imparts considerable stability to the hydrazine linkage; acylation is used as a means of reducing other groups in the presence of the nitrogen to nitrogen bond [1, 13]. An example of the conditions used for the hydrogenolysis of acylhydrazines is seen in the palladium on carbon hydrogenation of 1-(1,2,3,4-tetrahydroquinoline-6-carboxylic)hydrazide to 1,2,3,4-tetrahydroquinoline-6-carboxamide at 80° and 50 atm pressure [14].

In most instances the method of Ainsworth [15a], refluxing the acylhydrazide in alcohols with about 10 parts of Raney nickel, is very useful for obtaining the corresponding amide. Examples are seen in the conversion of *trans*-2-cyclohexyloxycyclopropanecarboxhydrazide to the carboxamide [15b] and in that of diethyl 4,5,6,7-tetrahydroindazol-3-one-5,5-dicarboxylate, which can be envisioned as an acylated cyclic hydrazide, to 4,4-dicarbethoxycyclohexylamine-2-carboxamide [15c] in a hydrogenolysis which was found to proceed stereospecifically [15d].

There is a hydrogenolysis of a related cyclic hydrazine which was carried out by conventional means, the reduction of 4-amino-3-pyrazolidinone, but the amount of Raney nickel and pressure conditions were not revealed [16].

Selective Reductions

In most instances selective reduction among hydrazines and hydrazones containing other reducible groups is aimed at retaining an intact nitrogen to nitrogen bond. There are a few examples where preferential hydrogenolysis of phenylhydrazine-type compounds took place without affecting a carbonyl group as in the nickel reduction of 2-hydrazino-7-methyl- and 6,7-dimethyl-2,4,6-cycloheptatrienone to the 2-aminocycloheptatrienones [17a] and without removing an aryl type chloride in the hydrogenations of chlorohydrazinopyridazines with Raney nickel [9b, 17b, c]. Reductions of the latter type may not be completely selective with palladium or platinum because of the affect of the resulting amino group on dehalogenation (Chapter XX).

PEROXIDES AND HYDROPEROXIDES

Most hydrogenolyses of peroxides and hydroperoxide-containing compounds have been carried out with palladium or platinum catalysts. At times an inactivated catalyst may be necessary to prevent further hydrogenolysis and/or saturation of the double bond when reduction leads to an allylic alcohol. The use of a lead-poisoned palladium catalyst proved advantageous in the hydrogenation of a steroidal peroxide where the resulting allylic system remained intact [18a, b]. Nickel and platinum reductions of styrene peroxide were inhibited by the presence of organic base [18c]. It did not prevent reduction of the double bond but it did prevent loss of the benzylic hydroxyl group to give 1-phenylethyleneglycol.

NONAROMATIC CARBOCYCLIC RINGS

Cyclopropanes

The cyclopropane ring is reasonably stable to catalytic hydrogenolysis and does not undergo cleavage too readily unless it is activated by conjugation or when it is a part of a highly strained system.

In general hydrogenolysis of cyclopropyl compounds with nickel requires elevated temperature and pressure. Ring opening has been effected under mild conditions with palladium and platinum catalysts but the ease of hydrogenation is dependent on the structure of the substrate, on substituents on the ring, and on the reaction medium, a point not too well investigated.

In most instances, however, the chemist can carry out hydrogenations of other reducible functions in the presence of a cyclopropane ring without causing rupture of the ring.

Reduction of relatively simple unconjugated cyclopropanes usually results in cleavage between the least substituted bonds. In reductions of 1,1-dialkyl and other *gem*-disubstituted unconjugated cyclopropanes, cleavage takes place at the bond opposite the quaternary carbon atom [19a]. This finding is well substantiated.

The difficulty in the hydrogenolysis of 1,1-diethylcyclopropane with palladium or platinum catalyst at low pressure is surprising in view of the number of successful ring openings in cyclopropanes containing a quaternary carbon atom. 1,1-Diethylcyclopropane was finally cleaved by reduction with nickel on kieselguhr in the absence of solvent, at 180° and 135 atm, to give 3,3-dimethylpentane [19a].

In contrast, spiropentane was reduced with comparative ease with platinum oxide at 2–3 atm to give 94% of 1,1-dimethylcyclopropane, some neopentane, and a small amount of isopentane [19b] (Eq. 23.5).

$$\text{spiropentane} \xrightarrow{\text{PtO}_2} \underset{94\%}{\text{1,1-dimethylcyclopropane}} + \underset{4.1\%}{(CH_3)_4C} \qquad (23.5)$$

Other examples of hydrogenolysis among cyclopropanes containing a quaternary carbon atom are seen in the reductions of 2-cyclopropanoadamantane to 2,2-dimethyladamantane and of 1-(1- and 2-adamantyl)-1-methylcyclopropane to 1- and 2-t-butyladamantane with platinum oxide in acetic acid at 50° and 3 atm [19c].

It would be of interest to attempt hydrogenolysis of 1,1-diethyl- or other 1,1-dialkylcyclopropanes with palladium or platinum catalyst in the presence of acid to determine whether they are still resistant to hydrogenolysis under mild conditions.

Conjugation by means of a carbonyl group [20a], a benzenoid [20b, c] or a furanoid system [21], or a vinyl group [22] increases the facility of hydrogenolysis. An alkyl carboxylate, which does not impart significant conjugation between it and the ring, has no effect on ring cleavage [22].

Cleavage caused by the promoting groups occurs at the bond adjacent to it when the ring is unsubstituted [20a, b, 21]. When there is also a substituent or substituents in 2-position, the 1,2-bond is cleaved [20c, 21, 22]. When 2- and 3-positions are occupied, cleavage may be complex, difficult, or may not take place.

Palladium catalyst appears to give the best results in the promoted cleavages. When it was employed for the hydrogenation of methyl cyclopropyl ketone absorption of hydrogen stopped at one molar equivalent to give 2-pentanone [20a]. Reduction with platinum gave 2-pentanol and some methyl cyclopropyl carbinol.

In examples of palladium reductions of aryl cyclopropanes, 1,2-diphenylcyclopropane yielded 1,3-diphenylpropane [20c] but 1,1-diphenylcyclopropane did not react (cf. the attempted hydrogenolysis of 1,1-diethylcyclopropane [19a]). 2-Cyclopropylfuran yielded 2-propylfuran, and 2-(2-methylcyclopropyl)furan gave 2-n-butylfuran [21]. The hydrogenolysis of 2-(2-cyclopropylcyclopropyl)furan to 2-(6-hexylfuran) is of interest (Eq. 23.6) because of the manner of cleavage. It is likely that the suggested

(23.6)

stepwise reduction is correct. If the middle ring was reduced first the loss of conjugation should make the terminal cyclopropane ring difficult to reduce. Cis- or trans-2-phenylcyclopropanecarboxylic acid underwent cleavage and saturation of the benzene ring during reduction with platinum oxide in alcohol at 60°; reaction gave 4-cyclohexylbutyric acid [23]. The hydrogenation of 2-cyclohexylcyclopropanecarboxylic acid under similar conditions is of interest because despite the absence of activating groups it too yielded 4-cyclohexylbutyric acid.

When vinylcyclopropanes are hydrogenated, cleavage if it occurs does not proceed through the saturated alkylcyclopropane. This is illustrated by a comparison of the reduction of ethyl 2-vinylcyclopropane-1,1-dicarboxylate with platinum oxide in methyl alcohol to ethyl n-butylmalonate which took place with ease while hydrogenolysis of the corresponding 2-ethylcyclopropane compound would not take place [22] (Eq. 23.7). In another

$$CH_2{=}CH{-}\underset{COOC_2H_5}{\overset{COOC_2H_5}{\triangle}} \xrightarrow[PtO_2]{2H_2} C_4H_9CH\underset{COOC_2H_5}{\overset{COOC_2H_5}{\diagdown}}$$

$$CH_3CH_2{-}\underset{COOC_2H_5}{\overset{COOC_2H_5}{\triangle}} \quad (23.7)$$

example, isopropenylcyclopropane could be converted to 2-methylpentane by reduction with palladium while isopropylcyclopropane would not undergo hydrogenolysis with either palladium or platinum [24].

In general vinylcyclopropanes, unless substituted in both 2- and 3-positions in the ring, absorb two equivalents of hydrogen without a break in the curve. While there is much discussion about whether cleavage takes place by 1,3-, 1,4-, or 1,5-addition of hydrogen, there is little doubt that the ring is reduced first when cleavage does take place in the hydrogenation of reducible vinylcyclopropanes.

Interrupted reduction usually gives cleavage product and starting material. However, when one molar equivalent of hydrogen was absorbed in the palladium-catalyzed reduction of 1-cyclopropyl-1-phenylethylene, 2-phenyl-2-pentene was obtained in quantitative yield by apparent cleavage and rearrangement [25] (Eq. 23.8).

$$\triangle{-}C(C_6H_5){=}CH_2 \xrightarrow[Pd \text{ on } C]{H_2} CH_3C(C_6H_5){=}CHC_2H_5 \quad (23.8)$$

When 2,3-substituted vinylcyclopropanes are hydrogenated, saturation of the double bond takes place without rupture of the ring [22]. In many cases when methylenecyclopropanes are reduced methylcyclopropanes can be obtained. When ring opening takes place it does so in the same manner as in the hydrogenation of vinylcyclopropanes—at the ring bond adjacent to the double bond through a related intermediate [26a] and not through the alkylcyclopropane. Both points are illustrated in the reduction of 2-methylenecyclopropane with copper chromite at high pressure to n-butane and in the failure to cleave methylcyclopropane under the same conditions [26b].

The hydrogenations of hypoglycin A, 2-amino-3-(2-methylenecyclopropyl)propionic acid with platinum oxide in methyl alcohol [27a] or in acetic acid [27b] show that saturation of the methylene group was the major reaction and that ring opening occurred on either side of the double bond (Eq. 23.9).

$$CH_2=\triangle-CH_2\overset{NH_2}{C}HCOOH \xrightarrow[PtO_2]{>H_2}$$

$$H_3C\triangle-CH_2\overset{NH_2}{C}HCOOH + C_2H_5\overset{CH_3}{C}H\overset{NH_2}{C}H_2\overset{}{C}HCOOH + C_5H_{11}\overset{NH_2}{C}HCOOH \quad (23.9)$$

$$\text{ratio} = 10 \quad : \quad 3.5 \quad : \quad 1$$

When methylenecyclopropane-1,1-dicarboxylic acid was reduced in ethyl acetate with platinum oxide (prereduced before use) n-propylmalonic acid was obtained in 68% yield [26a]; when it was hydrogenated over 10% palladium on carbon n-propylmalonic acid and methylcyclopropane-1,1-dicarboxylic acid were obtained in a ratio of two parts to one (Eq. 23.10).

$$CH_2=\triangle\overset{COOH}{\underset{COOH}{}} \begin{array}{c} \xrightarrow{Pt} C_3H_7CH(COOH)_2 \\ \\ \xrightarrow{Pd\ on\ C} C_3H_7CH(COOH)_2 + CH_3\triangle\overset{COOH}{\underset{COOH}{}} \end{array} \quad (23.10)$$

When a palladium reduction of an allenylcyclopropane, 1-(2-methylpropenylidene)-2-phenylcyclopropane, was carried out, interruption of the reaction could not prevent ring opening; 5-methyl-1-phenylhexane was obtained in 89% yield [28] (Eq. 23.11). Whether activation is due to the

$$C_6H_5\triangle=C=C(CH_3)_2 \xrightarrow[Pd]{3H_2} C_6H_5(CH_2)_4CH(CH_3)_2 \quad (23.11)$$

phenyl or allenyl group is a matter for further study.

Methylene-2,3-substituted cyclopropanes react in the same manner as vinyl-2,3-substituted cyclopropanes—saturation of the double bond takes place although chromatography may show that some hydrogenolysis does occur.

trans-3-Methylene-1,2-dicarboxylic acid or methyl ester yielded essentially *trans*-3-methyl-1,2-dicarboxylic acid or ester when hydrogenated over palladium on carbon in ethyl acetate [26a]. Some 2-ethylsuccinic acid and methylglutaric acid were shown to be present after reduction of the acid, while in the hydrogenation as ester a 4:1 ratio of dimethyl 3-methylcyclopropane-1,1-dicarboxylate to ethyl 2-ethylsuccinate was obtained.

Cyclopropenes

There are too few examples of reductions among cyclopropenes to determine the factors which favor saturation of the double bond or ring opening. Available examples for the most part suggest that double-bond reduction is the principal reaction. The hydrogenation of 2,2-dimethylcyclopropene in alcohol with palladium on carbon at 0° gave a 95% yield of 1,1-dimethylcyclopropane [29a]. Reduction of 1,2,2-trimethylcyclopropene under similar conditions yielded two principal components in a 3:1 ratio; the major one being 1,2,2-trimethylcyclopropane. Hydrogenations of sterculic acid, 8-(2-*n*-octylcycloprop-1-enyl)octanoic acid, and of malvalic acid, 7-(2-*n*-octylcycloprop-1-enyl)heptanoic acid, with palladium on calcium carbonate only resulted in saturation of the double bond [29b].

Among cyclopropen-3-ones the presence of substituents in conjugation with the double bond promoted ring cleavage but, from the author's suggestion, hydrogenolysis followed saturation of the double bond. This is shown in hydrogenation of 1,2-diphenylcyclopropen-3-one with platinum oxide in which a slight excess of two molar equivalents was absorbed to give dibenzyl ketone; absorption of one equivalent gave starting material and dibenzyl ketone [30a] (Eq. 23.12).

$$\underset{C_6H_5}{\overset{O}{\triangle}}C_6H_5 \xrightarrow{H_2}{Pt} \left[\underset{C_6H_5}{\overset{O}{\triangle}}C_6H_5 \right] \xrightarrow{H_2} C_6H_5CH_2\overset{O}{\overset{\|}{C}}CH_2C_6H_5 \qquad (23.12)$$

The manner in which cleavage takes place in the hydrogenation of 1,2-substituted cyclopropen-3-ones may be due to the catalyst. When 1,2-dipropylcyclopropene-3-one was reduced with platinum oxide, after one equivalent absorption cleavage occurred at the 1,2-bond (after double-bond reduction) to yield di-n-butyl ketone [30b] whereas hydrogenation in the presence of palladium on carbon yielded 2-propyl-2-hexenal from partial reduction, cleavage occurring at the 1,3-bond (Eq. 23.13).

$$\underset{C_3H_7 \quad C_3H_7}{\overset{O}{\triangle}} \xrightarrow[\text{Pd on C}]{H_2} C_3H_7CH{=}C(C_3H_7)CHO \qquad (23.13)$$

Cyclobutanes and Larger Cycloalkanes

Because cycloalkanes larger than cyclobutanes undergo catalytic hydrogenolysis with considerable difficulty, only that of cyclobutane will be discussed. Cyclobutanes are cleaved with more difficulty than cyclopropanes. As a rule elevated temperature and pressure conditions are necessary.

Ring strain increases the ease of hydrogenolysis as it does in cyclopropanes. Activating groups which aided hydrogenolysis among cyclopropanes may also have the same effect among cyclobutanes but there are too few examples for substantiation.

There is an illustration of the effect of a phenyl group but it may not be pertinent since the effect was not studied on cyclobutane itself. 1,2-Diphenylbenzocyclobutene was hydrogenated to 1,2-dibenzylbenzene with palladium on carbon in alcohol at room temperature [31a]. When the parent benzocyclobutene was reduced in a very slow reaction only the benzene ring was affected; bicyclo[4,2,0]octane was obtained [Ref. 14 in 31b] (Eq. 23.14).

(23.14)

The reduction of the following substituted cyclobutenone is of interest because it is one of the few instances where the cyclobutane ring is cleaved in preference to reduction of a functional group. When 3-ethoxy-4,4-dimethyl-2-butenone was hydrogenated with palladium on carbon at 60° and 100 atm, a 3:1 mixture of 3-ethoxy-4,4-dimethylcyclobutanone and 3-ethoxy-2,2-dimethylbutyraldehyde was obtained [31c]. Evidence that double-bond reduction preceded ring opening is shown in the hydrogenation of the saturated compound which gave a 3:1 ratio of recovered cyclobutanone and the substituted butyraldehyde (Eq. 23.15).

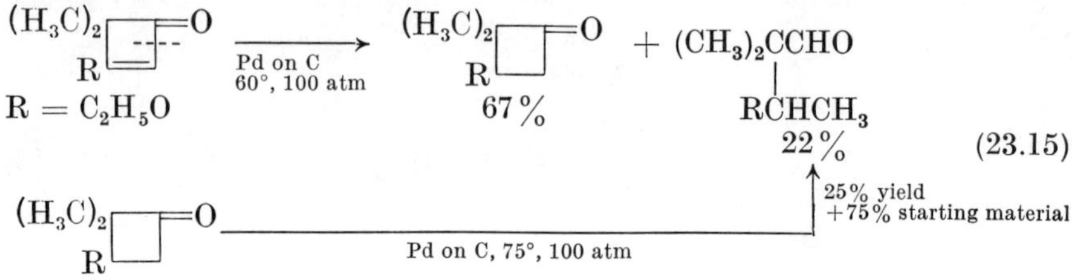

$$(H_3C)_2 \underset{R}{\overset{=O}{\boxed{}}} \xrightarrow[60°, 100 \text{ atm}]{\text{Pd on C}} (H_3C)_2 \underset{R}{\overset{=O}{\boxed{}}} + (CH_3)_2CCHO$$
$$R = C_2H_5O \qquad\qquad 67\% \qquad\qquad \underset{22\%}{\overset{|}{R\dot{C}HCH_3}}$$

$$(H_3C)_2 \underset{R}{\overset{=O}{\boxed{}}} \xrightarrow{\text{Pd on C, 75°, 100 atm}} \quad \uparrow 25\% \text{ yield} \atop +75\% \text{ starting material}$$

(23.15)

NONAROMATIC N-HETEROCYCLIC RINGS

Azirenes

From the few available examples an azirene, a 3-membered cyclic imine, is not hydrogenolyzed through an intermediate aziridine (ethyleneimine). When ethyl 2-methyl or 2,3-dimethylazirene-3-carboxylate was hydrogenated in tetrahydrofuran with palladium on carbon or with palladium on barium carbonate, ethyl 3-amino- or 3-amino-2-methyl-2-butenoate was obtained after absorption of one equivalent of hydrogen [32a], probably through cleavage of the 1,3-bond and accompanying rearrangement (Eq. 23.16). There was no evidence that the reduction was acid catalyzed. Hydrogenolysis

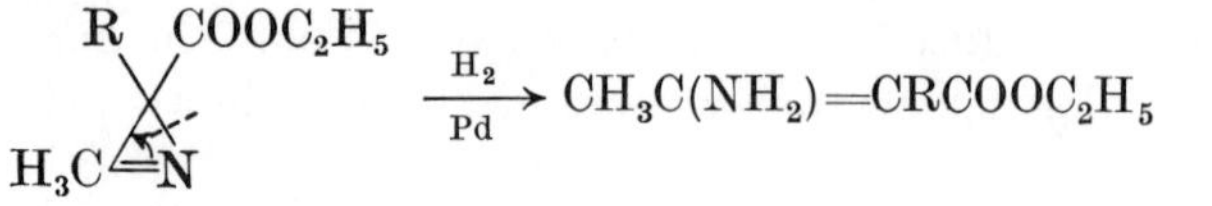

$$\underset{H_3C \overset{\diagdown}{\underset{N}{\diagup}}}{\overset{R \quad COOC_2H_5}{\diagdown \diagup}} \xrightarrow[Pd]{H_2} CH_3C(NH_2)=CRCOOC_2H_5 \qquad (23.16)$$

R = H or CH₃

of the corresponding aziridines did not take place under comparable conditions.

An earlier example is of considerable interest because the 1,3-bond in the strained ring system was cleaved in preference to reduction of aromatic nitro

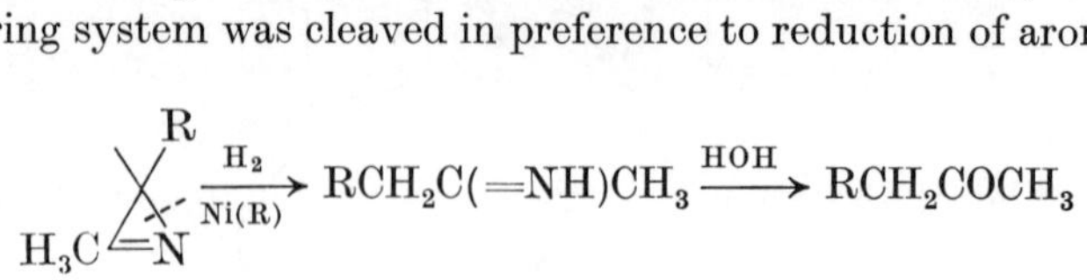

$$\underset{H_3C \overset{\diagdown}{\underset{N}{\diagup}}}{\overset{R}{\diagdown \diagup}} \xrightarrow[Ni(R)]{H_2} RCH_2C(=NH)CH_3 \xrightarrow{HOH} RCH_2COCH_3$$

R = 2,4-dinitrophenyl (23.17)

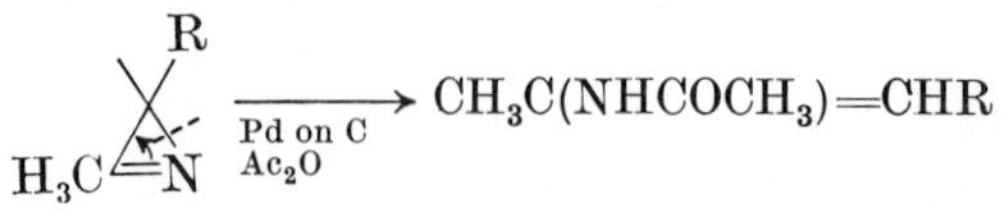

$$\underset{H_3C \overset{\diagdown}{\underset{N}{\diagup}}}{\overset{R}{\diagdown \diagup}} \xrightarrow[Ac_2O]{\text{Pd on C}} CH_3C(NHCOCH_3)=CHR$$

groups. The hydrogenation of 2-methyl-3-(2,4-dinitrophenyl)azirene with alcohol-free Raney nickel in dioxane yielded 2,4-dinitrophenylacetone (through hydrolysis of 2-imino-1-(2,4-dinitrophenyl)propane), and reduction with palladium on carbon in pyridine and acetic anhydride gave isomeric acetylated enamines, 2-acetylamino-1-(2,4-dinitrophenyl)-1-propenes [32b] (Eq. 23.17).

Aziridines

Aziridines or ethyleneimines are reasonably stable to hydrogenolysis. In general the more readily reducible functions can be hydrogenated without accompanying opening of the aziridine ring.

Catalysts used for hydrogenolysis include Raney nickel, platinum, and palladium, probably the best one. Although nickel hydrogenolyses are usually carried out at higher temperature and pressure than either palladium- or platinum-catalyzed ones, it was used at 60° and 4 atm to reduce 2,2-dimethylaziridine to t-butylamine in good yield [33].

The direction of ring opening is not always predictable except to state that it takes place at the bond adjacent to the ring nitrogen. In the hydrogenation of 1-azabicyclo[3,1,0]hexane with palladium on carbon in hexane a mixture of piperidine and 2-methylpyrrolidine, 1 part to 2, was obtained [34] (Eq. 23.18).

$$\text{structure} \xrightarrow[\text{Pd on C}]{\text{H}_2} \text{piperidine} + \text{2-methylpyrrolidine} \quad (23.18)$$

However, certain groups appear to direct which adjacent bond will be cleaved. Examples of reductions of aziridines containing a phenyl or a methylidene group on the carbon adjacent to the nitrogen indicate cleavage at the bond between those two atoms: for example, 2-phenylaziridine in heptane yielded phenethylamine on reduction with palladium black (preferred) or with platinum black [35a]; 2-phenyl-1-(4-tolylsulfonyl)-aziridine was hydrogenated with palladium on barium sulfate to N-(4-tolylsulfonyl)phenethylamine [35b]; 1,2-dimethyl-3-phenylaziridine gave N-methyl-1-phenylisopropylamine on reduction with palladium on carbon [35c]; and 2-phenylaziridin-3-carboxamide was very slowly hydrogenated over platinum oxide to 2-amino-3-phenylpropionamide (β-phenylalaninamide) [35d]. 1-(2-Methylene-1-aziridenyl)-3-buten-2-ol was reduced with platinum oxide in alcohol to 1-propylamino-2-butanol [35e]. The two types are illustrated in Eq. 23.19).

As previously pointed out most reducible functions exclusive of aromatic rings will be hydrogenated in the presence of an aziridene ring. There is an example showing ring opening in the presence of an N-benzyl group. 1-(1-Butylamino-2-methyl-1-phenylisopropyl)aziridine was reduced to N^1-butyl-N^2-ethyl-1-phenyl-2-methyl-1,2-propanediamine with palladium on carbon

$$\underset{H}{\overset{C_6H_5}{N}} \xrightarrow{H_2} C_6H_5CH_2CH_2NH_2$$

$$\underset{CH_2CHOHCH=CH_2}{\overset{=CH_2}{N}} \xrightarrow{3H_2} C_3H_7NHCH_2CHOHC_2H_5$$
(23.19)

in alcohol [36] (Eq. 23.20). However, there are good reasons for the results that were obtained. Even though palladium was used for the reaction, reduction as base probably inactivated the catalyst for debenzylation. In

$$\underset{NHC_4H_9}{C_6H_5CH-\overset{N\triangledown}{\underset{|}{C}}(CH_3)_2} \xrightarrow[\text{Pd on C}]{H_2} \underset{NHC_2H_5}{C_4H_9NHCH(C_6H_5)\overset{|}{C}(CH_3)_2} \quad (23.20)$$

addition benzylamines of the type ArCHRNH-alkyl, where R is not hydrogen, are not too readily cleaved (see Chapter XIX).

Azetidines

There is little information on the hydrogenolysis of azetidines or trimethylenimines, $(CH_2)_3NH$. Azetidine was obtained from the palladium-catalyzed hydrogenolysis of 1-benzylazetidine possibly accompanied by ring rupture, but (4-tolylsulfonyl)azetidine was unaffected by hydrogenation with nickel at 100° [37a]. Cleavage did occur in the hydrogenation of 6-azabicyclo[3,1,1]-heptane with an undisclosed amount of palladium to yield cyclohexylamine; no 2-pipecoline was reported [37b] (Eq. 23.21).

$$\xrightarrow{H_2}{Pd} \bigcirc NH_2 \quad (23.21)$$

In most instances 2-azetidinones are stable to hydrogenolysis. In a few instances rupture between the N—C4 bond has taken place when a phenyl group was in 4-position.

Refluxing some 4-phenylazetidin-2-ones in solvent with Raney nickel yielded 3-phenylpropionamides [37a, p. 947]. When 1-aryl-3,4-diphenyl-azetidin-2-one, where the aryl group was phenyl or 2-methoxyphenyl, was

Nonaromatic N-Heterocyclic Rings

hydrogenated at room temperature and atmospheric pressure with Raney nickel the substituted propionamides, $ArCHCH(C_6H_5)CONHC_6H_5$, were obtained [37c]. There was no reduction when Ar was 2-naphthyl or when palladium was used for reductions.

Other N-Heterocycles

Larger saturated heterocycles such as pyrrolidine and piperidine are not readily cleaved. Ring rupture of saturated heterocycles containing two nitrogen atoms which are separated by at least two carbon atoms, as in piperazine, is also difficult. Saturated heterocycles which contain a 1,1-diamino system, N—C—N, undergo cleavage between one carbon and nitrogen bond with comparative ease.

The instability of the 1,1-diamino system is well known. It is the reason that 2-aminopyridine cannot be catalytically hydrogenated to 2-aminopiperidine; piperidine and ammonia are obtained. Pyrimidines cannot be reduced to hexahydropyrimidines for the same reason. 1,2,5,6-Tetrahydropyrimidines can be obtained but further reduction results in cleavage between the 1,2-bond.

The instability of tetrahydropyrimidine to catalytic reduction was employed to obtain a variety of aminoalcohols from the hydrogenation of substituted tetrahydropyrimidines in water with Raney nickel [38]. In the absence of water diamines should be expected but in these reductions, of which 2,2,4,6,6-pentamethyl-1,2,5,6-tetrahydropyrimidine is an example, ring opening and hydrolysis of the intermediate aminoketimine (by water at high temperature) to aminoketone followed by further hydrogenation, yielded aminoalcohol (Eq. 23.22). Other examples of hydrogenolysis in

$$[(CH_3)_2CHNH_2 + (CH_3)_2C(NH_2)CH_2COCH_3] \xrightarrow{H_2} (CH_3)_2C(NH_2)CH_2CHOHCH_3 \quad (23.22)$$

N—C—N systems include the platinum-catalyzed hydrogenation of 1,4a,9-trimethyl-2,3,4,4a,9,9a-hexahydro-1H-pyrido[2,3-b]indole in acetic acid to 1,3-dimethyl-3-(3-methylaminopropyl)indoline [39a] (Eq. 23.23), and the reduction of 1,7-dimethyl-1,7-diaza-spiro[5,5]undecane to 1-methyl-2-(4-methylaminobutyl)piperidine, and those of 1,6-dialkyl- or dibenzyl-1,6-diaza-spiro[4,4]nonanes in acetic acid with platinum oxide or in alcohol with

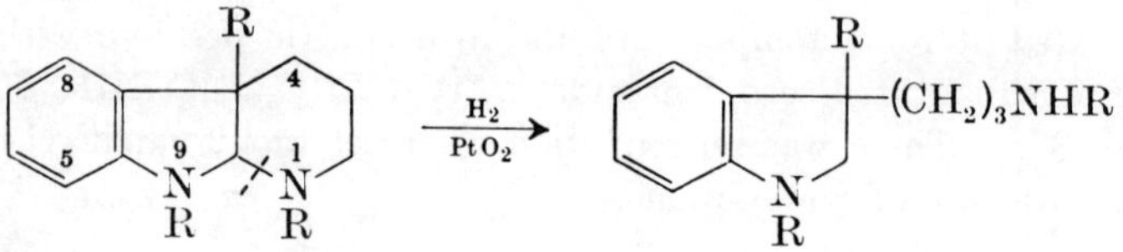

Miscellaneous Hydrogenolyses

$$\text{[structure]} \xrightarrow[\text{PtO}_2]{\text{H}_2} \text{[structure]}-(CH_2)_3NHR \quad (23.23)$$

R = CH₃

Raney nickel at 50° and 100 atm, which gave 1-alkyl-or benzyl-2-(3 alkyl-or benzylaminopropyl)pyrrolidines in good yield [39b] (Eq. 23.24).

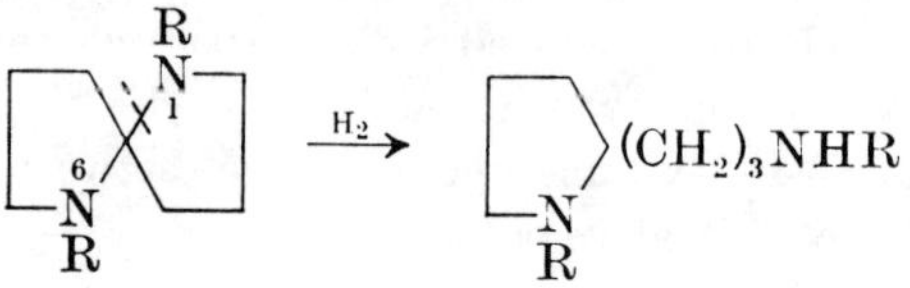

$$\text{[structure]} \xrightarrow{H_2} \text{[structure]}(CH_2)_3NHR \quad (23.24)$$

R—1 = R—6 = CH₃, C₃H₇, or C₆H₅CH₂—

It is of interest that, when R was benzyl, ring opening in this high-pressure reaction took place without removal of the *N*-benzyl group. The *N*-benzyl group was not affected during the hydrogenolysis of 3-benzyl-3,4,5,6-tetrahydro-2*H*-1,5-methanobenzo[*d*][1,3]-diazacine with palladium oxide to 3-benzylaminomethyl-1-methyl-1,2,3,4-tetrahydroquinoline in 93% yield [39c]. In a related reduction where the 2-substituent was phenyl, the benzyl group in the side chain remained intact but the benzyl group on the tetrahydroquinoline nitrogen (resulting from ring rupture) was removed to give 3-benzylaminomethyl-1,2,3,4-tetrahydroquinoline (Eq. 23.25). The possibility does exist in the last example that 1,2-cleavage took place not 2,3-.

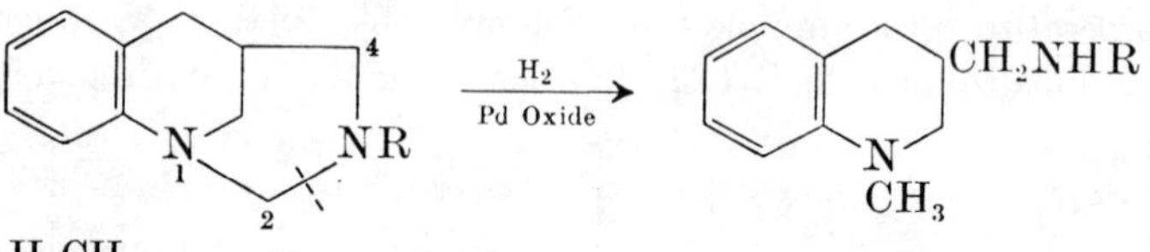

$$\text{[structure]} \xrightarrow[\text{Pd Oxide}]{H_2} \text{[structure]}CH_2NHR$$

R = C₆H₅CH₂— (23.25)

$$\text{[structure]} \rightarrow \left[\text{[structure]}CH_2NHR \right] \rightarrow \text{[structure]}CH_2NHR$$

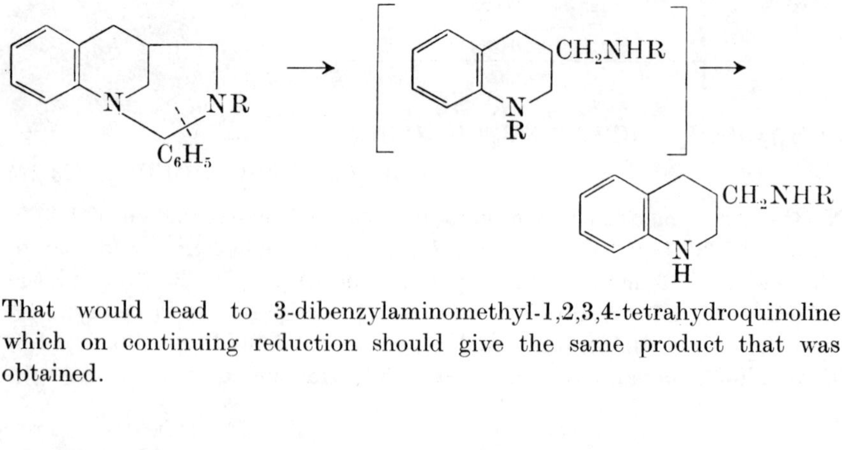

That would lead to 3-dibenzylaminomethyl-1,2,3,4-tetrahydroquinoline which on continuing reduction should give the same product that was obtained.

NONAROMATIC O,N-HETEROCYCLES

3- and 4-Membered Rings

There is very little information on catalytic hydrogenolyses of the smaller O,N-containing rings.

2-Cyclohexyl-3-phenyloxazirane absorbed two equivalents of hydrogen during reduction with platinum oxide to yield N-benzylcyclohexylamine [40a] but the manner of cleavage was not investigated.

There is little mention of the hydrogenolysis of 1,2-oxazetines except among N-substituted oxazetinones. In these reductions the N-substituent dictated the direction of cleavage. When 2-R-4,4-diphenyl-1,2-oxazetin-3-ones (R = methyl or cyclohexyl) were hydrogenated with palladium on carbon in alcohol under 3.3 atm pressure, cleavage took place between the O—C4 bond to yield N-substituted diphenylacetylhydroxamic acids; when the N-substituent was aryl, 1,2-cleavage took place to yield benzilanilides [40b] (Eq. 23.26).

$$(C_6H_5)_2 \begin{array}{c} {=}O \\ O{-}N{-}R \end{array} \xrightarrow{H_2} (C_6H_5)_2CHCON(R)OH$$

R = aliphatic (23.26)

$$(C_6H_5)_2 \begin{array}{c} {=}O \\ O{+}N{-}Ar \end{array} \xrightarrow{H_2} (C_6H_5)_2C(OH)CONHAr$$

5- and 6-Membered Rings

1,2-O,N-Heterocycles. The hydrogenolyses of nonaromatic 1,2-O,N-5- or 6-membered rings result in rupture between the oxygen and nitrogen atom. Among 5-membered rings, hydrogenolyses of isoxazolidines have been carried out with palladium or platinum catalysts and hydrogen at atmospheric pressure and at several atmospheres with Raney nickel to give good yields of 1,3-aminoalcohols [41a]. An isoxazoline, 3,5-diphenyl-3,4-dihydroxazole, was reduced in methyl alcohol and hydrochloric acid with 10% platinum on barium sulfate to yield a mixture of isomers of 3-amino-1,3-diphenylpropanol [41b]. 3,4-Dihydroisoxazole-3-carboxylic acid underwent rapid hydrogenolysis with 2% palladium on carbon to give high yield of 2-amino-4-hydroxybutyric acid [41c]. Other related compounds were prepared from the corresponding dihydroisoxazolecarboxylic acids.

Among isoxazolidinones, 4-benzoylamino-3-methylisoxazolidin-5-one gave 70–80% yields of 3-amino-2-benzoylaminobutyric acid on reduction with platinum oxide [42a] and the palladium on carbon hydrogenolysis of 4-benzylidene-2-phenyloxazolidin-3,5-dione in tetrahydrofuran was extremely rapid to give 72% yield of 2-carboxy-3-phenylpropionanilide [42b].

Hydrogenolysis of 6-membered 1,2-O,N-heterocycles give 1,4-aminoalcohols. If reaction conditions are too severe, cyclodehydration may take place to yield substituted pyrrolidines.

There is little available information on the hydrogenolysis of tetrahydro-1,2-oxazine. The reduction of 3,6-dihydro-2H-1,2-oxazine through the tetrahydro compound as intermediate with Raney nickel at 45° and 150 atm to 4-aminobutanol [43a] illustrates one method that can be employed. In a more complex system, in the hydrogenation of 3',4'-dihydrospiro[cyclohexane-1,6'(7'H)-[5H-2,4a]ethano(2H)pyrrolo[1,2-b]-1,2-oxazine-5',7'-dione, cleavage took place under 2 atm pressure with Raney nickel or with palladium on carbon [43b] (Eq. 23.27). Reduction with nickel at 195° and 150 atm led

$$\text{(23.27)}$$

85%

to conversion of the ketone function to yield a diol.

A complex tetrahydro-1,2-oxazine, containing an α,β-unsaturated double bond and an allylic alcohol, on hydrogenation with a large amount of prereduced platinum in acetic acid gave three compounds [43c]. The oxazine ring in one of these compounds was still intact but further hydrogenation of that product with prereduced platinum in acetic acid did result in ring opening.

The other two compounds were bases formed through a common ring-cleaved intermediate containing the allylic hydroxyl group which in one instance underwent hydrogenolysis and double-bond reduction and in the other instance reduction with retention of the hydroxyl group (Eq. 23.28).

1,3-O,N-Heterocycles. When most 5-membered 1,3-O,N-heterocycles undergo cleavage it takes place at the O—C2 bond to give N-substituted ethanolamines.

2-Oxazolines follow this pattern through reduction to an intermediate oxazolidine and subsequent ring opening to ethanolamines. However, when an amino grouping is present in 2-position only one equivalent is absorbed and cleavage is shown to have taken place at the O—C5 bond, to give a substituted urea. Such an example is seen in the palladium reduction of

Nonaromatic O,N-Heterocycles 537

2-amino-4-methyl-5-phenyl-2-oxazoline which yielded N-(1-phenylisopropyl)-urea [44] (Eq. 23.29).

$$\begin{array}{c} \underset{O}{\overset{N}{\diagup}}R \xrightarrow{2H_2} \left[\underset{O}{\overset{NH}{\diagup}}R \right] \longrightarrow HOCH_2CH_2NHR \end{array}$$

(23.29)

$$\underset{H_5C_6}{\overset{H_3C}{\diagdown}}\underset{O}{\overset{N}{\diagup}}NH_2 \xrightarrow{H_2} C_6H_5CH_2CH(CH_3)NHCONH_2$$
98%

Oxazolidines, which are isomeric with Schiff bases prepared from ethanolamines and carbonyl compounds, are cleaved at the O—C2 bond following reduction. The ease of hydrogenolysis has been reported to be dependent on whether the oxazolidines are derived from aldehydes or ketones [45a]. However, the structure of the ethanolamine can also be a factor [1].

Oxazolidines prepared from aldehydes, whether aliphatic or aromatic, were hydrogenolyzed under mild conditions with platinum oxide [45a, b, c] or with palladium [45b, d]. Nickel catalyst can be used [1, 45b] but is usually less efficient in low-pressure hydrogenolyses [1]. Oxadolidines obtained from ketones generally required more vigorous conditions for cleavage [1, 45a, 46].

The reduction of oxazolidinones does not appear to have been studied too widely. When there is a phenyl group in 5-position and the oxo group is in 2-position, the ring may be considered as a carbobenzyloxy group and subject to cleavage and loss of carbon dioxide. The procedure has found use in the preparation of compounds related to benzylpenicillins. An example of cleavage of such a system is shown in the hydrogenolysis of 5-phenyl-oxazolidine-2,4-dione-3-N-phenylacetamide with palladium on carbon in dioxane which gave 91% yield of phenylaceturic acid anilide [47] (Eq. 23.30).

$$\underset{91\%}{\text{[5-phenyloxazolidine-2,4-dione-3-N-phenylacetamide]} \xrightarrow{H_2} C_6H_5CH_2CONHCH_2CONHC_6H_5} \quad (23.30)$$

Cleavage in the hydrogenolysis of tetrahydro-2H-1,3-oxazines takes place at the O—C2 bond to give N-substituted 3-aminopropanols. Raney nickel has been used at 75–100° and 70 atm [48a]; platinum oxide has been employed under low-pressure conditions but with about 5% by weight of catalyst [48b, c].

Miscellaneous O,N-Heterocycles. The hydrogenation of 3-substituted-4-phenyl-1,2,4-oxadiazole-5-ones with platinum oxide yielded benzamidines through 1,2-cleavage and loss of carbon dioxide [49a] (Eq. 23.31).

$$\text{[3-substituted-4-phenyl-1,2,4-oxadiazole-5-one]} \xrightarrow[PtO_2]{H_2} C_6H_5C\begin{smallmatrix}NHR\\NH\end{smallmatrix} \quad (23.31)$$

R = H, CH$_3$, or C$_6$H$_5$

Nonaromatic O,N-Heterocycles

The hydrogenation of 5-acetoxy-2,2-pentamethylene-5-phenyl-Δ3-1,3,4-oxadiazoline with platinum oxide in alcohol followed two paths when one equivalent of hydrogen was absorbed—reduction of the nitrogen to nitrogen double bond to yield the corresponding 1,3,4-oxadiazolidine and cleavage at the O—C2 bond and rearrangement to give 1-benzoyl-2-cyclohexylidenehydrazine [49b]. When two equivalents of hydrogen were absorbed, 1-benzoyl-2-cyclohexylhydrazine was obtained. It could have been derived from either of the products from the first reduction step (Eq. 23.32).

$$\text{(23.32)}$$

When the 1,1-inner ether of 8,8'-dimethoxy-6,7,6',7'-bis-dioxymethylene-1,2,3,4,1',2',3',4'-octahydro-2,2'-bi-isoquinolyl (from its structure a 1,3,4-oxadiazolidine) was reduced with palladium on carbon, cleavage took place on each side of the oxygen giving 92% yield of the substituted 1,2,3,4,1',2',3',4'-octahydro bi-isoquinolyl [49c] (Eq. 23.33). It is likely that cleavage

$$\text{(23.33)}$$

took place simultaneously otherwise a 1-hydroxy compound would have been obtained.

The hydrogenation of 6-membered O,O,N-heterocycle, a 1,3,4-dioxazine, is of interest. Spiro[cyclohexane-1',3-9-carbamyl-3(H)-5,6,7,8-tetrahydrobenzo-1,2,4-dioxazine] on reduction with platinum oxide gave 2-carbamyl-2-hydroxydicyclohexylamine [50]. Reaction apparently took place through loss of one oxygen to a 1,3-oxazolidine which underwent the expected ring rupture (Eq. 23.34).

REFERENCES

[1] M. Freifelder, unpublished results.
[2] a. A. S. Dey and M. M. Joullie, *J. Heterocyclic Chem.*, **2**, 113 (1954); b. H. Paulsen, I. Sangster, and K. Heyns, *Ber.*, **100**, 802 (1967).
[3] a. H. M. Walborsky and M. E. Baum, *J. Org. Chem.*, **21**, 538 (1956); b. B. Belleau and P. Cooper, *J. Med. Chem.*, **6**, 579 (1963).
[4] a. M. F. Huber, *J. Am. Chem. Soc.*, **77**, 112 (1955); b. H. K. Nagy, A. J. Tomson, and J. P. Horwitz, *ibid.*, **82**, 1609 (1960); c. V. S. Kuznetsov and L. S. Efros, *J. Org. Chem. USSR*, **1**, 1479 (1965) (Eng. trans.).
[5] a. M. Ohno, S. Torimitsu, N. Naruse, M. Okamoto, and I. Saki, *Bull. Chem. Soc. (Japan)*, **39**, 1129 (1966); b. L. Krbechek and H. Takimoto, *J. Org. Chem.*, **29**, 1150 (1964).
[6] a. M. L. Wolfrom, F. Shafizadeh, J. O. Wehrmüller, and R. K. Armstrong, *J. Org. Chem.*, **23**, 571 (1958); b. I. Ziderman and E. Dimant, *ibid.*, **31**, 223 (1966).
[7] G. F. Chittenden and R. D. Guthrie, *J. Chem. Soc. (C)*, **1966**, 1508.
[8] a. H. E. Carter, F. R. Van Abeele, and J. W. Rothrock, *J. Biol. Chem.*, **178**, 325 (1949); b. A. N. Kost, G. A. Golubeva, and R. G. Stepanov, *J. Gen. Chem. USSR*, **32**, 2207 (1962) (Eng. trans.); c. N. Rabjohn, H. R. Havens, and J. L. Rutter, *J. Heterocyclic Chem.*, **3**, 413 (1966); d. M. Nagata, *Yakugaku Zasshi*, **86**, 608 (1966); through *Chem. Abstr.*, **65**, 15356 (1966).
[9] a. M. Pesson and D. Richer, *Compt. Rend. Ser. C*, **262**, 1719 (1966); b. W. D. Guither, D. G. Clark, and R. N. Castle, *J. Heterocyclic Chem.*, **2**, 67 (1965).
[10] a. J. H. Biel, W. K. Hoya, and H. A. Leiser, *J. Am. Chem. Soc.*, **81**, 2527 (1959); b. D. J. Cram and J. S. Bradshaw, *ibid.*, **85**, 1108 (1963); c. C. G. Overberger, L. C. Palmer, B. S. Marks, and N. R. Bird, *ibid.*, **77**, 4100 (1955).
[11] H. Stetter and K. Findeisen, *Ber.*, **98**, 3228 (1965).
[12] E. F. Elslager, D. F. Worth, N. F. Haley, and S. C. Perricone, *J. Heterocyclic Chem.*, **5**, 609 (1968).

[13] P. H. Bentley and J. S. Morley, *J. Chem. Soc.* (C), **1965**, 60.
[14] J. Piechaczek and H. Bojarska-Dahlig, *Acta. Polon. Pharm.*, **23**, 7 (1966); through *Chem. Abstr.*, **64**, 14164 (1966).
[15] a. C. Ainsworth, *J. Am. Chem. Soc.*, **76**, 5774 (1954); b. J. Finkelstein, E. Chiang, F. M. Vane, and J. Lee, *J. Med. Chem.*, **9**, 319 (1966); c. D. Škarić, V. Škarić, and V. Turjak-Zebić, *Croat. Chem. Acta*, **35**, 143 (1965); d. *ibid.*, p. 267.
[16] N. K. Kochetkov, E. I. Budovskii, and C. Chih-p'ing, *J. Gen. Chem. USSR*, **31**, 3072 (1961) (Eng. trans.).
[17] a. E. Zbiral, F. Takacs, and F. Wessely, *Monatsh.*, **95**, 402 (1964); b. T. Kuraishi and R. N. Castle, *J. Heterocyclic Chem.*, **1**, 42 (1964); c. G. A. Gerhardt and R. N. Castle, *ibid.*, p. 247.
[18] a. G. D. Laubach, E. C. Schreiber, E. J. Agnello, and K. J. Bruning, *J. Am. Chem. Soc.*, **78**, 4746 (1956); b. E. J. Agnello, R. Pinson, Jr., and G. D. Laubach, *ibid.*, p. 4756; c. G. A. Russell and F. R. Mayo, U.S. Patent 2,794,055 (1957).
[19] a. R. W. Shortridge, R. A. Craig, K. W. Greenlee, J. M. Derfer, and C. E. Boord, *J. Am. Chem. Soc.*, **70**, 946 (1948); b. V. A. Slabey, *ibid.*, **69**, 475 (1947); c. C. W. Woodworth, V. Buss, and P. v. R. Schleyer, *Chem. Commun.*, **1968**, 569.
[20] a. B. A. Kazansky, M. Yu. Lukina, and L. G. Salnikova, *Bull. Acad. Sci. USSR, Div. Chem. Sci.*, **1957**, 1422 (Eng. trans.); b. B. A. Kazansky, M. Yu. Lukina, and I. L. Safonova, *ibid.*, **1958**, 95; c. B. A. Kazansky, M. Yu. Lukina, and I. L. Safonova, *Proc. Acad. Sci., USSR, Sect. Chem.*, **130**, 55 (1960) (Eng. trans.).
[21] N. I. Shuikin, A. D. Petrov, V. G. Glukhovtsev, and R. A. Karakhanov, *Bull. Acad. Sci. USSR, Div. Chem. Sci.*, **1963**, 467 (Eng. trans.).
[22] R. W. Kierstead, R. P. Linstead, and B. C. L. Weedon, *J. Chem. Soc.*, **1952**, 3610.
[23] G. M. Badger, B. J. Christie, H. J. Rodda, and J. M. Pryke, *J. Chem. Soc.*, **1958**, 1179.
[24] B. A. Kazansky, M. Yu. Lukina, and L. G. Salnikova, *Proc. Acad. Sci. USSR, Sect. Chem.*, **1957**, 711.
[25] S. Sarel and E. Bruer, *J. Am. Chem. Soc.*, **81**, 6522 (1959).
[26] a. E. F. Ullman, *J. Am. Chem. Soc.*, **81**, 5386 (1959); b. J. T. Gragson, K. W. Greenlee, J. M. Derfer, and C. E. Boord, *ibid.*, **75**, 3344 (1953).
[27] a. R. S. DeRopp, J. C. VanMeter, E. C. Renzo, K. W. McKerns, C. Pidacks, P. H. Bell, E. F. Ullman, S. R. Safir, W. J. Fanshawe, and S. B. Davis, *J. Am. Chem. Soc.*, **80**, 1004 (1958); b. S. Wilkinson, *Chem. Ind.*, **1958**, 17.
[28] a. H. D. Hartzler, *J. Am. Chem. Soc.*, **83**, 4990 (1961).
[29] a. G. L. Closs and L. E. Closs, *J. Am. Chem. Soc.*, **85**, 99 (1963); b. F. L. Carter and V. L. Frampton, *Chem. Rev.*, **64**, 497 (1964).
[30] a. R. Breslow, T. Eicher, A. Krebs, R. A. Peterson, and J. Possner, *J. Am. Chem. Soc.*, **87**, 1320 (1965); b. R. Breslow, L. J. Altman, A. Krebs, E. Mohacsi, I. Murata, R. A. Peterson, and J. Possner, *ibid.*, p. 1326.
[31] a. F. R. Jensen and W. E. Coleman, *J. Am. Chem. Soc.*, **80**, 6149 (1958); b. M. P. Cava and R. Pohlke, *J. Org. Chem.*, **28**, 1012 (1963); c. R. H. Hasek, P. G. Gott, and J. C. Martin, *ibid.*, **29**, 2510 (1964).
[32] a. G. R. Harvey and K. W. Potts, *J. Org. Chem.*, **31**, 3907 (1966); b. D. J. Cram and M. J. Hatch, *J. Am. Chem. Soc.*, **75**, 33 (1953).
[33] K. N. Campbell, A. H. Sommers, and B. K. Campbell, *J. Am. Chem. Soc.*, **68**, 140 (1946).
[34] P. G. Gassman and A. Fentiman, *J. Org. Chem.*, **32**, 2388 (1967).
[35] a. S. V. Zotova, L. V. Loza, and M. Yu. Lukina, *Proc. Acad. Sci. USSR, Sect. Chem.*, **164**, 1016 (1965) (Eng. trans.); b. M. S. Kharasch and H. M. Priestley,

J. Am. Chem. Soc., **61,** 3425 (1939); c. R. Haberl, Monatsh., **89,** 814 (1958); d. Y. Yukawa and S. Kimura, Chem. Zentr., **128,** 10992 (1957); e. A. T. Bottini, V. Dev, and M. Stewart, J. Org. Chem., **28,** 156 (1963).

[36] C. L. Stevens, M. E. Munk, C. H. Chang, K. G. Taylor, and A. L. Schy, J. Org. Chem., **29,** 3146 (1964).

[37] a. J. A. Moore, *The Chemistry of Heterocyclic Compounds*, Interscience, New York, 1964, Vol. 19, Part 2, Ch. VII, p. 903; b. J. v. Braun, W. Haensel, and F. Zobel, Ann., **462,** 283 (1928); c. Al. Spasov and P. Panaiotova, J. Org. Chem. USSR, **1,** 1109 (1965) (Eng. trans.).

[38] V. E. Hawry, U.S. Patent 2,497,548 (1950).

[39] a. S. Yamada, T. Hino, and K. Ogawa, Chem. Pharm. Bull., **11,** 674 (1963); b. F. Korte, A. K. Bocz, and K. H. Buchel, Ber., **99,** 736 (1966); c. T. Kametani and K. Kigasawa, Chem. Pharm. Bull., **14,** 566 (1966).

[40] a. L. Horner and E. Jurgens, Ber., **90,** 2184 (1957); b. T. Sheradsky, U. Reichman, and M. Frankel, J. Org. Chem., **33,** 3619 (1968).

[41] a. N. A. LeBel, M. E. Post, and J. J. Whang, J. Am. Chem. Soc., **86,** 3759 (1964); b. A. Skita, W. Stühmer, and W. Heinrich, Ger. Patent 985,948 (1953); through Chem. Abstr., **52,** 12918 (1958); c. G. Drefahl and H. H. Hörhold, Ber., **97,** 159 (1964).

[42] a. N. K. Kochetkov, R. M. Khomutov, E. L. Budovskii, M. Ya. Karpeiskii, and E. S. Severin, J. Gen. Chem. USSR, **29,** 4030 (1959) (Eng. trans.); b. K. Michel, H. Gerlach-Gerber, Ch. Vogel, and M. Matter, Helv. Chim. Acta, **48,** 1973 (1965).

[43] a. D. Klamann, P. Weyerstahl, M. Fligge, and K. Ulm, Ber., **99,** 561 (1966); b. D. V. Nightingale, J. E. Johnson, and D. N. Heintz, J. Org. Chem., **33,** 360 (1968); c. T. Nakaro, S. Terao, K. H. Lee, Y. Saeki, and L. J. Durham, ibid., **31,** 2274 (1966).

[44] G. I. Poos, J. R. Carson, J. D. Rosenau, A. P. Roszkowski, N. M. Kelley, and J. McGowin, J. Med. Chem., **6,** 266 (1963).

[45] a. E. M. Hancock and A. C. Cope, J. Am. Chem. Soc., **66,** 1738 (1944); b. C. H. Bolen, U.S. Patent 2,865,925 (1958); c. E. Gil-Av, J. Am. Chem. Soc., **74,** 1346 (1952); d. M. C. Wani and S. G. Levene, J. Org. Chem., **31,** 2564 (1966).

[46] E. L. Engelhardt, F. S. Crossley, and J. M. Sprague, J. Am. Chem. Soc., **72,** 2718 (1950).

[47] J. C. Sheehan and G. D. Laubach, J. Am. Chem. Soc., **73,** 4752 (1951).

[48] a. M. Senkus, J. Am. Chem. Soc., **72,** 2967 (1950); b. W. R. Vaughan and R. S. Klonowski, J. Org. Chem., **26,** 147 (1961); c. Z. Horii, T. Inoi, S-W. Kim, Y. Tamura, A. Suzuki, and H. Matsumoto, Chem. Pharm. Bull., **13,** 1151 (1965).

[49] a. G. D'Alo and P. Grunanger, Farmaco, Ed. Sci., **221,** 346 (1966); b. R. W. Hoffman and H. J. Luthardt, Tetrahedron Letters, **1966,** 411; c. D. Korbonits and K. Harsányi, Ber., **99,** 273 (1965).

[50] a. A. E. McKay, J-M. Billy, and E. J. Tarlton, J. Org. Chem., **29,** 291 (1964).

XXIV

AROMATIC RING SYSTEMS

As a rule saturation of the various aromatic carbocyclic and heterocyclic ring systems is best carried out in the presence of platinum metal catalysts. Platinum as platinum oxide and rhodium on a support usually require mild conditions for these hydrogenations. At times palladium catalysts may be employed under similar conditions but more often elevated temperature and pressure is necessary. Ruthenium as dioxide or on a support is an excellent catalyst for ring reductions at 70–100° and 70–100 atm pressure. A rhodium-platinum oxide combination [1] prepared in the same manner as Adams' catalyst appears to be promising for ring reductions, particularly for those systems containing phenolic hydroxyl groups, ethers, benzylic alcohols, and esters [1, 2] where hydrogenolysis is to be avoided.

Nickel catalysts may also be useful but, unless they are especially active, will not cause saturation of the ring under mild hydrogenation conditions. When higher temperatures and pressures are employed, conditions necessary for ring reduction are usually much more vigorous than when a ruthenium catalyst is used [3].

In general acidic conditions are necessary for ring reduction when a platinum catalyst or rhodium-platinum oxide is employed. Acidic conditions can lead to an increase in hydrogen uptake in palladium-catalyzed reactions but is usually unnecessary in those employing rhodium or ruthenium. Indeed it was shown that the addition of hydrochloric acid had an adverse effect on the reduction of toluene or benzoic acid in methanol with 5% rhodium on carbon [4]. On the other hand, sulfuric acid had neither a retarding nor an enhancing effect on those reductions [3], as did the addition of weaker acids to the same hydrogenations.

CARBOCYCLES

Benzenes

The ease of reduction is dependent on substituents and their number. Compared with benzene the presence of a substituent on the ring increases the difficulty of reduction, except when the substituent is a phenolic hydroxyl group. In general phenols are the easiest of the substituted benzenes to hydrogenate. Alkylbenzenes and anilines are usually more difficult to reduce than benzene. The difficulty in reducing alkylbenzenes usually increases with the size of the alkyl chain and with branching [3]. The presence of a carboxyl group, either as acid or ester, usually has a definite retarding effect on hydrogen uptake. Among disubstituted benzenes, whether the substituents are alike or dissimilar, the ease of reducibility is 1,4- > 1,3- > 1,2- [3, 5].

It has been suggested that hydrogenation of a benzene ring proceeds in stages [6]. However, reduction of intermediate dihydro and tetrahydrobenzenes proceeds so rapidly that it is not usually possible to isolate them. In some instances partial reduction of certain substituted benzenes has been noted. These will be described in their proper categories.

Alkyl Benzenes. A useful method of converting alkylbenzenes to alkylcyclohexanes consists of reduction in methyl or ethyl alcohol with 20% by weight of 5% rhodium on alumina or rhodium on carbon at 25° and 2–3 atm [4]. Reaction time can be decreased by raising the temperature [3]. Reductions may be carried out at low pressure with platinum oxide in acidic medium. Hydrogenations may also be performed with or without solvent at 70–100° and 100 atm in the presence of ruthenium dioxide [3]. This method usually gives very high yield of alkyl substituted cyclohexanes.

Alkoxybenzenes. Platinum catalysts should not be considered for the hydrogenation of alkoxybenzenes in view of the extensive hydrogenolysis noted from the use of platinum oxide in the catalytic reduction of a number of methoxybenzenes [7]. There are many examples in the literature supporting this finding. Hydrogenolysis also often occurs during nickel-catalyzed hydrogenations of these compounds.

The results of Smith and Thompson [7], who found that rhodium on alumina caused far less hydrogenolysis than platinum, suggest that this catalyst should be useful in the hydrogenation of alkoxybenzenes. In examples, 2-(2-hydroxyethyl)anisole was converted to 2-(2-methoxycyclohexyl)ethanol in 78% yield by the use of rhodium on alumina in acetic acid [8] and hydrogenation of some 2-, 3-, and 4-alkoxyanilines (O-alkyl = methoxy to butoxy) in ethyl alcohol with 5% rhodium on alumina at 60° and

Carbocycles 545

3.5 atm gave fairly good yields of alkoxycyclohexylamines [9]. Hydrogenolysis to cyclohexylamine was of low order except during reduction of the 4-alkoxy compounds.

Hydrogenolysis which took place in those hydrogenations [9] probably occurred during an intermediate reduction stage. Attempted hydrogenolyses of the 2-, 3-, or 4-alkoxycyclohexylamines under the same hydrogenation conditions resulted in complete recovery of starting materials [3]. The presence of cyclohexylamine after reduction of the alkoxyanilines may be attributed to the formation of an allyl-type ether which underwent cleavage upon further absorption of hydrogen.

1,4-Addition of hydrogen to intermediate 2- or 3-alkoxy-2,5-cyclohexadienylideneamines followed by rearrangement to 2-alkoxy-1,2-dihydroaniline and 3-alkoxy-3,4-dihydroaniline respectively, both allylic ethers, should account for the formation of cyclohexylamine during the reduction of 2- and 3-alkoxyanilines. The largest amounts of cyclohexylamine found after hydrogenation of the 4-derivatives must be due to ease of formation of the allyl-type ether through the most unhindered pathway, 1,2-addition, or by a combination of this and 1,4-addition and rearrangement, which also will give an allyl ether (Eq. 24.1).

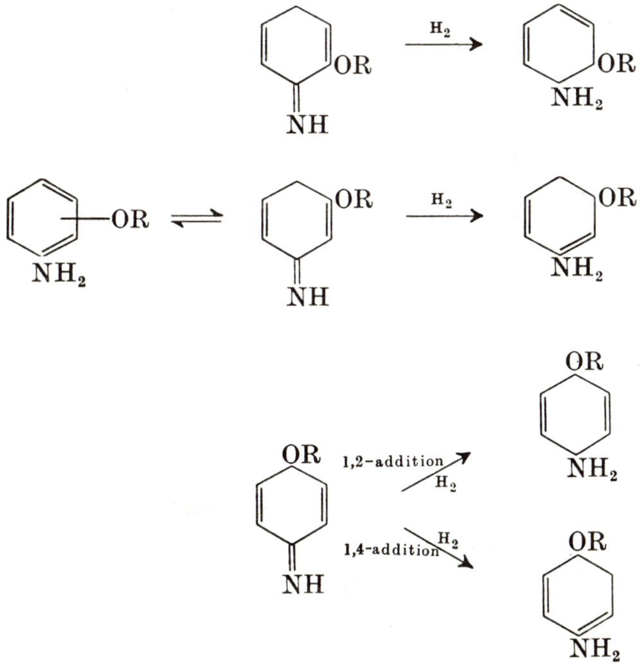

(24.1)

Another catalyst which has enabled hydrogenation of the ring to take place without hydrogenolysis of the ether group is rhodium-platinum oxide [1, 2]. It may be worth further study although it did produce varying amounts of methylcyclohexane (4–26%) during the reduction of some m-cresyl ethers in 96% acetic acid [10]. The largest amount of cleavage took place when the alkoxy group was methoxy or t-butoxy, the smallest when it was n-decyloxy.

It was claimed that hydrogenolysis did not take place when ruthenium on carbon was used to hydrogenate anisole and some dimethoxy-benzenes in aqueous sodium hydroxide [11]. In another example the reduction of 3-methyl-2-(4-methoxyphenyl)morpholine to 2-(4-methoxycyclohexyl)-3-methylmorpholine was successfully carried out in 86% yield over ruthenium dioxide in alcohol at 60–100° and 42–63 atm [12]. However, employment of ruthenium dioxide in the hydrogenation of 2-, 3-, and 4-alkoxyanilines in alcohol at 90–100° and 70–80 atm [13] did lead to a considerable amount of hydrogenolysis especially when methoxyanilines were reduced. Hydrogenolysis decreased with the increase in the size of the alkoxy group.

A comparison of the reductions of alkoxyanilines with rhodium and with ruthenium shows that rhodium on a support is preferred for these particular reductions. It may well be the catalyst of choice for the hydrogenation of alkoxybenzenes in general.

Other Ethers. The catalysts employed for successful hydrogenations of alkoxybenzenes should also be useful for the reduction of phenyl ethers without causing undue hydrogenolysis. Rhodium-platinum oxide gave fairly good yield of cyclohexyl ether (71%) from reduction of phenyl ether at room temperature and 125 atm [2]. About 10% of cyclohexanol was obtained. Hydrogenation under 1 atm pressure gave lower yield of cyclohexyl ether and more cyclohexanol.

Other rhodium and ruthenium catalysts gave high yield of cyclohexyl ether on reduction of phenyl ether at 85° and 80–100 atm [14].

Rhodium on alumina was employed for the hydrogenation of 2-methyl-2,3-dihydrobenzofuran in methyl alcohol to *cis*-2-methylperhydrobenzofuran without causing opening of the cyclic ether [8] (Eq. 24.2). In another

$$\text{benzofuran-O-CH}_3 \xrightarrow[\text{Rh on Al}_2\text{O}_3]{3\text{H}_2} \text{cyclohexane-O-CH}_3 \quad (24.2)$$
$$73\%$$

example dibenzofuran was hydrogenated in cyclohexane containing acetic acid with 5% rhodium on carbon at 100° and 50 atm also without causing ring opening [15].

Benzyl Alcohols, Esters, and Ethers. Rhodium on alumina catalyzed the low-pressure hydrogenation of a number of benzylic alcohols and ethers to the corresponding cyclohexyl compounds in high yields in ethyl alcohol containing a small amount of acetic acid with little or no accompanying hydrogenolysis except in case of methyl benzyl ether [16]. In the latter instance there was a tendency for hydrogen uptake to continue beyond three molar equivalents. Some of the compounds reduced included DL-mandelic acid, α-methylbenzyl alcohol, and benzhydrol.

In another example ethyl 4-hydroxy-4-phenylpyrrolidine-1-carboxylate was reduced with rhodium on alumina in alcohol under 3.5 atm pressure to yield ethyl 4-cyclohexyl-4-hydroxypyrrolidine-1-carboxylate [17] and 3-methyl-2-phenylmorpholine, which may be considered as a benzyl ether, was hydrogenated in acetic acid under 40–100 atm pressure with the same catalyst to produce 75–90% yield of the 2-cyclohexyl-3-methylmorpholine [12] (Eq. 24.3).

$$\underset{\substack{\\}}{\overset{H}{\underset{O}{\bigcirc}}\!\!\!\!\!\!\!\!\!\!\!\!\!\!\!\underset{C_6H_5}{\overset{N}{\diagdown}CH_3}} \xrightarrow[\text{Rh on Al}_2\text{O}_3]{3H_2} \underset{\substack{\\ 75\text{–}90\%}}{\overset{H}{\underset{O}{\bigcirc}}\!\!\!\!\!\!\!\!\!\!\!\!\!\!\!\underset{C_6H_{11}}{\overset{N}{\diagdown}CH_3}} \qquad (24.3)$$

Ruthenium dioxide catalyzed the hydrogenation of 1-phenylpropane-1,3-diol, which contains a secondary benzylic hydroxyl group, in alcohol at 100° and 70–140 atm initial pressure to give 1-cyclohexylpropane-1,3-diol [18]. When Raney nickel was used loss of the benzylic hydroxyl group took place. However, in another instance when ruthenium dioxide was employed under similar conditions in a reduction where a primary benzyl alcohol group was present, considerable hydrogenolysis occurred [19] (Eq. 24.4).

$$\underset{\substack{\\ R = CH_2OH}}{\underset{CH_3}{\bigcirc}\!\!\!\!\!\!\!\!\!\!\!R\text{—}R}\;\xrightarrow[\text{RuO}_2]{3H_2}\;\underset{\substack{\\ 26\%}}{\underset{CH_3}{\bigcirc}\!\!\!\!\!\!\!\!\!\!\!R\text{—}R}\;+\;\underset{\substack{\\ 40\%}}{\underset{CH_3}{\bigcirc}\!\!\!\!\!\!\!\!\!\!\!R\text{—}CH_3}\;+\;\underset{\substack{\\ }}{\underset{CH_3}{\bigcirc}\!\!\!\!\!\!\!\!\!\!\!H_3C\text{—}CH_3}\qquad(24.4)$$

Ruthenium catalyst may be useful in a number of instances, particularly when a secondary benzylic type alcohol is present, but the temperature necessary for reduction with this catalyst may cause hydrogenolysis when labile benzylic functions are present.

It appears likely that reductions among compounds containing benzylic alcohols, esters, and ethers may best be carried out with rhodium on a carrier or with the rhodium-platinum oxide of Nishimura [1, 2], although there is some indication that platinum oxide in ethyl alcohol containing a trace of acetic acid [20] or in t-butyl or t-amyl alcohol containing acetic acid [21] may reduce benzyl alcohols to cyclohexylcarbinols without accompanying hydrogenolysis.

Benzoic Acids, Esters, and Related Compounds. The hydrogenation of benzoic and substituted acids and derivatives to the corresponding cyclohexanes can be carried out under mild conditions with platinum oxide in acid media. However, when the ring contains a phenolic hydroxyl group or an ether linkage, reductions in the presence of platinum catalyst will often lead to cleavage of those functions.

Reductions with nickel or ruthenium catalyst require elevated temperature and pressure. Hydrogenations of esters and amides are usually carried out in alcohol while acids are best reduced in alkaline solution with nickel or in aqueous or aqueous alkaline solution with ruthenium. In either case the use of ruthenium generally gives better results under more moderate conditions and often in a shorter reaction period [3]. The difference in activity is more pronounced among benzenedicarboxylic acids as shown when sodium isophthalate was reduced at 116° and 40–90 atm in the presence of ruthenium dioxide while no uptake occurred with Raney nickel under similar conditions [22]. This example also pointed out the need of maintaining a pH of at least 8.0 in these ruthenium reductions; decarboxylation took place at a lower pH.

In this author's opinion the catalyst of choice for hydrogenations of benzoic acids and related compounds is rhodium on carbon or rhodium on alumina. Reductions of esters are carried out in alcohols, amides in alcohol, or water and acids in aqueous medium. The ease of reduction of the acids in aqueous solution is dependent on the solubility of the cyclohexanecarboxylic acids.

The reducibility of benzoic acid to cyclohexanecarboxylic acid with rhodium has been shown [4]. Other examples include the hydrogenations of phthalic, isophthalic, and terephthalic acids and of pyromellitic acid (benzene-1,2,4,5-tetracarboxylic acid) in water at 60–70° and 3 atm pressure [23]. When reductions were carried out at 70–100° and 70 atm, uptake of hydrogen was rapid and was complete by the time the reaction temperature was reached [3].

Rhodium oxide has been employed for the hydrogenation of benzenedicarboxylic acids in acetic acid or in water at 60–70° and 100 atm to give high yields or cyclohexanedicarboxylic acids [24], but this catalyst is ineffective under low-pressure conditions because of the difficulty in reducing it to a form in which it will catalyze hydrogenations [3].

PARTIAL REDUCTION. Partial reduction of benzoic acid by catalytic means has not been carried out, but terephthalic acid has been hydrogenated over 5% ruthenium on carbon in aqueous solution at 140° and 150 atm to yield

1,2,3,6-tetrahydroterephthalic and *cis*-1,2,3,4-tetrahydroterephthalic acids after absorption of two equivalents of hydrogen [25] (Eq. 24.5).

$$\underset{\text{COOH}}{\underset{|}{\text{C}_6\text{H}_4}}\text{COOH} \xrightarrow[\text{Ru on C}]{2\text{H}_2} \text{[tetrahydro isomer]} + \text{[tetrahydro isomer]} \quad (24.5)$$

10%   72%

Phenylalkanoic Acids and Derivatives. Catalysts and conditions useful for the hydrogenation of benzoic acids and related compounds should be applicable for the reduction of phenylalkanoic acids. A number of phenylalkanoic acids, $C_6H_5(CH_2)_{1-5}COOH$, were readily hydrogenated in acetic acid over platinum oxide under several atmospheres pressure [26]. A longer chain acid, phenylundecanoic acid, however, required much more drastic conditions [27].

Ethyl phenylacetate was reduced to ethyl cyclohexylacetate in 85% yield in alcohol containing a small amount of concentrated sulfuric acid with platinum oxide at 50° and 3 atm [3].

In another example mandelic acid was reduced to the cyclohexylhydroxyacetic acid with platinum oxide in acetic acid at low pressure without hydrogenolysis of the benzylic hydroxyl group [28].

Hydrogenations of phenylalkanoic acids with nickel require elevated temperatures and pressures, 150–200° and 175–200 atm [3, 27], while ruthenium-catalyzed reductions usually can be carried out at much lower temperature and pressure, 100° and 100 atm [3]. Ethyl phenylacetate was reduced without solvent with 2% by weight of ruthenium dioxide under those conditions in 4 hr; hydrogenation in solvent required much less time.

Hydrogenation with rhodium under low-pressure conditions should proceed readily although there is an example where benzenediacetic acids did not reduce readily at low pressure but required higher temperature and pressure [29]. In that instance the *cis-trans* distribution was changed from a predominance of *cis* isomer to one of *trans* because of the more vigorous reaction conditions.

Rhodium-platinum oxide [1] may also be useful in these reductions under low-pressure conditions. It was employed for the hydrogenations of 3-(3-acetylaminophenyl)propionic and 4-(3-acetylaminophenyl)butyric acids in acetic acid at 60° and 3 atm [30]. The main disadvantage in these reductions is the high ratio of combined metals to substrate, which might be lowered by reduction under higher pressure.

Other Acids. In general, a benzene ring containing an acid group other than carboxyl attached to the ring will not undergo reduction except in the case of phosphorous acids, where the valence of phosphorous is 5. Phenyl- and 2- and 4-(alkyl phenyl)phosphonic acids were readily hydrogenated in alcohol with rhodium on alumina under 4 atm pressure to the corresponding cyclohexyl compounds [31]. Phosphanilic acid and diphenylphosphinic acid, $(C_6H_5)_2P(=O)OH$, were reduced in aqueous ammonia.

The ring of phenylphosphinic acid could not be hydrogenated, probably because the compound was reduced to a phosphine which is a known catalyst poison. Phenylarsonic acid also could not be reduced to cyclohexylarsonic acid because the arsono group was reduced to a poisoning trivalent form of arsenic.

Phenol and Substituted Phenols. When phenols undergo hydrogenation, the predominant reduction product may be a cyclohexanol, a cyclohexanone, or a cyclohexane depending on the catalyst and reaction conditions and on the substrate. At times total uptake of hydrogen will depend on substituents and their steric effects. For example, in the nickel high-pressure hydrogenation of t-butyl cresols [32], partial hydrogenation was not possible when one position *ortho* to the phenol was unsubstituted, even if a t-butyl group occupied the other *ortho* position. However, when one *ortho* position was occupied by t-butyl and the other was methyl, either the cyclohexanone or cyclohexanol could be obtained. When t-butyl groups were in each *ortho* position, it was possible to add one or two equivalents of hydrogen but not three. This is exemplified in the hydrogenations of 2,6-di-t-butyl-4-methylphenol which yielded 2,6-di-t-butyl-4-methylcyclohexanone upon uptake of two equivalents and, upon interruption of the reaction at one equivalent, gave a product which was suggested to be 2,6-di-t-butyl-4-methyl-2-cyclohexen-1-one (Eq. 24.6).

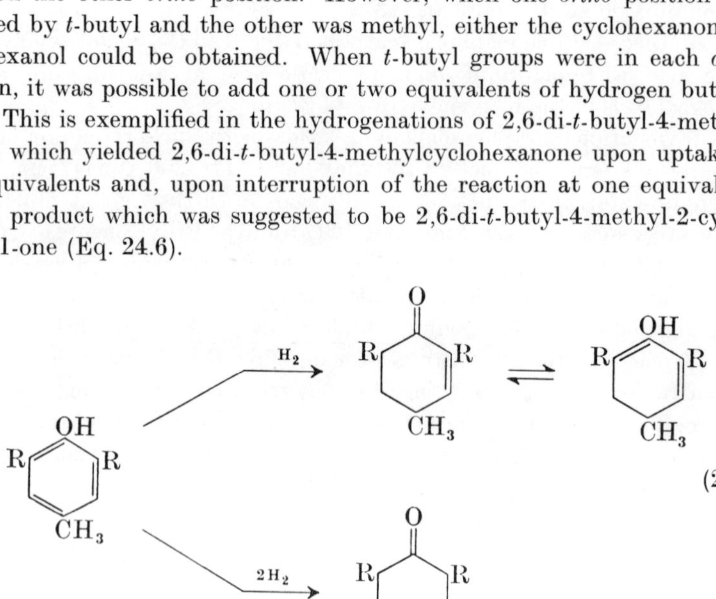

(24.6)

R = t-butyl

It might also be pointed out that rapid reaction favors the conversion of phenols to cyclohexanols over hydrogenolysis to cyclohexanes. This is often seen in hydrogenations run under elevated pressure as opposed to reactions carried out under mild conditions.

Reduction of a phenol proceeds through an intermediate cyclohexanone which in many instances can be isolated. However, it is believed that the true intermediate is the unstable 1-hydroxy-1-cyclohexene, the tautomeric form of cyclohexanone, which undergoes hydrogenolysis to cyclohexene and water or hydrogenation to cyclohexanol [33] (Eq. 24.7). In most instances

(24.7)

platinum-catalyzed reductions of phenolic compounds result in substantial loss of the hydroxyl group [33].

Raney nickel reductions of phenol and alkylphenols gave the corresponding cyclohexanols in high yield when sodium phenoxide (0.4 mole %) was present in hydrogenations carried out at 100° and 130–150 atm [34]. Uptake of hydrogen was very rapid in the presence of phenoxide ion, very much slower in its absence. The use of a small amount of 40% aqueous sodium hydroxide solution also aided reduction but not to the same extent as the sodium phenoxide. The authors reported that polar solvent such as methanol increased the reaction time somewhat, but it did not appear to make much difference when 4-ethylphenol was hydrogenated with Raney nickel in analytical reagent-grade methanol containing sodium methylate [3]. In none of these reductions was hydrogenolysis noted.

In another example, alkylphenols were hydrogenated with nickel in alcohol in the absence of sodium ion to give high yields of alkylcyclohexanols [35].

The nickel-catalyzed reduction of 4-trimethylsilylphenol, which may be considered as related to t-butylphenol, is of interest [36]. In this example reaction was carried out with alcohol-free Raney nickel at 90–100° and 130 atm pressure of hydrogen to yield 4-trimethylsilylcyclohexanol. Traces of alcohol in the catalyst caused cleavage of the trimethylsilyl group. If the reaction temperature was allowed to rise above 120°, rearrangement took place (Eq. 24.8).

Phenols have been hydrogenated over rhodium on carbon at low pressure without measurable hydrogenolysis [37]. In many instances lower catalyst

$$\underset{\mathrm{Si(CH_3)_3}}{\mathrm{C_6H_4(OH)}} \xrightarrow[90-100°,\ 130\ \mathrm{atm}]{3H_2} \underset{\mathrm{Si(CH_3)_3}}{\mathrm{C_6H_{10}(OH)}} \left(\mathrm{at}\ 120° \quad \mathrm{C_6H_{10}(OSi(CH_3)_3)} \right)$$

(24.8)

ratios can be employed and more rapid hydrogen uptake will take place when reactions are carried out under high pressure [3].

The value of rhodium for the reduction of phenols may be seen in the hydrogenation of 2-hydroxyacetophenone in alcohol to 2-(1-hydroxyethyl)-cyclohexanol at 50° and 3.3 atm [38] in which the readily cleaved benzylic alcohol function as well as the phenolic hydroxyl group remained intact.

Rhodium-platinum oxide, 10% by weight of substrate, was employed successfully for the hydrogenation of phenol in acetic acid to cyclohexanol at normal temperature and pressure [1]; much less catalyst was employed, 2% by weight, when the reduction was carried out at 130 atm where 91% yield of cyclohexanol was obtained in a very short reaction period [2].

Ruthenium catalyst should also be considered for the reduction of phenols. As dioxide it was reported to be useful for the hydrogenation of phenol in water or aqueous sodium hydroxide under both low- and high-pressure conditions [11]. It gave good yield of 2-hydroxycyclohexylmethanol when 2-hydroxybenzyl alcohol was reduced at 100° and 150 atm [3]. Hydrogen uptake was complete when the temperature reached 75–80° but under more moderate conditions, 120° and 50 atm, reduction was still incomplete after 6–7 hr. In instances where a benzylic alcohol or a related group is also present it may be more advisable to employ rhodium to prevent debenzylation. In a reaction related to the above reduction, that of 2,6-*bis*-(hydroxymethyl)-4-methylphenol [9], there was no loss of the phenolic hydroxyl group but there was considerable cleavage of the benzylic hydroxyl functions.

Phenolic Acids and Esters. Of the catalysts used for hydrogenation among this group, palladium on strontium carbonate has been found useful for the conversion of ethyl 4-hydroxybenzoate to ethyl 4-hydroxycyclohexanecarboxylate but only at high pressure where reaction was rapid and quantitative yield was obtained [39]. High-pressure conditions, 160–220° and 240–270 atm, with Raney nickel in alcohol containing 0.3 mole % of sodium ethoxide gave 75–85% yields of ethyl 2-, 3-, and 4-hydroxycyclohexanecarboxylates and 83–89% yields of ethyl (4-hydroxycyclohexane)-acetate and butyrate [40].

Probably the preferred methods for reducing phenolic acids and esters without accompanying hydrogenolysis involve the use of rhodium or ruthenium catalyst.

4-(3-Hydroxyphenyl)butyric acid was hydrogenated at 50° and 3.3 atm over rhodium on alumina to give 4-(3-hydroxycyclohexyl)butyric acid [38]. 4-Hydroxybenzoic acid was reduced in acetic acid with the same catalyst under 3 atm pressure to the lactone of cis-4-hydroxycyclohexanecarboxylic acid [41]. Some 4-alkyl-3-hydroxybenzoic acids were similarly hydrogenated [42]. When the alkyl group was t-butyl, high yield of the lactone was obtained (Eq. 24.9). When the alkyl group was methyl, ethyl, or isopropyl

$$\underset{\substack{C(CH_3)_3}}{\underset{OH}{\bigodot}{COOH}} \xrightarrow[\text{Rh on Al}_2\text{O}_3]{3H_2} \underset{\substack{C(CH_3)_3 \\ 85\%}}{\bigodot{\underset{O}{\overset{C=O}{|}}}} \qquad (24.9)$$

yield was lower; some hydrogenolysis of the hydroxyl group was observed in each case.

When ruthenium dioxide was employed for the hydrogenation of ethyl 3-hydroxybenzoate in alcohol at 90–100° and at an initial pressure of 130 atm high yield of ethyl 3-hydroxycyclohexanecarboxylate resulted [43]. 4-Hydroxybenzoic acid was reduced in water under similar conditions to give crude lactone but the yield was not noted.

There appear to be few comparisons concerning the usefulness of rhodium versus ruthenium for the reduction of phenolic acids except in the hydrogenation of pyrogallic acid (3,4,5-trihydroxybenzoic acid) to 3,4,5-trihydroxycyclohexanecarboxylic acid. Reduction with rhodium was carried out in 95% alcohol (absolute alcohol led to esterification) at 90–100° and an initial pressure of about 150 atm for 8–12 hr [44]. Pressures below 120 atm led to incomplete reduction. Hydrogen uptake was too slow at lower temperature; esterification was favored at 125°. Supported ruthenium and other catalysts including Raney nickel were less satisfactory or completely ineffective.

Rhodium-platinum oxide may also be useful in carrying out hydrogenations of phenolic acids with a minimum of hydrogenolysis if reduction is carried out at room temperature and elevated pressure. Hydrogenation of 4-hydroxybenzoic acid in acetic acid at low pressure was slower and more hydrogenolysis took place (8 hr, 23% cyclohexanecarboxylic acid) than at room temperature and 110–140 atm where reduction was complete in 0.5 hr, yield of 4-hydroxycyclohexanecarboxylic acid was 82% and only 15% of cyclohexanecarboxylic acid was obtained [2].

Di- and Polyhydric Phenols. Although platinum catalyst usually causes hydrogenolysis when phenols are hydrogenated, it was surprising to find that catechol was reduced to 1,2-cyclohexanediol in 95% yield in alcohol containing a small amount of aqueous sodium hydroxide under 3 atm pressure [28].

There are a number of examples where reductions of dihydric phenols are carried out at elevated temperature and pressure in alcohol containing sodium hydroxide with Raney nickel or in alcohol alone with W-7 Raney nickel, an alkaline containing active catalyst [45–47]. Hydrogenations can be and have been carried out in the absence of alkali since most nickel catalysts contain occluded alkali.

The addition of a small amount of sodium metal (0.2 mole%) to absolute alcoholic solutions of catechol, resorcinol, and hydroquinone reduced with about 15% by weight of Raney nickel resulted in fairly rapid reaction (0.5–2.0 hr) and quantitative yields of the corresponding cyclohexanediols when the reductions were carried out at 100–120° and 130 atm [3]. Hydrogenations in the absence of sodium usually took somewhat longer and required higher temperature and pressure. Reduction of resorcinol in alcohol with ruthenium catalyst was reported to give the diol and a small amount of cyclohexanol [43].

The catalyst of choice for the hydrogenation of polyhydric phenols to cyclohexane polyols may be rhodium on alumina. The reduction of resorcinol in ethanol over 5% rhodium on alumina gave 93% of the cyclohexane-1,3-diol and 7% of cyclohexanol [48].

It was surprising to find that the use of rhodium on carbon resulted in considerable hydrogenolysis to cyclohexanol (30%) in the same reduction. A mixed rhodium-palladium on carbon caused even more hydrogenolysis. Hydrogenations with palladium or platinum on carbon yielded 76–79.5% of cyclohexanol.

Hydroquinone as well as resorcinol were rapidly reduced to the corresponding diols with rhodium on alumina in alcohol or acetic acid at room temperature and 3.5 atm [49].

Pyrogallol was hydrogenated very rapidly in alcohol with 5% rhodium on alumina at 55–60° and 200 atm to give cyclohexane-1,2,3-triol [50] but there was no indication whether such pressure was necessary.

Phenols to Cyclohexanones. In general, attempts to partially reduce phenol or any unhindered monohydric phenol do not give appreciable amounts of cyclohexanones except when palladium on a support is used for hydrogenation. While it is fairly well known that hydrogenation in the presence of a strong acid leads to considerable hydrogenolysis of the phenolic hydroxyl group, it is interesting to note that phenol was reduced in very dilute hydrochloric acid with palladium on barium sulfate to yield 66% of cyclohexanone [51].

Molten phenol hydrogenated at 140–150° and pressures up to 20 atm with 0.1% by weight of 5% palladium on carbon gave 70–75% of cyclohexanone, 1–4% of cyclohexanol, and recovered phenol [52]. Reductions of phenol in aqueous alcohol promoted by sodium as a salt (0.4–0.7% based on phenol) gave 97% of cyclohexanone upon hydrogenation with a small amount of palladium on carbon (0.1% by weight) at 185° and 4.5 atm [53]. Without sodium promotion the yield of cyclohexanone was 91%. The use of excess sodium produced 80% cyclohexanol.

In palladium-catalyzed hydrogenations of 2-cyclohexyl- or 2-branched alkylphenols high yields of cyclohexanones (84–98%) were obtained without the aid of sodium [54]. 2,4-Dimethylphenol, neat or in methylcyclohexane, was reduced directly to cis-2,4-dimethylcyclohexanone with 1–3% by weight of palladium on carbon at 125–175° and 6.6–10 atm [55, 56].

Partial reduction of dihydric phenols does take place, but only with resorcinol can a dihydro compound be obtained. Hydrogenations of catechol and hydroquinone with rhodium interrupted after absorption of one equivalent of hydrogen were shown to give 2- and 4-hydroxycyclohexanone respectively [33] (Eq. 24.10).

$$\text{(24.10)}$$

Dihydroresorcinol was obtained in 87% yield during reduction of resorcinol in aqueous sodium hydroxide with rhodium on carbon under 3 atm pressure [33]. Hydrogen uptake stopped of its own accord. Neutralization of the solution with acid gave the enolone which is isomeric with cyclohexane-1,3-dione. Reduction with rhodium on alumina under similar conditions gave the same yield [50]. Hydrogenation under alkaline conditions with palladium on carbon gave lower yields, 55–60% [3, 57].

Raney nickel reductions of resorcinol and alkylresorcinols in water containing a molar equivalent of sodium hydroxide carried out at 50° and 50–100 atm have resulted in high yields of dihydroresorcinols [3, 58–60].

Among trihydric phenols, pyrogallol was partially hydrogenated with nickel to yield 2,3-dihydrocyclohex-2-en-1-one, dihydropyrogallol [61].

The presence of a group in 4- or 5-position appeared to have an effect on partial reduction. The partial hydrogenation of gallic acid was not too successful [62]; 4-alkylpyrogallol could not be partially reduced [61].

Anilines. The catalytic reduction of anilines gives cyclohexylamines. The hydrogenation of aniline itself to cyclohexylamine is a very important industrial process because of the large amounts of cyclohexylamine used in the production of sodium and calcium cyclohexylsulfamate, artificial sweeteners.

The pathway of reduction is through an enamine or imine through which formation of secondary amine, dicyclohexylamine, as well as conversion to primary amine takes place in the same manner as in the reduction of nitriles (see Chapter XII) (Eq. 24.11).

$$\text{C}_6\text{H}_6 \xrightarrow{2\text{H}_2} \left[\text{C}_6\text{H}_{10}\text{NH}_2 \rightleftharpoons \text{C}_6\text{H}_{10}\text{=NH} \right] \xrightarrow{\text{H}_2} \text{C}_6\text{H}_{11}\text{NH}_2 \quad (24.11)$$

Another side reaction product in the hydrogenation of aniline is 4-phenylcyclohexylamine which is formed by the addition of aniline to the intermediate imine and subsequent reduction of the addition product (Eq. 24.12).

$$\text{C}_6\text{H}_{10}\text{=NH} + \text{C}_6\text{H}_5\text{NH}_2 \longrightarrow$$

$$\text{C}_6\text{H}_{10}(\text{NH}_2)(\text{NHC}_6\text{H}_5) \xrightarrow[-\text{NH}_3]{\text{H}_2} \text{C}_6\text{H}_{11}\text{NHC}_6\text{H}_5 \quad (24.12)$$

It is essential in the hydrogenation of aniline that the substrate be dry since the intermediate imine may be hydrolyzed to cyclohexanone, which can lead to cyclohexylideneaniline or cyclohexylidenecyclohexylamine and subsequently to 4-phenylcyclohexylamine or dicyclohexylamine (Eq.24.13). All of these products including cyclohexanol have been found in the hydrogenation of aniline to cyclohexylamine [63].

Various means of inhibiting formation of dicyclohexylamine by mass action techniques have not been too successful. In one instance addition of 15% by weight of dicyclohexylamine to a feed stock of aniline did have some effect [64]. In other cases there was little inhibition of secondary amine [3, 65]. Reduction in the presence of ammonia was also ineffective in suppressing secondary amine formation or raising the yield of cyclohexylamine [65]. A technique that could eliminate secondary amine formation would necessitate removal of cyclohexylamine as rapidly as it formed.

Such a procedure is described in a hydrogenation of aniline with palladium on carbon at 155° and 1 atm [66]. However, the process, which consists of

$$\text{C}_6\text{H}_{10}=\text{NH} + \text{HOH} \longrightarrow$$

$$\text{C}_6\text{H}_5\text{NH}_2 \longrightarrow \text{C}_6\text{H}_{10}=\text{NC}_6\text{H}_5 \xrightarrow{\text{H}_2} \text{C}_6\text{H}_{11}\text{NHC}_6\text{H}_5$$

$$\text{C}_6\text{H}_{10}=\text{O}$$

(24.13)

$$\text{C}_6\text{H}_{11}\text{NH}_2 \longrightarrow \text{C}_6\text{H}_{10}=\text{NC}_6\text{H}_{11} \xrightarrow{\text{H}_2} \text{dicyclohexylamine}$$

sweeping hydrogen saturated with water through the reaction mixture, not only distills out cyclohexylamine but also aniline, making extraction, drying, and fractional distillation necessary. This may make the procedure unattractive economically.

The preferred method for the hydrogenation of aniline to cyclohexylamine involves reduction without solvent at 45–50° and 2 atm with rhodium on alumina [67]. Very high yield of cyclohexylamine (96.8%) was obtained with only 3% of dicyclohexylamine. The use of alcohol as solvent decreases the reaction time but it is difficult to distill off solvent without losing cyclohexylamine [3].

Ruthenium dioxide reduction of aniline at 80° and 100 atm was very rapid with a 1% catalyst ratio and gave high yield of cyclohexylamine [3]. By raising the temperature to 150–165° hydrogenation was complete in less than 1 hr with a 0.05% by weight of catalyst to substrate.

A 7:3 rhodium-platinum oxide catalyst was also effective in reducing aniline (in alcohol) to cyclohexylamine in 92% yield at 50° and 80–90 atm [68]. Hydrogenation was not carried out in the absence of solvent. The use of acetic acid or addition of it to reductions generally caused an increase in the amount of dicyclohexylamine.

Although aniline may be reduced over nickel at high pressure, conditions are usually more severe than with ruthenium or with rhodium-platinum oxide and more secondary amine is also produced [3]. It is also well known that platinum-catalyzed reductions generally must be carried out under acidic condition and yield as much as and even more dicyclohexylamine than cyclohexylamine.

558   Aromatic Ring Systems

In general rhodium reductions of anilines give good yields of cyclohexylamines. For example, N-methylaniline and N-ethylamine were converted to the corresponding N-alkylcyclohexylamine in quantitative yield [3]. One of the advantages of rhodium is that N,N-dialkylanilines can be hydrogenated without cleavage of the tertiary amino group. When N,N-dimethylaniline was reduced at 50–60° and 2–3 atm pressure, quantitative yield of N,N-dimethylcyclohexylamine was obtained [3]. In another instance hydrogenation of ethyl 4-dimethylaminobenzoate in acetic acid with rhodium on alumina at 70 atm gave 80% yield of ethyl 4-dimethylaminocyclohexanecarboxylate [69].

In some instances substituents on the ring may limit the effectiveness of supported rhodium catalysts in low-pressure reductions [3]. There did not appear to be any difficulty in the hydrogenations of 4-n-octylaniline or 4-n-octylacetanilide with rhodium on alumina at 56–67° and 4 atm [70]. However, a larger substituent on the ring did necessitate the application of more vigorous conditions. When 4-dodecyaniline (purified by pretreatment with Raney nickel) was hydrogenated in alcohol or acetic acid with 40% by weight of 5% rhodium on carbon at 60° and 3 atm hydrogen absorption was still incomplete after 18 hr [3]. Reduction at 100° and 100 atm was successful with the same amount of supported rhodium catalyst or with about 3.5% by weight of ruthenium dioxide [3].

Despite the usefulness of rhodium for low-pressure reductions of anilines it does have some failings. It is generally ineffective for hydrogenations of benzenoid compounds containing several amino groups [3]. Despite a report that phenylenediamines and 4,4'-diaminodiphenylmethane were reduced over a supported rhodium catalyst [71], this author was never able to completely hydrogenate phenylenediamines to diaminocyclohexanes under mild conditions with rhodium catalysts except when the amino groups were acylated. The hydrogenation of 4,4'-diaminodiphenylmethane at 60° and 3 atm in alcohol or alcohol containing acetic acid was too slow to be worth considering. Reductions of this compound and of phenylenediamines employing ruthenium catalyst at elevated temperature and pressure gave very good yields of cyclohexyl compounds [3, 72]. Rhodium on a support should probably be effective under similar hydrogenation conditions (100° and 100–125 atm).

Hydrogenation of a number of nuclear-substituted anilines in methyl or ethyl alcohol with ruthenium dioxide at 90–100° and 70 atm gave high yields of cyclohexylamines [73] except when dialkylamino or methoxy, ethoxy, and propoxy groups were attached to the ring. Hydrogenolysis of methoxy and ethoxy groups have been minimized by carrying out reductions in anhydrous t-butyl alcohol in the presence of ruthenium hydroxide and hydrated lithium hydroxide [74]. A rhodium oxide catalyst prepared by fusion of rhodium trichloride and lithium nitrate also gave high yields of methoxy and ethoxy-cyclohexylamines with somewhat less hydrogenolysis under high-pressure reduction conditions [75] but ruthenium reduction was reported to be preferred [74].

Possibly solvent may be a factor in controlling hydrogenolysis during reduction of N,N-dialkylanilines. 4-Diethylamino- and other N,N-disubstituted phenylenediamines were hydrogenated with ruthenium dioxide, 5% by weight of substrate, in dioxane at 100° and 100–167 atm to give high yields of the corresponding N,N-dialkylcyclohexanediamines except with the N,N-dibutyl compound [76]. In contrast hydrogenations of N,N-dimethyl-aniline [3] and of N,N-dimethyl, N,N-diethyl, and N,N-diethyl-2-methyl-1,4-phenylenediamines in alcohol over ruthenium dioxide, 2% by weight, at 70–100° and 70–80 atm caused considerable hydrogenolysis [73]. Where hydrogenolysis took place, it must have occurred during an intermediate stage of reduction because none of the 4-dialkylaminocyclohexylamines or N,N-dimethylcyclohexylamine were effected when they were subjected to hydrogenation at 100° and 100 atm for about 5 hr [3].

Secondary amine formation does not usually occur to any great extent during ruthenium-catalyzed hydrogenations of anilines. However, it was the major reaction product when 4-(2-pentyl)aniline was reduced, only 11.5% of 4-(2-pentyl)cyclohexylamine was obtained [73] (Eq. 24.14).

$$R\text{—}C_6H_4\text{—}NH_2 \xrightarrow{RuO_2} R\text{—}C_6H_{10}\text{—}NH_2 + [R\text{—}C_6H_{10}\text{—}]_2 NH \quad (24.14)$$

R = 2-pentyl   11.5%   63%

Salts of phenylsulfamic acid may be reduced to cyclohexylsulfamates over either rhodium on alumina at 25° and 2 atm [77] or with ruthenium dioxide at 75–80° and 20 atm, or ruthenium on alumina at 80° and 80 atm [78] (Eq. 24.15). This process to obtain artificial sweeteners is greatly dependent on

$$C_6H_5\text{—}NHSO_3NH_4 \xrightarrow{3H_2} C_6H_{11}\text{—}NHSO_3NH_4 \quad (24.15)$$

reducing a salt of phenylsulfamic acid as soon as it is prepared [3].

Among other substituted anilines, aminophenols were readily hydrogenated in alcohol to aminocyclohexanols with ruthenium dioxide at 100° and 70–100 atm with little if any loss of hydroxyl function [3]. Reductions under similar conditions with rhodium on a carrier were somewhat slower.

Hydrogenation as N-acetyl compounds with ruthenium or with rhodium usually resulted in more rapid reaction. As N-acetyl compound the inhibiting effect of the basic nitrogen atom on rhodium catalyst was so changed that

560   Aromatic Ring Systems

reduction could be carried out at 25° and 3 atm with 25% by weight of 5% rhodium on carbon in about 4 hr [3].

Hydrogenation as the N-acetyl derivatives apparently aided the nickel-catalyzed reductions of 2-, 3-, and 4-acetylaminophenol at 180° and 70 atm initial pressure giving over 90% yields of the corresponding acetylamino-cyclohexanols [79]. Reduction as aminophenol under the same conditions could not be accomplished as readily [3]; often loss of hydroxyl function occurred. Reaction as acetyl derivatives also aided hydrogenations of phenylenediamines. Indeed reduction of 1,2-phenylenediamine could not be accomplished over ruthenium dioxide at 150° and 150 atm, whereas N^1,N^2-diacetylphenylene-1,2-diamine gave very good yield (75–80%) of 1,2-diacetylaminocyclohexane by hydrogenation in methylcellosolve at 100° and 100 atm [3]. In another example of the effect of N-acetylation on rhodium reduction, hydrogenation of the same diacetyl compound was achieved with 5% rhodium on alumina at 55–60° and 3 atm to give even better yield, 91.5% [3].

When reduction of the 1,2-diacetyl compound was attempted in alcohol over Raney nickel under high-pressure conditions, little uptake of hydrogen occurred until the temperature was raised to 200°. Reaction at that temperature and 140 atm pressure yielded only 20% of a mixture of isomers of 1,2-diacetylaminocyclohexane and some 2-methylbenzimidazole (Eq. 24.16).

(24.16)

Ruthenium catalyst was found to be useful for the reduction of 4-aminobenzoic acid and methyl 4-aminobenzoate [73]. It was also possible to use platinum oxide in the reduction of 4-aminobenzoic acid in neutral solvent but in general either high-pressure conditions, 70 atm and/or a large amount of catalyst were necessary for complete hydrogen uptake [3].

In view of the ability of supported rhodium catalysts to reduce anilines and benzoic acids under mild conditions, either rhodium on carbon or on alumina should catalyze the hydrogenation of aminobenzoic acids and esters to the corresponding cyclohexanes.

However, there was some difficulty experienced in the reduction of ethyl anthranilate with rhodium on a support. When the hydrogenation of ethyl anthranilate was attempted in alcohol with 60% by weight of 5% rhodium on alumina at room temperature and 4 atm, only two equivalents of hydrogen were absorbed to yield ethyl 3,4,5,6-tetrahydroanthranilate and a high boiling component [80] (Eq. 24.17). Reduction at 25° and 70 atm also

$$\underset{NH_2}{\bigcirc}^{R} \xrightarrow{2H_2} \underset{-NH_2}{\bigcirc}^{R} + \underset{\underset{H}{N}}{\bigcirc}^{R}\underset{}{\bigcirc}^{R} \quad (24.17)$$

R = COOC$_2$H$_5$ 41%

yielded the same products. However hydrogenation at 85° and 33 atm caused the reaction of ethyl anthranilate to go to completion to give 70% yield of *cis* ethyl 2-aminocyclohexanecarboxylate. It is likely that reaction as the *N*-acetyl compound would overcome inactivation of the catalyst and allow the reduction to go to completion under low-pressure conditions especially if the reaction was warmed.

The hydrogenation of some acetylaminophenylalkanoic acids to the corresponding cyclohexanes was successfully carried out at low pressure (60°, 3 atm) with 9% by weight of rhodium-platinum oxide in glacial acetic acid [81]. In view of the large amount of this catalyst usually necessary for low-pressure reductions, it might be worthwhile to try rhodium on a carrier for hydrogenation of similar compounds. A much smaller metal to substrate ratio might be attained.

Phenylalkylamines. Phenylalkylamines have been converted to cyclohexylalkylamines in good yield with platinum catalysts at low pressure under acidic conditions but in general a large amount of catalyst was necessary and reaction time was often long [3, 82, 83]; in certain instances a change in rotation was noted when optically active compounds were reduced [3, 83].

Hydrogenations of these compounds with Raney nickel require elevated temperatures and superatmospheric pressure. Benzylamine, 1-phenethylamine, and tertiary 2-phenethylamines were reduced to the corresponding cyclohexylalkylamines at 140–160° and 50–100 atm [84]. Higher temperatures were necessary for ring reduction when the amino group in 2-phenethylamines was primary or secondary or when the amino group was further removed from the ring. Those conditions usually resulted in deamination [3, 84].

The method of choice for this group of compounds probably consists of reduction of the base in alcohol or of a salt in water with a 2% ratio of ruthenium dioxide at 90° and 70 atm although the temperature may be raised to 120–125° [85] (Eq. 24.18). In general reaction time was short

$$\text{C}_6\text{H}_5\text{CH}_2\text{CH}(\text{CH}_3)\text{NHCH}_3 \xrightarrow[\substack{\text{RuO}_2 \\ 90°,\ 70\ \text{atm}}]{3\text{H}_2} \text{C}_6\text{H}_{11}\text{CH}_2\text{CH}(\text{CH}_3)\text{NHCH}_3$$

(24.18)

(unless a branched chain was attached to the ring), yields were high, and no change in rotation was noted in the hydrogenation of optically active compounds. In a number of instances 30% by weight of 5% ruthenium on alumina (equal to a 1.5% metal to substrate ratio) was successfully substituted for ruthenium dioxide with the same good results as reported in Ref. 85 [3].

Attempted low-pressure reductions of most of the phenylalkylamines as bases or salts with rhodium on carriers were unsuccessful [3]. The hydrogenation of N-methylbenzylamine as acetate in alcohol did go to completion in 18 hr at 60° and 3 atm [3].

Stereochemistry. There are so many variables associated with the hydrogenation of polysubstituted benzenes to cyclohexanes that only a few general statements can be made about stereochemistry resulting from their reductions.

Hydrogenations of disubstituted benzenes to cyclohexanes under mild conditions is said to give mainly *cis* isomers [86]. Usually some *trans* material is also obtained. However, substituents on the ring, the catalyst, and reaction conditions have an effect on isomer distribution.

cis-Isomers predominated when phthalic and terephthalic acids were hydrogenated over rhodium in neutral solvent [23]. This predominance was generally seen in the reduction of other 1,2- and 1,4-disubstituted benzenes [3]. In one instance in the platinum-catalyzed hydrogenation of 1,2-diacetylaminobenzene in glacial acetic acid under 3 atm pressure, the *cis-trans* ratio was 4.5:5.5 [3]. When 1,3-disubstituted benzenes were reduced under mild conditions often nearly as much *trans* as *cis* isomer was obtained [3, 23, 87].

Reduction at elevated temperature and pressure usually leads to the most stable stereoisomer which, in 1,2- and 1,4-disubstituted cyclohexanes, is *trans* and in 1,3-disubstituted ones is the *cis* form [86]. This situation applies more to nickel hydrogenations than to those catalyzed by ruthenium. In general conditions used for saturation of the benzene ring with ruthenium (90–100°, 100 atm) are much milder than those employed in nickel reductions and still produced a predominance of *cis* isomer in the reduction of 1,2- and 1,4-disubstituted benzenes. For example, in the hydrogenation of 4-aminobenzoic acid about 9 parts of *cis*-4-aminocyclohexanecarboxylic acid to 1 part

Carbocycles 563

of *trans* acid were obtained [73]. In the hydrogenation of 1,2-diacetylaminobenzene, a 4:1 ratio of *cis* to *trans* diacetylaminocyclohexane resulted [3]; the same ratio was obtained upon reduction of phenylene-1,4-diamine under the same conditions (90–100°, 100 atm).

When phenylene-1,3-diamine or the diacetyl derivative was hydrogenated with ruthenium, the *cis-trans* ratio was similar to that obtained in rhodium reductions under mild conditions, about 1:1.

When more than two substituents are attached to a benzene ring isomer distribution after reduction can become complex because of the reaction conditions necessary for ring saturation.

Selective Reductions

PRESENCE OF FUNCTIONAL GROUPS. It is unusual to find that a benzene ring is reduced in preference to a functional group. In many instances such occurrences result from inaccessibility of the functional group to the catalyst sites. For example, in the hydrogenation of ethyl 3-(2-nitrobutyl)indole-2-carboxylate in acetic acid with 30% palladium on carbon the benzene ring was saturated while the aliphatic nitro group was unaffected [88]. However, when the nitro group was less hindered in the corresponding 3-(2-nitropropyl) compound, it was preferentially hydrogenated (Eq. 24.19).

$$(24.19)$$

In an example which seems rather unusual, the carbonyl function in (2-methoxyphenyl)acetone was left untouched when the compound was reduced in acetic acid with 5% rhodium on alumina to yield *cis* (2-methoxycyclohexyl)acetone [8]. In methanol the hydrogenation was less selective; the carbonyl group was partly reduced.

There is an example in which a pendent benzene ring was hydrogenated while aryl chloride contained in the same substrate remained intact [89]

(Eq. 24.20). Since base is usually necessary for hydrogenative dehalogenation, the acidity of the substrate promoted ring reduction and prevented loss

$$\text{Cl-Ar(R)(S(O)_2NH)-N=CH-CH_2C_6H_5} \xrightarrow[\text{100°, 70 atm, 10 hr}]{\text{PtO}_2\text{ 10% by wt,}} \text{Cl-Ar(R)(S(O)_2NH)-N=CH-CH_2C_6H_{11}} \quad (24.20)$$

R = NH$_2$SO$_2$—   50%

of halogen. In another example the benzene ring in 1-phenyl-3-(2-bromoethyl)-4,4-dicarbethoxyazetidine-2-one was preferentially reduced with an almost equal amount of platinum oxide in acetic acid containing hydrochloric acid [90] (Eq. 24.21). This result is not entirely unexpected since alkyl

$$\text{[azetidinone with CH}_2\text{CH}_2\text{Br, R, R, N-C}_6\text{H}_5\text{]} \xrightarrow{\text{PtO}_2} \text{[azetidinone with CH}_2\text{CH}_2\text{Br, R, R, N-C}_6\text{H}_{11}\text{]} \quad (24.21)$$

R = COOC$_2$H$_5$

halides are not readily hydrogenolyzed, particularly under acidic conditions (see Chapter XX).

POLYPHENYL COMPOUNDS. Selective hydrogenation of one phenyl ring in the presence of another can often be achieved. Success will depend on control of the amount of hydrogen absorbed, on the catalyst, on reaction conditions, on substituents on the ring, on the geometry of the substrate and at times on the solvent. In instances where the lack of steric effects make reduction of several rings possible it may be advisable to avoid the use of ruthenium [3].

At times results are unpredictable. A catalyst that works for one substrate may not produce selectivity in the half-hydrogenation of a related compound. For example, in the reduction of biphenyl to phenylcyclohexane in cyclohexane or acetic acid at 100° and 70 atm with noble metals on a support, palladium on carbon gave the best results, 87–97% yield, with little or no bicyclohexyl [91]. On the other hand it was ineffective for the attempted reduction of diphenylmethane to phenylcyclohexylmethane.

In another comparison diphenylacetic acid, (C$_6$H$_5$)$_2$CHCOOH, was hydrogenated almost quantitatively to phenylcyclohexylacetic acid in acetic acid with platinum oxide under low-pressure conditions while a similar reduction of benzilic acid, (C$_6$H$_5$)$_2$C(OH)COOH, was less successful giving starting material, dicyclohexylglycolic acid, and phenylcyclohexylglycolic acid in a ratio of 1:1:3.7 [92]. Low-pressure half-hydrogenations of diphenylmethane and 1,1-diphenylethane under the above conditions were reported to yield essentially pure phenylcyclohexylmethane and 1-phenyl-1-cyclohexylethane respectively but the selective reductions of biphenyl and 1,2-diphenylethane were less satisfactory.

In certain instances the presence of a substituent on the ring can determine which ring will reduce first. For example it was found that when aminoesters of 4-methyl and 3,5-dimethylbenzilic acids were hydrogenated the unsubstituted ring was selectively reduced [93]. In another example the phenolic-containing ring was preferentially hydrogenated as expected when 4-hydroxybiphenyl was reduced to 4-phenylcyclohexanol with nickel at 140–170° and 180 atm in the presence of a small amount of sodium metal [94].

The hydrogenation of ethyl 4′-hydroxybiphenyl-4-carboxylate over Raney nickel at elevated temperature and pressure [95] is an example of the preferential ring reduction of a phenol over a carboxyl-containing ring. That of 4′-hydroxybiphenyl-4-acetic acid in aqueous potassium hydroxide [96] is another example of the preferential saturation of a phenolic ring (Eq. 24.22).

$$\text{HO-C}_6\text{H}_4\text{-C}_6\text{H}_4\text{-CH}_2\text{COOH} \xrightarrow[\text{Ni(R)}\\ 240°,\ 170\ \text{atm}]{\text{in aq KOH}\\ 3\text{H}_2} \text{HO-C}_6\text{H}_{10}\text{-C}_6\text{H}_4\text{-CH}_2\text{COOH}$$

(24.22)

In some of these compounds there is such a difference in the rate of uptake between reduction of each ring, due probably to steric effects, that there is little difficulty in hydrogenating one selectively. This was observed in the reduction of some 3-hydroxy-2,3-diphenylalkanoic acids. When 3-hydroxy-2,3-diphenylpropionic acid was reduced in acetic acid with platinum oxide at 60° and 3 atm absorption of 6 equivalents was complete in 18–24 hr, but there was a sharp break in the curve for 3 equivalents after 4 hr [3]. Interruption of a hydrogenation after 3 equivalents yielded 84% of a product which was essentially a mixture of isomers of 3-cyclohexyl-3-hydroxy-2-phenylpropionic acid. In other related compounds one ring could always be selectively reduced.

In one reduction, that of 3-hydroxy-2,3,3-triphenylpropionic acid, it was not possible to saturate all three rings. Only two could be reduced. Attempts to reduce the sodium salt in aqueous solution with Raney nickel at 150° and 150 atm resulted in cleavage to yield cyclohexylacetic acid and other products which were not identified.

566   Aromatic Ring Systems

In attempted reductions of diphenylmethanols, $(C_6H_5)_2C(OH)R$ where R was 3- or 4-piperidyl, no uptake of hydrogen took place during reaction in acidic medium with platinum oxide at low pressure or in alcohol with 30% by weight of 5% rhodium on alumina at 85° and 70 atm [3].

An examination of framework molecular orbital models of these compounds seems to offer an explanation. If the position of the molecule is one in which contact of the piperidine ring with the catalyst surface is favored, then steric effects interfere with the approach of either benzene ring to the catalyst.

If a second carbon atom separates the piperidine ring and the benzene rings, then one ring can be saturated under acidic conditions but not both. When 4-(2-hydroxy-2,2-diphenyl)ethyl-1-methylpiperazine dihydrochloride was reduced in aqueous solution (pH 1.0) with platinum oxide, one ring was affected [97] (Eq. 24.23). Successful hydrogenation of one ring could also be

$$\begin{array}{c} C_6H_5 \\ \diagdown \\ C(OH)CH_2N\bigcircNCH_3 \cdot 2\,HCl \xrightarrow[PtO_2]{3H_2} \\ \diagup \\ C_6H_5 \end{array}$$

$$\begin{array}{c} C_6H_5 \\ \diagdown \\ C(OH)CH_2N\bigcircNCH_3 \qquad (24.23)\\ \diagup \\ C_6H_{11} \end{array}$$

achieved with rhodium on carbon or on alumina when the pH of the solution was adjusted to 4–5 [3, 98].

Both rings could be saturated only when hydrogenation was carried out as the N^1 quaternary salt [98]. Successful reduction of both rings is apparently due to the removal of the poisoning effect of the nitrogen atom in 1-position by complete shielding of that atom. This can make the favored position of the molecule one in which both phenyl rings are in the same plane and each can approach the catalyst.

The failure to saturate both rings when reduction was carried out as acid salt can be due to the inability to completely shield the same nitrogen atom. Reversibility of the reaction, acid + base $\rightleftharpoons$ salt, leaves free base which favors contact of the piperazine ring and only one benzene ring with the catalyst. The same reasoning may also apply to the nonreducibility of the 3- and 4-piperidyl diphenylmethanols noted earlier.

It was reported that the reduction of related tertiary amino 1,1-diphenylalkanols, $(C_6H_5)_2C(OH)(CH_2)_{2-5}N(R)_2$, yielded monocyclohexyl compounds [99] although it was not stated whether both rings could be saturated.

The effect of geometry on the saturation of one phenyl ring is evident in the hydrogenation of 3,3-diphenyl-2-(1-piperidinomethyl- or 4-morpholinomethyl)cyclopropane-cis-1-methanol in acetic acid with platinum oxide

[100] (Eq. 24.24). Hydrogen uptake stopped of its own accord after one ring

$$\underset{\underset{H}{|}}{\overset{H}{\underset{C_6H_5}{\overset{C_6H_5}{\diagdown}}}}\!\!\!\!\!\!\!\!\!\!\!\!\overset{CH_2N\langle\ \rangle}{\diagup}\!\!\!\!\!\!\!\!\!\!\!\!\!\!\underset{CH_2OH}{\diagdown} \quad\xrightarrow[PtO_2]{3H_2}\quad \underset{\underset{H}{|}}{\overset{H}{\underset{C_6H_5}{\overset{C_6H_{11}}{\diagdown}}}}\!\!\!\!\!\!\!\!\!\!\!\!\overset{CH_2N\langle\ \rangle}{\diagup}\!\!\!\!\!\!\!\!\!\!\!\!\!\!\underset{CH_2OH}{\diagdown} \quad (24.24)$$

was reduced. It seems obvious that only one ring could be saturated because the other was out of plane and could not make contact with the catalyst. It was assumed that the ring *trans* to the substituents in 1- or 2-position was reduced. This may be a correct assumption based on the failure of either ring to reduce when a methyl group was substituted for the hydrogen in 1-position.

As a rule fused rings appear to be reduced stepwise. In hydrogenation of systems containing two fused rings, one ring can generally be preferentially reduced over Raney nickel or the platinum metals under less drastic conditions than are necessary for perhydrogenation. However, there is a distinct tendency for ruthenium-catalyzed reductions to be less selective or even nonselective [3].

Indenes

Saturation of both rings is readily achieved by reduction with nickel at elevated temperature and pressure. Very likely ruthenium or rhodium catalysts would give better results under less strenuous conditions. Platinum oxide hydrogenations in acetic acid at low pressure will also result in saturation of both rings [101].

Absorption of one equivalent of hydrogen yields indanes. Indene itself was converted to indane by hydrogenation with nickel on kieselguhr at 100° and 70 atm [102]. Raney nickel gave equally good yield under the same conditions [3]. In a 4.0 mole experiment at 80° and 200 atm, uptake for one equivalent was complete in about 0.5 hr.

From certain references in the literature, of which the reduction of sodium indene-1-carboxylate to sodium indane-1-carboxylate with Raney nickel is one [103], it might be assumed that partial hydrogenation could be carried out under low-pressure conditions with this catalyst. In most cases pressure conditions are rather vaguely described.

Some 3-alkyl-2-(4-anisyl)-6-methoxyindenes were hydrogenated to the corresponding indanes in methyl alcohol over palladium on carbon at room temperature and atmospheric pressure [104]. A supported palladium catalyst may well be the one of choice for the reduction of indenes to indanes. Complete saturation of both rings is rather difficult with palladium even under high-pressure conditions.

Platinum oxide was found to be very useful in the hydrogenation of indenes to indanes. These reductions can be carried out in neutral solvent without danger of attacking the benzene ring. For example, a very small amount of catalyst was used to reduce indene to indane at 25° and 2–3 atm [105]. As previously noted, hydrogenation under acidic conditions can lead to hydrindanes [101].

Naphthalenes

Hydrogenation of naphthalenes proceeds stepwise to tetrahydronaphthalene and subsequently to decahydronaphthalene. Hydrogenation to decahydro compounds can be carried out at low pressure with platinum catalyst in acidic medium but reaction time is often long. Reduction of naphthalenes to decahydronaphthalenes with Raney nickel requires elevated temperature and pressure. In most instances hydrogenation with rhodium on a support will produce decahydro compounds under the same conditions used with nickel to prepare a tetrahydronaphthalene. An example is seen in comparisons of the reduction of 2-isopropylamino-1-(2-naphthyl)ethanol with 10% by weight of Raney nickel and with 5% by weight of 5% rhodium on carbon—equal to 0.25% metal to substrate [106] (Eq. 24.25).

(24.25)

Hydrogenation with ruthenium catalysts should also produce a decahydro compound. It has been this author's experience that ruthenium reductions of substituted naphthalenes are difficult to control at the tetrahydro stage. Decahydro compound was always present in substantial amounts in those reductions.

In general, tetrahydronaphthalenes are readily obtained from naphthalenes by hydrogenation with copper chromite at elevated temperature and pressure, with Raney nickel under high pressure or mild conditions, or with platinum oxide in acidic medium or with palladium on a support in neutral solvent under low pressure. Palladium-catalyzed reductions probably can be carried out under more vigorous conditions without danger of saturating the second ring if the hydrogenation of naphthalene with 5% palladium on carbon at 115–120° and 70 atm may be taken as an example [91]. Reaction stopped spontaneously at the tetralin stage.

It has been postulated that the unsubstituted ring should be preferentially reduced [107] but this may apply only where the substituent is an alkyl group as seen in the hydrogenations of 1- and 2-alkylnaphthalene with copper chromite which gave 5-alkyltetralins [108] and 6-alkyltetralins [109], respectively. In the reduction of other substituted naphthalenes, the nature and position of the substituent and the reaction conditions and perhaps the catalyst help to determine which ring will be affected.

It is difficult to predict results of partial reduction among naphthols. Partial hydrogenation of 1- or 2-naphthol with copper chromite favored saturation of the oxygenated ring [110]. Raney nickel reduction of 2-naphthol also yielded alicyclic alcohol but in the nickel reduction of 1-naphthol the ratio of phenolic compound, 5,6,7,8-tetrahydronaphthol, to alcohol was 2:1.

A low-pressure reduction of 1-naphthol in alcohol with 5% rhodium on alumina also yielded 5,6,7,8-tetrahydro-1-naphthol, 69% [111]. Some decalol was obtained but no 1,2,3,4-tetrahydro compound.

When 1-ethyl-2-naphthol or 2-ethyl-1-naphthol was hydrogenated over nickel, each oxygenated ring was saturated. In contrast, when N-(2-hydroxy-1-naphthylmethyl)piperidine was reduced with 5% palladium on carbon at 60° and 3.3 atm, 5,6,7,8-tetrahydro-1-methyl-2-naphthol was obtained in 60% yield [112] (Eq. 24.26).

$$\text{(24.26)}$$

R = 1-piperidyl   60%

Stork [107] studied the partial hydrogenation of 2-naphthol under neutral and acidic conditions with Raney nickel and also investigated the effect of added base. He found that neutral and acid conditions favored saturation of the unsubstituted ring while the addition of inorganic base gave a 6:1 ratio of alcohol to phenol. It was shown in another study that the addition of base, sodium hydroxide, or triethylamine, shifted the course of reduction of 2-naphthol with a number of nickel catalysts from the production of phenols to a predominance of alcohols [113].

When naphthols are to be converted to decalols, hydrogenation with rhodium on carbon or rhodium on alumina in alcohol under 3–4 atm pressure appears to be the method of choice [111, 114] producing high yields of decalols along with some decalones. Other catalysts, except ruthenium have been shown to be inferior, generally causing hydrogenolysis [114].

Ruthenium dioxide catalyzed the hydrogenation of 1-naphthol in alcohol at 110° and 110 atm to give 92% yield of a mixture of stereoisomeric 1-decalols [115] and yielded a mixture of *cis, cis-* and *cis, trans*-2-decalols with little accompanying hydrogenolysis in the reduction of 2-naphthol at 75° and 165 atm initial pressure [116]. However, there is a report which indicates some hydrogenolysis with this catalyst. When it was employed for the hydrogenation of 1,6-dihydroxynaphthalene at 75° and 165 atm initial pressure, good results were obtained, but in a larger size run the temperature rose because of an exothermic reaction causing lower yield of decalin-2,6-diol [117].

Rhodium-platinum oxide was also reported to give high yields of 1- and 2-decalols by reduction of the corresponding naphthols in acetic acid at room temperature and 100–140 atm [2].

In general study on the hydrogenation of naphthyl ethers has not been too extensive. High-pressure nickel reductions of 1-alkoxynaphthalenes gave 5-alkoxytetralines while those of 2-alkoxynaphthalenes resulted in saturation of the oxygenated ring giving 71–79% yields of 2-alkoxytetralins [110].

When 2-methoxynaphthalene was hydrogenated over Raney nickel in alcohol containing a small amount of acetic acid, 6-methoxytetralin was obtained but when alkali was added to another nickel-catalyzed hydrogenation, hydrogenolysis to tetralin was the major reaction [107].

The effect of other catalysts on the ring that will be affected during partial hydrogenation has not been studied except in the platinum reduction of 2-methoxynaphthalene in acetic acid which resulted in good yield of 6-methoxytetralin [118].

Results with rhodium catalyst cannot be predicted except to note, based on reductions of alkoxybenzenes, that loss of the alkoxy group should not take place if the oxygenated ring is saturated.

Different results have been obtained during partial reduction of 1-naphthoic acid as sodium salt and as free acid with Raney nickel. Reduction as ester gave about the same results as that of the acid. Hydrogenation as sodium salt at 75° and 165 atm initial pressure yielded 45% of 1,2,3,4-tetrahydro- and 28% of 5,6,7,8-tetrahydronaphthoic acids [119]. Reduction as acid gave 57% of 5,6,7,8-tetrahydroacid. Hydrogenation of 1-naphthoic acid in acetic acid under low-pressure conditions with platinum also gave a predominance of the same isomer.

Reduction as ester with W-6 Raney nickel under 2–3 atm pressure gave about a 2:1 ratio of 5,6,7,8-tetrahydronaphthoic acid to 1,2,3,4-tetrahydro acid [120].

When 2-naphthoic acid was partially hydrogenated with Raney nickel at 150° and 225 atm in dioxane containing a small amount of sodium hydroxide or enough to carry out reaction as sodium salt, only 5,6,7,8-tetrahydro-2-naphthoic acid was obtained [107]. Low-pressure reduction of the ethyl ester with W-6 Raney nickel gave practically the same results [120].

Perhydrogenation of 1-naphthoic acid with platinum oxide in acetic acid at low pressure yielded 85% of *cis,cis*-decahydro-1-naphthoic acid [121]. Reduction of 2-naphthoic acid under similar conditions gave 65% of *cis,cis*-decahydro-2-naphthoic acid [122]. Reaction times were not noted but it was pointed out that the presence of tetrahydro acid made the isolation of pure decahydro acid difficult [122]. Perhydrogenation of these acids might be more complete upon reduction in the presence of rhodium on a support. Employment of ruthenium under mild high-pressure conditions, 100° and 100 atm, might also be useful.

Perhydrogenation of 6-hydroxy-1- and 2-naphthoic acids with Raney nickel and with platinum oxide under a variety of conditions gave decahydro compounds accompanied by considerable hydrogenolysis [119]. Successful hydrogenation of 6-hydroxy-1-naphthoic acid or ethyl ester was carried out at 60–80° at 140 atm over W-4 Raney nickel promoted with sodium ethoxide [123]. Results appeared dependent on the age of the catalyst. Freshly prepared W-4 nickel gave quantitative yield of perhydro ester. Older catalysts required higher reaction temperature and gave a mixture of tetrahydro compounds, also without accompanying loss of hydroxyl function. Perhydrogenations of hydroxynaphthoic acids should be studied with rhodium or ruthenium catalysts.

Hydrogenation of 1- and 2-naphthylamines with nickel catalyst require temperatures above 150° at 100–120 atm unless the catalyst is especially active. Reaction under these conditions gives a mixture of 1,2,3,4- and 5,6,7,8-tetrahydronaphthylamines [3]. Reductions are best carried out in cyclohexane or methyl cyclohexane. Reduction in alcohols, unless they are secondary or tertiary, usually give *N*-alkylated tetrahydronaphthylamines as contaminants which are difficult to remove [3].

When tetrahydrogenations were attempted with ruthenium dioxide under milder conditions, 100° and 70–100 atm, *N*-alkylation did not take place but no matter how closely the reaction was watched it was not possible to prevent the formation of decahydronaphthylamine. Indeed raising the temperature to 125° and/or increasing the pressure to 125 atm gave 85–90% yields of decahydronaphthylamines in less than 1 hr [3]. This method can be the one of choice for production of decahydronaphthylamines although reductions with rhodium on a carrier might also be useful at lower pressure.

In general, as in the benzene series, ring reduction very seldom precedes that of a functional group. In one instance in the reduction of

2-acetyl-3,4-diacetoxy-1-naphthol with platinum oxide in acetic acid, some 2-acetyl-3,4-diacetoxy-5,6,7,8-tetrahydro-1-naphthol was obtained along with 3,4-diacetoxy-2-ethyl-5,6,7,8-tetrahydro-1-naphthol [124]. Apparently steric effects prevented complete conversion of the 2-acetyl group.

Phenanthrenes, Anthracenes, and Other Fused Ring Systems

The catalytic hydrogenation of unsaturated fused ring carbocycles containing more than two rings has not been studied too extensively because of the complexities associated with their reduction.

When phenanthrene or anthracene is hydrogenated, initial saturation takes place at the 9,10-bond (Eq. 24.27). This has been accomplished by the use of

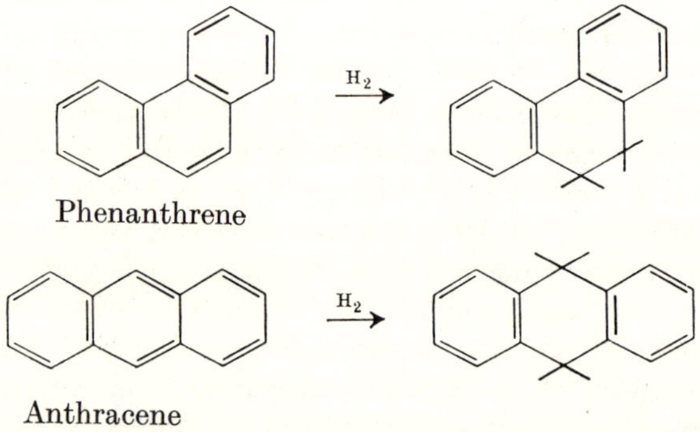

Phenanthrene

Anthracene

(24.27)

copper chromite at 150° and 150 atm pressure [125, 126]. At higher temperature migration of the double bond apparently takes place to give 1,2,3,4-tetrahydro compounds [125, 126].

Raney nickel is less selective for the preparation of dihydro compounds. A dihydroanthracene was obtained when 1,5,9-triacetoxy-2,6-dimethoxyanthracene was hydrogenated in dioxane over palladium on barium sulfate at 95° and 120 atm but it was accompanied by loss of the 9-acetoxy group to yield 1,5-diacetoxy-2,6-dimethoxy-9,10-dihydroanthracene [127].

Nickel was useful for the hydrogenation of phenanthrene to tetrahydro and octahydrophenanthrenes [128]. In general, rather vigorous conditions are necessary to obtain tetradecahydrophenanthrene.

Among anthracenes 2,6-dihydroxyanthracene was converted to 1,2,3,4,5,6,-7,8-octahydroanthracene-2,6-diol in 82% yield by hydrogenation with W-4 Raney nickel in the presence of potassium hydroxide at 150° and 280 atm [129]. It was not possible to obtain the perhydro compound by further reduction of the octahydroanthracenediol until it was purified.

Neither rhodium nor ruthenium catalysts have been employed for the hydrogenation of phenanthrenes or anthracenes. Rhodium might be useful for the preparation of tetrahydro and octahydro compounds. Ruthenium should be less useful for tetrahydrogenation but could lead to octahydro compounds. It may be the catalyst of choice for perhydrogenation.

In other fused ring systems the complexity of the molecule can make reduction extremely difficult. In certain instances what is actually taking place is hydrogenation of a polysubstituted benzene ring which is difficult to saturate because of steric effects.

HETEROCYCLES (ONE HETERO ATOM)

Furans

Furans can be readily hydrogenated to tetrahydrofurans without ring opening in the presence of palladium catalyst (usually on a support) or supported rhodium catalyst under low-pressure conditions. For work under higher pressure, reductions with ruthenium give excellent results. Ring opening does not occur, as a rule, because of the moderate reaction temperature [3]. Platinum oxide, useful in the reduction of benzenoid compounds in acetic acid, can cause ring opening in attempted saturations of furans in that acid [130]. In contrast the use of 70–30 rhodium-platinum oxide, for example, in the reduction of furfuryl alcohol in acetic acid at 100–120 atm gave 88% yield of tetrahydrofurfuryl alcohol [1, 2].

Raney nickel has catalyzed the hydrogenation of furan to tetrahydrofuran in 93% yield under 2–4 atm pressure [131]. However, ring opening did occur in the low-pressure reduction of 3-furoic acid in aqueous sodium hydroxide to yield 30% of tetrahydrofuran-3-carboxylic acid and the lactone from ring opening [132] (Eq. 24.28).

(24.28)

Nickel can be employed under elevated temperature and pressure without causing ring opening if reaction temperature is not too high. For example,

2-methylfuran was hydrogenated in the absence of solvent to 2-methyl-tetrahydrofuran in 80% yield with 10% by weight of Raney nickel at 100° and 160 atm, but in a larger run with 6.5% by weight of catalyst and gradual addition of hydrogen at 130 atm, an exothermic reaction, which carried the temperature to 140°, resulted in lower yield (65%) due to some cleavage [3].

In most instances ruthenium is superior to nickel because reaction conditions for the hydrogenation of furans to tetrahydrofurans can be carried out at 70–100° and 70 atm [3]. For example, when 2-furfurylamine was hydrogenated without solvent with Raney nickel, reaction at temperatures up to 150° and 70 atm required 30 hr [133] whereas reduction with ruthenium dioxide at 100° and 80 atm, also without solvent, was complete in 10 min to give tetrahydrofurfurylamine in high yield with only mechanical loss [3].

Some substituted furans have been converted to tetrahydrofurans with palladium black [134] but in most cases palladium on a support is employed. Palladium on carbon gave 93–97% yields of *cis*-tetrahydrofuran-2,5-dicarboxylic acid from the reduction of furan-2,5-dicarboxylic acid in water or when sodium hydroxide was added to pH 6.5 [135]. Very rapid reduction [0.5 hr] took place at 100° and 100 atm without ring cleavage. In another reaction at 100° and 140 atm palladium on carbon catalyzed the reduction of 2-hydroxymethylfuran 2-carboxylic acid without ring opening [136] but esterification did take place to give 97% of ethyl 2-hydroxymethyltetra-hydrofuran-2-carboxylate.

Palladium catalyst also performed satisfactorily in slightly acidic solution in the hydrogenations of (2-furyl)alkylamines to the corresponding tetrahydro compounds [137] (Eq. 24.29).

$$\underset{O}{\bigcirc}CH_2CHRNH_2 \xrightarrow[Pd(H^+)]{2H_2} \underset{O}{\bigcirc}CH_2CHRNH_2 \quad (24.29)$$

R = H or CH$_3$

5% Rhodium on carbon, 15% by weight, was employed for the reduction of a related *N*-ethyl-(2-furyl)alkylamine in neutral solvent at room temperature and 1.5 atm [3]. Hydrogen absorption was complete in 2–3 hr, good yield of tetrahydrofuran was obtained.

Selective Reductions

IN THE PRESENCE OF UNSATURATED CARBOCYCLES. The furan ring is reduced more readily than a benzene ring. It is unusual except under vigorous conditions with some catalysts to find accompanying saturation of a benzene ring when a substrate also contains a furan ring. For example, rhodium, which has the ability to readily saturate the benzene ring, had no effect on that ring in the hydrogenation of *N*-ethyl-1-(2-furyl)-2-phenethyl-amine to *N*-ethyl-1-(tetrahydro-2-furyl)-2-phenethylamine [3]. Hydrogen absorption stopped spontaneously at two equivalents. Raney nickel was employed in the hydrogenation of some aralkylamines, $C_6H_5CHRNHCH_3$ and $C_6H_5CH_2CHRNHCH_3$, where R = 2-furyl, at 100 atm to yield aralkylamines, R = tetrahydro-2-furyl [138].

The furan ring in benzofurans can always be reduced selectively particularly if reactions are carried out in neutral solvent with palladium on carbon as in the hydrogenation of 4-methoxybenzofuran in alcohol at 25° and 3 atm which yielded 96% of 4-methoxy-2,3-dihydrobenzofuran [139]. Some perhydrogenation occurred when 7-methoxy-2,3-dimethylbenzofuran was reduced with palladium on carbon in acetic acid [140]. Palladium on carbon in neutral solvent also catalyzed the hydrogenation of a naphthofuran in alcohol without affecting the naphthalene nucleus [141].

Raney nickel was employed for the reduction of 2-acetylbenzofuran to the benzo-2,3-dihydrofuran but it was not possible to prevent conversion of the carbonyl group to carbinol [3] (Eq. 24.30).

$$\text{CH}_3\text{O-benzofuran-COCH}_3 \xrightarrow[25°, 2.5 \text{ atm}]{2\text{H}_2, \text{ NiR}} \text{CH}_3\text{O-dihydrobenzofuran-CHOHCH}_3$$

(24.30)

IN THE PRESENCE OF FUNCTIONAL GROUPS. In most instances functional groups should be hydrogenated preferentially in the presence of a furan ring. Any tendency toward ring reduction may be prevented by adding ammonia [3].

An exception to the general rule is seen in the hydrogenation of 2-(2-furyl)-3-phenylacrylonitrile with 5% palladium on carbon which yielded 2-(2-furyl)-3-phenylpropionitrile and 2-(tetrahydro-2-furyl)-3-phenylpropionitrile [142] (Eq. 24.31). In this instance selective reduction of the furan

$$C_6H_5CH=C(\text{CN})\text{-furyl} \xrightarrow{\text{Pd on C}} C_6H_5CH_2CH(CN)\text{-furyl} + C_6H_5CH_2CH(CN)\text{-tetrahydrofuryl} \quad (24.31)$$

ring is probably due to steric effects which prevent approach of the nitrile group to the catalyst. Another factor is the general inability of palladium catalysts to promote hydrogenation of the nitrile function in neutral medium.

IN THE PRESENCE OF HYDROGENOLYZABLE GROUPS. Selective reduction of the furan ring in the presence of halogen is dependent on the activity of the halogen toward hydrogenolysis, on reaction conditions, and on the catalyst.

576   *Aromatic Ring Systems*

The furan ring in some aralkylamines, $ClC_7H_4CH_2CH(R)NHCH_3$ where R = 2-furyl, was saturated without removing halogen during high-pressure reduction with Raney nickel [138]. This may be expected since nickel is not especially reactive for removal of halogen.

In another hydrogenation, that of 4-chloro-1-oxo-1,2-dihydrofuro[3,4-*d*]-pyridazine with palladium on carbon, removal of the halogen from the pyridazine ring accompanied or probably preceded saturation of the furan ring [143] (Eq. 24.32). Loss of halogen in this case was due to the reactivity

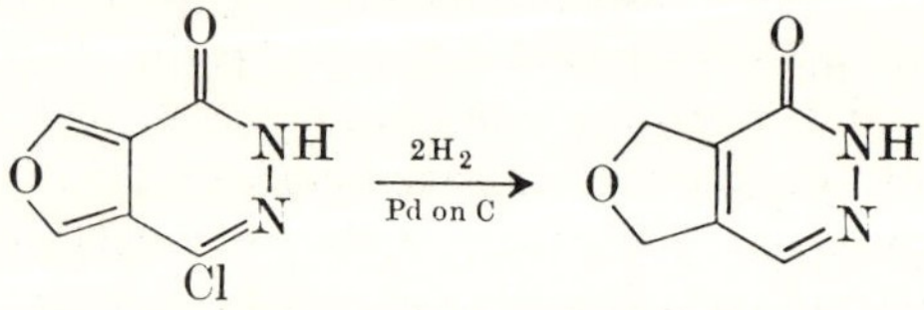

$$\text{(structure with Cl)} \xrightarrow[\text{Pd on C}]{2H_2} \text{(reduced structure)} \quad (24.32)$$

of the chlorine atom and the penchant of palladium catalysts for dehalogenation.

Examples of reduction of a furan ring in the presence of an *O*-benzyl function are rare. The hydrogenation of 6-benzyloxy-2-(2-hydroxy-2-propyl)benzofuran with Raney nickel at 3.3 atm yielded 6-hydroxy-2,3-dihydro-2-(2-hydroxy-2-propyl)benzofuran [144a]. Hydrogenolysis may have been due to the catalyst ratio, 60% [144b].

An example of selective reduction of the furan ring in the presence of an *N*-benzyl group is seen in the hydrogenation of *N*-methyl (α-2-furyl)-benzylamine with Raney nickel [138].

It should be pointed out that the system —RCH$_2$N⟨ where R is furyl is structurally equivalent to an *N*-benzyl function and may undergo hydrogenolysis during reduction with palladium. This was seen in the attempted

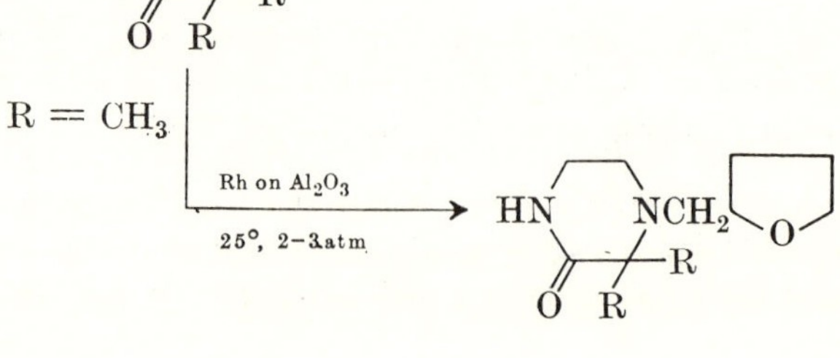

$$\text{(reaction scheme 24.33)} \quad (24.33)$$

R = CH$_3$, Rh on Al$_2$O$_3$, 25°, 2-3 atm

hydrogenation of 4-(2-furyl)-3,3-dimethylpiperazin-2-one with 5% palladium on carbon in alcohol which underwent cleavage to yield 3,3-dimethylpiperazin-2-one [3]. Success was attained by the use of 5% rhodium on alumina (Eq. 24.33). Some secondary 2-furylmethylamines were also hydrogenated successfully to the corresponding tetrahydro-2-furylmethyl secondary amines with either rhodium on carbon or rhodium on alumina [3].

The use of Raney nickel for low-pressure reductions of 2-furylmethyl tertiary amines to the corresponding tetrahydrofuryl compounds generally gave poor results [3]. Hydrogenolysis did not take place but reaction was too slow apparently because of the effect of the nitrogen bases on the catalyst. Better results would have been obtained by the use of more vigorous conditions exemplified by the hydrogenation of N-(5-methyl-2-furyl)methylpiperidine in methyl alcohol to N-(5-methyltetrahydro-2-furyl)methylpiperidine in 81% yield with Raney nickel at 150° and 115 atm [145].

Pyrroles. Pyrroles are among the more difficult heterocyclic rings to saturate. Reduction appears to proceed stepwise [146] but in most instances interruption of hydrogen absorption at one equivalent leads to mixtures. There is, however, an example of partial reduction of a pyrrole which does not have the double bonds at C-2 and C-4. Hydrogenation of 3-ethyl-2,4-dimethyl-5H-pyrrole, cryptopyrrole, in acidified alcohol over platinum oxide gave the corresponding 1-pyrroline [147] (Eq. 24.34). N-Substitution allows

$$\text{H}_3\text{C}\underset{\underset{N}{}}{\overset{}{\boxed{}}}\overset{\text{C}_2\text{H}_5}{\underset{\text{CH}_3}{}} \xrightarrow{\text{H}_2} \text{H}_3\text{C}\underset{\underset{N}{}}{\overset{}{\boxed{}}}\overset{\text{C}_2\text{H}_5}{\underset{\text{CH}_3}{}} \qquad (24.34)$$

hydrogenation of pyrroles to proceed more readily than that of an N-unsubstituted one because the ability of the formed basic pyrrolidine nitrogen atom to bond with the catalyst and cause poisoning is impaired.

Raney nickel has been employed for reduction of pyrroles to pyrrolidines but, in general, high temperatures and pressures are required and yields are not always high. Far better results are obtained with ruthenium catalysts under much milder conditions. For example, pyrrole was reduced without solvent over ruthenium dioxide (1% by weight of substrate) at 100° and 80 atm to give a quantitative yield of pyrrolidine [3]. As little as a 0.1% catalyst could be employed at 200° and 110 atm. When N-methylpyrrole was hydrogenated with a 1% by weight of ruthenium dioxide at 100° and 90 atm, uptake was 90% complete by the time the reaction temperature was reached.

The catalyst of choice for low-pressure hydrogenations of pyrroles may be rhodium on a carrier. Before the appearance of this catalyst, platinum oxide was widely used and gave good results in many instances [147–150]. However, there are disadvantages accompanying its use. Reductions of pyrroles with it as a rule must be carried out in acidic medium and a rather high catalyst ratio is necessary at times for complete reduction [149, 150]. There is also a report that aeration of the catalyst was required or reaction would not proceed to completion [151].

Rhodium reductions of pyrroles have also been carried out in acetic acid [146, 152, 153] but the advantage of rhodium on a support over platinum for these reductions is in the very low metal to substrate ratio, usually less than 0.5%.

The presence of acids, particularly weaker acids, does aid those hydrogenations with rhodium which lead to basic compounds. However, reduction of the pyrrole nucleus may easily be carried out in alcohols as seen in the hydrogenation of 3H-pyrrolizine or 1-methyl-3H-pyrrolizine [154] (Eq. 24.35).

$$\text{pyrrolizine} \xrightarrow[5\% \text{ Rh on C}]{3H_2} \text{pyrrolidine} \quad (24.35)$$

R = H or CH$_3$

Reductions may also be carried out in the absence of solvent with rhodium although reaction time is usually longer than when solvent is present. N-Methylpyrrole was hydrogenated in the absence of solvent with 10% by weight of 5% rhodium on carbon at 25° and 3 atm [3]. Doubling the amount of catalyst decreased reaction time from 14 to 4 hr. Raising the temperature to 50–60° resulted in much more rapid hydrogen uptake.

Selective Reductions

PRESENCE OF OTHER RING SYSTEMS. The pyrrole ring is more difficult to saturate than other aromatic ring systems. Exceptions may be dependent on the catalyst. Such an example is seen in the hydrogenation of 3-(2-pyrryl)pyridine (nicotyrine) with palladium on carbon in acetic acid in which the pyrrole ring was selectively reduced [155]. In contrast hydrogenation over platinum oxide gave 3-(2-pyrryl)piperidine [156] (Eq. 24.36).

If a phenyl or a phenylalkyl group is attached to the pyrrole nucleus, it is virtually impossible to saturate the pyrrole ring without concomitant reduction of the benzene ring. Indeed that ring is generally preferentially reduced. However, when an aromatic ring is fused with a pyrrole ring, such as indole or isoindole, the pyrrole portion can be preferentially hydrogenated.

Indole was converted to indoline, 2,3-dihydroindole, by reduction with Raney nickel at 90–100° and 70–100 atm [157]. Potassium indole-3-acetate was hydrogenated in water under similar reduction conditions followed by acetylation to give a 70% yield of N-acetylindoline-3-acetic acid [158]. Copper chromite has catalyzed the hydrogenation of indole to indoline [159].

Early work on reductions of indole with colloidal platinum in acetic acid gave mixtures of indole, indoline, and octahydroindole [160]. It has been

(24.36)

suggested that in acidic media the benzene ring of indole is preferentially hydrogenated but examples to substantiate this consist of indoles which have large groups on the pyrrole ring. In contrast indole itself was reduced to indoline with platinum oxide in 1:1 ethyl alcohol-aqueous fluoroboric acid [161].

Low-pressure hydrogenation employing palladium may be a useful method for obtaining indolines. 1-Acetyl-5-nitroindole was reduced in alcohol with 5% palladium on carbon under 2 atm pressure to give 93% of 1-acetyl-5-aminoindoline [162]. It is very possible that success was due to the presence of the acetyl group which rendered the indoline nitrogen nonbasic and thereby prevented catalyst inhibition.

Available references on the reduction of isoindoles cite the use of Raney nickel for selective reduction to isoindoline. For example, 2-phenylisoindole and 1-methyl-2-phenylisoindole were converted to the corresponding isoindolines in very high yield by hydrogenation over Raney nickel [163].

PRESENCE OF FUNCTIONAL GROUPS. In most instances functional groups, as in the benzene series, should be reduced without affecting the pyrrole ring. A few exceptions exist. The pyrrole ring in 3-(2-nitropropyl)indole was reduced while the hindered aliphatic nitro group remained intact. Hydrogenation, carried out with prereduced platinum in methyl alcohol containing hydrochloric acid, yielded unchanged material, 3-(2-nitropropyl)octahydroindole, and 3-(2-nitropropyl)indoline [164] (Eq. 24.37). A second hydrogenation carried out over a longer period gave unchanged starting material (12%), 27% of the nitropropyloctahydroindole, and only 3% of the nitropropylindoline.

A ketone group is usually selectively reduced in the presence of a pyrrole ring. However, when there was a carbon bridge between the ring and the

580 *Aromatic Ring Systems*

carbonyl group, as in 1-(1-methylpyrryl)acetone and in 1,3-*bis*(1-methylpyrryl)acetone, hydrogenation on glacial acetic acid over platinum oxide

[indole]-CH$_2$CH(NO$_2$)CH$_3$ $\xrightarrow{Pt, H^+}$ [indoline]-CH$_2$CH(NO$_2$)CH$_3$ +
                                                                   39%

[octahydroindole]-CH$_2$CH(NO$_2$)CH$_3$ + starting material   (24.37)
            12%

gave the corresponding pyrrolidinylacetones [165, 166]. In each instance hydrogen uptake came to a halt after ring reduction.

Pyrans and Pyrones

Pyrans have been hydrogenated over nickel at elevated temperature and pressure to tetrahydropyrans, often with very good results, but reaction can yield alkanediols as a result of ring opening [3]. In addition, as has been shown in the reduction of dihydropyrans [167], there is a wide variation in the susceptibility of pyrans to hydrogenation. A method reported for the rapid reduction of ethyl 2,3-dihydro-4H-pyran-2,5-dicarboxylate, in acetic acid over 5% rhodium on carbon at room temperature and atmospheric pressure, to the corresponding tetrahydropyran could be a general one for saturation of the pyran ring. A supported palladium catalyst (10% Pd on asbestos) was far less efficient but this could have been due to the much lower ratio of metal to substrate. Better results might also have been obtained with palladium on carbon since it is usually the most active of the supported palladium catalysts [3]. The medium appeared to have some effect in this study. Hydrogen uptake was slower when reduction was run in neutral solvent (ethyl alcohol).

2-Pyrones, upon hydrogenation under mild conditions, yield tetrahydropyrones. The resulting saturated lactone is usually resistant to further attack during the course of reaction. 4,6-Dimethyl-2-pyrone, ethyl 4,6-dimethyl-2-pyrone-5-carboxylate, and methyl 2-pyrone-6-carboxylate were reduced in ether or methyl alcohol with 10–20% by weight of 5% palladium on carbon to give very good yields of the corresponding tetrahydro-2-pyrones [168]. Some ring opening did take place in the hydrogenation of coumalic acid, 2-pyrone-5-carboxylic acid, with a smaller amount of catalyst to yield tetrahydro compound and 2-methylglutaric acid (Eq. 24.38).

Partial ring reduction took place in the palladium-catalyzed hydrogenation of dehydroacetic acid (3-acetyl-4-hydroxy-6-methyl-2-pyrone) in ethyl acetate. The 3-acetyl group was converted to 3-ethyl as a 5,6-dihydro-2-pyrone was obtained [169]. Partial reduction of the ring in this instance was probably due to substitution at 3- and 4-position.

$$\underset{\text{HOOC}}{\text{HOOC}}\underset{\text{O}}{\bigcirc}=O \xrightarrow[\text{3-4 atm}]{2H_2} \underset{\text{HOOC}}{\text{HOOC}}\underset{\text{O}}{\bigcirc}=O + \underset{\text{COOH}}{\overset{\text{COOH}}{\underset{|}{\text{CHCH}_3}}}\underset{|}{\overset{|}{(\text{CH}_2)_2}} \quad (24.38)$$
$$57.5\%$$

A 5,6-dihydropyran also resulted when ruthenium, generally an inefficient catalyst for low-pressure hydrogenations, was employed to reduce 6-methyl-4-(4-morpholino)-2-pyrone in methyl alcohol [170] (Eq. 24.39).

$$\underset{R}{\bigcirc}\underset{O}{=}O \xrightarrow[\text{low pressure}]{H_2} \underset{R}{\bigcirc}\underset{O}{=}O \quad (24.39)$$

R = 4-morpholino

When 4-pyrones are subjected to hydrogenation, either tetrahydropyrones or tetrahydropyranols may be obtained, dependent on catalyst and reaction conditions. Reductions with nickel are difficult to control at the tetrahydropyrone stage [3]. Attempts to poison a specially prepared Raney nickel catalyst in the hydrogenation of 2-phenyl-4-pyrone by the addition of chloroform (11% by weight of substrate) made it possible to obtain dihydropyrone. A tetrahydropyrone could not be produced under any of the experimental conditions; 4-hydroxy-2-phenyltetrahydropyran was always obtained [171]. Reductions of 4-pyrones with platinum also have yielded 4-hydroxytetrahydropyrans [172].

A study of the absorption curves in the hydrogenation of 4-pyronecarboxylic acids and esters at atmospheric pressure with palladium catalysts and with Raney nickel gave no evidence of selectivity [173]. Reduction tended to proceed to 4-hydroxytetrahydropyrans although it was possible to obtain partially hydrogenated products in poor yield by interrupting the reaction after absorption of one or two equivalents.

However, palladium catalysts have led to good yields of tetrahydropyrones when 4-pyrone and 2,6-dimethyl-4-pyrone were hydrogenated [174–176]. In the last-mentioned reduction [176] almost 15% of 4-hydroxytetrahydropyran was also obtained when 10% by weight of 10% palladium on carbon was employed. This result coincides with work of this author who found that hydrogenation with 10% by weight of 5% palladium on carbon made control of the reaction difficult. The use of a 3.5% catalyst ratio made it possible to prevent further reduction of the tetrahydropyrone [3].

582   *Aromatic Ring Systems*

In another example, the use of palladous oxide in the hydrogenation of 2,6-dimethyl-4-pyrone in ethyl alcohol under 2–3 atm pressure yielded about 60% of 2,6-dimethyltetrahydropyrone and 35% of 2,6-dimethyldihydro-4-pyrone [177]. It is of interest that in this same study a reduction with Raney nickel containing palladium, triethylamine, and chloroform which was interrupted after 2.2 moles of hydrogen yielded only tetrahydropyranol and starting material.

Selective Reductions. Examples are lacking on the selective reduction of the pyran ring in the presence of other unsaturated heterocycles but selectivity should be expected over saturation of a benzene ring. The hydrogenation of 2-phenyl-4-pyrone to 2-phenyltetrahydro-4-pyrone with nickel is an example [171]. When an unsaturated carbocyclic ring is fused with a pyran ring as in benzopyrans or benzopyrones there should be little danger of reducing the carbocyclic ring with most catalysts unless conditions are too vigorous. For example, coumarin was reduced to dihydrocoumarin with Raney nickel at 100° and 100–200 atm [178] (Eq. 24.40).

$$\text{coumarin} \xrightarrow{\text{H}_2,\ \text{Ni R}} \text{dihydrocoumarin} \qquad (24.40)$$

In most instances catalysts and conditions useful for hydrogenations of pyrans should be applicable for reduction of the double bonds in a pyran ring fused to an unsaturated carbocycle. The hydrogenation of ethyl 5,6-benzocoumarin-3-carboxylate under 4 atm pressure with 5% rhodium on alumina [179] should not be taken as an exception because 150% by weight of catalyst was employed and reduction was allowed to continue for 39 hr until uptake of hydrogen stopped. This reduction was a deliberate attempt at ring rupture and hydrogenation of the carbocyclic rings to obtain a decalylalkanoic acid.

Pyridines

Catalysts. The preferred catalysts for saturation of the pyridine ring are platinum oxide and rhodium on a support for low-pressure hydrogenations and ruthenium as dioxide or on a support for reductions at higher temperature and pressure.

In general when Raney nickel is employed, temperature and pressure for the conversion of pyridines to piperidines are much higher than when ruthenium is used. In addition, because of high temperatures, N-alkylation often occurs when the solvent is one of the lower primary alcohols [180]. Another side reaction resulting from the high temperature is seen in the hydrogenation of N,N-dimethyl-3-phenyl-3-(2-pyridyl)propylamine in methyl alcohol [181]. Not only did N-methylation take place at the piperidine nitrogen but the methyl groups were cleaved from the side chain nitrogen, Eq. 24.41 (this was noted previously in the benzene series). In another nickel-catalyzed reduction, that of 3-methyl-4-propoxypyridine at 150° and 150 atm initial pressure in the absence of solvent, only 26% yield of 3-methyl-4-propoxypiperidine was obtained [182], apparently because of loss of the propoxy group.

$$\underset{R = C_6H_5}{\underset{N}{\bigcirc}} CHR(CH_2)_2N(CH_3)_2 \xrightarrow[\substack{NiR \\ \text{init. press., 70 atm,} \\ 170°}]{3H_2 \text{ in MeOH}} \underset{CH_3}{\underset{N}{\bigcirc}} CHR(CH_2)_2N(CH_3)_2 + \underset{CH_3}{\underset{N}{\bigcirc}} CHR(CH_2)_2NH_2 \quad (24.41)$$

$$ 45\% 34\%$$

Side reaction in nickel hydrogenations of 2-(2-hydroxyethyl)pyridine led to low yields, below 50%, of 2-(2-hydroxyethyl)piperidine, [3, 183]. The contaminant, 1-[2-(2-piperidyl)ethyl]-2-(2-hydroxyethyl)piperidine, was apparently formed through dehydrogenation of the starting material to pyridine-2-acetaldehyde followed by reductive alkylation of 2-(2-hydroxyethyl)piperidine and subsequent hydrogenation (Eq. 24.42). In contrast, reduction of 2-(2-hydroxyethyl)pyridine with ruthenium could be carried out at 70–100° and 100 atm to give 94% yield of the corresponding piperidine [184].

Conditions necessary for successful hydrogenations of pyridines with nickel are dependent for the most part on the reactivity of the catalyst. The very reactive W-7 Raney nickel catalyzed the reduction of 2-dodecyl and 2-pentadecylpyridine at room temperature and atmospheric pressure [185]. However, as noted in other sections of this book, this catalyst must be freshly prepared since it loses its high activity in a short time.

Palladium on carbon has shown some utility both at 3–4 atm pressure and at elevated pressure. Methyl isonicotinate was reduced with it in methyl alcohol at 80° and 4 atm pressure to yield methyl isonipecotate [186]. Hydrogenation in the absence of solvent increased reaction time from 3.5 to 17 hr. Some phenyl pyridyl carbinols have been reduced in acetic acid to the corresponding phenyl piperidyl carbinols with 10% palladium on carbon at 70–80° and 3–4 atm [187], more conveniently, according to the author, than with platinum oxide. Palladium on carbon was also employed at 75° and 105–110 atm in the hydrogenation of dimethyl 4-methoxypyridine-2,6-dicarboxylate in methyl alcohol [188] and in the reduction of dilauryl 3-pyridyl carbinol in isopropyl alcohol at 115° and 110 atm [189]. In each instance very good yield of the corresponding piperidine was obtained.

The conditions used in Refs. 188 and 189 are in the range of those employed for reductions with ruthenium. The good results warrant further study on the use of palladium on carbon for pyridine hydrogenations under moderate high-pressure conditions.

Hydrogenation of a number of pyridines has been highly successful with ruthenium dioxide under moderate high-pressure conditions, (70–100° and 100 atm); yields were very high in most reductions [184]. In no instance was there evidence of N-alkylation during reaction in ethyl or methyl alcohol. Hydrogen uptake was much slower than usual when primary aminoalkylpyridines were hydrogenated, due apparently to the combined inhibitory effect of two basic nitrogen atoms. When the nitrogen atom in the side chain was secondary or tertiary, reduction was far more rapid. In this instance steric effects decreased the ability of the side chain nitrogen atom to inhibit the activity of the catalyst. In other reductions, except for that of 2,4,6-trimethylpyridine, hydrogen absorption was fairly rapid; in some cases uptake was complete by the time reaction temperature was reached. The catalyst could not be used to reduce the pyridine ring selectively in the presence of a benzene ring. 2- and 4-Benzylpyridines yielded cyclohexylmethylpiperidines. Some loss of an N-benzyl group as well as saturation of the benzene ring took place when 4-(2-benzylaminoethyl)pyridine was reduced; 4-(2-aminoethyl)piperidine (11%) was obtained along with the major component, 4-2-(cyclohexylmethylamino)ethyl piperidine. Decarboxylation occurred during hydrogenation of nicotinic acid but this was averted by the addition of an equivalent of sodium bicarbonate to an aqueous solution of the acid before reduction.

Other investigators have employed ruthenium with good results. Of interest is the hydrogenation of 4-alkoxypyridines at 140° and an initial pressure of 150 atm with ruthenium dioxide in 50% aqueous methyl alcohol which gave 75–85% yields of 4-alkoxypiperidines [190]. Another reduction of interest with ruthenium dioxide is the conversion of 3- and 4-trifluoromethylpyridines to the corresponding piperidines in 69 and 77% yields respectively by hydrogenation in tetrahydrofuran at 115° and 110 atm [191].

Most reductions of pyridines with platinum oxide must be carried out as an acid salt or as base under acidic conditions in order to prevent poisoning of the catalyst [192]. The hydrogenation of picolinic or isonicotinic acid can be carried out in neutral solvent [193] because the carboxyl group neutralizes the catalyst-inhibiting effect of the ring nitrogen. Reduction of quaternary pyridine compounds can also be run in neutral solvent.

Platinum oxide has been widely used for hydrogenations under acidic conditions and in general gives good yields of piperidines. It was useful in the reduction of 3-(dimethylaminomethyl)pyridine which can be considered as related to an N-tertiary benzylamine and subject to hydrogenolysis. Hydrogenation in acetic acid with little over 1% by weight of platinum oxide under 4 atm pressure gave 75% yield of 3-(dimethylaminomethyl)piperidine [194]. In contrast, reduction of 3-(diethylaminomethyl)pyridine under similar conditions resulted in considerable hydrogenolysis [3].

The main disadvantage of the use of platinum oxide lies in the fact that reductions with it, except in a few instances, must be carried out under acidic conditions. This is a distinct disadvantage when the resulting piperidine must be isolated as base and it is very soluble in water. Often yield is low unless continuous extraction is resorted to after basification [195].

Hydrogenations of pyridines with rhodium can be carried out in neutral media. A number of pyridines were reduced in alcohol at 25° and 2 atm pressure with 40% by weight of 5% rhodium on carbon [195]. In this series rhodium on alumina was not as effective. Reaction time was often lengthy. In another reduction, hydrogen absorption during the conversion of 2-methyl-2-(4-pyridyl)-1,3-dioxolane to the corresponding piperidine was more rapid in the presence of rhodium on alumina but higher yield resulted from the use of rhodium on carbon, 77 to 58% [196].

While the amount of catalyst employed in Ref. 195 appears excessive, it only represents a 2% metal to substrate ratio. This is often a much lesser amount of metal to substrate compared with that in many platinum reductions. Raising the temperature to 60° increases the speed of hydrogen absorption and often allows the use of a lower catalyst ratio [3].

The activity of the catalyst is affected, however, by the presence of a basic nitrogen-containing side chain. Hydrogenation of 4-(aminomethyl)pyridine and 4-(2-benzylaminoethyl)pyridine in neutral solvent took 12–13 hr before uptake was complete [195].

The presence of acids, in particular organic acids, has aided the reduction of a similar type compound. In another series, hydrogenation of 2-(2-isopropylaminoethyl)pyridine in acetic acid was complete in 3 hr with a 20% ratio of 5% rhodium on alumina but the reduction of other 2- and 4-aminoethylpyridines required additional catalyst under the same conditions [197]. It is possible that the difference in results was due to a difference in purity of the other substrates.

Effect of the Position of Substituents. The suggestion has been offered that piperidine is the poisoning agent in the catalytic reduction of pyridine and that the presence of a substituent in 2- or 6-position allows such pyridine to be more readily hydrogenated in neutral solvent than the parent pyridine or one with a similar substituent in another position [180]. The effect of position is shown in other instances. 2-Benzylpyridine was reduced with rhodium on carbon in one-third the time required to hydrogenate the 4-benzyl compound [195]. The same difference was noted in a ruthenium reduction [184]. This *ortho* effect is also seen in reductions with palladium on carbon in which 3- and 4-substituted pyridines were hydrogenated in acetic acid, while the corresponding 2-derivative could be reduced in ethyl acetate [187].

The effect of a substituent adjacent to the ring nitrogen appears to apply only to reductions in neutral solvent. This effect can be attributed to interference of the resulting piperidine nitrogen's ability to bond strongly to the catalyst.

When there are substituents in both 2- and 6-positions reduction becomes more difficult because the substituents interfere with the approach of the pyridine nitrogen to the catalyst surface. For example, although it was shown that 2,6-dimethylpiperidine did not act as an inhibitor in the reduction of cyclohexene and piperidine completely prevented hydrogen uptake [180], nevertheless, hydrogenations of pyridine to piperidine, in neutral solvent or in the absence of solvent, over rhodium or ruthenium go to completion far more rapidly than similar ones of 2,6-dimethylpyridine to 2,6-dimethylpiperidine [3].

Pyridine Acids, Esters, and Amides. Before the advent of rhodium and ruthenium catalysts, the standard method for reducing pyridine acids to the corresponding piperidines was through reduction in acidic solution or as an acid salt with platinum oxide. This made isolation of the reduced free acid difficult. Only piperidine-2-carboxylic acid could be isolated as the free acid after hydrogenation of picolinic acid in glacial acetic acid [198]. In all other examples of reductions in acid media the piperidine acid was isolated as an acid salt. More recently acidic media were found unnecessary for reductions of pyridines containing acid functions.

PYRIDINECARBOXYLIC ACIDS. Successful conversions of picolinic and isonicotinic acids to piperidine-2- and -4-carboxylic acids were carried out in very high yield by low-pressure reduction in water with 2% by weight of platinum oxide [193] and earlier with a 40% ratio of 5% rhodium on carbon [195]. The use of a 20% ratio, equal to a 1% metal to substrate ratio, was as effective as hydrogenation over 2–2.5% by weight of platinum oxide [3]. Reduction of picolinic acid was also successful with 5% palladium on carbon [3] but reaction time was longer (18 hr) than in hydrogenations with platinum or rhodium. The use of ruthenium at 70–100° and 100 atm also gave good results in reductions of picolinic and isonicotinic acids in water [184].

Hydrogenations of nicotinic acid with platinum, rhodium, and ruthenium in water resulted in extensive decarboxylation (palladium was not used). This was eliminated in the ruthenium reduction by carrying out reaction as sodium salt.

The method of choice for this reduction is hydrogenation as ammonium salt in water with 20% by weight of 5% rhodium on carbon [199]. Concentration of the solution after removal of the catalyst gave 88.5% yield of nipecotic acid. When platinum oxide was employed in a similar reaction, no uptake of hydrogen took place. The use of ammonia was not tried in the ruthenium reduction but should be successful.

Nicotinic esters and amides were readily hydrogenated without decarboxylation over rhodium [195] and over ruthenium [184]. Picolinic and isonicotinic esters and amides were also readily reduced in neutral solvent with the same catalyst. The comparative reductions of nicotinamide and N,N-diethylnicotinamide are of interest. Hydrogenation with ruthenium dioxide under pressure showed very little difference in time required for complete reduction. On the other hand there was a pronounced difference with rhodium catalyst. Nicotinamide was reduced eight times more rapidly than N,N-diethylnicotinamide. Steric effects are not apparent from models of starting or final reduction products but can be seen in models of the intermediate dihydro- or tetrahydronicotinic diethylamide. The ease with which the disubstituted amide was reduced over ruthenium was attributed to the reaction conditions which force the intermediates into a favorable position for contact with the catalyst whereas under the mild conditions of the rhodium reduction there is insufficient energy to accomplish the same thing.

PYRIDINEALKANOIC ACIDS. These acids can be reduced in water with platinum oxide or with rhodium on a support but, as in the hydrogenation of nicotinic acid, the method of choice is reduction in aqueous ammonia over rhodium on carbon or rhodium on alumina [200]. Concentration of the ammoniacal solution gives the piperidinealkanoic acid. In one instance, after the hydrogenation of 3-(2-pyridine)propionic acid, it was necessary to concentrate the reduction solution by freeze-drying to avoid complete loss of product that otherwise resulted from decarboxylation during concentration. 3- and 4-Piperidineacetic acids and 3-(2,3- and 4-piperidine)propionic acids were prepared in good yield by the above hydrogenation procedure. The reduction of 2-pyridineacetic acid was not attempted because it is known to decarboxylate in aqueous solution.

3-(3-Piperidine)propionic acid has been prepared by reduction of the pyridineacrylic acid in water over ruthenium [201], presumably under high-pressure conditions.

The reduction of esters, such as the related pyridinemalonic esters, can lead to cyclized products if the side chain is long enough and is in 2-position. Some examples include the hydrogenation of ethyl 2-pyridylmethylene-malonate with platinum oxide at 4 atm pressure which gave ethyl octahydro-indolizin-3-one-2-carboxylate [202] and that of ethyl 2-carbethoxy-4-(2-pyridyl)butyrate in alcohol and acetic acid to give the corresponding piperidine which underwent cyclization on distillation to ethyl 1,2,3,4,6,7,8,9-octahydroquinolizin-4-one-3-carboxylate [203] (Eq. 24.43). The same ester

$$\text{Pyridine-}(CH_2)_2CH(COOC_2H_5)_2 \xrightarrow{3H_2} \left[\text{piperidine-NH, CH(COOC_2H_5)_2} \right] \rightarrow \text{octahydroquinolizinone-COOC_2H_5} \quad (24.43)$$

was converted directly to the cyclic structure by hydrogenation with Raney nickel at 140–145° and 100 atm [204].

PYRIDINESULFONIC AND ALKANESULFONIC ACIDS. 2-(2-Pyridine and 4-pyridine)ethanesulfonic acids were converted to the (piperidine)ethanesulfonic acids [205] by the rhodium-aqueous ammonia method previously described but was not successful in the hydrogenation of pyridine-3-sulfonic acid. Piperidine-3-sulfonic acid was obtained by reduction of the free acid in water with 6–7% by weight of platinum oxide, plus a second portion of catalyst, at 60° and 27 atm pressure. 3-Pyridinehydroxymethanesulfonic acid could not be reduced by either method. Instead, as an aldehyde addition product, it underwent reversal with release of sulfur dioxide which was reduced to hydrogen sulfide, poisoning the catalyst immediately (Eq. 24.44).

$$RCH(OH)SO_3H \rightleftharpoons RCHO + H_2O + [SO_2 + 3H_2] \rightarrow H_2S + 2H_2O \quad (24.44)$$

R = 3-pyridyl

Hydroxypyridines and Derivatives. Of the three hydroxypyridines only the 3-derivatives resemble a phenol, the 2- and 4-compounds are considered as pyridones. Hydrogenation of 2-pyridones stops at the piperidone stage; the resultant cyclic amide is resistant to further attack except under vigorous hydrogenation conditions. There is no record of reductions of 4-hydroxypyridines (4-pyridones) that stop after absorption of two equivalents [180]. Hydrogen uptake continues to yield 4-hydroxypiperidines.

3-Hydroxypyridine was reduced over Raney nickel at 125° and 100 atm to give 60% yield of 3-hydroxypiperidine [206]. It was also hydrogenated over ruthenium in water (probably at lower temperature) to yield 82% of the hydroxypiperidine [43]. 3-Hydroxy-6-propylpyridine was hydrogenated to the corresponding piperidine in 96% yield in glacial acetic acid with a large amount of platinum oxide [207]. A quantitative yield of 3-hydroxypiperidine was reported from the platinum-catalyzed reduction of 3-hydroxypyridine hydrochloride in alcohol [208] but a repetition of this work by another group gave only 30–40% yield plus a considerable amount of piperidine [209]. Hydrogenolysis was suggested to take place through an allylic type alcohol—HN⟨⟩-OH.

It might be assumed, from hydrogenolysis occurring in platinum-catalyzed hydrogenations of phenols and the relationship of 3-hydroxypyridine to phenol, that hydrogenolysis should be expected in the presence of platinic oxide. In another example, that of the reduction of a betaine, platinum oxide caused considerable hydrogenolysis [210].

The preferred catalyst for the hydrogenation of 3-hydroxypyridines may be rhodium on a support. Although it has not been employed for the reduction of 3-hydroxypyridine itself, a suggestion for its general use is made valid by the comparison of rhodium on carbon with platinum oxide in the reduction of the following betaine (Eq. 24.45).

Further indications of the ability of rhodium catalysts to reduce 3-hydroxypyridines without loss of hydroxyl function is seen in hydrogenations of 2- and 2,6-tertiary aminoalkyl-3-hydroxypyridines with a 5% rhodium on alumina at pressures up to 130 atm [211, 212].

4-Hydroxypiperidine was obtained in 74% yield from hydrogenation of the 4-pyridone in water with ruthenium dioxide [43]. Reductions with

Raney nickel at 100 atm did not take place until the temperature reached 200° [180]. Reactions in alcohols resulted in N-alkylation.

$$\text{(24.45)}$$

N-Alkyl-4-pyridones have been successfully reduced to N-alkyl-4-hydroxypyridines with Raney nickel at elevated temperature and pressure. 1-Ethyl-4-pyridone yielded 85% of 1-ethyl-4-hydroxypiperidine after hydrogenation at 125° and 130 atm [213]. Others were reduced with nickel on silica under similar high-pressure conditions when low-pressure hydrogenations with palladium or platinum in acetic acid failed [214]. Some hydrogenolysis has taken place during nickel reductions because of high temperature [215].

The success of rhodium in hydrogenation of 3-hydroxypyridines suggest that it should be studied in the reduction of 4-pyridones and 1-alkyl-4-pyridones.

Among 2-pyridones, excellent yields of 2-piperidones were obtained during reduction with platinum oxide in acetic acid at 2–3 atm [216, 217] and with nickel at 50° and 150 atm [218].

Among alkoxypyridines, when 2-alkoxy compounds are reduced they undergo cleavage to piperidine and alkanol [219]. 3-Methoxypyridines were converted to 3-methoxypiperidines by high-pressure hydrogenation over Raney nickel without apparent hydrogenolysis [214], and with platinum oxide in acetic acid at 3 atm also with good results [220]. A series of 4-alkoxypyridines were hydrogenated to 4-alkoxypiperidines in good yield over ruthenium dioxide at 140° and 150 atm [190]; the use of Raney nickel gave only 25% yield in reaction at 150° and 150 atm [182], due probably to hydrogenolysis. Rhodium catalysts, which also should be of value for reductions of 3- and 4-alkoxypyridines, have not been investigated.

Aminopyridines. Of the aminopyridines only the 3-amino compound has aromatic character and behaves like aniline on reduction. 2- and 4-Aminopyridine can exist in tautomeric forms; their reduction is less straightforward.

3-Aminopyridine has been hydrogenated at low pressure with platinum oxide in dilute hydrochloric acid to provide good yields of 3-aminopiperidine as dihydrochloride [221, 222]. Neither rhodium nor ruthenium, catalysts preferred for aniline reductions, have not been employed for 3-aminopyridines; their use should be studied.

The hydrogenation of 4-aminopyridines has been meagerly investigated. 4-Aminopyridine was hydrogenated with difficulty over platinum oxide in alcoholic hydrogen chloride under 80 atm pressure to give only 16.5% of 4-aminopiperidine dihydrochloride [223]. On the other hand a tertiary aminopyridine, 4-(4-morpholino)pyridine, was readily hydrogenated in water in the presence of ruthenium dioxide at elevated temperatures and pressure [184] (Eq. 24.46).

$$N\text{—}\bigcirc\text{—}N\text{—}\bigcirc\text{—}O \xrightarrow[\text{RuO}_2]{3H_2} HN\text{—}\bigcirc\text{—}N\text{—}\bigcirc\text{—}O \quad (24.46)$$
$$90\text{–}100°, 100\ \text{atm}$$

2-Aminopyridine is initially reduced to the tetrahydro stage. Further reduction leads to piperidine and ammonia [219]. Hydrogenation of 2-aminopyridine over platinum in acetic acid-acetic anhydride was reported to absorb three equivalents to yield 1-acetyl-2-acetylaminopiperidine in 80% yield [224]. However, before this result is to be considered valid, more rigorous proof of structure is required beyond the meager analytical data offered.

Mono- and disubstituted 2-aminopyridines absorb only two hydrogen equivalents during reduction. 2-Benzylaminopyridine, reduced in acetic acid with palladium oxide, gave 2-benzylamino-3,4,5,6-tetrahydropyridine [225]. 2-Methylaminopyridine yielded the corresponding 3,4,5,6-tetrahydro compound upon hydrogenation with 40% by weight of 5% rhodium on alumina [226]; platinum oxide was less efficient. When 2-dimethylaminopyridine was reduced, although incapable of tautomerizing, it too absorbed only two equivalents to give 2-dimethylamino-3,4,5,6-tetrahydropyridine (Eq. 24.47).

$$\bigcirc_{N}\text{—}N(C_2H_5)_2 \xrightarrow[\text{5% Rh on Al}_2\text{O}_3]{2H_2,\ \text{AcOH}} \bigcirc_{N}\text{—}N(C_2H_5)_2 \quad (24.47)$$

None of the results is too surprising. All the tetrahydropyridines possess an amidine linkage, $-N\!\!=\!\!C\!\!-\!\!N\diagdown$. The $N\!\!=\!\!C$ bond in amidines is difficult to reduce. If hydrogenation does take place it results in cleavage at one of the $\diagdown N\!\!-\!\!C\!\!-\!\!N\diagup$ bonds.

Such an example is seen in the hydrogenation of 1-(2-hydroxy-2-phenyl)-ethyl-2-iminopyridine with about 10% by weight of 5% rhodium on carbon in alcohol [227]. The intermediate 1-substituted-2-iminopiperidine underwent further reduction to give 1-(2-hydroxy-2-phenyl) ethylpiperidine in quantitative yield. Deamination was prevented by reducing the iminopyridine as a hydrochloride salt (Eq. 24.48).

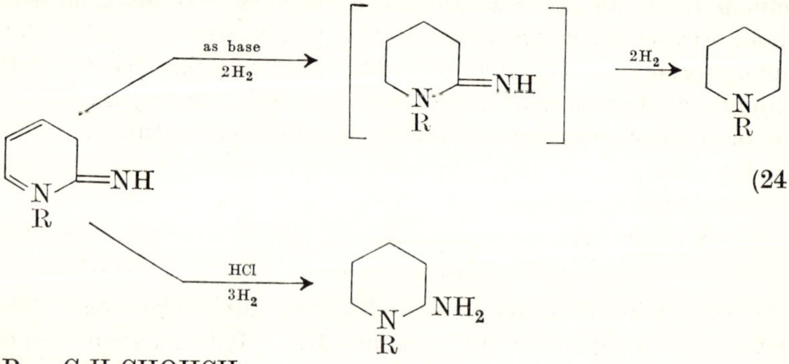

(24.48)

$R = C_6H_5CHOHCH_2$

7-Azaindole and 1-methyl-7-azaindole, both of which can be considered as substituted 2-aminopyridines, were reduced in hydrochloric acid solution with platinum oxide to amidine structures [228], similar to those in 2-mono- and dialkylamino-3,4,5,6-tetrahydropyridines (Eq. 24.49). However, when

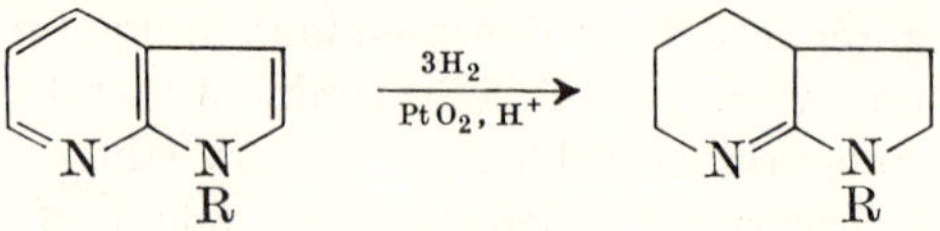

(24.49)

$R = H$ or CH_3

R was methyl and the compound was reduced as the methyl pyridinium iodide, cleavage occurred at the pyrrolidine nitrogen to yield 1-methyl-3-(2-methylaminoethyl)piperidine (Eq. 24.50).

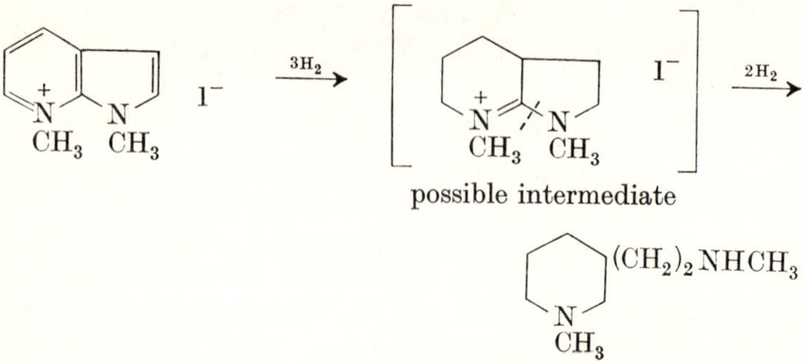

possible intermediate

(24.50)

Pyridylkylamines. Raney nickel is usually of little value for reduction of pyridines containing basic side chains because its activity is grossly affected by the basicity of the substrate and more so by the basicity of the reduction product. Conditions necessary to induce hydrogen absorption too often result in side reactions [180]. It was surprising that the pyridine ring of nicotine, 3-(1-methyl-2-pyrrolidinyl)pyridine, which may be viewed as a pyridylalkylamine, could be hydrogenated under such mild conditions with it, 100–140° and 15–20 atm [229].

Reductions with platinum oxide, which must be carried out under acidic conditions, often give very good results [3, 180, 230]. In a number of instances a larger than normal amount of catalyst was necessary or reduction would not go to completion [3, 197, 231]. At times it was necessary to use pressures up to 70–100 atm to foster increased hydrogen absorption [3].

Acylation of the basic side-chain nitrogen when it is primary or secondary can aid platinum-catalyzed hydrogenations of pyridylalkylamines (it should also be useful in rhodium and ruthenium reductions). An example is seen in the reduction of 2-(aminomethyl)pyridine in acetic acid at 50° and 3 atm with 3% by weight of platinum oxide plus additional catalyst; hydrogen uptake was complete in 30 hr [232]. In contrast, 2-(carbethoxyaminomethyl)pyridine was hydrogenated in acetic acid with 1% by weight of catalyst in 12 hr.

Rhodium has been employed for these reductions in neutral solvent [195] and under acidic conditions [197] with mixed results. The best method, if pressure equipment is available, may be hydrogenation with ruthenium dioxide in neutral solvent at 90–100° and 100 atm [184]. The catalyst appears to be less poisoned than the others by the substrate or reduction product although there was some inhibition when the amino group in the side chain was primary.

Miscellaneous Pyridines. This author favors the use of rhodium on a support for the low-pressure reductions of most other pyridines in neutral solvent or at elevated pressure with ruthenium. Rhodium was not very successful in the reduction of a dipyridyls under neutral or acidic conditions. The method of choice appears to be reduction with platinum oxide in water or alcohol containing a moderate excess of hydrochloric acid [233].

Quaternary Salts. Quaternizing the pyridine nitrogen as an alkyl, aryl, or an aralkylpyridinium compound prevents catalyst poisoning. Reductions, giving N-substituted piperidines in high yield, have been carried out under a variety of conditions with Raney nickel, palladium, or rhodium on a support and most frequently with platinum oxide. Hydrogenation is generally straightforward. However, it is possible to obtain tetrahydropyridines. This will be discussed in the following section.

Partial Reduction. Partial reduction of nonquaternized pyridines is limited to those containing certain functions. The effect of an amino group in 2-position has already been discussed. The hydrogenation of 3-acetylpyridine over 5% palladium on carbon in alcohol gave a 70% yield of 3-acetyl-1,4,5,6-tetrahydropyridine [234]. The formation of tetrahydro compound was postulated to take place by 1,4-addition of hydrogen followed by preferential reduction of the isolated double bond over the conjugated 2,3-bond (Eq. 24.51).

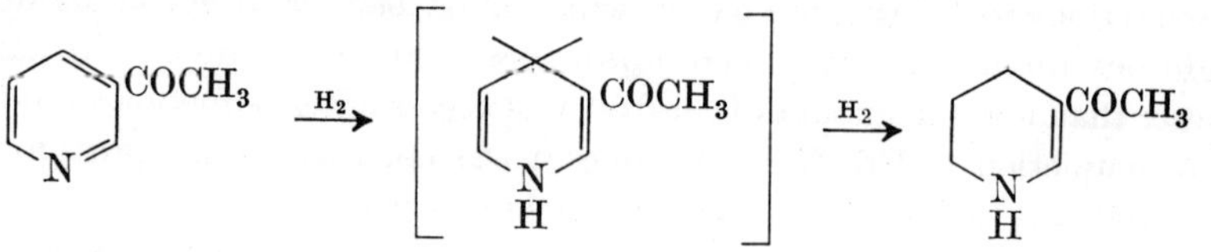

(24.51)

Partial hydrogenation did not take place in the presence of other catalysts, nor could the tetrahydropyridine be reduced further with palladium under acidic conditions.

Other pyridines with carbonyl-containing functions in 3-position were also converted to 1,4,5,6-tetrahydro compounds by reduction with palladium on carbon in neutral solvent. They include ethyl 3-nicotinylpropionate, ethyl nicotinate, and nicotinamide [235] and *t*-butyl nicotinate [236].

The alkaloid, myosmine, 3-(1-pyrrolin-2-yl)pyridine, when hydrogenated with palladium on carbon in 50% aqueous acetic acid also absorbed two molar equivalents to yield some 3-(1-pyrrolin-2-yl)1,4,5,6-tetrahydropyridine [235]. The authors suggested that partial ring reduction took place through an open chain aminoketone followed by cyclization (Eq. 24.52).

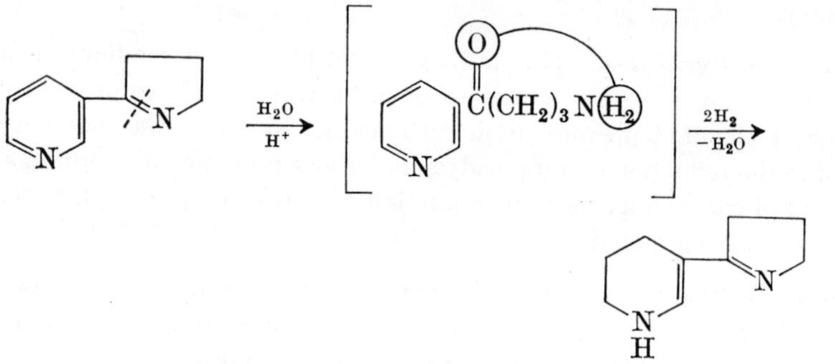

(24.52)

Partial hydrogenation of quaternized pyridines to tetrahydro compounds is not restricted to the use of palladium although it may be best for this purpose. Tetrahydrogenation of 3-substituted quaternized pyridines has the same limitation that applies to 3-substituted nonquaternized pyridines, the

Heterocycles (One Hetero Atom)

presence of unsaturated electron-withdrawing groups. A number of such compounds with aldehyde, amide, ester, ketone, and cyano groups in 3-position have been reduced with palladium on carbon and excess triethylamine to 2-piperideines, 1,4,5,6-tetrahydropyridines [236, 237]. In one instance when a methyl group was in 2-position, tetrahydrogenation did not take place; reduction of methyl 2-methylnicotinate as methotosylate produced the methyl 1,2-dimethylnipecotate exclusively [237]. An example of the use of platinum oxide is seen in the hydrogenation of 3-cyano-1,5-dimethylpyridinium bromide in aqueous sodium bicarbonate which yielded 1,5-dimethyl-1,4,5,6-tetrahydronicotinamide, the nitrile group undergoing conversion to amide during reduction [238].

It was also possible to obtain some tetrahydro compound with Raney nickel when methyl nicotinate methiodide was hydrogenated in its presence with excess triethylamine at 80° and 120 atm [239]. The formation of tetrahydropyridine was made possible by the tendency of iodide ion to inhibit catalyst activity (but not enough to prevent formation of over 50% of the piperidine).

Tetrahydrogenation of 4-substituted quaternized pyridines appears to be limited to isonicotinic esters and dependent to a great extent on reaction conditions, on the amount and activity of the catalyst, and on the halide ion.

Methyl isonicotinate methiodide was converted in part to methyl 1-methyltetrahydroisonicotinate by hydrogenation over platinum oxide in methyl alcohol [240]; continued hydrogen uptake yielded the piperidine. Formation of tetrahydro compound, despite the high catalyst ratio (15% by weight), attests to the effect of iodide ion in these reductions. The use of a 1.5% catalyst ratio in a similar reaction produced only tetrahydro compound [241]. The hydrogenation of crude quaternary compound with previously used catalyst resulted in slow hydrogen uptake to again yield methyl 1-methyltetrahydroisonicotinate [242]. The inhibiting effect of iodide ion in these reactions was emphasized when reduction of the corresponding bromide gave only methyl 1-methylpiperidine-4-carboxylate.

Tetrahydrogenation of the methiodide of the esters in 4-position was shown to yield 1-methyl-1,2,5,6-tetrahydropyridines [242]. They were apparently formed by 1,2-addition of hydrogen followed by preferential attack on the unhindered enamine double bond by another mole of hydrogen (Eq. 24.53).

$$\underset{\underset{CH_3}{\overset{+}{N}}}{\bigcirc}^{COOR} I^{(-)} \xrightarrow{H_2} \left[\underset{\underset{CH_3}{N}}{\bigcirc}^{COOR} \right] \xrightarrow{H_2} \underset{\underset{CH_3}{N}}{\bigcirc}^{COOR} \qquad (24.53)$$

Selective Reductions

PRESENCE OF FUNCTIONAL GROUPS. In most instances functional groups attached to the pyridine nucleus will be selectively reduced, but certain exceptions do exist. For example, the pyridine ring in 1-methyl-2-(6-methoxy-2-phenyl-3,4-dihydronaphthyl)pyridinium iodide was reduced in preference to the double bond in the naphthalene ring [243] (Eq. 24.54).

$$\text{(structure)} \xrightarrow[\text{PtO}_2]{3\text{H}_2} \text{(structure)} \quad (24.54)$$

R = C₆H₅ 53%

This reaction may not be entirely unexpected since quaternization aids saturation of the pyridine ring and the unattacked double bond is tetrasubstituted and should be difficult to reduce because of steric effects.

The other exceptions pertain to keto and cyano groups. Selective reduction of a pyridine ring when a ketone function is adjacent to it has taken place when an acyl group was in 3-position. Examples are seen in the previously noted hydrogenation of 3-acetylpyridine in alcohol with palladium on carbon [234], in the reduction of 3-acetyl-1-[2-(3-indolyl)ethyl]pyridinium bromide, also with palladium on carbon [244], in the hydrogenation of 3-benzoylpyridine hydrochloride with platinum oxide which yielded 3-benzoylpiperidine [245], and in the reduction of 3-benzoylpyridine methiodide with the same catalyst which gave 3-benzoyl-1-methylpiperidine as the major product but was accompanied by the formation of carbinols.

When N-methyl-4-acetylpyridinium iodide was hydrogenated in alcohol and triethylamine with palladium oxide, the ring and the keto group were both reduced [196]. There did not appear to be any attempt for selectivity. However, when N-benzyl-4-acetylpyridinium chloride was hydrogenated with either platinum or palladium, some 4-acetyl-1-benzylpiperidine was obtained along with the corresponding carbinol; they were inseparable by distillation.

If one or more methylene groups separate the ring and the ketone and hydrogenation is carried out under acidic conditions or the compound is reduced as a quaternary salt, examples show preferential ring reduction when the side chain is in 2-position.

1-(2-Pyridyl)-2-propanone in acetic acid was selectively reduced to the piperidylpropanone with platinum oxide [246]. When a number of 4-(2-pyridyl)-2-butanones were hydrogenated in aqueous or aqueous hydrochloric acid, ring closure took place to yield substituted quinolizidines

Heterocycles (One Hetero Atom) 597

[247, 248]. Reactions apparently took place as shown in Eq. 24.55. The pyridine ring was preferentially reduced when 1-(2-pyridyl)-2-propanone as

$$\text{(24.55)}$$

methosulfate was hydrogenated over platinum [249]. In another example, the methiodide of 2-(2-pyridyl)methyleneindan-1-one was reduced in water with platinum oxide to give 2-(1-methyl-2-piperidylmethyl)indan-1-one [250] (Eq. 24.56).

$$\text{(24.56)}$$

There is also a record of selective reduction of the ring when the side chain containing the keto group was in 4-position. 4-Phenacylpyridine in alcohol containing excess hydrochloric acid gave 95% yield of 4-phenacylpiperidine after hydrogenation at 30° and 33 atm [250a].

Conflicting results have been obtained from the reduction of 1-phenacylpyridinium compounds. In one instance hydrogenation of 1-phenacylpyridinium bromide in water with platinum oxide at low pressure gave 1-phenacylpiperidine [251]. In another reduction at 60 atm the ketone was selectively reduced [252].

Selectivity was reported to be governed by substitution on the pyridine ring. When 4-alkyl-1-phenacylpyridinium bromides were reduced with platinum oxide, the pyridine ring was reduced preferentially only when the alkyl group was methyl or ethyl; when it was larger the ketone group was selectively hydrogenated [253]. When 4-hexyl-1-(4-nitrophenacyl)pyridinium bromide was reduced, despite the large group on the pyridine nucleus, the ring was saturated and the ketone group was still intact; 4-hexyl-1-(4-aminophenacyl)piperidine was obtained in 74% yield [254].

Except for the reduction of methyl 3-cyanopyridinium iodide to methyl 3-cyanopiperidine [237] and that of the bromide in aqueous bicarbonate where ring reduction was accompanied by hydration of the nitrile function

598 Aromatic Ring Systems

[238], when a cyano group is attached to the pyridine nucleus it is reduced in preference to saturation of the ring. However, when there is an alkylene bridge between them, the ring can be selectively reduced.

2-(2-Cyanoethyl)pyridine was converted to indolizidine by reduction with platinum oxide in acidified aqueous alcohol (255). There is little doubt that ring reduction preceded cyclization and that 2-(2-cyanoethyl)piperidine was the intermediate through which cyclization took place (Eq. 24.57) although there is some question as to the manner in which it proceeded [180].

$$(24.57)$$

When 3-cyanomethylpyridine was reduced in dilute hydrochloric acid over platinum oxide or in acetic acid containing sulfuric acid, the cyano group was preferentially attacked. However, when the compound was hydrogenated as the methiodide the ring was selectively saturated to give 84.5% yield of 3-cyanomethyl-1-methylpiperidine with either 5% rhodium in carbon or with platinum oxide [256]. 1-Benzyl-3-cyanomethyl and 3-cyanoethyl-1-methylpiperidines were prepared in a similar manner from the corresponding quaternized cyanoalkylpyridines.

PRESENCE OF HYDROGENOLYZABLE GROUPS. The pyridine ring containing nuclear halogen cannot be saturated without loss of halogen. It can be reduced selectively in acidic medium in the presence of an aryl chloride or bromide (and probably in the presence of aryl iodide) with platinum oxide [3]. The hydrogenation at 10 atm of 3-(4-chlorobenzyl)-1,2,3,4-tetrahydroquinolizinium chloride, in alcohol contaiinng sodium acetate, with Raney nickel to 3-(4-chlorobenzyl)octahydroquinolizine is another example of the same selectivity [257] (Eq. 24.58).

$$(24.58)$$

Ring reduction should also take place in the presence of other halides which are not especially reactive to dehalogenation. This is exemplified in the retention of a secondary halide when 3-bromo-2-hydroxy-1,2,3,4-tetrahydroquinolizinium bromide was hydrogenated in alcohol over platinum oxide to 3-bromo-2-hydroxyoctahydroquinolizine [258] (Eq. 24.59).

$$\text{quinolizinium} \xrightarrow[\text{PtO}_2]{3\text{H}_2} \text{octahydroquinolizine} \quad (24.59)$$

When a substrate contains a pyridine ring and a benzyl type group, hydrogenolysis does occur occasionally during saturation of the pyridine ring, but this usually is dependent on the catalyst, on the amount employed and reaction conditions. For example, (α-4-pyridyl)benzhydrol, containing a tertiary benzylic hydroxyl group, and related compounds were reduced with platinum oxide, 3–5% by weight, in acetic acid to give high yields of the corresponding piperidines [3] (Eq. 24.60).

$$(C_6H_5)_2C(OH)\text{-pyridyl} \xrightarrow[\text{PtO}_2, \text{AcOH}]{3\text{H}_2} (C_6H_5)_2C(OH)\text{-piperidyl} \quad (24.60)$$

In contrast, reduction of a related compound (α-3-pyridyl)-4,4'-dimethylbenzhydrol in alcoholic hydrogen chloride with a large amount of catalyst resulted in loss of the benzylic alcohol to give 3-(4,4'-diphenylmethyl)piperidine [259].

Loss of the benzylic hydroxyl group in these reactions can be avoided if excess strong acid is not employed. The hydrogenation of 2-(1,2-dihydroxyethyl)pyridine hydrochloride, with a secondary benzylic type hydroxyl group, in water, alcohol, or acetic acid with platinum oxide or rhodium on carbon gave quantitative yield of the corresponding piperidine [260]. Reduction of the related 4-derivative as hydrochloride in alcohol containing additional hydrochloric acid under low pressure gave only 64% of 4-(1,2-dihydroxyethyl)piperidine and some 4-(2-hydroxyethyl)piperidine from loss of the secondary hydroxyl group [261].

2-Hydroxymethylpyridine, a primary alcohol related to benzyl alcohol, was easily reduced to 2-hydroxymethylpiperidine with rhodium on carbon in alcohol [3].

Selective reductions of pyridines containing N-benzyl linkages or pyridylmethyl secondary and tertiary amines (also subject to hydrogenolysis) may best be carried out as salts in the presence of platinum oxide [3]. Hydrogenations as base with 5% rhodium on carbon may also be selective. Some debenzylation did take place during the reduction of 4-(2-benzylaminoethyl)pyridine with 40% by weight of catalyst when uptake of hydrogen exceeded the theoretical amount [195]. The use of somewhat less catalyst should give better selectivity. Both catalysts were selective in the hydrogenations of benzyl pyridinium halides to 1-benzylpiperidines [210, 220, 256].

PRESENCE OF OTHER RINGS. A pyridine ring is generally reduced in preference to a benzene ring. When a phenyl ring is attached to a pyridine ring, most hydrogenations yield phenylpiperidines except when ruthenium is employed [3]; with that catalyst both rings are reduced. Steric effects can change or reverse selectivity. When 2,6-diphenylpyridine hydrochloride was reduced with platinum oxide all three rings were saturated; when 2,4,6-triphenylpyridine was hydrogenated only the pendant phenyl rings were reduced [262].

The use of platinum oxide gives excellent results in these selective reductions and may be the catalyst of choice. When employed for the reduction of 1-methyl-2-phenylpyridinium iodide, it gave very high yield of 1-methyl-2-phenylpiperidine [263]. A note of caution was voiced concerning overhydrogenation. It did not apply to saturation of the phenyl ring but rather to ring opening of the resulting piperidine because of its relationship to an N-benzyl tertiary amine (Eq. 24.61).

$$\underset{\underset{I^{(-)}}{CH_3}}{\overset{+}{N}}C_6H_5 \xrightarrow{3H_2}{PtO_2} \underset{CH_3}{N}C_6H_5 \xrightarrow{3H_2} C_6H_5(CH_2)_5NHCH_3 \quad (24.61)$$

Platinum oxide was very useful in the reductions of a number of pyridines of the type $(C_6H_5)_2C(R)Y$, where R was H, OH, or CH_3 and Y was 2-, 3-, or 4-pyridyl [3]. Reactions were carried out in acetic acid with 2–5% by weight of catalyst at 25–50° and 2–3 atm pressure. It worked equally as well in reductions of the type $(C_6H_5)_2CRCH_2Y$. It also was selective as far as competition between the pyridine and benzene rings, but did cause loss of the benzylic hydroxyl group when 2-phenyl-1,3-*bis*(4-pyridyl)-2-propanol was hydrogenated in 10% hydrochloric acid [250a], obviously from the use of strong acid. When 10% rhodium on carbon, 20% by weight, was employed for the same reduction in alcohol at 90–96° and an initial pressure of 100 atm, 2-phenyl-1,3-*bis*(4-piperidyl)-2-propanol was obtained in 98% yield.

Rhodium on carbon employed for the reduction of 4-benzylpyridine under low-pressure conditions gave 4-benzylpiperidine almost exclusively but did cause some benzene ring saturation during hydrogenation of 2-benzylpyridine to give 2-benzylpiperidine containing 8% of 2-(cyclohexylmethyl)-piperidine [195].

In general the pyridine ring is also preferentially reduced in the presence of other unsaturated heterocycles. Selective reduction leading to piperidine has occurred when 2-(2-pyridyl)- and 3-(2-pyridyl)pyrroles were hydrogenated over platinum oxide in alcohol containing an equivalent of hydrochloric acid [262]. However, when the pyrrole nitrogen was substituted both rings were reduced. Cleavage also occurred when 1-methyl-2-(2-pyridyl)pyrrole was hydrogenated to yield 3-(4-aminobutyl)piperidine.

When 3-substituted-2-methyl-4-(pyridyl)pyrroles were reduced with platinum in acetic acid only the pyridine ring was saturated. In the absence of a 3-substituent both rings were reduced [156].

Numerous references indicate that the pyridine ring is reduced in preference to the pyrrole ring in indoles. The following examples are those of (indolylalkyl)pyridines in which the pyridine nitrogen is quaternized; reductions are carried out in alcohol with platinum oxide [264, 265] or with supported rhodium [266]. One platinum-catalyzed reduction was carried out at 65° and 27 atm without causing attack on the pyrrole ring of the indole [267]. Reductions as hydrochloride salts were carried out in alcohol at 60 atm with platinum oxide to yield (indolylalkyl)piperidines [268].

Preferential saturation of the pyridine ring also took place in the hydrogenations of α-(3-indolyl)pyridine-2-methanol [269] and the corresponding 4-methanol [270] in alcohol containing excess acetic acid. However, in both instances hydrogenolysis of the allylic-like alcohol occurred to yield mainly 2- and 4-skatylpiperidines (Eq. 24.62).

$$\text{(24.62)}$$

The pyridine ring has also been reduced in the presence of a furan ring with platinum oxide [271] and in preference to an oxazole ring with palladium on carbon [272], each in neutral solvent. When 5-(2-, 3- and 4-pyridyl)-1,2,3,4-tetrazoles were hydrogenated in glacial acetic acid with platinum oxide very high yields of the 2-, 3-, and 4-piperidyltetrazoles were obtained [273].

Quinolines and Isoquinolines

When quinolines or isoquinolines are hydrogenated, the pyridine portion of the molecule is usually selectively saturated. Substituents on the carbocyclic ring aid *py*-tetrahydrogenation but substituents on the heterocyclic portion can change the course of reduction and produce 5,6,7,8-tetrahydro compounds.

The effect of alkyl and phenyl groups has been studied in nickel reductions among quinolines [274–276]. Their effect was reported to be dependent on the position of the substituent, the number of substituents and, in the 2-alkyl series, on the bulk of the substituent. Little effect was seen from the 2-methyl group, but hydrogenation of 2-propylquinoline resulted in 35% of 2-propyl-5,6,7,8-tetrahydroquinoline and 65% of 1,2,3,4-tetrahydro compound. Reductions of 3- and 4-substituted compounds gave mixtures of *py*- and *bz*-tetrahydroquinolines. When 2-phenylquinoline was hydrogenated only 2-phenyl-1,2,3,4-tetrahydroquinoline was obtained but in a similar reaction of the 3-phenyl derivative the ratio of *bz*-tetrahydro to *py*-tetrahydroquinoline was 2:1.

In these studies reductions of 2,3- or 3,4-disubstituted quinolines showed a tendency toward 5,6,7,8-tetrahydrogenation; trialkyl substitution favored almost complete conversion to it. It is of interest that when methyl 2,3-dimethylquinoline-4-carboxylate was hydrogenated only 24% of 5,6,7,8-tetrahydro compound was obtained against 70% of the *py*-tetrahydro derivative [277].

In a few instances amino and hydroxyl groups attached to the heterocyclic ring of quinoline or isoquinoline affected the course of tetrahydrogenation. When 3-hydroxyquinoline-8-carboxylic acid was reduced in aqueous sodium carbonate with Raney nickel only the 5,6,7,8-tetrahydroquinoline was obtained [278]. This is of interest since the reduction of 3-hydroxyquinoline with palladium in neutral solvent yielded the corresponding *py*-tetrahydro compound [279]. It is difficult to see how the carboxyl group caused the change since substituents on the benzene ring aid *py*-tetrahydrogenation.

In isoquinoline when the hydroxyl group was *meta* to the ring nitrogen, seen in the hydrogenation of 1-benzyl-7-chloro-4-hydroxyisoquinolinium chloride in acetic acid with 10% palladium on carbon at 90° and atmospheric pressure, debenzylation and dehalogenation accompanied selective reduction of the benzene ring to give 4-hydroxy-5,6,7,8-tetrahydroisoquinoline [280].

When isocarbostyril, the tautomer of 1-hydroxyisoquinoline, was hydrogenated in glacial acetic acid with platinum oxide, 1-hydroxy-5,6,7,8-tetrahydroisoquinoline was obtained [281]. Similar reductions of 1-ethoxy and 1-isopropoxyisoquinoline also gave *bz*-tetrahydro compounds.

The hydrogenations of 3-aminoquinoline and 4-aminoisoquinoline, each with the amino group *meta* to the ring nitrogen, provided different results. The reduction of 3-aminoquinoline in neutral solvent with Raney nickel at 55° and 90 atm gave 3-amino-1,2,3,4-tetrahydroquinoline and a bridged compound, N^1N^2-*bis*(1,2,3,4-tetrahydro-3-quinolyl)hydrazine [282]. When 4-aminoisoquinoline was reduced in acetic acid over platinum oxide or Raney nickel, only the benzene portion was affected [283]. When the 4-amino group was acylated, hydrogenations under similar conditions gave 1,2,3,4-tetrahydro compounds. Another change took place when reduction of the 4-acylamino compounds was carried out in acetic acid containing sulfuric

Heterocycles (One Hetero Atom)   603

acid; 5,6,7,8-tetrahydro compounds were obtained (Eq. 24.63). Hydrogenation of 1-aminoisoquinoline in a similar manner (strongly acidic conditions)

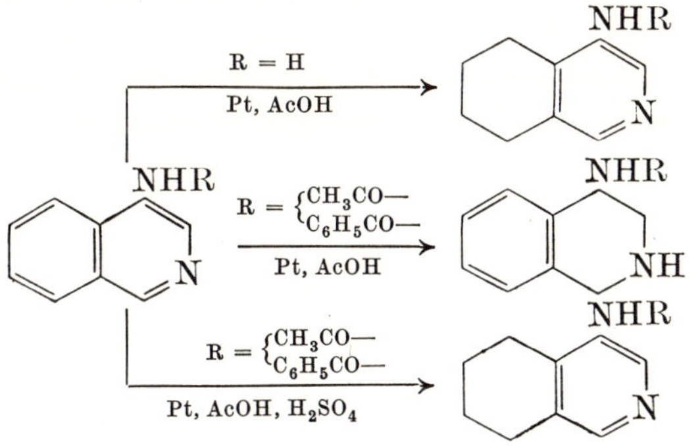

(24.63)

also yielded the 5,6,7,8-tetrahydro compound (65%) along with 1-aminodecahydroisoquinoline [281].

Further study is necessary in the hydrogenation of *py*-amino and hydroxy quinolines and isoquinolines to learn whether acidic conditions favors *bz*-tetrahydrogenation or whether it is peculiar to those substituted isoquinolines. The effect of reductions of the same isoquinolines in neutral solvent or under less acid conditions should also be studied.

In another example of 5,6,7,8-tetrahydrogenation, 5-dichloromethyl-5-methyl-[5H]8-quinolone was reduced in neutral solvent over platinum oxide or palladium on carbon to yield 8-hydroxy-5-dichloromethyl-5-methyl-5,6,7,8-tetrahydroquinoline [284]. This is not an unexpected result. A pyridine ring is not generally reduced with these catalysts, particularly platinum oxide, in neutral solvent under low-pressure conditions. In addition, the carbocyclic portion of this molecule is not truly aromatic. It is more logical to view this reaction as one in which the double bond and the carbonyl group in an α,β-unsaturated ketone are reduced in preference to saturation of the pyridine ring (Eq. 24.64).

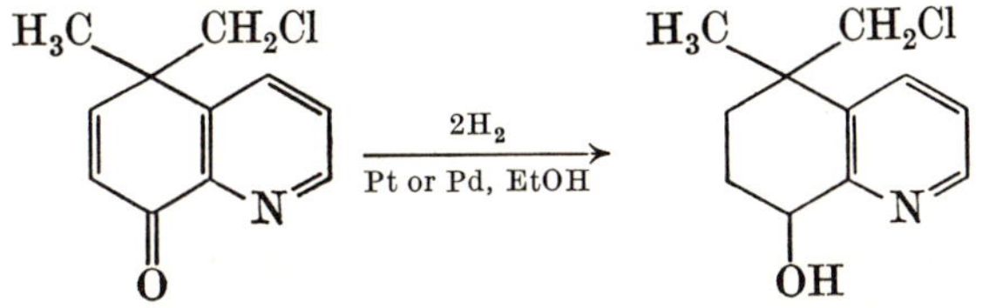

(24.64)

Catalysts. Tetrahydrogenations have been successful with nickel and copper chromite but the vigorous reaction conditions associated with the use of these catalysts are generally unnecessary. Hydrogenations in alcohols with them usually give *N*-alkylated products.

The method of choice is likely hydrogenation in acidic media with platinum oxide [192] because continuing uptake to saturate the benzene ring drops off very sharply after absorption of the first two equivalents. Overhydrogenation is less of a problem with isoquinolines since the benzene ring in 1,2,3,4-tetrahydroisoquinolines is rather difficult to saturate. It may be reasoned that py-tetrahydroisoquinolines resemble phenethylamines which have been shown to reduce with difficulty except with ruthenium catalysts at 70–100° and 100 atm [85].

Reactions to obtain decahydro compounds with platinum require a long reaction period [262] and usually a large amount of catalyst and the presence of strong acid [285, 286].

As in the pyridine series, hydrogenations in this series with platinum oxide cannot be carried out in neutral media unless there is some shielding effect from the bulk of a group adjacent to the ring nitrogen exemplified in the reduction of methyl quinoline-2-carboxylate in alcohol which yielded the corresponding 1,2,3,4-tetrahydro compound or if the substrate contains a strongly acidic group as in the rapid hydrogenation of quinoline-6-sulfonamide in alcohol with a 5% ratio of platinum oxide which gave very high yield of 1,2,3,4-tetrahydroquinoline-6-sulfonamide [3].

Rhodium on a support, which gave good results in pyridine hydrogenations in neutral solvent, has not been studied too extensively for reductions of quinolines and isoquinolines. There is a tendency toward overhydrogenation from its use.

When quinoline was reduced in the presence of 5% rhodium on alumina in alcohol at 60° and 2.5 atm, the resulting product was py-tetrahydroquinoline containing 10% of decahydro compound [3]. The same catalyst caused overhydrogenation when isoquinoline was reduced.

5% Rhodium on carbon, 20% by weight, seemed of value in the reduction of quinoline-6-carboxylic acid in aqueous ammonia to 1,2,3,4-tetrahydroquinoline-6-carboxylic acid in 75% yield [3]. On the other hand, when the carboxyl group was in 2-position, hydrogen uptake proceeded without any noticeable break in the rate until four equivalents were absorbed. The result may be attributed to the dual effect of the carboxyl group in preventing deactivation of the catalyst. It not only neutralized the inhibiting effect of the basic nitrogen atom on catalyst activity but also had a shielding effect because of its proximity.

Rhodium may be better suited for perhydrogenation.

Ruthenium, useful for the reduction of pyridines at 70–100° and 100 atm [184], should not be employed for tetrahydrogenation. Attempts to prepare 1,2,3,4-tetrahydroquinoline or isoquinoline resulted in substantial overhydrogenation in each experiment [3]. Indeed, where pressure equipment is available, ruthenium-catalyzed hydrogenations at 100–125° and 100 atm for quinoline and 120–140° and 100–150 atm for isoquinoline is the method of choice for direct conversion to decahydro compounds [3].

There are only a few examples describing the use of palladium. In one, ethyl quinoline-6-carboxylate was reduced in alcohol with a 5% ratio of 10% palladium on carbon at 80° and 50 atm for 4 hr to give 70% yield of ethyl 1,2,3,4-tetrahydroquinoline-6-carboxylate [287]. It was used exclusively in the hydrogenation of some hydroxyquinolines, benzoate esters, and benzoxy compounds [279] with varying success.

Substituted Quinolines and Isoquinolines. Reductions where the substituent is on the carbocyclic portion of the ring need no comment. In general, unless the heterocyclic ring contains several substituents *py*-tetrahydrogenation will take place.

ACIDS, ESTERS, AND AMIDES. No difficulty should be anticipated in the reductions of any of these compounds with the proper catalyst. Quinoline-2-carboxylic acid was hydrogenated with platinum in acetic acid to yield 1,2,3,4-tetrahydroquinoline-2-carboxylic acid [288]. It should be possible to reduce the 2-acid in neutral solvent in the same manner as the methyl ester described earlier. The 3-acid as sodium salt and the ethyl ester were reduced over Raney nickel at 100° and 110 atm and the corresponding 3-diethylamide was hydrogenated with palladium on carbon at 60° and 90 atm to give the corresponding *py*-tetrahydro compounds [282]. There are very few examples of isoquinolines containing these substituents in 1-, 3-, or 4-positions. In one instance when isoquinoline-1-carbocyclic acid in acetic acid was reduced with platinum oxide, uptake for the necessary two equivalents took 13 days [289].

AMINO AND HYDROXY COMPOUNDS. These reductions were discussed in a previous section. The benzyl ester of 3-hydroxyquinoline gave the corresponding *py*-tetrahydro compound on hydrogenation in neutral solvent with palladium; the 4-benzoate underwent hydrogenolysis to 4-hydroxyquinoline [279]. The reduction of the benzoate of 2-hydroxyquinoline and 1-hydroxyisoquinoline resulted in conversion to the 3,4-dihydro compounds and some *N*-benzoyl-3,4-dihydrocarbostyril and isocarbostyril. The hydrogenation of carbostyril (2-hydroxyquinoline) and isocarbostyril (1-hydroxyisoquinoline) gave only 3,4-dihydro compounds. This is expected as both starting products are amides, not hydroxyquinolines.

Methyl isocarbostyril-5-carboxylate was reduced in acetic acid with platinum oxide to the 3,4-dihydro compound [290]. Since carbostyril and isocarbostyril contain nonbasic nitrogen atoms, it should be possible to carry out hydrogenations in neutral solvent with platinum. An example is seen in the reduction of methyl 1-methylcarbostyril-4-acetate in alcohol with it to yield the corresponding 3,4-dihydro compound in 86% yield [291].

Reduction with Raney nickel or Raney cobalt at 100° and 80 atm for 4 hr gave 92% yield of 2-(2-hydroxyethyl)-3,4-dihydroisocarbostyril from 2-(2-hydroxyethyl)isocarbostyril [292].

Quaternary Compounds. Tetrahydrogenation of quaternized quinolines and isoquinolines very seldom presents any difficulties. As in the case of pyridines, the most commonly used catalyst is platinum oxide although others may also be satisfactory. Since reductions of quaternized quinolines and isoquinolines take place more readily than those of the parent compound, there may be a greater tendency toward overhydrogenation. It should be more pronounced in the case of quinolines, less so with isoquinolines since they are more difficult to perhydrogenate.

Perhydrogenation. Hydrogenation of quinolines and isoquinolines to decahydro compounds can be difficult. On preceding sections ruthenium at elevated temperature and pressure and rhodium at low pressure were shown to be useful for this purpose. In general those reductions give *cis*-decahydro compounds. Reductions with nickel usually require rather vigorous conditions and, as a result of the high temperature, produce *trans* isomers.

Difficulties have been encountered in attempts to obtain decahydro compounds directly from quinolines and isoquinolines by reduction with platinum oxide in acetic acid [285, 286]. It may be more advisable to carry out a two-step reduction involving acetylation of the tetrahydro compound [293]. When successful, reductions with platinum lead to a predominance of *cis*-compounds.

Acylation might have overcome the difficulty due to the poisoning effect of the basic nitrogen on ring saturation noted in the hydrogenation of ethyl 1,2,3,4-tetrahydroisoquinoline-3-carboxylate with 5% rhodium on alumina [294]. Reaction at 50° and 50 atm could not bring about ring reduction but did result in conversion of the ester to alcohol. On the other hand hydrogenation as the carboxylic acid under milder conditions, 50–100° and 15 atm, yielded decahydroisoquinoline-3-carboxylic acid because the carboxyl group in neutralizing the effect of the basic nitrogen allowed the normally expected reaction to proceed.

Although quaternization does facilitate reductions of isoquinolines, the difficulty in perhydrogenation of them enabled investigators to obtain 1,2,3,4,5,6,7,8-octahydro compounds. The platinum-catalyzed reduction of 2-benzyl 1-(4-substituted benzyl)-5,6,7,8-tetrahydroisoquinolinium bromide in alcohol and sodium ethylate gave only octahydro compound [295], the double bond at the ring juncture was unreduced. Reduction of 1-methyl-5-benzyl-5,6,7,8-tetrahydroisoquinolinium iodide [296] and the corresponding 8-benzyl compound [297] also resulted in octahydro compounds. When Raney nickel was employed in the hydrogenation of the corresponding

8-(4-methoxybenzyl), derivative, octahydro and decahydro compounds were obtained [298].

Partial Reduction. Partial hydrogenation of the heterocyclic ring of isoquinolines by catalytic means has not been recorded. Dihydrogenation among quinolines is limited to those containing carboxamide or carboxylic esters group in 3-position.

3-Acetylquinoline was reduced at atmospheric pressure in alcohol with Raney nickel to yield 40% of 3-acetyl-1,4-dihydroquinoline [299]. The authors concluded that failure to proceed beyond the dihydro stage was due to interaction between the imino nitrogen and the carbonyl group through the double bond.

When ethyl quinoline-3-carboxylate was hydrogenated in a mixture of ethyl alcohol and acetic acid with less than 2% by weight of platinum oxide, uptake also stopped at the dihydro stage [300]. The resulting compound was assumed to be ethyl 1,4-dihydroquinoline-3-carboxylate. The 1,4-dihydro structure was later proved when the methyl and ethyl esters and the 3-carboxamide were similarly reduced [301]. It is of interest that reduction of the carboxamide as hydrochloride salt in water yielded only tetrahydro compound. In contrast, hydrogenation of the 3-carboxamide as methiodide and benzochloride in water with 2.5% by weight of platinum oxide yielded 1-substituted 1,4-dihydroquinolines.

There is also a notation of the reduction of methyl quinoline-3-carboxylate with palladium on carbon at 60–65° and 90 atm which yielded a dihydro compound [282].

Selective Reductions

PRESENCE OF FUNCTIONAL GROUPS. As in the pyridine series most reducible functions are hydrogenated in preference to formation of *py*-tetrahydro compounds. Exceptions appear to exist only when a ketone group is present and reductions are carried out as base in neutral solvent. As previously noted, 3-acetylquinoline was reduced in alcohol over Raney nickel to 3-acetyl-1,4-dihydroquinoline [299] but in very many instances the carbinol is formed prior to or concurrent with saturation of the nitrogen-containing ring. This is seen in the hydrogenations of 2-, 4-, and 8-benzoylquinolines [302]. However, 6-benzoylquinoline did yield 6-benzoyl-1,2,3,4-tetrahydroquinoline after reduction with Raney nickel in alcohol at atmospheric pressure. In another study involving related compounds, other 6-(substituted benzoyl)quinolines and di(6-quinolyl)ketone yielded *py*-tetrahydro compounds with the carbonyl group intact after hydrogenation with active W-6 Raney nickel at 20° and 1 atm [303]. It is difficult to correlate the

Aromatic Ring Systems

effect of the benzoyl group in 6-position on preferential saturation of the nitrogen-containing ring.

Another example of selective ring reduction among carbonyl-containing quinolines and isoquinolines is seen in the hydrogenation of 1-phenacylisoquinolinium bromide with platinum oxide at 60° and 2 atm to yield 1-phenyl-1,2,3,4-tetrahydroisoquinoline [304]. In contrast, when 2-phenacylquinoline was reduced with platinum oxide in alcohol containing excess hydrochloric acid, the major product was the quinolyl carbinol; only 9% of 2-phenacyl-1,2,3,4-tetrahydroquinoline was obtained [305].

The result of the reduction in Ref. 304 is not entirely unexpected, although there are conflicting reports on the hydrogenations of the related 1-phenacylpyridinium bromide [251, 252].

PRESENCE OF HALOGEN. As in reduction among pyridines, when halogen is attached to the nitrogen-containing ring of quinoline or isoquinoline, it will be removed during py-tetrahydrogenation. If the halogen is attached to the benzene nucleus or is in a side chain, py-hydrogenation should be accomplished without hydrogenolysis except when the halide is an active one.

Azepines

In general, saturation of the azepine ring by catalytic hydrogenation takes place under mild conditions although this method is not often employed to obtain hexamethyleneimine or substituted hexamethyleneimines. Some examples are seen in the reduction of dimethyl 2,6-dimethyl-4H-azepine-3,6-dicarboxylate in purified cyclohexane over platinum oxide [306] and of the corresponding 1-methyl compound with palladium on carbon [307]. Hydrogen uptake in each case was rather slow (2–3 days) due, no doubt, to steric

(24.65)

effects. In the platinum-catalyzed reduction the first equivalent of hydrogen was absorbed rapidly and the corresponding 4,5-dihydro-1H-azepine could be isolated.

Reaction was not carried out in alcohol since in the above reactions the use of methanol was reported to cause loss of methylamine through a 1,6-transannular reaction to yield dimethyl 2,3-dimethylterephthalate (Eq. 24.65).

Other examples include the hydrogenation of 2,3-dihydro-1H-azepine and the 1-methyl derivative in methyl alcohol with platinum oxide to hexamethyleneimine and 1-methylhexamethyleneimine respectively and that of 2,3-dihydro-1H-azepin-2-one in ethyl acetate to caprolactam [308].

Among benzazepines, the heterocyclic portion of the molecule is selectively reduced in the same manner as in quinolines. Dimethyl 1-methyl-4,5-benzazepine-2,7-dicarboxylate was hydrogenated over palladium on barium sulfate to yield the corresponding 2,3,6,7-tetrahydrobenzazepine [309].

HETEROCYCLES (TWO OR MORE HETERO ATOMS)

Oxazoles

Although oxazoles undergo cleavage to aminoalcohols after conversion to oxazolidines, reductions with platinum catalyst in alcohol or acetic acid can be controlled to stop at the oxazolidine stage [3, 310]. Disubstitution may impede hydrogen uptake, for example, 2,5-dimethyl and 2,5-dimethyl-4-alkoxyoxazoles resisted reduction over active platinum and palladium catalysts [310]. However, 2,4-diphenyl-5-(benzylideneamino)oxazole was readily reduced to 2,4-diphenyl-5-benzylaminooxazolidine with Raney nickel at 65° and 60 atm [311].

Pyrazoles and Related Compounds

Pyrazoles on reduction either yield pyrazolines or pyrazolidines. Since the resultant products are cyclic hydrazides there is always the danger of cleavage to 1,3-diamines if conditions are too vigorous. Pyrazole itself was hydrogenated to 3-pyrazoline in alcohol-acetic acid over palladium on barium sulfate at 18° [312]. Similarly 1-phenylpyrazine absorbed one equivalent of hydrogen at room temperature and two equivalents at 70–80°.

1,2-Dimethyl-3-phenyl-3-pyrazoline was hydrogenated in acetic acid at 50° and 2.7 atm over palladium on carbon to yield 73% of 1,2-dimethyl-3-phenylpyrazolidine [313].

The platinum reductions of some 2-pyrazolines in methyl alcohol gave fairly good yields of pyrazolidines but the products contained 10–15% of 1,3-diamines [314].

Acylation of one of the ring nitrogens, a procedure known to increase the resistance of hydrazines to cleavage, or reduction in acetic anhydride should minimize or prevent hydrogenolysis. Both techniques were used successfully in the hydrogenation of 3,3,5-trimethyl- and 1-acetyl-5-phenyl-3-pyrazolines over Raney nickel at 150° and 100 atm [315]. In contrast when 5-cyclohexyl-3,4-cyclohexano-2-pyrazoline was reduced under similar reaction conditions in methanol, only cyclohexylamine was obtained (Eq. 24.66).

$$\text{[structure]} \xrightarrow[\text{150°, 100 atm}]{\text{2H}_2 \atop \text{Ni(R)}} 2\text{C}_6\text{H}_{11}\text{NH}_2 \qquad (24.66)$$

In general, when phenyl groups are attached to the nucleus they are not affected during hydrogenation of pyrazoles or pyrazolines. However, in indazoles (4,5-benzopyrazoles) the benzene ring is selectively reduced. This is exemplified in the hydrogenations of indazole and 1- and 2-methylindazoles in acetic acid with platinum which yielded 4,5,6,7-tetrahydro compounds [315a]. When palladium on barium sulfate and cobalt or nickel catalysts were employed in the same reactions, no reduction took place.

Imidazoles and Related Compounds

The imidazole ring is remarkably resistant to catalytic hydrogenation. It has been reported in the reduction of benzimidazoles that the benzene ring was saturated without affecting the heterocycle ring [316]. When 2,4,5-triphenylimidazole was hydrogenated, partial reduction of the imidazole ring accompanied saturation of the three benzene rings [317] (Eq. 24.67).

$$\text{[structure]} \xrightarrow{\text{Pt, AcOH}} \text{[structure]} \qquad (24.67)$$

$$X = C_6H_5, \ Y = C_6H_{11}$$

Imidazole has been reduced in one instance by hydrogenation in acetic anhydride in the presence of 50% by weight of platinum oxide to yield 80% of 1,3-diacetylimidazolidine [318]. Using the same technique, the imidazole ring in benzimidazole was selectively reduced to yield 86% of 1,3-diacetylbenzimidazoline.

2-Imidazolones can be reduced in acetic acid with platinum oxide [319], or with palladium on carbon [320], or with nickel under more vigorous conditions [321]. When a phenyl group is present, low-pressure reductions with palladium on carbon should be carried out in order to obtain the phenylimidazolidinone. Palladium oxide may also be employed for this purpose but hydrogen uptake is rather slow unless a large amount of catalyst is employed [3].

Among imidazol-4-ones, 5-ethyl-5-phenyl- and the corresponding 1-methyl-4-imidazolones have been reduced to imidazolidones in alcohol with 5% palladium on carbon [322] and 1-[2-(4-pyridyl)ethyl]-1,3-diazaspiro[4,5]dec-2-en-4-one was hydrogenated over platinum on carbon [323] (Eq. 24.68).

$$\underset{\underset{(CH_2)_2R}{|}}{\text{structure}} \xrightarrow[\text{Pd on C}]{H_2} \underset{\underset{(CH_2)_2R}{|}}{\text{structure}} \quad (24.68)$$

R = 4-pyridyl

Pyridazines and Related Compounds

There is little information available on the hydrogenation of pyridazines (1,2-diazines), but dihydro- and tetrahydropyridazines have been converted to hexahydropyridazines in excellent yield by reduction under mild conditions with platinum oxide in alcohol [324], or with palladium on strontium carbonate, or with Raney nickel [325]. Vigorous conditions may cause rupture between the two nitrogen atoms.

Among pyridazinones and pyridazinediones, 1-[2-(4-piperonyl)ethyl]-4,5-dihydropyridazin-6-one was reduced with platinum in alcohol containing acetic acid to give 70% yield of the corresponding tetrahydropyridazinone [326], and 1,2-dimethyl- and 1-methyl-2-phenylpyridazine-3,6-dione were reduced over palladium on calcium carbonate and platinum oxide respectively to give the corresponding saturated compounds [327]. In the latter example since reduction was carried out in alcohol the phenyl ring was unattacked.

Cleavage was caused during hydrogenation of 1-substituted-3-carboxy-4,5-dihydropyridazin-6-ones with palladium on carbon in water at 70 atm pressure to yield glutamines [328] (Eq. 24.69). Reduction of 3-carboxy-4,5-

$$\underset{\underset{R}{|}}{\text{structure with COOH}} \xrightarrow[\text{Pd on C}]{2H_2} \text{RNH}\overset{O}{\overset{\|}{C}}(CH_2)_2CH(NH_2)COOH \quad (24.69)$$

dihydropyridazin-6-one with platinum oxide in water at 60° and 1.6 atm also resulted in cleavage [3].

612   Aromatic Ring Systems

The hydrogenation of cinnolines (5,6-benzopyridazines) has been meagerly studied. Catalytic reduction has been reported to yield dihydrocinnoline and indole and in the case of 4-phenylcinnoline to give the 1,2,3,4-tetrahydro compound [329].

Recently the hydrogenation of cinnoline has been investigated in reactions over noble metal catalysts in neutral and acidic media at 4 atm and at 125° and 150–200 atm [330]. Low-pressure reductions with rhodium, ruthenium, or palladium catalysts gave 1,4-dihydrocinnoline and 1,1',4,4'-tetrahydro-4,4'-bicinnoline. In general absorption of hydrogen stopped at one equivalent or less. Reaction with platinum oxide in alcohol also stopped before one equivalent was absorbed but, on hydrogenation in alcohol containing hydrochloric acid, two equivalents were absorbed to yield 90% of 1,2,3,4-tetrahydrocinnoline along with 2-aminophenethylamine, indole, and 2,3-dihydroindole. When almost four equivalents were absorbed, the yield of tetrahydro compound dropped to 64% and the yield of aminophenethylamine rose from 5% (at $2H_2$) to 22%. Also obtained were more indole (4.5%), 2,3-dihydroindole (8%), and some octahydroindole (Eq. 24.70).

(24.70)

2-Aminophenethylamine was obtained either through (a) cleavage of tetrahydrocinnoline or (b) that of 1,4-dihydrocinnoline to 2-aminophenylacetaldimine and subsequent reduction. The pathway to indole has been suggested to go through (c) 1-aminoindole (formed by rearrangement of the dihydrocinnoline) followed by loss of ammonia [331]. Another possibility is reaction (d) in which 2-aminophenylacetaldimine either loses ammonia or in the presence of water and hydrochloric acid is hydrolyzed to 2-aminophenylacetaldehyde which in turn undergoes cyclization (Eq. 24.71).

(a) [structure] $\xrightarrow{H_2}$ [2-(2-aminophenyl)ethylamine structure with (CH$_2$)$_2$NH$_2$ and NH$_2$]

(b) [structure] $\xrightarrow{H_2}$ [structure with CH$_2$CH=NH and NH$_2$]

(c) ↓ [indole-type structure with NH$_2$ on N] $\xrightarrow{H_2, -NH_3}$ Indole

(24.71)

(d) [structure with CH$_2$CH=NH$_3$ and NH$_2$] → Indole

$\xrightarrow{HOH}$ [structure with CH$_2$CHO and NH$_2$] ↑

In general hydrogenations under the higher pressure conditions yielded octahydroindole as the major component if reactions were allowed to continue until no further hydrogen uptake was noted.

Catalytic hydrogenation of phthalazine (4,5-benzopyridazine) to 1,2,3,4-tetrahydrophthalazine has not been reported. However, from a study of the hydrogenative hydrogenolyses of phthalazine to 1,2-xylylenediamine [322], it appears likely that 1,2,3,4-tetrahydrophthalazine could be obtained. During Raney nickel reduction with 40% by weight of catalyst at 55–90° and 3.3 atm, slightly more than two equivalents were absorbed. Hydrogenation with 20% palladium on carbon (3% by weight) stopped at two equivalents after 18 hr. Although the intermediate was not isolated or identified, it seems certain that it was the tetrahydrophthalazine. The diamine sought by the investigators was obtained upon further hydrogenations with Raney nickel (Eq. 24.72).

$$\underset{\substack{\\}}{\bigcirc\!}} \xrightarrow[\text{Pd on C}]{2\,H_2} \left[\underset{\substack{\\}}{\bigcirc\!}} \right] \xrightarrow[\text{Ni R}]{H_2} \underset{\substack{\\}}{\bigcirc\!}}$$

(24.72)

Pyrimidines, Quinazolines, and Related Compounds

Pyrimidines, which may be viewed as cyclic amidines, tend to yield tetrahydro derivatives on hydrogenation. Excluding pyrimidones most reductions are carried out under acidic conditions in the presence of palladium, platinum, or rhodium catalysts although nickel has been employed in neutral solvent to yield tetrahydro compound.

There is a report that 4-methylpyrimidine, hydrogenated in alcohol with palladium on carbon, and 2-methylpyrimidine, hydrogenated under neutral and basic conditions with palladium on carbon or with Raney nickel, yielded dihydropyrimidines [333].

Attempts to obtain hexahydro compounds generally result in ring opening to 1,3-diamines. This is to be expected because of the 1,1-diamine structure at C-2 and the known instability of 1,1-diamines to hydrogenation.

It might be possible to obtain hexahydropyrimidines by the acylative reduction procedure employed to convert imidazole to 1,3-idacetylimidazolidine [318]. There is an indication of this in the reduction of 5-acetylaminopyrimidine in hot acetic anhydride with palladium which yielded 5-acetylamino-1,3-diacetyl-1,2,3,4 tetrahydropyrimidine and which on further hydrogenation with platinum oxide was reported to give a hexahydro derivative [334]. The latter structure was indicated by acid hydrolysis to formaldehyde and propane-1,2,3-triamine.

The product of reduction of pyrimidines is dependent on the substituents, on their position on the ring and, at times, on reaction conditions. Pyrimidine itself was converted to 1,4,5,6-tetrahydropyrimidine by reduction in acid solution with palladium on carbon [335]. Pyrimidines with alkyl and alkoxy substituents also yielded 1,4,5,6-tetrahydro derivatives on hydrogenation. 4,6-Dichloro-2-propylpyrimidine, reduced with Raney nickel in butyl alcohol containing barium oxide as the halogen acceptor, gave 2-methyl-1,4,5,6-tetrahydropyrimidine [336]. 1,4,5,6-Tetrahydropyrimidines were also obtained from palladium-catalyzed reductions of 2-, 4-, and 5-methylpyrimidines in acid solution [337] and from similar hydrogenations of 2,4,6-trimethyl-, 2-methoxy-, and 5-methoxypyrimidines [334]. Reduction of the 2-methoxypyrimidine was carried out in methanolic acetic acid to forestall hydrolysis. Despite the precaution, a mixture of 2-hydroxy- and 2-methoxy-tetrahydropyrimidine was obtained. On the other hand, hydrogenation of the 5-methoxy compound in dilute mineral acid went smoothly to yield the corresponding methoxytetrahydropyrimidine.

Hydrogenations of 2-, 4-, and 5-aminopyrimidines under acidic conditions were reported to yield dihydro compounds exclusively [337]. Different results were presented by other investigators.

2-Amino and 2-acylaminopyrimidines were reduced in aqueous acid over palladium on carbon to yield 1,4,5,6-tetrahydro compounds [338]. 4-Aminopyrimidine gave unstable products under similar conditions and 5-aminopyrimidine yielded 5,5-dihydroxy-1,4,5,6-tetrahydropyrimidine. In the latter reduction only one equivalent was absorbed giving a 5-amino-dihydropyrimidine (probably by 1,4-addition). The resultant enamine or the tautomeric imine underwent hydrolysis to a ketone and subsequent hydration to give the *gem*-diol (Eq. 24.73). When the 5-acetylamino

$$ (24.73) $$

compound was reduced in aqueous acid, 5-acetylamino-1,4,5,6-tetrahydropyrimidine was obtained. Reduction of the 5-acetylamino compound in acetic anhydride to give 5-acetylamino-1,3-diacetyl-1,2,3,4-tetrahydropyrimidine was noted earlier.

Aft and Christensen [334] also reported the formation of tetrahydropyrimidines from reductions of 2- and 4-aminopyrimidines in acid solution with palladium on carbon or on barium sulfate or with platinum oxide. The latter catalyst was ineffective when nuclear chlorine was present.

Tetrahydropyrimidines were also obtained when 2-amino-4-methylpyrimidine and 3-(2-amino-5-pyrimidyl)-2-aminopropionic acid (lathyrine) were hydrogenated over platinum oxide in methyl alcohol containing hydrochloric acid [340].

When hydroxypyrimidines, which exist primarily as pyrimidones, are hydrogenated reactions are usually carried out in neutral solvent in the presence of palladium, platinum, and rhodium under low-pressure conditions with absorption of two molar equivalents of hydrogen. For example,

2-hydroxypyrimidine hydrochloride was reduced in water over palladium on carbon to 2-hydroxy-1,4,5,6-tetrahydropyrimidine [334]. 2-Hydroxypyrimidine was also reduced in water with 5% rhodium on alumina to give the same saturated cyclic ureide in 80% yield [341].

It was assumed that reduction of 2-hydroxypyrimidine proceeded by 1,2- or 1,4-addition to give a 5,6-dihydropyrimidine [341]. Another group [342] suggested instead that reaction proceeded by 3,4- or 3,6- addition (different forms of 1,2- and 1,4-addition). To clarify the point the authors studied the hydrogenation of 1-methyl-2-oxopyrimidine. Their findings supported by NMR indicated that a 3,6-dihydro intermediate was obtained (at least with platinum oxide).

In general 2-pyrimidones do not undergo reductive cleavage on further hydrogenation after absorption of two equivalents. For example, some 2-pyrimidones with double bonds between C-5 and C-6 were successfully reduced at elevated temperature and pressure with Raney nickel [343, 344] and with copper chromite [344] without ring rupture.

Although 2-hydroxypyrimidine was readily reduced with rhodium on alumina [341] or with platinum oxide [342], 4-hydroxypyrimidine was resistant to hydrogenation under similar conditions. It did undergo slow reduction with 50% by weight of rhodium on carbon to give an 82% yield of 4-hydroxy-1,2,3,6-tetrahydropyrimidine [342]. When the ring nitrogen in 3-position was substituted so that tautomerism could not take place, 3-methyl-4-oxopyrimidine was rapidly reduced to 4-hydroxy-3-methyl-1,2,3,6-tetrahydropyrimidine.

Unlike the hydrogenations of 2-hydroxypyrimidines or 2-pyrimidones, those of 4-hydroxypyrimidines or 4-pyrimidones may lead to or be accompanied by ring opening. 2-Methyl and 2,6-dimethyl-4-hydroxypyrimidines upon hydrogenation in alcohol with platinum oxide yielded first the corresponding tetrahydropyrimidines and subsequently the corresponding 3-ethylamino amides, $C_2H_5NHCHRCHRCONH_2$. The reductions of other 4-hydroxypyrimidines in the series also yielded amides. In another example when 4-pyrimidone-6-carboxylic acid was hydrogenated in aqueous solution in the presence of 5% rhodium on alumina three major components were obtained after absorption of one equivalent of hydrogen [346]. They consisted of formiminoaspartic acid (from the highly unstable dihydropyrimidone), methyleneasparagine (the open-chain tautomer of the unstable tetrahydro compound), and asparagine (assumed to come from the dihydropyrimidone or from methyleneasparagine) (Eq. 24.74).

The palladium-catalyzed hydrogenation of 5-hydroxypyrimidine in dilute hydrochloric acid did not yield the expected 5-hydroxytetrahydropyrimidine [334]. Instead 5,5-dihydroxy-1,4,5,6-tetrahydropyrimidine was obtained after absorption of one equivalent of hydrogen in a manner in part similar to the reduction of 5-aminopyrimidine to 5,5-dihydroxy-1,4,5,6-tetrahydropyrimidine [334]. In this instance after formation of 5-hydroxydihydropyrimidine, tautomerization of it and hydration gave the *gem*-diol.

$$\begin{array}{c} \text{(structures)} \end{array} \xrightleftharpoons{} \text{H}_2\text{C}\!\!=\!\!\text{NCHRCH}_2\text{CONH}_2 \\ \text{Methyleneasparagine}$$

(24.74)

R = COOH

$$\xrightarrow{\text{HOH}} \text{H}_2\text{NCH}\!\!=\!\!\text{NCHRCH}_2\text{COOH} \\ + \text{H}_2\text{NCHRCH}_2\text{CONH}_2 \\ \text{Formiminoaspartic acid}$$

The hydrogenation of uracils and substituted uracils is relatively uncomplicated. Reductions have been carried out under high pressure with nickel or copper chromite [347]. Hydrogen adsorption in low-pressure reductions with platinum oxide [348, 349] and with palladium on carbon [350] was rather slow in neutral solvent. Reduction in acetic acid with platinum oxide was far more rapid [351]. In each instance [348–351] the catalyst ratios were much higher than normal.

Uracil and thymine were reduced to dihydrouracil and dihydrothymine respectively by hydrogenation in water with an equal weight of 5% rhodium on alumina [352]. The same catalyst had been previously employed in reducing pyrimidine nucleosides and nucleotides to the corresponding dihydropyrimidine derivatives [353] and is now widely used for this purpose because it separates from the solution rapidly and because it does not become colloidal [354].

It is difficult to state that supported rhodium catalysts are the ones of choice for the hydrogenation of uracil and related compounds, exclusive of nucleosides and nucleotides. In every reference on the reductions of uracils large amounts of noble metal catalysts are employed. In no instance is there a comparison of one catalyst with another.

When an amino group is present in 4-position in uracil and uracil nucleosides and nucleotides, deamination often accompanies formation of 5,6-dihydropyrimidines. This has been observed in rhodium reductions of cytosine and related compounds [355] but could be prevented by carrying out reactions at 5°C.

Deamination in these instances is not too surprising since reduction yields a hydrogenolyzable 1,1-diamine system. The mixture of products obtained from the palladium on carbon-hydrogenation of 4-aminopyrimidine [338] probably was a result of deamination.

The hydrogenation of quinazolines (5,6-benzopyrimidines) gives only 3,4-dihydroquinazolines. Quinazoline in 96% ethyl alcohol was first reduced over platinum oxide to 3,4-dihydroquinazoline in 75% yield [356]. It was also reduced over 5% palladium on carbon in alcohol to give the same product [357].

Substituents in 4-position are known to increase the difficulty of obtaining 3,4-dihydroquinazolines. Where quinazoline was reduced in 45 min with palladium on carbon, hydrogenation of the 4-methyl was incomplete in 10 hr and the 4-ethyl derivative was reduced in 16 hr [357]. 4-Alkoxyquinazolines also resisted reduction in alcohol in this instance with palladium on calcium carbonate [358].

When a chlorine atom was in 4-position the solvent appeared to be the controlling factor. Reduction in hydroxylated solvent proceeded rapidly with absorption of two equivalents of hydrogen to yield 3,4-dihydroquinazoline [358]. Hydrogenation in a nonhydroxylated solvent resulted only in dehalogenation unless an acid acceptor was present.

The difficulty in reducing 4-substituted quinazolines appears to have been overcome by the use of water and acid exemplified in the palladium or platinum on carbon-hydrogenation of 7-methoxy-4-methylquinazoline [359]. In the absence of acid, the reaction was apparently poisoned. In the absence of water, even though acid was present no uptake of hydrogen occurred during hydrogenations with palladium, platinum, rhodium, or ruthenium on carbon or with palladium on calcium carbonate, or with platinum oxide in ethanol, ethanolic hydrogen chloride, or acetic acid.

Pyrazines and Related Compounds

Pyrazines have been reduced catalytically to piperazines with little difficulty although at times the type substituent, their number, and position may only allow di- or tetrahydrogenation. For example, there was no difficulty in reducing 2,3,5,6-tetramethylpyrazine hydrochloride in alcohol containing acetic acid [360] or in hydrogenating 1,2,3,5,6-pentamethylpyrazinium methiodide in alcohol containing some water [361] with platinum oxide under mild conditions. In each case the corresponding piperazine was obtained.

No difficulty was encountered in the hydrogenation of pyrazine-2-carboxylic acid in dilute potassium hydroxide in the presence of 10% palladium on carbon [362], but reduction of dimethyl pyrazine-2,3-dicarboxylate with the same catalyst under mild conditions yielded only the tetrahydropyrazine [363]. When tetramethyl pyrazine-2,3,5,6-tetracarboxylate was hydrogenated with 5% platinum on alumina under more vigorous conditions (100° and 100 atm), only the 1,4-dihydropyrazine was obtained [364] (Eq. 24.75). It is of interest that reductions of pyrazine-2,3-dicarboxylic

$$\text{pyrazine} \xrightarrow[\text{Pd on C}]{2H_2} \text{dihydropyrazine}$$

R = COOCH₃ (24.75)

$$\text{tetrasubstituted pyrazine} \xrightarrow[\substack{\text{Pt on Al}_2\text{O}_3 \\ 100°, 100\text{ atm}}]{H_2} \text{1,4-dihydropyrazine}$$

acid and the 2,3-diamide with palladium on carbon at 50° and atmospheric pressure did give the corresponding piperazines [362].

In many instances low-pressure hydrogenations in this series with palladium or platinum can be carried out in the absence of acid. Rhodium appeared to be poisoned by the formation of the resulting piperazine [3]. 2-Butylpyrazine was converted to 2-butylpiperazine by reduction with palladium on carbon in alcohol [365] and 2-(2-hydroxyethyl)pyrazine was reduced over platinum oxide [366]. It is likely that the presence of acid would speed up reaction since in the latter reduction about 25% by weight of catalyst was employed and reaction time was 20 hr. The effect of acid may be seen in the attempted reduction of ethyl 4-(2-pyrazinomethyl)-aminobenzoate with platinum oxide which failed in ethyl alcohol but succeeded upon hydrogenation in acetic acid [367]. In another example when 2-(2-hydroxy-2,2-diphenyl)ethylpyrazine was hydrogenated with platinum oxide in alcohol containing one equivalent of hydrogen chloride (this only neutralizes the effect of one nitrogen atom), reaction time was 7 hr. In contrast, another reduction containing an equivalent of hydrogen chloride plus additional acetic acid was complete in 3 hr [3].

In certain instances reductions have been carried out under elevated temperature and pressure with nickel and with supported palladium or rhodium catalysts [368], apparently in order to overcome the inhibiting effect of the basic piperazine nitrogen atoms on the activity of the catalyst. For such a purpose reduction with ruthenium at 70–100° and 70 atm may be the method of choice [3].

Quinoxalines (5,6-benzopyrazines) are also readily reduced, the heterocyclic portion being preferentially saturated. Palladium on carbon was employed for the hydrogenation of 2-methylquinoxaline [369] and for that of 3-phenyl-1,2-dihydroquinoxaline [370].

2-Methylquinoxaline was also reduced over platinum oxide in alcohol [371]. However, yield of 2-methyl-1,2,3,4-tetrahydroquinoxaline was only 10% but was raised to 93% by hydrogenation in acetic acid. Quinoxaline and derivatives were reduced in pure benzene with platinum oxide at 3–5 atm pressure after pretreatment with Raney nickel [372]. Reaction with platinum at 50–60 atm was very rapid and yields were almost quantitative [3]. Reductions with Raney nickel in alcohol or benzene or with copper chromite in alcohol at 100° and 150° and 150 atm respectively were also successful, yielding 1,2,3,4-tetrahydroquinoxalines [3]. It might be pointed out that hydrogenations with ruthenium catalyst at 100° and 100 atm are difficult to control at the tetrahydro stage [3]. There is a tendency for reduction to proceed to decahydroquinoxalines. Indeed it may be the catalyst of choice for the preparation of decahydroquinoxalines. While 5% rhodium on alumina also catalyzed the conversion of quinoxaline to decahydroquinoxaline by reduction at 100° and 136 atm [373], the investigators reported that the catalyst was sensitive to poisoning and that as a result only tetrahydroquinoxaline was obtained at times.

Very few low-pressure reductions of quinoxalines with Raney nickel have been recorded. 3-(2-quinoxalinyl)propenoic and propanoic acids have been hydrogenated in aqueous sodium hydroxide with an equal weight of Raney nickel to yield the corresponding tetrahydro compounds [374]. The use of less than an equal weight of catalyst gave unchanged quinoxalinylpropionic acids.

Oxazepines, Diazepines, and Their Benzo Derivatives

The double bond in 5,5,7-trimethyl-2,3,4,5-tetrahydro-1,4-oxazepine was reduced over palladium on carbon in alcohol to yield the hexahydrooxazepine [375]. However, when 4-ethyl-5,7-dimethyl-5-phenyl-2,3,4,5-tetrahydro-1,4-oxazepine was hydrogenated, two equivalents were absorbed. Here with a phenyl group in 5-position the compound acted as a tertiary N-benzylamine and underwent hydrogenolysis (Eq. 24.76). In the latter instance a catalyst less active for debenzylation such as Raney nickel might be useful.

In a related example in a benzoxazepine, the 2,3-double bond in 3-phenyl 6,7-benzo-1,4-oxazepin-5-one was very quickly saturated over palladium on carbon [376]. Cleavage should not be expected in this example because the benzylic nitrogen system is that of an N-benzylamide which is usually resistant to hydrogenolysis.

The 6,7-double bond in 7-phenyl-1,2,3,4-tetrahydro-1,4-diazepine was reduced with palladium on carbon to give 39% yield of the corresponding hexahydro compound [377]. The poor yield may have been due to cleavage of the resultant hexahydro compound since it contained a secondary N-benzyl moiety. In contrast, when the related 7-methyl compound was

$$\text{(diazepine structure)} \xrightarrow{H_2, \text{Pd on C}} \text{(reduced structure)} \qquad (24.76)$$

$$\text{(diazepine structure)} \xrightarrow{2H_2} C_6H_5CH(CH_3)CH_2CH(CH_3)OCH_2CH_2NH_2C_2H_5$$

hydrogenated over platinum oxide (it would not undergo hydrogenolysis after reduction), an 80% yield of the hexahydrodiazepinone was obtained. It is likely that the use of palladium on carbon would have also led to good results.

Among benzodiazepines, when 2-methylamino-10-chloro-5-phenyl-6,7-benzo-3H-1,4-diazepine was hydrogenated in acetic acid over prereduced platinum oxide only the 4,5-double bond was affected [378]. The aryl chloride remained intact as did the 1,2-double bond because of its amidine-like structure. Of interest in the same work was the reduction of the corresponding 4-N-oxide in which reaction over palladium on carbon resulted in saturation of the 4,5-double bond as well as removal of oxygen and chlorine at absorption of three equivalents of hydrogen and no loss of halogen after absorption cf two equivalents.

Triazoles, Benzotriazoles, and Triazines

There is little available literature on the hydrogenation of triazoles. There is little likelihood of ring reduction among 1,2,4-triazoles because of the cyclic amidine structure of any of the possible isomeric forms. Partial or complete saturation of the isomeric 1,2,3-triazoles might be possible if cleavage of the nitrogen to nitrogen double bonds could be avoided.

In one example of hydrogenation among 1,2,3-triazoles, cleavage was the aim in the reduction of 1-(2,4-dichlorophenyl)-3-(2-hydroxyethyl)-2,3-dihydro-1,2,3-triazole and its sulfate ester with Raney nickel at 70° and 60 atm [379] (Eq. 24.77).

In a reduction of a benzotriazole the fused benzene ring in 2-(2-aminophenyl)-4,5-benzo-1,2,3-triazole was saturated when the hydrogenation was

$$\text{R} \overset{\text{H}}{\underset{\underline{\quad\quad}}{\text{N}\diagdown\text{N}\diagup\text{N}(\text{CH}_2)_2\text{OH}}} \xrightarrow[\text{N(R), 70°, 60 atm}]{3\text{H}_2} \text{RNH}(\text{CH}_2)_2\text{NH}(\text{CH}_2)_2\text{OH} \quad (24.77)$$

R = 2,4-dichlorophenyl

carried out in acetic acid in the presence of palladium on carbon at 2.7 atm [380]. In a related example, that of pyrido[1,2-c]-1,2,4-triazole-5-one, the 6-membered fused ring was reduced with the same catalyst in alcohol to give the corresponding piperidinotriazolone [381].

In the triazine series there do not appear to be examples of reductions of 1,2,3-triazines. There are reports that 1,3,5-triazines cannot be hydrogenated because the s-triazine system is a powerful poison for palladium and platinum catalysts [382, 383]—the formation of s-triazine was claimed as the reason for the failure to dehalogenate cyanoric chloride [382]. In another example 2,4-dihydroxy-1,3,5-triazine would not reduce over platinum oxide; it and the 6-carboxy derivative were found to act as poisons to prevent the reduction of the double bond in trimethoxycinnamic acid [383].

In contrast to the above experiments on hydrogenations among 1,3,5-triazines, when 6-(α,α-dimethylbenzyl)-1,3,5-triazine-2,4-dione was subjected to reduction over palladium oxide (usually not a very active hydrogenation catalyst) in glacial acetic acid for 24 hr at 50° and 3 atm, one equivalent of hydrogen was absorbed to indicate saturation of the 5,6-double bond [3]. Although rigorous proof of structure was not obtained (NMR was not available), other methods indicated a single substance (not starting material) which gave an excellent elemental analysis for the dihydro compound.

It is possible that the use of a catalyst less active than those employed in Refs. 382 and 383 allowed hydrogen absorption to proceed, albeit slowly, because it was less sensitive to poisoning [384]. The other possible reason for success could be ascribed to the effect of the bulky group in 6-position which interfered with the ability of the triazine nucleus to bond strongly with the active portions of the catalyst surface.

Successful hydrogenations of 1,2,4-triazines have been accomplished. In reductions under neutral or alkaline conditions only one equivalent of hydrogen was absorbed, while two equivalents were absorbed in the presence of acid [385]. In certain instances, because of the catalyst employed and because of the presence of phenyl groups on carbon atoms adjacent to the ring nitrogen (reduction in the latter case leads to formation of N-benzyl type amines), rupture of the ring system was the major reaction. An example of cleavage is seen in the hydrogenations of some 1,2,4-triazines with W-6 Raney nickel, a very active catalyst, which resulted in 6–30% yields of products suggested to be 1,2-dihydro compounds [386]. The extraordinarily active catalyst probably induced cleavage between the 1,2-NH-NH bonds and also caused ring rupture from N-debenzylation in those examples where a phenyl group was attached to a carbon atom adjacent to the ring nitrogen. In a reaction of a similar C-phenyl-1,2,4-triazine, hydrogenolysis took place when ethyl 5,6-diphenyl-1,2,4-triazine-3-carboxylate (in 80% alcohol containing hydrochloric acid) was reduced with an equal weight of palladium on carbon. No hydrogenated triazine was found; only ethyl 3,4-diphenyl-pyrazole-5-carboxylate was obtained [387]. While the manner in which the small ring was formed was not explained, it appears likely that a benzyl-type secondary amine was formed during reduction and was cleaved, the resulting intermediate undergoing recyclization with loss of ammonia to form the corresponding pyrazole.

When the possibility of cleavage exists during the reduction of C-phenyl-1,2,4-triazines, catalysts which are not especially active for N-debenzylation should be employed. For example, 3-methoxy-5,6-diphenyl-1,2,4-triazine and its 4,5-dihydro analog were hydrogenated in glacial acetic acid with platinum oxide without incident to yield 3-methoxy-5,6-diphenyl-1,4,5,6-tetrahydro-1,2,4-triazine [385]. When reduction of the triazine was carried out in alcohol, one equivalent was absorbed and the corresponding 4,5-dihydro compound was obtained. The successful use of normally active Raney nickel as well as platinum oxide for the reduction of related phenyl, 1,2,4-triazines is illustrated in the hydrogenations of 3-hydroxy-5,6-diphenyl-1,2,4-triazine in aqueous alkaline solution which gave the corresponding 4,5-dihydro derivative [388].

In another series, 1,2,4-triazine-3,5-dione and 6-methyl-1,2,4-triazine-3,5-dione and their 2- and 4-alkyl derivatives compounds (6-azauracils and 6-azathymines) upon hydrogenation with platinum oxide in water, alcohol, or aqueous alcohol gave high yields of the 1,6-dihydro compounds [389]. In certain instances reduction only proceeded in water. 2,6-Dimethyl- and 2,4,6-trimethyl-1,2,4-triazine-2,4-diones could not be reduced under any of the described conditions, presumably because of steric effects.

MISCELLANEOUS HETEROCYCLES (TWO OR MORE HETERO ATOMS IN SEPARATE RINGS OR IN ONE RING WITH A BRIDGEHEAD NITROGEN)

In many instances it may be difficult to find specific examples in systems which contain several hetero atoms in separate rings or where they are in one ring and a bridgehead nitrogen is involved. In most cases it should be possible to reduce such compounds partially or completely based on the reduction of related systems already discussed in this chapter.

Azaindoles

The hydrogenation of these compounds may be viewed as examples in which the pyridine portion of the molecule is saturated in preference to reduction of the fused pyrrole ring. Although the hydrogenation of 7-azaindole [228] to 4,5,6,7-tetrahydroazaindole was cited as such an example earlier in this chapter, 5-azaindole would not undergo reaction under similar conditions [390]. However, in the hydrogenation of a 6-azaindole, that of 3-(2-carboxy-5-oxo-6-azaindol-3-yl)propionic acid, over palladium on carbon two molar equivalents were absorbed to give the 4,5,6,7-tetrahydro compound [391].

Imidazo[1,2-a]pyridines

Reductions in this system (Eq. 24.78) show that preferential saturation of

$$\text{(structure)} \xrightarrow{2H_2} \text{(structure)} \qquad (24.78)$$

the pyridine portion of the molecule takes place as expected since the reaction can be viewed as that of a pyridine ring over a difficultly reducible imidazole system.

Examples consist of the platinum-catalyzed hydrogenation of the 5,8-dimethyl compound in acidified alcohol [392], the palladium on carbon reduction of the 2-dimethylaminomethyl compound also in acidified alcohol [393], and the Raney nickel reduction of the 2-phenyl compound in alcohol [394]. All yielded 5,6,7,8-tetrahydroimidazopyridines.

Naphthpyridines

When naphthpyridines are reduced by catalytic means either tetrahydro or decahydro compounds can be obtained, depending on reaction conditions. Hydrogenations when the nitrogen atoms were in symmetrical positions as in 1,5-, 1,8-, and 2,7-naphthpyridine with palladium on calcium carbonate under neutral conditions yielded only one tetrahydro compound [395]. A reduction of 1,5-naphthpyridine in alcohol with platinum oxide was slow but gave a quantitative yield of tetrahydro-1,5-naphthpyridine [396]. Hydrogenation of 1,6- and 1,7-naphthpyridines, either of which might form a 5,6,7,8- as well as a 1,2,3,4-tetrahydro derivative, gave one isomer upon reduction of the 1,6-compound and 98% of 1,2,3,4-tetrahydro and 2% of 5,6,7,8-tetrahydro derivative after reduction of 1,7-naphthpyridine [395]. The same investigators showed that a substituent on one ring favored saturation of the unsubstituted ring.

Later work by another investigator with palladium on carbon in alcohol [397] confirmed that the hydrogenations of 1,5- and 1,8-naphthpyridines gave only 1,2,3,4-tetrahydro compounds while that of 1,7-naphthpyridine yielded 57% and 43% of 1,2,3,4- and 5,6,7,8-tetrahydro-1,7-naphthpyridines, respectively.

Reduction of 1,5-naphthpyridine under acidic conditions with platinum oxide gave a separable 2:1 mixture of *cis* and *trans*-decahydronaphthpyridines [397]. The reduction of other naphthpyridines to decahydro compounds was not studied. In view of the penchant of ruthenium catalysts for perhydrogenation they should be valuable in the preparation of decahydronaphthpyridines under moderate high-pressure conditions, 70–100° and 70–100 atm. Low-pressure reduction with rhodium under conditions employed for the hydrogenation of pyridines [195] might also yield decahydronaphthpyridines.

Pteridines

The hydrogenation of pteridines, except for those of 2-hydroxy- and 2,7-dihydroxypteridines, follows the expected course, selective saturation of the pyrazine portion of the molecule (Eq. 24.79).

$$\text{pteridine} \xrightarrow{2H_2} \text{5,6,7,8-tetrahydropteridine} \qquad (24.79)$$

Examples are seen in the hydrogenations of 2,4-diactyelamino-6,7-dimethylpteridine in alcohol with platinum oxide [398], of 4-amino-7-methyl-2-phenylpteridine with Raney nickel in alcohol—very slow [399], of 2-amino-4-hydroxypteridines in trifluoroacetic acid with platinum oxide or with rhodium on a carrier [400], and of 2-amino-4-hydroxy-6,7-dimethylpteridine hydrochloride with palladium on carbon in alcohol [401], all of which gave 5,6,7,8-tetrahydropteridines.

2-Hydroxypteridine resisted reduction in neutral or acidic solution over Raney nickel or platinum oxide but yielded a 3,4-dihydro compound in alkaline solution in a Raney nickel or palladium on carbon hydrogenation [402]. If the pyrazine ring of 2-hydroxypteridine was partially reduced, further reaction under hydrogen gave only 2-hydroxy-5,6,7,8-tetrahydropteridine.

A study of the reduction of other mono- and dihydroxypteridines showed that only 2,7-dihydroxypteridine yielded a 3,4-dihydro compound [403]. Others upon hydrogenation with palladium or with platinum yielded 5,6- or 7,8-dihydro- or 5,6,7,8-tetrahydropteridines.

MISCELLANEOUS POLYNUCLEAR HETEROCYCLES

Three Rings with a Common Junction Point

The course of reduction of polynuclear heterocycles depends on the structure of the substrate. In compounds where three rings have a common junction point the most readily reduced ring system will be affected. In the hydrogenation of 2-methyl-1-oxaphenalene, the oxygen-containing ring was saturated as should be expected to yield the 2,3-dihydro compound [404] (Eq. 24.80).

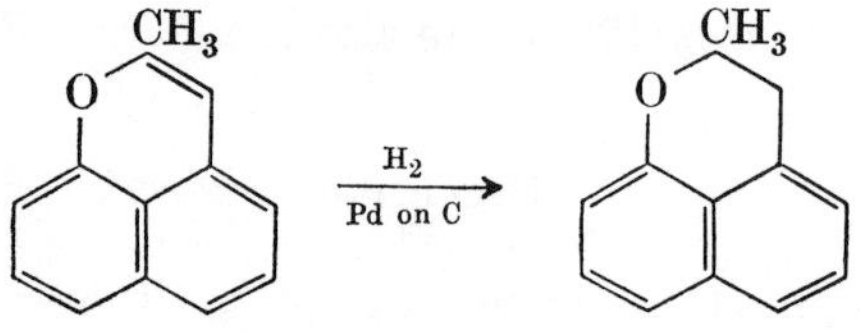

(24.80)

Unsymmetrically Fused Rings

In many instances hydrogenations among these systems can be likened to a related one so that results can often be predicted. This is readily seen in the reductions of unsymmetrically fused polynuclear heterocycles, such as phenanthridines and phenanthrolines. In the hydrogenation of phenanthridine, selective saturation of the pyridine ring took place in the same manner as in the reduction of a quinoline or isoquinoline; 5,6-dihydrophenanthridine was obtained [405] (Eq. 24.81).

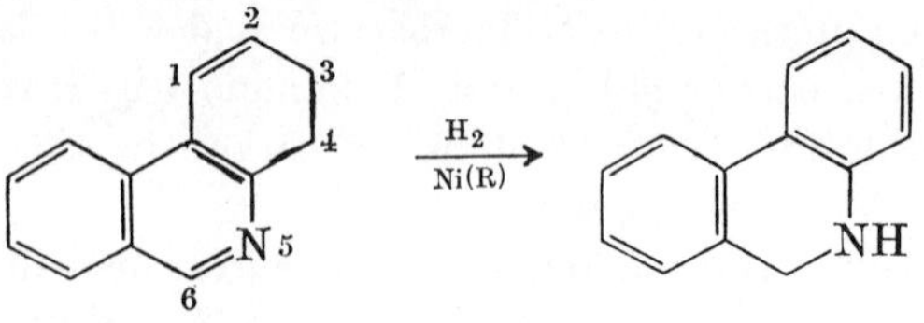

(24.81)

In nickel-catalyzed reductions of phenanthrolines preferential saturation of the pyridine ring also took place to yield 1,2,3,4-tetrahydrophenanthrolines under mild conditions and 1,2,3,4,5,6,7,8-octahydro compounds under more vigorous conditions [406] (Eq. 24.82). However, these reactions yield small amounts of 9,10-dihydrophenanthrolines. They should be readily removed as contaminants by treatment of tetrahydro or octahydrophenanthrolines with very dilute acetic acid in which they are soluble and the 9,10-dihydro compound is very poorly soluble [3].

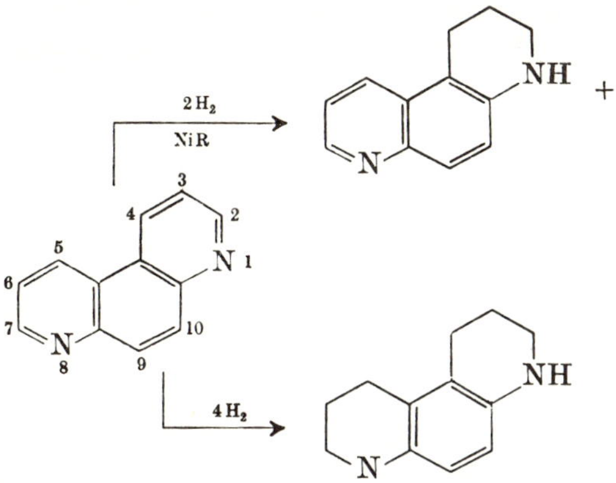

(24.82)

When one of the heterocyclic rings in a phenanthroline contains substituents, the unsubstituted ring can be selectively tetrahydrogenated.

Symmetrically Fused Rings

The product (or products) of reduction of compounds containing unsaturated fused rings on each side of a heterocyclic ring is dependent not only on reaction conditions but on the rings involved.

If the middle ring is 5-membered and the adjoining rings are 6-membered, tetrahydro- or octahydro compounds are obtained; the bonds at the ring junctions usually remain intact. For example, when dibenzofuran was hydrogenated in glacial acetic acid with platinum, 1,2,3,4-tetrahydrodibenzofuran was obtained in quantitative yield.

When one of the side rings is also heterocyclic, unless it is a difficultly reducible system, it will be hydrogenated in preference to an unsaturated carbocycle fused on the other side of the middle ring. An example is seen in the reduction of some alkyl indolo[2,3-d]-2-pyrones in tetrahydrofuran with palladium on carbon [408] (Eq. 24.83). If the middle ring is 6-membered

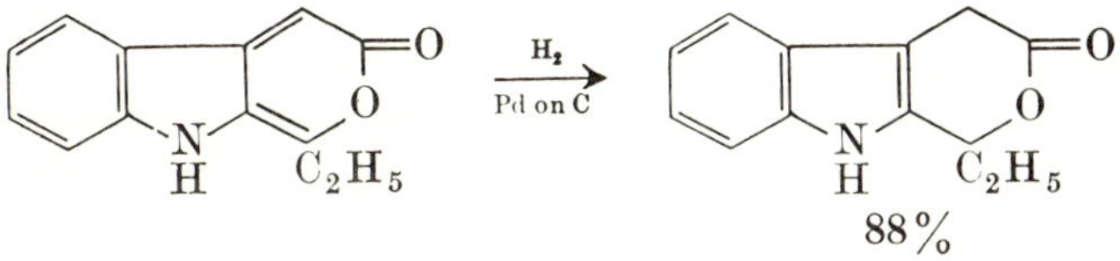

(24.83)

88%

628   *Aromatic Ring Systems*

and contains nitrogen as in acridine or phenazine a 9,10-dihydro compound is first obtained.

Carbazoles. Reductions of carbazoles can at times be controlled to yield only tetrahydrocarbazoles. This is seen in the hydrogenation of highly purified carbazole with copper chromite at 230° and 240–300 atm [158]. Reduction in water adjusted to pH 12.0 with potassium hydroxide with prereduced and stabilized nickel on kieselguhr under more moderate conditions, 200° and 33–70 atm, gave 87% yield of 1,2,3,4-tetrahydrocarbazole [409].

It was also reported [409] that hydrogenation in decalin with ruthenium on carbon at 250° and 17 atm yielded tetrahydrocarbazole (53%). This author has found it extremely difficult to control reductions of carbazoles with ruthenium catalysts although decalin was never used as solvent. There was always a tendency toward formation of octahydrocarbazole or dodecahydrocarbazole. In a reduction at 70 atm pressure over two equivalents of hydrogen were absorbed when the temperature reached 70–95°. There was little break in the rate curve until a total of four equivalents was taken up within 0.5 hr; the products of such reactions consisted of 10% of starting material, 11% of tetrahydrocarbazole, about 40% of octahydrocarbazole, and 30% of dodecahydrocarbazole.

$$\text{carbazole-N-CH}_3 \xrightarrow[\text{Pt O}_2]{3\,\text{H}_2} \text{hexahydrocarbazole-N-CH}_3 \qquad (24.84)$$

Hexahydrocarbazoles have been prepared in low yields during high-pressure reductions of carbazoles with Raney nickel or copper chromite [158]. Some 9-methyl-1,2,3,4,4a,9a-hexahydro compounds were prepared in 85% yield or better by hydrogenation of the corresponding 1,2,3,4-tetrahydrocarbazoles with platinum oxide under acidic conditions [410] (Eq. 24.84) but prolonged hydrogenation led to dodecahydrocarbazoles.

Perhydrogenations of carbazoles are readily carried out with rhodium, ruthenium, or palladium on carbon or with nickel in organic or aqueous media at 200° and 70 atm or with rhodium on carbon in aqueous acid at 50–100° and 55 atm [409]. As previously noted, prolonged hydrogenation of carbazoles with platinum oxide also will yield dodecahydrocarbazoles [410].

Pyridoindoles (Carbolines). The pyridoindoles (carbolines) are related to carbazoles, but reductions are more complex because of the presence of the nitrogen atom in ring A. Their catalytic hydrogenation has been meagerly studied and appears to be limited to 9H-pyrido[3,4-b]indoles or β-carbolines (Eq. 24.85).

From their relationship to carbazoles we should expect and we do obtain tetrahydro compounds upon reduction. However, the expected preferential tetrahydrogenation of ring A does not always take place. 1,2,3,4-Tetrahydro

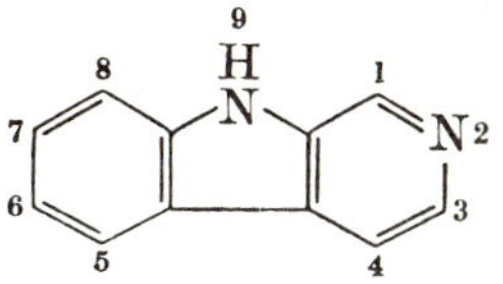

(24.85)

9H-Pyrido[3,4-b]indole (β-carboline)

compounds are obtained from reductions of 3,4-dihydropyridoindoles [411] but a study of reductions of completely unsaturated pyrido[3,4-b]indoles shows that the product of reaction is dependent on the medium.

In neutral solvent the bases were stable to hydrogenation. Despite the ease with which quaternized pyridines are reduced to piperidines in neutral solvent, it is of interest that pyrido[3,4-b]indoles quaternized in ring A were stable to hydrogenation under neutral conditions. Py-tetrahydro compounds were obtained only if hydrogenations of quaternized pyridoindoles were carried out with platinum oxide (other catalysts were not employed) in the presence of a strong base [411, p. 102]. In these instances anhydro bases were formed which were susceptible to reduction.

When hydrogenations were carried out in glacial acetic acid with platinum oxide, tetrahydrogenation took place in ring C.

Acridines. From the structure of acridine, which may be viewed as a 2,3-benzoquinoline, one could expect reduction to take place first at the pyridine ring. One molar equivalent of hydrogen was absorbed during the reductions of acridine, 1-, 2-, 3-, and 4-aminoacridine with Raney nickel at atmospheric pressure to yield 9,10-dihydroacridines [412] (Eq. 24.86).

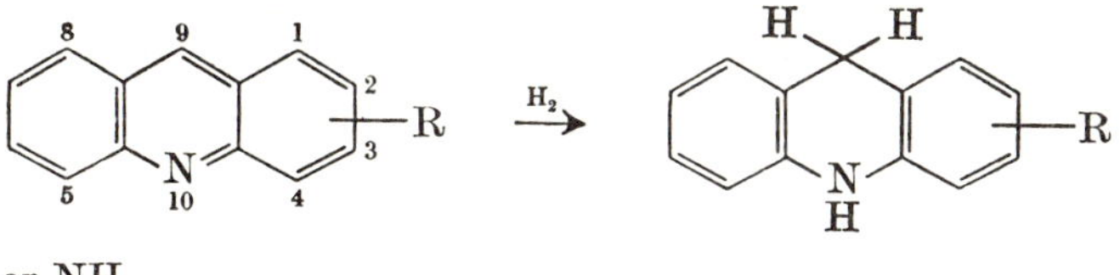

(24.86)

R = H or NH$_2$

In this series the length of reaction increased with increasing basicity of the substrate. 2,9-Diaminoacridine because of its basicity apparently caused inactivation of the catalyst and did not undergo hydrogenation.

Acridine was also converted to 9,10-dihydroacridine by hydrogenation over Raney nickel at elevated pressure or with copper chromite at elevated temperature and pressure [158]. Further hydrogenation or hydrogenation under more drastic conditions led to mixtures of 1,2,3,4,5,6,7,8- and 1,2,3,4,

630　Aromatic Ring Systems

4a,9,9a,10-octahydroacridines, dodecahydro-, and tetradecahydroacridines with nickel and high yield of 1,1a,2,3,4,4a,9,10-octahydroacridine and dodecahydroacridine with copper chromite. In all instances 9,10-dihydroacridine was assumed to have been formed first.

Neither tetrahydro- nor hexahydroacridine has been obtained by hydrogenation of acridine.

In general noble metal catalysts have found little use in the hydrogenation of acridines. A mixture of palladium and platinum on carbon was employed to hydrogenate 1,2,3,4-tetrahydroacridines to the corresponding *as*-octahydroacridines [413]; palladium on carbon or platinum on carbon were ineffective. There are no reports on the use of rhodium or ruthenium either of which might be useful for reductions of acridines to perhydroacridines at moderate high temperature and pressure conditions.

Phenazines. Phenazines, which are weak bases, are readily hydrogenated at low pressure to dihydrophenazines (Eq. 24.87). The parent phenazine has

$$\text{(structure)} \xrightarrow[\text{Pd Oxide}]{H_2} \text{(structure)} \quad (24.87)$$

R = 2-CH$_3$,1-COOCH$_3$,1- and 2-CONH$_2$

been converted to 5,10-dihydrophenazine by reduction in alcohol with palladium on carbon, rhodium on alumina, or ruthenium on carbon at 100° and 55 atm [415].

Tetrahydrophenazines have not been prepared by catalytic hydrogenation of phenazines, nor have octahydrophenazines. Octahydrophenazine has been obtained by reducing tetrahydrophenazine catalytically under 6–7 atm pressure of hydrogen over palladium on carbon in alcohol, or with palladium on carbon in acetic acid (this method gave quantitative yield), or with nickel from nickel hydroxide [416]. The product of reaction was *cis*-1,2,3,4,4a,5,10,10a-octahydrophenazine (Eq. 24.88).

$$\text{(structure)} \xrightarrow{2H_2} \text{(structure)} \quad (24.88)$$

Passing the tetrahydro compound with hydrogen over nickel from nickel nitrate at 180° yielded 1,2,3,4,6,7,8,9-octahydrophenazine. It is of interest that when the catalyst was prepared from nickel hydroxide, reduction gave *trans*-1,2,3,4,4a,5,10,10a-octahydrophenazine.

The isomeric perhydrophenazines have been obtained by hydrogenation of 1,2,3,4,6,7,8,9-octahydrophenazines with platinum in acetic acid at 6–7 atm [417] and by reduction of phenazine in alcohol with palladium on carbon at 180° and 50 atm [415]. It is of interest that neither rhodium nor ruthenium was employed for perhydrogenation in the latter example since either is usually more satisfactory for ring reductions than palladium.

REFERENCES

[1] S. Nishimura, *Bull. Chem. Soc. Japan*, **33**, 566 (1960).
[2] S. Nishimura and H. Taguchi, *Bull. Chem. Soc. Japan*, **36**, 353 (1963).
[3] M. Freifelder, unpublished results.
[4] M. Freifelder, *J. Org. Chem.*, **26**, 1835 (1961).
[5] H. A. Smith, *Annals N.Y. Acad. Sci.*, **145**, 72 (1967).
[6] S. Siegel and M Dunkel, *Advances in Catalysis*, Vol. IX, Academic Press, New York, 1957, p. 15.
[7] H. A. Smith and R. G. Thompson, *Advances in Catalysis*, Vol. IX, Academic Press, New York, 1957, p. 727.
[8] S. E. Cantor and D. S. Tarbell, *J. Am. Chem. Soc.*, **86**, 2902 (1964).
[9] M. Freifelder, Y. H. Ng, and P. F. Helgren, *J. Org. Chem.*, **30**, 2485 (1965).
[10] F. Zymalkowski and T. Yupraphat, *Arch. Pharm.*, **300**, 969 (1967).
[11] P. N. Rylander and J. F. Kreidl, U.S. Patent 3,193,584 (1965).
[12] M. J. Kalm, *J. Med. Chem.*, **7**, 427 (1964).
[13] M. Freifelder and G. R. Stone, *J. Org. Chem.*, **27**, 3568 (1962).
[14] Y. Takagi, T. Naito, and S. Nishimura, *Bull. Chem. Soc. Japan*, **38**, 2119 (1965).
[15] P. N. Rylander and D. R. Steele, *Engelhard Ind. Tech. Bull.*, **7**, 153 (1967).
[16] J. H. Stocker, *J. Org. Chem.*, **27**, 2288 (1962).
[17] Y-H. Wu, W. A. Gould, W. G. Lobeck, Jr., H. R. Roth, and R. F. Feldkamp, *J. Med. Chem.*, **5**, 752 (1962).
[18] H. W. Arnold, U.S. Patent 2,555,912 (1951).
[19] C. E. Frank, U.S. Patent 2,478,261 (1949).
[20] S. Nishimura, *Bull. Chem. Soc. Japan*, **32**, 1158 (1959).
[21] Y. Ichinohe and H. Ito, *Bull. Chem. Soc. Japan*, **37**, 887 (1964).
[22] L. L. Ferstandig and W. A. Pryor, U.S. Patent 2,828,335 (1958).
[23] M. Freifelder, D. A. Dunnigan, and E. J. Baker, *J. Org. Chem.*, **31**, 3438 (1966).
[24] H. Maegawa, Jap. Patent **27**, 245 (1964).
[25] P. N. Rylander and N. F. Rakoncza, U.S. Patent 3,162,679 (1964).
[26] H. A. Smith, D. M. Alderman, and F. W. Nadig, *J. Am. Chem. Soc.*, **67**, 272 (1945).
[27] N. L. Smith and F. L. Schmehl, *J. Org. Chem.*, **13**, 859 (1948).
[28] F. F. Blicke and W. K. Johnson, *J. Am. Pharm. Assoc.*, **45**, 437 (1956).
[29] J. P. Schaeffer, L. S. Endres, and M. D. Moran, *J. Org. Chem.*, **32**, 3963 (1967).
[30] I. D. Pletneva, R. S. Muramova, I. V. Pervukhina and I. V. Shkhiyants, *J. Org. Chem. USSR*, **1**, 2020 (1965) (Eng. trans.).
[31] L. D. Freeman, G. O. Doak, and E. L. Petit, *J. Am. Chem. Soc.*, **77**, 4262 (1958).
[32] A. C. Whitaker, *J. Am. Chem. Soc.*, **69**, 2414 (1947).
[33] H. A. Smith and B. L. Stump, *J. Am. Chem. Soc.*, **83**, 2739 (1961).
[34] H. E. Ungnade and D. E. Nightingale, *J. Am. Chem. Soc.*, **66**, 1218 (1944).

[35] C. V. Banks, D. T. Hooker, and J. J. Richards, *J. Org. Chem.*, **21**, 547 (1956).
[36] R. J. Fessenden, K. Seeler, and M. Dagani, *J. Org. Chem.*, **31**, 2483 (1966).
[37] G. Gilman and G. Cohn, *Advances in Catalysis*, Vol. 9, Academic Press, New York, 1957, p. 733.
[38] I. A. Kaye and R. S. Matthews, *J. Org. Chem.*, **28**, 325 (1963).
[39] R. H. Levin and J. H. Prendergrass, *J. Am. Chem. Soc.*, **69**, 2436 (1947).
[40] H. E. Ungnade and F. V. Morriss, *J. Am. Chem. Soc.*, **70**, 1898 (1948).
[41] D. S. Noyce, G. L. Woo, and B. R. Thomas, *J. Org. Chem.*, **25**, 260 (1960).
[42] D. S. Noyce and L. J. Dolby, *J. Org. Chem.*, **26**, 1732 (1961).
[43] H. K. Hall, Jr., *J. Am. Chem. Soc.*, **80**, 6413 (1958).
[44] A. W. Burgstahler and Z. J. Birthos, *Org. Syn.*, **42**, 62 (1962).
[45] J. English, Jr. and G. W. Barber, *J. Am. Chem. Soc.*, **71**, 3310 (1949).
[46] W. J. Bailey and W. B. Lawson, *J. Am. Chem. Soc.*, **79**, 1444 (1957).
[47] C. A. Greb and W. Baumann, *Helv. Chim. Acta*, **38**, 594 (1955).
[48] P. N. Rylander and N. Himelstein, *Engelhard Ind. Tech. Bull.*, **5**, 43 (1964).
[49] J. C. Circar and A. I. Meyers, *J. Org. Chem.*, **30**, 3206 (1965).
[50] A. W. Burgstahler and Z. J. Bithos, *J. Am. Chem. Soc.*, **82**, 5466 (1960).
[51] R. Kuhn and H. J. Haas, *Ann.*, **611**, 57 (1958).
[52] G. G. Joris and J. Vitrone, Jr., U.S. Patent 2,829,166 (1958).
[53] R. J. Duggan, E. J. Murray, and L. O. Winstrom, U.S. Patent 3,076,810 (1963).
[54] L. J. Dankert and D. A. Permoda, U.S. Patent 3,124,614 (1964).
[55] F. Johnson, N. A. Starkovsky, A. C. Paton, and A. A. Carlson, *J. Am. Chem. Soc.*, **86**, 118 (1964).
[56] Details of experiment in Ref. 55 supplied by L. G. Duquette of the same laboratory.
[57] B. Esch and H. J. Schaeffer, *J. Am. Pharm. Assoc.*, **49**, 786 (1960).
[58] R. B. Thompson, *Org. Syn.*, **27**, 21 (1947).
[59] C. A. Grob and H. R. Kieger, *Helv. Chim. Acta*, **48**, 799 (1965).
[60] H. J. Teuber, D. Cornelius, and U. Wölche, *Ann.*, **696**, 116 (1966).
[61] B. Pecherer, L. M. Jampolsky, and H. M. Wuest, *J. Am. Chem. Soc.*, **70**, 2587 (1948).
[62] W. Mayer, R. Bachman, and F. Kraus, *Ber.*, **88**, 316 (1955).
[63] Personal communication from R. M. Robinson and F. van Munster, Abbott Laboratories, North Chicago, Ill.
[64] G. M. Illich, Jr. and R. M. Robinson, Brit. Patent 836,951 (1960).
[65] H. Greenfield, *J. Org. Chem.*, **29**, 3082 (1964).
[66] R. J. Duggan, U.S. Patent 3,117,992 (1964).
[67] R. M. Robinson, U.S. Patent 3,196,179 (1965).
[68] S. Nishimura and H. Taguchi, *Bull. Chem. Soc. Japan*, **36**, 873 (1963).
[69] F. J. Villani and C. A. Ellis, *J. Org. Chem.*, **29**, 2585 (1964).
[70] F. G. Abraham and C. Lamb, U.S. Patent 3,228,975 (1966).
[71] E. C. Shokal and H. A. Newey, U.S. Patent 2,817,644 (1957).
[72] G. M. Whitman, U.S. Patent 2,606,925 (1952).
[73] M. Freifelder and G. R. Stone, *J. Org. Chem.*, **27**, 3568 (1962).
[74] S. Nishimura and H. Yoshino, *Bull. Chem. Soc. Japan*, **42**, 499 (1969).
[75] S. Nishimura, H. Uchino, and H. Yoshino, *Bull. Chem. Soc. Japan*, **41**, 2194 (1968).
[76] J. E. Kirby, U.S. Patent 2,606,926 (1952).
[77] M. Freifelder, B. Meltzner, G. M. Illich, and R. M. Robinson, Brit. Patent 882,952 (1961).
[78] M. Freifelder, U.S. Patent 3,082,247 (1963).

[79] J. H. Billman and J. A. Buehler, *J. Am. Chem. Soc.*, **75**, 1345 (1953).
[80] K. J. Liska, *J. Pharm. Sci.*, **53**, 1427 (1964).
[81] I. D. Pletneva, R. S. Muromova, I. V. Pervukhina, and I. V. Shkhiyants, *J. Org. Chem. USSR*, **1**, 2020 (1965) (Eng. trans.).
[82] B. L. Zenitz, E. B. Macks, and M. L. Moore, *J. Am. Chem. Soc.*, **69**, 1117 (1947).
[83] A. LaManna, V. Ghislandi, P. M. Scopes, and R. J. Swann, *Farmaco Sci. Ed.*, **20**, 842 (1965).
[84] M. Metayer, *Bull. Soc. Chim. France*, **1952**, 276.
[85] M. Freifelder and G. R. Stone, *J. Am. Chem. Soc.*, **80**, 5270 (1958).
[86] R. L. Burwell, Jr., *Chem. Rev.*, **57**, 895 (1957).
[87] B. Rickborn and J. Quartucci, *J. Org. Chem.*, **29**, 3185 (1964).
[88] D. V. Young and H. R. Snyder, *J. Am. Chem. Soc.*, **83**, 3160 (1961).
[89] C. W. Whitehead, J. J. Traverso, F. J. Marshall, and D. E. Morrison, *J. Org. Chem.*, **26**, 2809 (1961).
[90] A. K. Bose and M. S. Manhas, *J. Org. Chem.*, **27**, 1244 (1962).
[91] P. N. Rylander and D. R. Steele, *Engelhard Ind. Tech. Bull.*, **5**, 113 (1965).
[92] H. A. Smith, D. M. Alderman, Jr., C. D. Shacklett, and C. M. Welch, *J. Am. Chem. Soc.*, **71**, 3772 (1949).
[93] H. A. Smith, C. A. Buehler, T. A. Magee, K. V. Nayak, and D. M. Glenn, *J. Org. Chem.*, **24**, 1301 (1959).
[94] H. E. Ungnade, *J. Org. Chem.*, **13**, 361 (1948).
[95] A. L. Wilds and C. H. Shunk, *J. Am. Chem. Soc.*, **72**, 2388 (1950).
[96] W. H. Linnell and H. J. Smith, *J. Chem. Soc.*, **1959**, 557.
[97] H. E. Zaugg, R. J. Michaels, H. J. Glenn, L. R. Swett, M. Freifelder, G. R. Stone, and A. W. Weston, *J. Am. Chem. Soc.* **80**, 2763 (1955).
[98] M. Freifelder, *J. Org. Chem.*, **29**, 979 (1964).
[99] D. W. Adamson and S. Wilkinson, U.S. Patent 2,682,543 (1954).
[100] R. Baltzly, N. B. Mehta, P. B. Russell, R. E. Brooks, E. M. Grivsky, and A. M. Steinberg, *J. Org. Chem.*, **26**, 3669 (1961).
[101] W. G. Dauben and J. Jiu, *J. Am. Chem. Soc.*, **76**, 4426 (1954).
[102] W. M. Kutz, J. E. Nickels, J. J. McGovern, and B. B. Corson, *J. Am. Chem. Soc.*, **70**, 4026 (1948).
[103] W. Wunderlich, *Arch. Pharm.*, **286**, 512 (1953).
[104] U. V. Solmssen and E. Wenis, *J. Am. Chem. Soc.*, **70**, 4197 (1948).
[105] N. L. Allinger and J. L. Coke, *J. Am. Chem. Soc.*, **82**, 2553 (1960).
[106] R. Howe, L. H. Smith, and J. S. Stephenson, U.S. Patent 3,255,249 (1966).
[107] G. Stork, *J. Am. Chem. Soc.*, **69**, 576 (1947).
[108] H. F. Hipsher and P. H. Wise, *J. Am. Chem. Soc.*, **76**, 1747 (1954).
[109] A. S. Bailey, J. C. Smith, and C. M. Stavely, *J. Chem. Soc.*, **1956**, 2731.
[110] D. M. Musser and H. Adkins, *J. Am. Chem. Soc.*, **60**, 664 (1938).
[111] M. Freifelder and G. R. Stone, *J. Pharm. Sci.*, **53**, 1134 (1964).
[112] R. L. Hull, *J. Am. Chem. Soc.*, **77**, 6376 (1955).
[113] H. Adkins and G. Krsek, *J. Am. Chem. Soc.*, **70**, 412 (1948).
[114] A. I. Meyers, W. Beverung, and G. Garcia-Munoz, *J. Org. Chem.*, **29**, 3427 (1964).
[115] A. C. Cope, R. J. Cotter, and G. G. Roller, *J. Am. Chem. Soc.*, **77**, 3594 (1955).
[116] O. R. Rodig and L. C. Ellis, *J. Org. Chem.*, **26**, 2197 (1961).
[117] W. S. Johnson, D. S. Allen, R. R. Hindersinn, G. N. Sausen, and R. Pappo, *J. Am. Chem. Soc.*, **84**, 2181 (1962).
[118] W. E. Bachmann and J. Controulis, *J. Am. Chem. Soc.*, **73**, 2636 (1951).
[119] W. G. Dauben, C. F. Hiskey, and A. H. Markhart, Jr., *J. Am. Chem. Soc.*, **73**, 393 (1951).

[120] H. Adkins and E. E. Burgoyne, *J. Am. Chem. Soc.*, **71**, 3528 (1949).
[121] W. G. Dauben, R. C. Tweit, and C. Mannerskantz, *J. Am. Chem. Soc.*, **76**, 4420 (1954).
[122] W. G. Dauben and E. Hoerger, *J. Am. Chem. Soc.*, **73**, 1504 (1951).
[123] H. E. Ungnade and F. V. Morriss, *J. Am. Chem. Soc.*, **72**, 2112 (1950).
[124] D. J. Cram, *J. Am. Chem. Soc.*, **71**, 3953 (1949).
[125] J. R. Durland and H. Adkins, *J. Am. Chem. Soc.*, **60**, 1501 (1938).
[126] E. A. Garlock, Jr. and E. Mosettig, *J. Am. Chem. Soc.*, **67**, 2255 (1945).
[127] P. Boldt and P. Paul, *Ber.*, **99**, 2337 (1966).
[128] J. R. Durland and H. Adkins, *J. Am. Chem. Soc.*, **59**, 135 (1937).
[129] R. L. Clarke and W. S. Johnson, *J. Am. Chem. Soc.*, **81**, 5706 (1959).
[130] H. A. Smith and J. F. Fusek, *J. Am. Chem. Soc.*, **71**, 415 (1949).
[131] D. S. Tarbell and C. Weaver, *J. Am. Chem. Soc.*, **63**, 2939 (1941).
[132] V. Boekelheide and G. C. Morrison, *J. Am. Chem. Soc.*, **80**, 3905 (1958).
[133] J. B. Tindall, U.S. Patent 2,739,159 (1956).
[134] A. Windaus and O. Delber, *Ber.*, **53**, 2304 (1920).
[135] B. W. Lew, U.S. Patent 3,225,066 (1965).
[136] B. W. Lew, U.S. Patent 3,225,069 (1965).
[137] W. C. McCarthy and R. J. Kahl, *J. Org. Chem.*, **21**, 1118 (1956).
[138] R. L. Clarke, U.S. Patent 3,194,818 (1965).
[139] S. D. Darling and K. D. Wills, *J. Org. Chem.*, **32**, 2794 (1967).
[140] D. P. Brust, D. S. Tarbell, S. M. Hecht, E. C. Hayward, and D. L. Colebrook, *J. Org. Chem.*, **31**, 2192 (1966).
[141] J. S. Moffatt, *J. Chem. Soc.*, (C) **1966**, 734.
[142] M. Pesson, S. Dupin, M. Antoine, D. Humbert, and M. Joannic, *Bull. Soc. Chim. France*, **1965**, 2262.
[143] M. Robba and M-C. Zaluski, *Compt. Rend.*, Serie C, **1966**, 429.
[144] a. W. W. Epstein, P. Gerike, and W. J. Horton, *Tetrahedron Letters*, **1965**, 3991;
b. Conditions of experiment and amount of catalyst obtained through personal communication from Dr. Epstein.
[145] A. C. Cope and E. E. Schweizer, *J. Am. Chem. Soc.*, **81**, 4577 (1959).
[146] C. G. Overberger, L. C. Palmer, B. S. Marks, and N. R. Byrd, *J. Am. Chem. Soc.*, **77**, 4100 (1955).
[147] J. H. Atkinson, R. Grigg, and A. W. Johnson, *J. Chem. Soc.*, **1964**, 893.
[148] O. Dann and W. Dimmling, *Ber.*, **86**, 1383 (1953).
[149] H. Rapoport, C. G. Christian, and G. Spencer, *J. Org. Chem.*, **18**, 840 (1954).
[150] L. R. Kray and M. G. Reinecke, *J. Org. Chem.*, **32**, 225 (1967).
[151] L. H. Andrew and S. M. McElvain, *J. Am. Chem. Soc.*, **51**, 887 (1929).
[152] R. Adams, S. Miyano, and M. D. Nair, *J. Am. Chem. Soc.*, **83**, 3323 (1961).
[153] J. M. Patterson, J. Brasch, and P. Drenchko, *J. Org. Chem.*, **27**, 1652 (1962).
[154] E. E. Schweizer and K. R. Light, *J. Org. Chem.*, **31**, 870 (1966).
[155] E. Späth and F. Kuffner, *Ber.*, **68**, 494 (1935).
[156] E. Ochiai, K. Tsuda, and S. Ikuma, *Ber.*, **69**, 2238 (1936).
[157] F. E. King, J. A. Barltrop, and R. J. Walley, *J. Chem. Soc.*, **1945**, 277.
[158] H. E. Johnson and D. G. Crosby, *J. Org. Chem.*, **28**, 2794 (1963).
[159] H. Adkins and H. L. Coonradt, *J. Am. Chem. Soc.*, **63**, 1563 (1941).
[160] R. Willstatter and D. Jacquet, *Ber.*, **51**, 767 (1918).
[161] A. Smith and J. H. P. Utley, *Chem. Commun.*, **1965**, 427.
[162] A. E. Hydorn, *J. Org. Chem.*, **32**, 4100 (1967).
[163] G. Wittig, G. Closs, and F. Mindermann, *Ann.*, **594**, 89 (1955).
[164] A. Cohen and B. Heath-Brown, *J. Chem. Soc.*, **1965**, 7179.

[165] F. Sorm, *Coll. Czechoslov. Chem. Communs.*, **12**, 245 (1947).
[166] H. Rapoport and E. Jorgensen, *J. Org. Chem.*, **14**, 664 (1949).
[167] H. C. Silberman, *J. Org. Chem.*, **25**, 151 (1960).
[168] R. H. Wiley and A. J. Hart, *J. Am. Chem. Soc.*, **77**, 2340 (1955).
[169] G. N. Walker, *J. Am. Chem. Soc.*, **78**, 3201 (1956).
[170] R. H. Hasek, P. G. Gott, and J. C. Martin, *J. Org. Chem.*, **29**, 3513 (1964).
[171] R. Cornubert, M. Real, and P. Thomas, *Bull. Soc. Chim. France*, **1954**, 534.
[172] W. Borsche and R. Frank, *Ber.*, **59**, 237 (1926).
[173] J. Attenburrow, J. Elks, D. F. Elliott, B. A. Hems, J. O. Harris, and C. I. Broderick, *J. Chem. Soc.*, **1945**, 571.
[174] S. R. Cawley and S. G. P. Plant, *J. Chem. Soc.*, **1938**, 1214.
[175] S. A. Ballard, R. T. Holm, and P. H. Williams, *J. Am. Chem. Soc.*, **72**, 5734 (1950).
[176] C. A. Grob and V. Krasnobajew, *Helv. Chim. Acta*, **47**, 2145 (1964).
[177] J. J. de Vrieze, *Rec. Trav. Chim.*, **78**, 91 (1959).
[178] P. L. de Benneville and R. Connor, *J. Am. Chem. Soc.*, **62**, 283 (1940).
[179] K. J. Liska and L. Salerni, *J. Org. Chem.*, **25**, 124 (1960).
[180] M. Freifelder, *Advances in Catalysis*, Vol. 14, Academic Press, New York, 1963, p. 203.
[181] N. Sperber, D. Papa, E. Schwenk, M. Sherlock, and R. Fricano, *J. Am. Chem. Soc.*, **73**, 5752 (1952).
[182] E. Profft and G. Schulz, *Arch. Pharm.*, **294**, 292 (1961).
[183] M. Freifelder, Y. H. Ng, and G. R. Stone, *J. Org. Chem.*, **30**, 1319 (1965).
[184] M. Freifelder and G. R. Stone, *J. Org. Chem.*, **26**, 3805 (1961).
[185] D. E. Ames and R. E. Bowman, *J. Chem. Soc.*, **1952**, 1057.
[186] J. A. Pianfetti, U.S. Patent 3,192,220 (1965).
[187] G. N. Walker, *J. Org. Chem.*, **27**, 2966 (1962).
[188] E. A. Steck, L. T. Fletcher, and R. P. Brundage, *J. Org. Chem.*, **28**, 2233 (1963).
[189] K. Hoffman and E. Sury, U.S. Patent 3,153,046 (1965).
[190] K. Stach, M. Thiel and F. Bickelhaupt, *Monatsh.*, **93**, 1090 (1962).
[191] H. K. Hall, Jr., *J. Org. Chem.*, **29**, 3539 (1964).
[192] T. S. Hamilton and R. Adams, *J. Am. Chem. Soc.*, **50**, 2260 (1928).
[193] M. Freifelder, *J. Org. Chem.*, **27**, 4046 (1962).
[194] F. Haglid and I. Wellings, *Acta. Chem. Scand.*, **17**, 1735 (1963).
[195] M. Freifelder, R. M. Robinson, and G. R. Stone, *J. Org. Chem.*, **27**, 284 (1962).
[196] A. T. Nielsen, D. W. Moore, J. H. Mazur, and K. H. Berry, *J. Org. Chem.*, **29**, 2898 (1964).
[197] L. E. Brady, M. Freifelder, and G. R. Stone, *J. Org. Chem.*, **26**, 4757 (1961).
[198] F. Sorm, *Coll. Czechoslov. Chem. Communs.*, **13**, 57 (1948).
[199] M. Freifelder, *J. Org. Chem.*, **28**, 1135 (1963).
[200] M. Freifelder, *J. Org. Chem.*, **28**, 602 (1963).
[201] H. K. Hall, Jr., *J. Am. Chem. Soc.*, **82**, 1209 (1960).
[202] R. J. Mohrbacher, U.S. Patent 3,245,991 (1966).
[203] S. Ohki and I. Matuo, *Chem. Pharm. Bull.* (Japan), **7**, 892 (1959).
[204] J. Ratusky and F. Sorm, *Coll. Czechoslov. Chem. Communs.*, **19**, 340 (1954).
[205] M. Freifelder and H. B. Wright, *J. Med. Chem.*, **7**, 664 (1964).
[206] J. H. Biel, U.S. Patent 2,802,007 (1957).
[207] L. Marion and W. F. Cockburn, *J. Am. Chem. Soc.*, **71**, 3402 (1949).
[208] C. H. Kao, *J. Chem. Eng. China*, **15**, 80 (1948); through *Chem. Abstr.*, **44**, 3993 (1950).
[209] J. H. Biel, H. L. Friedman, H. A. Leiser, and E. P. Sprengler, *J. Am. Chem. Soc.*, **74**, 1485 (1952).

[210] S. L. Shapiro, K. Weinberg, T. Bazga, and L. Freedman, *J. Am. Chem. Soc.*, **81**, 5146 (1959).
[211] J. H. Biel and F. F. Blicke, U.S. Patent 3,051,715 (1962).
[212] J. H. Biel and C. E. Aiman, U.S. Patent 3,310,567 (1967).
[213] S. B. Coan, B. Jaffe, and D. Papa, *J. Am. Chem. Soc.*, **78**, 3701 (1956).
[214] K. N. Campbell, J. F. Ackerman, and B. K. Campbell, *J. Org. Chem.*, **15**, 337 (1950).
[215] T. Ishii, *J. Pharm. Soc. Japan*, **71**, 1097 (1951); through *Chem. Abstr.*, **46**, 5042 (1952).
[216] N. J. Leonard and E. Barthel, Jr., *J. Am. Chem. Soc.*, **71**, 3098 (1949).
[217] N. J. Leonard and F. P. Hauck, Jr., *J. Am. Chem. Soc.*, **79**, 5279 (1957).
[218] K. H. Büchel, A. K. Bocz, and F. Korte, *Ber.*, **99**, 724 (1966).
[219] T. Grave, *J. Am. Chem. Soc.*, **46**, 1460 (1924).
[220] B. R. Baker and F. J. McEvoy, *J. Org. Chem.*, **20**, 136 (1955).
[221] H. Nienburg, *Ber.*, **70**, 635 (1937).
[222] G. N. Walker and M. A. Moore, *J. Org. Chem.*, **26**, 432 (1961).
[223] L. N. Yakhontov, S. V. Yatsenko, and M. V. Rubstov, *J. Gen. Chem. USSR*, **28**, 3146 (1958) (Eng. trans.).
[224] A. V. Kirsanov and Y. N. Ivaschenko, *Bull. Soc. Chim. France*, **1936**, 2279.
[225] L. Birkhofer, *Ber.*, **75**, 429 (1942).
[226] M. Freifelder, R. W. Mattoon, and Y. H. Ng, *J. Org. Chem.*, **29**, 3730 (1964).
[227] S. L. Shapiro, H. Soloway, and L. Freedman, *J. Org. Chem.*, **26**, 818 (1961).
[228] M. M. Robison, F. P. Butler, and B. L. Robison, *J. Am. Chem. Soc.*, **79**, 2573 (1957).
[229] E. C. Britton and L. H. Horsley, U.S. Patent 2,834,784 (1958).
[230] W. F. Minor, J. B. Hoekstra, D. Fisher, and J. Sam, *J. Med. Chem.*, **5**, 96 (1962).
[231] H. L. Cohen and L. M. Minsk, *J. Am. Chem. Soc.*, **79**, 1759 (1957).
[232] K. Winterfeld and H. Schüler, *Arch. Pharm.*, **293**, 203 (1960).
[233] C. R. Smith, *J. Am. Chem. Soc.*, **50**, 1936 (1928).
[234] M. Freifelder, *J. Org. Chem.*, **29**, 2895 (1964).
[235] P. M. Quan and L. D. Quin, *J. Org. Chem.*, **31**, 2487 (1966).
[236] E. Wenkert, K. G. Dave, and F. Haglid, *J. Am. Chem. Soc.*, **89**, 5461 (1965).
[237] E. Wenkert, K. G. Dave, F. Haglid, R. G. Lewis, T. Oishi, R. V. Stevens, and M. Terashima, *J. Org. Chem.*, **33**, 747 (1968).
[238] Personal communication from Dr. R. E. Lyle, University of New Hampshire, Durham, N.H.
[239] C. A. Grob and F. Ostermayer, *Helv. Chim. Acta*, **45**, 1119 (1963).
[240] J. V. Supniewski and M. Serafinowna, *Arch. Chem. Farm.*, **3**, 109 (1936); through *Chem. Abstr.*, **33**, 7301 (1939).
[241] R. E. Lyle and G. G. Lyle, *J. Am. Chem. Soc.*, **76**, 3536 (1954).
[242] R. E. Lyle, E. F. Perlowski, H. J. Troscianiec, and G. G. Lyle, *J. Org. Chem.*, **20**, 1761 (1955).
[243] D. Lednicer, S. C. Lyster, and G. W. Duncan, *J. Med. Chem.*, **10**, 78 (1967).
[244] E. Wenkert and B. Wickberg, *J. Am. Chem. Soc.*, **87**, 1580 (1965).
[245] R. E. Lyle and G. H. Warner, *J. Med. Chem.*, **3**, 597 (1961).
[246] J. P. Wibaut, C. C. Kloppenburg, and M. G. J. Beets, *Rec. Trav. Chim.*, **63**, 134 (1944).
[247] V. Boekelheide and E. J. Agnello, *J. Am. Chem. Soc.*, **72**, 5005 (1950).
[248] V. Boekelheide and S. Rothschild, *J. Am. Chem. Soc.*, **71**, 879 (1949).
[249] J. P. Wibaut and C. C. Kloppenburg, *Rec. Trav. Chim.*, **65**, 100 (1946).
[250] J. Sam, J. D. England, and D. W. Alwani, *J. Med. Chem.*, **7**, 732 (1964).

[250a] M. E. Derieg, B. Brust, and R. I. Fryer, *J. Heterocyclic Chem.*, **3**, 165 (1966).
[251] F. Kröhnke and K. Fasold, *Ber.*, **67**, 656 (1934).
[252] B. Riegel and H. Wittcoff, *J. Am. Chem. Soc.*, **68**, 1805 (1946).
[253] P. Truitt, B. Bryant, W. E. Goode, and B. Arnwine, *J. Am. Chem. Soc.*, **74**, 2179 (1952).
[254] P. Truitt, B. Hall, and B. Arnwine, *J. Am. Chem. Soc.*, **74**, 4552 (1952).
[255] V. Boekelheide, W. J. Linn, P. O'Grady, and M. Lamborg, *J. Am. Chem. Soc.*, **75**, 3243 (1953).
[256] M. Freifelder, *J. Pharm. Sci.*, **55**, 535 (1966).
[257] I. Matsuo and S. Oki, Japan Patent 111,1967; through *Chem. Abstr.*, **66**, 75918b (1967).
[258] K. W. Wischmann, A. Logan, and D. M. Stuart, *J. Org. Chem.*, **26**, 2794 (1961).
[259] Neth. Appl. 6,515,775 (1966); through *Chem. Abstr.*, **65**, 15339 (1966).
[260] W. R. Hardie, J. Hidalgo, I. F. Halverstadt, and R. E. Allen, *J. Med. Chem.*, **9**, 127 (1966).
[261] H. S. Aaron, O. O. Owens, P. D. Rosenstock, S. Leonard, S. Elkin, and J. I Miller, *J. Org. Chem.*, **30**, 1331 (1965).
[262] J. Overhof and J. P. Wibaut, *Rec. Trav. Chim.*, **50**, 957 (1931).
[263] D. Lednicer and C. R. Hauser, *J. Am. Chem. Soc.*, **79**, 4459 (1957).
[264] A. P. Gray and W. L. Archer, *J. Am. Chem. Soc.*, **79**, 3354 (1957).
[265] R. C. Elderfield, B. Fischer, and J. M. Lagowski, *J. Org. Chem.*, **22**, 1376 (1957).
[266] R. N. Castle and C. W. Whittle, *J. Org. Chem.*, **24**, 1189 (1959).
[267] A. P. Gray and H. Kraus, *J. Org. Chem.*, **26**, 3368 (1961).
[268] J. King, Brit. Patent 1,023,781 (1966).
[269] H. Bader and W. Oroshnik, *J. Am. Chem. Soc.*, **79**, 5686 (1957).
[270] H. Bader and W. Oroshnik, *J. Am. Chem. Soc.*, **81**, 163 (1959).
[271] F. J. McCarty, C. H. Tilford, and M. G. Van Campen, Jr., *J. Am. Chem. Soc.*, **79**, 472 (1957).
[272] D. G. Ott, F. N. Hayes, E. Hansbury, and V. N. Kerry, *J. Am. Chem. Soc.*, **79**, 5448 (1957).
[273] J. M. McManus and R. M. Herbst, *J. Org. Chem.*, **24**, 1462 (1959).
[274] J. von Braun, W. Gmelin, and A. Schultheiss, *Ber.*, **56**, 1338 (1923).
[275] J. von Braun, W. Gmelin, and A. Petzold, *Ber.*, **57**, 382 (1924).
[276] J. von Braun, A. Petzold, and J. Seeman, *Ber.*, **55**, 3779 (1922).
[277] J. von Braun and G. Lemke, *Ann.*, **478**, 176 (1930).
[278] E. Ochiai, C. Kaneko, I. Shimada, Y. Murata, T. Kosuye, and C. Kawasaki, *Chem. Pharm. Bull., Tokyo*, **8**, 126 (1960).
[279] C. J. Cavallito and T. H. Haskell, *J. Am. Chem. Soc.*, **66**, 1166 (1946).
[280] G. Grethe, H. L. Lee, M. Uskoković, and A. Brossi, *J. Org. Chem.*, **33**, 494 (1968).
[281] E. Ochiai and Y. Kawazoe, *Pharm., Bull. (Tokyo)*, **5**, 606 (1957).
[282] S. Chiaverelli and G. B. Marini-Betal, *Gazz. Chim. Ital.*, **82**, 86 (1952).
[283] S. Shitani, K. Sakai, and K. Mitsuhashi, *Yakugaku Zasshi*, **87**, 547 (1967); through *Chem. Abstr.*, **67**, 54022k (1967).
[284] K. Isogai, *J. Chem. Soc. Japan*, **81**, 1594 (1960); through *Chem. Abstr.*, **56**, 2420 (1962).
[285] B. Witkop, *J. Am. Chem. Soc.*, **70**, 2617 (1948).
[286] W. L. F. Armstrong, *J. Chem. Soc.*, (C), **1967**, 377.
[287] J. Piechazek and H. Bojarska-Dahlig, *Acta Polon. Pharm.*, **23**, 7 (1966); through *Chem. Abstr.*, **64**, 14164 (1966).
[288] H. Wieland, O. Hettche, and T. Hoshina, *Ber.*, **61**, 2371 (1928).
[289] W. Solomon, *J. Chem. Soc.*, **1947**, 129.

[290] E. Wenkert, D. B. R. Johnston, and R. G. Dave, *J. Org. Chem.*, **29**, 2534 (1964).
[291] T. Kametani, K. Kigasawa, and M. Hiiragi, *Chem. Pharm. Bull., Tokyo*, **13**, 1220 (1965).
[292] W. Schneider and B. Müller, *Ber.*, **93**, 1579 (1960).
[293] R. B. Woodward and W. E. Doering, *J. Am. Chem. Soc.*, **67**, 860 (1945).
[294] R. Rapala, E. R. Lavagnino, E. R. Shepard, and E. Farkas, *J. Am. Chem. Soc.*, **79**, 3770 (1957).
[295] A. Brossi and O. Schneider, *Helv. Chim. Acta*, **39**, 1376 (1956).
[296] N. Sugimoto, S. Ohshiro, H. Kugita, and S. Saito, *Chem. Pharm. Bull., Tokyo*, **5**, 62 (1957).
[297] N. Sugimoto and H. Kugita, *Chem. Pharm. Bull., Tokyo*, **5**, 70 (1957).
[298] N. Sugimoto and H. Kugita, *Chem. Pharm. Bull., Tokyo*, **6**, 432 (1958).
[299] R. B. Woodward and E. C. Kornfeld, *J. Am. Chem. Soc.*, **70**, 2508 (1948).
[300] F. F. Blicke and J. F. Gearien, *J. Am. Chem. Soc.*, **76**, 3586 (1954).
[301] J. F. Munshi, Doctoral Thesis, Part II, *An Investigation of the Reduction of Some Quinoline Derivatives*, University of Pennsylvania, Philadelphia, Pa. (1965).
[302] H. de Diesbach, A. Pugin, F. Morard, W. Nowacinski, and J. Dessibourg, *Helv. Chim. Acta*, **35**, 2322 (1952).
[303] H. Kühnis and H. de Diesbach, *Helv. Chim. Acta*, **41**, 894 (1958).
[304] W. E. Goode, *J. Am. Chem. Soc.*, **70**, 3946 (1948).
[305] P. E. Wright and W. E. McEwen, *J. Am. Chem. Soc.*, **76**, 4540 (1954).
[306] M. Anderson and A. W. Johnson, *J. Chem. Soc.*, **1965**, 2411.
[307] R. F. Childs and A. W. Johnson, *J. Chem. Soc., C*, **1966**, 1950.
[308] E. Vogel, R. Erb, G. Lenz, and A. A. Bothner-By, *Ann.*, **682**, 1 (1965).
[309] K. Dimroth and H. Freyschlag, *Angewandte Chem.*, **68**, 518 (1956).
[310] R. C. Elderfield, *Heterocyclic Compounds*, Vol. 5, John Wiley, New York, 1957, p. 329.
[311] J. Lichtenberger and J. P. Fleury, *Bull. Soc. Chim. France*, **1955**, 1320.
[312] H. Thomas and J. Schnupp, *Ann.*, **434**, 296 (1923).
[313] R. L. Hinman, R. D. Ellefson, and R. D. Campbell, *J. Am. Chem. Soc.*, **82**, 3988 (1960).
[314] R. J. Crawford, A. Mishra, and R. J. Dummel, *J. Am. Chem. Soc.*, **88**, 3959 (1966).
[315] A. N. Kost and G. A. Golubeva, *J. Gen. Chem. USSR*, **33**, 240 (1963) (Eng. trans.)
[315a] L. C. Behr, *The Chemistry of Heterocyclic Compounds*, Interscience, New York, 1967, Vol. 22, Chapter 10, p. 318.
[316] K. Hoffman, *Imidazole and Its Derivatives*, Interscience, New York, 1953, p. 16.
[317] E. Waser and A. Gratsos, *Helv. Chim. Acta*, **11**, 944 (1928).
[318] H. Bauer, *J. Org. Chem.*, **26**, 1649 (1961).
[319] R. Duschinsky and L. A. Dolan, *J. Am. Chem. Soc.*, **67**, 2079 (1945).
[320] R. Duschinsky, L. A. Dolan, L. O. Randall, and G. Lehmann, *J. Am. Chem. Soc.*, **69**, 3150 (1947).
[321] H. McKennis, Jr. and V. Du Vigneaud, *J. Am. Chem. Soc.*, **68**, 832 (1946).
[322] E. Schipper and E. Chinery, *J. Org. Chem.*, **26**, 4480 (1961).
[323] E. Schipper and E. Chinery, *J. Org. Chem.*, **26**, 3597 (1961).
[324] M. Rink, S. Mehta, and K. Grabowski, *Arch. Pharm.*, **292**, 225 (1959).
[325] P. Baranger and J. Levisalles, *Bull. Soc. Chim. France*, **1957**, 704.
[326] S. Sugasawa and K. Kohno, *Pharm. Bull. Tokyo*, **4**, 477 (1956); through *Chem. Abstr.*, **51**, 13866 (1956).
[327] K. Eichenberger, A. Staehlin, and J. Druey, *Helv. Chim. Acta*, **37**, 837 (1954).
[328] G. B. Kline, U.S. Patent 2,873,294 (1959).
[329] R. Elderfield, *Heterocyclic Compounds*, Wiley, New York, 1957, Vol. 6, p. 159.

References 639

[330] J. D. Westover, *Hydrogenation of Cinnoline*, University Microfilms, Inc., Ann Arbor, 1966.
[331] D. I. Haddlesey, P. A. Mayor, and S. S. Szinai, *J. Chem. Soc.*, **1964**, 5269.
[332] E. F. Elslager, D. F. Worth, N. F. Haley, and S. C. Perricone, *J. Heterocyclic Chem.*, **5**, 609 (1968).
[333] W. J. Haggerty, Jr., R. H. Springer, and C. C. Cheng, *J. Heterocyclic Chem.*, **2**, 1 (1965).
[334] R. F. Evans and J. S. Shannon, *J. Chem. Soc.*, **1965**, 1406.
[335] D. J. Brown and R. F. Evans, *J. Chem. Soc.*, **1962**, 527.
[336] H. R. Henze and S. O. Winthrop, *J. Am. Chem. Soc.*, **79**, 2230 (1957).
[337] V. H. Smith and B. E. Christiensen, *J. Org. Chem.*, **20**, 829 (1955).
[338] R. F. Evans, *J. Chem. Soc.*, **1964**, 2450.
[339] H. Aft and B. E. Christiensen, *J. Org. Chem.*, **27**, 2170 (1962).
[340] B. J. Whitlock, S. H. Lipton, and F. M. Strong, *J. Org. Chem.*, **30**, 115 (1963).
[341] J.J. Fox and D. Van Praag, *J. Am. Chem. Soc.*, **82**, 486 (1960).
[342] V. Škarić, B. Gašpert, and D. Škarić, *Croat. Chem. Acta*, **36**, 87 (1964).
[343] H. Gault and M. Suquet, *Compt. Rend.*, **233**, 180 (1951).
[344] K. Folkers and T. B. Johnson, *J. Am. Chem. Soc.*, **56**, 1180 (1934).
[345] S. David and P. Sinay, *Bull. Soc. Chim. France*, **1965**, 2301.
[346] H. Kny and B. Witkop, *J. Am. Chem. Soc.*, **81**, 6245 (1959).
[347] J. C. Ambelang and T. B. Johnson, *J. Am. Chem. Soc.*, **61**, 74 (1939).
[348] G. E. Hilbert, *J. Am. Chem. Soc.*, **54**, 2076 (1932).
[349] R. D. Batt, J. K. Martin, J. McT. Ploeser, and J. Murray, *J. Am. Chem. Soc.*, **76**, 3663 (1954).
[350] S. Y. Wang, *J. Am. Chem. Soc.*, **80**, 6196 (1959).
[351] F. J. DiCarlo, A. S. Schultz, and A. M. Kent, *J. Biol. Chem.*, **199**, 333 (1952).
[352] M. Green and S. S. Cohen, *J. Biol. Chem.*, **225**, 397 (1957).
[353] W. E. Cohn and D. G. Doherty, *J. Am. Chem. Soc.*, **78**, 2863 (1956).
[354] Private communications from a number of biochemists.
[355] M. Green and S. S. Cohen, *J. Biol. Chem.*, **228**, 601 (1957).
[356] E. B. Marr and M. T. Bogert, *J. Am. Chem. Soc.*, **57**, 729 (1935).
[357] W. L. F. Armarego and J. I. C. Smith, *J. Chem. Soc.*, **1965**, 5360.
[358] R. C. Elderfield, T. A. Williamson, W. J. Gensler, and C. B. Kremer, *J. Org. Chem.*, **12**, 405 (1947).
[359] E. R. H. Jones, *J. Chem. Soc.*, **1964**, 5911.
[360] F. B. Kipping, *J. Chem. Soc.*, **1929**, 2889.
[361] F. B. Kipping, *J. Chem. Soc.*, **1932**, 1336.
[362] E. Felder, S. Maffei, S. Pietra, and D. Pitré, *Chimia*, **13**, 263 (1959).
[363] H. I. X. Magers and W. Berends, *Rec. Trav. Chim.*, **78**, 109 (1959).
[364] H. I. X. Magers and W. Berends, *Rec. Trav. Chim.*, **76**, 28 (1957).
[365] J. D. Behun and R. Levine, *J. Org. Chem.*, **26**, 3379 (1961).
[366] E. F. Rogers and H. J. Becker, U.S. Patent 3,281,423 (1966).
[367] M. P. Mertes and N. R. Patel, *J. Med. Chem.*, **9**, 868 (1966).
[368] J. J. Scigliano, U.S. Patent 2,843,589 (1958).
[369] M. Munk and H. P. Schultz, *J. Am. Chem. Soc.*, **74**, 3433 (1952).
[370] J. Figueras, *J. Org. Chem.*, **31**, 803 (1966).
[371] P. Schuyler, F. D. Popp, A. C. Noble, and B. R. Masters, *J. Med. Chem.*, **9**, 704 (1966).
[372] J. C. Cavagnol and F. Y. Weislogle, *J. Am. Chem. Soc.*, **69**, 795 (1947).
[373] H. S. Broadbent, E. L. Allred, L. Pendleton, and C. W. Whittle, *J. Am. Chem. Soc.*, **82**, 189 (1960).

[374] E. C. Taylor and A. McKillop, *J. Am. Chem. Soc.*, **87**, 1984 (1965).
[375] R. D. Dillard and N. R. Easton, *J. Org. Chem.*, **31**, 122 (1966).
[376] K. Schenker, *Helv. Chim. Acta*, **51**, 413 (1968).
[377] C. M. Hoffman and S. R. Safir, *J. Org. Chem.*, **27**, 3565 (1962).
[378] L. H. Sternbach and E. Reeder, *J. Org. Chem.*, **26**, 111 (1961).
[379] R. Mohr and H. Hertel, Ger. Patent 1,089,392 (1960); through *Chem. Abstr.*, **56**, 1391 (1962).
[380] R. A. Carboni, U.S. Patent 3,197,475 (1965).
[381] G. Pallazzo and L. Baiocchi, *Ann. Chim.* (Rome), **56**, 199 (1966).
[382] C. Grundmann and A. Kreutzberger, *J. Am. Chem. Soc.*, **77**, 44 (1955).
[383] H. Brandenburger and R. Schyzer, *Helv. Chim. Acta*, **38**, 1396 (1955).
[384] This has been noted by this author in a number of instances.
[385] M. Polonovski, M. Pesson, and P. Rajzman, *Bull. Soc. Chim. France*, **1955**, 1171.
[386] R. Metze and G. Scherowsky, *Ber.*, **92**, 2481 (1959).
[387] E. Hayer and R. Gompper, *Ber.*, **92**, 564 (1959).
[388] M. Polonovski, M. Pesson, and P. Rajzman, *Bull. Soc. Chim. France*, **1955**, 1166.
[389] J. Gut, M. Prytaš, J. Jonáš, and F. Šorm, *Coll. Czechoslov. Chem. Commun.*, **26**, 974 (1964).
[390] S. Okuda and M. M. Robison, *J. Org. Chem.* **24**, 1008 (1959).
[391] B. Frydman, M. F. Despuy, and H. Rapoport, *J. Am. Chem. Soc.*, **87**, 3530 (1965).
[392] A. M. Roe, *J. Chem. Soc.*, **1963**, 2195.
[393] J. G. Lombardino, *J. Org. Chem.*, **30**, 2403 (1965).
[394] L. M. Werbel and M. L. Zamora, *J. Heterocyclic Chem.*, **2**, 290 (1965).
[395] N. Ikekawa, *Chem. Pharm. Bull.*, **6**, 408 (1958).
[396] H. Rapoport and A. D. Batcho, *J. Org. Chem.*, **28**, 1753 (1963).
[397] W. L. F. Armarego, *J. Chem. Soc.*, (C), **1967**, 377.
[398] R. A. Archer and H. S. Mosher, *J. Org. Chem.*, **32**, 1378 (1967).
[399] I. J. Pechter and J. Weinstock, U.S. Patent 3,159,627 (1964).
[400] A. Bobst and M. Viscontini, *Helv. Chim. Acta*, **49**, 875 (1966).
[401] H. I. X. Mager, R. Addink, and W. Berends, *Rec. Trav. Chim.*, **86**, 833 (1967).
[402] A. Albert and S. Matsuura, *J. Chem. Soc.*, **1961**, 5131.
[403] R. C. Elderfield, *Heterocyclic Compounds*, Vol. 9, Wiley, New York, 1967, p. 32.
[404] A. J. Birch, M. Salahud-Din, and D. C. C. Smith, *J. Chem. Soc.*, (C), **1966**, 523.
[405] C. P. Huttrer, *J. Am. Chem. Soc.*, **71**, 4147 (1949).
[406] W. O. Kermack and J. E. McKail, *Heterocyclic Chemistry*, R. C. Elderfield, ed., Vol. 7, Wiley, 1961, p. 354.
[407] N. M. Cullinane and H. J. H. Padfield, *J. Chem. Soc.*, **1935**, 1131.
[408] H. Plieninger, W. Müller, and K. Weinerth, *Ber.*, **97**, 667 (1964).
[409] H. Dressler and M. E. Baum, *J. Org. Chem.*, **26**, 102 (1961).
[410] K. H. Bloss and C. E. Timberlake, *J. Org. Chem.*, **28**, 267 (1963).
[411] R. A. Abramovitch and I. D. Spenser, *Advances in Heterocyclic Chemistry*, Vol. 3, Academic Press, New York, 1964, p. 102.
[412] A. Albert and B. Ritchie, *J. Chem. Soc.*, **1943**, 458.
[413] E. Hayashi and T. Nagao, *Yakugaku Zasshi*, **84**, 198 (1964); through *Chem. Abstr.*, **61**, 3071 (1964).
[414] L. Birkhofer, *Ber.*, **85**, 1023 (1952).
[415] S. Maffei and S. Pietra, *Gazz. Chim. Ital.*, **88**, 556 (1958).
[416] G. R. Clemo and H. McIlwain, *J. Chem. Soc.*, **1936**, 258.
[417] G. R. Clemo and H. McIlwain, *J. Chem. Soc.*, **1936**, 1698.

XXV

HYDROGENATION OF SULFUR-CONTAINING COMPOUNDS

Of all the elements, sulfur is the one most capable of inhibiting the activity of hydrogenation catalysts. As described in Chapter IV its effect is due to the unshared electron pairs in the octet surrounding it through which strong bonding to the active sites of the catalyst takes place. Its bonding power is dependent on its accessibility to the catalyst surface. There is also an effect dependent on the number of unshared electron pairs, that is, whether a substrate contains a divalent, tetravalent, or hexavalent sulfur atom. This difference in reducibility can be seen in the comparative hydrogenations of 4-nitrophenyl sulfone, 4-nitrophenyl sulfoxide, and 4-nitrophenyl sulfide, to the corresponding amines with palladium on carbon. The order of reducibility was sulfone (no unshared pairs), 1 hr, sulfoxide (one unshared pair), 3–4 hr, and sulfide (two unshared pairs), 6–7 hr [1]. In another comparison it was shown that unsaturated aliphatic sulfones were reduced far more rapidly than the corresponding sulfides [2].

COMPOUNDS CONTAINING HEXAVALENT SULFUR

Reductions of functional groups among sulfones, sulfonamides, sulfonic acids, and related sulfur-containing compounds should proceed in normal fashion unless the sulfonyl system is converted to a poisoning form during the reaction. This does not generally occur except under vigorous hydrogenation conditions although it may take place in systems, RSO_2R' where R and R' are aliphatic. Such an example is seen in the reduction of some unsaturated aliphatic sulfones where thiols were formed and hydrogen uptake became markedly slow [2].

In a few instances an aromatic ring has been saturated in the presence of a sulfonyl-containing system but in general examples appear to be an exception not a rule. Ammonium phenylsulfamate, $C_6H_5NHSO_3NH_4$, was converted to ammonium cyclohexylsulfamate by hydrogenation with rhodium and with ruthenium (see Refs. 77 and 78, Chapter XXIV). 3-Pyridinesulfonic acid and 2-(2- and 4-pyridine)ethanesulfonic acids were also reduced to the corresponding cyclohexyl compounds (see Ref. 205, Chapter XXIV). Of these compounds, only in 3-pyridinesulfonic acid was the sulfur-containing group attached to the aromatic ring.

Attempts to reduce the aromatic ring in benzenesulfonic acids and amides or in phenyl sulfones under a variety of conditions resulted in failure [1].

COMPOUNDS CONTAINING TETRAVALENT SULFUR

There are too few examples of reductions among tetravalent sulfur-containing compounds to suggest too many conclusions. However, some conclusions may be drawn about specific types.

Despite the availability of an unshared electron pair in a sulfoxide, the hydrogenation of 4-nitrophenyl sulfoxide to 4-aminophenylsulfoxide [1] suggests that successful hydrogenations in this group will be dependent on the reducibility of the function undergoing attack and on the stability of the sulfinyl group in the system. Poisoning, which occurred during the reduction of some allyl sulfones [2], suggests that it should also take place (probably more readily) in the case of sulfoxides to give mercaptans, RSH, which are potent catalyst inhibitors. The hydrogenation of alkyl phenyl sulfoxide clearly illustrates this; hydrogen uptake was very slow due to formation of thiophenol [2].

Among sulfinic acids, reducible functions should be difficult to hydrogenate. The geometry of the substrate (indicated by construction of a model of benzenesulfinic acid) suggests that catalyst inhibition should occur because of proximity of the sulfur atom to the catalyst. Indeed, benzenesulfinic acid was shown to be a potent inhibitor when present in the Raney nickel reduction of cyclohexene [3].

Poisoning, which occurs during reduction of tetravalent sulfur compounds, may be overcome at times by the use of larger than normal amounts of catalyst. For example, sulfilimines have been hydrogenolyzed to substituted benzenesulfonamides and the corresponding sulfides by hydrogenation in alcohol with an equal weight of 5% palladium on carbon [4] (Eq. 25.1). In other reductions where the system SXY was 2,6-dimethylthiamorpholine

$$H_3C\langle\bigcirc\rangle SO_2N=S{<}^X_Y \xrightarrow[Pd\ on\ C]{H_2} H_3C\langle\bigcirc\rangle SO_2NH_2 + XSY$$

X = alkyl, Y = aryl and aralkyl (25.1)

hydrochloride, it was necessary to use a 250–400% catalyst ratio to achieve hydrogenolysis [1].

COMPOUNDS CONTAINING DIVALENT SULFUR

Despite the inhibiting effect of a divalent sulfur atom, compounds codṇaining it can be and have been reduced. Success will depend on the ease of reducibility of the system undergoing attack by hydrogen, on the accessibility of the poisoning atom to the surface of the catalyst, on the catalyst employed for reduction (often on the amount used), and on the energy applied to obtain uptake of hydrogen during the course of reaction.

In this series, construction of molecular models can be especially useful. Examination of them can indicate the relative ease or difficulty in the approach of the poisoning atom to the catalyst surface and on this basis often can enable the chemist to decine on the amount of catalyst and on the severity of the reaction conditions to be employed.

Disulfides

Hydrogenolysis to Mercaptans. One might expect hydrogenolysis of disulfides, RS—SR, to mercaptans to be difficult because of the contact between the catalyst and the poisoning atoms in the disulfides and in the reduction products. Successful reactions have been carried out by employing rather high catalyst ratios as exemplified in the conversion of cystine to cysteine and in the reduction of related compounds to the corresponding thiols [5] and in the hydrogenation of the methyl ester of cystine to the methyl ester of cysteide [6]. In each case about 25% by weight of palladium oxide was used.

Much less catalyst was employed when cystine was hydrogenated in dilute hydrochloric acid in the presence of palladium on polyvinyl alcohol [7]. However, the ratio of precious metal to substrate was still an abnormal one (5%) indicative of the anticatalytic effect of an unhindered divalent sulfur atom. This effect was further emphasized by the long reaction time (45 hr). Somewhat better results were obtained when 40% by weight of 10%

644   Hydrogenation of Sulfur-Containing Compounds

palladium on carbon was employed for the hydrogenolysis of L-cystinyl-di-L-phenylalanine; reaction was complete in 12 hr [8].

Ruthenium dioxide, generally unsatisfactory for low-pressure hydrogenations, was surprisingly effective in the hydrogenolysis of 2,2'-diamino-5,5'-disulfamylbenzenedisulfide at 75° and 2.7 atm with 0.5% by weight of catalyst [9].

Hydrogenation of Reducible Functions. Attempts to remove the protecting N-carbobenzyloxy groups from cystine with palladium or platinum blacks resulted in no uptake of hydrogen [10]. The more labile 4-nitrocarbobenzyloxy group was removed from cystine with 40% by weight of 10% palladium on carbon in dilute sodium hydroxide [8]. A very modest amount of the same catalyst, about 3.5% by weight, was employed for removal of the protecting group from di-4-nitrocarbobenzyloxy-L-cystinyl-diglycine when the reaction was carried out in alcohol. While the catalyst was effective in removing the N-protecting groups, it is of interest that the readily reducible aromatic nitro group was not converted to an amino group when 6 or 7 moles of hydrogen were absorbed: only the hydroxylamino compound was obtained (Eq. 25.2). The incomplete reduction of the nitro

$$\left[\begin{array}{c}-SCH_2CHCOOH\\|\\NHCOCH_2\!-\!\!\!\bigcirc\!\!\!-R\\\|\\O\end{array}\right]_2 \quad\begin{array}{c}\xrightarrow[6\,hr]{6H_2}\\\\\xrightarrow[17\,hr]{7H_2}\end{array}\quad\begin{array}{l}\left[-SCH_2CHCOOH\atop |\atop NH_2\right]_2 + H_3C\!-\!\!\!\bigcirc\!\!\!-NHOH\\[2em]HSCH_2CHCOOH + H_3C\!-\!\!\!\bigcirc\!\!\!-NHOH\\\quad\;\;|\\\quad\;NH_2\end{array} \qquad(25.2)$$

$R = NO_2$

group in each instance could be attributed to the effect of the divalent sulfur atom in the disulfide or in the mercaptoamino acid.

The inhibiting effect of the sulfur atoms in a disulfide was evident in the attempted reduction of the azomethine group in some compounds, $[RC_6H_4CH{=}NC_6H_4S]_2$, with a 10:1 ratio of Raney nickel to substrate; no uptake of hydrogen was observed except when R was 4-chloro [10]. When larger amounts of catalyst were employed, desulfurization resulted to yield N-aralkylanilines.

Mercaptans

Attempts to reduce functional groups in substrates containing an SH group usually fail because of the poisoning effect of that group. Attempts to convert the readily reducible aromatic nitro group in thiophenols to an amino group with palladium, platinum, or Raney nickel were never successful [1]; little or no hydrogen uptake was observed in low-pressure reductions at 60–70° with excess catalyst.

It is possible that success might be attained in the hydrogenation of nitrothiophenols with platinum sulfide or other noble metal sulfided catalysts under higher pressure since they are capable of reducing aromatic nitro groups. Although they have not been studied for such hydrogenations, the fact that they are already sulfided may make them resistant to sulfur poisoning and worth investigating for the reduction of divalent sulfur-containing substrates.

Sulfides, RSR-

Hydrogenation of Reducible Functions. The ease or difficulty in hydrogenating reducible functions in sulfides will depend on the groups that R and R' represent.

WHERE R AND R' ARE ALIPHATIC. When R and R' are both aliphatic some difficulty should be expected. Not only is there good contact between the sulfur atom in the sulfide and the catalyst, but hydrogenolysis can also occur to yield a more potent inhibitor, a thiol, to cause a marked slowdown in hydrogen absorption as noted in the reductions of some allyl sulfides [2].

Hydrogenations can be successful but will require high catalyst ratios. Examples are seen in the hydrogenolysis of (N-carbobenzyloxy-L-seryl)-L-methionine to L-seryl-L-methonine in 90% acetic acid with ten times the weight of 5% palladium on barium sulfate [11], in the reduction of crotyl 2-acetylaminoethyl sulfide, $CH_3CH=CHC(=O)S(CH_2)_2NHCOCH_3$, to butyryl 2-acetylaminoethyl sulfide with about ten times the weight of 30% palladium on carbon [12], and in the hydrogenations of 1-n-butylmercapto- and 1-t-butylmercapto-1-chlorobutadiene [13]. In the latter instance the ratio of 5% palladium on barium sulfate was 250–400%. The lower catalyst ratio in contrast to that employed in Refs. 11 and 12 may be due to the effect of the chlorine atom which interfered with the approach of the sulfur atom to the catalyst.

WHERE R IS ALIPHATIC AND R' IS AROMATIC. Reduction of functional groups among sulfides, RSR' where R is aliphatic and R' is aromatic, may be more successful than in those among aliphatic sulfides. The bulk of the aromatic group has some effect on contact between the sulfur atom and the catalyst. In addition these compounds are generally fairly stable and do not undergo cleavage to mercaptans.

Hydrogenations among this group are dependent on the relative reducibility of the function undergoing attack. The reduction of 1-phenylmercapto-1-chlorobutadiene required the same high ratio of catalyst employed for that of the corresponding 1-butylmercapto compound [13]. Where an aromatic nitro group is to be converted to an amino group, more moderate amounts of catalyst may often be used. For example about 75% by weight of Raney nickel was employed in the reduction of 4-amino-6-methyl-2-methylmercapto-5-nitropyrimidine to the diaminopyridine at low pressure [14]. The rapid absorption (1 hr) suggests that less catalyst could be used. About 25% by weight of palladium on carbon catalyzed the reduction of 3,5-dinitro-2-methylmercaptopyridine to the diamine [15]. A rather large amount of platinum oxide, 20% by weight, was employed to convert 4-methoxy-5-nitro-6-pyrimidinemercaptoacetic acid to 4-methoxy-7H-pyrimido[4,5-b][1,4]thiazine-6(5H)-one [16] (Eq. 25.3). However, the

$$\underset{\substack{\text{PtO}_2 \\ \text{2 atm, 2 hr}}}{\xrightarrow{3H}}$$

fairly rapid absorption of hydrogen under low-pressure conditions suggests that less catalyst could also give good results. Indeed the same nitro compound was reduced in alcohol to the cyclized product with 20–25% by weight of commercially available Raney nickel under 2 atm pressure within 2–3 hr [1]. The same procedure also was employed to reduce a number of 2-(nitroarylmercapto)acetic acids (NO$_2$-ArS)CH$_2$COOH (or alkyl) where Ar was an unsaturated N-heterocycle [1].

In some instances excessive amounts of catalyst were employed (in the following examples it was Raney nickel) to reduce nitro groups in N-(2-chloroethyl)-2-nitro-5-mercaptoethylaniline [17], in 1,1-bis(4-nitrophenylthio)butane [18], and in 4-nitrophenyl 2-diethylaminoethyl sulfide [19]. In the last reduction the authors stated that excess catalyst was employed to counteract the poisonous effect of the sulfur atom.

In a few instances higher pressures have been employed to overcome possible poisoning effects of the sulfur atom in alkyl nitroaryl sulfides.

Examples are seen in the hydrogenation of 1(2-nitrophenylmercapto)-acetone with nickel on kieselguhr [20], in the reduction of methyl 2-nitrophenyl sulfide with palladium on alumina under 10 atm pressure, and in the Raney nickel catalyzed hydrogenation of longer chain alkyl 2-nitrophenyl sulfides under 50 atm pressure [21]. In the last mentioned reference palladium appeared to be superior to Raney nickel because reductions with nickel were carried out under higher pressure. In contrast Raney nickel gave better results than palladium as palladium on carbon when each was used in alcohol, at 25–40° and 2–3 atm in the hydrogenations of 3-(6-methyl-3-nitrophenylmercapto)propionic acid [1]. In the palladium reductions it was necessary to filter the reaction mixture and add fresh catalyst each time for four additions before hydrogen uptake was complete in 24 hr. Better results with nickel were also obtained when the methyl ester was hydrogenated in methyl cellosolve.

WHERE R IS ALIPHATIC AND R' IS ARALKYL. Some difficulties might be expected in reductions among this group such as are seen when R and R' are aliphatic. However, they should be of a lesser degree. When a readily reducible function, such as an aromatic nitro group, is to be hydrogenated little difficulty should be encountered. The reductions of 2-(2-nitrophenylmercapto)acetic acid and 3-(2-nitrophenylmercapto)propionic acid with 16.5% by weight of 10% palladium on carbon at room temperature and 4–7 atm pressure of hydrogen were accomplished with apparent ease [22]. The first reaction yielded a mixture of the amino and cyclized product (Eq. 25.4). The second reaction yielded 3-(2-aminophenylmercapto)-

$$\underset{CH_2SCH_2COOH}{\bigcirc NO_2} \xrightarrow[\text{4–6 atm}]{\underset{\text{Pd on C}}{3H_2}}$$

$$\underset{CH_2SCH_2COOH}{\bigcirc NH_2} + \text{[structure]} \quad (25.4)$$

propionic acid exclusively in 84% yield.

WHERE R AND R' ARE AROMATIC. Molecular models of aryl sulfides suggest that the sulfur atom in these compounds cannot approach the catalyst as easily as it can in other sulfur-containing systems. Therefore, it should be possible to reduce functional groups more or less readily depending on their susceptibility to hydrogenation, although examples appear to be limited to nitro groups.

Some difficulty may be encountered during reductions because of incomplete removal of impurities from the preparation of the sulfides (they are often prepared from thiophenols) but this can often be remedied by pretreatment of a solution of the substrate with Raney nickel.

Nitro groups in these compounds have been reduced under low-pressure conditions with palladium on carbon [1] and with platinum oxide [23]. More often Raney nickel has been employed possibly because it may be less sensitive to sulfur poisoning than the noble metals. Examples consist of the fairly rapid conversion of 4-methyl-4'-nitrophenyl sulfide to the amine with 7–8% by weight of Raney nickel at 1–3 atm [24], of the reduction of 4-nitronaphthyl phenyl sulfide with nickel in alcohol at 4 atm pressure [25], and of that of the methyl ester of 2-carboxyphenyl 2'-methyl-5'-nitrophenyl sulfide [26].

In other examples heat and increased pressures have aided reduction. This is seen in the hydrogenation of 4-methyl-4'-nitrophenyl sulfide with Raney nickel in methyl alcohol containing a small amount of acetic acid at 50–60° and 20 atm [27] and in the reduction of 2-amino-2',4'-dinitrophenyl sulfide with 9% by weight of 25% nickel on filtercel at 87–107° and 33 atm [28].

The sulfur atom in an aryl sulfide is not completely without poisoning effect. The difference in the reaction rate in the reductions of nitrophenyl sulfide, sulfoxide, and sulfone previously described [1] attests to the fact that it can and does inhibit catalytic activity. Evidence of its effect is also suggested in the incomplete hydrogenations of the aromatic nitro group in 4-acetyl-4'-nitrophenyl sulfide with Raney nickel in which only 22% of 4-acetyl-4'-aminophenyl sulfide was obtained along with 4,4'-di-(4-acetylphenylmercapto)azoxybenzene and in that of 4-nitro-4'-(1-hydroxyiminoethyl)phenyl sulfone in which aromatic amine and azoxy compound were obtained [29]. It is of interest that in these reductions neither the carbonyl nor the hydroxyimine group was affected (Eq. 25.5).

$$O_2N\!-\!\!\bigcirc\!\!-\!S\!-\!\!\bigcirc\!\!-\!R \xrightarrow{3H_2}$$

R = COCH$_3$ or C(CH$_3$)=NOH

$$H_2N\!-\!\!\bigcirc\!\!-\!S\!-\!\!\bigcirc\!\!-\!R \;+\; \begin{matrix} R\!-\!\!\bigcirc\!\!-\!S\!-\!\!\bigcirc\!\!-\!\overset{+}{N}\!-\!O^- \\ \| \\ R\!-\!\!\bigcirc\!\!-\!S\!-\!\!\bigcirc\!\!-\!N \end{matrix} \qquad (25.5)$$

22%; R = COCH$_3$
35%; R = C(CH$_3$)=NOH

WHERE R IS AROMATIC AND R' IS ARALKYL. From molecular models one might suspect that the sulfur atom in aryl aralkyl sulfides should have a greater anticatalytic activity than the sulfur atom in aryl sulfides. Nevertheless, good results have been obtained in the reduction of aromatic nitro groups in this series. 8-(3-Nitrobenzylmercapto)adanine was hydrogenated with 50% by weight of 10% palladium on carbon to yield the amine in a 12 hr reaction period under 2–3 atm [30] (Eq. 25.6). When the methyl ester

$$\text{purine-SCH}_2\text{-C}_6\text{H}_4\text{-NO}_2 \xrightarrow{3H_2} \text{purine-SCH}_2\text{-C}_6\text{H}_4\text{-NH}_2$$

(25.6)

of 2-carboxybenzyl 2'-nitrophenyl sulfide was reduced with a 10% ratio of 5% palladium on carbon in methyl cellosolve under 2 atm pressure, 75% yield of the corresponding amine was obtained after 18 hr [1]. However, hydrogenation as the free acid in alcohol or as sodium salt in water with the same amount of catalyst was complete in 2–3 hr; a very rapid and exothermic reaction took place when the reduction was carried out under a hydrogen pressure of 40 atm. When the reduction of 4-acetylaminophenacyl 4'-nitrophenyl sulfide was carried out, Raney nickel, 75% by weight, was employed [31]. The nitro group was converted to amino but the carbonyl group was unaffected.

$$R\text{-C}_6\text{H}_4\text{-COCH}_2\text{S-C}_6\text{H}_4\text{-NO}_2 \xrightarrow[\substack{\text{NiR} \\ 2\ \text{hr}}]{3H_2} R\text{-C}_6\text{H}_4\text{-COCH}_2\text{S-C}_6\text{H}_4\text{-NH}_2 \quad (25.7)$$

R = CH₃CONH   73%

WHERE RSR' IS A RING SYSTEM *Thiophenes.* Although thiophene is a rather potent catalyst poison functional groups in thiophene compounds have undergone hydrogenation, in most cases by the use of large amounts of catalyst or by the application of heat and pressure.

2-Nitro-3,5-diphenylthiophene was rapidly reduced in acetic acid and acetic anhydride (isolated as the acetylated amine) by reaction under 2.6 atm of hydrogen with excess Raney nickel [32]. 2-Nitrothiophene-4-sulfonamide was hydrogenated in 6 hr with an equal weight of Raney nickel [33]. Not only were the nitro groups reduced in 2,5-dibromo-3,4-dinitrothiophene but debromination also occurred during hydrogenation with an enormous amount of palladium on carbon prepared *in situ* [34].

The acetylenic linkage in 5-phenyl-2-(2-phenylethinyl)thiophene was completely saturated during reduction with an undisclosed amount of palladium on calcium carbonate and converted to an ethylenic linkage when Lindlar catalyst was employed [35].

Side chain ethylenic linkages in 2-butenyl and 2-pentenylthiophenes were saturated by hydrogenation with a 30–40% by weight of 0.5% palladium on alumina at 80–100° and 100 atm [36]. A good result was obtained when 2-(2-carboxyphenyl)vinylthiophene was reduced with a 5% ratio of 10% palladium on carbon under similar conditions for 4–6 hr [37]; yield of 2-(2-carboxyphenethyl)thiophene was 92%.

Attempts to reduce the azomethine linkage in thiophenes,

$$Ar—2—CH\!=\!NHR$$

where Ar was a thiophene ring and R was 2-pyridyl, failed under low-pressure conditions with Raney nickel or palladium on carbon [1]; the substrate had been pretreated with Raney nickel. However, success was reported where R was 4-hydroxyphenyl and 4-methoxyphenyl when reductions were carried out over rhenium heptasulfide in dioxane at 150–170° and 130 atm [38] (Eq. 25.8). Hydrogenation where R was 2-pyridyl would

$$\underset{S}{\bigcirc}\!-\!CH\!=\!NR \xrightarrow[150-170°,\ 130\ atm]{H_2,\ Re_2S_7} \underset{S}{\bigcirc}\!-\!CH_2NHR \qquad (25.8)$$

R = 4-hydroxyphenyl or 4-methoxyphenyl

probably succeed under similar conditions.

Carbonyl functions, which are less readily reduced than aromatic nitro groups or double or triple bonds, have been hydrogenated when part of a thiophene system in reactions over rhenium heptasulfide or palladium sulfide at 170° and 130 atm [39] and in the presence of cobalt polysulfide and sulfur at 200–225° and an initial pressure of 100 atm [40]. In neither case was any alcohol obtained. Thiophene 2 carboxaldehyde and 2-acetylthiophene were converted to 2-methyl and 2-ethylthiophene respectively [39] and all the acylthiophenes reduced over the cobalt catalyst mixture were converted to alkylthiophenes and in some cases to alkylthiophanes [40]. In the latter reactions a nitro group attached to the ring was reduced to amino and a nuclear bromine atom was completely removed.

While there are no reports concerning poisoning of the reaction apparatus with rhenium or palladium sulfide, the use of cobalt polysulfide and sulfur caused damage to the metal reactor [41].

In an attempted reduction with an unsulfided catalyst, that of 2 acetyl-thiophene in alcohol in the presence of ruthenium dioxide at 150° and 140–185° atm, little or no 2-(1-hydroxyethyl)thiophene was found [42]. The products of reduction were not determined except to note that about 5% of 2-(1-hydroxyethyl)tetrahydrothiophene was formed.

Reductive amination of 2 acetylthiophene was partially successful in the presence of rhenium heptasulfide at 160° and 130 atm [39]. Because of competition between reduction of the carbonyl function and amination, almost equal parts of 2-ethylthiophene and 2-(1-aminoethyl)thiophene were obtained along with secondary amine (Eq. 25.9). Poor results were obtained

$$\underset{S}{\bigcirc}COCH_3 + NH_3 \xrightarrow[160°, 130 \text{ atm}]{H_2, Re_2S_7}$$

$$\underset{S}{\bigcirc}C_2H_5 + \underset{S}{\bigcirc}CH(CH_3)NH_2 + \left[\underset{S}{\bigcirc}CH(CH_3)\right]_2 NH \quad (25.9)$$

in amination in the presence of palladium sulfide.

Attempts to aminate 2 acetylthiophene under low-pressure conditions with noble metal catalysts failed; no uptake of hydrogen was observed at any time [1].

Benzo- and Dibenzothiophenes. Examples of reductions of functional groups in benzo- and dibenzothiophenes are limited to groups attached to the benzene rings.

The hydrogenation of 5-nitrobenzothiophene proceeded rather smoothly and rapidly [1 hr] to 5-aminobenzothiophene by hydrogenation with 10% by weight of 5% palladium on carbon under 2 atm pressure [43]. Low-pressure reduction with Raney nickel was also reported [44].

Reductions among dibenzothiophenes should be less troublesome since these compounds are related to phenyl sulfides and hydrogenation of functional groups in that series was not too difficult. In examples among dibenzothiophenes, 1-nitrodibenzothiophene was reduced to the amine within 1 hr with Raney nickel under 3 atm of hydrogen [45] and methyl 4-(4-dibenzothiophenyl)-3-pentenoate was hydrogenated to the corresponding pentanoate with palladium on carbon [46] (Eq. 25.10). It is of interest that,

$$\text{dibenzothiophene-}C(CH_3)=CHCH_2COOCH_3 \xrightarrow{H_2, \text{ Pd on C}}$$

$$\text{dibenzothiophene-}CH(CH_3)(CH_2)_2COOCH_3 \quad (25.10)$$

in contrast to the ease of reduction of 1-nitrodibenzothiophene, reduction of 3-nitrodibenzothiophene required more Raney nickel and took 18 hr for complete uptake of hydrogen [47]. Molecular models suggest that there is little difference in contact between the sulfur atom and the catalyst in either compound. The difference may have been one of purity of the substrate and/or age of the catalyst.

In another reduction among dibenzothiophenes, 2-acetyldibenzothiophene was hydrogenated over copper chromite at 130° and 135 atm for 5 hr to give high yield of 2-(1-hydroxyethyl)dibenzothiophene [48].

Thiazoles and related compounds. Since thiazoles are structurally related to thiophenes, the use of larger than normal amounts of catalyst should be expected for reductions of functional groups in substrates containing a thiazole ring. In addition reaction time may be long and it may be necessary to apply elevated temperature and pressure to achieve success.

5-Methyl-2-nitrothiazole was hydrogenated fairly rapidly in acetic anhydride with an undisclosed amount of Raney nickel (probably with a high catalyst ratio) to yield 2-acetylamino-5-methylthiazole [49]. In other Raney nickel reductions where the nitro group was attached to a benzene ring in a thiazole-containing substrate, as in 4- and 5-(nitrophenyl)thiazoles [50, 51], the amount of catalyst also was not noted. However, the hydrogenation of 2-amino- and 2 acetylamino-5-thiazolyl 4-nitrophenyl sulfones with an equal weight of Raney nickel [52] suggests that excess catalyst was probably employed in Refs. 50 and 51.

Reduction of an aromatic nitro group in thiazole-containing substrates with noble metal catalysts also required excess catalyst. About 40% by weight of 10% palladium on carbon was employed for the hydrogenation of 2-amino-4-(4-nitrophenyl)thiazole [53] and 5% by weight of platinum oxide was used for that of 2-(4-methyl-3-nitrobenzoyl)aminothiazole 54] to the corresponding amines. Complete hydrogen uptake in the latter example took 20 hr. When platinum on carbon was employed in the reduction of a (nitrophenyl)thiazole, the reaction was carried out in the presence of 50% by weight of catalyst for 24 hr [55] (Eq. 25.11).

$$CH_3O-C_6H_3(NO_2)(NHCOR) \xrightarrow[\text{Pt on C}]{3H_2}$$

R = 4-thiazolyl

$$\left[CH_3O-C_6H_3(NH-CR=O)(N(H_2)) \right] \longrightarrow C_6H_3(NH)(N)=CR \quad (25.11)$$

In the reduction of other thiazole-containing substrates, 4-cyanomethylthiazole was converted to 4-(2-aminoethyl)thiazole by hydrogenation in ammoniacal ethyl alcohol with 50% by weight of Raney nickel at 75° and 115 atm [56], and 4-pyridyl-2-thiazolyl ketone was reduced to the carbinol over Raney nickel [57]. Reductive alkylation of N^1,N^1-diethyl-N^2-2-thiazolyethylenediamine with benzaldehyde and Raney nickel at 70° and 13 atm pressure of hydrogen was successful yielding N^1-benzyl-N^1-2-thiazolyl-N^2,N^2-diethylethylenediamine [58]. Debrominations among thiazoles were also successful with Raney nickel and acid acceptors [59–61].

The removal of a nitrogen-protecting carbobenzyloxy group in the presence of a thiazolidine ring also required a large amount of catalyst. Almost twice the weight of 30% palladium on barium carbonate was employed to convert 6-(α-carbobenzyloxyamino)phenylacetylaminopenicillanic acid to 6-(α-amino)phenylacetylaminopenicillanic acid [62] (Eq. 25.12).

$$\begin{array}{c} O \quad H \\ \parallel \\ RCH\overset{}{C}NH \\ | \\ RH_2COCNH \\ \parallel \\ O \end{array} \underset{H}{\overset{S}{\underset{N}{\bigsqcup}}} (CH_3)_2 \quad \xrightarrow[30\% \text{ Pd on BaCO}_3]{H_2}$$

$R = C_6H_5$

$$\begin{array}{c} O \quad H \\ \parallel \\ RCH\overset{}{C}NH \\ | \\ NH_2 \end{array} \underset{O}{\overset{S}{\underset{N}{\bigsqcup}}} \qquad (25.12)$$

In another reaction the azide grouping in a related penicillin compound was reported to be readily hydrogenolyzed with only 25% by weight of Raney nickel under 3 atm pressure of hydrogen [63] to yield the same product seen in Eq. 25.12. It was also stated that palladium, platinum, or rhodium catalysts could be employed.

Benzothiazoles. Reduction of functional groups in benzothiazole compounds probably also requires high catalyst ratios although in a few examples the amount employed was not reported. 2-(2-Diethylaminoethyl)-6-nitrobenzothiazole was hydrogenated over Raney nickel [64] as was 2-benzothiazolyl 4-pyridyl ketone [65]. An equal weight of 5% palladium on carbon was used to reduce 2-benzothiazolylsulfonylethylene [66] and probably a similar amount of 5% palladium on barium sulfate catalyzed the hydrogenations of some 1-benzothiazolyl-2-(phenyl, substituted phenyl and 2-furyl)-ethylenes [67].

In one instance, however, a moderate amount of catalyst, 20% by weight of 5% palladium on barium sulfate, was employed to convert 12-(2-benzothiazolyl)-12-oxododecanoic acid in glacial acetic acid containing a small amount of perchloric acid to 12-(2-benzothiazolyl)dodecanoic acid [68]. In this example the long side chain may have interfered with the approach of the sulfur atom to the catalyst and minimized its inhibitory effect.

Thiadiazoles. In the few available examples among this group of sulfur-containing heterocycles, reductions proceeded as in the thiazole series, usually with an abnormal amount of catalyst. 2-(4-Nitrobenzenesulfonamido)-5-R-1,3,4-thiadiazole, R=OCH$_3$ and SO$_2$CH$_3$, was reduced to the corresponding amine in 22 hr with about 50% by weight of Raney nickel [69]. When 5-bromo-1,2,4-thiadiazole was dehalogenated in the presence of triethylamine and Raney nickel, about 60% by weight of catalyst was employed [70]. When 2-bromo-1,3,4-thiazole underwent hydrogenolysis in the presence of triethylamine and platinum oxide the catalyst had to be replaced three times. In the case of 2-bromo-5-methyl-1,3,4-thiadiazole the catalyst was replaced only once, perhaps because the methyl group in 5-position did not allow as firm contact between the catalyst and the sulfur atom as took place during the reduction of 2-bromo-1,3,4-thiadiazole. It is of interest that supported palladium, generally the most effective catalyst for dehalogenation, was not employed in the described debrominations. The large amount of palladium catalyst that was reported necessary for the dehalogenation of bromothiophenes [34] may have deterred the investigators from studying the effect of palladium on bromothiadiazoles.

Phenothiazines. Since the sulfur atom in phenothiazine (as in phenyl sulfide and dibenzothiophene) does not approach the catalyst with the same ease as it does in other sulfur-containing systems, more normal amounts of catalyst usually may be employed for reduction although, depending on the function to be attacked, higher temperature and pressure may be necessary at times.

Some 1- and 3-nitro-10*H*-phenothiazines were reduced in acetic acid with platinum oxide under 3 atm pressure [71]. Hydrogenations of some 10-(2-cyanoethyl)phenothiazines were carried out under elevated temperature and pressure, 60–65° and 25 atm [72] and 106–115° and 45 atm [73], with Raney nickel and ammonia. Substituting secondary amines for ammonia, employment of similar conditions yielded 10-(3-tertiaryaminopropyl)phenothiazines [74].

Reductive alkylation of methylamine with 1-[3-(10-phenothiazinyl)-propyl]-4-piperidone and reductive amination of that ketone were successfully carried out over Raney nickel under mild conditions to give good yields of the 1-substituted-4-aminopiperidines [75] (Eq. 25.13).

$$R(CH_2)_3N\diagup\!\!\!\bigcirc\!\!\!\diagdown=O + R'NH_2 \xrightarrow{H_2} R(CH_2)_3N\diagup\!\!\!\bigcirc\!\!\!\diagdown NHR'$$

R = 10-phenothiazinyl
R' = H or CH$_3$

(25.13)

Only a moderate amount of 10% palladium on carbon (6% by weight) was needed for the dehalogenations of 3- and 4-chloro-1,2-diaza-10H-phenothiazine [76].

Other sulfur-containing ring systems. Attempts to reduce the carbonyl group in 4-oxobenzothiazine-3-acetic acid with Raney nickel or with palladium on carbon failed apparently because of catalyst poisoning. This was suggested by construction of a molecular model which indicated that the sulfur atom would have no difficulty reaching the catalyst.

SATURATION OF SULFUR-CONTAINING RINGS. In general saturation of bonds in cyclic sulfur compounds does not take place readily unless a large amount of catalyst or vigorous reaction conditions are employed. In many instances, despite the use of excess catalyst, hydrogenation may fail. The effect of the sulfur atom in a ring is seen in the reduction of 1,6-dithiacyclodeca-3,8-diyne in ethyl acetate with ten times its weight of 10% palladium on carbon [77]. Uptake of hydrogen for conversion of diyne to diene was never complete. Furthermore, the isolated *cis*-3,*cis*-8-diene resisted attempts at saturation (Eq. 25.14).

$$S\begin{matrix}CH_2C{\equiv}CCH_2\\ \\ CH_2C{\equiv}CCH_2\end{matrix}S \xrightarrow[10\% \text{ Pd on C}]{2H_2} S\begin{matrix}CH_2CH{=}CHCH_2\\ \\ CH_2CH{=}CHCH_2\end{matrix}S \qquad (25.14)$$

Thiophenes. Thiophene and substituted thiophenes have been hydrogenated to the corresponding tetrahydro compounds with 5 to 25 times their weight of 5% palladium on carbon or palladium on barium sulfate under 2–3 atm pressure [34]. Reductions have also been carried out with 10% molybdenum sulfide on alumina under an initial hydrogen pressure of 225 atm at 230° [78]. Reaction took about 150 hr, probably because of the low catalyst ratio (3.2%). It was reported that nickel and tungsten sulfide could also be used under similar conditions.

Some dihydrothiophenes were rapidly converted to tetrahydrothiophenes by low-pressure hydrogenations with palladium on carbon [2], but with 400–500% by weight of catalyst.

Thiopyrans. While there do not appear to be any references on the tetrahydrogenation of thiopyrans, it is likely that, as in the case of thiophenes, it could be accomplished by the use of a large amount of catalyst.

Among dihydrothiopyrans, the hydrogenation of 5,6-dihydro-4H-thiopyran with three times its weight of palladium on carbon was very rapid whereas reduction of 5,6-dihydro-2H-thiopyran under similar conditions was slow [2] (Eq. 25.15).

The difference in the rate of reduction between the two compounds is readily seen upon examination of molecular models of each. The double bond

$$\text{(25.15)}$$

in the first compound can make contact with the catalyst without difficulty as can the sulfur atom. However, since a large amount of catalyst was employed there were enough active sites for reduction to take place rapidly. In the second compound, since the sulfur atom is the strongest bonding force, the 3,4-double bond is generally not in position to make good contact with the catalyst.

Hydrogenation of 2,3-dimethyl-5,6-dihydro-4H-thiopyran under the conditions reported above was very slow (10 days) due to a combination of steric effects and the inhibiting action of the sulfur atom.

Thiazoles. Catalytic reduction of thiazoles have not been reported. Conversion to a saturated system, if it could be accomplished, would probably require a large amount of catalyst as in the hydrogenation of thiophenes. The 4,5-bond in 2-methyl-3-phthaloylglycyl-2,3-dihydrothiasole was reduced but required two large portions of palladium on carbon and a 4-day reaction period [79].

Miscellaneous sulfur-containing rings. 3,5-Diphenyl-1,4-thiazine was reported to absorb two equivalents of hydrogen during reduction over platinum oxide to yield 3,5-diphenylthiamorpholine [80]. In contrast the hydrogenations of 3,5-*bis*(phenyl and 4-substituted phenyl)-1,4-thiazines with platinum oxide or with a mixture of palladium on carbon and platinum oxide were found to give 2-methyl-2,4-*bis*(phenyl and substituted phenyl)-2,5-dihydrothiazoles [81]. The authors' suggested mechanism involves hydrogenolysis of the starting material followed by ring closure and rearrangement (Eq. 25.16).

Saturation of the sulfur-containing ring in benzo-1,4-dithin was rapid in methyl alcohol containing sulfuric acid when eight times its weight of 30%

palladium on carbon was employed during hydrogenation under 3 atm pressure [82] (Eq. 25.17).

REFERENCES

[1] M. Freifelder, unpublished results.
[2] L. Bateman and F. W. Shipley, *J. Chem. Soc.*, **1958**, 2888.
[3] L. Horner, H. Reuter, and E. Hermann, *Ann.*, **660**, 1 (1962).
[4] M. A. McCall, D. S. Tarbell, and M. A. Havill, *J. Am. Chem. Soc.*, **73**, 4476 (1951).
[5] M. Bergmann and G. Michalis, *Ber.*, **63**, 987 (1930).
[6] L. Zervas and D. M. Theodoropoulos, *J. Am. Chem. Soc.*, **78**, 1359 (1956).
[7] K. E. Kavanaugh, *J. Am. Chem. Soc.*, **64**, 2721 (1942).
[8] C. Berse, R. Boucher, and L. Piché, *J. Org. Chem.*, **22**, 805 (1957).
[9] W. H. Vinton, U.S. Patent 2,483,447 (1949).
[10] D. Froehling and R. Pohloudek, *Arch. Pharm.*, **298**, 617 (1965).
[11] K. Hofmann, A. Jöhl, A. E. Furlenmeier, and H. Koppler, *J. Am. Chem. Soc.*, **79**, 1636 (1957).
[12] J. C. Sheehan and C. W. Beck, *J. Am. Chem. Soc.*, **77**, 4875 (1955).
[13] W. E. Parham and S. H. Groen, *J. Org. Chem.*, **29**, 2214 (1964).
[14] R. J. Prasad, C. W. Noell, and R. K. Robns, *J. Am. Chem. Soc.*, **81**, 193 (1959).
[15] P. Tomasik and Z. Skrowaczewska, *Rocz. Chem.*, **41**, 275 (1967); through *Chem. Abstr.*, **67**, 11404u (1967).
[16] E. C. Taylor and E. E. Garcia, *J. Org. Chem.*, **29**, 2121 (1964).

[17] P. Clarke and A. Moorhouse, *J. Chem. Soc.*, **1963**, 4763.
[18] I. Kh. Feldman and T. I. Gurevich, *J. Gen. Chem.*, **21**, 1689 (1951) (Eng. trans.).
[19] J. Buchi, J. Enezian, G. Enezian, G. Vallette, and C. Pattani, *Helv. Chim. Acta*, **43**, 1971 (1960).
[20] W. H. Strain and J. B. Dickey, U.S. Patent 2,381,935 (1945).
[21] F. Gialdi and A. Baruffini, *Farmaco* (Pavia), *Ed. Sci.*, **12**, 206 (1957); through *Chem. Abstr.*, **51**, 12850 (1957).
[22] M. Uskovic, G. Grethe, J. Iacobelli, and W. Wenner, *J. Org. Chem.*, **30**, 3113 (1965).
[23] V. Hach and M. Protiva, *Coll. Czechoslov. Chem. Commun.*, **23**, 1941 (1958).
[24] H. Gilman and H. S. Broadbent, *J. Am. Chem. Soc.*, **69**, 2053 (1947).
[25] E. F. Elslager, D. B. Capps, and L. M. Werbel, *J. Med. Chem.*, **7**, 658 (1964).
[26] E. F. Elslager, M. Maienthal, and D. R. Smith, *J. Org. Chem.*, **21**, 1528 (1956).
[27] W. Wenner, *J. Org. Chem.*, **22**, 1508 (1957).
[28] W. V. Wirth and S. E. Krahler, U.S. Patent 2,765,341 (1956).
[29] H. H. Szmant and D. A. Irwin, *J. Am. Chem. Soc.*, **78**, 4386 (1956).
[30] B. R. Baker and J. A. Kozma, *J. Med. Chem.*, **11**, 652 (1968).
[31] I. Kh. Feldman and N. G. Prein, *J. Gen. Chem.*, **21**, 1811 (1951) (Eng. trans.).
[32] W. E. Parham, I. Nicholson, and V. J. Traynelis, *J. Am. Chem. Soc.*, **78**, 850 (1956).
[33] H. Y. Lew and C. R. Noller, *J. Am. Chem. Soc.*, **72**, 5715 (1950).
[34] R. Mozingo, S. A. Harris, D. E. Wolf, C. E. Hoffhine, Jr., N. R. Easton, and K. Folkers, *J. Am. Chem. Soc.*, **67**, 2092 (1945).
[35] K. E. Schultz, J. Reisch, and L. Horner, *Ber.*, **95**, 1943 (1962).
[36] H. Pines, B. Kvetinskas, J. A. Vesely, and E. Baclawski, *J. Am. Chem. Soc.*, **73**, 5173 (1951).
[37] J. M. Bastian, A. Ebnöther, E. Jucker, E. Rissi, and A. P. Stoll, *Helv. Chim. Acta* **49**, 214 (1966).
[38] M. A. Ryashentseva, Kh. M. Minachev, O. A. Kalinovskii, and Ya. L. Goldfarb, *Zh. Organ. Khim.*, **1**, 1104 (1965); through *Chem. Abstr.*, **63**, 11474 (1965).
[39] M. A. Ryashentseva, O. A. Kalinovskii, Kh. M. Minachev, and Ya. L. Goldfarb, *Khim. Geterotsiki Soedin*, **1966**, 694; through *Chem. Abstr.*, **66**, 54772j (1967).
[40] E. Campaigne and J. L. Diedrich, *J. Am. Chem. Soc.*, **73**, 5240 (1951).
[41] A personal communication from Dr. Campaigne, Dept. of Chemistry, Indiana University, Bloomington, Indiana noted that the hydrogenation bomb was pitted and had to be rebored and refitted to make it leakproof.
[42] T. L. Cairns and B. C. McKusick, *J. Org. Chem.*, **15**, 790 (1950).
[43] F. G. Bordwell and H. Stange, *J. Am. Chem. Soc.*, **77**, 5939 (1955).
[44] K. Rabindran, A. V. Sunthankar, and B. D. Tilak, *Proc. Indian Acad. Sci.*, **36A**, 405 (1952); through *Chem. Abstr.*, **47**, 11189 (1953).
[45] H. Gilman and G. R. Wilder, *J. Am. Chem. Soc.*, **76**, 2906 (1954).
[46] W. Carruthers and H. N. M. Stewart, *J. Chem. Soc.*, **1965**, 6221.
[47] H. Gilman and J. F. Nobis, *J. Am. Chem. Soc.*, **71**, 274 (1949).
[48] R. G. Flowers and L. W. Flowers, *J. Am. Chem. Soc.*, **71**, 3102 (1949).
[49] H. v. Babo and B. Prijs, *Helv. Chim. Acta*, **33**, 306 (1950).
[50] H. Erlenmeyer, C. Becker, E. Sorkin, H. Bloch, and E. Suter, *Helv. Chim. Acta* **30**, 2058 (1947).
[51] W. Vögtli, E. Sorkin, and H. Erlenmeyer, *Helv. Chim. Acta*, **33**, 1297 (1950).
[52] T. V. Gortinskaya, V. G. Samolovova, and M. N. Shchikina, *J. Gen. Chem. USSR*, **27**, 2027 (1957) (Eng. trans.).
[53] G. De Stephens and V. P. Arya, *J. Org. Chem.*, **29**, 2064 (1964).
[54] A. Adams, J. N. Ashley, and H. Bader, *J. Chem. Soc.*, **1956**, 3739.

[55] D. J. Tocco, R. P. Buhs, H. D. Brown, A. R. Matsuk, H. E. Mertel, R. E. Herman, and N. R. Trenner, *J. Med. Chem.*, **7**, 399 (1964).
[56] H. Erlenmeyer and M. Müller, *Helv. Chim. Acta*, **28**, 922 (1945).
[57] V. G. Ermolaeva and M. N. Shchukina, *J. Gen. Chem. USSR*, **32**, 2623 (1962); Eng. trans.
[58] H. H. Fox and W. Wenner, *J. Org. Chem.*, **16**, 225 (1951).
[59] H. Erlenmeyer and H. Kiefer, *Helv. Chim. Acta*, **28**, 985 (1945).
[60] A. v. Wartburg, *Helv. Chim. Acta*, **32**, 1097 (1949).
[61] H. Erlenmeyer and Ch. J. Morel, *Helv. Chim. Acta*, **28**, 362 (1945).
[62] F. P. Doyle, G. R. Fosker, J. H. C. Nayler, and H. Smith, *J. Chem. Soc.*, **1962**, 1440.
[63] B. O. H. Sjoberg and B. A. Ekstrom, U.S. Patent 3,228,930 (1966).
[64] E. Hoggarth, *J. Chem. Soc.*, **1949**, 3311.
[65] V. V. Avidon and M. N. Shchukina, *J. Gen. Chem. USSR*, **34**, 3002 (1964); Eng. trans.
[66] E. A. Kuznetsova, S. V. Zhuravlev, and T. N. Stepanova, *J. Org. Chem. USSR*, **1**, 769 (1965((Eng. trans.).
[67] W. Ried and S. Hinsching, *Ann.*, **600**, 47 (1956).
[68] E. Graef, J. M. Frederickson, and A. Burger, *J. Org. Chem.* **11**, 257 (1946).
[69] M. L. Sassiver and R. G. Shepherd, *J. Med. Chem.*, **9**, 541, (1966).
[70] J. Goerdeler, J. Ohm, and O. Tegtmeyer, *Ber.*, **89**, 1534 (1956).
[71] A. B. Sen and R. C. Sharma, *J. Indian Chem. Soc.*, **35**, 202 (1958).
[72] Brit. Patent 731,016 (1955).
[73] K. Fujui, *J. Pharm. Soc. Japan*, **76**, 640 (1956); through *Chem. Abstr.*, **51**, 424 (1957).
[74] K. Fujii, K. Okumura, J. Arita, H. Yoshikawa, and H. Watanabe, Japan. Patent 5171 (1957); through *Chem. Abstr.*, **52**, 10219 (1958).
[75] K. Stack, M. Thiel, and F. Bickelhaupt, *Monatsh.*, **93**, 1090 (1962).
[76] F. Yoneda, T. Ohtaka, and Y. Nitta, *Chem. Pharm. Bull.*, **13**, 580 (1965).
[77] G. Eglinton, I. A. Hardy, R. A. Raphael, and G. A. Sim, *J. Chem. Soc.*, **1964**, 1154.
[78] G. B. Hatch, U.S. Patent 2,648,675 (1953).
[79] J. C. Sheehan, C. W. Beck, K. R. Henery-Logan, and J. J. Ryan, *J. Am. Chem. Soc.*, **78**, 4478 (1956).
[80] K. Fujii, *Yakugaku Zasshi*, **77**, 359 (1957); through *Chem. Abstr.*, **51**, 12103 (1957).
[81] D. Sica, C. Santacroce, and R. A. Nicolaus, *Gazz. Chim. Ital.*, **98**, 488 (1968).
[82] W. E. Parham, T. M. Roder, and W. R. Hasek, *J. Am. Chem. Soc.*, **75**, 1647 (1953).